Microglia in Health and Disease

Marie-Ève Tremblay • Amanda Sierra
Editors

Microglia in Health and Disease

Springer

Editors
Marie-Ève Tremblay
Axe Neurosciences, Centre de recherche du CHU de Québec, and Department of Molecular Medicine
Université Laval
Québec, QC, Canada

Amanda Sierra
Achucarro Basque Center for Neuroscience
University of the Basque Country EHU/UPV
Ikerbasque Foundation
Zamudio, Bizkaia, Spain

ISBN 978-1-4939-1428-9 ISBN 978-1-4939-1429-6 (eBook)
DOI 10.1007/978-1-4939-1429-6
Springer New York Heidelberg Dordrecht London

Library of Congress Control Number: 2014949136

Printed on acid-free paper

Springer is part of Springer Science+Business Media (www.springer.com)

To our families

Preface

These past few years have witnessed a revolution in our understanding of microglial cells, since their roles in the healthy central nervous system have just started to be revealed. These resident macrophages were shown to actively contribute to maintaining health, in cooperation with neurons, glial cells, and other types of immune cells, throughout the life-span, prompting reinterpretation of their long known involvement with diseases.

To share with the broader scientific community the recent discoveries on some of the most relevant topics in microglia research, from a diverse perspective, we propose a collection of 19 chapters from 52 specialists, working in 11 countries (Australia, Canada, Chile, China, France, Germany, Japan, Spain, Switzerland, the UK, and the USA) across 5 continents (Asia, Australia, Europe, North America, South America), and presently at the graduate, postdoctoral, assistant, associate, or full professor stages of their career.

To set microglia on the stage, we begin by explaining briefly who they are and what they do: their origin, history, physiology, and immune functions, the recent development of noninvasive methodologies to study microglia, and the ongoing controversy about their neurotoxic versus neuroprotective implication in disease.

In the first section, we describe in more detail their physiological roles in the maturation, function, and plasticity of the central nervous system, across normal development, adolescence, adulthood, and aging. Doing so, we also address their crucial involvement in neuropathic pain and drug addiction.

In the second section, we discuss their implication in pathologies impacting on the quality of life: neurodevelopmental and neuropsychiatric disorders, AIDS, and multiple sclerosis; and their contribution to leading causes of death: ischemia and stroke, neurodegenerative diseases, as well as trauma and injury.

The chapters strictly discriminate between experimental data, hypotheses, and speculations, so that the open questions are presented clearly to newcomers in this young and vibrating field. Outdated terminologies such as "resting" and "activated" microglial cells are replaced by a more thorough description of their actual phenotype, comprising concerted changes in morphology, gene expression, and functions. The distinction between in vivo and in vitro data is also emphasized throughout the book, considering the importance of studying microglia in their normally prevailing behavior, without inadvertently causing their reactive transformation during experimental procedures. Contextual differences between central nervous system regions and stages of the lifespan are also covered, whenever data is available.

As a result of this book, hopefully engaging the scientific community in a lively discussion about microglial involvement in many more contexts of health and disease, we can envision an explosion of discoveries that will translate into the development of better targeted and more efficient therapies in the near future.

We are particularly grateful to our editors at Springer, Simina Calin and Gina Kahn, all the collaborators, our institutions and lab members, and most of all, our families for supporting this grand endeavor.

Québec, QC, Canada Marie-Ève Tremblay
Zamudio, Bizkaia, Spain Amanda Sierra

Contents

Contributors

Jessica K. Alexander, Ph.D. Program in Neurosciences and Mental Health, Hospital for Sick Children, Toronto, ON, Canada

Simon Beggs, Ph.D. Program in Neurosciences and Mental Health, Hospital for Sick Children, Toronto, ON, Canada

Rommy von Bernhardi, M.D., Ph.D. Department of Neurology, School of Medicine, Pontificia Universidad Católica de Chile, Santiago, Chile

Alain Bessis, Ph.D. Institut de Biologie de l'Ecole Normale Supérieure, Inserm, Paris, France

Knut Biber, Ph.D. Department of Psychiatry and Psychotherapy, Molecular Psychiatry, University Hospital Freiburg, Freiburg, Germany

Department of Neurosciences, University Medical Center Groningen, The Netherlands

Michelle L. Block, Ph.D. Department of Anatomy and Neurobiology, Virginia Commonwealth University, Richmond, VA, USA

Department of Anatomy & Cell Biology Indiana University School of Medicine and The Stark Neuroscience Research Institute, Indianapolis, IN

Astrid E. Cardona, Ph.D. Department of Biology and South Texas Center for Emerging Infectious Diseases, University of Texas at San Antonio, San Antonio, TX, USA

Sandra M. Cardona, Ph.D. Department of Biology, University of Texas at San Antonio, San Antonio, TX, USA

Raffaela Cipriani, Ph.D. Department of Neurosciences, CIBERNED and Achucarro Basque Center for Neuroscience, University of Basque Country EHU/UPV, Leioa, Spain

Thomas A. Cochran, B.A. Department of Psychology and the Center for Neuroscience, University of Colorado at Boulder, Boulder, CO, USA

Dimitrios Davalos, Ph.D. Gladstone Institute of Neurological Disease, University of California, San Francisco, San Francisco, CA, USA

Samuel David, Ph.D. Centre for Research in Neuroscience, The Research Institute of the McGill University Health Center, Montreal, QC, Canada

María Domercq, Ph.D. Department of Neurosciences, CIBERNED and Achucarro Basque Center for Neuroscience, University of Basque Country EHU/UPV, Leioa, Spain

Betsi Flores Department of Biological Sciences, Universidad Andrés Bello, Santiago, Chile

Martin Fuhrmann, Ph.D. Department of Neuroimmunology and Imaging, German Center for Neurodegenerative Diseases (DZNE), Bonn, Germany

Harris A. Gelbard, M.D., Ph.D. Department of Neurology, Center for Neural Development and Disease, University of Rochester Medical Center, Rochester, NY, USA

Diego Gomez-Nicola, Ph.D. Centre for Biological Sciences, Southampton General Hospital, University of Southampton, Southampton, UK

Richa Hanamsagar, B. Pharm., Ph.D. Department of Psychology and Neuroscience, Duke University, Durham, NC, USA

Uwe-Karsten Hanisch, Ph.D. Institute of Neuropathology, University of Göttingen, Göttingen, Germany

G. Jean Harry, Ph.D. National Toxicology Program Laboratory, National Institute of Environmental Health Sciences/National Institutes of Health, Research Triangle Park, NC, USA

Mark R. Hutchinson, Ph.D. Discipline of Physiology, School of Medical Sciences, University of Adelaide, Adelaide, SA, Australia

Lee-Way Jin, M.D., Ph.D. Department of Pathology and Laboratory Medicine, School of Medicine, University of California Davis Medical Center, Sacramento, CA, USA

Takahiro A. Kato, M.D., Ph.D. Department of Neuropsychiatry, Graduate School of Medical Sciences, Kyushu University, Fukuoka, Japan

Tammy Kielian, Ph.D. Departments of Pathology and Microbiology, University of Nebraska Medical Center, Omaha, NE, USA

Shao-Ming Lu, Ph.D. Department of Neurology, Center for Neural Development and Disease, University of Rochester Medical Center, Rochester, NY, USA

Izumi Maezawa, Ph.D. Department of Pathology and Laboratory Medicine, School of Medicine, University of California Davis, M.I.N.D. Institute, Sacramento, CA, USA

Michel Mallat, M.D., Ph.D. Institut du Cerveau et de la moelle épinière (ICM), Hôpital de la Salpêtrière, Paris, France

José L. Marín-Teva, Ph.D. Departamento de Biología Celular, Universidad de Granada, Granada, Spain

Daniel F. Marker, Ph.D. Department of Neurology, Center for Neural Development and Disease, University of Rochester Medical Center, Rochester, NY, USA

Carlos Matute, Ph.D. Department of Neurosciences, CIBERNED and Achucarro Basque Center for Neuroscience, University of Basque Country EHU/UPV, Leioa, Spain

Yoshito Mizoguchi, M.D., Ph.D. Department of Psychiatry, Faculty of Medicine, Saga University, Saga, Japan

Akira Monji, M.D., Ph.D. Department of Psychiatry, Faculty of Medicine, Saga University, Saga, Japan

Hiroshi Nakanishi, Ph.D. Department of Aging Science and Pharmacology, Faculty of Dental Sciences, Kyushu University, Fukuoka, Japan

Julio Navascués, Ph.D. Departamento de Biología Celular, Universidad de Granada, Granada, Spain

Rosa C. Paolicelli, Ph.D. Division of Psychiatry Research, University of Zurich, Schlieren, Switzerland

V. Hugh Perry, Ph.D. Centre for Biological Sciences, Southampton General Hospital, University of Southampton, Southampton, UK

Jennifer M. Pocock, Ph.D. Department of Neuroinflammation, University College London Institute of Neurology, London, UK

Lorena Pont-Lezica, Ph.D. Institut de Biologie de l'Ecole Normale Supérieure, Inserm, Paris, France

Phillip G. Popovich, Ph.D. Department of Neuroscience, The Ohio State University Wexner Medical Center, Center for Brain and Spinal Cord Repair, Columbus, OH, USA

Payam Rezaie, Ph.D. Brain & Behavioural Sciences Discipline, Department of Life, Health & Chemical Sciences, Faculty of Science, The Open University, Milton Keynes, UK

Michael W. Salter, M.D., Ph.D. Program in Neurosciences and Mental Health, Hospital for Sick Children, Toronto, ON, Canada

Department of Physiology, Faculty of Medicine, Centre for the Study of Pain, University of Toronto, Toronto, ON, Canada

Amanda Sierra, Ph.D. Achucarro Basque Center for Neuroscience, University of the Basque Country EHU/UPV, Ikerbasque Foundation, Zamudio, Bizkaia, Spain

Ana Sierra Departamento de Biología Celular, Universidad de Granada, Granada, Spain

Yoshifumi Sonobe, Ph.D. Department of Neuroimmunology, Research Institute of Environmental Medicine, Nagoya University, Nagoya, Japan

Beth Stevens, Ph.D. Department of Neurology, Harvard Medical School, Boston Children's Hospital, F.M. Kirby Neurobiology Center, Boston, MA, USA

Akio Suzumura, M.D., Ph.D. Department of Neuroimmunology, Research Institute of Environmental Medicine, Nagoya University, Nagoya, Japan

Marie-Ève Tremblay, Ph.D. Axe Neurosciences, Centre de recherche du CHU de Québec, and Department of Molecular Medicine, Université Laval, Québec, QC, Canada

Hiroaki Wake, M.D., Ph.D. National Institute for Physiological Sciences, Okazaki, Japan

Xiaohui Wang, Ph.D. Chemical Biology Laboratory, Changchun Institute of Applied Chemistry, Chinese Academy of Sciences, Changchun, Jilin, China

Department of Psychology and the Center for Neuroscience, Department of Chemistry and Biochemistry and BioFrontiers Institute, University of Colorado at Boulder, Boulder, CO, USA

Linda R. Watkins, Ph.D. Department of Psychology and the Center for Neuroscience, University of Colorado at Boulder, Boulder, CO, USA

Jyoti J. Watters, Ph.D. Department of Comparative Biosciences, University of Wisconsin, Madison, Madison, WI, USA

Hang Yin, Ph.D. Department of Chemistry and Biochemistry and BioFrontiers Institute, University of Colorado at Boulder, Boulder, CO, USA

Part I
The General Setting

Chapter 1
Introduction

Amanda Sierra and Marie-Ève Tremblay

Abstract Microglia are taking a central role in the Neuroscience arena nowadays, following a series of recent discoveries challenging their traditional portrayal as simple "brain macrophages". Indeed, microglia have unique properties compared with other cells of the immune system, including their origin from the embryonic yolk sac, their capacity for self-renewal, and the extreme motility of their processes which closely interact with all the neighboring elements of the nervous parenchyma. In parallel, we have just begun to unravel novel roles for microglia during normal physiological conditions that are impacting our understanding of their crucial participation in pathological situations. Throughout the chapters of this book, some of the most intriguing open questions in microglial research today are transversally discussed. How is microglial population homeostasis maintained? Are there heterogeneous subpopulations of microglia? Can microglia respond to neuronal activity, and alter it? Do microglia interact with other types of immune cells? Are microglia beneficial or detrimental in diseases of the central nervous system? Are there novel tools to specifically manipulate microglia in a non-invasive manner? We do not have the answers yet for most of these questions, but in the next few years their investigation will will undoubtly continue to shape the future of Neuroscience in unexpected manners.

A. Sierra, Ph.D. (✉)
Achucarro Basque Center for Neuroscience, University of the Basque Country EHU/UPV, Ikerbasque Foundation, Bizkaia Science and Technology Park, Building 205, Zamudio, Bizkaia 48116, Spain
e-mail: a.sierra@ikerbasque.org

M.-È. Tremblay, Ph.D. (✉)
Axe Neurosciences, Centre de Recherche du CHU de Québec, and Department of Molecular Medicine, Université Laval, Québec, QC, Canada
e-mail: tremblay.marie-eve@crchudequebec.ulaval.ca

M.-È. Tremblay, A. Sierra (eds.), *Microglia in Health and Disease*,
DOI 10.1007/978-1-4939-1429-6_1

Keywords Microglia • Origin • Motility • Homeostasis • Heterogeneity • Neuronal activity • Central nervous system • Diseases • Experimental tools

Microglia have been traditionally perceived simply as brain macrophages controlling a mostly detrimental inflammatory response and have been for ages confined to the darkest and most pejorative corner of morphological, observational descriptions. With the recent development of non-invasive tools to visualize microglial dynamics and manipulate their gene expression, without inducing their pathological transformation in response to experimental procedures, microglia have taken the center stage in neuroscience. High impact papers have just started pouring in, discovering new roles for microglial cells in various contexts of health and disease. At the core of this turmoil has been the realization that microglia are not, indeed, like other macrophages.

Microglia do share many characteristics with circulating monocytes and macrophages residing in other tissues, including the expression of surface markers, receptors for pathogen and danger-associated molecules, intracellular signalling pathways involved with innate immunity, and above all, the capacity to phagocytose cellular debris and release various types of pro-inflammatory mediators. In fact, this similarity has hindered the understanding of microglia, and for a long time, macrophage behaviors were simply extrapolated to microglia. However, microglia have a distinctive origin, coming from primitive precursors in the yolk sac that invade the developing brain during early embryonic stages, whereas most hematopoietic lineages are generated later in the embryonic fetal liver. Furthermore, monocytes and macrophages are continuously replaced throughout adulthood, from bone marrow-derived precursors, whereas microglia are a long-lived population that self-renews in the healthy central nervous system (CNS) parenchyma, at least when the blood–brain barrier remains intact. Microglia also display unique properties that render them exquisitely adapted to their CNS environment, most notably the extreme dynamism of their processes, which interact intimately with all the other parenchymal elements, including neuronal bodies, processes, and synapses, astrocytes, neural progenitors, oligodendrocytes, and blood vessels, during normal physiological conditions.

The recent discoveries have opened many functional possibilities for microglia, beyond merely passive immune surveillance, but have also raised many more questions about their homeostasis and heterogeneity, regulation by neuronal activity, interactions with the other types of CNS and immune cells, and their active versus passive contributions to diseases, as well as serious concerns regarding the specificity of the tools commonly used for studying microglia.

Microglial homeostasis. Upon invasion of the CNS during embryonic development, microglial cells are well known to differentiate, migrate, and occupy nonoverlapping territories, but the molecular signals recruited to establish and maintain their territorial organisation are still undetermined. Little is known also regarding their homeostasis, at the cellular and population levels, including proliferation, migration, lifespan, and dynamic changes in phenotype and organisation, within different contexts of health and disease. Additionally, how the CNS copes with inflammatory barbarians from the periphery which reach the parenchyma upon disruption of the blood–brain barrier is not yet understood.

Microglial heterogeneity. Microglia are viewed as a single cell category, but different subsets most likely coexist, with each one being tightly adapted to the regional needs of the CNS, across the various stages of the lifespan and contexts of health and disease, in terms of structure and function, extracellular signalling, neurotransmitters and neuromodulators, blood–brain barrier permeability, energetic constraints, etc. Similarly, during pathological states, several 'activated' microglial phenotypes were recently described, comprising qualitative and quantitative differences in a wide range of microglial properties. Throughout the book, a strong effort is put into systematically describing the contextual particularities of microglia, such as CNS region, stage of the lifespan, and state of health, and their actual phenotype at all the levels: morphological, transcriptional, and functional.

Response to neuronal activity. Microglia express a plethora of receptors for neurotransmitters and neuromodulators in vitro, but whether these receptors are also present and serving any purpose in the healthy CNS is still undetermined. Furthermore, we have just started to unravel to which extent microglial functions such as surveillance, interactions with synapses, phagocytosis, and secretion of cytokines, extracellular matrix proteins, and growth factors, etc. depend on the levels of neuronal activity and sensory experience. This bidirectional communication between neurons and microglia will hold the key to ultimately understand microglial contribution to normal CNS development and function, as much as their consequences on essential cognitive processes, such as information processing, learning, memory, and decision making.

Interactions with the other CNS and immune cells. An outstanding question in most pathologies concerns the relative contribution of microglia versus their close cousins from the periphery, monocytes and macrophages, and the other resident cells which can also contribute to the production of pro-inflammatory mediators, such as astrocytes, endothelial cells, perivascular, and meningeal macrophages, as well as parenchymal mast cells. Furthermore, microglia may engage in some forms of crosstalk with lymphocytes, but it remains to be determined whether this interaction is physical and related to antigen presentation.

Microglial contribution to diseases. Our understanding of microglial implication in pathological conditions is now being revisited in light of their emerging roles in the healthy brain. A game-changer concept is that microglia might actually contribute to the disease pathogenesis by becoming dysfunctional or deregulated, rather than simply "responding" to the pathological damage. The chicken-or-the-egg kind of question about which particular cell type initiates the disease is complicated by the potential feedback processes through which diseased neurons affect microglia, and vice versa, creating an ever-growing vicious circle of pathological damage. Beyond the prevention of inflammatory responses, novel therapeutic approaches might be specifically targeted at curing the diseased microglia.

Tools for studying microglia. Some of the pharmacological manipulations widely claimed to inhibit microglial 'activation', including the antibiotics minocycline and doxycycline, are far from being specific for these cells, and in fact broadly target inflammatory responses. Other strategies commonly used to study microglial function involve the use of several transgenic-based or liposome-based methods to deplete microglia, but in this case, the bystander effects of replacing healthy microglia by corpses must be seriously considered when interpreting the results. Finally, the hot star CX3CR1 transgenic mice, which include the GFP and newcomers Cre and CreER mice, have proven to be incredibly valuable for our understanding of the fractalkine receptor, which in the healthy brain is only expressed by microglia. Nonetheless, the CX3CR1 locus is replaced by different transgenic constructs, and even the heterozygous mice were recently revealed to have several alterations at the functional and behavioural levels. Undoubtedly, the field will strongly benefit from the development of novel, cell-type specific tools to visualise and manipulate microglia in a non-invasive manner.

Some of these open questions that will certainly shape our future understanding of the roles of microglia in health and disease, in synergy with the other types of CNS and immune cells, are discussed transversally across the chapters of this book.

Chapter 2
Historical Context

Payam Rezaie and Uwe-Karsten Hanisch

Abstract The term 'microglia' was first introduced into the scientific literature almost a century ago. The various eras of microglia research have not only been defined by the number of reports subsequently generated but, more critically, by the concepts that have shaped our present-day views and understanding of microglia. Key methods, technologies, and models as well as seminal discoveries made possible through their deployment have enabled breakthroughs, and now pave the way for lines of investigation that could not have been anticipated even a decade ago. Advances in our understanding of microglial origin, forms, and functions have relied fundamentally on parallel developments in immunology. As *the* 'neuro-immune' cells of the brain, microglia are now under the spotlight in various disciplines. This chapter surveys the gradual processes and precipitous events that helped form ideas concerning the developmental origin of microglia and their roles in health and disease. It covers first the dawning phase during which the early pioneers of microglia research discovered cellular entities and already assigned functions to them. Following a recess period, the 1960s brought a renaissance of active interest, with a development of tools and models—and fundamental notions on microglial contributions to central nervous system pathologies. These seminal efforts laid the fundament for the awakening of a sweeping research era beginning with 1980s and being spurred on by a blast of immunological discoveries. Finally, the chapter stresses the advancement of molecular, genetic as well as imaging approaches to the study of microglia with the turn of the millennium, enabling insights into virtually all facets of microglial physiology. Moving forward it is clear that the future holds substantial promise for further discoveries. The next epoch in the history of microglia research has just begun.

Keywords Microglia • Developmental origin • Roles • Physiology • Pathology • Methods • Technologies • Models

---

P. Rezaie, Ph.D. (✉)
Brain & Behavioural Sciences Discipline, Department of Life, Health & Chemical Sciences, Faculty of Science, The Open University, Milton Keynes, UK
e-mail: p.rezaie@open.ac.uk

U.-K. Hanisch, Ph.D. (✉)
Institute of Neuropathology, University of Göttingen, Robert-Koch-Straße 40, Göttingen 37075, Germany
e-mail: ukhanisch@med.uni-goettingen.de

M.-È. Tremblay, A. Sierra (eds.), *Microglia in Health and Disease*,
DOI 10.1007/978-1-4939-1429-6_2

Bullet Points

- This chapter surveys the historical context that surrounds microglial cells, including their discovery, developmental origin, and functional roles in health and disease.
- We describe the concepts that have shaped our present-day views and understanding of microglia.
- We also present the key methods, technologies, and models that made possible the seminal discoveries across the various eras of microglia research.
- Lastly, we discuss the recent advancement of molecular, genetic, and imaging approaches, holding substantial promise for further discoveries.

2.1 Introduction

What relevance does the historical context have to today's research on microglia? With the pace of research and scientific discovery moving so rapidly in the present day, it is perhaps inevitable that those who are new to the field would focus on the latest outputs and deem work that has been published beyond the last few years as already 'dated'. Those with more than a passing interest in the neurosciences, however, find greater appreciation and deeper insight through a realisation of the concepts, the controversies, and breakthroughs that have shaped today's understanding of microglia. Indeed, it is remarkable to note the extent to which today's research is still directed at questions raised almost a century ago, and the evidence, in some areas, that merely confirms those pioneering observations (Rezaie and Male 2002a). Among the non-neural elements that constitute the nervous system, these enigmatic cells—the microglia—have arguably one of the most widely debated and contentious of historical perspectives (Rezaie and Male 2002a). In this chapter, we will examine the early history in more detail, beginning with pioneers of the late nineteenth and early twentieth centuries, and their contributions. We will thereafter journey through the renaissance and modern eras of research arriving at the new millennium. Throughout, we will consider major concepts and technical breakthroughs that have shaped present-day understanding of the biology (origin, form, function) of microglia and their responses to central nervous system (CNS) injury and disease. Figure 2.1 provides an overview of the history of research on microglia, from the pioneers, major discoveries, technical developments, and fundamental concepts across one and half centuries (up to 2013).

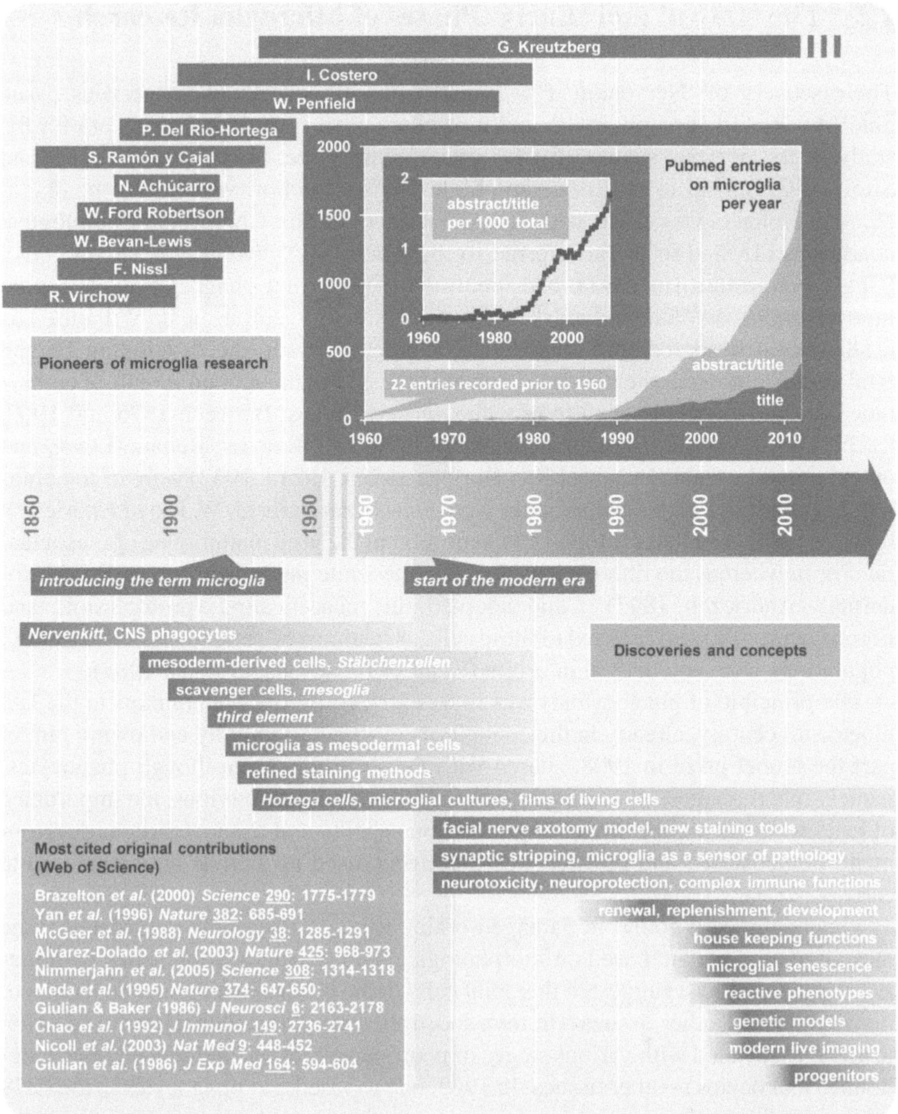

Fig. 2.1 Historical overview of research on microglia. Pioneers who contributed to the early discoveries are indicated with their biographic dates. Some major discoveries, technical developments, and fundamental concepts are shown below the timeline. The graphs (*inset*) illustrate the growth in research on microglia, according to the number of publications per year, carrying the term 'microglia' in the abstract and/or title (based on PubMed entries). The steady increase of microglia-related work is also reflected when normalised for the total record (number of microglial publications per 1000 total entries). The box lists the 10 most cited original contributions (by the topic microglia, based on the Web of Science)

2.2 The 'Dawn' and 'Early' Phases of Microglia Research

The discovery of 'Nervenkitt' ('*neuroglia*', literally translated as 'nervous tissue glue') by Rudolf Virchow in the mid-nineteenth century heralded a new era for the study of the nervous system (for historical reviews see: Ferrer 1973; Theele and Streit 1993; Barron 1995; Rezaie and Male 2002a). Virchow was among the first to recognise phagocytes filled with fatty substances in the CNS under pathological conditions (1851–1867). He referred to these cells as '*Schaumzelle*' (foam cells). Gluge, before him (in 1841), had identified mesodermal phagocytic cells within injured brains as '*Entzündungskugeln*' (inflammatory corpuscles). Wilhelm His (1888–1890) observed that mesodermal elements invade the developing human fetal spinal cord from the pia mater, towards the end of the second month of embryonic life, and distribute equally in white and grey areas. Between 1875 and 1900, Lachi, Duval, Gadow, Eichhorst, Schwalb, Friedman, Ranvier, Renaut, Lowe, and others (cited by Cajal 1925; del Río-Hortega 1920a, 1921a, 1932) were of the opinion that neuroglia in the white matter were mesoderm-derived. W. Lloyd Andriezen, however, proposed that protoplasmic neuroglia in the grey matter to be of mesoblastic origin, whereas the fibrous cells found in the white matter were considered ectodermal (Andriezen 1893). Cajal adopted this classification (protoplasmic and fibrous neuroglia) and referred to these cells as 'astrocytes', but he insisted that both populations were ectodermal in origin (Cajal 1913; see also Somjen 1988).

The principle of phagocytosis was first described by Ilja Metchnikoff in the late nineteenth century, already in the context of cell-based immunity and owing him in part the Nobel prize in 1908, shared with Paul Ehrlich. Even though phagocytes, namely macrophages, exhibit a much broader spectrum of activities, the importance of cells with the ability of clearing infectious agents and tissue debris in development, homeostasis, and defence was thus recognised as well as anticipated quite early.

Phagocytic cells found in damaged brains were on the whole considered to be mesodermal in origin, based on morphological and structural similarities with leukocytes. Franz Nissl suggested that glial cells in the brain adopt similar functions to macrophages in other tissues. He introduced the term "Stäbchenzellen" (rod cells) in 1899, associated with various stages of neurodegeneration, and related their functions to leukocytes in other tissues. In 1903, he proposed that phagocytes in the CNS originated from adventitial connective tissue of intracerebral vessels and mesenchyme, referring to these cells in 1904 as "*Gitterzellen*" (compound granular corpuscles). He considered them to be derived from mesoderm and not from blood cells (see Rezaie and Male 2002a). Like his contemporary Andriezen (1893), Bevan-Lewis (1899, 1906) was of the opinion that neuroglia were of different types, but unlike Andriezen (who had in fact been describing protoplasmic and fibrous astrocytes), Bevan-Lewis proposed a phagocytic 'scavenging' function for neuroglia in '*clearing out of fatigue products*' within the CNS (Bevan-Lewis 1906). Marinesco, before him, had reported dying neurons being removed by glial cells through phagocytosis. Forster, Marchand, and Pick later concluded that phagocytes could originate from neuroglia, and in 1910, Merzbacher was to include the term "*Abräumzelle*" (scavenger cell) in this group (see Rezaie and Male 2002a).

A significant breakthrough came from a somewhat unexpected and relatively inconspicuous source around the turn of twentieth century. Through developing a new technique to stain cellular elements within the CNS (a modification of the Golgi method using platinous oxide) (Robertson 1899), the Scottish pathologist William Ford Robertson who had been studying the neuroglia in some depth (Robertson 1897, 1898) showed that 'neuroglia' were not in fact a homogeneous grouping of cells, but were actually composed of different sets of elements. In 1899, using this new 'platinum method', the 32-year old pathologist to the Scottish Asylums described a novel group of cells in the canine and human CNS, referring to these as 'mesoglia'. He considered them to be derived from mesodermal elements (based on their staning properties), with morphologies that were quite distinct from other neuroglia (Robertson 1899, 1900a, b). Robertson later stated that the 'mesoglia' were able to act as phagocytes in certain pathological conditions (Robertson 1900a, b, see also Rezaie and Male 2002a). With the public announcement of his findings, Robertson became the first to demonstrate that the neuroglia were in fact composed of different cellular elements. John Turner was among the first to investigate Robertson's mesoglia. He was able to verify these cells in tissue sections stained with methylene blue dye (Turner 1905).

It was around this time that Santiago Ramón y Cajal, the founding father of the Spanish School of Histology and Histopathology in Madrid (awarded the Nobel prize in 1906, joint with Camillo Golgi, for his contributions to the 'neuron doctrine') started to take a serious interest in the neuroglia. One of his young and talented students (and later colleague whom he held in great esteem), Nicolás Achúcarro, had studied with Emil Kraepelin and Alois Alzheimer in Munich and developed a keen interest in the neuroglia. Between 1908 and 1910, Achúcarro embarked on a series of investigations into these cells, focusing specifically on astrocytes and on *Stäbchenzellen* first reported by Nissl in 1898 in cerebral palsy. He began by employing modifications of the reduced silver nitrate method used by Cajal and others and went on to develop a more selective tannin and ammoniacal silver nitrate method to describe *Stäbchenzellen* (microglia 'rod' cells; see Graeber and Mehraein 1994). Achúcarro considered *Stäbchenzellen* to 'adapt to degenerating pyramidal neurons' and noted that their peculiar elongated configuration conformed to the structure of the nervous tissue, aligning with the dendrites (Achúcarro 1908). He studied experimental wounds and local inflammation within the hippocampus, showing morphological transformation of these 'rod' cells and their phagocytic activity, with high lipid content in the vicinity of necrotic foci. Achúcarro proposed that *Stäbchenzellen* were phagocytic cells of mesodermal origin (specialised leukocytes) that adapted mechanically to the form and direction of neuronal processes and were capable of engulfing the decay products of neurons in inflammatory processes. He considered these cells to belong to the category of phagocytic '*Abräumzellen*' (Achúcarro 1908, 1909, 1910).

In 1911, Achúcarro presented his new technique (the method of tannin and ammoniacal silver) to the Spanish Society of Biology (Achúcarro 1911). Achúcarro used this method in a number of investigations on the cerebral cortex in 'general paralysis' (syphilitic encephalopathy) and to describe changes in neuroglia in senile dementia. Cajal referred to Achúcarro's work, as well as his own gold-sublimate method for neuroglia, extensively in his subsequent publications on the neuroglia (Cajal 1913, 1920a, b, 1925). 1911–1913 was a period of intense focus on neuroglia research (see Garcia-Segura 2002; García-Marín et al. 2007). Using formol uranium nitrate and gold chloride sublimate methods, Santiago Ramón y Cajal described what he termed a 'new class of cells that appeared to lack processes' in 1913 (Cajal 1913). He referred to these 'adendritic', 'apolar', and 'dwarf' cells as the 'third element' within the CNS, distinguishing them from neurons and astrocytes (see García-Marín et al. 2007). Cajal described these cells within the white matter, and also occurring as perineuronal and perivascular satellites, considering them to be mesodermal in origin. He believed that the cells identified in the central white matter were analogous to Schwann cells in the peripheral nervous system (Cajal 1913). At the time, he was not aware of Robertson's earlier work and only chanced across this later through a publication by Cerletti (lacking illustrative detail), which he considered in more detail in his 1920 and 1925 papers on the subject (Cajal 1920a, 1925). Achúcarro's successes were tragically cut short, as the first symptoms of a fatal illness began to take hold of him just 3 years following this appointment, forcing him to abandon his research activities. Initially considered to be tuberculosis (a disease that had afflicted his brother), Achúcarro recognised the symptoms as a form of Hodgkin's disease, which he self-diagnosed. Achúcarro's death at the age of 37, on 23 April 1918, came as a tremendous blow to Ramón y Cajal (see Cajal et al. 1968; Bustamante 1982; Andres-Barquin 2002).

Pío del Río-Hortega is probably the most prominent figure of the 'Spanish School' after Cajal (Gonzalez 1971; Andres-Barquin 2002). A pupil and later friend of Achúcarro, del Río-Hortega, had studied Medicine at the University of Valladolid, obtaining his license in 1905 and his doctorate in histology from the University of Madrid in 1909. Del Río-Hortega applied to work with Cajal, and after a brief spell with Tello (Jorge Francisco Tello-Muñoz), whose work at this time focused on regeneration of the nervous system—see Garcia-Segura 2002), in the newly established Laboratorio de Investigaciones Biológicas (Andres-Barquin 2002), joined Achúcarro's laboratory in the Museum of Natural History. This was later merged with Cajal's Institute and named the 'Laboratorio de Histopatología (Histologia Normal y Patológica) de la Junta para Ampliación de Estudios'. Here began his independent training, learning silver impregnation techniques and developing an interest in neuroglia. In 1916, his curiosity spurred by Cajal's 'third element', del Río-Hortega began to seek more stable variations of Cajal and Achúcarro's methods. Del Río-Hortega's diligence in pursuit of the third element led to his first publications on neuroglia in 1916 and 1917. He had also begun to doubt the accuracy of Cajal's concept that the 'third element' was composed solely of "corpuscles without processes" (see Cannon 1949, p. 239). Following closely in Achúcarro's footsteps, del Río-Hortega developed a new modification of Achúcarro's ammoniacal silver method (Achúcarro 1911), using silver carbonate (del Río-Hortega 1918), and identified distinct elements, initially the 'microglia', and what he called 'interfascicular cells' (del Río-Hortega 1919a, b, 1920a) which, after further modifications, he was able to define as 'oligodendroglia' (del Río-Hortega 1921b). He proposed that these cells were distinct populations; that the oligodendroglia belonged to the class of 'neuroglia' cells; and that microglia represented the true 'third element' on account of their mesodermal (as opposed to ectodermal) origin. He also alluded to the fact that Cajal had been unable to see the full extent of the morphology (fine cellular processes) of these cells in his 'third element' paper of 1913, due to the limitations of the technique he had used. Although Cajal eventually came to accept the existence of microglia as a separate class of cell (Cajal 1920a, 1925), he was not convinced regarding the oligodendroglia. Furthermore, Cajal considered Robertson's mesoglia and del Río-Hortega's microglia to be one and the same (Cajal 1920a, b). This contention was at the centre of a dispute that developed between Cajal and Hortega and led to tensions in their professional relationship.

Achúcarro, by this time, had become gravely ill (Jelliffe 1919). Consequently, del Río-Hortega's technical breakthrough and exemplary work were marred by the tragic illness and subsequent death of Nicolás Achúcarro in 1918. Del Río-Hortega succeeded Achúcarro in directing the Histopathology Laboratory and was later also appointed Histopathologist at the Provincial Hospital of Madrid. Going against Cajal's advice, del Río-Hortega proceeded to present his work to the Society of Biology. He published his work on microglia in four parts under the heading of 'the third element of the nervous system' (del Río-Hortega 1919a, b; see Fig. 2.2). Not only did this echo Cajal's own 1913 publication, but the work deigned to 'criticise and correct' it. Publishing his work without Cajal's approval had placed del Río-Hortega in a delicate position and tensions continued to grow (del Río-Hortega 1986), forcing his departure from the laboratory of Cajal in October 1919, and transfer to the Students' Residence (La Residencia de Estudiantes) located on the outskirts of Madrid in 1920, where he established a new laboratory.

Del Río-Hortega's follow-up paper on microglia (del Río-Hortega 1920a) related microglia with the different morphological forms of phagocytic cells found in the CNS and made the direct 'connection' with *Stäbchenzellen*. Cajal published a critique in the same journal volume (dated 15 December 1920), attributing del Río-Hortega's discovery to Robertson who, according to Cajal, had named these elements 'mesoglia' two decades prior to their 'rediscovery' as microglia by Hortega (Cajal 1920a). In a second (companion) paper directly following his critique, Cajal published his own method to detect mesoglia/microglia and their derivatives, using "a simple modification of the method of Bielschowsky" (Cajal 1920b, p. 136). He referred to mesoglia/microglia as the "cells of Robertson and of del Río". In his paper, Cajal compared the mesoglia of Robertson and microglia of del Río-Hortega to the interstitial corpuscles that Achúcarro had described earlier in 1910, referring also to the method he had published in 1911 (another variation of the Bielschowsky method) (Achúcarro 1910, 1911). Perhaps the problem also stemmed from inconsistencies and confusion surrounding published methods (and their citation), which were based on numerous modifications of existing methods, as well as a lack of clarity regarding the 'ownership' of these methods (see also Penfield 1924a, p. 437). For example, del Río-Hortega, having been inspired by Cajal and Achúcarro's

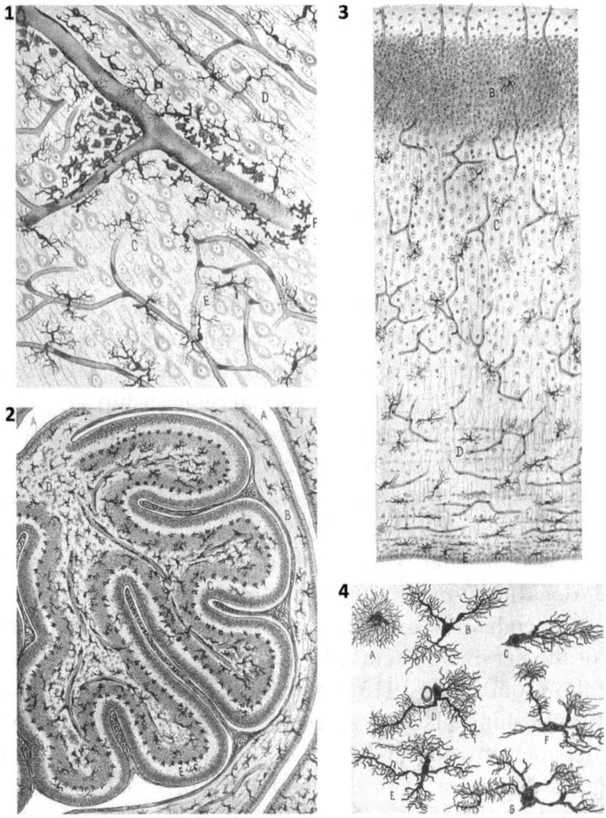

Fig. 2.2 Original microglia drawings from Pío del Río Hortega. Reprinted with permission from Memorias de la Real Sociedad Española de Historia Natural (del Río-Hortega 1921a) [In Spanish]. Original figure legends are transcribed. (**1**) Vertical section of the 4-day old rabbit brain protuberance. (*A*) blood vessel; (*B*, *F*) perivascular microglia with tuberosities and pseudopods; (*C*) ramified microglia; (*D*) cell with branches that follow the direction of the nervous fibres; (*E*) vascular satellites. (**2**) Cerebella microglia in the 4-day-old rabbit. (*A*) fourth ventricle; (*B*) medullary velum; (*C*) cerebellar folium; (*D*) microglia migrating through the white matter; (*E*) microglial cell reaching the grey substance. (**3**) Cerebral circumvolution of the 1-day-old rabbit. (*A*) molecular layer; (*B*, *C*), cortical layers; (*D*) white matter, from which microglia ascend to the inferior and medial layers of the cortex; (*E*) ependyma. (**4**) Simple ramified forms of the cerebral white substance in the 9-day-old rabbit. (*A*) hairy microglia; (*B*, *C*, *D*) mono- and tripolar cells, with penniform appendices; (*E*, *F*, *G*) multipolar cells with dendritic tufts

approaches, had undertaken a number of trials using the Bielschowsky ammoniacal silver method (the reagent used by Achúcarro), and by substituting carbonate for nitrate (Cajal had used a reduced silver nitrate method), had obtained finer micelles that were capable of impregnating neuroglia and revealing these cells in much greater detail (see Ortíz Picón 1971). Del Río-Hortega published this method in February 1918, and in November of that year presented his initial findings to the Spanish Society of Biology "On the true significance of neuroglial 'ameboid' cells".

This paper noted the appearance of ameboid cells in inflammatory and destructive processes. It provided evidence that astrocytes did not participate in the formation of rod cells, or granuloadipose corpuscles (Ortíz Picón 1971). In fact, del Río-Hortega published a number of variants of his method (see for example del Río-Hortega 1919c), which he reviewed shortly before his death, in a four-part publication entitled "The method of silver carbonate. General review of the technique and its application in Normal and Pathological Histology". This collection was published in the Archives of Normal and Pathological Histology, Buenos Aires (del Río-Hortega 1943a, b, 1944a, b).

In his 1921 paper on the histogenesis, normal development, exodus, and distribution of microglia (del Río-Hortega 1921a; see Fig. 2.2), del Río-Hortega provided a full translation (in Spanish) of Robertson's works (Robertson 1898, 1900b), emphasising that what Robertson had described as 'mesoglia' in fact corresponded to the type of neuroglia del Río-Hortega had himself described in some detail and had referred to as 'oligodendroglia' (see del Río-Hortega 1921b). In the footnote on the last page, he invited Cajal to examine the original preparations for himself (del Río-Hortega 1921a). It appears from this tail note that del Río-Hortega had secured some of Robertson's original samples (see also Ortíz Picón 1983). It is unlikely that Cajal ever took del Río-Hortega up on this offer, but Wilder Penfield who visited his laboratory 3 years later did review "an original preparation of Robertson" and referred to this in his 1924 paper (Penfield 1924a, p. 433). Del Río-Hortega renamed the new laboratory in La Residencia de Estudiantes the 'Laboratory of Normal and Pathological Histology' to provide greater scope to the work that was undertaken there (Palmero and del Río-Hortega 2004).

The impact of the work conducted in that laboratory was extraordinary, attracting international scientists to learn the new methods for detecting the neuroglia. Among those who visited del Río-Hortega was Wilder Penfield. Having closely followed the protocols of the Spanish School, including del Río-Hortega's published methods, Penfield had been attempting to stain microglia and oligodendroglia in brain tissue sections between 1921 and 1923. He wanted, in particular, to investigate these cells in epilepsy. He had started this work at Columbia College of Physicians and Surgeons, and continued with his attempts by looking at brain tissue scars later at the Presbyterian Hospital Laboratories in New York (Gill and Binder 2007), convinced that glial cells were central to understanding the causes of epilepsy, as well as tissue repair processes. For that purpose, Penfield had already made numerous attempts using Cajal's and del Río-Hortega's published methods for the neuroglia, but these had been disappointing and he had been unable to interpret the results. In his eagerness to set off, Penfield had dashed a letter to del Río-Hortega and set sail for Madrid before having received a reply. This finally arrived, forwarded by telegram from New York, while Penfield was halfway across the Atlantic Ocean. It was a single word 'venga' (come) (Penfield 1977). He arrived in April 1924 and remained in the lab until July. For independent accounts of Penfield's stay at La Residencia de Estudiantes, see Obrador (1975), García-Albea (2004), and Gill and Binder (2007).

In his 1924 paper, Penfield noted that the method of Robertson "was specific when successful but it was so capricious" and that it had not been possible to reproduce the descriptions provided by the Scottish pathologist. Likewise, del Río-Hortega's method had also given variable results in different hands, typically staining one or the other cell type more prominently—rather than providing uniform and selective labelling. Penfield went on to devise further modifications (Penfield 1924a, b, 1925) and a more 'reliable' method for combined staining of these distinct classes of cells, which he published in 1928. This was itself a further modification of Globus's method, trialled at length for consistency of staining by Penfield's assistant, Fulstow, over a period of 1 year whilst working within Penfield's laboratory in New York (Penfield 1928). Penfield confirmed del Río-Hortega's findings and disseminated these across Europe, North America, and Canada. During his visit to La Residencia in 1924, he submitted two influential papers: the first published in Brain (1924), the other in the first edition of the American Journal of Pathology (1925), in which he confirmed the existence of microglia, and after a further modification of del Río-Hortega's stain, the oligodendroglia. In his 1924 paper 'Oligodendroglia and its relation to classical neuroglia', he agreed with del Río-Hortega's initial assumptions that the apolar and adendritic cellular elements described by Cajal in 1913 were microglia and oligodendrocytes which appeared to be devoid of cellular processes because of the limitations of the technique employed, i.e. an imperfect technique producing incomplete staining of the cells. He also pointed out that microglia represented a small proportion of perineuronal and perivascular satellite cells (Penfield 1924a), although perivascular cells and parenchymal microglia were not yet recognised as distinct populations at this time (see Graeber and Streit 1990b; Streit and Graeber 1993). Penfield also believed that Robertson's mesoglial cells corresponded to del Río-Hortega's oligodendrocytes (Penfield 1924a). The paper he submitted to the American Journal of Pathology (1925) on phagocytosis of microglia in gliomas was also published in French in Cajal's own laboratory journal the previous year (Penfield 1924b). In this paper Penfield provides a rather general summary of del Río-Hortega's earlier work along with his own (in human material and in cats, rabbits, and mice). He referred also to the work by Collado who had been studying microglia in cases of rabies, and to Metz and Spatz (1924) who had confirmed the morphological characteristics of these cells, among other investigators (Penfield 1925). He further described del Río-Hortega's 'fountains' of microglia within the developing nervous system, areas where these cells were encountered in greatest densities and thought to be centres for their genesis. Moreover, he stressed the migratory and phagocytic activities of microglia following experimental injury, before looking at two cases of brain tumours that he termed 'gliomas'– the first study of its kind. He proposed that microglia acted as scavenger cells to clear away the products of degeneration in the area surrounding the glioma and performed 'dendrophagocytosis' within the tumour itself. His impression was that microglia transported the ingested substances to the outer surface of blood vessels and discharged these into the vessel lumen or perivascular space, based on the assumption that "*the presence of scavenger cells about vessels indicates transfer and delivery of ingested substance rather than new formation of these cells…*" (Penfield 1925). During his visit, he moved from normal histology to neuropathology, examining these glial cells in experimentally induced tissue injury, specifically in response to the formation of scars (del Río-Hortega and Penfield 1927). The 1927 paper 'Cerebral cicatrix—the reaction of neuroglia and microglia to brain wounds' described morphological changes (in microglia and astocytes) that accompanied scar formation in response to puncture wounds induced in rabbit brains, and more extensive wounds generated in canine cerebral hemispheres after craniotomy, followed over a period of several months (del Río-Hortega and Penfield 1927). The paper describes stages in glial scar formation clearly, with initial microglial activation and phagocytosis, later reorganisation of astrocytes surrounding the wound, and the laying down of connective tissue over time (del Río-Hortega and Penfield 1927).

Cajal however, while coming to accept microglia as a distinct class of cells, was not convinced that interfascicular glia or 'oligodendroglia' were a component of his 'third element', but that adendritic cells, including perivascular and satellite cells and dwarf cells of the white matter, were the 'real third element' (Cajal 1920a). More than two decades later, some authors still clung to the idea that microglia and oligodendroglia did not represent the complete picture, and that Cajal's 'adendritic' cells should be classed separately as 'adendroglia' (Andrew and Ashworth 1945). In turn, Cajal had considered that perhaps some of the mesoglia cells described by Robertson were in fact incompletely stained microglia (Cajal 1925). Having read the influential work of the German neurohistologists Metz and Spatz (1924), and that of Penfield (1924a, b), who had independently reproduced del Río-Hortega's findings and confirmed his theories using his silver carbonate method, as well as other investigators at the time (e.g. see for example Gans 1923; Bailey and Hiller 1924; Rezza 1925), Cajal later publically acknowledged microglia as a distinct entity, in his extended treatise on the neuroglia, published in 1925. This work was also published in French in order to reach a wider audience (Cajal 1925). Through his 'official recognition', Cajal had given del Río-Hortega merit for the discovery of 'normal' microglia, and Robertson the credit for discovery of the cells del Río-Hortega referred to as oligodendroglia. Returning to the Presbyterian Hospital in New York in September 1924, Penfield inaugurated the Laboratory of Neurocytology and met his research associate William Cone, who subsequently wrote a paper on 'Acute Pathological Changes in Neuroglia and Microglia' published in 1928. Penfield was to influence a number of other colleagues and associates who worked with him in New York and later on at the Montreal Neurological Institute, instilling in them his own enthusiasm and keen interest in pursuing research into the microglia (and oligodendroglia). Among these were Dorothy Russell, John Kershman, and Webb Haymaker who had worked with Alpers in Pennsylvania, and Carmichael at the London National Hospital. It was around this time that Penfield founded the idea to write a textbook on the general principles of neuropathology (without describing specific diseases); he was looking for a 'mechanistic description of disease pathophysiology' (Gill and Binder 2007). He wrote to 27 eminent investigators around the world, del Río-Hortega among them, and managed to secure their contribution to his three-volume collection: 'Cytology and Cellular Pathology of the Nervous System' (see Penfield 1932).

Dorothy Russell's contribution to research on microglia was made through investigating the functions of these cells by means of intravital dyes (trypan blue) administered intraperitoneally and examining cellular responses to aseptic cerebral puncture wounds in rabbits (Russell 1929). Russell showed that trypan blue could be taken up by all transitional forms of microglia from typical (normal) forms to *Gitterzellen* (compound granular corpuscles) and by ameboid or spindle-shaped cells lying in the adventitia of cerebral vessels. She proposed that such intravital staining was related to phagocytic function. She also identified these cells with the rest of the reticulo-endothelial system, which supported Hortega's contention that microglia was a mesodermal element. There was no evidence for uptake of dye by the neuroglia (Russell 1929).

In Germany, Creutzfeldt and Metz (1926), following up on the earlier work by Metz and Spatz (1924), charted the involvement of microglia ('Hortegazellen') in acute as well as chronic degenerative processes, progressive paralysis, in tuberculous meningitis, in presenile dementia, and the specific association of these cells with senile plaques (Creutzfeldt and Metz 1926; see also Timmer 1925 for an early discussion on the relationship between microglia and senile plaques). By 1930, the concepts of microglia and of oligdendroglia were firmly established, as can be read in the Society Transactions of the 11th Annual International Neurologic Assembly, held in Paris (Winkler et al. 1931). The mesodermal nature of microglia, however, was not universally accepted. Others (e.g. Polderman 1926; Schaffer 1926; Pruijs 1927), while accepting that microglia were a class apart from astrocytes (the 'macroglia'), questioned whether these cells could in fact be derived from the neuropeithelium (see Winkler et al. 1931).

Meanwhile, a series of influential studies were emerging from North American laboratories. Key among these were papers by William Cone (Penfield's coworker from The Department of Surgery and Pathology at the Presbyterian Hospital in New York), Bernard Alpers from Philadelphia (and the Laboratorio de Histologia Normal y Patologica, Madrid), and Carl Rand and Cyril Courville from the Neurological Service and Neuropathological Laboratory of the Los Angeles County General Hospital. Cone described acute changes in neuroglia and the transformation of microglia (rod cells and compound granular corpuscles) in routine human autopsy material (poliomyelitis and cerebrospinal meningitis), including responses to 10-day-old surgical needle tracks (ventricular punctures) (Cone 1928). Alpers, who had studied with del Río-Hortega, described reactions of cells within the CNS to experimental intoxication with urea in rabbits (Alpers 1930). He described profound changes within the cerebral white matter affecting fibrous and perivascular astrocytes (transforming to 'ameboid cells' with degeneration of vascular end-feet), with only minor changes affecting the microglia—typically fusiform or nodular swellings scattered at intervals along microglial processes in more severe cases. These changes varied with the degree of toxicity (Alpers 1930). Rand and Courville undertook a detailed histological examination of cellular changes in the human brain in 24 cases of fatal head injury (with survival ranging from a few hours to several months following injury). They used Penfield's 'combined method' for detecting microglia and oligodendroglia in tissue fixed with formaldehyde (Penfield 1928).

These authors reported that changes in the microglia depend directly on the severity of the contusion, its age, and proximity of individual cells to the point of greatest injury. They noted that the degree and timing of morphological changes in the microglia depended on the severity of the injury. Transitional forms could be found within the first 24 h following injury and fully developed compound granular corpuscles (foamy macrophages) around 4 days' survival time, with the development of vacuoles and swelling within these cells corresponding to their phagocytic activity (Rand and Courville 1932). They did not agree with the views of Penfield and Cone that rod cells were a typical 'stage' in the transformation between normal microglia and formation of compound granular corpuscles (rod cells were only noted in one case with 'dementia paralytica') (Rand and Courville 1932). They concluded that microglia react to local destruction and disintegration of tissue and persist as compound granular corpuscles as long as the products of disintegration were present.

Del Río-Hortega published the first of two definitive synopses on microglia in the English language, in 1932, in Penfield's 'Cytology and Cellular Pathology of the Nervous system' (del Río-Hortega 1932). The second publication was in The Lancet, in 1939, while he was at Oxford (del Río-Hortega 1939). In a four-part publication, he provided a thorough review of his silver carbonate method and its various modifications (del Río-Hortega 1943a, b, 1944a, b). Del Río-Hortega died in Buenos Aires, from a malignant cancer in 1945. His work, however, continued to live on through researchers whom he had influenced and inspired and was integrated within systematic investigations which took place later on.

The earliest accounts of microglia in tissue culture were provided by Marinesco and Minea (Winkler et al. 1931, pp. 663–664), Wells and Carmichael (1930) as well as Costero (1930a, b, c, 1931). Of these, the most definitive studies were those reported by Isaac Costero, who was the first to culture human microglia and to record their activities using time-lapse cinemicroscopy. AQ Wells and Edward Arnold Carmichael's work published in 1930 formed part of a long-standing collaboration between the Strangeways Research Laboratory in Cambridge and St Bartholomew's Hospital in London (see Wilson 2005). Carmichael, who had already started researching microglia and published a paper in 1929 looking at the microglial response to intracerebral injection of blood in rabbits (Carmichael 1929), joined up with Wells to examine cultured cells derived from the embryonic fowl nervous system (pons and medulla) and the retina in solid and fluid media, using the 'coverslip technique' that had been established at the time, along with standard 'static' staining and photographic methods (Wells and Carmichael 1930). Wells and Carmichael showed that cells 'resembling normal microglia' derived from nervous system tissues displayed selective affinity for silver impregnation methods, took up vital dyes (trypan blue), and were similar in their properties to 'wandering cells' or 'histiocytes' (i.e. macrophages; see Maximow 1928) cultured from periosteum and limb buds that belonged to the reticulo-endothelial system. On this basis, they considered that microglia also belonged to this system and were therefore of mesoblastic origin (Wells and Carmichael 1930). They described 'remarkable activity' in these cells: "*Protopasmic protrusions were constantly thrown out and then withdrawn. Mitochondria and cell granules were in constant violent movement. Highly refractile fat droplets were present in varying amounts*." (Wells and Carmichael 1930). Webb Haymaker later adapted del Rio-Hortega's silver impregnation technique to the staining of tissue cultures, in a paper read at the meeting of the International Association of Medical Museums, New York, April 1935, and published in the scientific apparatus and laboratory methods section of Science (Haymaker and Sánchez-Pérez 1935; see also von Mihálik 1935). Costero, having spent a number of years studying under del Río-Hortega (1922–1931), had developed quite a deep interest in microglia, writing his own paper in support of del Rio-Hortega's findings concerning the 'third element' when he was just 22 years old (Costero 1925). A scholarship in 1929 enabled him to travel to Germany, to the Paul Ehrlich Institute in Frankfurt, where he trained in tissue culture techniques (1929–1930) under Drs Caspari and Vollmar (Palmero 2005). There, he made the first cinefilm in the world documenting the experimental behaviour of microglia isolated in culture, using time-lapse cinemicrography (Fernández-Guardiola 1997). This material formed the basis of Costero's first manuscript demonstrating the experimental behaviour of microglia under normal and pathological conditions (Costero 1930a), accepted for publication by Dr. Wilhelm Kolle, Director of the Paul Ehrlich Institute (Fernández-Guardiola 1997). Costero became the first to thoroughly investigate the *dynamic* nature of these enigmatic cells 'live' under cell culture conditions. His cinefilm was apparently shown by del Río-Hortega on his visits to various universities and institutions around Europe to much acclaim and played an important part in establishing del Río-Hortega's concepts concerning the microglia (Fernández-Guardiola 1997). Costero returned to Spain, where he wrote two further papers on the cultivation and study of microglia 'in vitro' (Costero 1930b, c). Costero's findings at this time are particularly noteworthy, since they reflect a conscious anticipation of the motile behaviour of microglia that was to be reported 75 years later using in vivo imaging techniques in transgenic mice with fluorescent protein-expressing microglia (Davalos et al. 2005; Nimmerjahn et al. 2005).

Del Río-Hortega referred to Costero's work in his treatise on microglia, published in the second volume of Penfield's 'Cytology and Cellular Pathology of the Nervous System' in 1932: "*Recent investigations by Costero (*1930a, b, c*) clearly demonstrate the existence of ameboid movements and the capacity of migration of the microglia. This investigator has succeeded in obtaining cultures of microglia in vitro and he shows that the motile elements seen by other investigators in explants of nervous tissue are microgliocytes. In cultures, the microgliocytes show shapes ranging from the globose types characteristic of the fat granule cells to the bipolar and branched forms. The former occur near the cultivated tissue, while the bipolar and branched cells are found in the plasma. The observations of Costero on the activity of the microglia are conclusive since all stages, from the moment the microgliocytes enter into motion and send off prolongations with pseudopodia (lateral spines) to phagocytosis of erythrocytes, tissue detritus and carmine granules, are present in the cultures. In explants of nervous tissue and in pure cultures of microglia we find a strong argument in favour of its inclusion among the elements of the reticulo-endothelial system*." (del Río-Hortega 1932, p. 518).

The concept that microglia belonged to the 'reticulo-endothelial system' (of Aschoff 1924), later renamed the 'mononuclear phagocyte system' (see Rezaie and Male 2002a), was established by the 1940s. Del Río-Hortega had referred to his own work and that of his students Jiménez de Asúa (del Río-Hortega and Jiménez de Asúa 1921; Jiménez de Asúa 1927) and Isaac Costero (1930a, b, 1931) who made significant advances to his own studies on two fronts: (1), the concept that microglia belonged to the reticulo-endothelial system proposed by Aschoff (Jiménez de Asúa 1927; Aschoff 1924; Russell 1929; Visintini 1931; Belezky 1931, 1932; Bolsi 1936; and others); and (2) the morphological characteristics, motile behaviour, and phagocytic functions of microglia cultivated in vitro (Costero 1930a, b, 1931). Henry Dunning and his colleagues Lewis Stevenson and Jacob Furth gave a clear account of the evidence supporting Jiménez de Asúa's proposal. These were based on properties (similarities in morphology, uptake of colloidal dyes, behaviour and functions in vitro) that microglia shared with mononuclear phagocytes resident within other tissues, for example the liver, spleen and kidney (which were called 'histiocytes' at the time; see Maximow 1928; Dunning and Stevenson 1934; Dunning and Furth 1935). These investigators also referred to a study carried out by Lebowich (1934), who observed that microglia were capable of phagocytosing bacteria. Von Mihálik had concluded that there were no differences between macrophages cultured from the brain, liver, and subcutaneous tissue (von Mihálik 1935). Further studies on the ontogeny of microglia and their relationship with the reticulo-endothelial system were carried out by Belezky (1931, 1932), von Sántha (1932), Juba (1933, 1934), von Sántha and Juba (1933), and by Dougherty (1944) (see also Cardona 1937).

John Kershman, Assistant Professor of Neurology at McGill University, was among the first to chart the development and differentiation of microglia within the human fetal nervous system, significantly extending the initial studies presented by Rydberg (1932), von Sántha (1932), and Juba (1933), in his exemplary study published in 1939 (Kershman 1939). The detail and accuracy of the descriptions provided make this one of the key landmark papers on microglia. Penfield's influence is acknowledged by Kershman therein. He separately confirmed observations in mammals by a number of previous researchers, showing that microglia appear within the human brain, with the first evidence of vascularisation (Kershman 1939).

From 1941 to 1960, only 14 articles appear listed on NCBI PubMed (using 'microglia' as the search term entered under 'all fields'). These are studies by Vazquez-Lopez (1942) on microglia in the neurohypophysis, Kurobane (1950) on microglia in gliomas, Meo (1950) on cerebral changes in leukemia, Costero (1951, 1952) on microglia in rheumatism, Mazzi (1952) who looked at the brain in teleosts, Herrera (1953) who examined microglial genesis, Imamura (1954) focusing on gliomas, Field (1955) who focused on the development of microglia and the influence of cortisone, Pickering and Vogel (1956) examining demyelinating lesions, Jufe (1957) on rabies, Koenig (1958) addressing nucleic acid and protein turnover, Wolman (1958) who examined the mechanism of selective impregnation, and Tsypkin (1959) examining microglia in senile dementia and their role in the structural genesis of senile plaques. There are other articles that are not currently listed on PubMed, for example Dougherty (1944) and Bullo (1945). So it is clear that the above list does not represent the complete picture, but it does provide an indication of the relative paucity of research on microglia that followed the death of del Río-Hortega in 1945, and for more than a decade after the end of the second world war. Glees' publication in 1953 provides a good point of reference on the neuroglia up to that period (Glees 1953).

We end this section with the words of del Río-Hortega. Summing up two decades of work on microglia in a lecture delivered at Oxford University in 1938, del Río-Hortega provided a concise account of the state of knowledge with respect to the origins, forms, and functions of the microglia up to that period. This was published in The Lancet the following year (del Río-Hortega 1939). In his conclusion he noted: "*Nerve tissue contains a type of cell with properties corresponding to those of the reticulo-endothelial system. The chief function of these cells (the microglia) is the phagocytosis of waste products, and, if need be, of red blood-cells….. Even when they are in apparent rest, microglial cells remain capable of migration, and under pathological conditions they are mobilised to undertake phagocytic activity. Phagocytosis in nerve centres is a specific function of microglia; despite opinions to the contrary, neither the astrocytes nor the oligodendrocytes take part in it. Microglia cells take up the products of disintegration of the nerve tissue, digesting and elaborating them, and making them finally disappear. These ideas have been developed in my papers from 1919 to 1921, and though they have been discussed in many subsequent papers by other workers, almost nothing new has been added to our knowledge of this subject. Today these ideas are widely accepted.*" (del Río-Hortega 1939, p. 1026).

At the time this statement was made, it was perhaps unlikely to envisage that it would be decades before research on microglia would 'take off' once again. Yet the modern era is indebted to these pioneering achievements, and progress has been made through building on these earlier discoveries.

2.3 The 'Renaissance' Period

Much of the credit for the revival of interest in microglia and our present-day knowledge can be attributed to Georg Kreutzberg who was studying the mechanisms of axonal regeneration in the 1960s (Kreutzberg 1963, 1967, 1968, 1972, 1996; Kreutzberg and Barron 1978), and those whom he inspired and influenced, including Manuel Graeber, Wolfgang Streit, Helmut Kettenmann, and Richard Banati, among others. Kreutzberg began investigating the facial nerve transection model through a series of light and electron microscopic and autoradiographic studies (see Moran and Graeber 2004 for a review on this model). He observed that excision of the facial nerve (distal to the facial motor nucleus within the brainstem) produced characteristic responses in astrocytes and microglia associated with motor neuronal cell bodies, which enabled the peripheral nerve to regenerate (Blinzinger and Kreutzberg 1968; see also the work of Cammermeyer 1965a, b). Blinzinger and Kreutzberg first described the phenomenon of 'synaptic stripping' (Blinzinger and Kreutzberg 1968), whereby microglia in the facial motor nucleus initially displaced synapses from traumatised neuronal somata, followed by ensheathment of these cell bodies by astrocytes, preventing further synaptic input, allowing the injured neurons to conserve energy, recuperate, and the peripheral nerve to regenerate over time. Until then, it had been generally assumed that the sole function of microglia was to carry out phagocytosis and 'clearance' (e.g. of red blood corpuscles, dead cells, or pathogens like bacteria), and the expectation was that these cells would simply engulf, assimilate, and thus remove the damaged neuronal cell bodies. The novel findings therefore came both as a surprise and a revelation—this was the first evidence for a 'neuroprotective' function of microglia; their involvement in neuronal regeneration. While the discovery of synaptic stripping was just one of the seminal contributions of this school, it was a fundamental addition to the known spectrum of microglia functions. Synaptic 'nursing' and 'pruning' are essential elements of a portfolio which critically involves microglia in processes of CNS development, maintenance, and recovery (Kettenmann et al. 2013) (for further reading, see Chap. 9).

Later (in the 1980s and 1990s), Kreutzberg and his colleagues showed, using this model, that the microglial response within the CNS was tailored to the extent of consequent 'damage' inflicted to the neuron (Kreutzberg and Barron 1978; Graeber and Streit 1990a; Graeber et al. 1988, 1993, 1998; Streit et al. 1989; Haas et al. 1993). Two key insights were gleaned from these studies: (1), the blood–brain barrier within the brainstem nucleus remained intact in this model, so that both 'synaptic stripping' and phagocytosis were functions carried out by microglia alone, without recruitment of mononuclear phagocytes from the periphery, and (2), microglia were shown to be capable of proliferating following axotomy, and therefore had the capacity to 'self-renew' their population without recruitment of progenitors from the blood (Graeber et al. 1988, 1993, 1998; Graeber 1993; Haas et al. 1993; Moran and Graeber 2004; Ajami et al. 2007). More recently, it has been argued that 'synaptic stripping' cannot be extended more generally to models of chronic neurodegeneration, and that degeneration of synapses and envelopment of a degenerating terminal are neuron autonomous events in which microglial involvement is merely 'guilt by association' (Perry and O'Connor 2010). Nevertheless, there is growing evidence that microglia are indeed able to remove synaptic terminals, and in this respect contribute to remodelling and plasticity of the nervous system in development, with implications for neurodevelopmental disorders, even including autism and Rett syndrome (Boggio et al. 2010; Tremblay et al. 2010; Paolicelli et al. 2011; Paolicelli and Gross 2011; Derecki et al. 2012; Hughes 2012; Schafer et al. 2012).

2.4 The Period of 'Awakening'

The 1980s saw an 'awakening' in the understanding of microglial 'immune' functions, the processes of neuroinflammation, and the concept of 'innate immunity' inspiring large-scale research on microglia in vitro and in vivo, with a particular focus on Alzheimer's disease. Neuroglial cells moved into the spotlight, and immunological concepts and approaches fuelled research in the neurosciences. It was at this time also that the myeloid nature of microglia gained broader attention, supported and influenced by immunological work on mononuclear phagocytes, and a distinction between perivascular macrophages and perivascular microglia was clarified.

The morphological transformation and phagocytic function of microglia, and their role in glial scar formation (in response to stab injury, for example) had already gained widespread recognition following del Rio-Hortega's publication in Penfield's 'Cytology and Cellular Pathology of the Nervous System' (reprinted by Hafner in 1965) (del Río-Hortega 1932), and in his Lancet paper (del Río-Hortega 1939). However, evidence for the concept of microglia as 'immunocompetent' cells that were capable, like their macrophage counterparts within other tissues, of presenting antigen to T lymphocytes (through expression of Major Histocompatibility Complex (MHC) class II antigen), of secreting pro- and anti-inflammatory cytokines, as well as cytotoxic factors, of signalling to cerebral endothelium, and initiating an immune response within the CNS, was first drawn together in the 1980s by Streit, Graeber, and Kreutzberg (Streit et al. 1988). These authors went on to propose that microglia represent 'networks' of immune surveillance within the CNS (Graeber and Streit 1990a; Gehrmann et al. 1993, 1995). In 1996, Kreutzberg coined the term 'pathological sensor' to describe the inherent activity of the microglia, as occurring in a 'heightened' or 'alert' state, monitoring the environment, and ready to initiate a rapid response to insult or injury. The article is still one of the most cited in the field. The view of microglia as the 'innate immune system' of the CNS later led to the concept of 'neuroinflammation' (Streit et al. 2004a; Mrak 2009) in which microglia were the key cellular mediators in acute and chronic inflammatory responses associated with infectious (e.g. human immunodeficiency virus (HIV)) and chronic neurodegenerative diseases such as multiple sclerosis, Alzheimer's disease, and prion diseases (see McGeer and McGeer 2011 for a history of innate immunity in neurodegenerative disorders). Microglia were found to play roles in both innate (or 'cell-mediated') and adaptive immunity within the CNS.

The dichotomy in the nature of microglial functions (neurotoxicity versus neuroprotection) is now well-recognised (Czeh et al. 2011; Zhang et al. 2011), and reference to microglia functioning in this regard as 'a double-edged sword' and a 'pathological sensor' can be traced back to Georg Kreutzberg (Kreutzberg 1995, 1996). The concept of 'neuroinflammation', however, is being re-examined in the light of the continuum of microglial activation and multi-faceted responses in both healthy and pathological states (Hanisch and Kettenmann 2007; Graeber 2010; Graeber et al. 2011).

A separate focus of work that began during this period was a revival in research into the origin and cell lineage of microglia. Eng Ang Ling and co-workers, pursuing the autoradiographic work of Imamoto and Leblond (1978) and that of Kitamura (1969, 1973), were particularly interested in defining the origins of microglia, which had remained contentious for decades (Fujita and Kitamura 1975; Ling 1976a, b;

Boya et al. 1979; Fujita et al. 1981). Ling and colleagues described two forms of microglia in the neonatal and postnatal rodent brain—ameboid cells within white matter tracts, with special reference to the corpus callosum, and ramified cells in the gray matter (Ling and Tan 1974; Ling 1979; Leong and Ling 1992). They observed that ameboid cells were taking up carbon particles injected intravenously, and that a subpopulation of these cells could transform into the resident (ramified) microglial population in adult animals (Ling 1979; Kaur et al. 1985; Wu et al. 1992). The literature on the origin of microglia, namely their relationship with the mononuclear phagocyte system and the initial view that microglia were derived from circulating blood monocytes, is summarised in the review by Kaur et al. published in 2001 (for related reviews, see also Ling and Wong 1993; Theele and Streit 1993; Cuadros and Navascués 1998, 2001; Rezaie 2003, 2007; Rezaie et al. 1999; Rezaie and Male 2002a, b).

Bone marrow (BM) reconstitution experiments by Hickey and Kimura in the late 1980s, and later by a number of other investigators, showed that donor BM-derived stem cells transferred to irradiated hosts (rodents) could subsequently be traced as 'microglia' within the brain parenchyma (Hickey and Kimura 1988; Lassmann et al. 1993; Lassmann and Hickey 1993; Flügel et al. 2001; Priller et al. 2001; Simard and Rivest 2004). Likewise gender-mismatched BM-transplant recipients were found to have a small number of donor-derived cells located within the human brain, mainly around vasculature (perivascular positions), at postmortem (Unger et al. 1993). However, the fact that BM-derived elements primarily took up residence within the brain at perivascular positions and were largely absent in the parenchyma proper (Hickey and Kimura 1988; Unger et al. 1993) was intriguing. Observations that these cells were able to differentiate into 'microglia-like cells' with immunological properties resurfaced questions concerning the relationship between perivascular macrophages and parenchymal microglia that had been posed two decades earlier (see Kitamura and Hattori 1972). Manuel Graeber, Wolfgang Streit and colleagues re-examined this relationship, offered clarification for differences between perivascular cells (perivascular macrophages) and perivascular microglia, and demonstrated that perivascular cells were preferentially located within the basement membrane (glia limitans) surrounding cerebral vessels (Graeber et al. 1992; Graeber and Streit 1990b; Streit and Graeber 1993). While perivascular cells appeared to be repopulated by BM-derived progenitors, with a fairly consistent high turnover, and served as immunoregulatory 'go-betweens' connecting the nervous system and the peripheral immune system (Williams et al. 2001), microglia in contrast were capable of self-renewal with little need for recruitment from the BM, under non-pathological conditions (see Ajami et al. 2007; Mildner et al. 2007). In fact, a criticism levied against irradiation experiments was that the consequent damage to (opening of) the blood–brain barrier and additional impact on CNS cells, facilitated 'engraftment' of donor BM progenitors, whereas an intact 'healthy' barrier and unchallenged tissue would restrict this (see Ajami et al. 2007; Mildner et al. 2007; Ginhoux et al. 2010).

2.5 The Turn of the Twenty-first Century and the New Millennium

Around the turn of the century, genetic models in combination with imaging techniques offered new options and precipitated research activity on a massive scale, focusing on responses, reactive phenotype diversity, origin, turnover and replenishment, and interactions with both resident CNS and infiltrating immune cells. Microglia were no longer considered solely as 'destructive' phagocytes. The first house-keeping functions were revealed, ending the concept of a 'resting' or 'dormant' cell. Contributions to normal development (e.g. synaptic 'pruning'/remodeling) as well as to neurological diseases were further unraveled, venturing into neuropsychiatric disorders (Bilbo and Schwarz 2009; Chen et al. 2010). Neuron–glial interactions, the roles of microglia in autoimmunity, chronic pain, in brain tumours and aging were demonstrated. Regional specifications of microglia throughout the CNS and long-term outcomes of microglial challenges began to be examined.

Dana Giulian and colleagues were among the first to culture microglia from rodent brains. They investigated the properties of these cells during the mid-1980s (Giulian 1987; Giulian and Baker 1985, 1986; Giulian and Ingeman 1988; Giulian et al. 1988). Studies soon focused on the cultivation and characterisation of adult and fetal human microglia (Hayes et al. 1988; Grenier et al. 1989; Lee et al. 1992; Williams et al. 1992; Lauro et al. 1995). Cell cultures paved the way for extensive characterisation of microglial morphology and phenotype (Grenier et al. 1989; Rieske et al. 1989; Sedgwick et al. 1991; Giulian et al. 1995), identification of their secretory profiles, cellular responses, and interactions with other CNS cell types (e.g. Giulian 1987; Suzumura et al. 1987; Bocchini et al. 1988; Raivich et al. 1993; Sawada et al. 1993; Hanisch 2002; Pocock and Kettenmann 2007). This led to a greater understanding of the neurotoxic and neurotrophic repertoire of these cells (Banati et al. 1993; Banati and Graeber 1994; Nakajima and Kohsaka 1993a, Nakajima and Kohsaka 1993b, 1998, 2001, 2004; Giulian et al. 1994; Hanisch 2002), among these their expression of growth factors and mitogen receptors (Raivich et al. 1993), neurotransmitter receptors (Pocock and Kettenmann 2007), and cytokine and receptor profiles (Hanisch 2002). Microglia were found to ramify in response to a number of factors, most prominently trophic factors released by astrocytes in cocultures (Suzumura et al. 1990, 1991). The proteomic and genomic expression patterns of microglia have been profiled (Duke et al. 2004; Moran et al. 2004; Glanzer et al. 2007; Parakalan et al. 2012; Veremeyko et al. 2012) and their electrophysiological properties examined (Kettenmann et al. 1993; Draheim et al. 1999; Prinz et al. 1999), including those of the ramified variety maintained in vitro (Eder et al. 1999) and in tissue slices (Boucsein et al. 2000). Helmut Kettenmann and colleagues first drew attention to the fact that cultured microglia possess a distinct pattern of membrane ion channels (including an inward-rectifying potassium channel) that distinguished these cells from other mononuclear phagocytes, including peritoneal macrophages, a unique property that was shared with a small subpopulation of progenitors located within the bone marrow (Kettenmann et al. 1990; Banati et al. 1991; Kettenmann 1994).

Attention soon turned towards deciphering the motile nature of these cells, much as it had done earlier with Costero's studies (Costero 1930a, b, 1931). This time, however, with more sophisticated (light, confocal, and time-lapse) microscopic imaging techniques. Microglial cell dynamics and morphological 'plasticity' were investigated in slice cultures (Hailer et al. 1996; Dailey and Waite 1999; Stence et al. 2001; Grossmann et al. 2002; Petersen and Dailey 2004; Kurpius et al. 2006; Grinberg et al. 2011), in isolated cells as well as co-cultures (Ward et al. 1991; Glenn et al. 1992; Rezaie et al. 2002; Ohsawa and Kohsaka 2011), and 'real-time' microglial activation could now be examined (e.g. see Rangroo Thrane et al. 2012). A number of studies also began to examine the behaviour of ramified cells (Ward et al. 1991; Booth and Thomas 1991; Kloss et al. 1997; Eder et al. 1999; Rosenstiel et al. 2001; Rezaie et al. 2002). The morphological plasticity of microglia was demonstrated even in their fully ramified forms and, with the later confirmation of these findings in vivo (Davalos et al. 2005; Nimmerjahn et al. 2005), it was clear that microglia were highly dynamic cells capable of motility, constantly on the move even at 'rest'. In this respect, microglia could no longer be regarded as 'resting' in their ramified forms in the normal, healthy brain (see Rezaie 2007). These observations impacted also long-held definitions of the 'activation' of these cells (Hanisch and Kettenmann 2007), which were in fact not 'step-wise' but occurring along a dynamic 'continuum', and offered further support for their rapid morphological and phenotypic response to insult or injury. These features—dynamic cell motility and plasticity—remain unique to this cell type within the CNS.

Areas that had until recently remained relatively less well-explored included the functional role(s) played by microglia in the normal, healthy adult brain, and during development. It was clear that the functional status of microglia depended on their immediate environment (Neumann 2001). Signalling from, and direct contact with, neurons and astrocytes are essential to their maintenance within the healthy nervous system (e.g. see Biber et al. 2007). Neuronal fractalkine (Harrison et al. 1998; Maciejewski-Lenoir et al. 1999) and CD200 as well as their corresponding receptors present on microglia (Hoek et al. 2000), and normal electrical activity of neurons, were discovered to maintain microglia in their 'resting' states—the number of ligand-receptor pairs and conditions with activity control still being incomplete (Kettenmann et al. 2011, 2013; Hanisch 2013a, b, c) (for further reading, see Chaps. 3 and 9). Diffusible factors (including the trophic factors MCSF (macrophage colony stimulating factor) and GMCSF (granulocyte-macrophage colony stimulating factor)), released from astrocytes, and direct contact with these cells were likewise found to be important for maintaining microglia in their ramified (presumed 'non-activated') states (Suzumura et al. 1990, 1991; Liu et al. 1994; Sievers et al. 1994; Fujita et al. 1996; Tanaka and Maeda 1996; Eder et al. 1999; Schilling et al. 2001; see Rezaie and Male 2002b for review). Discussion of potential roles of these cells in the healthy brain, including in development, was the subject of a recent symposium (Tremblay et al. 2011).

The normal 'housekeeping' activities of microglia in the healthy nervous system (see Kettenmann 2007; Raivich 2005), with roles in surveillance (policing/patrolling) and phagocytosis (scavenging/clearance), are now firmly established

(Nimmerjahn et al. 2005; Davalos et al. 2005). Key insight into their function in maintaining a healthy nervous system was derived through their expression of functional TREM-2 (triggering receptor expressed on myeloid cells-2) (Schmid et al. 2002), required for the regular clearance of 'debris' within the nervous system (including apoptotic neurons) without inducing an inflammatory response, via a process termed 'controlled phagocytosis' (Neumann and Takahashi 2007; Takahashi et al. 2005, 2007). More recent evidence supports the existence of subsets of microglial cells, with distinct phenotypes (and potential immunological diversity), whose functions may be more diverse than originally thought, and further specialised to support the brain regions they subserve (de Haas et al. 2008; Olah et al. 2011, 2012; Scheffel et al. 2012; Hanisch 2013b, c). For in-depth reviews on our progressive understanding of the physiological functions of microglial cells, see Färber and Kettenmann (2005), Kettenmann (2006), Hanisch and Kettenmann (2007), Ransohoff and Perry (2009), and Kettenmann et al. (2011) (for further reading, see Chap. 3).

Views on the origins and lineage of fetal and adult mononuclear phagocytes were also changing at the turn of the twenty-first century, with emerging concepts for differences in lineage, form, and functions between fetal and adult macrophages (see Chan et al. 2007). Tissue-resident mononuclear phagocytes in particular were now considered to belong to the fetal macrophage lineage which developed independent of monocyte-derived macrophages (Shepard and Zon 2000). A number of studies had shown that microglia were already present during embryonic and fetal stages in the rodent and human CNS (see Dalmau et al. 2003; Rezaie et al. 1997, 1999; Rezaie 2003; Chan et al. 2007), which precluded the BM as the main source of microglial progenitors. The yolk sac was proposed as the 'developmental' source of microglial progenitors (Alliot et al. 1999; see Chan et al. 2007), later confirmed by Miriam Merad's group using fate-mapping experiments (Ginhoux et al. 2010). The fact that microglia were specifically of myeloid lineage had already been indicated, given their expression of PU.1 (Walton et al. 2000), a transcription factor essential for the development of all myeloid cells including monocyte-derived macrophages (Greaves and Gordon 2002; see Chan et al. 2007). Frederic Geissmann's group (Geissmann et al. 2010; Gomez Perdiguero et al. 2013; Schulz et al. 2012) confirmed that microglia, Langerhans cells of the skin and Kupffer cells of the liver develop from Myb-independent, FLT3-independent, but PU.1-dependent precursors that express CSF-1 receptors. Wang et al. (2012) proposed interleukin 34 (IL-34) as the tissue-restricted ligand of CSF-1 receptors, required for the development of both Langerhans cells and of microglia. Most recently, additional factors and steps for microgliogenesis have been added, confirming PU.1 and introducing also interferon regulatory factor 8 (Irf8) (Kierdorf et al. 2013). For recent, relevant reviews on the origin and lineage of microglia see Saijo and Glass (2011), Greter and Merad (2013), and Ginhoux et al. (2013).

2.6 The Technical Bases

Del Rio-Hortega's ammoniacal silver method and the 'improved' methods of Penfield (1928), Penfield and Cone (1937), and others (e.g. McCarter 1940; Fain 1963; Gallyas 1963; Naoumenko and Feigin 1963; Malik 1964) were still being used, albeit selectively within histopathology laboratories in Europe and elsewhere around the world in the 1950s, 1960s, and 1970s, to detect microglia and to examine their responses in various pathological conditions. The techniques proved to be too capricious for routine use, however, with varying results obtained in different laboratories (and even in the same labs, on different samples). Electron microscopy was in its heyday during the 1960s and 1970s, as a tool to investigate the cellular cytoarchitecture of the brain, and a number of investigators turned their attention to examining the ultrastructural morphology of microglia in more detail (e.g. Blinzinger and Hager 1962; Yasuzumi et al. 1964; Hager 1968; Mori and Leblond 1969; Blakemore 1975; Boya 1975). There had been some who questioned the existence of these cells, given difficulties in identifying them (see Vaughn and Peters 1968; Kitamura 1969; Cammermeyer 1970). Cammermeyer (1970) reviewed this debate in greater depth and provided a good coverage of the state of knowledge (and controversies) during this period.

Silver impregnation methods (Gallyas 1963; Naoumenko and Feigin 1963) and morphological studies were replaced by histochemical methods (Mladenov and Gerebtzoff 1967; Kraśnicka and Renhawek 1969, 1970; Ibrahim et al. 1974) as a new tool for investigating mononuclear phagocytes (including lysosomal enzymes found in macrophages) and microglia (see Rezaie and Male 2002a). Enzyme histochemical methods, however, much like the prior silver impregnation techniques, also proved to lack specificity for microglia. They were technically demanding procedures, requiring tissues to be prepared in a specific way, combined with skilled training and a 'good eye' to be able to reliably discriminate cells (Esiri and Booss 1984).

Histochemical staining with plant lectins introduced in the late 1970s and 1980s (Streit and Kreutzberg 1987) was soon replaced by the use of antibodies and immunohistochemical methods to detect surface and intracellular antigens. These enabled the cellular phenotype and distribution of microglia and other populations of mononuclear phagocytes associated with the CNS (e.g. perivascular cells, supraependymal macrophages, and Kolmer cells) to be examined for the first time by Hugh Perry, Siamon Gordon, and others (Oehmichen et al. 1979; Hume et al. 1983; Perry et al. 1985; Perry and Gordon 1988; Imamura et al. 1990; Lawson et al. 1990, 1992). Factors that modulated the microglial phenotype could be examined (Perry 1994). Complex immunophenotypic characterisation was proposed as a way to differentiate between so-called 'resting' and 'activated' cells, based on constitutive 'high' or 'low' levels of expression of CD45 and CD11b/c antigens by microglia (Sedgwick et al. 1991; Ford et al. 1995). Subsets of activated microglia continue to be selectively discriminated by their cellular phenotype, e.g. differential expression of MHC class II, CD68, CD11b/c and CD45 using light microscopy and more sophisticated fluorescence and confocal microscopic imaging techniques. A more reliable method for detecting all forms of microglia *in situ* and in vitro was developed in Shinichi Kohsaka's lab by Imai and colleagues raising a polyclonal antibody against a novel protein called 'Iba1' (ionised calcium adapter-binding molecule-1) (Imai et al. 1996; Ito et al. 1998). Iba1 is encoded by a gene located within a segment of the MHC class III region, between the Bat2 and tumour necrosis factor (TNF) alpha genes (Imai et al. 1996), and expressed by mononuclear phagocytes, including microglia. Immunohistochemistry with Iba1 antibody detected all morphological and functional forms of microglia in both rodent and human preparations allowing direct comparisons to be made in translational studies, and between species (Ito et al. 1998; Hirasawa et al. 2005). Although a considerable step forward, this tool still does not allow recently recruited/infiltrating macrophages to be discriminated from CNS-resident microglia (see Kettenmann et al. 2011 for an overview of methods for microglia detection).

Over the last two decades advances in imaging techniques (light, fluorescence, confocal, time-lapse, and electron microscopy), combined with immunoassays (e.g. ELISA and Elispot), electrophysiological methods, proteomic and molecular techniques (e.g. 2D gel electrophoresis, in situ hybridisation, RT-PCR, Western blotting), laser capture microdissection, and gene profiling methods, have rapidly become powerful 'everyday' tools to examine the physiological functions of microglia under varied conditions, from living cells (single isolated cells, microglial cocultures, slice culture preparations) to histological preparations (frozen and fixed tissue sections) obtained at postmortem. Each and every study has had its limitations, the greatest of which has always been the unanswered query: To what extent can the multitude of observations and experimental findings in vitro, ex-vivo, and in situ (in tissue slices for example) be extrapolated to microglial functions in vivo? This has been the ultimate drive—understanding microglial functioning within the intact CNS environment, and perhaps the most exciting period in the history of these cells lies ahead of us, but there have already been fundamental advances on several fronts over the last decade; each allowing us a glimpse of the 'real world' of the microglia (inside the living and intact nervous system).

The first is in a clinical setting: imaging 'activated' microglia using positron emission tomographic (PET) scanning of the human brain—a methodological breakthrough developed by Richard Banati and coworkers, which relies on intravenous injection of a radiolabelled ligand (PK11195) that selectively binds to peripheral benzodiazepine-binding sites, found on all mononuclear phagocytes, including microglia (Banati et al. 1997), and subsequent PET imaging of the brain to visualise the bound ligand (Banati 2002, 2003; Banati et al. 2000). Using this procedure, the in vivo imaging of microglia was first demonstrated, and quantitated, in patients with multiple sclerosis (in which microglial activation is associated with lesions or 'plaques' located within the white matter), as a measure of disease severity (Banati et al. 2000). Microglial activation has subsequently been visualised in a number of neurodegenerative conditions (Cagnin et al. 2001, 2002, 2007). Although the selectivity and specificity of the technique have been questioned (peripheral benzodiazepine receptor/18KDa translocator protein ligands may not bind exclusively to microglia, but may additionally detect activated astrocytes—see Lavisse et al. 2012), use of the radioactive ligand may modify microglial functions and induce bystander effects on microglial activity (see Choi et al. 2011), and resident 'non-activated' microglia which have limited (or lack such) binding sites cannot be detected using this method, it is still a significant step forward in being able not only to identify, but to pinpoint and quantitate activated microglial involvement in situ in pathological disease progression. Newer and more selective ligands are being developed (see for example, Imaizumi et al. 2008; Politis et al. 2012). These imaging techniques have important value not only for charting the clinical progression and anatomical localisation of disease (which could act as an aid to diagnosis), but also in determining the clinical efficacy and/or responses to drug treatments aimed at slowing down or arresting disease progression, including those that target and modify microglial functions directly (Aldskogius 2001; Skaper 2011; Harry and Kraft 2012).

The second is within an experimental domain: 'real-time' in vivo imaging of microglia in transgenic and knock-in mice engineered to express fluorescent mononuclear phagocytes (for example, by tagging a fluorophore such as enhanced Green Fluorescent Protein (e-GFP) to the fractalkine receptor gene or Iba1 gene), and two-photon fluorescence microscopy (Davalos et al. 2005; Nimmerjahn 2012; Nimmerjahn et al. 2005) (for further reading, see Chap. 4). The combination of genetic engineering and fluorescence microscopy has allowed microglia within the upper 100–200 μm of the mouse cerebral cortex (cortical layers 1, 2 and upper layer 3) to be visualised in vivo and in 3D, through an 'optical window' or thinned skull, and therefore in their intact and undisturbed environment, for the first time (Nimmerjahn et al. 2005). These studies have confirmed the resting motility of microglia and charted and confirmed the characteristic microglial responses to local brain injury in vivo (chemotaxis, morphological transformation, phagocytosis), including localised responses to laser-induced damage to cerebral microvessels (Davalos et al. 2005). This combined technology (genetic engineering and in vivo imaging) promises to offer further insights into the behaviour of microglia, the dynamic interaction between microglia and synapses being one such contemporary focus, in the developing and aged brain, and in a variety of disease models. For example, crossing mice that express green fluorescent protein tagged to the fractalkine receptor (CX3CR1-GFP mice) with Thy1-YFP mice (expressing yellow fluorescent protein in neurons) has already enabled a more discrete visualisation of microglia interactions with dendritic spines directly in vivo (see Tremblay 2011; Tremblay et al. 2010). With new animal models becoming available for conditional targeting of microglia, more and more sophisticated options are available for visualisation, tracing, and manipulation of these cells (Goldmann et al. 2013; Yona et al. 2013). The combined application of modern technologies and genetic approaches is now a significant driver pushing research forward (Prinz et al. 2011).

2.7 Looking to the Future

Where will the field move from here? While much has been learned concerning the form and functions of microglia from cell culture and tissue slice work, the physiology of these cells (their cellular interactions with neurons, macroglia, and endothelium) in their natural environment in vivo still needs to be investigated. Understanding the versatility of these cells and dynamic control over their phenotype in health and in disease, facilitating their phagocytic capabilities to clear senile plaques in Alzheimer's disease and modulating their reactive functions in other instances (for example in brain tumour progression and in chronic pain), their roles in development, potential for engraftment, conditional gene expression, and as therapeutic and pharmacological targets within the CNS, all remain to be explored. Looking back over more than a century of work (see Fig. 2.1), it is clear that these are exciting times to be involved in research on microglia. The next epoch in the history of microglia research has just begun.

References

Achúcarro N (1908) Sur la formation de cellules à bâttonnet (Stäbchenzellen) et d'autres elements similaires dans le système nerveux central. Trab Lab Invest Biol Univ Madrid 6:97–122 [In French]

Achúcarro N (1909) Cellules allongées et Stäbchenzellen, cellules névrologique et cellules granulo-adipeuses à la corne d'Ammon du lapin. Trab Lab Invest Biol Univ Madrid 7:201–215 [In French]

Achúcarro N (1910) Algunos datos relativos a la naturaleza de las células en bastoncito de la corteza cerebral humana obtenidos con el método de Cajal. Trab Lab Invest Biol Univ Madrid 8:8 [In Spanish]

Achúcarro N (1911) Neuroglía y elementos intersticiales patológicos del cerebro impregnados por los métodos de reducción de la plata o por sus modificaciones. Trab Lab Invest Biol Univ Madrid 9:161–179 [In Spanish]

Ajami B, Bennett JL, Krieger C, Tetzlaff W, Rossi FM (2007) Local self-renewal can sustain CNS microglia maintenance and function throughout adult life. Nat Neurosci 10:1538–1543

Aldskogius H (2001) Regulation of microglia—potential drug targets in the CNS. Expert Opin Ther Targets 5:655–668

Alliot F, Godin I, Pessac B (1999) Microglia derive from progenitors, originating from the yolk sac, and which proliferate in the brain. Brain Res Dev Brain Res 117:145–152

Alpers BJ (1930) The reaction of the central nervous system to experimental urea intoxication. Arch Neurol Psychiatry 24:492–508

Andres-Barquin PJ (2002) Santiago Ramón y Cajal and the Spanish school of neurology. Lancet Neurol 1:445–452

Andrew W, Ashworth CT (1945) The Adendroglia. A new concept of the morphology and reactions of the smaller neuroglial cells. J Comp Neurol 82:101–127

Andriezen WL (1893) The neuroglia elements in the human brain. Br Med J 2:227–230

Aschoff L (1924) Reticulo-endothelial system. Lectures on pathology. Hoeber, New York, pp 1–33

Bailey P, Hiller G (1924) The interstitial tissues of the central nervous system: a review. J Nerv Ment Dis 59:337–361

Banati RB (2002) Visualising microglial activation in vivo. Glia 40:206–217

Banati RB (2003) Neuropathological imaging: in vivo detection of glial activation as a measure of disease and adaptive change in the brain. Br Med Bull 65:121–131

Banati RB, Hoppe D, Gottmann K, Kreutzberg GW, Kettenmann H (1991) A subpopulation of bone marrow-derived macrophage-like cells share a unique ion channel pattern with microglia. J Neurosci Res 30:593–600

Banati RB, Gehrmann J, Schubert P, Kreuzberg GW (1993) Cytotoxicity of microglia. Glia 7:111–8

Banati RB, Graeber MB (1994) Surveillance, intervention and cytotoxicity: is there a protective role of microglia Dev Neurosci 16:114–27

Banati RB, Myers R, Kreutzberg GW (1997) PK ('peripheral benzodiazepine')—binding sites in the CNS indicate early and discrete brain lesions: microautoradiographic detection of [3H] PK11195 binding to activated microglia. J Neurocytol 26:77–82

Banati RB, Newcombe J, Gunn RN, Cagnin A, Turkheimer F, Heppner F, Price G, Wegner F, Giovannoni G, Miller DH, Perkin GD, Smith T, Hewson AK, Bydder G, Kreutzberg GW, Jones T, Cuzner ML, Myers R (2000) The peripheral benzodiazepine binding site in the brain in multiple sclerosis: quantitative in vivo imaging of microglia as a measure of disease activity. Brain 123:2321–2337

Barron KD (1995) The microglial cell. A historical review. J Neurol Sci 134(Suppl):57–68

Belezky WK (1931) Die PyridinsodamethodezurImprägnation der Mesoglia (Hortegazellen, Oligodendroglia, Drenagzellen) und Reticuloendothel-zellen (fürGelatin und Celloidinschnitte). Virchows Arch Pathol Anat Physiol Klin Med 282:214–224 [In German]

Belezky WK (1932) Über die Histogenese der Mesoglia. Virchows Arch Pathol Anat Physiol Klin Med 284:295–311 [In German]

Bevan-Lewis W (1899) Textbook of mental diseases, 2nd edn. Charles Griffin and Company, London

Bevan-Lewis W (1906) The neuron theory—fatigue, rest and sleep. J Ment Sci 52:661–673

Biber K, Neumann H, Inoue K, Boddeke HW (2007) Neuronal 'on' and 'off' signals control microglia. Trends Neurosci 30:596–602

Bilbo SD, Schwarz JM (2009) Early-life programming of later-life brain and behavior: a critical role for the immune system. Front Behav Neurosci 3:14

Blakemore WF (1975) The ultrastructure of normal and reactive microglia. Acta Neuropathol Suppl 6:273–278

Blinzinger K, Hager H (1962) Electron microscope studies on the fine structure of resting and progressive microglia cells in the mammalian brain. Beitr Pathol Anat Allg Pathol 127: 173–192 [In German]

Blinzinger K, Kreutzberg H (1968) Displacement of synaptic terminals from regenerating motoneurons by microglial cells. Z Zellforsch Mikrosk Anat 85:145–157, (journal was later renamed Cell & Tissue Research) [In English]

Bocchini V, Artault JC, Rebel G, Dreyfus H, Massarelli R (1988) Phagocytosis of polystyrene latex beads by rat brain microglia cell cultures is increased by treatment with gangliosides. Dev Neurosci 10:270–276

Boggio EM, Lonetti G, Pizzorusso T, Giustetto M (2010) Synaptic determinants of Rett syndrome. Front Synaptic Neurosci 2:28, 11 p

Bolsi D (1936) Il problema della origine della microglia. Riv Patol Nerv Ment 48:1 [In Italian]

Booth PL, Thomas WF (1991) Evidence for motility and pinocytosis in ramified microglia in tissue culture. Brain Res 548:163–171

Boucsein C, Kettenmann H, Nolte C (2000) Electrophysiological properties of microglial cells in normal and pathologic rat brain slices. Eur J Neurosci 12:2049–2058

Boya J (1975) Contribution to the ultrastructural study of microglia in the cerebral cortex. Acta Anat 92:364–375

Boya J, Calvo J, Prado A (1979) The origin of microglial cells. J Anat 129:177–186

Bullo J (1945) Contribuciones de la escuela de Cajal sobre histopatologia de la neuroglia y microglia. Arch Histol Norm Patol 2:425–445 [In Spanish]

Bustamante M (1982) Nicolás Achúcarro: the man and his work. Arch Neurobiol 45:55–72 [In Spanish]
Cagnin A, Brooks DJ, Kennedy AM, Gunn RN, Myers R, Turkheimer FE, Jones T, Banati RB (2001) In vivo measurement of activated microglia in dementia. Lancet 358:461–467
Cagnin A, Gerhard A, Banati RB (2002) In vivo imaging of neuroinflammation. Eur Neuropsychopharmacol 12:581–586
Cagnin A, Kassiou M, Meikle SR, Banati RB (2007) Positron emission tomography imaging of neuroinflammation. Neurotherapeutics 4:443–452
Cajal SR (1913) Contribución al conocimiento de la neuroglía del cerebro humano. Trab Lab Invest Biol Univ Madrid 11:55–315 [In Spanish]
Cajal SR (1920a) Algunas consideraciones sobre la mesoglia de Robertson y Rio-Hortega. Trab Lab Invest Biol Univ Madrid 18:109–127 [In Spanish]
Cajal SR (1920b) Una modificación del método de Bielschowsky para la impregnación de la neuroglía común y mesoglía y algunos consejos acerca de la técnica del oro-sublimado. Trab Lab Invest Biol Univ Madrid 18:129–141 [In Spanish]
Cajal SR (1925) Contribution a la connaissance de la névroglie cérébrale et cérébelleuse dans la paralysie générale progressive. Avec quelques indications techniques sur l'imprégnation argentique du tissu nerveux pathologique. Trab Lab Invest Biol Univ Madrid 23:157–216 [In French]
Cajal SR, Marañón G, Gasset JO (1968) Nicolás Achúcarro (1880–1918): su vida y su obra. Taurus Ed, Madrid
Cammermeyer J (1965a) Juxtavascular karyokinesis and microglia cell proliferation during retrograde reaction in the nouse facial nucleus. Ergeb Anat Entwicklungsgesch 38:1–22 [In German]
Cammermeyer J (1965b) Histiocytes juxtavascular mitotic cells and microglia cells during retrograde changes in the facial nucleus of rabbits of varying age. Ergeb Anat Entwicklungsgesch 38:195–229 [In German]
Cammermeyer J (1970) The life history of the microglial cell: a light microscopic study. Neurosci Res 3:43–129
Cannon DF (1949) Explorer of the human brain: the life of Santiago Ramón y Cajal (1852–1934). H. Schuman, New York
Cardona F (1937) Sui quadri iso-patologici della microglia cerebrale umana. Riv Patol Nerv Ment 50:461–472 [In Italian]
Carmichael EA (1929) Microglia: an experimental study in rabbits after intracerebral injection of blood. J Neurol Psychopathol 9:209–216
Chan WY, Kohsaka S, Rezaie P (2007) The origin and cell lineage of microglia: new concepts. Brain Res Rev 53:344–354
Chen SK, Tvrdik P, Peden E, Cho S, Wu S, Spangrude G, Capecchi MR (2010) Hematopoietic origin of pathological grooming in Hoxb8 mutant mice. Cell 141:775–785
Choi J, Ifuku M, Noda M, Guilarte TR (2011) Translocator protein (18kDa)/peripheral benzodiazepine receptor specific ligands induce microglia functions consistent with an activated state. Glia 59:219–230
Cone W (1928) Acute pathologic changes in neuroglia and in microglia. Arch Neurol Psychiatry 20:34–72
Costero I (1925) Conocimientos actuales sobre el tercer elemento de los centros nerviosos. Clin Lab 29:386–406 [In Spanish]
Costero I (1930a) Studien an Mikrogliazellen (sogenanten Hortegazellen) in Gewebskulturen von Gehirn. Arb Staatinstit Exp Ther Frankf 23:27–37 [In German]
Costero I (1930b) Cultivo "in vitro" de microglía. Bol R Soc Esp Hist Nat 30:165–171 [In Spanish]
Costero I (1930c) Estudio del comportamiento de la microglía "in vitro". Mem Real Soc Esp Hist Nat 14:125–127 [In Spanish]
Costero I (1931) Experimenteller Nachweis der morphologischen und funktionellen Eigenschaften und des mesodermischen Charakters der Mikroglia. Z Gesamte Neurol Psychiatr 132:371–406 [In German]
Costero I (1951) Reaction of cerebral microglia in rheumatism. Gac Med Mex 81:49–70 [In Spanish]

Costero I (1952) Contribution to the study of reactions of microglia of the central nervous system in acute rheumatism. Arch Mal Coeur Vaiss 45:143–146 [In French]
Creutzfeldt HG, Metz A (1926) Űber Gestalt und Tätigkeit der Hortegazellen bei Pathologischen Vorgängen. Z Gesamte Neurol Psychiatr 106:18–53 [In German]
Cuadros MA, Navascués J (1998) The origin and differentiation of microglial cells during development. Prog Neurobiol 56:173–189
Cuadros MA, Navascués J (2001) Early origin and colonization of the developing central nervous system by microglial precursors. Prog Brain Res 132:1–9
Czeh M, Gressens P, Kaindl AM (2011) The yin and yang of microglia. Dev Neurosci 33:199–209
Dailey ME, Waite M (1999) Confocal imaging of microglial cell dynamics in hippocampal slice cultures. Methods 18:222–230
Dalmau I, Vela JM, González B, Finsen B, Castellano B (2003) Dynamics of microglia in the developing rat brain. J Comp Neurol 458:144–157
Davalos D, Grutzendler J, Yang G, Kim JV, Zuo Y, Jung S, Littman DR, Dustin ML, Gan WB (2005) ATP mediates rapid microglial response to local brain injury in vivo. Nat Neurosci 8:752–758
de Haas AH, Boddeke HW, Biber K (2008) Region-specific expression of immunoregulatory proteins on microglia in the healthy CNS. Glia 56:888–894
del Río-Hortega P (1918) Noticia de un nuevo fácil método para la coloración de la neuroglía y del tejido conectivo. Trab Lab Invest Biol Univ Madrid 15:367–368 [In Spanish]
del Río-Hortega P (1919) El "tercer elemento" de los centros nerviosos. I. La microglía en estado normal. II. Intervención de la microglía en los procesos patológicos (células en bastoncito y cuerpos gránulo-adiposos). III. Naturaleza probable de la microglía. Bol Soc Esp Biol 69–120 [In Spanish]
del Río-Hortega P (1919b) Poder fagocitario y movilidad de la microglía. Bol Soc Esp Biol 9: 154–166 [In Spanish]
del Río-Hortega P (1919c) Coloración rápida de tejidos normales y patológicos con carbonato de plata amoniacal. Bol Soc Esp Biol 9:7–14 [In Spanish]
del Río-Hortega P (1920a) Estudios sobre la neuroglía. La microglíay su transformación en células en bastoncito y cuerpos gránulo-adiposos. Trab Lab Invest Biol Univ Madrid 18:37–82 [In Spanish]
del Río-Hortega P (1920b) Coloración de tejidos normales y patológicos con carbonato de plata amoniacal. Trab Lab Invest Biol Univ Madrid 18 [In Spanish]
del Río-Hortega P (1921a) Histogénesis y evolucíon normal, éxodo y distribución regional de la microglía. Mem Real Soc Esp Hist Nat 11:213–268 [In Spanish]
del Río-Hortega P (1921b) La glía de escasas radiaciones (oligodendroglía). Bol Real Soc Esp Hist Nat 21:63–92 [In Spanish]
del Río-Hortega P (1932) Microglia. In: Penfield W (ed) Cytology and cellular pathology of the nervous system, vol 2. Hoeber, New York, pp 483–534
del Río-Hortega P (1939) The microglia. The Lancet 233:1023–1026
del Río-Hortega P (1943a) El método del carbonato argéntico. Revisión general de sus técnicas y aplicaciones en histología normal y patológica I. Arch Hist Normal Patol 1:165–206 [In Spanish]
del Río-Hortega P (1943b) El método del carbonato argéntico. Revisión general de sus técnicas y aplicaciones en histología normal y patológica II. Arch Hist Normal Patol 1:329–363 [In Spanish]
del Río-Hortega P (1944a) El método del carbonato argéntico. Revisión general de sus técnicas y aplicaciones en histología normal y patológica III. Arch Hist Normal Patol 2:231–244 [In Spanish]
del Río-Hortega P (1944b) El método del carbonato argéntico. Revisión general de sus técnicas y aplicaciones en histología normal y patológica IV. Arch Hist Normal Patol 2:577–604 [In Spanish]

del Río-Hortega P (1986) El Maestro y yo. Consejo Superior del Investigaciones Científicas (CSIC), Madrid, 1986 [In Spanish]
del Río-Hortega P, Jiménez de Asúa F (1921) Sobre la fagocitosis en los tumores y en otros procesos patológicos. Arch Cardiol Hematol 2:161–220 [In Spanish]
del Río-Hortega P, Penfield WG (1927) Cerebral cicatrix. The reaction of neuroglia and microglia to brain wounds. Bull John Hopkins Hosp 41:278–303
Derecki NC, Cronk JC, Lu Z, Xu E, Abbott SB, Guyenet PG, Kipnis J (2012) Wild-type microglia arrest pathology in a mouse model of Rett syndrome. Nature 484:105–109
Dougherty TF (1944) Studies on the cytogenesis of microglia and their relation to cells of the reticulo-endothelial system. Am J Anat 74:61–97
Draheim H, Prinz M, Weber JR, Weiser T, Kettenmann H, Hanisch UK (1999) Induction of K^+ channels in mouse brain microglia: cells acquire responsiveness to pneumococcal cell wall components during late development. Neuroscience 89:1379–1390
Duke DC, Moran LB, Turkheimer FE, Banati R, Graeber MB (2004) Microglia in culture: what genes do they express? Dev Neurosci 26:30–37
Dunning HS, Furth J (1935) Studies on the relation between microglia, histiocytes and monocytes. Am J Pathol 11:895–914
Dunning HS, Stevenson L (1934) Microglia-like cells and their reaction following injury to the liver, spleen and kidney. Am J Pathol 10:343–348
Eder C, Schilling T, Heinemann U, Haas D, Hailer N, Nitsch R (1999) Morphological, immunophenotypical and electrophysiological properties of resting microglia in vitro. Eur J Neurosci 11:4251–4261
Esiri MM, Booss J (1984) Comparison of methods to identify microglial cells and macrophages in the human central nervous system. J Clin Pathol 37:150–156
Fain NG (1963) On the method of impregnating microglia in celloidin sections. Arkh Patol 25: 52–54 [In Russian]
Färber K, Kettenmann H (2005) Physiology of microglial cells. Brain Res Rev 48:133–143
Fernández-Guardiola A (1997) Isaac Costero Tudanca. In: Las Neurociencias en el Exilio Español en México. La Ciencia para Todos 153; Fondo de Cultura Económica (FCE), Mexico [In Spanish]
Ferrer D (1973) Microglia (mesoglia or Rio Hortega cells) yesterday and today. Rev Med Univ Navarra 17:145–161 [In Spanish]
Field EJ (1955) Observations on the development of microglia together with a note on the influence of cortisone. J Anat 89:201–208
Flügel A, Bradl M, Kreutzberg GW, Graeber MB (2001) Transformation of donor-derived bone marrow precursors into host microglia during autoimmune CNS inflammation and during the retrograde response to axotomy. J Neurosci Res 66:74–82
Ford AL, Goodsall AL, Hickey WF, Sedgwick JD (1995) Normal adult ramified microglia separated from other central nervous system macrophages by flow cytometric sorting. Phenotypic differences defined and direct ex vivo antigen presentation to myelin basic protein-reactive CD4+ T cells compared. J Immunol 154:4309–4321
Fujita S, Kitamura T (1975) Origin of brain macrophages and the nature of the so-called microglia. Acta Neuropathol Suppl 6:291–296
Fujita S, Tsuchihashi Y, Kitamura T (1981) Origin, morphology and function of the microglia. Prog Clin Biol Res 59A:141–169
Fujita H, Tanaka J, Toku K, Tateishi N, Suzuki Y, Matsuda S, Sakanaka M, Maeda N (1996) Effects of GM-CSF and ordinary supplements on the ramification of microglia in culture: a morphometric study. Glia 18:269–281
Gallyas F (1963) Silver impregnation method for microglia. Acta Neuropathol 3:206–209
Gans A (1923) Del Rio-Hortega's Derde Element in het centrale zenuwstelsel. Ned Tijdschr Geneeskd 67:1024–1027
García-Albea E, Wilder G (2004) Penfield en la Residencia de Estudiantes (Madrid, 1924). Rev Neurol 39:872–878 [In Spanish]

García-Marín V, García-López P, Freire M (2007) Cajal's contributions to glia research. Trends Neurosci 30:479–487

Garcia-Segura LM (2002) Cajal and glial cells. Prog Brain Res 136:255–260

Gehrmann J, Banati RB, Kreutzberg GW (1993) Microglia in the immune surveillance of the brain: human microglia constitutively express HLA-DR molecules. J Neuroimmunol 48:189–198

Gehrmann J, Matsumoto Y, Kreutzberg GW (1995) Microglia: intrinsic immuneffector cell of the brain. Brain Res Rev 20:269–287

Geissmann F, Manz MG, Jung S, Sieweke MH, Merad M, Ley K (2010) Development of monocytes, macrophages, and dendritic cells. Science 327:656–661

Gill AS, Binder DK (2007) Wilder Penfield, Pio del Rio-Hortega and the discovery of oligodendroglia. Neurosurgery 60:940–948

Ginhoux F, Greter M, Leboeuf M, Nandi S, See P, Gokham S, Mehler MF, Conway SJ, Ng LG, Stanley ER, Samokhvalov IM, Merad M (2010) Fate mapping analysis reveals that adult microglia derive from primitive macrophages. Science 330:841–845

Ginhoux F, Lim S, Hoeffel G, Low D, Huber T (2013) Origin and differentiation of microglia. Front Cell Neurosci 7:45, 14 p

Giulian D (1987) Ameboid microglia as effectors of inflammation in the central nervous system. J Neurosci Res 18:155–171

Giulian D, Baker TJ (1985) Peptides released by ameboid microglia regulate astroglial proliferation. J Cell Biol 101:2411–2415

Giulian D, Baker TJ (1986) Characterisation of ameboid microglia isolated from developing mammalian brain. J Neurosci 6:2163–2178

Giulian D, Ingeman JE (1988) Colony-stimulating factors as promoters of ameboid microglia. J Neurosci 8:4707–4717

Giulian D, Young DG, Woodward J, Brown DC, Lachman LB (1988) Interleukin-1 is an astroglial growth factor in the developing brain. J Neurosci 8:709–714

Giulian D, Li J, Leara B, Keenen C (1994) Phagocytic microglia release cytokines and cytotoxins that regulate the survival of astrocytes and neurons in culture. Neurochem Int 25:227–33

Giulian D, Bartel S, Broker J, Li X, Kirkpatrick JB (1995) Cell surface morphology identifies microglia as a distinct class of mononuclear phagocyte. J Neurosci 15:7712–7726

Glanzer JG, Enose Y, Wang T, Kadiu I, Gong N, Rozek W, Liu J, Schlautman JD, Ciborowski PS, Thomas MP, Gendelman HE (2007) Genomic and proteomic microglial profiling: pathways for neuroprotective inflammatory responses following nerve fragment clearance and activation. J Neurochem 102:627–645

Glees P (1953) Neuroglia: morphology and function. Charles C. Thomas, Springfield, IL

Glenn JA, Ward SA, Stone CR, Booth PL, Thomas WE (1992) Characterisation of ramified microglial cells: detailed morphology, morphological plasticity and proliferative capability. J Anat 180:109–118

Goldmann T, Wieghofer P, Müller PF, Wolf Y, Varol D, Yona S, Brendecke SM, Kierdorf K, Staszewski O, Datta M, Luedde T, Heikenwalder M, Jung S, Prinz M (2013) A new type of microglia gene targeting shows TAK1 to be pivotal in CNS autoimmune inflammation. Nat Neurosci 16:1618–1626

Gomez Perdiguero E, Schulz C, Geissmann F (2013) Development and homeostasis of 'resident' myeloid cells: the case of the microglia. Glia 61:112–120

Gonzalez C (1971) El merecido homenaje a Don Pío del Río Hortega. Anal Real Acad Med Cir Valladolid 9:70–73 [In Spanish]

Graeber MB (1993) Microglia, macrophages and the blood–brain barrier. Clin Neuropathol 12:296–297

Graeber MB (2010) Changing face of microglia. Science 330:783–788

Graeber MB, Mehraein P (1994) Microglia rod cells. Neuropathol Appl Neurobiol 20: 178–18014

Graeber MB, Streit WJ (1990a) Microglia: immune network in the CNS. Brain Pathol 1:2–5

Graeber MB, Streit WJ (1990b) Perivascular microglia defined. Trends Neurosci 13:366

Graeber MB, Tetzlaff W, Streit WJ, Kreutzberg GW (1988) Microglial cells but not astrocytes undergo mitosis following rat facial nerve axotomy. Neurosci Lett 85:317–321
Graeber MB, Streit WJ, Büringer D, Sparks DL, Kreutzberg GW (1992) Ultrastructural location of major histocompatibility complex (MHC) class II positive perivascular cells in histologically normal human brain. J Neuropathol Exp Neurobiol 51:303–311
Graeber MB, Bise K, Mehraein P (1993) Synaptic stripping in the human facial nucleus. Acta Neuropathol 86:179–181
Graeber MB, López-Redondo F, Ikoma E, Ishikawa M, Imai Y, Nakajima K, Kreutzberg GW, Kohsaka S (1998) The microglia/macrophage response in the neonatal rat facial nucleus following axotomy. Brain Res 813:241–253
Graeber MB, Li W, Rodriguez ML (2011) Role of microglia in CNS inflammation. FEBS Lett 3585:3798–3805
Greaves DR, Gordon S (2002) Macrophage-specific gene expression: current paradigms and future challenges. Int J Hematol 76:6–15
Grenier Y, Ruijs TC, Robitaille Y, Olivier A, Antel JP (1989) Immunohistochemical studies of adult human glial cells. J Neuroimmunol 21:103–115
Greter M, Merad M (2013) Regulation of microglia development and homeostasis. Glia 61:121–127
Grinberg YY, Milton JG, Kraig RP (2011) Spreading depression sends microglia on Lévy flights. PLoS One 6:e19294. doi:10.1371/journal.pone.0019294
Grossmann R, Stence N, Carr J, Fuller L, Waite M, Dailey ME (2002) Juxtavascular microglia migrate along brain microvessels following activation during early postnatal development. Glia 37:229–240
Haas CA, Donath C, Kreutzberg GW (1993) Differential expression of immediate early genes after transection of the facial nerve. Neuroscience 53:91–99
Hafner (1965) refers to the New York-based publishers who reprinted Wilder Penfield's original (1932) set of three volumes for Cytology and Cellular Pathology of the Nervous system in 1965. 652–653 http://www.worldcat.org/title/cytology-cellular-pathology-of-the-nervous-system/oclc/187412121
Hager H (1968) Ultrastructural pathology of the macro- and microglia. Acta Neuropathol Suppl 4:86–97 [In German]
Hailer NP, Jarhult JD, Nitsch R (1996) Resting microglial cells in vitro: analysis of morphology and adhesion molecule expression in organotypic hippocampal slice cultures. Glia 18: 319–331
Hanisch UK (2002) Microglia as a source and target of cytokines. Glia 40:140–155
Hanisch UK (2013a) Factors controlling microglial activation. In: Kettenmann H, Ransom B (eds) Neuroglia, 3rd edn. Oxford University Press, New York, pp 614–625, Chapter 48
Hanisch UK (2013b) Functional diversity of microglia—how heterogeneous are they to begin with? Front Cell Neurosci 7:65, 18 p
Hanisch UK (2013c) Proteins in microglial activation—inputs and outputs by subsets. Curr Protein Pept Sci 14:3–15
Hanisch UK, Kettenmann H (2007) Microglia: active sensor and versatile effector cells in the normal and pathologic brain. Nat Neurosci 10:1387–1394
Harrison JK, Jiang Y, Chen S, Xia Y, Maciejewski D, McNamara RK, Streit WJ, Salafranca MN, Adhikari S, Thompson DA, Botti P, Bacon KB, Feng L (1998) Roles for neuronally-derived fractalkine in mediating interactions between neurons and CX3CR1-expressing microglia. Proc Natl Acad Sci U S A 95:10896–10901
Harry GJ, Kraft AD (2012) Microglia in the developing brain: a potential target with lifetime effects. Neurotoxicology 33:191–206
Hayes GM, Woodroofe MN, Cuzner ML (1988) Characterisation of microglia isolated from adult human and rat brain. J Neuroimmunol 19:177–189
Haymaker W, Sánchez-Pérez M (1935) Rio-Hortega's double silver impregnation technique adapted to the staining of tissue cultures. Science 82:355–356

Herrera JM (1953) Studies on the problem of microglia genesis. I. Reaction capacity of the microglia of cadaveric encephalon. Arch Med Panamenos 2:161–182 [In Spanish]
Hickey WF, Kimura H (1988) Perivascular microglial cells of the CNS are bone marrow-derived and present antigen in vivo. Science 239:290–292
Hirasawa T, Ohsawa K, Imai Y, Ondo Y, Akazawa C, Uchino S, Kohsaka S (2005) Visualization of microglia in living tissues using Iba1-EGFP transgenic mice. J Neurosci Res 81:357–362
Hoek RM, Ruuls SR, Murphy CA, Wright GJ, Goddard R, Zurawski SM, Blom B, Homola ME, Murphy CA, Wright GJ, Goddard R, Zurawski SM, Blom ME, Streit WJ, Brown MH, Barclay AN, Sedgwick JD (2000) Down-regulation of the macrophage lineage through interaction with OX2 (CD200). Science 290:1768–1771
Hughes V (2012) Microglia: the constant gardeners. Nature 485:570–572
Hume DA, Perry VH, Gordon S (1983) Immunohistochemical localization of a macrophage-specific antigen in developing mouse retina: phagocytosis of dying neurons and differentiation of microglial cells to form a regular array in the plexiform layers. J Cell Biol 97:253–257
Ibrahim MZ, Khreis Y, Kosahyan DS (1974) The histochemical identification of microglia. J Neurol Sci 22:211–233
Imai Y, Ibata I, Ito D, Ohsawa K, Kohsaka S (1996) A novel gene iba1 in the major histocompatibility complex class III region encoding an EF hand protein expressed in a monocytic lineage. Biochem Biophys Res Commun 224:855–862
Imaizumi M, Briard E, Zoghbi SS, Gourley JP, Hong J, Fujimura Y, Pike V, Innis RB, Fujita M (2008) Brain and whole-body imaging in non-human primates of [11C]PBR28, a promising PET radioligand for peripheral benzodiazepine receptors. Neuroimage 39:1289–1298
Imamoto K, Leblond CP (1978) Radioautographic investigation of gliogenesis in the corpus callosum of young rats. II. Origins of microglial cells. J Comp Neurol 180:139–163
Imamura S (1954) Microglia in gliomas. Folia Pyshciatr Neurol Jpn 8:99–126
Imamura K, Ito M, Suzumura A, Asai J, Takahashi A (1990) Generation and characterisation of monoclonal antibodies against rat microglia and ontogenic distribution of positive cells. Lab Invest 63:853–861
Ito D, Imai Y, Ohsawa K, Nakajima K, Fukuuchi Y, Kohsaka S (1998) Microglia-specific localisation of a novel calcium binding protein, Iba1. Mol Brain Res 57:1–9
Jelliffe SE (1919) Nicolas Achúcarro. Obituary J Nerv Ment Dis 49:271–272
Jiménez de Asúa F (1927) Die Mikroglia (Hortegasche Zellen) und das Reticulo-endotheliale System. Z Neurol Psychiatr 109:354–379 [In German]
Juba A (1933) Untersuchungen über die Entwicklung der Hortegaschen Mikroglia des Menschen. Arch Psychiatr Nervenkr 101:577–592 [In German]
Juba A (1934) Das erste Erscheinen und die Urformen der Hortegaschen Mikroglia im Zentralnervensystem. Arch Psychiatr Nervenkr 102:225–232 [In German]
Jufe R (1957) Changes in the microglia of the plexiform ganglion in rabies. Arch Histol Norm Patol 7:51–59 [In Spanish]
Kaur C, Ling EA, Wong WC (1985) Transformation of ameboid microglial cells into microglia in the corpus callosum of the postnatal rat brain—an electron microscopical study. Arch Histol Jpn 48:17–25
Kaur C, Hao AJ, Wu CH, Ling EA (2001) Origin of microglia. Microsc Res Tech 54:2–9
Kershman J (1939) Genesis of microglia in the human brain. Arch Neurol Psychiatr 41:24–50
Kettenmann H (1994) Electrophysiological behavior of microglia. Neuropathol Appl Neurobiol 20:177–178
Kettenmann H (2006) Triggering the brain's pathology sensor. Nat Neurosci 9:1463–1464
Kettenmann H (2007) Neuroscience: the brain's garbage men. Nature 446:987–9
Kettenmann H, Hoppe D, Gottmann K, Banati R, Kreutzberg G (1990) Cultured microglial cells have a distinct pattern of membrane channels different from peritoneal macrophages. J Neurosci Res 26:278–287
Kettenmann H, Banati R, Walz W (1993) Electrophysiological behavior of microglia. Glia 7:93–101

Kettenmann H, Hanisch U-K, Noda M, Verkhratsky A (2011) Physiology of microglia. Physiol Rev 91:461–553

Kettenmann H, Kirchhoff F, Verkhratsky A (2013) Microglia: new roles for the synaptic stripper. Neuron 77:10–18

Kierdorf K, Erny D, Goldmann T, Sander V, Schulz C, Perdiguero EG, Wieghofer P, Heinrich A, Riemke P, Holscher C, Muller DN, Luckow B, Brocker T, Debowski K, Fritz G, Opdenakker G, Diefenbach A, Biber K, Heikenwalder M, Geissmann F, Rosenbauer F, Prinz M (2013) Microglia emerge from erythromyeloid precursors via Pu.1- and Irf8-dependent pathways. Nat Neurosci 16:273–80

Kitamura T (1969) Re-examination of the concept of microglia. Shinkei Kenkyu No Shimpo 13:355–364 [In Japanese]

Kitamura T (1973) The origin of brain macrophages—some considerations on the microglia theory of del Rio-Hortega. Pathol Int 23:11–26

Kitamura T, Hattori H (1972) Re-examination of the origin, function and morphology of microglia. 3. Role of perivascular cells in relation to macrophages. Shinkei Kenkyu No Shimpo 16: 447–451 [In Japanese]

Kloss CU, Kreutzberg GW, Raivich G (1997) Proliferation of ramified microglia on an astrocyte monolayer: characterization of stimulatory and inhibitory cytokines. J Neurosci Res 49: 248–254

Koenig H (1958) An autoradiographic study of nucleic acid and protein turnover in the mamalian neuraxis. J Biophys Biochem Cytol 4:785–792

Kraśnicka Z, Renhawek K (1969) Morphology and histochemistry of microglia in tissue culture studied in normal and pathological conditions. Neuropatol Pol 7:73–90 [In Polish]

Kraśnicka Z, Renhawek K (1970) Morphology and histochemistry of microglia cultured in normal and pathological conditions. Pol Med J 9:96–212 [In Polish]

Kreutzberg GW (1963) Changes of coenzyme (TPN) diaphorase and TPN-linked dehydrogenase during axonal reaction of the nerve cell. Nature 199:393–394

Kreutzberg GW (1967) Autoradiographic study on incorporation of leucine-H3 in peripheral nerves during regeneration. Experientia 23:33–34

Kreutzberg GW (1968) Autoradiographic studies on perineuronal microgliocytes. Acta Neuropathol Suppl 4:141–145 [In German]

Kreutzberg GW (1972) Experimental studies on nerve regenerarion. Fortschr Med 90:910–912 [In German]

Kreutzberg GW (1995) Microglia, the first line of defence in brain pathologies. Arzneimittelforschung 45:357–360 [In German]

Kreutzberg GW (1996) Microglia: a sensor for pathological events in the CNS. Trends Neurosci 19:312–318

Kreutzberg GW, Barron KD (1978) 5′-Nucleotidase of microglial cells in the facial nucleus during axonal reaction. J Neurocytol 7:601–610

Kurobane T (1950) Microglia in gliomas; a contribution to the study of microglia. Folia Psychiatr Neurol Jpn 4:123–131

Kurpius D, Wilson N, Fuller L, Hoffman A, Dailey ME (2006) Early activation, motility and homing of neonatal microglia to injured neurons does not require protein synthesis. Glia 54:58–70

Lassmann H, Hickey WF (1993) Radiation bone marrow chimeras as a tool to study microglia turnover in normal brain and inflammation. Clin Neuropathol 12:284–285

Lassmann H, Schmied M, Vass K, Hickey WF (1993) Bone marrow derived elements and resident microglia in brain inflammation. Glia 7:19–24

Lauro GM, Babilono D, Buttarelli FR, Starace G, Cocchia D, Ennas MG, Sogos V, Gremo F (1995) Human microglia cultures: a powerful model to study their origin and immunoreactive capacity. Int J Dev Neurosci 13:739–752

Lavisse S, Guillermier M, Hérard AS, Petit F, Delahaye M, Van Camp N, Ben Haim L, Lebon V, Remy P, Dollé F, Delzescaux T, Bonvento G, Hantraye P, Escartin C (2012) Reactive astrocytes

overexpress TSPO and are detected by TSPO positron emission tomography imaging. J Neurosci 32:10809–10818
Lawson LJ, Perry VH, Dri P, Gordon S (1990) Heterogeneity in the distribution and morphology of microglia in the normal adult mouse brain. Neuroscience 39:151–170
Lawson LJ, Perry VH, Gordon S (1992) Turnover of resident microglia in the normal adult mouse brain. Neuroscience 48:405–415
Lebowich RJ (1934) Phagocytic behavior of interstitial cells of brain parenchyma of adult rabbit toward colloidal solutions and bacteria. Arch Pathol 18:50–71
Lee SC, Liu W, Brosnan CF, Dickson DW (1992) Characterisation of primary human fetal dissociated cultures with an emphasis on microglia. Lab Invest 67:465–476
Leong SK, Ling EA (1992) Ameboid and ramified microglia: their interrelationship and response to brain injury. Glia 6:39–47
Ling EA (1976a) Electron microscopic identification of ameboid microglia in the spinal cord of newborn rats. Acta Anat 96:600–609
Ling EA (1976b) Some aspects of ameboid microglia in the corpus callosum and neighbouring regions of neonatal rats. J Anat 121:29–45
Ling EA (1979) Transformation of monocytes into ameboid microglia in the corpus callosum of postnatal rats, as shown by labelling monocytes by carbon particles. J Anat 128:847–858
Ling EA, Tan CK (1974) Ameboid micriglial cells in the corpus callosum of neonatal rats. Arch Histol Jpn 36:265–280
Ling EA, Wong WC (1993) The origin and nature of ramified and ameboid microglia: a historical review and current concepts. Glia 7:9–18
Liu W, Brosnan CF, Dickson DW, Lee SC (1994) Macrophage colony-stimulating factor mediates astrocyte-induced microglial ramification in human fetal central nervous system culture. Am J Pathol 145:48–53
Maciejewski-Lenoir D, Chen S, Feng L, Maki R, Bacon KB (1999) Characterisation of fractalkine in rat brain cells: migratory and activation signals for CX3CR1-expressing microglia. J Immunol 163:1628–1635
Malik V (1964) Impregnation of microglia and oligodendroglia in paraffin sections. Cesk Morfol 12:214–219 [In Czech]
Maximow AA (1928) The macrophage or histiocytes. In: Cowdry EV (ed) Special cytology, vol 1. Hoeber, New York, pp 427–484
Mazzi V (1952) Macrophages and microglia in the brain of teleosts. Arch Ital Anat Embriol 57:330–348 [In Italian]
McCarter JC (1940) A silver carbonate staining method for oligodendrocytes and microglia for routine use. Am J Pathol 16:233–235
McGeer PL, McGeer EG (2011) History of innate immunity in neurodegenerative disorders. Front Pharmacol 2:77
Meo L (1950) Cerebral changes and microglioblastic reactions in leukemia, with several considerations on the histogenesis of microglia. Riv Anat Patol Oncol 3:1060–1088 [In Italian]
Metz A, Spatz H (1924) Die Hortegaschen Zellen (das sogenannte 'dritte element') und über ihre funktionelle Bedeutung. Z Neurol Psychiatr 89:138–170 [In German]
Mildner A, Schmidt H, Nitsche M, Merkler D, Hanisch UK, Mack M, Heikenwalder M, Brück W, Priller J, Prinz M (2007) Microglia in the adult brain arise from Ly-6ChiCCR2+ monocytes only under defined host conditions. Nat Neurosci 10:1544–1553
Mladenov S, Gerebtzoff MA (1967) Histochemical research methods on reactive microglia in the rat brain. C R Séances Soc Biol ses Fil 161:1467–1469
Moran LB, Graeber MB (2004) The facial nerve axotomy model. Brain Res Rev 44:154–178
Moran LB, Duke DC, Turkheimer FE, Banati RB, Graeber MB (2004) Towards a transcriptome definition of microglial cells. Neurogenetics 5:95–108
Mori S, Leblond CP (1969) Identification of microglia in light and electron microscopy. J Comp Neurol 135:57–80
Mrak RE (2009) Neuropathology and the neuroinflammation idea. J Alzheimers Dis 18:473–481

Naoumenko J, Feigin I (1963) A modification for paraffin sections of silver carbonate impregnation for microglia. Acta Neuropathol 3:402–406
Nakajima K, Kohsaka S (1993a) Functional roles of microglia in the brain. Neureosci Res 17:187–203
Nakajima K, Kohsaka S (1993b) Characterization of brain microglia and the biological significance in the central nervous system. Adv Neurol 60:734–43
Nakajima K, Kohsaka S (1998) Functional roles of microglia in the central nervous system. Hum Cell 11:141–55
Nakajima K, Kohsaka S (2001) Microglia: activation and their significance in the central nervous system. J Biochem 130:169–75
Nakajima K, Kohsaka S (2004) Microglia: neuroprotective and neurotrophic cells in the central nervous system. Curr Drug Targets Cardiovasc Haematol Disord 4:65–84
Neumann H (2001) Control of glial immune function by neurons. Glia 36:191–199
Neumann H, Takahashi K (2007) Essential role of the microglial triggering receptor expressed on myeloid cells-2 (TREM2) for central nervous system tissue immune homeostasis. J Neuroimmunol 184:92–99
Nimmerjahn A (2012) Two-photon imaging of microglia in the mouse cortex in vivo. Cold Spring Harb Protoc 2012(5). pii: pdb.prot069294. doi:10.1101/pdb.prot069294
Nimmerjahn A, Kirchhoff F, Helmchen F (2005) Resting microglial cells are highly dynamic surveillants of brain parenchyma in vivo. Science 308:1314–1318
Obrador S (1975) Penfield and the school of Cajal. Penfield and his relation to Cajal and Rio-Hortega. Rev Esp Otoneurooftalmol Neurocir 33:39–53 [In Spanish]
Oehmichen M, Wiethölter H, Greaves MF (1979) Immunological analysis of human microglia: lack of monocytic and lymphoid membrane differentiation antigens. J Neuropathol Exp Neurol 38:99–103
Ohsawa K, Kohsaka S (2011) Dynamic motility of microglia: purinergic modulation of microglial movement in the normal and pathological brain. Glia 59:1793–1799
Olah M, Biber K, Vinet J, Boddeke HW (2011) Microglia phenotype diversity. CNS Neurol Disord Drug Targets 10:108–118
Olah M, Amor S, Brouwer N, Vinet J, Eggen B, Biber K, Boddeke HW (2012) Identification of a microglia phenotype supportive of remyelination. Glia 60:306–321
Ortíz Picón JM (1971) La obra neurohistológica del doctor Pío del Río Hortega (1882–1945). Arch Neurobiol 34:39–70 [In Spanish]
Ortíz Picón JM (1983) Pío del Río Hortega. Un estudio biográfico. Arch Neurobiol 46:209–226 [In Spanish]
Palmero JR (2005) Pío del Río Hortega en la Universidad de Valladolid. ARBOR Cien Pensamento Cultura 181:181–197 [In Spanish]
Palmero JR, del Río-Hortega J (2004) Pío Del Río-Hortega y la Institucionalización de la ciencia en España. In: Actas VIII Congreso de la Sociedad Española de Historia de las Ciencias y de las Técnicas, pp 161–199 [In Spanish]
Paolicelli RC, Gross CT (2011) Microglia in development: linking brain wiring to brain environment. Neuron Glia Biol 7:77–83
Paolicelli RC, Bolasco G, Pagani F, Maggi L, Scianni M, Panzanelli P, Giustetto M, Ferreira TA, Guiducci E, Duman L, Ragozzino D, Gross CT (2011) Synaptic pruning by microglia is necessary for normal brain development. Science 333:1456–1458
Parakalan R, Jiang B, Nimmi B, Janani M, Jayapal M, Lu J, Tay SS, Ling EA, Dheen ST (2012) Transcriptome analysis of ameboid and ramified microglia isolated from the corpus callosum of rat brain. BMC Neurosci 13:64
Penfield W (1924a) Oligodendroglia and its relation to classical neuroglia. Brain 47:430–452
Penfield W (1924b) Microglie et son rapport avec la dégénération névrogliale dans un gliome. Trav Lab Rech Biol Univ Madrid 22:277–293 [In French]
Penfield W (1925) Microglia and the process of phagocytosis in gliomas. Am J Pathol 1:77–89
Penfield W (1928) A method of staining oligodendroglia and microglia (combined method). Am J Pathol 28:153–157

Penfield W (1932) Neuroglia: normal and pathological. In: Penfield W (ed) Cytology and cellular pathology of the nervous system, vol 2. Hoeber, New York, pp 423–479
Penfield W (1977) No man alone. A neurosurgeon's life. Little, Brown Company, Boston
Penfield W, Cone W (1937) Neuroglia and microglia (the metallic methods). In: McClung CE (ed). Paul B. Hoeber, New York, p 489–521.
Perry VH (1994) Modulation of microglia phenotype. Neuropathol Appl Neurobiol 20:177
Perry VH, Gordon S (1988) Macrophages and microglia in the nervous system. Trends Neurosci 11:273–277
Perry VH, O'Connor V (2010) The role of microglia in synaptic stripping and synaptic degeneration: a revised perspective. ASN Neuro 2:e00047. doi:10.1042/AN20100024
Perry VH, Hume DA, Gordon S (1985) Immunohistochemical localisation of macrophages and microglia in the adult and developing mouse brain. Neuroscience 15:313–326
Petersen MA, Dailey ME (2004) Diverse microglial motility behaviors during clearance of dead cells in hippocampal slices. Glia 46:195–206
Pickering JE, Vogel FS (1956) Demyelinization in the brains of monkeys by means of fast neutrons; pathogenesis of the lesion and comparison with the lesions of multiple sclerosis and Schilder's disease. J Exp Med 104:435–442
Pocock JM, Kettenmann H (2007) Neurotransmitter receptors on microglia. Trends Neurosci 30:527–535
Polderman H (1926) Die Entdeckung der Mikroglia und ihre Bedeutung für die Neurogliafrage. Nederland Tijdschr v Geneesk 70:537–549 [In German]
Politis M, Su P, Piccini P (2012) Imaging of microglia in patients with neurodegenerative disorders. Front Pharmacol 3:96
Priller J, Flügel A, Wehner T, Boentert M, Haas CA, Prinz M, Fernánddez-Klett F, Prass K, Bechmann I, deBoer BA, Kreutzberg GW, Persons DA, Dirnagl U (2001) Targeting gene-modified hematopoietic cells to the central nervous system: use of green fluorescent protein uncovers microglial engraftment. Nat Med 7:1356–1361
Prinz M, Kann O, Draheim HJ, Schumann RR, Kettenmann H, Weber JR, Hanisch UK (1999) Microglial activation by components of gram-positive and -negative bacteria: distinct and common routes to the induction of ion channels and cytokines. J Neuropathol Exp Neurol 58: 1078–1089
Prinz M, Priller J, Sisodia SS, Ransohoff RM (2011) Heterogeneity of CNS myeloid cells and their roles in neurodegeneration. Nat Neurosci 13:1227–1235
Pruijs WM (1927) Über Mikroglia, ihre Herkunft, Funktion und ihr Verhältnis zu anderen Glialelementen. Z Neurol Psychiatr 108:298–331 [In German]
Raivich G (2005) Like cops on the beat: the active role of resting microglia. Trends Neurosci 28: 571–573
Raivich G, Gehrmann J, Moreno-Floros M, Kreutzberg GW (1993) Microglia: growth factor and mitogen receptors. Clin Neuropathol 12:293–295
Rand CW, Courville CB (1932) Histologic changes in the brain in cases of fatal injury to the head. III. Reaction of microglia and oligodendroglia. Arch Neurol Psychiatry 27:605–644
Rangroo Thrane V, Thrane AS, Chanag J, Alleluia V, Nagelhus EA, Nedergaard M (2012) Real-time analysis of microglial activation and motility in hepatic and hyperammonemic encephalopathy. Neuroscience 220:247–255
Ransohoff RM, Perry VH (2009) Microglial physiology: unique stimuli, specialised responses. Annu Rev Immunol 27:119–145
Rezaie P (2003) Microglia in the human nervous system during development. Neuroembryology 2:18–31
Rezaie P (2007) Microglia: from origins to differentiation, distribution and dynamics. In: Dheen ST, Ling EA (eds) Trends in glial research—basic and applied. Research Signpost, Kerala, pp 1–41
Rezaie P, Male D (2002a) Mesogia and microglia. A historical review of the concept of mononuclear phagocytes within the central nervous system. J Hist Neurosci 11:325–374

Rezaie P, Male D (2002b) Differentiation, ramification and distribution of microglia within the central nervous system examined. Neuroembryology 1:29–43

Rezaie P, Cairns NJ, Male DK (1997) Expression of adhesion molecules on human fetal cerebral vessels: relationship to microglial colonisation during development. Dev Brain Res 104: 175–189

Rezaie P, Patel K, Male DK (1999) Microglia in the human fetal spinal cord—patterns of distribution, morphology and phenotype. Dev Brain Res 115:71–81

Rezaie P, Trillo-Pazos G, Greenwood J, Everall IP, Male DK (2002) Motility and ramification of human fetal microglia in culture: an investigation using time-lapse video microscopy and image analysis. Exp Cell Res 274:68–82

Rezza A (1925) Le cellule di Del Rio Hortega. Quistioni intorno alla nevroglia. Rass Studi Psichiatr 14;307–335 [In Italian]. http://www.cbt.biblioteche.provincia.tn.it/oseegenius/resource?uri=6253037&v=l&dcnr=8

Rieske E, Graeber MB, Tetzlaff W, Czlonkowska A, Streit WJ, Kreutzberg GW (1989) Microglia and microglia-derived brain macrophages in culture: generation from axotomized rat facial nuclei, identification and characterisation in vitro. Brain Res 492:1–14

Robertson WF (1897) The normal histology and pathology of neuroglia (in relation specially to Mental Diseases). J Ment Sci 43:733–752

Robertson WF (1898) The normal histology and pathology of neuroglia (in relation specially to mental diseases). Edinb Hosp Rep 5:240–271

Robertson WF (1899) On a new method of obtaining a black reaction in certain tissue-elements of the central nervous system (platinum method). Scottish Med Surg J 4:23–30

Robertson WF (1900a) A microscopic demonstration of the normal and pathological histology of mesoglia cells. J Ment Sci (Br J Psychiatr) 46:724

Robertson WF (1900b) A textbook of pathology in relation to mental diseases. W.F. Clay, Edinburgh

Rosenstiel P, Lucius R, Deuschl G, Sievers J, Wilms H (2001) From theory to therapy: implications from an in vitro model of ramified microglia. Microsc Res Tech 54:18–25

Russell DS (1929) Intravital staining of microglia with trypan blue. Am J Pathol 5:451–458

Rydberg E (1932) Cerebral injury in new-born children consequent on birth trauma; with an inquiry into the normal and pathological analtomy of the neuroglia. Acta Pathol Microbiol Scand Suppl 10:1–247

Saijo K, Glass CK (2011) Microglial cell origin and phenotypes in health and disease. Nat Rev Immunol 11:775–787

Sawada M, Itoh Y, Suzumura A, Marunouchi T (1993) Expression of cytokine receptors in cultured neuronal and glial cells. Neurosci Lett 160:131–134

Schafer DP, Lehrman EK, Kautzman AG, Koyama R, Mardinly AR, Yamsaki R, Ransohoff RM, Greenberg M, Barres BA, Stevens B (2012) Microglia sculpt postnatal neural circuits in an activity and complement-dependent manner. Neuron 74:691–705

Schaffer K (1926) Über die Hortegasche Mikroglia. Z Anat Entwicklungsgesch 81:715–719 [In German]

Scheffel J, Regen T, Van Rossum D, Seifert S, Ribes S, Nau R, Parsa R, Harris RA, Boddeke HW, Chuang HN, Pukrop T, Wessels JT, Jürgens T, Merkler D, Brück W, Schnaars M, Simons M, Kettenmann H, Hanisch UK (2012) Toll-like receptor activation reveals developmental reorganization and unmasks responder subsets of microglia. Glia 60:1930–1943

Schilling T, Nitsch R, Heinemann U, Haas D, Eder C (2001) Astrocyte-released cytokines induce ramification and outward K+ channel expression in microglia via distinct signalling pathways. Eur J Neurosci 14:463–473

Schmid CD, Sautkulis LN, Danielson PE, Cooper J, Hasel KW, Hilbush BS, Sutcliffe JG, Carson MJ (2002) Heterogeneous expression of the triggering receptor expressed on myeloid cells-2 on adult murine microglia. J Neurochem 83:1309–1320

Schulz C, Gomez Perdiguero E, Chorro L, Szabo-Rogers H, Cagnard N, Kierdorf K, Prinz M, Wu B, Jacobsen SE, Pollard JW, Frampton J, Liu KJ, Geissmann F (2012) A lineage of myeloid cells independent of Myb and hematopoietic stem cells. Science 336:86–90

Sedgwick JD, Schwender S, Imrich H, Dörries R, Butcher GW, terMeulen V (1991) Isolation and direct characterization of resident microglial cells from the normal and inflamed central nervous system. Proc Natl Acad Sci U S A 88:7438–7442
Shepard JL, Zon LI (2000) Developmental derivation of embryonic and adult macrophages. Curr Opin Hematol 7:3–8
Sievers J, Parwaresch R, Wottge HU (1994) Blood monocytes and spleen macrophages differentiate into microglia-like cells on monolayers of astrocytes: morphology. Glia 12:245–258
Simard AR, Rivest S (2004) Bone marrow stem cells have the ability to populate the entire central nervous system into fully differentiated parenchymal microglia. FASEB J 18:998–100
Skaper SD (2011) Ion channels on microglia: therapeutic targets for neuroprotection. CNS Neurol Disord Drug Targets 10:44–56
Somjen JS (1988) Nervenkitt: notes on the history of the concept of neuroglia. Glia 1:2–9
Stence N, Waite M, Dailey ME (2001) Dynamics of microglial activation: a confocal time-lapse analysis in hippocampal slices. Glia 33:256–266
Streit WJ, Graeber MB (1993) Heterogeneity of microglial and perivascular cell populations: insights gained from the facial nucleus paradigm. Glia 7:68–74
Streit WJ, Kreutzberg GW (1987) Lectin binding by resting and reactive microglia. J Neurocytol 16:249–260
Streit WJ, Graeber MB, Kreutzberg GW (1988) Functional plasticity of microglia: a review. Glia 1:301–307
Streit WJ, Graeber MB, Kreutzberg GW (1989) Peripheral nerve lesion produces increased levels of major histocompatibility complex antigens in the central nervous system. J Neuroimmunol 21:117–123
Streit WJ, Mrak RE, Griffin SE (2004a) Microglia and neuroinflammation: a pathological perspective. J Neuroinflammation 1:14
Streit WJ, Sammons NW, Kuhns AJ, Sparks DL (2004b) Dystrophic microglia in the aging human brain. Glia 45:208–212
Suzumura A, Mezitis SG, Gonatas NK, Silberberg DH (1987) MHC antigen expression on bulk isolated macrophage-microglia from newborn mouse brain: induction of Ia antigen expression by gamma-interferon. J Neuroimmunol 15:263–278
Suzumura A, Sawada M, Yamamoto H, Marunouchi T (1990) Effects of colony stimulating factors on isolated microglia in vitro. J Neuroimmunol 30:111–120
Suzumura A, Marunouchi T, Yamamoto H (1991) Morphological transformation of microglia in vitro. Brain Res 545:301–306
Takahashi K, Rochford CD, Neumann H (2005) Clearance of apoptotic neurons without inflammation by microglial triggering receptor expressed on myeloid cells-2. J Exp Med 201:647–657
Takahashi K, Prinz M, Stagi M, Chechneva O, Neumann H (2007) TREM2-transduced myeloid precursors mediate nervous tissue debris clearance and facilitate recovery in an animal model of multiple sclerosis. PLoS Med 4:e124
Tanaka J, Maeda N (1996) Microglial ramification requires nondiffusible factors derived from astrocytes. Exp Neurol 137:367–375
Theele DP, Streit WJ (1993) A chronicle of microglial ontogeny. Glia 7:5–8
Timmer AP (1925) Der Anteil der Mikroglia und Makroglia am Aufbau der Senilen Plaques. Z Neurol Psychiatr 98:43–58 [In German]
Tremblay M-E (2011) The role of microglia at synapses in the healthy CNS: novel insights from recent imaging studies. Neuron Glia Biol 7:67–76
Tremblay M-E, Lowery R, Majeewska AK (2010) Microglial interactions with synapses are modulated by visual experience. PLoS Biol 8:e1000527
Tremblay M-E, Stevens B, Sierra A, Wake H, Bessis A, Nimmerjahn A (2011) The role of microglia in the healthy brain. J Neurosci 31:16064–16069
Tsypkin LB (1959) Microglial reaction in senile dementia and the role of the microglia in the structural genesis of senile plaques. Zh Nevropatol Psikhiatr Im SS Korsakova 59:1325–1331 [In Russian]

Turner J (1905) A note concerning mesoglia cells. Rev Neurol Psychiatr 3:773–775
Unger ER, Sung JH, Manivel JC, Chenggis ML, Blazar BR, Krivit W (1993) Male donor-derived cells in the brains of female sex-mismatched bone marrow transplant receipients: a Y-chromosome specific in situ hybridization study. J Neuropathol Exp Neurol 52:460–470
Vaughn JE, Peters A (1968) A third neuroglial cell type. An electron microscopic study. J Comp Neurol 133:269–288
Vazquez-Lopez E (1942) The existence of microglia in the neurohypophysis. J Anat 76:178–186
Veremeyko T, Starossom SC, Weiner HL, Ponomarev ED (2012) Detection of microRNAs in microglia by real-time PCR in normal CNS and during neuroinflammation. J Vis Exp 20165:e4097. doi:10.3791/4097 [Video Journal Article]
Visintini F (1931) Sulla presenza di cellule ramificate simili alla microglia, nel cuore, nei muscoli volontari e nella vescica urinaria. Riv Patol Nerv Ment 37:36–47 [In Italian]
von Mihálik P (1935) Über die Nervengewebekulturen, mit besonderer Berücksichtigung der Neuronenlehre und der Mikrogliafrage. Arch Exp Zellforsck 17:119–176 [In German]
von Sántha K (1932) Untersuchungen über die Entwicklung der Hortegaschen Mikroglia. Arch Psychiatr Nervenkr 96:36 [In German]
von Sántha K, Juba A (1933) Weitere Untersuchungen über die Entwicklung der Hortegaschen Mikroglia. Arch Psychiatr Nervenkr 98:598–613 [In German]
Walton MR, Gibbons H, MacGibbon GA, Sirimanne E, Saura J, Gluckman PD, Dragunow M (2000) PU.1 expression in microglia. J Neuroimmunol 104:109–115
Wang Y, Szretter KJ, Vermi W, Gilfillian S, Rossini C, Cella M, Barrow AD, Diamond MS, Colonna M (2012) IL-34 is a tissue-restricted ligand of CSF1R required for the development of Langerhans cells and microglia. Nat Immunol 13:753–760
Ward SA, Ransom PA, Booth PL, Thomas WE (1991) Characterisation of ramified microglia in tissue culture: pinocytosis and motility. J Neurosci Res 29:13–28
Wells AQ, Carmichael EA (1930) Microglia: an experimental study by means of tissue culture and vital staining. Brain 53:1–10
Williams K, Bar-Or A, Ulvestad E, Olivier A, Antel JP, Yong VW (1992) Biology of adult human microglia in culture: comparisons with peripheral blood monocytes and astrocytes. J Neuropathol Exp Neurol 51:538–549
Williams K, Alvarez X, Lackner AA (2001) Central nervous system perivascular cells are immunoregulatory cells that connect the CNS with the peripheral immune system. Glia 36:156–164
Wilson D (2005) The early history of tissue culture in Britain: the interwar years. Soc Hist Med 18:225–243
Winkler A, de Bedo, Purves-Stewart J, Negro F (1931) Eleventh annual international neurologic assembly—society transactions. Arch Neurol Psychiatr 1931;25:653–666
Wolman M (1958) Studies on the impregnation of nervous tissue elements. IV. Mechanism of impregnation of oligodendroglia and microglia. Lab Invest 7:52–57
Wu CH, Wen CY, Shieh JH, Ling EA (1992) A quantitative and morphomteric study of the transformation of ameboid microglia into ramified microglia in the developing corpus callosum in rats. J Anat 181:423–430
Yasuzumi G, Tsubo I, Sugihara R, Nakai Y (1964) Analysis of the development of Japanese B Encephalitis (JBE) Virus. I. Electron microscope studies of microglia infected with JBE virus. J Ultrastruct Res 11:213–229
Yona S, Kim KW, Wolf Y, Mildner A, Varol D, Breker M, Strauss-Ayali D, Viukov S, Guilliams M, Misharin A, Hume DA, Perlman H, Malissen B, Zelzer E, Jung S (2013) Fate mapping reveals origins and dynamics of monocytes and tissue macrophages under homeostasis. Immunity 38:79–91
Zhang H, Wang FW, Yao LL, Hao AJ (2011) Microglia—friend or foe. Front Biosci 3:869–883

Chapter 3
Microglial Physiology

Jyoti J. Watters and Jennifer M. Pocock

Abstract In this chapter, we will review what is known about factors regulating microglial activities in the healthy CNS. Although microglia are considered to be quiescent or resting in the uninjured brain, they are not inactive but rather more in a surveillant mode. Indeed microglia perform many critical activities throughout all stages of life, including monitoring the health and function of neuronal synapses, and eliminating neuronal debris from normal synaptic remodeling and plasticity processes, as well as providing growth factors to maintain neuronal health. These microglial activities have been largely neglected because they are not obvious until an insult is present, and because appropriate in vivo experimental tools have only recently become available. As a result, comparatively little is known about routine, housekeeping functions of microglia in the healthy CNS, nor how basic microglial activities are shifted due to alterations in the normal CNS. Most information on microglial activities comes from in vitro studies, but in vivo results are becoming more available. Emphasis will be on physiologic processes such as development, aging, puberty, and learning that induce changes in microglial activities; pathological events are discussed at length in other chapters. In particular, we will focus on the modulation of microglial activities by hormones, peptides, neurotransmitters, and nucleotides, molecules whose abundance in the CNS vary with standard neural activities over the course of life. We will also highlight the critical regulation of microglial housekeeping activities by cross talk with neighboring neurons via CD200, CD300, and fractalkine interactions.

Keywords Microglia • Physiology • Inflammation • Hormones • Peptides • Neurotransmitters • Nucleotides • Growth factors • Neuron–microglia interaction • Estrous cycle • Puberty • Aging • Learning

---

J.J. Watters, Ph.D. (✉)
Department of Comparative Biosciences, University of Wisconsin,
2015 Linden Drive, Madison, WI 53706, USA
e-mail: jjwatters@wisc.edu

J.M. Pocock, Ph.D.
Department of Neuroinflammation, University College London Institute of Neurology,
1 Wakefield Street, London WC1N 1PJ, UK
e-mail: j.pocock@ucl.ac.uk

M.-È. Tremblay, A. Sierra (eds.), *Microglia in Health and Disease*,
DOI 10.1007/978-1-4939-1429-6_3

Bullet Points

- Much information is available on factors that regulate microglial inflammatory activities in vitro and in vivo.
- Little is understood about the routine housekeeping activities of microglia in the healthy CNS, nor about the factors that control them in vivo.
- Neurotransmitters and hormones can modulate microglial inflammatory factor production in vitro, but few studies have investigated these aspects in vivo.
- Microglia, especially their inflammatory activities, influence physiological processes including neurogenesis, aging, puberty, and learning, but little is known about how these processes influence microglia.
- If microglia are to be effective therapeutic targets to treat CNS disorders, a better understanding of their essential activities in the healthy CNS and how they are regulated by normal physiological processes and endogenous molecules is needed.

3.1 Introduction

Microglia play an important role in the healthy brain including synapse monitoring/pruning, proliferation, migration, and phagocytosis, and many of these activities will be detailed in other chapters. Microglia are considered to be "resting or quiescent" in the healthy central nervous system (CNS), and this term implies that they are inactive or passive. However, in vivo imaging studies demonstrate that resting microglia perform constant surveillance activities (Davalos et al. 2005; Nimmerjahn et al. 2005), physically monitoring all aspects of the CNS several times/day (Kettenmann et al. 2011). There are approximately 6,500 microglia per cubic millimeter of CNS tissue, and their processes maintain nonoverlapping regions of surveillance (Nimmerjahn et al. 2005; Beggs et al. 2012). Since the role of microglia in the injured CNS remains the best-studied aspect of microglia, the mechanisms involved in preventing microglial process overlap and regulating other microglial activities in the healthy, uninjured CNS remain poorly understood. The purpose of this chapter is therefore not only to generally review what is known about endogenous molecules that regulate microglial activities, but also to identify areas about which there is little information available. We will focus on what is known about the regulation of microglial activities by endogenous molecules whose levels naturally vary throughout the life of an organism. We will begin with how microglial activities are modulated by endogenous factors and then how these activities change with physiological processes. Molecules that regulate microglial activities include factors whose levels appear, change, or disappear in the CNS over the course of an animal's life (aging) or during normal physiological functions such as stress, pain, the estrous cycle, puberty, and learning. Because most studies in this regard have been performed in the context of an inflammatory stimulus or injury, most of what is known about the effects of hormones and neurotransmitters on microglial activities is related to modulating their inflammatory activities. However, we will point out information related to other, non-inflammatory activities of microglia when available. Each class of molecule, and its effects on microglial activities, will be discussed in its own section below.

Microglia express a plethora of neurotransmitter receptors both in vivo and in vitro and full reviews of this are published elsewhere with particular reference to the role of these receptors in neurodegeneration (Pocock and Kettenmann 2007; Kettenmann et al. 2011; Lee 2013). Recent evidence suggests that microglia can become activated in response to excessive neuronal activity during neurochemical or electrically induced seizures (Hung et al. 2010). Microglial neurotransmitter receptors can act as molecular switches to control microglial responses (Kettenmann et al. 2011) via changes in intracellular calcium signals (Seifert et al. 2011). It is likely that the repertoire of microglial neurotransmitter receptors expressed and responsive on a given cell is influenced by that cell's immediate surroundings and state of activation. Thus, microglia present in a brain region rich in dopaminergic neurotransmission such as the striatum may possess more of a propensity to express dopamine (DA) receptors and respond to DA neurotransmitters than, perhaps, a microglial cell in the hippocampus. In addition, the regular influence of local neurotransmitters may act to suppress microglial reactivity and promote ramification, while microglial reactivity may be promoted by high levels of neuronal activity. Given the evolutionary benign nature of microglia, it is to be supposed that the expression and activation of neurotransmitter receptors on microglia is beneficial [Pocock and Kettenmann (2007) for review]. While increasing evidence implicates the activity of microglia in neurodegenerative and chronic neurological conditions to be otherwise, the current review will limit the discussion to what microglial neurotransmitter receptors might do for "normal" and benign microglial function. Furthermore, we will also discuss how neurotransmitter receptors might be used to target microglia to promote regenerative "alternative" activation profiles.

In the healthy CNS, microglia produce very low levels of inflammatory cytokines (at the protein level), although they have considerable mRNA levels for several inflammatory cytokines such as tumor necrosis factor (TNF)-α. Eukaryotic cytokine mRNAs are typically short-lived, but mRNA half-lives vary widely among eukaryotes (Seko et al. 2006) and perhaps even among cell types depending upon the expression levels of requisite RNA-binding proteins (Schott and Stoecklin 2010). While cytokine mRNA stability has not been well-studied in microglia, these and other CNS cell types may maintain low levels of cytokine mRNAs to enable the rapid translation and secretion of cytokine proteins when an appropriate stimulus is experienced. In this regard, some cytokines, such as TNFα, are actually produced more highly by cells other than microglia in the healthy CNS (Nikodemova and Watters 2012), although this may change in the presence of an inflammatory stimulus such as lipopolysaccharide (LPS), where microglial production would likely dominate. However, it should be mentioned that several cytokines typically thought of as inflammatory molecules have dual functions in the CNS. Cytokines such as TNFα and interleukin (IL)-1β play roles in neuronal plasticity in the hippocampus for example (Pickering and O'Connor 2007), and basal levels of these molecules in the healthy CNS may therefore be involved in normal CNS functions instead of mediating inflammatory reactions. The neuroprotective activities of cytokines and the mechanisms involved in these beneficial activities have been reviewed in detail elsewhere (Pickering and O'Connor 2007; Spulber et al. 2009).

In contrast to inflammatory cytokines, in the healthy CNS, microglia produce neurotrophic proteins in the basal, unactivated state including insulin-like growth factor (IGF)-1 (Lalancette-Hébert et al. 2007; Suh et al. 2013) and brain-derived neurotrophic factor (BDNF) (Dougherty et al. 2000). Interestingly, following exposure to an injection injury, but not LPS, microglial production of vascular endothelial growth factor (VEGF) mRNA strongly increases (Nikodemova and Watters 2011), suggesting differential production of inflammatory and trophic molecules, and stimulus-specific production of neurotrophic factors. Although the purpose of this type of neurotrophin production/regulation by microglia is not yet known, it is reasonable to posit that neurotrophic factors produced by microglia perform the same types of neurosupportive/neuroprotective functions that are well-described for these molecules when produced by other cell types. However, specific studies investigating this have not yet been performed. Increased microglial production of BDNF in response to neuronal damage in vitro has also been reported (Lai and Todd 2008), suggesting that the microglial response to a potentially injurious insult may be to upregulate their release of neurotrophins to provide support to neurons. It remains unclear whether a single microglial cell is capable of producing neurotrophic factors at the same time as inflammatory factors.

3.2 Regulation of Microglia by Endogenous Molecules

3.2.1 Gonadal Hormones

Gonadal hormones (estrogens, progestins, androgens) are sex steroids that are made by the gonads (ovaries and testes) and adrenal glands. They can also be locally produced in tissues such as the CNS from precursor hormones. In addition to fluctuations over the monthly estrous cycle in females, circulating hormone levels also decrease with age in both males and females. In females, ovarian steroid hormone production eventually ceases and reproductive competence is lost (menopause). Gonadal hormones play critical roles in reproduction and in virtually every organ in the body via interactions with receptors that include members of the ligand-activated transcription factor family (i.e., nuclear receptors). In healthy microglia in vivo, estrogen receptor (ER)α and not ERβ are detectably expressed (Sierra et al. 2007; Crain et al. 2009), although ERβ is upregulated in microglia following ischemic injury (Takahashi et al. 2004). Glucocorticoid and mineralocorticoid receptors (GR and MR, respectively) are expressed in microglia isolated from adult brains, but androgen and progesterone receptors were undetectable (Sierra et al. 2008). These hormones are generally considered to be anti-inflammatory, but most studies have been done with estrogens and/or estrogen receptor ligands in some kind of injury or disease model. These areas have been comprehensively reviewed previously (DonCarlos et al. 2006; Pozzi et al. 2006; Vegeto et al. 2008; Crain and Watters 2010) and will not be the focus here. The effect of estrogens and progestins on the beneficial activities of microglia is less well understood than their effects on inflammatory parameters.

To the best of our knowledge, there have been no studies investigating the effects of estrogens (or any gonadal steroid) on neurotrophic factor production or release in microglia. However, a recent review article suggests the hypothesis that estrogen may regulate microglial production of IGF-1 (Sohrabji and Williams 2013), although this has not been directly tested. Progestins, like estrogens, are also anti-inflammatory to microglia in the context of an injury (Grossman et al. 2004; Pettus et al. 2005; Muller and Kerschbaum 2006; Wright et al. 2007), although their effects are not as well-studied as those of estrogens. It remains controversial whether the combination of estrogen and progesterone is beneficial to reducing cognitive decline during aging and in Alzheimer's disease (AD) (Henderson 2013; Lobo 2013), but studies on microglia in this context have been largely limited to considerations regarding decreasing their inflammatory activities.

The effects of androgens have also not been as thoroughly studied as those of estrogen in microglia. However, in general, they are anti-inflammatory after injury (García-Ovejero et al. 2002; Tomas-Camardiel et al. 2002). The ability of microglia to produce estrogens from androgen precursors has been shown in vitro (Jellinck et al. 2005, 2006, 2007; Gottfried-Blackmore et al. 2008), and in vivo, microglia express the requisite estrogen synthetic enzymes (Gottfried-Blackmore et al. 2008), suggesting that microglia may be a local source of estrogen in the CNS. Neurosteroids such as allopregenolone (derived from progesterone) are reported to reduce microglial activation (Chen et al. 2011), reduce amyloid burden in AD, and delay onset of seizures in a model of epilepsy (Biagini et al. 2009; Chen et al. 2011). However, again, little work has been done assessing the effects of these steroids, or any other, on microglial activities other than their inflammatory ones. It should also be mentioned that microglial expression of estrogen, progesterone, and androgen receptors is considerably less than glucocorticoid receptors, and most anti-inflammatory activities of these hormones have been studied in CNS tissue homogenates in which all cell types are present. Thus, it is possible that the anti-inflammatory effects of gonadal hormones may be mediated by other cell types in which these receptors are more plentiful (such as neurons and astrocytes).

3.2.2 *Adrenal Hormones*

The adrenal gland releases glucocorticoid and mineralocorticoid hormones (often in response to stress) that control important metabolic and immune processes, as well as salt and water balance. Adrenal hormone levels can vary over the course of an animal's daily experiences, and like gonadal hormones, they exert many of their effects by interactions with the nuclear receptor family members GR and MR.

In otherwise healthy animals, stress hormone levels can fluctuate rapidly in response to an acute stressor or stay elevated in situations of chronic stress. Glucocorticoids are the prototype of anti-inflammatory drugs, and like their effects on other cells of the immune system, they are also inhibitory to microglia. Microglia express both GR and MR in vivo, although GR is the most prevalent (Sierra et al. 2008). The anti-inflammatory effects of glucocorticoids have been recently reviewed (Bellavance and Rivest 2012) so they will not be addressed in detail here. Of note in microglia, GR and MR activation exert differential effects on neuroinflammatory gene regulation in vitro (Chantong et al. 2012). The MR is recruited at low doses of 11-dehydrocorticosterone which activates NFκB, and the GR is engaged at high doses and represses NFκB. Interestingly, 11-β hydroxysteroid dehydrogenase type 1 (the enzyme that converts 11-dehydrocorticosterone to the active metabolites corticosterone and cortisol) is also increased by both MR and GR activation in a microglial cell line (BV2) (Chantong et al. 2012). These data indicate a positive regulatory loop by adrenal hormones that would serve to exert biphasic effects on inflammatory gene expression. Whether this is also true in microglia in vivo is not yet known. Although microglial GR levels in vivo are downregulated by inflammatory challenge (Sierra et al. 2008), acutely, glucocorticoids have been shown to "prime" the inflammatory response of microglia to subsequent inflammatory stimuli (Frank et al. 2010, 2012), suggesting that stress may augment microglial responses to neuroinflammatory stimuli or neurodegenerative processes. However, in situations of chronic stress, when glucocorticoid levels are elevated for sustained periods, microglial GR levels decrease (Park et al. 2011), perhaps as a protective mechanism to prevent the exaggerated inflammatory responses observed with acute stress hormone exposure.

3.2.3 Hypothalamic-Acting Hormones

Corticotropin-releasing hormone (CRH) is a hormone produced by the hypothalamus that acts on the anterior pituitary gland to induce adrenocorticotropic hormone (ACTH) release that subsequently acts on the adrenal gland to promote corticosteroid release. Leptin is an adipokine produced by fat tissue that circulates in the blood and acts in the hypothalamus to inhibit appetite. Both CRH and leptin levels vary with animal activity. CRH is increased during times of stress and leptin is released during weight gain, two natural physiologic processes occurring in the absence of injury. Microglia are responsive to CRH in vitro via expression of the receptor CRH-R1, but not CRH-R2 (Wang et al. 2002). In vivo, microglia express both receptors, although CRH-R1 appears to be the dominant receptor signaling to cyclic AMP (cAMP) accumulation in primary microglial cultures (Stevens et al. 2003). CRH induces microglial proliferation and inflammatory mediator production in vitro (Wang et al. 2003c; Yang et al. 2005), and this corresponds to exacerbated neuronal damage following injury in vivo and in vitro (Stevens et al. 2003; Huitinga et al. 2004; Wang et al. 2004). Leptin plays a central role in feeding behavior by suppressing appetite. High fat diets promote hypothalamic inflammation thought to involve microglial production of pro-inflammatory cytokines including IL-6 and IL-1β (Pinteaux et al. 2007; Tang et al. 2007; Lafrance et al. 2010). Neonatal rats that are overnourished during development have higher numbers of activated microglia and higher IL-6 levels (Tapia-González et al. 2011). Hypothalamic neuroinflammation is believed to be due to long chain saturated fatty acid activation of Toll-like receptor 4 (TLR4) in the hypothalamus (Milanski et al. 2009). TLR4 is a key pattern recognition receptor whose activation promotes production of inflammatory molecules from microglia. Because most TLR4 in the CNS is expressed by microglia, the inflammatory effects of saturated fatty acids may be mediated by microglia (Milanski et al. 2009). Interestingly however, centrally administered leptin also has the ability to exert neuroprotective effects in injuries such as spinal cord injury where it increases neuroprotective gene expression and decreases microglial reactivity (Fernandez-Martos et al. 2012). Leptin also appears to be protective in a model of ischemia/stroke where leptin levels increase in microglia in the penumbra region (Valerio et al. 2009). Leptin increases levels of the anti-apoptotic protein Bcl-xl via an NFκB-dependent pathway in neurons in vitro (Valerio et al. 2009). Thus, leptin upregulation by microglia during stroke may be neuroprotective due its effects on preventing neuronal apoptosis.

3.2.4 Neuropeptides

Substance P (SP) is a peptide neurotransmitter that is important in transmitting pain signals. SP activates microglia in culture, induces their production of reactive oxygen species (Block et al. 2006), and augments their production of prostaglandins in the presence of bacteria (Rasley et al. 2004). However, following a spinal cord injury in vivo, SP is neuroprotective and promotes an M2-like reparative phenotype as assessed by their decreased production of the pro-inflammatory molecules iNOS and TNFα, increased production of the anti-inflammatory IL-10, and increased axonal growth through the lesioned spinal cord (Jiang et al. 2012, 2013). Whether the effects of SP are pro-inflammatory or neuroprotective may depend on the injury, as an antagonist of the SP receptor (NK1) was also effective at reducing inflammation and microglial activation in a model of Parkinson's disease (PD) (Thornton and Vink 2012).

Vasoactive intestinal peptide (VIP) and pituitary adenylyl cyclase-activating peptide (PACAP) are up-regulated in immune cells after injury and/or inflammation. They exert anti-inflammatory effects in vivo via high-affinity receptors expressed on microglia and peripheral inflammatory cells. PACAP is a pituitary hormone and neurotransmitter that is highly related to VIP. Both PACAP and VIP have potent neuroprotective and anti-apoptotic effects in the CNS (Fernandez-Martin et al. 2006; Kojro et al. 2006; Shioda et al. 2006; Kulkarni et al. 2011; Nakamachi et al. 2011; Szabadfi et al. 2012). The anti-inflammatory and neuroprotective effects of these related peptides have been recently reviewed

(White et al. 2010; Waschek 2013). Although VIP and PACAP inhibit microglial production and release of inflammatory mediators such as TNFα, little is known about the activities of these peptides in the absence of CNS injury. How the levels of these peptides vary with age, puberty, or over the estrous cycle in the CNS and how these alterations might influence microglial activities in the healthy or injured CNS are also poorly understood.

3.2.5 5-Hydroxytryptamine (5HT)

5HT acts as an anti-inflammatory signal on peripheral immune cells by influencing cytokine and chemokine secretion, migration, and T-cell priming capacity of dendritic cells in vitro and in vivo (Müller et al. 2009). To date, a number of subclasses of 5HT receptors have been identified ($5HT_1$, $5HT_2$, $5HT_3$, $5HT_4$, $5HT_5$, $5HT_6$ and $5HT_7$) as well as splice variants of each receptor [reviewed in Filip and Bader (2009)], and of these, a limited few have been shown to be expressed on microglia. The $5HT_7$ receptor has been shown to be expressed on a human microglial cell line, MC-3 (Mahé et al. 2005). Two splice variants of the $5HT_7$ receptor ($5HT_7$ (a/b)) were detected by RT-PCR and subsequently both 5-HT and the specific $5HT_7$ agonist, 5-carboxamidotryptamine (5-CT), were found to evoke concentration-dependent stimulation of cAMP accumulation in these cells. Following incubation with 5-CT, mRNA for the cytokine interleukin-6 (IL-6) was rapidly increased, in a manner sensitive to the $5HT_7$ receptor antagonist SB-269970. Although these data indicate that functional $5HT_7$ receptors are present in human microglial MC-3 cells, the actual secretion of IL-6 in response to 5-HT has not been confirmed, as stimulation of $5HT_7$ receptors might promote microglial priming. 5HT receptor antagonists may therefore promote a quiescent microglial phenotype.

Recent evidence suggests that 5HT may also act in concert with ATP signaling in murine microglia to enhance the chemotactic response of ATP, although no inherent ability of 5HT alone to promote chemotaxis has been reported (Krabbe et al. 2012). Furthermore, 5HT may decrease the phagocytic activity of microglia in early postnatal mice, as assessed by the uptake of microspheres, in areas of the brain in which amoeboid microglia are dominant and are actively migrating into the brain (corpus callosum). However, 5HT was unable to modify phagocytosis of microspheres in ramified microglia of adult mice in vivo or in vitro, suggesting that 5HT signaling through the particular subset of microglial 5HT receptors responsible here ($5HT_{2b}$) may be silenced with maturation (Krabbe et al. 2012). One function of such signaling may be to recruit microglia to developing synapses where they are known to play a direct role in the process of "synapse stripping" of inappropriately formed redundant synapses (Tremblay et al. 2010). It is possible therefore, although this remains to be proven, that the high concentrations of 5HT present and possibly released from these early synapses as well as from other non-synaptic sites may disappear during synapse maturation, promoting the silence of this particular microglial signaling cascade in the mature brain. In vitro, murine microglial $5HT_{2b}$ stimulation does not influence LPS-evoked IL-6 secretion (or LPS-evoked TNFα, nitric oxide (NO), or macrophage inflammatory protein (MIP)-1α production (Krabbe et al. 2012). Together, these data suggest that different 5HT receptors on microglia have distinct functions that may be activated at different developmental stages. The continued silence of specific 5HT receptors may only remain so if abnormal brain environments do not develop. Thus, the acute or chronic opening of the blood–brain barrier may allow sufficiently elevated brain concentrations of 5HT to reactivate the microglial $5HT_{2b}$ receptor.

3.2.6 Dopamine (DA)

Functional expression of dopamine receptors has been described in mouse and rat microglial cultures and brain slices using patch clamping assays (Färber et al. 2005). D1- and D2-like dopamine receptors were identified and found to trigger the inhibition of the constitutive potassium inward rectifier and activated potassium outward currents in a sub-population of microglia. Chronic dopamine receptor stimulation enhanced migratory activity and attenuated LPS-induced NO release in a manner similar to adrenergic receptor stimulation. Further, DA can modulate cytokine secretion in mouse peripheral and tissue-located immune cells, enhancing LPS-evoked cytokine secretion of TNFα, interferon (IFN)-γ or IL-10 in a highly tissue-specific manner (Matalka et al. 2011). These effects appeared to be mediated by DA receptors, although the exact subtype remains to be determined, i.e., D2/D3/D4.

Coculture of human elderly microglia with the differentiated dopaminergic neuronal cell line SH-SY_5Y resulted in elevated neuronal toxicity which was significantly enhanced in the presence of DA (Mastroeni et al. 2009). DA induces microglial chemotaxis, an effect that was inhibited by spiperone, a DA receptor antagonist with higher affinity for D2-like receptors (Färber et al. 2005). Cultured human elderly microglia express mRNAs and protein for D1–D4 but not D5 receptors. In situ PD brain microglia in substantia nigra and striatum tissue were also immunoreactive for D1–D4 but not D5 receptors (Mastroeni et al. 2009), and microglia are also thought to phagocytose degenerating DA neurons in rodent models (Cho et al. 2003). Together, these observations suggest that specific microglial DA receptor activation may play a role in the selective vulnerability of DA neurons in PD by chemotactically targeting DA neurons resulting in their phagocytosis. However, direct chemotactic effects of DA have not been demonstrated in vivo. The apparent disparities in microglial expression of the different receptors for DA may point to species differences, age-dependent, or disease-susceptible changes in the repertoire of microglial DA receptors expressed. Further studies are needed to define the functional role of DA receptors in microglia.

3.2.7 Norepinephrine (NE)

Microglia express α1, α2, as well as β2 adrenergic receptors (AR) that are Gi/Gq-coupled and Gs-coupled, respectively (Kohm and Sanders 2001; Mori et al. 2002). Patch-clamp analyses have revealed that both DA and NE can directly modulate membrane currents in primary neonatal microglial cells which could suppress cytokine release and inflammatory activation of microglia (Färber et al. 2005). Murine microglia (N9 cell line and primary cells) express β2AR, but not β1AR or β3AR (but see (Mori et al. 2002) for rat primary microglia which do express β1AR and are also immunosuppressive). β2 agonists inhibit microglial proliferation by enhancing intracellular cAMP (Fujita et al. 1998). In primary rat-cultured microglia, βAR agonists and α1AR agonists (e.g., phenylephrine) suppress IL-6 and TNFα expression and release, as well as NO production (Mori et al. 2002). Microglial gene expression of a range of inflammatory mediators (including TNFα, COX-2, IFNγ, CCL-2, IL-1β, and IL-8) and the secretion of chemokines and cytokines such as TNFα are suppressed by NE (McNamee et al. 2010), and by the non-selective beta-adrenergic agonist, isoproterenol, via increased cAMP accumulation (Heneka et al. 2010). The anti-inflammatory role of NE in inflammatory and neurodegenerative disease with regard to microglial–neuronal interactions has been reviewed further elsewhere (Carnevale et al. 2007). The induction of a neuroprotective phenotype in microglia in vitro and in vivo by NE is further supported by the finding that NE can regulate the expression of cytokine receptors such as IL-1 receptor antagonist (IL-1ra) and the release of a range of chemokines and cytokines (McNamee et al. 2010). In general, NE promotes a neuroprotective phenotype, although a recent study reports that β2AR stimulation increases microglial NADPH oxidase (nicotinamide adenine dinucleotide phosphate-oxidase; a membrane-bound enzyme complex that produces superoxide) activity and DA-induced neurotoxicity in rat primary mesencephalic neuronal-glial cultures (Qian et al. 2009). Furthermore, treatment with the βAR agonist propranolol following surgical injury enhances microglial pro-inflammatory cytokine production (Wang et al. 2003b). These disparate effects suggest that the context of microglial AR stimulation is important. The exact repertoire and function of NE receptors expressed on rodent as well as human microglia remain to be fully elucidated.

In vivo studies have demonstrated that NE deficiency enhances the deposition of amyloid-beta (Aβ) plaques in mouse models of AD, leading to the suggestion that NE signaling to microglia promotes phagocytosis (Kalinin et al. 2007; Heneka et al. 2010). In addition, the number of microglia recruited to Aβ plaques is reduced by NE depletion, suggesting that recruitment of microglia is disrupted in the absence of NE (Heneka et al. 2010). NE directly promotes Aβ 1–42 ($A\beta_{1-42}$) phagocytosis by murine microglia in vitro and downregulates $A\beta_{1-42}$-induced cytokine and chemokine production (Kalinin et al. 2007; Heneka et al. 2010), suggesting that NE can switch microglia from an M1 inflammatory, to an M2 neuroprotective phenotype. β2AR activation with NE or isoproterenol enhances the expression of the Aβ receptor mFPR2 in vivo and in vitro, the mouse homologue of the human formyl peptide receptor FPR2 (Kong et al. 2010). In addition, NE or isoproterenol induced the expression of the Aβ-degrading enzyme and enhanced the capacity of microglia to phagocytose and degrade $A\beta_{1-42}$ through mFPR2. Mechanistic studies indicate that isoproterenol induces mFPR2 expression in microglia, suggesting that NE acts as a link between neurons and microglia to orchestrate the host response to Aβ (Kong et al. 2010).

3.2.8 GABA and Glycine

Mouse microglia in vivo and in vitro express functional $GABA_B$ receptors (metabotropic receptors linked via G proteins to potassium channels) (and subunits 1 and 2) whose activation triggers an outwardly rectifying K^+ current and intracellular Ca^{2+} transients resulting in reduced interleukin secretion (Kuhn et al. 2004). Mouse microglia also express $GABA_A$ receptors (ligand gated ionotropic receptors) which modulate microglial functions such as motility and volume sampling (Fontainhas et al. 2011; Kettenmann et al. 2011) and may also promote an anti-inflammatory phenotype possibly by modulating Cl^- fluxes (Lee 2013). $GABA_A$ but not $GABA_B$ receptor activation on cultured rat microglia can induce NADPH oxidase activity and intracellular superoxide generation (Mead et al. 2012) which modulates downstream intracellular signaling cascades to promote a neuroprotective microglial phenotype. Human microglia express $GABA_A$ and $GABA_B$ receptors (Lee et al. 2011) which, upon activation with either a $GABA_A$ receptor agonist (muscimol), or a $GABA_B$ receptor agonist (baclofen), attenuate lipopolysaccharide (LPS)/interferon γ-mediated activation and neurotoxicity. Overall, the activation of GABA receptors on microglia appears to promote a protective phenotype.

Like GABA, glycine also has inhibitory effects on macrophages by reducing agonist-stimulated Ca^{2+} signals and attenuating the generation of reactive oxygen species, the production of pro-inflammatory cytokines, phagocytic activity, and cell proliferation [for review, see Zhong et al. (2003), Schilling and Eder (2004)]. Nevertheless, data for the expression of glycine receptors and L-type Ca^{2+} channels in microglia are so far unconvincing. Glycine alone can also enhance microglial intracellular calcium transients induced by 100 μM ATP or by 500 nM thapsigargin (Van den Eynden et al. 2011). However, since these responses are unaffected by the glycine receptor antagonist strychnine, and they are not mimicked by glycine receptor agonists such as taurine or β-alanine, glycine receptors per se may not be involved. Instead, these effects of glycine were found to be mediated by sodium-coupled neutral amino acid (glycine) transporters (Van den Eynden et al. 2011). Evidence in neuropathic pain suggests these transporters may be downregulated on microglia, thereby increasing extracellular glycine concentrations which then potentiate NMDA receptor activation (Cavaliere et al. 2007). Thus to date, there is little evidence of glycine receptor expression by microglia under physiological conditions, although this may not be true in pathological conditions.

3.2.9 Acetylcholine (ACh)

Overall, cholinergic pathways in the brain confer anti-inflammatory properties to microglia by acting at α7 nicotinic receptors (α7nAChRs) (Wang et al. 2003a; Giunta et al. 2004; Carnevale et al. 2007). Decreased cholinergic input to microglia in neurodegeneration is implicated in fuelling neuroinflammation by relieving this block on microglia (Carnevale et al. 2007; Kawamata and Shimohama 2011). Cultured rodent microglia express mRNA and protein for α7nAChR and activation of these receptors selectively upregulates COX-2 and prostaglandin E2 (Shytle et al. 2004), and suppresses LPS-evoked TNFα secretion (De Simone et al. 2005; Thomsen and Mikkelsen 2012). This is proposed to occur by modulating upstream phosphorylation of several signalling pathways including p44/42 and p38 mitogen-activated protein kinases (De Simone et al. 2005), JAK2/STAT3 (Kawamata and Shimohama 2011) and PLCC/$InsP_3$/Ca^{2+} signalling (Suzuki et al. 2006). Rhesus monkey retinal microglia express α6 and β4 nAChR subunits (Liu et al. 2009) and human embryonic-cultured microglia express mRNA for α3, α5, α7, and β4 subunits (Rock et al. 2008). Human-cultured microglia may also express functional muscarinic ACh receptors (mAChR) due to the induction of an intracellular calcium response by carbachol, which could be blocked by atropine (Whittemore et al. 1993; Zhang et al. 1998). To date, little else is known about mAChR function in microglia, and this remains a somewhat neglected area.

3.2.10 Excitatory Amino Acids

The role of microglial glutamate receptors in neurodegeneration has been reviewed previously (Pocock and Kettenmann 2007; Noda and Beppu 2013). In vitro, activation of the ionotropic glutamate receptors (AMPA and KA) on microglia is thought to promote a pro-inflammatory microglial phenotype (Lee 2013). Rat microglia have been shown to express mRNA for the AMPA receptor subunits GluR1, GluR2, GluR3, and GluR4 which function to evoke cationic currents that are inhibited by CNQX (6-cyano-7-nitroquinoxaline-2,3-dione), a competitive AMPA/kainate receptor antagonist, and which may modulate TNFα secretion (Hagino et al. 2004; Noda and Beppu 2013). In addition, activation of AMPA receptors by glutamate can promote membrane ruffling and chemotaxis by regulating the cytoskeleton (Liu et al. 2009) (also see review (Lee 2013)). Expression data for KA receptor subunits (GluR5, GluR6, GluR7, KA1, and KA2) and NMDA receptor subunits (NR1, NR2A, NR2B, and NR3) on microglia in the healthy CNS is currently lacking. However, application of KA to rat microglia induces Ca^{2+} currents (Noda and Beppu 2013) and promotes TNFα secretion from primary cultured rat microglia (Zhu et al. 2010). Increased KA receptor expression (GluA6 mRNA) has been reported in a proliferating rodent microglial cell line (GMIR1) (Yamada et al. 2006), suggesting cell cycle-dependent regulation of KA receptors. Functional studies suggest that

NMDA receptor activity on microglia might be pro-inflammatory because the NMDA receptor antagonist MK-801 prevented microglial reaction in rat hippocampus after forebrain ischemia (Streit et al. 1992; Hirayama and Kuriyama 2001). Further studies are needed to consolidate these findings.

Microglia also express the Group I metabotropic glutamate receptors mGluR5, but to date, not mGluR1 (Biber et al. 1999), which mediates intracellular signalling via PLC/IP3/Ca_2^+/PKC (Conn and Pin 1997). In vitro, mGluR5 activation downregulates microglial inflammatory activities in the presence of stimuli such as LPS or fibrinogen (Farso et al. 2009; Piers et al. 2011). Cultured microglia also express mRNA and protein for the Group II receptors, mGluRs 2 and 3, and activation of these receptors is negatively coupled to adenylate cyclase (Taylor et al. 2002). In particular, in vitro activation of mGluR2 promotes a neuroinflammatory and neurotoxic microglial phenotype mediated by TNF-α secretion and microglial-derived FAS ligand shedding (Taylor et al. 2002, 2005). A similar neurotoxic profile was observed in vivo (Kaushal and Schlichter 2008). Primary microglial cultures express mRNA and protein for Group III receptors mGluRs 4, 5, and 8, but not 7. These receptors are also coupled to adenylate cyclase activation, and their activation promotes a neuroprotective microglial phenotype (Taylor et al. 2003), which might be due to modulation of microglial stress pathways preventing the activation of Group II mGluRs, or by regulating NADPH oxidase-coupled signaling cascades (Mead et al. 2012).

Microglia can also release glutamate via several mechanisms: reversal of the X_c^- transporter in exchange for cysteine (Kingham et al. 1999; Barger and Basile 2001), a bafilomycin-sensitive pathway (Kingham et al. 1999) or by glutamate-permeable volume-regulated anion channels (VRACs) (Harrigan et al. 2008). Basal expression of glutamate transporters on microglia appears to be very low. However, expression can be enhanced upon microglial activation, as the glutamate transporters GLT1/EAAT-2 (SLC1A2) and GLAST/EAAT-1 are upregulated following brain injuries and in neurodegenerative diseases (Kanai and Hediger 2004; Kettenmann et al. 2011), suggesting a beneficial role for microglia in reducing excitotoxicity.

3.2.11 Nucleotides

Nucleotides (adenine- and uracil-containing di- and tri-phosphates) are released in the brain by several normal physiologic mechanisms (Dubyak and el-Moatassim 1993; Lazarowski et al. 1997; Lutz and Kabler 1997) including co-release with other neurotransmitters from neurons (Dunant and Ballmer-Hofer 1997; Bodin and Burnstock 2001; Kasai et al. 2001; Matsuka et al. 2001; Poelchen et al. 2001) and release from nearby cells such as astrocytes, through membrane channels or gap junctions (Cantiello 1997; Bodin and Burnstock 2001). There are 19 known receptors for nucleotides [thoroughly reviewed elsewhere (Burnstock 2007)]. In brief, receptors for nucleosides and monophosphate containing nucleotides (e.g., adenosine and AMP) interact with P_1 receptors (also called adenosine receptors), whereas di- and triphosphate containing nucleotides (e.g., ADP and ATP) interact with receptors of the P_2 family. The G protein-coupled P_1 receptor family includes the A_1, A_{2a}, A_{2b}, and A_3 receptors (Burnstock 2007). The P_2 receptor family is subdivided based upon agonist specificity and putative transmembrane topologies into two classes: P_2Y (hetero-trimeric G-protein-coupled) and P2X (ionotropic) receptors. P_2Y receptors include P_2Y_1, P_2Y_2, P_2Y_4, P_2Y_6, P_2Y_{11}, P_2Y_{12}, P_2Y_{13}, and P_2Y_{14}, and adenine, cytosine, and uridine containing nucleotides are ligands (Burnstock and Knight 2004). P_2X receptors are designated P_2X_1-P_2X_7 and all are activated by ATP (Abbracchio and Burnstock 1998). Phosphate groups are removed from nucleotides by the actions of four families of ectonucleotidases, extracellular enzymes that have complex enzymatic substrate specificities, cellular localizations, and regulation [(reviewed in Kukulski et al. (2011)]. While ATP is a ligand for most known P_2 receptors, ADP tends to prefer certain P_2Y receptors. UTP prefers P_2Y receptor subtypes, although it does have some affinity for P_2X_5 receptors (Burnstock and Knight 2004).

Of the 15 P_2 receptor for nucleotides, microglia express most (but not all) of them although their levels vary with age, sex, and whether or not detection is being performed in primary microglial cultures or in vivo (Crain et al. 2009). Nucleotides are an important regulator of both inflammatory and anti-inflammatory microglial cell functions in vitro, including promoting plasminogen (Inoue et al. 1998), TNFα (Hide et al. 2000), IL-1β (Ferrari et al. 1997b; Sanz and Di Virgilio 2000), IL-6 (Friedle et al. 2011), TGF-β (Wang et al. 2003a), CC-chemokine ligand (CCL)3 (Kataoka et al. 2009), and CCL2 (Morioka et al. 2013) production. Nucleotides also have cytotoxic (Ferrari et al. 1996, 1997a) and chemotactic (Honda et al. 2001) effects in vitro. While many of these effects are attributed to the actions of P_2X receptors, certain P_2Y receptors also influence these endpoints (Honda et al. 2001; Boucsein et al. 2003; Brautigam et al. 2005; Morioka et al. 2013). However, P_2 receptor activation is not always pro-inflammatory as it can reduce the effects of other inflammatory stimuli like LPS (Boucsein et al. 2003; Brautigam et al. 2005; Apolloni et al. 2013) and also promote the production of anti-inflammatory cytokines such as IL-10 (Seo et al. 2004, 2008). Although the effects of nucleotides on trophic factors produced by microglia have not been well described, P_2 receptor activation controls microglial transcription and release of BDNF in vitro and in vivo (Lai and Todd 2008; Ulmann et al. 2008; Trang et al. 2009), suggesting that the effects of nucleotides in microglia are not only restricted to modulation of their inflammatory activities. Together, these results suggest that activation of P_2X receptors may have both inflammatory and anti-inflammatory activities in microglia.

Whereas microglial P_2X_4 is best studied in the development of tactile allodynia and neuropathic pain (Tsuda et al. 2003; Trang et al. 2011; North and Jarvis 2013), P_2X_4, along with P_2Y_1 and P_2Y_{12}, also contribute to ATP-dependent chemotaxis of microglia within the CNS (Haynes et al. 2006; Ohsawa et al. 2007, 2010; De Simone et al. 2010). The receptors mediating exogenously administered ATP-induced microglial motility in vivo (Davalos et al. 2005) are not yet clear, but likely do not involve the actions of P_2Y_1 or P_2Y_{12} receptors, as ATP only weakly agonizes them. However, P_1 receptors may play a role as A_{2a} receptors regulate microglial process retraction (Orr et al. 2009) and A_{3a} receptors control process extension (Ohsawa et al. 2012). UTP/UDP activation of P_2Y_6 receptors also controls microglial phagocytosis in vitro and in vivo (Koizumi et al. 2007). Thus, nucleotides, acting through multiple P_1 and P_2 receptor subtypes, control many critical microglial functions.

3.2.12 *Fractalkine*

Fractalkine–CX_3CR_1 interactions attenuate microglial reactivity. CX_3CR_1 is a microglial receptor activated by the neuronal chemokine CX_3CL_1 (fractalkine), which controls key functions of microglial cells. For example, neutralizing fractalkine antibodies exacerbate neuroinflammation following central delivery of LPS (Zujovic et al. 2001). In cultured microglia, fractalkine treatment attenuates LPS-induced production of the inflammatory cytokines TNFα, IL-6, and IL-1β (Zujovic et al. 2000; Mizuno et al. 2003). CX_3CR_1 deficiency increases microglial IL-1β expression and subsequent neurotoxicity following repeated peripheral LPS injections (Cardona et al. 2006) suggesting that fractalkine–CX_3CR_1 interactions function to repress microglial inflammatory activities. The enhanced expression of fractalkine by developing neurons also acts as a microglial chemoattractant and indicates that microglia may be recruited to developing synapses to aid in synapse maturation (Hoshiko et al. 2012). However, fractalkine–CX_3CR_1 interactions can also have detrimental effects in the CNS. CX_3CR_1 activation induces p38 MAPK phosphorylation in microglia, which is a critical mediator of spinal neuronal–microglial communication in chronic pain (Staniland et al. 2010). Fractalkine–CX_3CR_1 interactions are implicated in pathological pain and in particular, hypersensitivity associated with neuropathic pain in the spinal dorsal horn (Scholz and Woolf 2007; Milligan et al. 2008; Ren and Dubner 2008; Staniland et al. 2010). Microglial CX_3CR_1 levels are chronically downregulated in the aging brain (Wynne et al. 2010), which may contribute to their exaggerated inflammatory responses. Thus, whereas fractalkine–CX_3CR_1 interactions promote beneficial microglial activities in the developing brain and contribute to microglial quiescence in the healthy adult brain, they may also be detrimental in chronic pathology (such as pain) and in the aging brain.

3.2.13 *CD200 and CD300 Interactions*

In addition to fractalkine–CX_3CR_1 interactions that maintain microglia in a surveillance and quiescent state, specific ligand–receptor pairs, for example via CD200–CD200R interactions, also contribute (Lynch 2013). CD200 is expressed on a number of cell types including neurons, while its receptor, CD200R, is expressed on myeloid cells, including microglia where it functions to suppress inflammatory responses in these cells in vivo (Denieffe et al. 2013). It has recently been suggested that the expression of CD200R on human macrophages is indicative of an M2, or alternatively activated phenotype, which is neuroprotective (Jaguin et al. 2013).

CD300 molecules are glycoproteins that comprise a family of transmembrane and soluble receptors that are important regulators of mammalian immune cell functions including differentiation, viability, cytokine and chemokine secretion, phagocytosis, and chemotaxis. CD300 is also known as "triggering receptor expressed on myeloid cells" (TREM) and TREM-like (TREML) receptors (Cannon et al. 2012). These receptors can inhibit (CD300a, CD300f) or trigger (CD300b, CD300e) microglial and monocyte cell functions (Clark et al. 2009; Simhadri et al. 2013), although some have as yet unknown functions in this regard (CD300c, CD300d). The overall function of these molecules may be determined by their ability to form homo- and heterodimeric signaling complexes (Martínez-Barriocanal et al. 2010). While specific ligands for most of the CD300 family members remain elusive, recent data indicate TREM and TREML families can recognize lipids, such as extracellular ceramide, phosphatidylserine, and phosphatidyl-ethanolamine that become flipped and exposed on the outer leaflet of the plasma membrane of apoptotic and activated immune cells (Cannon et al. 2012; Borrego 2013). CD300c is expressed on the cell surface of monocytes following in vitro differentiation, and on macrophages and dendritic cells (Simhadri et al. 2013). CD300c expression can be dynamically regulated by the TLR ligands LPS and flagellin, and activation of CD300c can potentiate LPS-stimulated production of inflammatory cytokines on human monocytes (Simhadri et al. 2013). Although it has not yet been specifically evaluated in microglia, CD300c may also be an activating receptor regulating microglial pro-inflammatory activities. In contrast, CD300a (also called TREM2) is an inhibitory receptor which binds to non-MHC class I ligands. Phosphatidylserine (PS) is a ligand for CD300a, and interactions between CD300a and PS expressed on apoptotic cells inhibit the uptake of apoptotic cells by phagocytic cells (Lankry et al. 2013). The activation of TREM2 promotes phagocytosis by microglia and inhibits their inflammatory responses (Hsieh et al. 2009). Recent data implicate mutations in TREM2 (which is expressed on microglia), as a risk factor for the development of late onset AD (Forabosco et al. 2013; Guerreiro et al. 2013), consistent with the notion that impaired microglial phagocytosis of Aβ may contribute to AD (Sierra et al. 2013).

3.3 Regulation of Microglia by Normal Physiological Processes

3.3.1 Neurogenesis and Learning

Neurogenesis occurs throughout life in the mammalian brain, particularly in the hippocampal dentate gyrus subgranular zone (SGZ) and the sub-ventricular zone (SVZ) (Lois and Alvarez-Buylla 1993; Abrous et al. 2005). Since the role of microglia in neurogenesis is extensively covered in Chap. 10, we will simply mention here that microglial inflammatory activities have been correlated with impaired neurogenesis in the SGZ (Ekdahl et al. 2003; Ekdahl 2012). However, more recently, microglial neuroprotective activities have been associated with increased neurogenesis through release of anti-inflammatory cytokines and growth factors (Aarum et al. 2003; Morgan et al. 2004; Battista et al. 2006; Butovsky et al. 2006; Ziv et al. 2006; Vukovic et al. 2012). It is necessary to point out that most microglial studies to date have been mostly correlative; direct causative or mechanistic studies have not been performed in vivo. However, an important role for surveillant/ramified microglia was recently identified in neurogenesis (Sierra et al. 2010). Surveillant microglia were shown to phagocytose the debris from newly created non-surviving neurons that undergo apoptosis, and this phagocytic process did not correlate with an increase in microglial markers indicative of activation to the pro-inflammatory phenotype (Sierra et al. 2010). Interestingly, although the rate of neurogenesis decreases in the aging CNS (Kuhn et al. 1996), and unactivated microglia become more pro-inflammatory with age (Sierra et al. 2007; Crain et al. 2013), the ability of aged microglia to phagocytose apoptotic neurons did not change (Sierra et al. 2010), suggesting that this microglial property is unaffected by age. These novel data underscore the importance of normal, healthy microglial contributions to proper CNS function and neurogenesis, even in the aging brain, and distinguish between their inflammatory and surveillant activities.

The vast majority of the literature in the realm of learning focuses on how microglial activities, predominantly their inflammatory activities, interfere with various types of learning and memory. Hence, scant information is available regarding how healthy, unactivated microglia influence learning (also see Chaps. 9 and 10 for further information). However, two recent reports have begun to shed light on how normal microglial activities contribute to learning (Rogers et al. 2011; Parkhurst Christopher et al. 2013). Microglia are maintained in the surveillant state in the healthy CNS due to the influence of several neuronal signals. One such signal, fractalkine (CX_3CL1), is produced by healthy neurons and acts on CX_3CR_1 receptors which are expressed on microglia in the CNS (Harrison et al. 1998). In the presence of fractalkine, microglial production of inducible nitric oxide synthase (iNOS), IL-1β, TNFα, and IL-6 is kept to a minimum (Cardona et al. 2006). However, when neurons are damaged, tonic fractalkine levels decrease, sustained CX_3CR_1 activation is interrupted, and microglia are permitted to synthesize pro-inflammatory molecules (Cardona et al. 2006). Therefore, chronic fractalkine–CX_3CR_1 interactions regulate normal microglial activities. In mice in which fractalkine–CX_3CR1 signaling has been disrupted, not only was neural progenitor cell proliferation and survival decreased (Bachstetter et al. 2011), but motor learning, cognitive function, and synaptic plasticity were also impaired (Rogers et al. 2011), suggesting that basal microglial activities contribute to effective learning and normal neurogenesis. Further, in CX_3CR1-deficient mice, environmental enrichment failed to augment synaptic plasticity (Maggi et al. 2011), indicating that microglial activities are also necessary for experience-dependent learning. Recently, it was shown that mice in which microglia had been depleted had deficits in many learning tasks and motor-dependent synapse formation, an effect that was recapitulated in mice possessing microglial-specific deletion of BDNF (Parkhurst Christopher et al. 2013). Further, these studies demonstrated that the absence of microglial BDNF altered synaptic protein levels, synaptic structure, and mouse learning. Significantly, these results show the importance of a non-inflammatory microglial molecule in learning and memory. Additional studies like these, focused on the role of non-inflammatory microglial activities on learning, as well as the converse set of studies—how learning influences microglial activities, are needed. To the best of our knowledge, there have been no studies where animals have been subjected to different learning paradigms and any microglial activity evaluated thereafter.

3.3.2 Estrous Cycle

Ovarian hormone levels and synaptic density vary over the estrous cycle. Estradiol levels peak at ovulation, just prior to onset of the luteal phase, during which time progesterone levels dominate until the onset of menses when progesterone levels fall (Staley and Scharfman 2005). Neuronal synapse density is highest when serum estrogen levels are highest (during proestrus) (Woolley and McEwen 1992), and the effects of estrogen on synapse density are counteracted by progesterone (Woolley and McEwen 1993), suggesting that both hormones work together to regulate synapses. In in vitro wounding in a dish coculture studies, the antagonistic effect of progesterone on neurite outgrowth was shown to require microglia, as the effects of progesterone were lost in the absence of microglia (Wong et al. 2009). Microglial molecules also vary over the estrous cycle. For example, phosphorylated levels of the BDNF receptor TrkB (indicating functional activation of the receptor) increase in hippocampal microglia during proestrus compared to males (Spencer-Segal et al. 2011), suggesting an important role for microglia in the healthy hippocampus. Microglial production of ApoE also varies throughout the estrus cycle, with levels being highest in proestrus (Stone et al. 1997). Interestingly, estrogen was not able to increase ApoE levels in microglial cultures in vitro, suggesting that the CNS milieu is important for this effect. ApoE in the CNS is thought to be important in scavenging cholesterol from cellular debris to facilitate axonal regeneration and remyelination (Mahley 1988). Cholesterol is enriched in plasma membranes, growth cones, and synapses, and it is an essential membrane component that in the CNS is typically synthesized locally by neurons themselves and astrocytes (Fünfschilling et al. 2012). Thus, microglial activities contributing to modulation of synapse density, cholesterol scavenging and responses to neuronal trophic factors such as BDNF vary during the estrus cycle, in concert with coordinated changes in serum estrogen and progesterone levels.

3.3.3 *Puberty*

Puberty, the time when adolescents become reproductively competent, involves hypothalamic gonadotropin-releasing hormone (GnRH) stimulation of the anterior pituitary gland gonadotrophs to produce luteinizing hormone (LH), necessary for triggering ovulation in females and testicular androgen production in males. Microglia play a role in normal ovulation and estrogen feedback to the hypothalamus for LH release (Cohen et al. 1999) in females, although little is known about the converse—how puberty affects microglia. Osteopetrotic (Op) mice have a mutation in the macrophage colony-stimulating factor (*M-CSF* or *CSF-1*) gene, and as a result, they have reduced macrophage numbers in many tissues including the CNS (Wegiel et al. 1998; Kondo et al. 2007; Kondo and Duncan 2009). Op females have extended estrus cycles and poor ovulation rates, and both sexes have reduced fertility. Puberty is also significantly delayed in Op females compared to littermate controls, and males have low circulating LH and testosterone levels. Importantly, these endpoints are normalized by administration of a GnRH agonist (Cohen et al. 2002), suggesting that Op mice have a defect at the level of the hypothalamus. Because microglia are the primary cell type expressing receptors for M-CSF in the hypothalamus, microglia likely influence hypothalamic GnRH neuron function. Further studies are necessary to identify the specifics of this interaction.

3.3.4 *Aging*

Microglial inflammatory activities have been proposed to be dysregulated with age (Franceschi et al. 2000; Dilger and Johnson 2008), an effect that is thought to contribute to or increase the propensity for neurodegeneration in aged animals (Mrak and Griffin 2005). With age, microglial inflammatory activities in vivo increase while their capacity to migrate and phagocytose Aβ decreases (Harry 2013). The increased pro-inflammatory nature of aged microglia is supported by increased expression of immune activation markers such as MHC class II, LCA, CD4, and ED1 in aged rat microglia (≥24 months) (Perry et al. 1993) and in postmortem human brain (Sheng et al. 1998). Basal microglial expression of cytokines also increases in vivo with age (Sierra et al. 2007; Crain et al. 2013), and in microglial cultures from the aged brain (Ye and Johnson 1999; Njie et al. 2012). Increased basal cytokine levels likely contribute to exacerbated neuronal damage and microglial responses to stimulation or injury (Sugama et al. 2003; Godbout et al. 2005; Sandhir et al. 2008; Wasserman et al. 2008). Aged microglia in the retina are slower to initiate an inflammatory response to injury or ATP, and their responses are prolonged, as they fail to appropriately inactivate (Damani et al. 2011), indicating that aging dysregulates the onset and resolution of microglial activities (see Chap. 13). Concurrent reductions in beneficial, neurosupportive microglial activities as a function of age are also observed in microglial dynamic behavior and ramification in the retina (Damani et al. 2011). Reductions in microglial surveillant properties may therefore not only influence the regulation of synapses (Wake et al. 2009; Schafer Dorothy et al. 2012) and neuronal activity (Li et al. 2012), but also their neuroprotective functions (Vinet et al. 2012).

Microglial support activities are also critical in the CNS, and they are hypothesized to decrease with age (Streit and Xue 2009). In support of this idea is a recent study reporting lower basal levels of BDNF release and glutamate uptake in microglial cultures from aged animals compared to young adults (Lai et al. 2013) even though inflammatory activities increase with age (Godbout et al. 2005; Sierra et al. 2007; Crain et al. 2009). Interestingly, however, a recent sequencing of the microglial transcriptome from brains of young (5 months-old) and aged (24 months-old) adult mice shows that the genes that were most upregulated in aged microglia belonged to pathways considered to be neuroprotective, whereas genes in pathways considered to be neurotoxic were downregulated (Hickman et al. 2013). Moreover in aged microglia, genes indicative of the M2 alternatively activated phenotype were upregulated as were genes encoding receptors that sense bacterial and fungal ligands, whereas receptors involved in sensing endogenous ligands from apoptotic neurons etc. were downregulated (Hickman et al. 2013). These results suggest that microglia may not be pro-inflammatory in the aged, healthy brain, but rather that their responses to pathogen challenge may be exaggerated in the geriatric CNS compared to the young adult. Whereas this conclusion is consistent with most aging studies that use an inflammatory stimulus to assess the activities of aged microglia, there may also be CNS region-specific effects of aging on microglial activities. Thus, microglia in the retina may behave differently from the whole brain microglia assessed in the study by Hickman and colleagues.

Young adult microglia produce many factors that are beneficial to neuron health and survival including neurotrophins such as BDNF and glial-derived neurotrophic factor (GDNF) (Beggs et al. 2012; Wang et al. 2013), IGF-1 (Kaur et al. 2006; Lalancette-Hébert et al. 2007), and VEGF (Nikodemova and Watters 2011; Crain et al. 2013). Basal IGF-1 levels appear to be upregulated in microglia following injury in vivo (Lalancette-Hébert et al. 2007) or activation in vitro (Kaur et al. 2006). Interestingly, the effects of IGF-1 appear to be biphasic. The absence of microglial IGF-1 exacerbates ischemic injury in the adult (Lalancette-Hébert et al. 2007), but delivery of IGF-1 to the developing brain in the presence of an acute inflammatory stimulus like LPS was severely detrimental (Pang et al. 2010). Importantly, how microglial production of these trophic molecules changes in the aging CNS in vivo is not yet known. However, a recent report indicates that the neurotrophic factors IGF-1 and fractalkine may also become dysregulated in aged microglia since there are increased levels of microRNAs in the aged CNS that function to reduce the translation of these trophic factors (Fenn et al. 2013).

In vitro, microglia produce BDNF and GDNF in response to conditioned medium from damaged neurons that sustained a hypoxic injury (Lai and Todd 2008). Although the severity of neuronal damage did not affect microglial production of these trophic molecules (they produced them regardless of magnitude of neuron injury), it did influence their production of inflammatory molecules. Only conditioned medium from mildly injured neurons promoted microglial production of inflammatory molecules. Interestingly, microglia exposed to conditioned medium from moderately injured neurons attenuated hypoxia-induced neuronal death in cocultures, suggesting neuroprotective activities of microglia. Whereas the role of BDNF release from microglia is well-established in mediating pain in the spinal dorsal horn (Trang et al. 2011) (also see Chap. 11), the physiological effects of BDNF or microglial TrkB receptor activation on the immune activities of microglia are not yet clear. However, in macrophages in vitro, BDNF treatment increases phagocytic activity and promotes IL-1β synthesis and release (Asami et al. 2006), suggesting similar effects may also be observed in microglia. To our knowledge, this has not been tested. In contrast, GDNF attenuates microglial inflammatory activities in vitro (Rocha et al. 2012), and centrally delivered VEGF in vivo reduces microglial activation and neuroinflammation in focal cerebral ischemia (Herz et al. 2012). Together, these data suggest that aging interferes with both normal inflammatory and neurotrophic factor responses in microglia, both aspects of which likely contribute to the hyper-reactivity of microglia in the aged CNS.

3.4 Conclusions/Summary

In this chapter, we have discussed the major classes of endogenous molecules regulating microglial activities. As described above, most studies of regulation in this regard focus on the inflammatory activities of microglia, typically during pathology. As a result, comparatively little is known about normal, housekeeping functions of microglia in the healthy CNS, nor how basic microglial activities are shifted due to alterations in the healthy CNS. These microglial activities have been largely neglected because they are not obvious until an insult is present, and because appropriate in vivo experimental tools have only recently become available. More studies are needed to understand how the healthy/beneficial activities of microglia are regulated by the local CNS milieu. In vitro studies in cell lines (immortalized most often with an oncogene) or primary microglia derived from neonatal animals are frequently used and have provided important insights, but interpretation of results from these studies comes with several caveats.

We have discussed many normal physiological processes (e.g., learning, neurogenesis, puberty, aging) and classes of molecules whose availability varies over the course of normal activity (e.g., neurotransmitters, peptides, hormones). Even though microglia have a multitude of important activities aside from their inflammatory actions, little remains understood about how these activities are impacted upon by these normal physiologic processes. Moreover, the studies that do address normal CNS physiology tend to focus on how microglia influence those processes (neurogenesis, puberty, etc.) and not the converse—how those processes influence microglia. This is likely due to the difficulty of performing these types of experiments and the paucity of microglial endpoints to study, but these are important questions that remain virtually untackled to date. If we are to target these cells for therapeutic benefit to control what goes awry during pathology, we must better understand their normal, healthy activities. Because microglia play a role in virtually all CNS functions and disorders, results from these studies will impact all aspects of neuroscience.

References

Aarum J, Sandberg K, Haeberlein SLB, Persson MAA (2003) Migration and differentiation of neural precursor cells can be directed by microglia. Proc Natl Acad Sci U S A 100: 15983–15988

Abbracchio MP, Burnstock G (1998) Purinergic signalling: pathophysiological roles. Jpn J Pharmacol 78:113–145

Abrous DN, Koehl M, Le Moal M (2005) Adult neurogenesis: from precursors to network and physiology. Physiol Rev 85:523–569

Apolloni S, Amadio S, Montilli C, Volonté C, D'Ambrosi N (2013) Ablation of P2X7 receptor exacerbates gliosis and motoneuron death in the SOD1-G93A mouse model of amyotrophic lateral sclerosis. Hum Mol Genet 22:4102–4116

Asami T, Ito T, Fukumitsu H, Nomoto H, Furukawa Y, Furukawa S (2006) Autocrine activation of cultured macrophages by brain-derived neurotrophic factor. Biochem Biophys Res Commun 344:941–947

Bachstetter AD, Morganti JM, Jernberg J, Schlunk A, Mitchell SH, Brewster KW et al (2011) Fractalkine and CX3CR1 regulate hippocampal neurogenesis in adult and aged rats. Neurobiol Aging 32:2030–2044

Barger SW, Basile AS (2001) Activation of microglia by secreted amyloid precursor protein evokes release of glutamate by cystine exchange and attenuates synaptic function. J Neurochem 76:846–854

Battista D, Ferrari CC, Gage FH, Pitossi FJ (2006) Neurogenic niche modulation by activated microglia: transforming growth factor β increases neurogenesis in the adult dentate gyrus. Eur J Neurosci 23:83–93

Beggs S, Trang T, Salter MW (2012) P2X4R+ microglia drive neuropathic pain. Nat Neurosci 15:1068–1073

Bellavance M-A, Rivest S (2012) The neuroendocrine control of the innate immune system in health and brain diseases. Immunol Rev 248:36–55

Biagini G, Longo D, Baldelli E, Zoli M, Rogawski MA, Bertazzoni G et al (2009) Neurosteroids and epileptogenesis in the pilocarpine model: evidence for a relationship between P450scc induction and length of the latent period. Epilepsia 50(Suppl 1):53–58

Biber K, Laurie DJ, Berthele A, Sommer B, Tölle TR, Gebicke-Härter P-J et al (1999) Expression and signaling of Group I metabotropic glutamate receptors in astrocytes and microglia. J Neurochem 72:1671–1680

Block ML, Li G, Qin L, Wu X, Pei Z, Wang T et al (2006) Potent regulation of microglia-derived oxidative stress and dopaminergic neuron survival: substance P vs. dynorphin. FASEB J 20: 251–258

Bodin P, Burnstock G (2001) Purinergic signalling: ATP release. Neurochem Res 26:959–969

Borrego F (2013) The CD300 molecules: an emerging family of regulators of the immune system. Blood 121:1951–1960

Boucsein C, Zacharias R, Farber K, Pavlovic S, Hanisch UK, Kettenmann H (2003) Purinergic receptors on microglial cells: functional expression in acute brain slices and modulation of microglial activation in vitro. Eur J Neurosci 17:2267–2276

Brautigam VM, Frasier C, Nikodemova M, Watters JJ (2005) Purinergic receptor modulation of BV-2 microglial cell activity: potential involvement of p38 MAP kinase and CREB. J Neuroimmunol 166:113–125

Burnstock G (2007) Purine and pyrimidine receptors. Cell Mol Life Sci 64:1471–1483
Burnstock G, Knight GE (2004) Cellular distribution and functions of p2 receptor subtypes in different systems. Int Rev Cytol 240:31–304
Butovsky O, Ziv Y, Schwartz A, Landa G, Talpalar AE, Pluchino S et al (2006) Microglia activated by IL-4 or IFN-γ differentially induce neurogenesis and oligodendrogenesis from adult stem/progenitor cells. Mol Cell Neurosci 31:149–160
Cannon J, O'Driscoll M, Litman G (2012) Specific lipid recognition is a general feature of CD300 and TREM molecules. Immunogenetics 64:39–47
Cantiello HF (1997) Nucleotide transport through the cystic fibrosis transmembrane conductance regulator. Biosci Rep 17:147–171
Cardona AE, Pioro EP, Sasse ME, Kostenko V, Cardona SM, Dijkstra IM et al (2006) Control of microglial neurotoxicity by the fractalkine receptor. Nat Neurosci 9:917–924
Carnevale D, De Simone R, Minghetti L (2007) Microglia–neuron interaction in inflammatory and degenerative diseases: role of cholinergic and noradrenergic systems. CNS Neurol Disord Drug Targets 6:388–397
Cavaliere C, Cirillo G, Rosaria Bianco M, Rossi F, De Novellis V, Maione S et al (2007) Gliosis alters expression and uptake of spinal glial amino acid transporters in a mouse neuropathic pain model. Neuron Glia Biol 3:141–153
Chantong B, Kratschmar D, Nashev L, Balazs Z, Odermatt A (2012) Mineralocorticoid and glucocorticoid receptors differentially regulate NF-kappaB activity and pro-inflammatory cytokine production in murine BV-2 microglial cells. J Neuroinflammation 9:260
Chen S, Wang JM, Irwin RW, Yao J, Liu L, Brinton RD (2011) Allopregnanolone promotes regeneration and reduces beta-amyloid burden in a preclinical model of Alzheimer's disease. PLoS One 6:e24293
Cho BP, Sugama S, Shin DH, DeGiorgio LA, Kim SS, Kim YS et al (2003) Microglial phagocytosis of dopamine neurons at early phases of apoptosis. Cell Mol Neurobiol 23:551–560
Clark GJ, Ju X, Azlan M, Tate C, Ding Y, Hart DNJ (2009) The CD300 molecules regulate monocyte and dendritic cell functions. Immunobiology 214:730–736
Cohen PE, Nishimura K, Zhu L, Pollard JW (1999) Macrophages: important accessory cells for reproductive function. J Leukoc Biol 66:765–772
Cohen PE, Zhu L, Nishimura K, Pollard JW (2002) Colony-stimulating factor 1 regulation of neuroendocrine pathways that control gonadal function in mice. Endocrinology 143:1413–1422
Conn PJ, Pin J-P (1997) Pharmacology and functions of metabotropic glutamate receptors. Annu Rev Pharmacol Toxicol 37:205–237
Crain JM, Watters JJ (2010) Estrogen and P2 purinergic receptor systems in microglia: therapeutic targets for neuroprotection. Open Drug Discov J 2:148–167
Crain J, Nikodemova M, Watters J (2009) Expression of P2 nucleotide receptors varies with age and sex in murine brain microglia. J Neuroinflammation 6:24
Crain JM, Nikodemova M, Watters JJ (2013) Microglia express distinct M1 and M2 phenotypic markers in the postnatal and adult central nervous system in male and female mice. J Neurosci Res 91:1143–1151
Damani MR, Zhao L, Fontainhas AM, Amaral J, Fariss RN, Wong WT (2011) Age-related alterations in the dynamic behavior of microglia. Aging Cell 10:263–276
Davalos D, Grutzendler J, Yang G, Kim JV, Zuo Y, Jung S et al (2005) ATP mediates rapid microglial response to local brain injury in vivo. Nat Neurosci 8:752–758
De Simone R, Ajmone-Cat M, Carnevale D, Minghetti L (2005) Activation of alpha7 nicotinic acetylcholine receptor by nicotine selectively up-regulates cyclooxygenase-2 and prostaglandin E2 in rat microglial cultures. J Neuroinflammation 2:4
De Simone R, Niturad CE, De Nuccio C, Ajmone-Cat MA, Visentin S, Minghetti L (2010) TGF-beta and LPS modulate ADP-induced migration of microglial cells through P2Y1 and P2Y12 receptor expression. J Neurochem 115:450–459
Denieffe S, Kelly RJ, McDonald C, Lyons A, Lynch MA (2013) Classical activation of microglia in CD200-deficient mice is a consequence of blood brain barrier permeability and infiltration of peripheral cells. Brain Behav Immun 34:86–97

Dilger RN, Johnson RW (2008) Aging, microglial cell priming, and the discordant central inflammatory response to signals from the peripheral immune system. J Leukoc Biol 84: 932–939

DonCarlos LL, Sarkey S, Lorenz B, Azcoitia I, Garcia-Ovejero D, Huppenbauer C et al (2006) Novel cellular phenotypes and subcellular sites for androgen action in the forebrain. Neuroscience 138:801–807

Dougherty KD, Dreyfus CF, Black IB (2000) Brain-derived neurotrophic factor in astrocytes, oligodendrocytes, and microglia/macrophages after spinal cord injury. Neurobiol Dis 7:574–585

Dubyak GR, el-Moatassim C (1993) Signal transduction via P2-purinergic receptors for extracellular ATP and other nucleotides. Am J Physiol 265:C577–606

Dunant N, Ballmer-Hofer K (1997) Signalling by Src family kinases: lessons learnt from DNA tumour viruses. Cell Signal 9:385–393

Ekdahl CT (2012) Microglial activation—tuning and pruning adult neurogenesis. Front Pharmacol 3:41

Ekdahl CT, Claasen J-H, Bonde S, Kokaia Z, Lindvall O (2003) Inflammation is detrimental for neurogenesis in adult brain. Proc Natl Acad Sci U S A 100:13632–13637

Färber K, Pannasch U, Kettenmann H (2005) Dopamine and noradrenaline control distinct functions in rodent microglial cells. Mol Cell Neurosci 29:128–138

Farso M, O'Shea R, Beart P (2009) Evidence group I mGluR drugs modulate the activation profile of lipopolysaccharide-exposed microglia in culture. Neurochem Res 34:1721–1728

Fenn AM, Smith KM, Lovett-Racke AE, Guerau-de-Arellano M, Whitacre CC, Godbout JP (2013) Increased micro-RNA 29b in the aged brain correlates with the reduction of insulin-like growth factor-1 and fractalkine ligand. Neurobiol Aging 34:2748–2758

Fernandez-Martin A, Gonzalez-Rey E, Chorny A, Martin J, Pozo D, Ganea D et al (2006) VIP prevents experimental multiple sclerosis by downregulating both inflammatory and autoimmune components of the disease. Ann N Y Acad Sci 1070:276–281

Fernandez-Martos CM, Gonzalez P, Rodriguez FJ (2012) Acute leptin treatment enhances functional recovery after spinal cord injury. PLoS One 7:e35594

Ferrari D, Villalba M, Chiozzi P, Falzoni S, Ricciardi-Castagnoli P, Di Virgilio F (1996) Mouse microglial cells express a plasma membrane pore gated by extracellular ATP. J Immunol 156: 1531–1539

Ferrari D, Chiozzi P, Falzoni S, Dal Susino M, Collo G, Buell G et al (1997a) ATP-mediated cytotoxicity in microglial cells. Neuropharmacology 36:1295–1301

Ferrari D, Chiozzi P, Falzoni S, Hanau S, Di Virgilio F (1997b) Purinergic modulation of interleukin-1 beta release from microglial cells stimulated with bacterial endotoxin. J Exp Med 185:579–582

Filip M, Bader M (2009) Overview on 5-HT receptors and their role in physiology and pathology of the central nervous system. Pharmacol Rep 61:761–777

Fontainhas AM, Wang M, Liang KJ, Chen S, Mettu P, Damani M et al (2011) Microglial morphology and dynamic behavior is regulated by ionotropic glutamatergic and GABAergic neurotransmission. PLoS One 6:e15973

Forabosco P, Ramasamy A, Trabzuni D, Walker R, Smith C, Bras J et al (2013) Insights into TREM2 biology by network analysis of human brain gene expression data. Neurobiol Aging 34:2699–2714

Franceschi C, BonafÈ M, Valensin S, Olivieri F, De Luca M, Ottaviani E et al (2000) Inflamm-aging: an evolutionary perspective on immunosenescence. Ann N Y Acad Sci 908:244–254

Frank MG, Miguel ZD, Watkins LR, Maier SF (2010) Prior exposure to glucocorticoids sensitizes the neuroinflammatory and peripheral inflammatory responses to E. coli lipopolysaccharide. Brain Behav Immun 24:19–30

Frank MG, Thompson BM, Watkins LR, Maier SF (2012) Glucocorticoids mediate stress-induced priming of microglial pro-inflammatory responses. Brain Behav Immun 26:337–345

Friedle SA, Brautigam VM, Nikodemova M, Wright ML, Watters JJ (2011) The P2X7-Egr pathway regulates nucleotide-dependent inflammatory gene expression in microglia. Glia 59:1–13

Fujita H, Tanaka J, Maeda N, Sakanaka M (1998) Adrenergic agonists suppress the proliferation of microglia through β2-adrenergic receptor. Neurosci Lett 242:37–40
Fünfschilling U, Jockusch WJ, Sivakumar N, Möbius W, Corthals K, Li S et al (2012) Critical time window of neuronal cholesterol synthesis during neurite outgrowth. J Neurosci 32:7632–7645
García-Ovejero D, Veiga S, García-Segura LM, Doncarlos LL (2002) Glial expression of estrogen and androgen receptors after rat brain injury. J Comp Neurol 450:256–271
Giunta B, Ehrhart J, Townsend K, Sun N, Vendrame M, Shytle D et al (2004) Galantamine and nicotine have a synergistic effect on inhibition of microglial activation induced by HIV-1 gp120. Brain Res Bull 64:165–170
Godbout JP, Chen J, Abraham J, Richwine AF, Berg BM, Kelley KW et al (2005) Exaggerated neuroinflammation and sickness behavior in aged mice after activation of the peripheral innate immune system. FASEB J 19(10):1329–1331
Gottfried-Blackmore A, Sierra A, Jellinck PH, McEwen BS, Bulloch K (2008) Brain microglia express steroid-converting enzymes in the mouse. J Steroid Biochem Mol Biol 109:96–107
Grossman KJ, Goss CW, Stein DG (2004) Effects of progesterone on the inflammatory response to brain injury in the rat. Brain Res 1008:29–39
Guerreiro R, Wojtas A, Bras J, Carrasquillo M, Rogaeva E, Majounie E et al (2013) TREM2 variants in Alzheimer's disease. New Eng J Med 368:117–127
Hagino Y, Kariura Y, Manago Y, Amano T, Wang B, Sekiguchi M et al (2004) Heterogeneity and potentiation of AMPA type of glutamate receptors in rat cultured microglia. Glia 47:68–77
Harrigan TJ, Abdullaev IF, Jourd'heuil D, Mongin AA (2008) Activation of microglia with zymosan promotes excitatory amino acid release via volume-regulated anion channels: the role of NADPH oxidases. J Neurochem 106:2449–2462
Harrison JK, Jiang Y, Chen S, Xia Y, Maciejewski D, McNamara RK et al (1998) Role for neuronally derived fractalkine in mediating interactions between neurons and CX3CR1-expressing microglia. Proc Natl Acad Sci U S A 95:10896–10901
Harry GJ (2013) Microglia during development and aging. Pharmacol Ther 139:313–326
Haynes SE, Hollopeter G, Yang G, Kurpius D, Dailey ME, Gan WB et al (2006) The P2Y12 receptor regulates microglial activation by extracellular nucleotides. Nat Neurosci 9:1512–1519
Henderson VW (2013) Alzheimer's disease: review of hormone therapy trials and implications for treatment and prevention after menopause. J Steroid Biochem Mol Biol 142:99–106
Heneka MT, Nadrigny F, Regen T, Martinez-Hernandez A, Dumitrescu-Ozimek L, Terwel D et al (2010) Locus ceruleus controls Alzheimer's disease pathology by modulating microglial functions through norepinephrine. Proc Natl Acad Sci U S A 107:6058–6063
Herz J, Reitmeir R, Hagen SI, Reinboth BS, Guo Z, Zechariah A et al (2012) Intracerebroventricularly delivered VEGF promotes contralesional corticorubral plasticity after focal cerebral ischemia via mechanisms involving anti-inflammatory actions. Neurobiol Dis 45:1077–1085
Hickman SE, Kingery ND, Ohsumi TK, Borowsky ML, LC W, Means TK et al (2013) The microglial sensome revealed by direct RNA sequencing. Nat Neurosci 16:1896–1905
Hide I, Tanaka M, Inoue A, Nakajima K, Kohsaka S, Inoue K et al (2000) Extracellular ATP triggers tumor necrosis factor-alpha release from rat microglia. J Neurochem 75:965–972
Hirayama M, Kuriyama M (2001) MK-801 is cytotoxic to microglia in vitro and its cytotoxicity is attenuated by glutamate, other excitotoxic agents and atropine: Possible presence of glutamate receptor and muscarinic receptor on microglia. Brain Res 897:204–206
Honda S, Sasaki Y, Ohsawa K, Imai Y, Nakamura Y, Inoue K et al (2001) Extracellular ATP or ADP induce chemotaxis of cultured microglia through Gi/o-coupled P2Y receptors. J Neurosci 21:1975–1982
Hoshiko M, Arnoux I, Avignone E, Yamamoto N, Audinat E (2012) Deficiency of the microglial receptor CX3CR1 impairs postnatal functional development of thalamocortical synapses in the barrel cortex. J Neurosci 32:15106–15111
Hsieh CL, Koike M, Spusta SC, Niemi EC, Yenari M, Nakamura MC et al (2009) A role for TREM2 ligands in the phagocytosis of apoptotic neuronal cells by microglia. J Neurochem 109:1144–1156

Huitinga I, Erkut ZA, van Beurden D, Swaab DF (2004) Impaired hypothalamus-pituitary-adrenal axis activity and more severe multiple sclerosis with hypothalamic lesions. Ann Neurol 55: 37–45

Hung J, Chansard M, Ousman SS, Nguyen MD, Colicos MA (2010) Activation of microglia by neuronal activity: results from a new in vitro paradigm based on neuronal-silicon interfacing technology. Brain Behav Immun 24:31–40

Inoue K, Nakajima K, Morimoto T, Kikuchi Y, Koizumi S, Illes P et al (1998) ATP stimulation of Ca2+-dependent plasminogen release from cultured microglia. Br J Pharmacol 123: 1304–1310

Jaguin M, Houlbert N, Fardel O, Lecureur V (2013) Polarization profiles of human M-CSF-generated macrophages and comparison of M1-markers in classically activated macrophages from GM-CSF and M-CSF origin. Cell Immunol 281:51–61

Jellinck PH, Croft G, McEwen BS, Gottfried-Blackmore A, Jones G, Byford V et al (2005) Metabolism of dehydroepiandrosterone by rodent brain cell lines: relationship between 7-hydroxylation and aromatization. J Steroid Biochem Mol Biol 93:81–86

Jellinck PH, Kaufmann M, Gottfried-Blackmore A, Croft G, Byford V, McEwen BS et al (2006) Dehydroepiandrosterone (DHEA) metabolism in the brain: identification by liquid chromatography/mass spectrometry of the delta-4-isomer of DHEA and related steroids formed from androstenedione by mouse BV2 microglia. J Steroid Biochem Mol Biol 98:41–47

Jellinck PH, Kaufmann M, Gottfried-Blackmore A, McEwen BS, Jones G, Bulloch K (2007) Selective conversion by microglia of dehydroepiandrosterone to 5-androstenediol-A steroid with inherent estrogenic properties. J Steroid Biochem Mol Biol 107:156–162

Jiang MH, Chung E, Chi GF, Ahn W, Lim JE, Hong HS et al (2012) Substance P induces M2-type macrophages after spinal cord injury. Neuroreport 23:786–792

Jiang MH, Lim JE, Chi GF, Ahn W, Zhang M, Chung E et al (2013) Substance P reduces apoptotic cell death possibly by modulating the immune response at the early stage after spinal cord injury. Neuroreport 24:846–851

Kalinin S, Gavrilyuk V, Polak PE, Vasser R, Zhao J, Heneka MT et al (2007) Noradrenaline deficiency in brain increases β-amyloid plaque burden in an animal model of Alzheimer's disease. Neurobiol Aging 28:1206–1214

Kanai Y, Hediger M (2004) The glutamate/neutral amino acid transporter family SLC1: molecular, physiological and pharmacological aspects. Pflugers Arch 447:469–479

Kasai Y, Ohta T, Nakazato Y, Ito S (2001) Release of dopamine and ATP from PC12 cells treated with dexamethasone, reserpine and bafilomycin A1. J Vet Med Sci 63:367–372

Kataoka A, Tozaki-Saitoh H, Koga Y, Tsuda M, Inoue K (2009) Activation of P2X7 receptors induces CCL3 production in microglial cells through transcription factor NFAT. J Neurochem 108:115–125

Kaur C, Sivakumar V, Dheen ST, Ling EA (2006) Insulin-like growth factor I and II expression and modulation in amoeboid microglial cells by lipopolysaccharide and retinoic acid. Neuroscience 138:1233–1244

Kaushal V, Schlichter LC (2008) Mechanisms of microglia-mediated neurotoxicity in a new model of the stroke penumbra. J Neurosci 28:2221–2230

Kawamata J, Shimohama S (2011) Stimulating nicotinic receptors trigger multiple pathways attenuating cytotoxicity in models of Alzheimer's and Parkinson's diseases. J Alzheimers Dis 24(Suppl 2):95–109

Kettenmann H, Hanisch U-K, Noda M, Verkhratsky A (2011) Physiology of microglia. Physiol Rev 91:461–553

Kingham PJ, Cuzner ML, Pocock JM (1999) Apoptotic pathways mobilized in microglia and neurones as a consequence of chromogranin A-induced microglial activation. J Neurochem 73:538–547

Kohm AP, Sanders VM (2001) Norepinephrine and β2-adrenergic receptor stimulation regulate CD4+ T and B lymphocyte function in vitro and in vivo. Pharmacol Rev 53:487–525

Koizumi S, Shigemoto-Mogami Y, Nasu-Tada K, Shinozaki Y, Ohsawa K, Tsuda M et al (2007) UDP acting at P2Y6 receptors is a mediator of microglial phagocytosis. Nature 446:1091

Kojro E, Postina R, Buro C, Meiringer C, Gehrig-Burger K, Fahrenholz F (2006) The neuropeptide PACAP promotes the alpha-secretase pathway for processing the Alzheimer amyloid precursor protein. FASEB J 20:512–514
Kondo Y, Duncan ID (2009) Selective reduction in microglia density and function in the white matter of colony-stimulating factor-1-deficient mice. J Neurosci Res 87:2686–2695
Kondo Y, Lemere CA, Seabrook TJ (2007) Osteopetrotic (op/op) mice have reduced microglia, no Abeta deposition, and no changes in dopaminergic neurons. J Neuroinflammation 4:31
Kong Y, Ruan L, Qian L, Liu X, Le Y (2010) Norepinephrine promotes microglia to uptake and degrade amyloid β peptide through upregulation of mouse formyl peptide receptor 2 and induction of insulin-degrading enzyme. J Neurosci 30:11848–11857
Krabbe G, Matyash V, Pannasch U, Mamer L, Boddeke HWGM, Kettenmann H (2012) Activation of serotonin receptors promotes microglial injury-induced motility but attenuates phagocytic activity. Brain Behav Immun 26:419–428
Kuhn H, Dickinson-Anson H, Gage F (1996) Neurogenesis in the dentate gyrus of the adult rat: age-related decrease of neuronal progenitor proliferation. J Neurosci 16:2027–2033
Kuhn SA, van Landeghem FKH, Zacharias R, Färber K, Rappert A, Pavlovic S et al (2004) Microglia express GABAB receptors to modulate interleukin release. Mol Cell Neurosci 25:312–322
Kukulski F, Lévesque SA, Sévigny J (2011) Impact of ectoenzymes on P2 and P1 receptor signaling. Adv Pharmacol 61:263–299
Kulkarni S, Zou B, Hanson J, Micci MA, Tiwari G, Becker L et al (2011) Gut-derived factors promote neurogenesis of CNS-neural stem cells and nudge their differentiation to an enteric-like neuronal phenotype. Am J Physiol Gastrointest Liver Physiol 301:G644–G655
Lafrance V, Inoue W, Kan B, Luheshi GN (2010) Leptin modulates cell morphology and cytokine release in microglia. Brain Behav Immun 24:358–365
Lai AY, Todd KG (2008) Differential regulation of trophic and proinflammatory microglial effectors is dependent on severity of neuronal injury. Glia 56:259–270
Lai AY, Dibal CD, Armitage GA, Winship IR, Todd KG (2013) Distinct activation profiles in microglia of different ages: a systematic study in isolated embryonic to aged microglial cultures. Neuroscience 254:185–195
Lalancette-Hébert M, Gowing G, Simard A, Weng YC, Kriz J (2007) Selective ablation of proliferating microglial cells exacerbates ischemic injury in the brain. J Neurosci 27:2596–2605
Lankry D, Rovis TL, Jonjic S, Mandelboim O (2013) The interaction between CD300a and phosphatidylserine inhibits tumor cell killing by NK cells. Eur J Immunol 43:2151–2161
Lazarowski ER, Homolya L, Boucher RC, Harden TK (1997) Direct demonstration of mechanically induced release of cellular UTP and Its implication for uridine nucleotide receptor activation. J Biol Chem 272:24348–24354
Lee M (2013) Neurotransmitters and microglial-mediated neuroinflammation. Curr Protein Pept Sci 14:21–32
Lee M, Schwab C, McGeer PL (2011) Astrocytes are GABAergic cells that modulate microglial activity. Glia 59:152–165
Li Y, Du X-F, Liu C-S, Wen Z-L, Du J-l (2012) Reciprocal regulation between resting microglial dynamics and neuronal activity in vivo. Dev Cell 23:1189–1202
Liu J, McGlinn AM, Fernandes A, Milam AH, Strang CE, Andison ME et al (2009) Nicotinic acetylcholine receptor subunits in rhesus monkey retina. Invest Ophthalmol Vis Sci 50: 1408–1415
Lobo RA (2013) Where are we 10 years after the women's health initiative? J Clin Endocrinol Metab 98:1771–1780
Lois C, Alvarez-Buylla A (1993) Proliferating subventricular zone cells in the adult mammalian forebrain can differentiate into neurons and glia. Proc Natl Acad Sci U S A 90:2074–2077
Lutz PL, Kabler S (1997) Release of adenosine and ATP in the brain of the freshwater turtle (Trachemys scripta) during long-term anoxia. Brain Res 769:281–286
Lynch MA (2013) The impact of neuroimmune changes on development of amyloid pathology; relevance to Alzheimer's disease. Immunology 141(3):292–301

Maggi L, Scianni M, Branchi I, D'Andrea I, Lauro C, Limatola C (2011) CX3CR1 deficiency alters hippocampal-dependent plasticity phenomena blunting the effects of enriched environment. Front Cell Neurosci 5:22

Mahé C, Loetscher E, Dev KK, Bobirnac I, Otten U, Schoeffter P (2005) Serotonin 5-HT7 receptors coupled to induction of interleukin-6 in human microglial MC-3 cells. Neuropharmacology 49:40–47

Mahley RW (1988) Apolipoprotein E: cholesterol transport protein with expanding role in cell biology. Science 240:622–630

Martínez-Barriocanal Á, Comas-Casellas E, Schwartz S, Martín M, Sayós J (2010) CD300 heterocomplexes, a new and family-restricted mechanism for myeloid cell signaling regulation. J Biol Chem 285:41781–41794

Mastroeni D, Grover A, Leonard B, Joyce JN, Coleman PD, Kozik B et al (2009) Microglial responses to dopamine in a cell culture model of Parkinson's disease. Neurobiol Aging 30: 1805–1817

Matalka KZ, Attallah LJ, Qinna NA, Alhussainy T (2011) Dopamine selectively modulates lipopolysaccharide-induced TNF-alpha, IFN-gamma and IL-10 within mice tissues. Neuro Endocrinol Lett 32:176–186

Matsuka Y, Neubert JK, Maidment NT, Spigelman I (2001) Concurrent release of ATP and substance P within guinea pig trigeminal ganglia in vivo. Brain Res 915:248–255

McNamee EN, Ryan KM, Kilroy D, Connor TJ (2010) Noradrenaline induces IL-1ra and IL-1 type II receptor expression in primary glial cells and protects against IL-1β-induced neurotoxicity. Eur J Pharmacol 626:219–228

Mead EL, Mosley A, Eaton S, Dobson L, Heales SJ, Pocock JM (2012) Microglial neurotransmitter receptors trigger superoxide production in microglia; consequences for microglial–neuronal interactions. J Neurochem 121:287–301

Milanski M, Degasperi G, Coope A, Morari J, Denis R, Cintra DE et al (2009) Saturated fatty acids produce an inflammatory response predominantly through the activation of TLR4 signaling in hypothalamus: implications for the pathogenesis of obesity. J Neurosci 29:359–370

Milligan ED, Sloane EM, Watkins LR (2008) Glia in pathological pain: a role for fractalkine. J Neuroimmunol 198:113–120

Mizuno T, Kawanokuchi J, Numata K, Suzumura A (2003) Production and neuroprotective functions of fractalkine in the central nervous system. Brain Res 979:65–70

Morgan SC, Taylor DL, Pocock JM (2004) Microglia release activators of neuronal proliferation mediated by activation of mitogen-activated protein kinase, phosphatidylinositol-3-kinase/Akt and delta–Notch signalling cascades. J Neurochem 90:89–101

Mori K, Ozaki E, Zhang B, Yang L, Yokoyama A, Takeda I et al (2002) Effects of norepinephrine on rat cultured microglial cells that express α1, α2, β1 and β2 adrenergic receptors. Neuropharmacology 43:1026–1034

Morioka N, Tokuhara M, Harano S, Nakamura Y, Hisaoka-Nakashima K, Nakata Y (2013) The activation of P2Y6 receptor in cultured spinal microglia induces the production of CCL2 through the MAP kinases-NF-κB pathway. Neuropharmacology 75:116–125

Mrak RE, Griffin WST (2005) Glia and their cytokines in progression of neurodegeneration. Neurobiol Aging 26:349–354

Muller E, Kerschbaum HH (2006) Progesterone and its metabolites 5-dihydroprogesterone and 5-3-tetrahydroprogesterone decrease LPS-induced NO release in the murine microglial cell line, BV-2. Neuro Endocrinol Lett 27:675–678

Müller T, Dürk T, Blumenthal B, Grimm M, Cicko S, Panther E et al (2009) 5-Hydroxytryptamine modulates migration, cytokine and chemokine release and T-cell priming capacity of dendritic cells in vitro and in vivo. PLoS One 4:e6453

Nakamachi T, Farkas J, Watanabe J, Ohtaki H, Dohi K, Arata S et al (2011) Role of PACAP in neural stem/progenitor cell and astrocyte—from neural development to neural repair. Curr Pharm Des 17:973–984

Nikodemova M, Watters JJ (2011) Outbred ICR/CD1 mice display more severe neuroinflammation mediated by microglial TLR4/CD14 activation than inbred C57Bl/6 mice. Neuroscience 190:67–74

Nikodemova M, Watters JJ (2012) Efficient isolation of live microglia with preserved phenotypes from adult mouse brain. J Neuroinflammation 9:147

Nimmerjahn A, Kirchhoff F, Helmchen F (2005) Resting microglial cells are highly dynamic surveillants of brain parenchyma in vivo. Science 308:1314–1318

Njie EG, Boelen E, Stassen FR, Steinbusch HWM, Borchelt DR, Streit WJ (2012) Ex vivo cultures of microglia from young and aged rodent brain reveal age-related changes in microglial function. Neurobiol Aging 33:195e1–12

Noda M, Beppu K (2013) Possible contribution of microglial glutamate receptors to inflammatory response upon neurodegenerative diseases. J Neurol Disord 1:131

North RA, Jarvis MF (2013) P2X receptors as drug targets. Mol Pharmacol 83:759–769

Ohsawa K, Irino Y, Nakamura Y, Akazawa C, Inoue K, Kohsaka S (2007) Involvement of P2X4 and P2Y12 receptors in ATP-induced microglial chemotaxis. Glia 55:604–616

Ohsawa K, Irino Y, Sanagi T, Nakamura Y, Suzuki E, Inoue K et al (2010) P2Y12 receptor-mediated integrin-beta1 activation regulates microglial process extension induced by ATP. Glia 58:790–801

Ohsawa K, Sanagi T, Nakamura Y, Suzuki E, Inoue K, Kohsaka S (2012) Adenosine A3 receptor is involved in ADP-induced microglial process extension and migration. J Neurochem 121: 217–227

Orr AG, Orr AL, Li XJ, Gross RE, Traynelis SF (2009) Adenosine A(2A) receptor mediates microglial process retraction. Nat Neurosci 12:872–878

Pang Y, Zheng B, Campbell LR, Fan L-W, Cai Z, Rhodes PG (2010) IGF-1 can either protect against or increase LPS-induced damage in the developing rat brain. Pediatr Res 67:579–584

Park J, Yoo K-Y, Lee C, Kim I, Shin B, Choi J et al (2011) Comparison of glucocorticoid receptor and ionized calcium-binding adapter molecule 1 immunoreactivity in the adult and aged gerbil hippocampus following repeated restraint stress. Neurochem Res 36:1037–1045

Parkhurst Christopher N, Yang G, Ninan I, Savas Jeffrey N, Yates Iii John R, Lafaille Juan J et al (2013) Microglia promote learning-dependent synapse formation through brain-derived neurotrophic factor. Cell 155:1596–1609

Perry VH, Matyszak MK, Fearn S (1993) Altered antigen expression of microglia in the aged rodent CNS. Glia 7:60–67

Pettus EH, Wright DW, Stein DG, Hoffman SW (2005) Progesterone treatment inhibits the inflammatory agents that accompany traumatic brain injury. Brain Res 1049:112–119

Pickering M, O'Connor JJ (2007) Pro-inflammatory cytokines and their effects in the dentate gyrus. Prog Brain Res 163:339–354

Piers TM, Heales SJ, Pocock JM (2011) Positive allosteric modulation of metabotropic glutamate receptor 5 down-regulates fibrinogen-activated microglia providing neuronal protection. Neurosci Lett 505:140–145

Pinteaux E, Inoue W, Schmidt L, Molina-Holgado F, Rothwell NJ, Luheshi GN (2007) Leptin induces interleukin-1beta release from rat microglial cells through a caspase 1 independent mechanism. J Neurochem 102:826–833

Pocock JM, Kettenmann H (2007) Neurotransmitter receptors on microglia. Trends Neurosci 30:527–535

Poelchen W, Sieler D, Wirkner K, Illes P (2001) Co-transmitter function of ATP in central catecholaminergic neurons of the rat. Neuroscience 102:593–602

Pozzi S, Benedusi V, Maggi A, Vegeto E (2006) Estrogen action in neuroprotection and brain inflammation. Ann N Y Acad Sci 1089:302–323

Qian L, Hu X, Zhang D, Snyder A, Wu H-M, Li Y et al (2009) β2 Adrenergic receptor activation induces microglial NADPH oxidase activation and dopaminergic neurotoxicity through an ERK-dependent/protein kinase A-independent pathway. Glia 57:1600–1609

Rasley A, Marriott I, Halberstadt CR, Bost KL, Anguita J (2004) Substance P augments Borrelia burgdorferi-induced prostaglandin E2 production by murine microglia. J Immunol 172: 5707–5713
Ren K, Dubner R (2008) Neuron-glia crosstalk gets serious: role in pain hypersensitivity. Curr Opin Anaesthesiol 21:570–579
Rocha SM, Cristovão AC, Campos FL, Fonseca CP, Baltazar G (2012) Astrocyte-derived GDNF is a potent inhibitor of microglial activation. Neurobiol Dis 47:407–415
Rock RB, Gekker G, Aravalli R, Hu S, Sheng W, Peterson P (2008) Potentiation of HIV-1 expression in microglial cells by nicotine: involvement of transforming growth factor-β1. J Neuroimmune Pharmacol 3:143–149
Rogers JT, Morganti JM, Bachstetter AD, Hudson CE, Peters MM, Grimmig BA et al (2011) CX3CR1 deficiency leads to impairment of hippocampal cognitive function and synaptic plasticity. J Neurosci 31:16241–16250
Sandhir R, Onyszchuk G, Berman NEJ (2008) Exacerbated glial response in the aged mouse hippocampus following controlled cortical impact injury. Exp Neurol 213:372–380
Sanz JM, Di Virgilio F (2000) Kinetics and mechanism of ATP-dependent IL-1 beta release from microglial cells. J Immunol 164:4893–4898
Schafer Dorothy P, Lehrman Emily K, Kautzman Amanda G, Koyama R, Mardinly Alan R, Yamasaki R et al (2012) Microglia sculpt postnatal neural circuits in an activity and complement-dependent manner. Neuron 74:691–705
Schilling T, Eder C (2004) A novel physiological mechanism of glycine-induced immunomodulation: Na+-coupled amino acid transporter currents in cultured brain macrophages. J Physiol 559:35–40
Scholz J, Woolf CJ (2007) The neuropathic pain triad: neurons, immune cells and glia. Nat Neurosci 10:1361–1368
Schott J, Stoecklin G (2010) Networks controlling mRNA decay in the immune system. Wiley Interdiscip Rev RNA 1:432–456
Seifert S, Pannell M, Uckert W, Färber K, Kettenmann H (2011) Transmitter- and hormone-activated Ca2+ responses in adult microglia/brain macrophages in situ recorded after viral transduction of a recombinant Ca2+ sensor. Cell Calcium 49:365–375
Seko Y, Cole S, Kasprzak W, Shapiro BA, Ragheb JA (2006) The role of cytokine mRNA stability in the pathogenesis of autoimmune disease. Autoimmun Rev 5:299–305
Seo DR, Kim KY, Lee YB (2004) Interleukin-10 expression in lipopolysaccharide-activated microglia is mediated by extracellular ATP in an autocrine fashion. Neuroreport 15: 1157–1161
Seo DR, Kim SY, Kim KY, Lee HG, Moon JH, Lee JS et al (2008) Cross talk between P2 purinergic receptors modulates extracellular ATP-mediated interleukin-10 production in rat microglial cells. Exp Mol Med 40:19–26
Sheng JG, Mrak RE, Griffin WST (1998) Enlarged and phagocytic, but not primed, interleukin-1α-immunoreactive microglia increase with age in normal human brain. Acta Neuropathol 95: 229–234
Shioda S, Ohtaki H, Nakamachi T, Dohi K, Watanabe JUN, Nakajo S et al (2006) Pleiotropic functions of PACAP in the CNS. Ann N Y Acad Sci 1070:550–560
Shytle RD, Mori T, Townsend K, Vendrame M, Sun N, Zeng J et al (2004) Cholinergic modulation of microglial activation by α7 nicotinic receptors. J Neurochem 89:337–343
Sierra A, Gottfried-Blackmore AC, McEwen BS, Bulloch K (2007) Microglia derived from aging mice exhibit an altered inflammatory profile. Glia 55:412–424
Sierra A, Gottfried-Blackmore A, Milner TA, McEwen BS, Bulloch K (2008) Steroid hormone receptor expression and function in microglia. Glia 56:659–674
Sierra A, Encinas JM, Deudero JJP, Chancey JH, Enikolopov G, Overstreet-Wadiche LS et al (2010) Microglia shape adult hippocampal neurogenesis through apoptosis-coupled phagocytosis. Cell Stem Cell 7:483–495
Sierra A, Abiega O, Shahraz A, Neumann H (2013) Janus-faced microglia: beneficial and detrimental consequences of microglial phagocytosis. Front Cell Neurosci 7:6. doi:10.3389/fncel.2013.00006, eCollection 2013

Simhadri VR, Mariano JL, Gil-Krzewska A, Zhou Q, Borrego F (2013) CD300c is an activating receptor expressed on human monocytes. J Innate Immun 5:389–400
Sohrabji F, Williams M (2013) Stroke neuroprotection: estrogen and IGF-1 interactions and the role of microglia. J Neuroendocrinol 25(11):1173–1181
Spencer-Segal JL, Waters EM, Bath KG, Chao MV, McEwen BS, Milner TA (2011) Distribution of phosphorylated TrkB receptor in the mouse hippocampal formation depends on sex and estrous cycle stage. J Neurosci 31:6780–6790
Spulber S, Bartfai T, Schultzberg M (2009) IL-1/IL-1ra balance in the brain revisited—evidence from transgenic mouse models. Brain Behav Immun 23:573–579
Staley K, Scharfman H (2005) A woman's prerogative. Nat Neurosci 8:697–699
Staniland AA, Clark AK, Wodarski R, Sasso O, Maione F, D'Acquisto F et al (2010) Reduced inflammatory and neuropathic pain and decreased spinal microglial response in fractalkine receptor (CX3CR1) knockout mice. J Neurochem 114:1143–1157
Stevens SL, Shaw TE, Dykhuizen E, Lessov NS, Hill JK, Wurst W et al (2003) Reduced cerebral injury in CRH-R1 deficient mice after focal ischemia: a potential link to microglia and atrocytes that express CRH-R1. J Cereb Blood Flow Metab 23:1151–1159
Stone DJ, Rozovsky I, Morgan TE, Anderson CP, Hajian H, Finch CE (1997) Astrocytes and microglia respond to estrogen with increased apoE mRNA in vivo and in vitro. Exp Neurol 143:313–318
Streit W, Xue Q-S (2009) Life and death of microglia. J Neuroimmune Pharmacol 4:371–379
Streit WJ, Morioka T, Kalehua AN (1992) MK-801 prevents microglial reaction in rat hippocampus after forebrain ischemia. Neuroreport 3:146–148
Sugama S, Yang L, Cho BP, DeGiorgio LA, Lorenzl S, Albers DS et al (2003) Age-related microglial activation in 1-methyl-4-phenyl-1,2,3,6-tetrahydropyridine (MPTP)-induced dopaminergic neurodegeneration in C57BL/6 mice. Brain Res 964:288–294
Suh HS, Zhao ML, Derico L, Choi N, Lee SC (2013) Insulin-like growth factor 1 and 2 (IGF1, IGF2) expression in human microglia: differential regulation by inflammatory mediators. J Neuroinflammation 10:37
Suzuki T, Hide I, Matsubara A, Hama C, Harada K, Miyano K et al (2006) Microglial α7 nicotinic acetylcholine receptors drive a phospholipase C/IP3 pathway and modulate the cell activation toward a neuroprotective role. J Neurosci Res 83:1461–1470
Szabadfi K, Danyadi B, Kiss P, Tamas A, Fabian E, Gabriel R et al (2012) Protective effects of vasoactive intestinal peptide (VIP) in ischemic retinal degeneration. J Mol Neurosci 48:501–507
Takahashi N, Tonchev A, Koike K, Murakami K, Yamada K, Yamashima T et al (2004) Expression of estrogen receptor-beta in the postischemic monkey hippocampus. Neurosci Lett 369:9–13
Tang CH, Lu DY, Yang RS, Tsai HY, Kao MC, Fu WM et al (2007) Leptin-induced IL-6 production is mediated by leptin receptor, insulin receptor substrate-1, phosphatidylinositol 3-kinase, Akt, NF-kappaB, and p300 pathway in microglia. J Immunol 179:1292–1302
Tapia-González S, García-Segura LM, Tena-Sempere M, Frago LM, Castellano JM, Fuente-Martín E et al (2011) Activation of microglia in specific hypothalamic nuclei and the cerebellum of adult rats exposed to neonatal overnutrition. J Neuroendocrinol 23:365–370
Taylor DL, Diemel LT, Cuzner ML, Pocock JM (2002) Activation of group II metabotropic glutamate receptors underlies microglial reactivity and neurotoxicity following stimulation with chromogranin A, a peptide up-regulated in Alzheimer's disease. J Neurochem 82:1179–1191
Taylor DL, Diemel LT, Pocock JM (2003) Activation of microglial group III metabotropic glutamate receptors protects neurons against microglial neurotoxicity. J Neurosci 23:2150–2160
Taylor DL, Jones F, Kubota ESFCS, Pocock JM (2005) Stimulation of microglial metabotropic glutamate receptor mGlu2 triggers tumor necrosis factor α-induced neurotoxicity in concert with microglial-derived Fas ligand. J Neurosci 25:2952–2964
Thomsen MS, Mikkelsen JD (2012) The α7 nicotinic acetylcholine receptor ligands methyllycaconitine, NS6740 and GTS-21 reduce lipopolysaccharide-induced TNF-α release from microglia. J Neuroimmunol 251:65–72

Thornton E, Vink R (2012) Treatment with a substance P receptor antagonist is neuroprotective in the intrastriatal 6-hydroxydopamine model of early Parkinson's disease. PLoS One 7:e34138

Tomas-Camardiel M, Sanchez-Hidalgo MC, Sanchez del Pino MJ, Navarro A, Machado A, Cano J (2002) Comparative study of the neuroprotective effect of dehydroepiandrosterone and 17beta-estradiol against 1-methyl-4-phenylpyridium toxicity on rat striatum. Neuroscience 109:569–584

Trang T, Beggs S, Wan X, Salter MW (2009) P2X4-receptor-mediated synthesis and release of brain-derived neurotrophic factor in microglia is dependent on calcium and p38-mitogen-activated protein kinase activation. J Neurosci 29:3518–3528

Trang T, Beggs S, Salter MW (2011) Brain-derived neurotrophic factor from microglia: a molecular substrate for neuropathic pain. Neuron Glia Biol 7:99–108

Tremblay M-È, Lowery RL, Majewska AK (2010) Microglial interactions with synapses are modulated by visual experience. PLoS Biol 8:e1000527

Tsuda M, Shigemoto-Mogami Y, Koizumi S, Mizokoshi A, Kohsaka S, Salter MW et al (2003) P2X4 receptors induced in spinal microglia gate tactile allodynia after nerve injury. Nature 424:778–783

Ulmann L, Hatcher JP, Hughes JP, Chaumont S, Green PJ, Conquet F et al (2008) Up-regulation of P2X4 receptors in spinal microglia after peripheral nerve injury mediates BDNF release and neuropathic pain. J Neurosci 28:11263–11268

Valerio A, Dossena M, Bertolotti P, Boroni F, Sarnico I, Faraco G et al (2009) Leptin is induced in the ischemic cerebral cortex and exerts neuroprotection through NF-κB/c-Rel-dependent transcription. Stroke 40:610–617

Van den Eynden J, Notelaers K, Brône B, Janssen D, Nelissen K, SahebAli S et al (2011) Glycine enhances microglial intracellular calcium signaling. A role for sodium-coupled neutral amino acid transporters. Pflügers Arch 461:481–491

Vegeto E, Benedusi V, Maggi A (2008) Estrogen anti-inflammatory activity in brain: a therapeutic opportunity for menopause and neurodegenerative diseases. Front Neuroendocrinol 29: 507–519

Vinet J, van Weering H, Heinrich A, Kalin R, Wegner A, Brouwer N et al (2012) Neuroprotective function for ramified microglia in hippocampal excitotoxicity. J Neuroinflammation 9:27

Vukovic J, Colditz MJ, Blackmore DG, Ruitenberg MJ, Bartlett PF (2012) Microglia modulate hippocampal neural precursor activity in response to exercise and aging. J Neurosci 32: 6435–6443

Wake H, Moorhouse AJ, Jinno S, Kohsaka S, Nabekura J (2009) Resting microglia directly monitor the functional state of synapses in vivo and determine the fate of ischemic terminals. J Neurosci 29:3974–3980

Wang W, Ji P, Riopelle RJ, Dow KE (2002) Functional expression of corticotropin-releasing hormone (CRH) receptor 1 in cultured rat microglia. J Neurochem 80:287–294

Wang C-M, Chang Y-Y, Sun SH (2003a) Activation of P2X7 purinoceptor-stimulated TGF-[beta]1 mRNA expression involves PKC/MAPK signalling pathway in a rat brain-derived type-2 astrocyte cell line, RBA-2. Cell Signal 15:1129

Wang H, Yu M, Ochani M, Amella CA, Tanovic M, Susarla S et al (2003b) Nicotinic acetylcholine receptor [alpha]7 subunit is an essential regulator of inflammation. Nature 421:384–388

Wang W, Ji P, Dow KE (2003c) Corticotropin-releasing hormone induces proliferation and TNF-alpha release in cultured rat microglia via MAP kinase signalling pathways. J Neurochem 84:189–195

Wang W, Solc M, Ji P, Dow KE (2004) Corticotropin-releasing hormone potentiates neural injury induced by oxygen-glucose deprivation: a possible involvement of microglia. Neurosci Lett 371:133–137

Wang J, Yang Z, Liu C, Zhao Y, Chen Y (2013) Activated microglia provide a neuroprotective role by balancing glial cell-line derived neurotrophic factor and tumor necrosis factor-alpha secretion after subacute cerebral ischemia. Int J Mol Med 31:172–178

Waschek JA (2013) VIP and PACAP: neuropeptide modulators of CNS inflammation, injury, and repair. Br J Pharmacol 169:512–523

Wasserman JK, Yang H, Schlichter LC (2008) Glial responses, neuron death and lesion resolution after intracerebral hemorrhage in young vs. aged rats. Eur J Neurosci 28:1316–1328

Wegiel J, Wisniewski HM, Dziewiatkowski J, Tarnawski M, Kozielski R, Trenkner E et al (1998) Reduced number and altered morphology of microglial cells in colony stimulating factor-1-deficient osteopetrotic op/op mice. Brain Res 804:135–139

White CM, Ji S, Cai H, Maudsley S, Martin B (2010) Therapeutic potential of vasoactive intestinal peptide and its receptors in neurological disorders. CNS Neurol Disord Drug Targets 9:661–666

Whittemore ER, Korotzer AR, Etebari A, Cotman CW (1993) Carbachol increases intracellular free calcium in cultured rat microglia. Brain Res 621:59–64

Wong AM, Rozovsky I, Arimoto JM, Du Y, Wei M, Morgan TE et al (2009) Progesterone influence on neurite outgrowth involves microglia. Endocrinology 150:324–332

Woolley C, McEwen B (1992) Estradiol mediates fluctuation in hippocampal synapse density during the estrous cycle in the adult rat [published erratum appears in J Neurosci 1992 Oct;12(10):following table of contents]. J Neurosci 12:2549–2554

Woolley CS, McEwen BS (1993) Roles of estradiol and progesterone in regulation of hippocampal dendritic spine density during the estrous cycle in the rat. J Comp Neurol 336:293–306

Wright DW, Kellermann AL, Hertzberg VS, Clark PL, Frankel M, Goldstein FC et al (2007) ProTECT: a randomized clinical trial of progesterone for acute traumatic brain injury. Ann Emerg Med 49:391–402, 402.e1–2

Wynne AM, Henry CJ, Huang Y, Cleland A, Godbout JP (2010) Protracted downregulation of CX3CR1 on microglia of aged mice after lipopolysaccharide challenge. Brain Behav Immun 24:1190–1201

Yamada J, Sawada M, Nakanishi H (2006) Cell cycle-dependent regulation of kainate-induced inward currents in microglia. Biochem Biophys Res Commun 349:913–919

Yang Y, Hahm E, Kim Y, Kang J, Lee W, Han I et al (2005) Regulation of IL-18 expression by CRH in mouse microglial cells. Immunol Lett 98:291–296

Ye S-M, Johnson RW (1999) Increased interleukin-6 expression by microglia from brain of aged mice. J Neuroimmunol 93:139–148

Zhang L, McLarnon JG, Goghari V, Lee YB, Kim SU, Krieger C (1998) Cholinergic agonists increase intracellular Ca2+ in cultured human microglia. Neurosci Lett 255:33–36

Zhong Z, Wheeler MD, Li X, Froh M, Schemmer P, Yin M et al (2003) L-Glycine: a novel antiinflammatory, immunomodulatory, and cytoprotective agent. Curr Opin Clin Nutr Metab Care 6:229–240

Zhu W, Zheng H, Shao X, Wang W, Yao Q, Li Z (2010) Excitotoxicity of TNFα derived from KA activated microglia on hippocampal neurons in vitro and in vivo. J Neurochem 114:386–396

Ziv Y, Ron N, Butovsky O, Landa G, Sudai E, Greenberg N et al (2006) Immune cells contribute to the maintenance of neurogenesis and spatial learning abilities in adulthood. Nat Neurosci 9:268–275

Zujovic V, Benavides J, Vigé X, Carter C, Taupin V (2000) Fractalkine modulates TNF-α secretion and neurotoxicity induced by microglial activation. Glia 29:305–315

Zujovic V, Schussler N, Jourdain D, Duverger D, Taupin V (2001) In vivo neutralization of endogenous brain fractalkine increases hippocampal TNFα and 8-isoprostane production induced by intracerebroventricular injection of LPS. J Neuroimmunol 115:135–143

Chapter 4
Lessons from In Vivo Imaging

Dimitrios Davalos and Martin Fuhrmann

Abstract The first line of defense against injury or disease in the central nervous system (CNS) is through microglia. In the adult brain, microglia were long believed to stay in a dormant/resting state, activated only in the event of an insult to the brain. This view changed dramatically with the development of modern imaging techniques that allowed the study of microglial behavior in the intact brain over time to reveal the dynamic nature of their responses. In vivo imaging studies using two-photon microscopy revealed a previously unknown function for microglia: they continuously screen the intact brain parenchyma with their fine processes on a timescale of minutes. By doing so, they contact neuronal cell bodies, axons, dendrites, and dendritic spines and are believed to play a central role in sculpting neuronal networks during development, adulthood, and the normal aging process. Following acute trauma, or in neurodegenerative or neuroinflammatory diseases, microglial responses range from protective to harmful, underscoring the need to better understand their diverse roles in different pathological conditions. In this chapter we will introduce two-photon microscopy and compare the in vivo and in vitro imaging approaches for studying microglia. We will also discuss relevant mouse models available for in vivo imaging studies of microglia and review how such studies are constantly reshaping our understanding of the multifaceted role of microglia in the healthy and diseased CNS.

Keywords Microglia • In vivo imaging • Imaging technologies and methods • Mouse models • Two-photon microscopy • In vitro experiments • Alzheimer's disease • Multiple sclerosis

---

D. Davalos, Ph.D. (✉)
Gladstone Institute of Neurological Disease, University of California, San Francisco, 1650 Owens Street, San Francisco, CA 94158, USA
e-mail: ddavalos@gladstone.ucsf.edu

M. Fuhrmann, Ph.D.
Department of Neuroimmunology and Imaging, German Center for Neurodegenerative diseases (DZNE), c/o Biomedical Center, Sigmund-Freud-Str. 25, Bonn 53175, Germany
e-mail: martin.fuhrmann@dzne.de

M.-È. Tremblay, A. Sierra (eds.), *Microglia in Health and Disease*,
DOI 10.1007/978-1-4939-1429-6_4

Bullet Points

- This chapter highlights how in vivo imaging has revolutionized our understanding of microglial biology in the healthy and diseased central nervous system.
- We provide a brief introduction to available imaging technologies, methods, and mouse models used to study microglia in vivo, with a focus on two-photon microscopy.
- We compare findings from in vitro and in vivo imaging experiments and discuss how different experimental conditions have generated distinct and often conflicting conclusions regarding microglial functions.

4.1 Introduction

Microglia are the resident immune cells of the brain and spinal cord. They were first identified by Nissl and Robertson at the end of the nineteenth century. In 1932, Pio del Rio Hortega called them "microglia" for the first time and introduced some basic characteristics of the cells that are still valid today ((Kaur et al. 2001; Kettenmann et al. 2011), and for more information about the discovery of microglia, please see Chap. 2). For many decades after the identification and first description of microglia, their importance for central nervous system (CNS) physiology was still largely underappreciated. Moreover, their developmental origin remained an area of active debate until recently, almost 90 years after their discovery. We now know that microglia colonize the brain from the yolk-sac at the early stages of embryonic development and, unlike all other glial cells in the brain, are of mesodermal origin (Chan et al. 2007; Ginhoux et al. 2010; Kierdorf et al. 2013). Once they enter the CNS, microglia populate every part of the brain and spinal cord, and they remain present throughout life. In addition, their developmental links with myeloid progenitors and the monocytic lineage, as well as their ability to perform most of the typical immune cell functions (e.g., phagocytosis of dead cells, secretion of pro-inflammatory cytokines and chemokines, and antigen presentation), has established their classification as the resident immune cells of the CNS (see Chap. 5 for more information).

The classic immunological functions of microglia have been thoroughly described in several disease, injury, and infection paradigms (Kreutzberg 1996; Ransohoff and Brown 2012; Aguzzi et al. 2013). Interestingly, microglia have been directly associated with a robust morphological transformation, defined as their "activated state". For many decades, however, microglia were essentially assumed to be inactive in the absence of a challenge, with minimal or no involvement in physiological brain functions. The role of microglia in physiological conditions was significantly re-evaluated with the development of in vivo imaging approaches, which allowed the study of microglia in the intact brain for the first time. In fact, in vivo imaging studies with subcellular resolution became possible with the advent of two-photon microscopy (Denk et al. 1990), a major technological advance that proved instrumental for revealing previously unknown functions for microglia in both the physiological and pathological CNS. While other in vivo imaging modalities like positron emission tomography (PET) and whole animal luminescent or fluorescent imaging allow monitoring of large areas of microglial activation throughout the brain, the spatial resolution of these techniques remains relatively low. In contrast, imaging individual microglial processes with two-photon microscopy uncovered the constant physical surveillance function of microglia, underscored their significance for the architecture and function of the brain, and thereby opened new avenues of exploration into the cellular neuroscience behind these dynamic behaviors. These new endeavors generated the need for new animal models for endogenous labeling of microglia with fluorescent proteins, as well as new in vivo imaging methods to allow access to previously unexplored areas of the CNS, such as the spinal cord. Microglia are also involved in pathological processes associated with essentially every CNS disease. Investigators are now taking advantage of these newly developed imaging resources and approaches to perform longitudinal in vivo imaging studies in various animal disease models, unveiling additional roles and distinct functions for microglia in neurodegenerative and neuroinflammatory diseases. In this chapter we will first make a brief and simplified introduction to available technologies, imaging methods, and mouse models that have made the in vivo imaging studies of microglia possible, with a particular focus on two-photon microscopy. We will then highlight some of the early attempts to image microglia in culture or in slice preparations and briefly discuss their advantages and limitations. Finally, we will discuss some of the key findings that were uncovered by in vivo imaging studies of microglia, underscoring how these findings challenged longstanding preconceptions about them and helped redefine our understanding of their role in both the physiological and pathological CNS.

4.2 In Vivo Imaging of Microglia: Technologies, Methods, and Mouse Models

Several in vivo imaging techniques such as magnetic resonance imaging (MRI), ultrasound imaging, PET, and whole animal luminescence or fluorescence microscopy allow the observation of biological phenomena in living organisms. Among these modalities, in vivo imaging specifically of microglial cells in the mammalian CNS has been performed with luminescence imaging, PET, and multi-photon or two-photon laser scanning microscopy. For example, luminescence imaging has allowed to study microglial activation in a model of transient ischemia using a transgenic mouse line with expression of both luciferase and green fluorescent protein (GFP) under the promoter of the toll-like receptor 2 (TLR2) (Lalancette-Hebert et al. 2009). This receptor, though not specific for microglia, is upregulated in these cells following certain brain insults. Bioluminescence imaging using a charged coupled device (CCD) camera showed TLR-2 upregulation indicative of microglial activation that was interestingly evident in the olfactory bulb a few hours before it was detected over the site of infarction (Lalancette-Hebert et al. 2009). Microglial activation in different brain regions can also be detected using PET imaging and the radioligand [11C]PK11195 (Venneti et al. 2013). [11C]PK11195 binds to the translocator protein (TSPO) which has been shown to be upregulated in activated microglia upon brain injury and neurodegenerative conditions. However, this ligand is not specific for microglia, as it detects to a smaller degree also activated astrocytes. Moreover, [11C]PK11195 cannot be used to image microglia under physiological conditions (Venneti et al. 2013). New tracers are therefore needed to specifically image microglia in physiological and pathological conditions, using non-invasive modalities like PET and MRI imaging. Developing better microglia-specific tracers for these imaging modalities would be an invaluable resource for in vivo studies of microglial responses across large brain regions, not only in animal models but also in human patients. On the other hand, though PET and MRI are the most widely used clinically applicable imaging modalities, they can still only accomplish a relatively limited spatial resolution.

Among all available in vivo imaging modalities, two-photon microscopy can accomplish the highest spatial resolution. It is therefore typically used to track individual cells or even cellular compartments such as dendritic spines and microglial processes, as they partake in complex biological processes in real time, within their natural microenvironment, in both health and disease. This is accomplished by the use of specialized lasers that emit low-energy infrared light that penetrates below the surface of living organs, such as the brain, with minimal photobleaching and photodamage (Helmchen and Denk 2005; Svoboda and Yasuda 2006).

4.2.1 Two-Photon Laser Scanning Microscopy: Introduction to the Technique

Fluorescence microscopy is largely based upon the ability of specific molecules called fluorophores to absorb and then emit light. In single-photon fluorescence microscopy, one photon delivers enough energy to excite an electron of a fluorophore to a higher energy state (Fig. 4.1a). Part of this absorbed energy is released by light emission that occurs when the excited electron returns to its basal state as the fluorophore relaxes. In 1931, Maria Göppert-Mayer first described the principle of multi-photon excitation (Göppert-Mayer 1931), such that if two or more photons arrive at a fluorophore simultaneously, their energies could be combined in a sum that, if sufficient, would excite an electron to its higher energy state, after which it can relax to its ground state and emit light, namely produce fluorescence (Fig. 4.1a). Sixty years later, Winfried Denk invented the first microscope equipped with a Titan-Sapphire laser, which experimentally accomplished two-photon excitation (Denk et al. 1990). The space and time requirements for two-photon excitation are very strict, as two photons must arrive at the same molecule within 10^{-18} s. To meet this time requirement, Titan-Sapphire lasers have undergone extensive

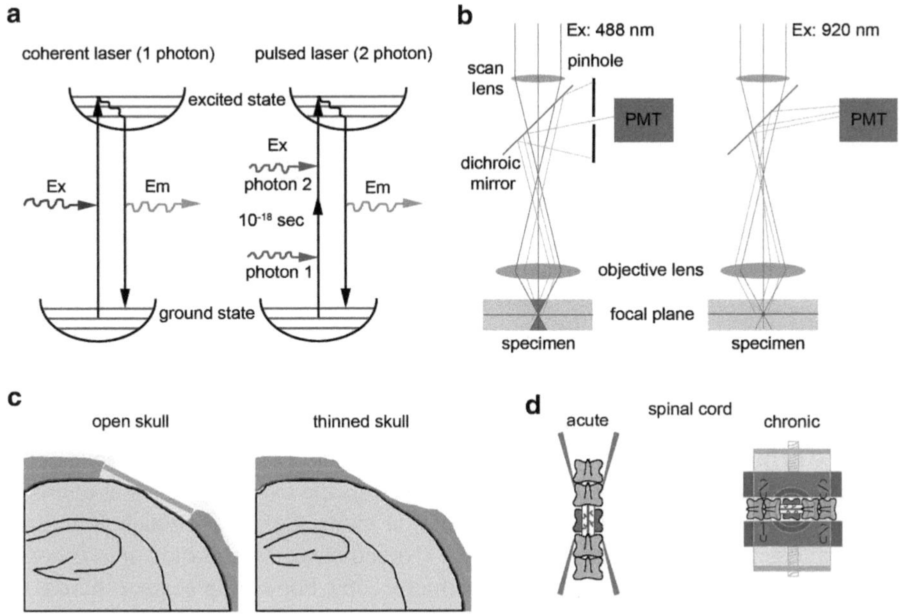

Fig. 4.1 Two photon in vivo imaging of the central nervous system. (**a**) Jablonski diagrams illustrating single versus two-photon excitation of a green fluorophore. *Left panel*: For single photon excitation, coherent laser light of higher energy and shorter wavelength (e.g., *blue* 488 nm) than the emission wavelength (*green* 530 nm) is used. *Right panel*: For two-photon excitation pulsed infrared laser light of lower energy and longer wavelength (e.g., *red* 920 nm) than the emission wavelength of the fluorophore (*green* 530 nm) is used. Two photons have to spatio-temporally coincide to combine their energy and elevate the electron to a higher energy state within 10^{-18} s. (**b**) Exemplary light paths of a confocal and a two-photon microscope. All fluorophores of the specimen within the light-cone of the objective are excited with coherent laser light in a confocal microscope. The pinhole in front of the photomultiplier detector (PMT) prevents out of focus light from reaching the PMT. Two-photon excitation occurs only at the focal point, preventing out of focus excitation of the fluorophore. Hence, no pinhole is necessary and all emitted photons reach the PMT. (**c**) The two widely used surgery techniques chronic window or "open skull" and "thinned skull". A circular coverslip is glued with dental acrylic as skull replacement (*left panel*). For thinned skull the bone stays intact and is thinned to about 50 μm (*right panel*). (**d**) Two options for two-photon in vivo imaging of the spinal cord. After performing a laminectomy on a single vertebra the vertebral column is stabilized by two external holders (*left panel*). For a chronic window, two metal bars laterally stabilize the vertebral column after the laminectomy. A glass coverslip is held in position by a placeholder (*right panel*)

development over the past two decades to pulse light with a repetition rate of 70–100 MHz and a pulse width of 70–100 fs (10^{-15} s). Additionally, to accomplish a high enough abundance of photons for two-photon excitation it is necessary to concentrate the laser light in space, namely increase the density of photons available to excite a fluorophore molecule. High numerical aperture (NA) objective lenses contribute to this end. In recent years, most major microscope manufacturers have constructed high NA (>0.95) objective lenses, with magnifications ranging from ×16 to ×60 and working distances of up to 8 mm. These are typically water-immersion lenses that have been optimized for infrared light and manufactured specifically for use with two-photon microscopy of living tissues or organs immersed in physiological solutions.

In single photon excitation, like that used in conventional confocal microscopy, specimens are illuminated with coherent laser light, leading to fluorescence generation throughout the depth of the specimen (Fig. 4.1b). To accomplish sufficient spatial resolution, the abundance of fluorescence generated from the tissue above and below the focal plane is prevented from reaching the light detector by the use of mechanical aperture regulators called pinholes. In contrast, the strict time and space requirements for successful two-photon excitation result in fluorescence generation only at the focal plane of the lens that is used to focus the excitation beam within the specimen (Fig. 4.1b). As a result, two-photon excitation only happens at the focal point, making pinholes redundant. This is also important, because less tissue is exposed to high amounts of energy, thereby reducing the amount of photobleaching and photodamage of the imaged tissue. Conversely, the use of pinholes in confocal microscopy results in slightly better resolution along the z-axis, which is a slight disadvantage of two-photon microscopes. Nonetheless, two-photon microscopes have numerous advantages over confocal microscopy. For example, since there is no out-of-focus light generated, two-photon microscopes are designed to collect all fluorescence-generated photons, which translates into enhanced detection sensitivity (Fig. 4.1b). Furthermore, the long-wavelength infrared light generated by two-photon lasers is scattered less by biological tissue, resulting in better depth of penetration than any other light microscopy approach. Collectively, these features suggest that two-photon microscopy is the most appropriate and advantageous technique for imaging microglia in animal models in vivo, especially at the single cell level, in both physiology and disease.

Even as such, since it still requires light penetration for both fluorescence generation and detection, light scattering within living tissues is the main limiting factor for two-photon microscopy, with a penetration limit of ~1 mm below an organ's surface. One way to overcome this limitation and image deeper into living tissues is to use fiber optics. Fiberscopes with small diameters of 1–2 mm may be inserted in living organs at depths of more than a millimeter, as needed (Flusberg et al. 2005). The technique of microendoscopy for in vivo imaging of deep brain structures such as the hippocampus was greatly advanced by Mark Schnitzer's laboratory that fabricated grin-lenses, suitable for two-photon microscopy (Barretto et al. 2011). The same group also built a miniaturized fluorescence microendoscope that can be implanted in the brain and thus allow in vivo imaging in freely moving animals, with single cell resolution (Flusberg et al. 2008; Ziv et al. 2013). Imaging in head-fixed mice has also been exploited as an approach to study cellular responses in awake animals, using two-photon microscopy (Dombeck et al. 2007; Lovett-Barron et al. 2014). Though most of these approaches have been used to study neuronal activity in vivo, expanding them to studies of microglia in deeper brain structures, and in animals free from the possible effects of anesthesia, remain interesting directions for the future.

4.2.2 In Vivo Imaging Methods for the Mouse Brain and Spinal Cord

The breakthrough of two-photon microscopy generated a need for developing new methods to apply this major technological advance to imaging cells in living organisms. In the past two decades, many laboratory animal species have been used for in vivo imaging, including *Drosophila* flies, zebrafish, *Caenorhabditis elegans* (*C. elegans*) worms, mice, rats, cats, and primates. Microglia, for example, have been imaged in vivo mostly in mice, both in the brain and, more recently, in the spinal cord. Proper exposure and immobilization of the imaged tissues are essential for acquiring motion-artifact free, high-resolution images. As a result, several approaches have been exploited for stabilizing the skull and the vertebral column while having optical access to image the cerebral cortex and the spinal cord, respectively (Fig. 4.1c, d; Holtmaat et al. 2005; Davalos et al. 2008).

Two main approaches have been used for in vivo imaging of the mouse cortex: the thinned skull and the chronic cranial window (Fig. 4.1c; Xu et al. 2007; Holtmaat et al. 2009). After skin removal and exposure of the skull, a high-speed dental drill is typically used to remove the upper layers of bone. Subsequently, the skull is very carefully thinned with a surgical blade to a thickness of 20–50 μm (Grutzendler et al. 2002), allowing sufficient light to get through for imaging up to a depth of ~300 μm. For the chronic cranial window, a circular part of the skull is removed using a dental drill. A glass coverslip with optimal optical characteristics replaces the removed skull, allowing imaging up to a depth of 600–700 μm in the cerebral cortex. Each approach has advantages and disadvantages that should be carefully considered depending on the experimental aims. The thinned skull procedure has the advantage of keeping the skull intact and the pial surface injury and inflammation to a minimum (Xu et al. 2007). This may be particularly important for studying acute microglial behavior in the brain as microglia quickly respond to tissue damage as will be discussed later in this chapter. On the other hand, depth penetration and the ability to repeatedly reimage through the same thinned window are limited. Since the thinned part of the bone regrows, the skull-thinning procedure must be repeated every time the same cortical area or individual cells need to be reimaged, over time-periods ranging from days to weeks or months. With every attempt to re-thin the bone, the possibility of causing pial damage and inflammation becomes greater, making it difficult to image through the same thinned window more than three to four times in total. Alternatively, the complete removal of a bone flap to generate a chronic cranial window enables unrestricted repetitive and deep imaging up to a period of a year or even longer, after an initial healing period that limits the applicability of this method for acute studies. This is usually a 2- to 3-week period, which is required for the increased amount of inflammation that this technique introduces to subside and allows for useful imaging data collection thereafter. To overcome these limitations, a combination of both techniques was recently described using a polished and reinforced thinned skull (PoRTS) or thinned-skull cortical window (TSCW) (Drew et al. 2010; Marker et al. 2010). This approach requires placing a coverslip on top of a thinned and polished area of the skull and, though more technically demanding, can significantly improve the accessibility and durability of thinned skull preparations for long-term imaging of the intact mouse brain.

In contrast to the skull, the spinal cord lies within the flexible structure of the vertebral column, close to the lungs and the heart. This location makes stabilization a major challenge for in vivo imaging of the spinal cord. The first attempt to image the spinal cord in vivo on a regular microscope stage required an elaborate experimental setup with several serial laminectomies, a perfusion system, and animal intubation that allowed breathing to be stopped for each image acquisition (Kerschensteiner et al. 2005; Misgeld et al. 2007). This approach also required extensive image post-processing to align frames that had shifted position during animal breathing intervals. This method was appropriate for imaging and realigning single axons in the spinal cord, but was limiting for use with more dense cellular populations, such as microglia. These limitations were overcome in a more recent method that accomplished stable in vivo imaging of a spinal cord segment using a spinal stabilization device to immobilize the vertebral column (Davalos et al. 2008; Davalos and Akassoglou 2012). In this method, after the paravertebral muscles are separated from the vertebral column, a laminectomy is carried out on a single vertebra. The spinal column, head, and tail base of the animal are then clamped in a way that provides space underneath the body for breathing movements (Fig. 4.1d; Davalos et al. 2008; Davalos and Akassoglou 2012). Since the exposed spinal cord is not sealed with a coverslip during this procedure, repetitive imaging in the spinal cord requires suturing and reopening the musculature over the exposed spinal cord window. This allows for only a limited number of reimaging attempts, as scar tissue can form over time. Recently, however, an implantable window was developed, which is placed over the exposed spinal cord segment with a coverslip that—at least to some degree—limits tissue regrowth (Fig. 4.1d; Farrar et al. 2012). This approach allows for repetitive and unrestricted access to the spinal cord over extended periods of time.

4.2.3 Mouse Models to Study Microglia by In Vivo Imaging

In vivo imaging has been facilitated tremendously by the availability of transgenic animals that express fluorescent proteins in specific cell types (Tsien 1998; Feng et al. 2000). This technology made possible the direct observation and in vivo tracking of many cell types—including microglia—under physiological and pathological conditions (Helmchen and Denk 2005; Germain et al. 2006; Misgeld and Kerschensteiner 2006; Svoboda and Yasuda 2006; Ishii and Ishii 2011; Kawakami et al. 2012). Using fluorescent proteins to image microglia has proven difficult, as microglia are derived from the monocytic lineage and, therefore, share a large number of markers with peripheral cells of the same origin. Additionally, a unique microglial promoter has yet to be identified, so there currently is not a transgenic or knockout mouse line available that can accomplish specific and exclusive genetic manipulation, including fluorescent protein expression, only in microglia. Nevertheless, since microglia are the only monocytes normally residing in the brain and spinal cord, there are a few mouse lines that have been used as fluorescent reporters of microglia.

The first and most widely used mouse line for in vivo imaging of microglia is the CX3CR1-GFP line that was generated by Steffen Jung in Dan Littman's laboratory (Jung et al. 2000). In this line, the coding sequence for the enhanced GFP is knocked into the *Cx3cr1* locus, partially replacing the *Cx3cr1* exon. Therefore, in homozygosity, this results in GFP expression and a loss of function of the *CX3CR1* protein. The *Cx3cr1* gene is expressed by myeloid precursors in the bone-marrow, circulating monocytes in the blood and dendritic cells throughout the body, but only by microglia in the physiological CNS. While homozygote mice are viable and have bright expression of GFP, they are typically used for studying the effects of *Cx3cr1* deletion. Heterozygote mice that bear one copy of both the *Cx3cr1* and the GFP gene were recently shown to have altered synaptic transmission, learning, and memory (Maggi et al. 2011; Rogers et al. 2011). Nevertheless, as a reporter line, the heterozygous CX3CR1-GFP mice have proven an invaluable tool for in vivo imaging studies of microglia in the CNS.

Two additional mouse lines that further exploit the *Cx3cr1* locus and may also be used as reporters for microglia were recently generated by two different groups. They are the CX3CR1-Cre and CX3CR1-CreER mouse lines (Parkhurst et al. 2013; Yona et al. 2013), in which the coding sequence for the Cre-recombinase is introduced into the *Cx3cr1* locus. The tamoxifen-inducible CX3CR1-CreER line offers the advantage of driving the Cre-recombinase expression in mice through the administration of tamoxifen. When these mice are bred to a reporter mouse line expressing floxed forms of a fluorescent protein, such as the Rosa26-YFP or others, the expression of that fluorescent protein is turned on specifically in CX3CR1-expressing cells. Additionally, using these mouse lines conditional targeted deletion of floxed genes in microglia can be carried out.

An additional microglial reporter line is the one expressing GFP under the *Iba1* promoter (Iba1-GFP) (Hirasawa et al. 2005). These mice were generated via pronucleus injection and carry a transgene that consists of the 5′ non-coding region upstream of the promoter of the ionized calcium-binding adapter molecule 1 (Iba1), exon 1 and the beginning of exon 2. The coding sequence of *Gfp* replaces most of the exon 2 sequence, but all interferon-γ response elements (IRE) are preserved and the genomic *Iba-1* locus also remains functional in this transgenic mouse line. Iba1-GFP mice exhibit GFP expression also outside the brain (e.g., macrophages, liver, and foreleg) and are not used as commonly as the CX3CR1-GFP mice.

Another transgenic mouse line uses the c-fms promoter, also named macrophage colony-stimulating factor receptor (M-CSFR or CSF-1R) or CD115, for GFP expression in monocytes and macrophages (Sasmono et al. 2003; Geissmann et al. 2010). M-CSFR-GFP mice express GFP in microglia and other mononuclear cells. Similar to Iba-1-GFP mice the transgene is randomly introduced into the genome leaving the *M-csfr* gene intact. They have been used to study microglia in the aging and prion-infected brain, as well as to study the role of microglia for neurogenesis

(Sierra et al. 2007, 2010; Ebert et al. 2009; Erblich et al. 2011; Gomez-Nicola et al. 2013). However, whether the expression of GFP is sufficient for two-photon in vivo imaging remains to be determined.

CD11b-DTR transgenic mice may also be used as a microglial reporter. In these mice the diphteria toxin receptor is expressed under the CD11b promoter together with GFP (Duffield et al. 2005). This line has GFP expression in all CD11b-expressing leukocytes, which include monocytes, neutrophils, natural killer cells, granulocytes, microglia and macrophages, a significant percentage of bone marrow cells, and a small percentage of spleen cells. This mouse was recently used to study neurotoxicity of microglia during development (Ueno et al. 2013). Interestingly, the depletion of all CD11b-positive cells, which is accomplished by administration of diphteria toxin, was found to be transient in this mouse model, implicating that the expression of GFP may have a similar time-sensitive expression pattern. Whether the amount of GFP expression is sufficient for two-photon in vivo imaging also remains an open question.

In summary, there are several reporter mouse lines that can be used to study different aspects of microglial function using in vivo two-photon microscopy. Heterozygous CX3CR1-GFP mice provide the best model to visualize microglial morphology, to date. In the future, the generation of a mouse reporter line that specifically and exclusively labels only microglia and not peripheral cells of the monocytic lineage will be instrumental for deciphering the specific roles of CNS-resident microglia in physiology, as well as in neurological disease.

4.3 Histological, In Vitro, and Ex Vivo Studies of Microglia

4.3.1 Histological Studies of Microglia

Prior to the development of the technological and genetic tools that made in vivo imaging possible, studies were performed either in fixed tissue using classic histological staining or immunohistochemical techniques, or by in vitro imaging of cultured cells or ex vivo preparations of CNS tissues. Since the invention of the light microscope, scientists have been compelled to describe the morphological and structural characteristics of cells and tissues and use this information to propose potential functions and cell–cell relationships in different organs. Using a modified silver carbonate impregnation technique to label cells in situ allowed Pio del Rio Hortega to draw microglial cells in their highly ramified morphology as commonly seen in the physiological brain. These drawings were impressively accurate representations of the cellular morphology of microglia in an unchallenged setting. Using his advanced staining technique, Pio del Rio Hortega managed to conduct and publish a number of studies between 1919 and 1927 which put him in a position to describe and propose some of the fundamental principles of microglial biology that remain unchallenged to date (Kettenmann et al. 2011). Interestingly, several studies of the same era also proposed that some brain glial cells were of mesodermal origin and infiltrated the brain at an early developmental stage (Kettenmann et al. 2011). This concept was only recently confirmed for microglia using elegant genetic fate mapping approaches (Chan et al. 2007; Ginhoux et al. 2010; Kierdorf et al. 2013) and for reviews see (Prinz et al. 2011; Saijo and Glass 2011).

In more recent decades, immunohistochemical studies of healthy and diseased tissue have identified several morphological phenotypes of microglia, which were further characterized as distinct functional states of the cells. The amoeboid microglia which are first seen upon infiltration and differentiation of pial macrophages and peripheral monocyte progenitors during the early developmental stages were shown to be the precursor form of adult microglia (Kaur et al. 2001; Zusso et al. 2012). In the adult CNS, under physiological conditions, microglia bear long processes with many finer branches that have given them the name "ramified" microglia (Fig. 4.2a, c; Kettenmann et al. 2011). Ramified microglia are ubiquitously deployed throughout the CNS, in a mostly regular manner, where they each cover their own territory (Lawson et al. 1990). Interestingly, their distribution varies considerably among different brain regions. For example, they are denser in the hippocampus, basal ganglia, and the substantia nigra than in fiber tracts, the cerebellum, and much of the brainstem. Additionally, they appear to be denser in gray rather than white matter (Lawson et al. 1990). Unlike ramified microglia, the reactive or activated microglia have enlarged, mostly rounded cell bodies and a few short or no processes (Fig. 4.2b, d; Kreutzberg 1996). Activated microglia are typically found around sites of traumatic injury and in most pathological CNS conditions. Some of these distinct microglial phenotypes have been shown to interconvert in vitro, in response to stimulation with cytokines, growth factors, astrocyte condition media, or other stimulating agents like lipopolysaccharide (LPS) and adenosine triphosphate (ATP) (Suzumura et al. 1991; Stence et al. 2001), implying the dynamic capacity of these cells to adapt to changes in their environment.

4.3.2 Studies of Microglia in Cell Culture Conditions In Vitro

Beyond the morphological characterization of cells in situ, however, histological analyses can only allow speculative assessments of cellular functions and cell–cell relationships. In order to confirm or analyze the specific function of a particular cell type, experimental observation and manipulation of living cells is required. The development and continuous improvement of cell culture conditions over the past few decades has allowed numerous studies of living cells in vitro, both in immortalized cell lines and in primary cultures. Although the isolation of primary microglia was already described in 1930, microglia-derived immortalized cell lines have only been established in the last three decades (Rodhe 2013).

Microglial cells in culture express several of the typical markers of the monocytic lineage like CD11b, CD68, FcR (Fragment crystallizable region Receptor), major histocompatibility complex (MHC) class I and II, and many others (Duke et al. 2004; Rodhe 2013). Studying microglial cells in culture has demonstrated some of

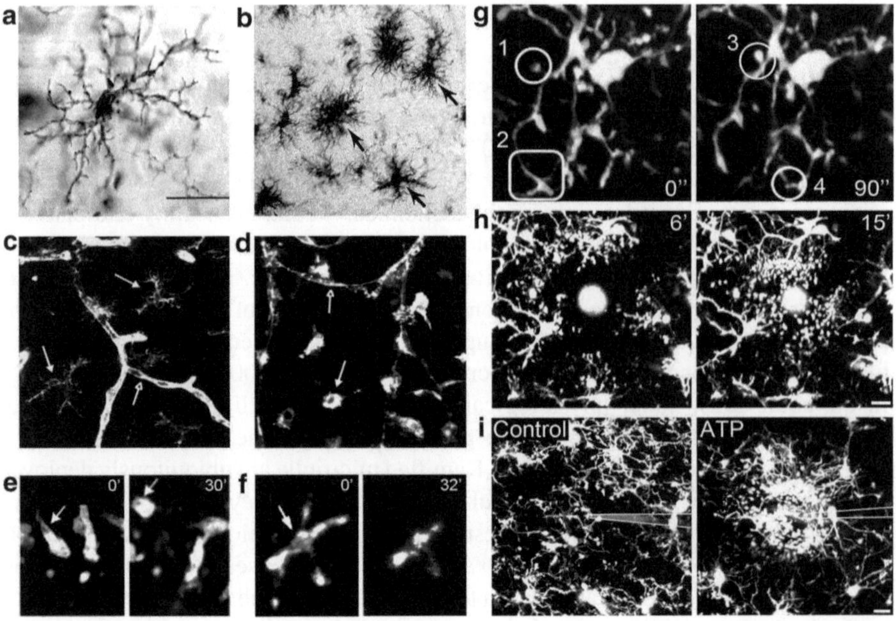

Fig. 4.2 Imaging of microglia in fixed tissue, in vitro, in brain slice cultures, and in the intact brain in vivo revealed different aspects of microglial biology. (**a**) Ramified microglial cell from a 64-year-old human brain and (**b**) activated microglia with a "bushy appearance" from a 41-year-old human brain, stained with the monoclonal antibody LN-3. Adapted from (Streit 2013) and reprinted with permission from Oxford University Press (**c**, **d**) Confocal microscopy images of hippocampal brain slices from a postnatal day 7 rat, stained with FITC-isolectin B4 (IB4), which labels both microglia and blood vessels. Highly ramified microglial cells (*solid arrows*), as well as blood vessels (*open arrows*), are seen when the slice was fixed, stained, and imaged immediately after preparation (**c**), whereas microglia appear rounded or amoeboid when the slice was first left in culture for 1 day (**d**); notice that the blood vessels have also begun to collapse at this point in culture. Adapted from (Dailey and Waite 1999) and reprinted with permission from Elsevier. (**e**) Live-cell imaging of IB4-labeled microglia in culture shows robust changes in shape and movement among different amoeboid cells, over a 30 min time-lapse. Adapted from (Dailey et al. 2006) and reprinted with permission from Springer. (**f**) Ex vivo confocal imaging of microglia in a postnatal day 5 mouse brain slice demonstrates the process of process retraction as cells in the slice get activated. The first time-point in the time-lapse (0 min) was collected within 45 min of sacrificing the animal; while half of the cell's processes are already retracted another 32 min later, all processes are absorbed in the cell body (*arrow*) within 1–2 h from the time of sacrifice (not shown). Adapted from (Stence et al. 2001) and reprinted with permission from Wiley and Sons. (**g**–**i**) In vivo imaging of GFP-expressing microglia in the cortex of an adult $CX3CR1^{GFP/+}$ mouse. (**g**) Time-lapse imaging of the same microglial branches demonstrated rapid extension and retraction of fine microglial processes over seconds. *Circles* and *rounded box* indicate four representative processes that change in length and shape over time. (**h**) Time-lapse in vivo imaging of microglial responses to a localized ablation (bright sphere, ~15 μm in diameter, in the *center*) delivered using the two-photon laser inside the mouse cortex. Nearby microglial processes responded immediately with bulbous termini and started extending within just a few minutes toward the ablation until they formed a spherical containment around it (not shown); at the same time, the same cells retracted those processes that lay in directions opposite to the site of injury. (**i**) Microglial response to local injection of adenosine triphosphate (ATP) at the 30 min time point after insertion of the injecting glass electrode (*white lines*) through a craniotomy. Whereas a control microelectrode containing artificial cerebrospinal fluid (ACSF) and rhodamine-dextran caused little or no microglial response (*left*), similar electrodes containing 10 mM ATP induced rapid extension of microglial processes towards the tip of the electrode (*right*). Panels g-i adapted from (Davalos et al. 2005)

their classic known abilities, such as their membrane ruffling upon stimulation, chemotactic migration towards a concentration gradient, activation and secretion of cytokines in the culture medium following stimulation, and phagocytic behaviors (Smith et al. 1998; Hide et al. 2000; Honda et al. 2001; Shigemoto-Mogami et al. 2001; Wollmer et al. 2001; Xiang et al. 2006; Ohsawa et al. 2007; Jeon et al. 2012). In vitro studies have the advantage that they can be performed under well-defined conditions, allowing the experimental manipulation of cells by transfection of exogenous genes and addition of stimulatory or inhibitory agents. As a result, in vitro cell culture studies have elucidated both cellular and molecular properties and capabilities of microglia under specific experimental conditions. However, as is the case with all cell types, both immortalized and primary microglial cultures have significant limitations with respect to the artificial environment that cells grow in, their viability, and their relevance to cells in their physiological environment.

4.3.3 Ex Vivo Studies of Microglia in Acute or Organotypic Slice Cultures

To overcome some of the in vitro limitations, microglia have been studied in their endogenous microenvironment, mostly within acutely extracted tissue from brain slices. In the field of neuroscience, brain slice preparations have been used in electrophysiological, morphological, biochemical, and pharmacological studies of brain structures. For example, imaging studies performed in brain slices have expanded our understanding of dendritic calcium dynamics and the biochemical signals regulated by neuronal activity (Svoboda et al. 1997; Nagerl et al. 2004; Lohmann et al. 2005; Yasuda et al. 2006; Muller et al. 2012). Brain slice preparations provide precise control of experimental conditions, such as temperature, pH, oxygen, carbon dioxide (CO_2), and ionic and drug concentrations, while maintaining cells in the context of the tissue that surrounds them. As such, these preparations have facilitated the examination of electrophysiological properties and metabolic parameters of intact neuronal and glial cells, which can easily be located, identified, and accessed within a brain slice. A technical advantage of brain slices over in vivo studies is that they can be performed without contamination from anesthetics, muscle relaxants, or intrinsic regulatory substances (Wang and Kass 1997). They also offer great stability for imaging purposes and electrophysiological recordings due to the lack of movement artifacts that are generated from the heartbeat and respiration pulsations.

On the other hand, brain slices also have inherent limitations. Although the architecture and connectivity of the spared tissue can be maintained and kept functional within a distance from the edges of a slice and for a certain period of time, brain slices lack inputs and outputs from distant connections that have been severed. In addition, both acute and organotypic brain slices have a limited lifespan, with the first being viable for a few hours and the latter for a few days to weeks, depending on the preparation and the goal of the study (Lossi et al. 2009). Moreover, since blood circulation through the vascular system cannot be maintained, even evolved artificial bathing media that have improved the viability of slices lack blood-derived factors and the physiological signaling that occurs between vascular, neuronal, and glial cells. The effects of decapitation ischemia produced by the loss of blood supply and the stress that the tissue undergoes until bathed in media are not well understood, and probably significantly underestimated (Mayevsky 1978). Most importantly for microglial studies, slicing an acutely extracted brain delivers extensive tissue injury, especially at the top and bottom surfaces of the slice. Although this may not appear to affect neuronal and other glial cell populations that may continue to function "physiologically" as long as they are located further away from the slice surfaces, the same is not true for microglia. Microglia have myeloid origins and are, thus, immune cells programmed to respond to tissue damage in the brain or the spinal cord (Thomas 1992; Kreutzberg 1996; Gonzalez-Scarano and Baltuch 1999; Raivich et al. 1999). As tissue injury is an unavoidable effect of the preparation procedure, brain slicing causes microglia to become activated and undergo a series of changes in their cellular morphology, behavior, and gene expression profile, the degree of which depends on the severity of the incurred injury (Fig. 4.2c, d; Davis et al. 1994; Stence et al. 2001; Petersen and Dailey 2004). Some of the most commonly observed functions of activated microglia include cell proliferation, migration, phagocytosis, and up-regulation of typical immunological markers and antigen-presenting cell capabilities (Kreutzberg 1996; Aloisi 2001; Carson 2002). Several of these functions were first observed and described before the advent of two-photon microscopy, by ex vivo imaging studies performed in brain slices using white light or fluorescence microscopy (see below).

4.4 Imaging Studies of Microglial Morphology and Activation In Vitro or Ex Vivo

The ability of microglia to perform highly active movements has been evident since the earliest imaging studies were performed in dissociated cell cultures (Fig. 4.2e; Booth and Thomas 1991; Ward et al. 1991; Haapaniemi et al. 1995; Tomita et al. 1996; Takeda et al. 1998). Imaging microglia by time-lapse confocal microscopy in acutely extracted mouse brain slices led to a detailed description of the morphological transformation that they undergo from their presumed sessile ramified phenotype in the unperturbed brain (Stoll and Jander 1999) to that of fully activated cells that migrate to sites of injury, typically towards the surfaces of the slices (Czapiga and Colton 1999; Stence et al. 2001). This transformation included almost complete withdrawal of microglial "resting" processes before activated microglia were able to migrate, proliferate, and phagocytose dead or injured cells and cellular debris (Fig. 4.2f; Stence et al. 2001). Interestingly, under acute slicing conditions, preexisting ramified branches were found to be incapable of extension and only underwent a slow retraction, sometimes until complete withdrawal, while new "reactive" processes had to form for activated microglia to phagocytose injured cells or debris

(Stence et al. 2001). Similarly, different microglial motility behaviors were described in mouse brain slices towards recently deceased versus preexisting dead cells (Petersen and Dailey 2004).

In another confocal time-lapse video microscopy study, microglial migratory patterns were studied in acute mouse brain slices in which a stab wound injury was delivered 1–7 days prior to slice preparation. As a result, microglia in the injured hemisphere had sufficient time to become activated by the injury in vivo, and their motility patterns were compared to those of cells in the contralateral, uninjured hemisphere (Carbonell et al. 2005). Microglia activated by the stab wound injury exhibited increased migratory activity for 1–4 h after acute brain slice preparation from mice sacrificed 1 and 3 days after injury. These motility patterns were not directed but instead appeared autologous and non-biased and were hence characterized as a "random walk" (Carbonell et al. 2005). The behavior of microglia in organotypic slice cultures was shown to vary significantly over time and with the preparation methods, as discussed above, but also between animal species and medium contents (Czapiga and Colton 1999).

The analysis of "snapshots" of biological events by classic histological approaches could hinder the full appreciation of the sequence or causality between ongoing cellular processes, responses to stimuli, and transient cell–cell interactions (Davalos and Akassoglou 2008). Although some of these phenomena can be studied ex vivo, the preparation and conditioning of slice cultures inherently activates microglia, the first responders to brain injury (Dailey and Waite 1999). Therefore, careful consideration of the physiological relevance of observed microglial responses in slice studies is required, as many of the cells are exposed and hence react to the massive tissue injury in their vicinity. In contrast, imaging microglia in their intact environment represents a reliable approach to investigate their interactions with the surrounding tissue, in the absence of underlying injury signals.

4.5 Redefining the Role of Microglia by Imaging Them In Vivo for the First Time

4.5.1 Baseline Process Dynamics

For many decades, the highly branched microglia in the unperturbed adult brain were considered to be largely inactive. As a result, ramified microglia were characterized as "resting", underscoring the limited functional role that was attributed to the cells under normal conditions compared to their activated state (Streit et al. 1988; Davis et al. 1994; Stoll and Jander 1999). The concept of "resting" microglia in the physiological brain was challenged by the first in vivo imaging studies of microglia in the mouse cortex, published in 2005. By performing transcranial two-photon imaging in $CX3CR1^{GFP/+}$ mice, Davalos et al. and Nimmerjahn et al. showed that the higher order processes of ramified microglia demonstrate a highly motile behavior, while their cell bodies and main processes remain relatively stationary (Fig. 4.2g; Davalos et al. 2005; Nimmerjahn et al. 2005). These real-time recordings of the continuous cycles of small extensions and retractions of the finer microglial processes were the first in vivo demonstration of the tissue surveillance function of microglia in the unperturbed mammalian brain, in real time. Moreover, this was a completely unexpected result that generated extensive interest in the physiological significance of microglia and is currently being investigated by many groups globally.

4.5.2 Rapid Response to Focal Brain Injury

The first imaging study of microglial behavior in a living organism was performed in the leech, examining their responses to massive nerve injury. By labeling only cell nuclei and using ultraviolet illumination for fluorescence video microscopy, invertebrate microglia were found capable of rapidly migrating along axonal tracks towards nerve lesions from distances often further than 200 μm away (McGlade-McCulloh et al. 1989). These migratory responses were similar to those described in mouse brain slices that were discussed above and seem to be typical for microglia challenged with massive neuronal injury. The ability of microglia to rapidly contain localized damage in their vicinity was also first described by in vivo two-photon microscopy in *CX3CR1*$^{GFP/+}$ mice.

By taking advantage of the focal properties of the two-photon laser, Davalos et al. and Nimmerjahn et al. introduced small focal injuries in the mouse cortex, delivered either to the brain parenchyma or on cortical blood vessels. In both cases, neighboring microglia exhibited very rapid responses by extending their processes directly towards the sites of injury and physically containing them (Davalos et al. 2005; Nimmerjahn et al. 2005). Microglial processes showed enlarged bulbous termini within seconds and began extending within only a few minutes (Fig. 4.2h). Interestingly, when challenged with two injuries very close to each other, individual microglial processes of even the same cell can differentially respond by extending towards the nearest injury site (Davalos et al. 2005). These studies showed that acute microglial responses to localized injury delivered in the living brain were significantly different from those previously described in ex vivo setups, thereby challenging the concept that a "reactive transformation" was required for microglia to respond to traumatic injuries. Injuries that can be contained locally do not necessarily require process retraction, migration, and proliferation of microglia, as described in brain slice experiments. Also, preexisting processes of the ramified microglia are capable of morphological activation within seconds after they are challenged and can individually respond to nearby damage without delay (Davalos and Akassoglou 2008). In line with these observations, it was recently shown in situ that microglial phagocytosis of apoptotic neurons in the physiological brain is fast and efficient, can be performed by ramified cell processes, and does not require pre-activation of the cells by an inflammatory or a traumatic challenge (Sierra et al. 2010). Finally, microglial process extension was always directed towards the sites of injury (Davalos et al. 2005; Nimmerjahn et al. 2005). Cell body migration was not observed over the first 10 h after laser-induced injury (Davalos et al. 2005), but likely occurs over longer periods, since an increase in microglial numbers was reported after reimaging in vivo the same sites 1–3 days after injury (Kim and Dustin 2006), unlike what was previously shown ex vivo following stab wound injury in mice (Carbonell et al. 2005).

4.5.3 Discrepancies Between In Vitro and In Vivo Imaging Studies of Microglial Behavior

Though similar approaches might have proven more reliable for studying other cell types, attempting to predict microglial behavior in the intact CNS based on what had previously been described either in culture or in brain slices proved largely unsuccessful. The conflicting findings likely represent the different cellular and molecular properties that microglial cells have to acquire under artificial conditions in vitro, compared to those in the intact CNS. Interestingly, dissimilarities in the findings from in vitro and in vivo studies of microglial biology are not only limited to imaging studies of cell morphology and behavior. Gene expression profiles in cultured microglia differ significantly among cells that were extracted using different isolation methods, and often more so if compared to microglia in vivo. A comprehensive and comparative transcriptional profiling following LPS stimulation of microglia in vitro and in vivo was done for 480 inflammation-related genes and analyzed by custom-spotted microarray analysis (Lund et al. 2006). The outcome of these studies was correlated with data from a proteomics analysis and genome spanning Affymetrix chips to evaluate differences between methodologies and to obtain additional information on the regulation of non-inflammatory genes and proteins (Lund et al. 2006). This thorough comparison of the effects of the most commonly used activating molecule for microglia showed that although most inflammatory genes induced in vitro were also upregulated in vivo, several discrepancies were identified between the two experimental settings. These differences may offer the molecular basis of the observed functional and behavioral differences for microglia in vivo and are likely due to regulatory factors and compensatory mechanisms present in the CNS but not in a dish. Nevertheless, the identification of such discrepancies and limitations does not discredit the observations made in vitro or ex vivo, but highlights the importance of fully acknowledging the applicability and limitations of each experimental design. Accurate interpretation of the results acquired under specific experimental conditions is imperative for placing study findings in the appropriate context, especially for cells as responsive to changes in their microenvironment as microglia.

4.6 Regulation of Microglial Process Dynamics In Vivo

The surprising features of the observed microglial behaviors in vivo created the need to study their molecular properties and signaling mechanisms in their natural habitat inside the living brain (Davalos and Akassoglou 2008). Among the first questions asked were about the mechanisms that regulate microglial tissue surveillance, and about the nature of the molecules involved in attracting microglial processes towards sites of damage. Several molecules have been shown to activate microglia, induce morphological changes, and attract them in a concentration gradient-dependent manner in vitro. However, most studies have addressed these issues on a time scale of several hours to days, in accordance with typical immunological response paradigms. The constant motility of microglial processes in unperturbed conditions implies a mechanism that is acutely regulated within the physiological brain. Moreover, the rapid directional convergence of microglial processes towards a localized injury implies the presence of a gradient of one or more highly diffusible and abundant molecules that can mediate this phenomenon.

4.6.1 Regulation of Rapid Microglial Responses to Localized Injury

Davalos et al. demonstrated that extracellular ATP is sufficient for mediating rapid microglial responses by directly injecting ATP in the living cortex and imaging microglial responses in vivo (Fig. 4.2i). Using several pharmacological inhibitors applied directly on the exposed cortex, they showed that ATP signaling through P2Y receptors is essential for mediating microglial responses to local ATP application, as well as to focal brain injury (Davalos et al. 2005). This result was genetically confirmed by another in vivo imaging study that used microglia lacking the P2Y12 receptor and showed that they are incapable of responding to laser-induced injury in vivo (Haynes et al. 2006). Indeed, ATP is a highly diffusible small molecule that is also highly concentrated intracellularly, but not present in the cerebrospinal fluid, making ATP a good candidate for explaining the rapid microglial process responses as they were imaged in real time in vivo, following tissue damage.

Since these responses required process extension lasting for 30–60 min until the injury was contained, how could a single blast of ATP from the injured cells sustain microglial responses to the site of injury? One possibility is that the surrounding tissue plays a role in preserving and regenerating the signals that fuel the observed microglial response to damage. Indeed, pharmacologically blocking ATP release from astrocytes in vivo was sufficient to arrest microglial process extension to laser injury, implying a role for astrocytes in mediating the rapid and widespread microglial response to focal brain injury (Davalos et al. 2005). The ability of ATP to act as a chemoattractant of microglia in vitro had previously been recognized (Honda et al. 2001), and there is extensive literature describing the activating effects of purinergic signaling on microglial cell lines, primary cells, and microglia in brain slices (Hide et al. 2000; Honda et al. 2001; Shigemoto-Mogami et al. 2001; Xiang et al. 2006; Ohsawa et al. 2007; Inoue 2008; Koizumi et al. 2013). However, the study of this phenomenon inside the intact brain, where all cellular constituents are present, allowed the identification of this intriguing cell–cell interaction between astrocytes and microglia and revealed their orchestrated responses to brain injury (Davalos et al. 2005). Follow-up studies done in acute brain slices were able to recapitulate aspects of the rapid microglial responses to ATP and characterize some of the molecular downstream players involved in them. For example, the generation of outward potassium currents, the phosphoinositide 3-kinase (PI3K) pathway (Wu et al. 2007), and the activation of integrin-β1 (Ohsawa et al. 2010) were shown to be involved in the directional process extension by microglia in brain tissue. The role of ATP signaling in microglia is discussed in further detail in Chap. 3.

A similar mechanism of ATP-dependent rapid microglial responses to focal laser injury was also described in white matter by in vivo imaging in the dorsal mouse spinal cord (Lee et al. 2008). This study revealed that the nitric oxide (NO) pathway is also involved in regulating rapid microglial responses to injury. Using a combination of local inhibition or activation experiments for both the NO and ATP pathways, Dibaj et al. demonstrated that blocking the NO pathway arrests microglial responses, which can be triggered by donors and downstream compounds of the NO pathway. Interestingly, though the ATP effect on microglia was found to be independent of the NO pathway, a substantial purinergic influence on the NO-mediated microglial attraction was evident (Lee et al. 2008). The involvement of the NO pathway in mediating microglial migration to neuronal injury had previously been demonstrated in the leech nerve cord, in an elegant study combining a similar array of pharmacological treatments and low-light fluorescence video microscopy to track individual Hoechst-stained microglia as they migrated to a crush site (Chen et al. 2000). Finally, volume-sensitive chloride channels were shown to be required for the rapid process outgrowth of microglia in response to laser-induced damage in brain slices (Hines et al. 2009). The same study showed that the microglial containment of injury sites prevented lesion expansion at least in the acute brain slice setting, suggesting a beneficial role for this rapid and localized microglial response following brain injury (Hines et al. 2009).

4.6.2 Regulation of Microglial Baseline Process Motility

The striking observation that microglial processes are so dynamic, even under physiological conditions, has attracted a lot of attention since it was first described by performing in vivo imaging in the mouse cortex. Similar baseline motility observations were made in vivo in the developing zebrafish embryo (Peri and Nusslein-Volhard 2008), as well as in mouse spinal cord and retinal explants under physiological conditions (Davalos et al. 2008; Lee et al. 2008). This implies a function of fundamental significance for all anatomical sites of the CNS. What is the biological significance of the "baseline motility" pattern of microglial processes? Are they simply performing surveillance of the extracellular space to ensure structural integrity of their immediate neighbors, or are they participating in physiological brain functions? What are the molecular signals and downstream mechanisms that mediate these small extension and retraction events? Do these events play different roles in the developing, adult, and diseased CNS? Such questions became the subject of several studies in postnatal development, adult neuronal plasticity, and neuronal circuit function and dysfunction. While several of these matters are discussed in depth in separate chapters of this book, we will highlight some of the findings that emerged from in vivo imaging studies.

One hypothesis is that microglial processes extend rapidly in response to neuronal synaptic activity in their vicinity. This is certainly a reasonable assumption, since microglial baseline process dynamics are very transient, similar to events occurring in nearby synaptic clefts. However, reducing neuronal activity with tetrodotoxin (TTX, a sodium channel blocker) had no significant effect on baseline microglial dynamics in vivo, while enhancing synaptic activity with a γ-aminobutyric acid (GABA)-receptor inhibitor had a slightly stimulating effect at increasing their sampled area (Nimmerjahn et al. 2005). In both cases, the velocity of microglial process changes was also unaffected, suggesting that microglial tissue surveillance may not be directly modulated by neuronal activity in their vicinity. This finding was in agreement with a study in acute hippocampal slices where simultaneous electrophysiological recordings and time-lapse confocal microscopy showed that glutamate, GABA, or long-term potentiation did not induce chemotaxis in microglia (Wu and Zhuo 2008).

Interestingly, sensory deprivation by binocular enucleation in mice induced a reduction of neuronal activity, which was followed by a prominent retraction of microglial processes in the visual cortex over a few hours (Wake et al. 2009). This study performed the first in vivo imaging analysis of the interaction between microglia and axonal boutons or dendritic spines, the sites of synaptic connections between neuronal cells. By crossing the Iba1–GFP mice with a Thy1-driven neuronal reporter line (the GFP-expressing M-line (Feng et al. 2000)), Wake et al. detailed the frequency and duration of direct contacts between microglial processes and pre- and post-synaptic neuronal structures. They found a clear reduction in the number of contacts with pre-synaptic axon terminals following a reduction of neuronal activity, implying that at least part of the microglial baseline surveillance activity is directed toward synapses, and may depend on their functional status (Wake et al. 2009). Interestingly, following an ischemic challenge, microglial processes made prolonged contacts with axon terminals that lasted for >60 min (versus ~5 min in normal conditions), and sometimes resulted in the disappearance of the neuronal structures (Wake et al. 2009).

The possibility that microglia may be involved with the removal of neuronal processes was explored in another elegant study that used in vivo two-photon microscopy in combination with immunocytochemical electron microscopy (EM) and serial section EM followed by tedious three-dimensional (3D) reconstruction analyses. Tremblay et al. crossed CX3CR1-GFP with the Thy1-YFP H-line

(Feng et al. 2000) to label microglial and neuronal processes with different fluorescent proteins and record their interactions over time, in the presence or absence of sensory experience in the visual cortex of juvenile mice (Tremblay et al. 2010). In normal conditions, microglial processes interacted mostly with small and transient dendritic spines, which were typically lost over 2 days. Following light deprivation, however, microglial processes became less motile and shifted their preference towards larger dendritic spines that persistently shrank (Tremblay et al. 2010). The EM analysis confirmed the extensive envelopment of synaptic elements and the presence of phagocytic structures sometimes with ultrastructural features of axon terminals and dendritic spines inside of microglial processes, particularly under conditions of sensory deprivation, in the juvenile mouse cortex (Tremblay et al. 2010). Interestingly, similar inclusions were also observed in the aged brain (Tremblay et al. 2012). Overall, these findings imply a role for microglia in synaptic stripping throughout life.

Microglia are indeed essential for sculpting the developing mammalian brain, by contributing to the elimination of newborn cells, neurons, and synapses, both in the early postnatal days and in adulthood (Marin-Teva et al. 2004; Sierra et al. 2010; Paolicelli et al. 2011). In accordance with the in vivo observations, another elegant set of studies in the developing retino-geniculate system has demonstrated in situ that sculpting of this network requires microglial engulfment and remodeling of synaptic connections in a complement C1q and C3/C3R-dependent manner (Stevens et al. 2007; Schafer et al. 2012). Interestingly, the initiation of the complement protein expression seems to be regulated by transforming growth factor β (TGFβ) (Bialas and Stevens 2013), and there is a clear link between the complement-mediated synaptic pruning by microglia and neuronal activity (Schafer et al. 2012). For further information about the role of microglia in development, please refer to Chaps. 7, 8, and 9. Overall, there are several indications linking the microglial baseline process motility with synaptic activity and neuronal network modulation, but we are only beginning to uncover the cellular and molecular mechanisms involved in these processes and the potential impact they may have for neuronal function.

4.7 Imaging Studies of Microglia in the Retina

The retina is considered a part of the CNS and it is often preferred by investigators as an experimental model due to its simple cellular architecture and accessibility, compared to the brain and spinal cord. Proper preparation of whole retinal explants seems to introduce less tissue damage to the retina itself than brain slicing, an attribute that is likely responsible for the recapitulation in these preparations of several microglial behaviors previously described in the cortex in vivo. For example, ex vivo imaging of microglia in retinal explants using time-lapse confocal imaging showed similar baseline extension and retraction process dynamics to the adult unperturbed cortex (Lee et al. 2008). In response to laser injury, ramified microglia responded by rapid processes extension, as well as cell migration (Lee et al. 2008).

Abolishing the fractalkine receptor (CX3CR1) in microglia diminished both the responses to injury and the baseline dynamics of microglial processes in retinal explants (Liang et al. 2009). This finding implies that these microglial responses may be regulated through the fractalkine ligand–receptor interaction expressed in neurons and microglia, respectively. Moreover, both the morphology of microglia and their process dynamics were differentially regulated by different modes of neurotransmission in retinal explants. Specifically, there was an increase in baseline motility by glutamatergic and a decrease by GABAergic neurotransmission (Fontainhas et al. 2011). However, similar to experiments in the mouse cortex, these neurotransmitters could not induce microglial responses directly; instead, they appeared to be mediated indirectly via extracellular ATP released in response to glutamatergic neurotransmission through probenecid-sensitive pannexin hemichannels (Fontainhas et al. 2011). Interestingly, the baseline dynamics of microglial processes were also reduced in the mouse cortex, in the presence of pharmacological inhibitors of purinergic receptors (Davalos et al. 2005). Overall, these observations imply that retinal explants might offer a good ex vivo alternative for studying microglia in a controlled setting that may still preserve some fundamental aspects of the in vivo condition. Nevertheless, since retinal explants are still ex vivo preparations devoid of blood circulation, they also only remain stable for limited periods of time.

The retina has also been imaged in vivo in CX3CR1-GFP mice using confocal scanning laser ophthalmoscopes (cSLO). Highly motile microglia were shown to migrate from distances up to 400 μm towards laser-induced injuries, starting from a few minutes to hours, and persisting at the injury sites for at least a few weeks (Eter et al. 2008; Paques et al. 2010). Similarly, longitudinal cSLO imaging was used to track microglial responses in models of retinal ganglion cell damage. Following optic nerve crush injury and acute elevation of intraocular pressure, microglia transformed from a ramified to an amoeboid morphology and significantly increased in numbers over the first week post-injury. These microglia gradually returned to the ramified morphology and near-baseline cell densities over 3 additional weeks (Liu et al. 2012). These results show that the massive microglial responses that were previously described in brain slice experiments also occur in vivo, but only under conditions of extensive tissue injury that require recruitment of distant cells to contain such widespread trauma. They also underscore the adaptability of the brain's resident immune cells to different types of damage as well as their ability to return to a nonactivated state once a particular insult has been addressed.

On the other hand, even microglia are limited in their abilities to address some challenges, especially those that are of a chronic nature, such as in the context of neurological disease or normal aging. For example, ex vivo imaging in a retinal explant identified a number of morphological and functional impairments in aged microglia that were also recently confirmed in vivo, in the mouse cortex (Hefendehl et al. 2014). Unchallenged cells presented with significantly smaller and less branched dendritic arbors and exhibited slower process motilities (Damani et al. 2011). In addition, while young microglia extended processes toward extracellularly applied ATP, aged microglia became less dynamic and ramified. In response to laser-induced injury, aged microglia demonstrated slower responses acutely, which were sustained for longer than those of microglia in younger retinas (Damani et al. 2011). The compromised responses of senescent microglia both in terms of tissue surveillance and in response to injury reveal potential roles for microglia in age-dependent neuroinflammatory degeneration of the CNS.

4.8 In Vivo Imaging Studies of Microglia in CNS Disease

Microglia have been implicated in the pathogenesis of neurodegenerative and neuroinflammatory diseases, such as Alzheimer's disease (AD), Huntington's disease (HD), age-related macular degeneration, multiple sclerosis (MS), amyotrophic lateral sclerosis (ALS), and several others (Perry et al. 2010). The role of microglia in MS and AD will be discussed in detail in Chaps. 16 and 18, respectively. In the following paragraph, we will primarily focus on imaging studies of microglia related to these diseases. The involvement of microglia in pathologies of both the brain and spinal cord can be acute or chronic, and depending on the disease or tissue context, microglia may exert harmful or beneficial effects. As discussed above, microglia can contain injuries and limit the damage to surrounding healthy tissue, by clearing cellular debris in the eye, the spinal cord, or the cortex. Their accumulation around lesions can also be found in most neurological diseases. For example, microglia reside around amyloid beta (Aβ) plaques in AD, and they are found in increased numbers in prion disease, in the substantia nigra in Parkinson's disease, in MS lesions, etc. However, whether their chronic activation in many of these pathologies is protective or damaging remains largely unknown. In recent years, animal models for most neurological diseases have been developed and tremendously improved, and although they still present limitations, some can recapitulate several aspects of the human conditions to a satisfactory degree. Given the very dynamic nature of microglial responses to their microenvironment, using two-photon microscopy to follow microglial responses over the course of disease in available animal models represents a powerful approach to further elucidate their role and understand how to treat them in the context of CNS disease.

4.8.1 In Vivo Imaging of Microglia in Neurodegeneration: Alzheimer's Disease

AD is characterized by the accumulation of Aβ in plaques and by the intracellular aggregation of hyper-phosphorylated protein tau (Haass and Selkoe 2007; Palop and Mucke 2010). In AD, increased numbers of activated microglia have been identified around Aβ plaques (Fig. 4.3a) and in proximity of hyper-phosphorylated tau-bearing neurons (Hanisch and Kettenmann 2007; Kettenmann et al. 2011), suggesting that AD exhibits a prominent neuroinflammatory component comprising the innate immune system. Two-photon in vivo imaging has had a significant impact on AD

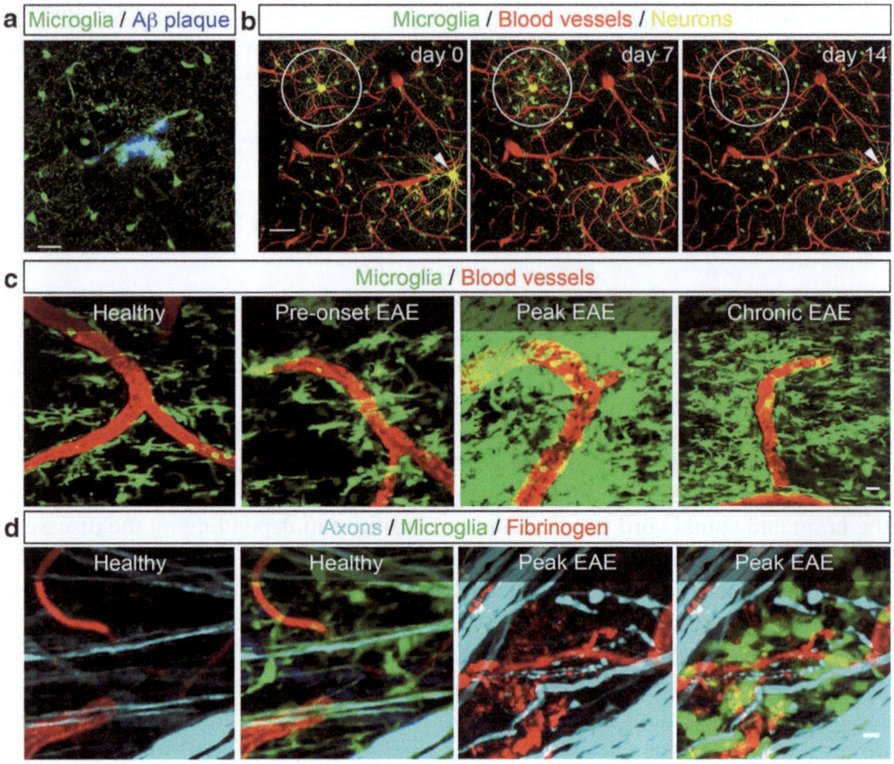

Fig. 4.3 In vivo imaging of microglia in mouse models of neurological disease. (**a**) Microglia (*green*) accumulate around amyloid beta plaques that are stained with MethoxyX04 (*blue*). (**b**) Microglia (*green*) increase in density around neurons that subsequently disappear (*yellow*) suggesting that they are involved in neuronal loss in a mouse model of Alzheimer's disease. Blood vessels (*red*) are labeled with an intravenous injection of Texas-red dextran. Adapted from (Fuhrmann et al. 2010). (**c**) Microglia (*green*) cluster around blood vessels (*red*, labeled as in **b**) at different stages of neuroinflammatory disease. Microglial perivascular clusters start forming before the onset of clinical signs, become very pronounced at the peak of experimental autoimmune encephalomyelitis (EAE), and continue forming and resolving even through the chronic phase of the disease. (**d**) In the spinal cord of healthy *CX3CR1*$^{GFP/+}$*Thy1-CFP*$^{+}$ mice ramified microglia (*green*) appear evenly distributed among healthy intact axons (*cyan*) and blood vessels (*red*, labeled by intravenous administration of fibrinogen solution). In contrast, at the peak of disease, blood-brain barrier disruption results in perivascular fibrin deposition in the perivascular space of leaky vessels; these areas of fibrin deposition define the sites of perivascular microglial clusters and the areas where neuronal damage (the main culprit of multiple sclerosis) is evident, in the form of swelling, bending, and fragmentation of axons. Adapted from (Davalos et al. 2012). Scale bars: 20 μm (**a**, **c**), 10 μm (**b**), 5 μm (**d**)

research and our understanding of the role of microglia in this disease. Meyer-Luehmann et al. investigated the formation of new plaques and showed in a semi-quantitative way that microglia were rapidly recruited to newly formed Aβ plaques, using in vivo labeling of Aβ with the blood-brain barrier (BBB)-permeant dye MethoxyX04 (Meyer-Luehmann et al. 2008). In another study,

Bolmont et al. (2008) observed plaques that increased or decreased in size over a 4-week period. In this study, the authors performed an in-depth in vivo imaging analysis of the relationship between microglia and plaques in an APP/PS1 transgenic mouse model of Aβ deposition crossed with the IBA1-GFP microglial reporter (Bolmont et al. 2008). They found a positive correlation between the size of amyloid plaques and the number of microglia associated with them. Moreover, they analyzed the migration of microglia in the proximity of Aβ plaques and found that while the majority of microglial cell bodies were sessile, a small proportion of cells were moving. Specifically, during a 24-h period, they observed microglia migrating towards Aβ plaques by first extending a main process and then subsequently pulling the soma along the axis of this process towards the plaque. In the proximity of Aβ plaques, Bolmont et al. found an increase in total microglial process length, which was significantly reduced for cells directly contacting the plaque. The authors found no difference with respect to the baseline motility of microglial processes in AD mice (Bolmont et al. 2008).

Interestingly, Bolmont et al. also examined whether microglia are involved with the phagocytosis of Aβ. They showed that microglial lysosomal-associated membrane protein 1 (LAMP-1)-positive lysosomes contained some MethoxyX04-labeled Aβ, suggesting that microglia can uptake Aβ, and thereby providing an in vivo assay to measure Aβ-phagocytosis (Bolmont et al. 2008). Liu et al. (2010) later demonstrated that microglia are unable to phagocytose fibrillar Aβ, since MethoxyX04-labeled plaques did not diminish in size over time, but instead found newly formed protofibrillar—presumably oligomeric—Aβ within microglial lysosomes. Moreover, they concluded that the microglial fractalkine receptor CX3CR1 is involved in regulating Aβ-uptake, since CX3CR1-deficient mice exhibit an increased amount of Aβ inside microglia and a diminished number of amyloid plaques. However, decreased plaque size (Aβ clearance) after microglial recruitment was not observed by Meyer-Luehmann et al., suggesting that microglial phagocytosis is not effective, or at least cannot compensate for continuous deposition. Overall, though the process of engulfment of Aβ has not been recorded in real time in the living brain to date, increasing the phagocytic capacity of microglia may represent a credible target for the treatment of AD. Indeed, microglia were involved in antibody-mediated clearing of Aβ in mouse models of AD, but their contribution was found to be limited (Garcia-Alloza et al. 2007; Koenigsknecht-Talboo et al. 2008). However, the treatment periods in these studies were limited to several days, and the time point of treatment may also be critical. Thus, whether boosting the phagocytic activity of microglia, without changing the inflammatory profile, represents a valuable strategy to treat AD remains unresolved.

Another method in which to target microglia for the treatment of AD may be to improve the fine-process motility of microglia in response to tissue damage, as this process was shown to be impaired in the proximity of Aβ plaques (Krabbe et al. 2013). While many studies have investigated the relationship of microglia and Aβ, there is only one study that has analyzed the relationship of microglia and neurons in a mouse model of AD, particularly the 3xTg-AD which display both Aβ accumulation and tau hyperphosphorylation. Herein, Fuhrmann et al. used two-photon in vivo imaging to show that microglia migrated towards neurons in advance of their elimination, in the somatosensory cortex still devoid of Aβ plaques at 4–6 months of age (as confirmed in situ using correlative histological analyses of the same neurons imaged in vivo), and subsequently displayed a reduction in fine process motility once the neurons were lost (Fig. 4.3b; Fuhrmann et al. 2010). Additionally, they found that the CX3CR1 was involved in this process, since CX3CR1-deficient mice were rescued from microglia-mediated neuronal loss (Fuhrmann et al. 2010). The mechanisms underlying this loss of neurons are still unclear and could involve microglial release of cytokines, chemokines, and neurotoxins.

Several in vivo imaging studies have now been performed using mouse models of AD to investigate Aβ plaque formation, stability and growth, the kinetics of dendritic spine loss in relation to the proximity of Aβ plaques, etc. However, the potential contribution of microglia to amyloid deposition, plaque formation and/or clearance, and the potential role of microglia in shaping neuronal networks in AD are still very poorly understood. Besides the classical aspects of AD like synapse-, neuron-loss, Aβ-deposition, tau-hyperphosphorylation and gliosis, future studies using in vivo imaging approaches may focus also on new risk factors identified for AD (Seshadri et al. 2010) and help design new strategies for patient treatments.

4.8.2 In Vivo Imaging of Microglia in Neuroinflammation

The involvement of microglia with neuroinflammatory diseases such as MS and ALS has been known for decades. However, in vivo imaging has only recently been employed to study the role of microglia in neuroinflammation at different stages of disease in available mouse models. In the experimental autoimmune encephalomyelitis (EAE) animal model of MS, mice are immunized with a myelin antigen, resulting in the development of autoimmune disease that simulates several of the pathological characteristics of human MS, such as demyelination and axonal damage in the CNS. Davalos et al. evaluated the progression of microglial responses at different stages of EAE by performing longitudinal in vivo imaging in the spinal cord of CX3CR1-GFP mice, using a previously developed imaging method (Davalos et al. 2008; Davalos and Akassoglou 2012). They found that microglia are the first cellular responders in EAE, as they begin accumulating around blood vessels several days before the onset of clinical signs of disease (Davalos et al. 2012). The number of microglial clusters was also shown to increase at the peak of disease and was correlated with disease severity (Fig. 4.3c). Longitudinal in vivo imaging of the same spinal cord areas in the same mice at different stages of EAE revealed that microglial cluster formation and resolution is an ongoing process; some microglial clusters grow in size and some eventually get resolved, while new ones appear around different vessels during various disease stages (Davalos et al. 2012). Herein, time-lapse imaging of forming clusters showed the directed microglial cell body migration and process extension towards blood vessels for cells that were previously further away, while those that were already close to a given vessel maintained their vascular association. Moreover, in vivo imaging of dye leakage demonstrated that microglial clusters specifically formed around vessels with a disrupted BBB and, as a consequence of BBB disruption, perivascular fibrin deposition was shown to be essential for microglial clustering in the perivascular space (Davalos et al. 2012). Finally, by crossing the CX3CR1-GFP reporter with a Thy1-CFP mouse line, they were able to record the interactions of GFP-labeled microglia with CFP-labeled axons in vivo, at the peak of EAE (Fig. 4.3d). These experiments showed abrupt physical manipulations of axons passing through perivascular microglial clusters, and showed for the first time axonal fragments still expressing CFP being detached and taken up by microglial cells (Davalos et al. 2012). By performing in vivo imaging and correlated histology in the same spinal cord areas, combined with genetic or pharmacological inhibition approaches, this study provided a mechanistic link between the leakage of the blood protein fibrinogen at sites of BBB disruption and the development of axonal damage through the perivascular activation of microglia at these sites, in a mouse model of MS.

A mutation in the superoxide dismutase-1 enzyme (SOD1-G93A) has been associated with the sporadic form of ALS in humans (Synofzik et al. 2010). A similar phenotype of progressive degeneration of motor neurons and pyramidal tracts was also seen in mice bearing the mutation ($SOD1^{G93A}$ mice). In this mouse model of ALS, Dibaj et al. examined the interactions between GFP-expressing microglia and YFP-expressing cortico-spinal tract axons in the lateral column of the spinal cord also by using in vivo two-photon microscopy. They found that microglia progressively reduced their ramification and baseline process motility velocity as a function of disease progression (Dibaj et al. 2011). The number of amoeboid microglia also increased in areas of axonal degeneration as the disease progressed to a more advanced stage with more prominent and widespread axonal pathology. In addition, when challenged with a focal axonal injury in their vicinity, microglia demonstrated a diverse profile of acute responses at different disease stages. At disease onset, they appeared hyper-activated and capable of significantly more robust responses towards the injury sites than healthy controls, while their responses were significantly reduced in advanced disease stages (Dibaj et al. 2011). Interestingly, in clinically unaffected areas of the spinal cord, microglial responses were comparable to control, irrespective of disease stage. This in vivo imaging study in a mouse model of ALS revealed that microglia undergo a transition to a progressively more activated phenotype with disease progression that not only influences their morphological appearance, but also determines their ability to perform tissue surveillance and contain injuries in the CNS. Overall, the application of in vivo imaging to studying microglial responses in animal models of neuroinflammatory disease has helped elucidate anatomical and structural characteristics of their activation and identify key pathological culprits and consequences of their sustained activation at different stages of CNS disease.

4.9 Conclusion

In the past decade, microglial biology has essentially been redefined, as a large number of unexpected physiological and pathological functions for these cells have emerged. In vivo imaging studies have been at the forefront of this dramatic expansion of our understanding of microglia. The constant development of new microscopy technologies, mouse models and imaging methodologies promises to further expand our access to uncharted territories of the CNS, such as image deeper brain structures or in awake animals, and shed more light on the extensive functional repertoire of microglia. Our ability to follow the natural defenders of the brain over the course of a disease is an important step towards characterizing and, eventually, modifying their behavior in therapeutic ways. In vivo imaging is essential for capturing the true potential of microglia, which is only beginning to be discovered.

References

Aguzzi A, Barres BA, Bennett ML (2013) Microglia: scapegoat, saboteur, or something else? Science 339:156–161

Aloisi F (2001) Immune function of microglia. Glia 36:165–179

Barretto RP, Ko TH, Jung JC, Wang TJ, Capps G, Waters AC et al (2011) Time-lapse imaging of disease progression in deep brain areas using fluorescence microendoscopy. Nat Med 17:223–228

Bialas AR, Stevens B (2013) TGF-beta signaling regulates neuronal C1q expression and developmental synaptic refinement. Nat Neurosci 16(12):1773–1782

Bolmont T, Haiss F, Eicke D, Radde R, Mathis CA, Klunk WE et al (2008) Dynamics of the microglial/amyloid interaction indicate a role in plaque maintenance. J Neurosci 28: 4283–4292

Booth PL, Thomas WE (1991) Dynamic features of cells expressing macrophage properties in tissue cultures of dissociated cerebral cortex from the rat. Cell Tissue Res 266:541–551

Carbonell WS, Murase S, Horwitz AF, Mandell JW (2005) Migration of perilesional microglia after focal brain injury and modulation by CC chemokine receptor 5: an in situ time-lapse confocal imaging study. J Neurosci 25:7040–7047

Carson MJ (2002) Microglia as liaisons between the immune and central nervous systems: functional implications for multiple sclerosis. Glia 40:218–231

Chan WY, Kohsaka S, Rezaie P (2007) The origin and cell lineage of microglia: new concepts. Brain Res Rev 53:344–354

Chen A, Kumar SM, Sahley CL, Muller KJ (2000) Nitric oxide influences injury-induced microglial migration and accumulation in the leech CNS. J Neurosci 20:1036–1043

Czapiga M, Colton CA (1999) Function of microglia in organotypic slice cultures. J Neurosci Res 56:644–651

Dailey ME, Manders E, Soll D, Terasaki M (2006) Confocal microscopy of live cells. In: Pawley JB (ed) Handbook of biological confocal microscopy, 3rd edn. Plenum, New York

Dailey ME, Waite M (1999) Confocal imaging of microglial cell dynamics in hippocampal slice cultures. Methods 18:222–230, 177

Damani MR, Zhao L, Fontainhas AM, Amaral J, Fariss RN, Wong WT (2011) Age-related alterations in the dynamic behavior of microglia. Aging Cell 10:263–276

Davalos D, Akassoglou K (2008) Imaging microglia in the central nervous system: past, present and future. In: Lane TE, Carson M, Bergmann C, Wyss-Coray T (eds) Central nervous system diseases and inflammation. Springer, New York, pp 45–57

Davalos D, Akassoglou K (2012) *In vivo* imaging of the mouse spinal cord using two-photon microscopy. J Vis Exp 59:e2760.
Davalos D, Grutzendler J, Yang G, Kim JV, Zuo Y, Jung S et al (2005) ATP mediates rapid microglial response to local brain injury *in vivo*. Nat Neurosci 8:752–758
Davalos D, Lee JK, Smith WB, Brinkman B, Ellisman MH, Zheng B et al (2008) Stable *in vivo* imaging of densely populated glia, axons and blood vessels in the mouse spinal cord using two-photon microscopy. J Neurosci Methods 169:1–7
Davalos D, Ryu JK, Merlini M, Baeten KM, Le Moan N, Petersen MA et al (2012) Fibrinogen-induced perivascular microglial clustering is required for the development of axonal damage in neuroinflammation. Nat Commun 3:1227
Davis EJ, Foster TD, Thomas WE (1994) Cellular forms and functions of brain microglia. Brain Res Bull 34:73–78
Denk W, Strickler JH, Webb WW (1990) Two-photon laser scanning fluorescence microscopy. Science 248:73–76
Dibaj P, Steffens H, Zschuntzsch J, Nadrigny F, Schomburg ED, Kirchhoff F et al (2011) *In vivo* imaging reveals distinct inflammatory activity of CNS microglia versus PNS macrophages in a mouse model for ALS. PLoS One 6:e17910
Dombeck DA, Khabbaz AN, Collman F, Adelman TL, Tank DW (2007) Imaging large-scale neural activity with cellular resolution in awake, mobile mice. Neuron 56:43–57
Drew PJ, Shih AY, Driscoll JD, Knutsen PM, Blinder P, Davalos D et al (2010) Chronic optical access through a polished and reinforced thinned skull. Nat Methods 7:981–984
Duffield JS, Forbes SJ, Constandinou CM, Clay S, Partolina M, Vuthoori S et al (2005) Selective depletion of macrophages reveals distinct, opposing roles during liver injury and repair. J Clin Invest 115:56–65
Duke DC, Moran LB, Turkheimer FE, Banati R, Graeber MB (2004) Microglia in culture: what genes do they express? Dev Neurosci 26:30–37
Ebert S, Weigelt K, Walczak Y, Drobnik W, Mauerer R, Hume DA et al (2009) Docosahexaenoic acid attenuates microglial activation and delays early retinal degeneration. J Neurochem 110:1863–1875
Erblich B, Zhu L, Etgen AM, Dobrenis K, Pollard JW (2011) Absence of colony stimulation factor-1 receptor results in loss of microglia, disrupted brain development and olfactory deficits. PLoS One 6:e26317
Eter N, Engel DR, Meyer L, Helb HM, Roth F, Maurer J et al (2008) *In vivo* visualization of dendritic cells, macrophages, and microglial cells responding to laser-induced damage in the fundus of the eye. Invest Ophthalmol Vis Sci 49:3649–3658
Farrar MJ, Bernstein IM, Schlafer DH, Cleland TA, Fetcho JR, Schaffer CB (2012) Chronic *in vivo* imaging in the mouse spinal cord using an implanted chamber. Nat Methods 9:297–302
Feng G, Mellor RH, Bernstein M, Keller-Peck C, Nguyen QT, Wallace M et al (2000) Imaging neuronal subsets in transgenic mice expressing multiple spectral variants of GFP. Neuron 28:41–51
Flusberg BA, Cocker ED, Piyawattanametha W, Jung JC, Cheung EL, Schnitzer MJ (2005) Fiber-optic fluorescence imaging. Nat Methods 2:941–950
Flusberg BA, Nimmerjahn A, Cocker ED, Mukamel EA, Barretto RP, Ko TH et al (2008) High-speed, miniaturized fluorescence microscopy in freely moving mice. Nat Methods 5:935–938
Fontainhas AM, Wang M, Liang KJ, Chen S, Mettu P, Damani M et al (2011) Microglial morphology and dynamic behavior is regulated by ionotropic glutamatergic and GABAergic neurotransmission. PLoS One 6:e15973
Fuhrmann M, Bittner T, Jung CK, Burgold S, Page RM, Mitteregger G et al (2010) Microglial Cx3cr1 knockout prevents neuron loss in a mouse model of Alzheimer's disease. Nat Neurosci 13:411–413
Garcia-Alloza M, Ferrara BJ, Dodwell SA, Hickey GA, Hyman BT, Bacskai BJ (2007) A limited role for microglia in antibody mediated plaque clearance in APP mice. Neurobiol Dis 28:286–292

Geissmann F, Manz MG, Jung S, Sieweke MH, Merad M, Ley K (2010) Development of monocytes, macrophages, and dendritic cells. Science 327:656–661
Germain RN, Miller MJ, Dustin ML, Nussenzweig MC (2006) Dynamic imaging of the immune system: progress, pitfalls and promise. Nat Rev Immunol 6:497–507
Ginhoux F, Greter M, Leboeuf M, Nandi S, See P, Gokhan S et al (2010) Fate mapping analysis reveals that adult microglia derive from primitive macrophages. Science 330:841–845
Gomez-Nicola D, Fransen NL, Suzzi S, Perry VH (2013) Regulation of microglial proliferation during chronic neurodegeneration. J Neurosci 33:2481–2493
Gonzalez-Scarano F, Baltuch G (1999) Microglia as mediators of inflammatory and degenerative diseases. Annu Rev Neurosci 22:219–240
Göppert-Mayer M (1931) Über Elemtarakte mit zwei Quantensprüngen. Ann Phys 401:273–294
Grutzendler J, Kasthuri N, Gan WB (2002) Long-term dendritic spine stability in the adult cortex. Nature 420:812–816
Haapaniemi H, Tomita M, Tanahashi N, Takeda H, Yokoyama M, Fukuuchi Y (1995) Non-amoeboid locomotion of cultured microglia obtained from newborn rat brain. Neurosci Lett 193:121–124
Haass C, Selkoe DJ (2007) Soluble protein oligomers in neurodegeneration: lessons from the Alzheimer‘s amyloid beta-peptide. Nat Rev Mol Cell Biol 8:101–112
Hanisch UK, Kettenmann H (2007) Microglia: active sensor and versatile effector cells in the normal and pathologic brain. Nat Neurosci 10:1387–1394
Haynes SE, Hollopeter G, Yang G, Kurpius D, Dailey ME, Gan WB et al (2006) The P2Y12 receptor regulates microglial activation by extracellular nucleotides. Nat Neurosci 9:1512–1519
Hefendehl JK, Neher JJ, Suhs RB, Kohsaka S, Skodras A, Jucker M (2014) Homeostatic and injury-induced microglia behavior in the aging brain. Aging Cell 13:60–69
Helmchen F, Denk W (2005) Deep tissue two-photon microscopy. Nat Methods 2:932–940
Hide I, Tanaka M, Inoue A, Nakajima K, Kohsaka S, Inoue K et al (2000) Extracellular ATP triggers tumor necrosis factor-alpha release from rat microglia. J Neurochem 75:965–972
Hines DJ, Hines RM, Mulligan SJ, Macvicar BA (2009) Microglia processes block the spread of damage in the brain and require functional chloride channels. Glia 57:1610–1618
Hirasawa T, Ohsawa K, Imai Y, Ondo Y, Akazawa C, Uchino S et al (2005) Visualization of microglia in living tissues using Iba1-EGFP transgenic mice. J Neurosci Res 81:357–362
Holtmaat A, Bonhoeffer T, Chow DK, Chuckowree J, De Paola V, Hofer SB et al (2009) Long-term, high-resolution imaging in the mouse neocortex through a chronic cranial window. Nat Protoc 4:1128–1144
Holtmaat AJ, Trachtenberg JT, Wilbrecht L, Shepherd GM, Zhang X, Knott GW et al (2005) Transient and persistent dendritic spines in the neocortex *in vivo*. Neuron 45:279–291
Honda S, Sasaki Y, Ohsawa K, Imai Y, Nakamura Y, Inoue K et al (2001) Extracellular ATP or ADP induce chemotaxis of cultured microglia through Gi/o-coupled P2Y receptors. J Neurosci 21:1975–1982
Inoue K (2008) Purinergic systems in microglia. Cell Mol Life Sci 65:3074–3080
Ishii T, Ishii M (2011) Intravital two-photon imaging: a versatile tool for dissecting the immune system. Ann Rheum Dis 70(Suppl 1):i113–i115
Jeon H, Kim JH, Kim JH, Lee WH, Lee MS, Suk K (2012) Plasminogen activator inhibitor type 1 regulates microglial motility and phagocytic activity. J Neuroinflammation 9:149
Jung S, Aliberti J, Graemmel P, Sunshine MJ, Kreutzberg GW, Sher A et al (2000) Analysis of fractalkine receptor CX3CR1 function by targeted deletion and green fluorescent protein reporter gene insertion. Mol Cell Biol 20:4106–4114
Kaur C, Hao AJ, Wu CH, Ling EA (2001) Origin of microglia. Microsc Res Tech 54:2–9
Kawakami N, Bartholomaus I, Pesic M, Mues M (2012) An autoimmunity odyssey: how autoreactive T cells infiltrate into the CNS. Immunol Rev 248:140–155
Kerschensteiner M, Schwab ME, Lichtman JW, Misgeld T (2005) *In vivo* imaging of axonal degeneration and regeneration in the injured spinal cord. Nat Med 11:572–577
Kettenmann H, Hanisch UK, Noda M, Verkhratsky A (2011) Physiology of microglia. Physiol Rev 91:461–553

Kierdorf K, Erny D, Goldmann T, Sander V, Schulz C, Perdiguero EG et al (2013) Microglia emerge from erythromyeloid precursors via Pu.1- and Irf8-dependent pathways. Nat Neurosci 16:273–280
Kim JV, Dustin ML (2006) Innate response to focal necrotic injury inside the blood-brain barrier. J Immunol 177:5269–5277
Koenigsknecht-Talboo J, Meyer-Luehmann M, Parsadanian M, Garcia-Alloza M, Finn MB, Hyman BT et al (2008) Rapid microglial response around amyloid pathology after systemic anti-Abeta antibody administration in PDAPP mice. J Neurosci 28:14156–14164
Koizumi S, Ohsawa K, Inoue K, Kohsaka S (2013) Purinergic receptors in microglia: functional modal shifts of microglia mediated by P2 and P1 receptors. Glia 61:47–54
Krabbe G, Halle A, Matyash V, Rinnenthal JL, Eom GD, Bernhardt U et al (2013) Functional impairment of microglia coincides with beta-amyloid deposition in mice with Alzheimer-like pathology. PLoS One 8:e60921
Kreutzberg GW (1996) Microglia: a sensor for pathological events in the CNS. Trends Neurosci 19:312–318
Lalancette-Hebert M, Phaneuf D, Soucy G, Weng YC, Kriz J (2009) Live imaging of Toll-like receptor 2 response in cerebral ischaemia reveals a role of olfactory bulb microglia as modulators of inflammation. Brain 132:940–954
Lawson LJ, Perry VH, Dri P, Gordon S (1990) Heterogeneity in the distribution and morphology of microglia in the normal adult mouse brain. Neuroscience 39:151–170
Lee JE, Liang KJ, Fariss RN, Wong WT (2008) Ex vivo dynamic imaging of retinal microglia using time-lapse confocal microscopy. Invest Ophthalmol Vis Sci 49:4169–4176
Liang KJ, Lee JE, Wang YD, Ma W, Fontainhas AM, Fariss RN et al (2009) Regulation of dynamic behavior of retinal microglia by CX3CR1 signaling. Invest Ophthalmol Vis Sci 50:4444–4451
Liu S, Li ZW, Weinreb RN, Xu G, Lindsey JD, Ye C et al (2012) Tracking retinal microgliosis in models of retinal ganglion cell damage. Invest Ophthalmol Vis Sci 53:6254–6262
Liu Z, Condello C, Schain A, Harb R, Grutzendler J (2010) CX3CR1 in microglia regulates brain amyloid deposition through selective protofibrillar amyloid-beta phagocytosis. J Neurosci 30:17091–17101
Lohmann C, Finski A, Bonhoeffer T (2005) Local calcium transients regulate the spontaneous motility of dendritic filopodia. Nat Neurosci 8:305–312
Lossi L, Alasia S, Salio C, Merighi A (2009) Cell death and proliferation in acute slices and organotypic cultures of mammalian CNS. Prog Neurobiol 88:221–245
Lovett-Barron M, Kaifosh P, Kheirbek MA, Danielson N, Zaremba JD, Reardon TR et al (2014) Dendritic inhibition in the hippocampus supports fear learning. Science 343:857–863
Lund S, Christensen KV, Hedtjarn M, Mortensen AL, Hagberg H, Falsig J et al (2006) The dynamics of the LPS triggered inflammatory response of murine microglia under different culture and *in vivo* conditions. J Neuroimmunol 180:71–87
Maggi L, Scianni M, Branchi I, D'Andrea I, Lauro C, Limatola C (2011) CX(3)CR1 deficiency alters hippocampal-dependent plasticity phenomena blunting the effects of enriched environment. Front Cell Neurosci 5:22
Marin-Teva JL, Dusart I, Colin C, Gervais A, van Rooijen N, Mallat M (2004) Microglia promote the death of developing Purkinje cells. Neuron 41:535–547
Marker DF, Tremblay ME, Lu SM, Majewska AK, Gelbard HA (2010) A thin-skull window technique for chronic two-photon *in vivo* imaging of murine microglia in models of neuroinflammation. J Vis Exp 43
Mayevsky A (1978) Ischemia in the brain: the effects of carotid artery ligation and decapitation on the energy state of the awake and anesthetized rat. Brain Res 140:217–230
McGlade-McCulloh E, Morrissey AM, Norona F, Muller KJ (1989) Individual microglia move rapidly and directly to nerve lesions in the leech central nervous system. Proc Natl Acad Sci U S A 86:1093–1097
Meyer-Luehmann M, Spires-Jones TL, Prada C, Garcia-Alloza M, de Calignon A, Rozkalne A et al (2008) Rapid appearance and local toxicity of amyloid-beta plaques in a mouse model of Alzheimer's disease. Nature 451:720–724

Misgeld T, Kerschensteiner M (2006) *In vivo* imaging of the diseased nervous system. Nat Rev Neurosci 7:449–463
Misgeld T, Nikic I, Kerschensteiner M (2007) *In vivo* imaging of single axons in the mouse spinal cord. Nat Protoc 2:263–268
Muller C, Beck H, Coulter D, Remy S (2012) Inhibitory control of linear and supralinear dendritic excitation in CA1 pyramidal neurons. Neuron 75:851–864
Nagerl UV, Eberhorn N, Cambridge SB, Bonhoeffer T (2004) Bidirectional activity-dependent morphological plasticity in hippocampal neurons. Neuron 44:759–767
Nimmerjahn A, Kirchhoff F, Helmchen F (2005) Resting microglial cells are highly dynamic surveillants of brain parenchyma *in vivo*. Science 308:1314–1318
Ohsawa K, Irino Y, Nakamura Y, Akazawa C, Inoue K, Kohsaka S (2007) Involvement of P2X4 and P2Y12 receptors in ATP-induced microglial chemotaxis. Glia 55:604–616
Ohsawa K, Irino Y, Sanagi T, Nakamura Y, Suzuki E, Inoue K et al (2010) P2Y12 receptor-mediated integrin-beta1 activation regulates microglial process extension induced by ATP. Glia 58:790–801
Palop JJ, Mucke L (2010) Amyloid-beta-induced neuronal dysfunction in Alzheimer's disease: from synapses toward neural networks. Nat Neurosci 13:812–818
Paolicelli RC, Bolasco G, Pagani F, Maggi L, Scianni M, Panzanelli P et al (2011) Synaptic pruning by microglia is necessary for normal brain development. Science 333:1456–1458
Paques M, Simonutti M, Augustin S, Goupille O, El Mathari B, Sahel JA (2010) *In vivo* observation of the locomotion of microglial cells in the retina. Glia 58:1663–1668
Parkhurst CN, Yang G, Ninan I, Savas JN, Yates JR 3rd, Lafaille JJ et al (2013) Microglia promote learning-dependent synapse formation through brain-derived neurotrophic factor. Cell 155:1596–1609
Peri F, Nusslein-Volhard C (2008) Live imaging of neuronal degradation by microglia reveals a role for v0-ATPase a1 in phagosomal fusion *in vivo*. Cell 133:916–927
Perry VH, Nicoll JA, Holmes C (2010) Microglia in neurodegenerative disease. Nat Rev Neurol 6:193–201
Petersen MA, Dailey ME (2004) Diverse microglial motility behaviors during clearance of dead cells in hippocampal slices. Glia 46:195–206
Prinz M, Priller J, Sisodia SS, Ransohoff RM (2011) Heterogeneity of CNS myeloid cells and their roles in neurodegeneration. Nat Neurosci 14:1227–1235
Raivich G, Bohatschek M, Kloss CU, Werner A, Jones LL, Kreutzberg GW (1999) Neuroglial activation repertoire in the injured brain: graded response, molecular mechanisms and cues to physiological function. Brain Res Brain Res Rev 30:77–105
Ransohoff RM, Brown MA (2012) Innate immunity in the central nervous system. J Clin Invest 122:1164–1171
Rodhe J (2013) Cell culturing of human and murine microglia cell lines. Methods Mol Biol 1041:11–16
Rogers JT, Morganti JM, Bachstetter AD, Hudson CE, Peters MM, Grimmig BA et al (2011) CX3CR1 deficiency leads to impairment of hippocampal cognitive function and synaptic plasticity. J Neurosci 31:16241–16250
Saijo K, Glass CK (2011) Microglial cell origin and phenotypes in health and disease. Nat Rev Immunol 11:775–787
Sasmono RT, Oceandy D, Pollard JW, Tong W, Pavli P, Wainwright BJ et al (2003) A macrophage colony-stimulating factor receptor-green fluorescent protein transgene is expressed throughout the mononuclear phagocyte system of the mouse. Blood 101:1155–1163
Schafer DP, Lehrman EK, Kautzman AG, Koyama R, Mardinly AR, Yamasaki R et al (2012) Microglia sculpt postnatal neural circuits in an activity and complement-dependent manner. Neuron 74:691–705
Seshadri S, Fitzpatrick AL, Ikram MA, DeStefano AL, Gudnason V, Boada M et al (2010) Genome-wide analysis of genetic loci associated with Alzheimer disease. JAMA 303:1832–1840

Shigemoto-Mogami Y, Koizumi S, Tsuda M, Ohsawa K, Kohsaka S, Inoue K (2001) Mechanisms underlying extracellular ATP-evoked interleukin-6 release in mouse microglial cell line, MG-5. J Neurochem 78:1339–1349

Sierra A, Encinas JM, Deudero JJ, Chancey JH, Enikolopov G, Overstreet-Wadiche LS et al (2010) Microglia shape adult hippocampal neurogenesis through apoptosis-coupled phagocytosis. Cell Stem Cell 7:483–495

Sierra A, Gottfried-Blackmore AC, McEwen BS, Bulloch K (2007) Microglia derived from aging mice exhibit an altered inflammatory profile. Glia 55:412–424

Smith ME, van der Maesen K, Somera FP (1998) Macrophage and microglial responses to cytokines *in vitro*: phagocytic activity, proteolytic enzyme release, and free radical production. J Neurosci Res 54:68–78

Stence N, Waite M, Dailey ME (2001) Dynamics of microglial activation: a confocal time-lapse analysis in hippocampal slices. Glia 33:256–266

Stevens B, Allen NJ, Vazquez LE, Howell GR, Christopherson KS, Nouri N et al (2007) The classical complement cascade mediates CNS synapse elimination. Cell 131:1164–1178

Stoll G, Jander S (1999) The role of microglia and macrophages in the pathophysiology of the CNS. Prog Neurobiol 58:233–247

Streit WJ (2013) Microglial cells. In: Kettenmann H, Ransom BR (eds) Neuroglia, 3rd edn. Oxford University Press, New York

Streit WJ, Graeber MB, Kreutzberg GW (1988) Functional plasticity of microglia: a review. Glia 1:301–307

Suzumura A, Marunouchi T, Yamamoto H (1991) Morphological transformation of microglia *in vitro*. Brain Res 545:301–306

Svoboda K, Denk W, Kleinfeld D, Tank DW (1997) *In vivo* dendritic calcium dynamics in neocortical pyramidal neurons. Nature 385:161–165

Svoboda K, Yasuda R (2006) Principles of two-photon excitation microscopy and its applications to neuroscience. Neuron 50:823–839

Synofzik M, Fernandez-Santiago R, Maetzler W, Schols L, Andersen PM (2010) The human G93A SOD1 phenotype closely resembles sporadic amyotrophic lateral sclerosis. J Neurol Neurosurg Psychiatry 81:764–767

Takeda H, Tomita M, Tanahashi N, Kobari M, Yokoyama M, Takao M et al (1998) Hydrogen peroxide enhances phagocytic activity of ameboid microglia. Neurosci Lett 240:5–8

Thomas WE (1992) Brain macrophages: evaluation of microglia and their functions. Brain Res Brain Res Rev 17:61–74

Tomita M, Fukuuchi Y, Tanahashi N, Kobari M, Takeda H, Yokoyama M et al (1996) Swift transformation and locomotion of polymorphonuclear leukocytes and microglia as observed by VEC-DIC microscopy (video microscopy). Keio J Med 45:213–224

Tremblay ME, Lowery RL, Majewska AK (2010) Microglial interactions with synapses are modulated by visual experience. PLoS Biol 8:e1000527

Tremblay ME, Zettel ML, Ison JR, Allen PD, Majewska AK (2012) Effects of aging and sensory loss on glial cells in mouse visual and auditory cortices. Glia 60:541–558

Tsien RY (1998) The green fluorescent protein. Annu Rev Biochem 67:509–544

Ueno M, Fujita Y, Tanaka T, Nakamura Y, Kikuta J, Ishii M et al (2013) Layer V cortical neurons require microglial support for survival during postnatal development. Nat Neurosci 16: 543–551

Venneti S, Lopresti BJ, Wiley CA (2013) Molecular imaging of microglia/macrophages in the brain. Glia 61:10–23

Wake H, Moorhouse AJ, Jinno S, Kohsaka S, Nabekura J (2009) Resting microglia directly monitor the functional state of synapses *in vivo* and determine the fate of ischemic terminals. J Neurosci 29:3974–3980

Wang T, Kass IS (1997) Preparation of brain slices. Methods Mol Biol 72:1–14

Ward SA, Ransom PA, Booth PL, Thomas WE (1991) Characterization of ramified microglia in tissue culture: pinocytosis and motility. J Neurosci Res 29:13–28

Wollmer MA, Lucius R, Wilms H, Held-Feindt J, Sievers J, Mentlein R (2001) ATP and adenosine induce ramification of microglia *in vitro*. J Neuroimmunol 115:19–27
Wu LJ, Vadakkan KI, Zhuo M (2007) ATP-induced chemotaxis of microglial processes requires P2Y receptor-activated initiation of outward potassium currents. Glia 55:810–821
Wu LJ, Zhuo M (2008) Resting microglial motility is independent of synaptic plasticity in mammalian brain. J Neurophysiol 99:2026–2032
Xiang Z, Chen M, Ping J, Dunn P, Lv J, Jiao B et al (2006) Microglial morphology and its transformation after challenge by extracellular ATP *in vitro*. J Neurosci Res 83:91–101
Xu HT, Pan F, Yang G, Gan WB (2007) Choice of cranial window type for *in vivo* imaging affects dendritic spine turnover in the cortex. Nat Neurosci 10:549–551
Yasuda R, Harvey CD, Zhong H, Sobczyk A, van Aelst L, Svoboda K (2006) Supersensitive Ras activation in dendrites and spines revealed by two-photon fluorescence lifetime imaging. Nat Neurosci 9:283–291
Yona S, Kim KW, Wolf Y, Mildner A, Varol D, Breker M et al (2013) Fate mapping reveals origins and dynamics of monocytes and tissue macrophages under homeostasis. Immunity 38:79–91
Ziv Y, Burns LD, Cocker ED, Hamel EO, Ghosh KK, Kitch LJ et al (2013) Long-term dynamics of CA1 hippocampal place codes. Nat Neurosci 16:264–266
Zusso M, Methot L, Lo R, Greenhalgh AD, David S, Stifani S (2012) Regulation of postnatal forebrain amoeboid microglial cell proliferation and development by the transcription factor Runx1. J Neurosci 32:11285–11298

Chapter 5
Roles in Immune Responses

Richa Hanamsagar, Sandra M. Cardona, Tammy Kielian, and Astrid E. Cardona

Abstract Microglia are best known as the mononuclear phagocytes of the central nervous system (CNS) parenchyma. As a resident glial cell population, microglia play key roles during the initiation, propagation, and/or resolution of inflammation. Recently, the discovery that microglial cells continuously survey their local CNS environment in vivo improved our understanding of their immune-surveillance properties in health and disease. Microglial interactions with other elements of the immune system and resident cells of the CNS define a fine balance between neuroprotection and irreparable tissue damage. In this chapter we highlight the innate immune properties of microglia, with a focus on events that initiate an inflammatory response within the brain proper including, Toll-like receptors, inflammasomes, cytokines, and chemokines, and their relationship to immune-mediated disease exacerbation or resolution.

Keywords Microglia • Phagocytosis • Toll-like receptor • Nod-like receptor • Inflammasome • Chemokine • Scavenging • Antigen presentation • Blood–brain barrier

---

R. Hanamsagar, B. Pharm., Ph.D.
Department of Psychology and Neuroscience, Duke University,
210 Research Dr, Room 3019, Durham, NC 27710, USA
e-mail: rmh45@duke.edu

S.M. Cardona, Ph.D.
Department of Biology, University of Texas at San Antonio,
West Campus/Tobin lab MBT 1.212, San Antonio, TX 78249, USA
e-mail: sandra.cardona@utsa.edu

T. Kielian, Ph.D. (✉)
Department of Pathology and Microbiology, University of Nebraska Medical Center,
985900 Nebraska Medical Center, Room 7029 DRCII, Omaha, NE 68198-5900, USA
e-mail: tkielian@unmc.edu

A.E. Cardona, Ph.D. (✉)
Department of Biology and South Texas Center for Emerging Infectious Diseases,
University of Texas at San Antonio, West Campus/Tobin lab MBT 1.216,
San Antonio, TX 78249, USA
e-mail: astrid.cardona@utsa.edu

M.-È. Tremblay, A. Sierra (eds.), *Microglia in Health and Disease*,
DOI 10.1007/978-1-4939-1429-6_5

Bullet Points

- Immune surveillance, antigen recognition, and phagocytosis are key functions of microglia during the initiation, propagation, and/or resolution of inflammation.
- Innate immune activation of microglia can be triggered by numerous pathways including pattern recognition receptors and chemokine receptors.
- Microglial cells participate actively during innate immune responses, but also activate pathways that contribute to adaptive immunity.
- The physiological importance of microglia in inflammasome activation during central nervous system (CNS) infection and neurodegeneration highlights potential venues for CNS pathogen clearance.
- Functions of microglia do not occur in isolation but are influenced by neighboring cells including astrocytes and neurons, and the blood–brain barrier.

5.1 Regulation of CNS Innate Immune Reactions

Microglia are the resident mononuclear phagocytes of the CNS, comprising approximately 10–15 % of the total cell population within the parenchyma. These cells share many phenotypic and functional characteristics with macrophages and, as such, are major players in the brain's innate immune responses. Developmental studies in mice have shown that microglia are derived from primitive yolk sac myeloid progenitors seeding the brain parenchyma between embryonic days 9 and 10 (Alliot et al. 1999; Ginhoux et al. 2010), indicating that microglia constitute a population of cells ontogenically distinct from tissue macrophages, which originate from bone marrow-derived monocytes (Fogg et al. 2006; Parwaresch and Wacker 1984).

5.1.1 Immune Surveillance

Historically, the CNS was considered an immune-privileged site based on its lack of lymphatic drainage, low levels of major histocompatibility complex (MHC) expression, which is essential for antigen presentation (see below), and restricted diffusion of molecules and cells from the periphery by the blood–brain barrier (BBB). However, the healthy CNS is now recognized as a site where immune surveillance occurs (Greenwood et al. 2011). Microglia share several markers with bone marrow-derived macrophages (Prinz et al. 2011), including the fractalkine receptor CX3CR1 and the lectin Siglec-H (Gautier et al. 2012; Harrison et al. 1998). It has been proposed that the expression of CX3CR1 and its ligand CX3CL1 (fractalkine or FKN) on microglia and neurons, respectively, assists in chemotactic migration of microglia to neurons during neuronal injury or inflammatory conditions (Chapman et al. 2000; Harrison et al. 1998; Mizutani et al. 2012; Streit et al. 2005) (see Chap. 9 for further reading on their roles in the healthy brain). This highlights the use of traditional immune chemotactic signals to facilitate specific interactions between CNS cellular constituents.

Immune surveillance within the CNS occurs during normal physiological conditions and microglia play a critical role in this process. Although microglia were previously considered to exist in a "resting" state characterized by a ramified morphology, elegant studies have demonstrated that in the intact adult brain, microglia are highly active and continuously surveying their microenvironment by extending and retracting their motile processes (Davalos et al. 2005; Nimmerjahn et al. 2005) (see Chaps. 2 and 4 for further reading). This attribute may be essential to elicit immediate responses to neuronal injury or invading pathogens.

Microglia exist in at least two morphologically distinct states based exclusively on their appearance; namely, amoeboid and ramified. The amoeboid state refers to cells with larger cell bodies as well as shorter and thicker processes, a phenotype which has been correlated with microglial "activation" in response to neuronal injury and inflammatory conditions. Conversely, "quiescent" microglia are generally described as ramified, exhibiting longer and thinner processes. Interestingly, microglia possess an amoeboid morphology during early stages of CNS development, transitioning into a ramified form during the late fetal and early neonatal periods (Boya et al. 1979). Ramified microglia transform again into the amoeboid-like morphology after CNS insults, such as injury or inflammation (Streit 2002; Xiang et al. 2006) (Fig. 5.1). However, in some models of inflammation the correlation between microglial morphology and function is not as strong. For instance, low doses of systemic lipopolysaccharide (LPS, a major component of the outer membrane of gram-negative bacteria) can induce the production of pro-inflammatory cytokines without any apparent changes in morphology (Sierra et al. 2007) (Table 5.1). Among other mechanisms, microglial remodeling, swelling, and process extension were shown to be controlled by chloride channels (Ducharme et al. 2007; Zierler et al. 2008) and the phosphorylation/dephosphorylation of cofilin, a member of the actin depolymerizing factor (ADF) family (Hadas et al. 2012), while the inflammatory response is regulated by the transcription factor NFκB (Wilms et al. 2003). Although it remains unclear whether these two pathways interact, disruption of the actin cytoskeleton has been shown to activate NFκB in cultured human intestinal epithelial cells, which supports the notion that cytoskeletal disruption may occur upstream of NFκB (Nemeth et al. 2004).

5.1.2 Antigen Recognition in the CNS

Microglial responses are often measured based on the array of cytokines, chemokines, and complement components produced during specific pathologies (Tables 5.1 and 5.2). Cytokines are small signaling molecules that circulate within and among organs, affecting nearly every biological process by orchestrating the degree of

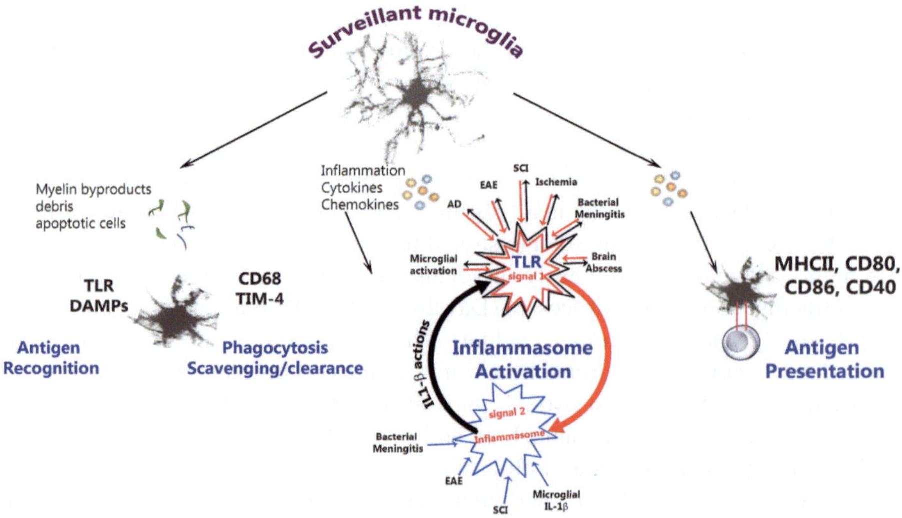

Fig. 5.1 Microglia and innate immunity. Antigen recognition, phagocytosis, inflammatory mediator release, and antigen presentation are among the most recognized functions of microglia in health and disease. Their Toll-like receptors (TLR) and danger-associated molecular patterns (DAMPs) receptors play an important role in recognizing antigens such as myelin byproducts, debris, or amyloid-beta (Aβ), making their engagement critical for sensing cellular damage. Membrane receptors such as CD68 and T cell immunoglobulin mucin-4 (TIM-4) are involved in the processes of phagocytosis and clearance of damaged cells. In general, antigens presented through MHC class II lead to adaptive responses that elicit Th1, Th2, or Th17 responses. In parallel, antigen recognition activates the inflammasome. The pro-inflammatory cytokine interleukin (IL)-1β is implicated in the pathophysiology of numerous neurodegenerative diseases, including multiple sclerosis/experimental autoimmune encephalomyelitis (EAE) and Alzheimer's disease (AD), and a number of central nervous system (CNS) infections ranging from bacterial meningitis, brain abscess, and human immunodeficiency virus (HIV)-associated dementia. The initial response upon TLR activation (signal 1) in certain pathologies with further inflammatory cues leads to IL-1β release and inflammasome activation (signal 2) that in turns activates TLR components creating a positive feedback regulation of the inflammatory cascade. Although the roles of inflammasomes are becoming increasingly well-defined, insights into inflammasome biology in the CNS are only just emerging. Interactions between TLRs and the inflammasome mediate responses to both pathogen-associated and endogenous triggers during CNS pathology

inflammatory reaction. Cytokines are broadly categorized as pro-inflammatory (e.g., interleukin (IL)-1β, IL-6, and tumor necrosis factor α (TNFα)) and anti-inflammatory (e.g., transforming growth factor β (TGFβ) and IL-4), and further stratified based on their capacity to promote (1) cellular immunity (i.e., the activation of phagocytes and antigen-specific T lymphocytes, such as helper (Th)1 cells, which produce mainly interferon γ (IFNγ) and IL-12); (2) antibody-mediated responses (mediated by Th2 cells and associated with cytokines such as IL-4, IL-10, and IL-13); and (3) responses mediated by Th17 cells (distinguished by their release of IL-17, associated with pathogen clearance and tissue inflammation in autoimmune diseases) (Dinarello 2007) (see Chap. 16 for further reading on their roles in multiple sclerosis). Chemotactic cytokines (or chemokines) participate in cellular

Table 5.1 Microglial cytokines[a]

Cytokine	Condition	References
IL-1α	In vitro, viral nucleocapsid, LPS	Lee et al. (1993), Lokensgard et al. (2001)
IL-1β	In vitro, viral nucleocapsid, LPS, Aβ	Lee et al. (1993), Lokensgard et al. (2001), Lue et al. (2001b)
IL-3	In vitro, IFNγ	Hanisch (2002)
IL-6	In vitro, viral nucleocapsid, LPS, Aβ, cytomegalovirus	Lee et al. (1993), Lokensgard et al. (2001), Lue et al. (2001b), Pulliam et al. (1995)
IL-8	In vitro, IFNγ, HIV	Hanisch (2002), Renner et al. (2012)
IL-10	In vitro, IFNγ/LPS,	Hanisch (2002), Ledeboer et al. (2002), Williams et al. (1996)
IL-12	In vitro, IFNγ/LPS, IL-12	Aloisi et al. (1997), Becher et al. (1996), Hanisch (2002), Stalder et al. (1997)
IL-15	In vitro, IFNγ	Hanisch (2002)
IL-16	In vivo, fetal tissues	Schwab et al. (2001)
IL-23	In vitro, IFNγ/LPS	Li et al. (2003)
M-CSF	In vitro, β-amyloid	Lue et al. (2001b)
TGFβ	In vitro, IL-1	da Cunha et al. (1997)
TNFα	In vitro, viral nucleocapsid, LPS, HIV	Koka et al. (1995), Lee et al. (1993), Lokensgard et al. (2001), Pulliam et al. (1995)

[a]Selected reports, influenced by activation stimuli, model, age, and species from which microglial cells were derived

activation by arresting and positioning relevant cells with spatiotemporal precision (Ransohoff 2009). The complement system also constitutes one of the first defense mechanisms by coating the surface of cells and pathogens, thereby assisting in the process of phagocytosis (see Chap. 9 for more information on its involvement with synaptic pruning). Therefore, the coordinated control of soluble mediators by microglial cells is considered a key determinant in the processes of antigen recognition in the CNS (Tables 5.1 and 5.2).

Innate immune activation in the CNS can be triggered by numerous pathways upon recognition of invading pathogens and/or tissue damage by pattern recognition receptors (PRRs). Within the last 10 years, much attention has been focused on Toll-like receptor (TLR) activation in several models of CNS infection, neurodegenerative disease, and injury (Hanamsagar et al. 2012). TLRs are stimulated by pathogen- or danger-associated molecular patterns of pathogen or host origin (PAMPs and DAMPs), respectively, leading to NFκB and MAPK activation, and the subsequent release of pro-inflammatory mediators. A total of 13 TLRs have been identified in human, all of which are expressed at the cell surface except TLRs 3, 7, and 9, which are exclusive to the endosomal compartments.

Microglia express a wide range of TLRs to varying degrees depending on their phenotype (Olson and Miller 2004). In the resting state, many TLRs cannot be detected in microglia; however, TLRs are rapidly induced upon cellular activation (Bsibsi et al. 2002) (Fig. 5.1). In microglia, TLRs are considered a crucial first line

Table 5.2 Microglial chemokines[a]

Cytokine	Condition	References
CCL1	Ex vivo, *S. aureus*	Kielian et al. (2001), Kielian et al. (2002), Kielian (2004)
CCL2	In vitro, viral Tat protein	D'Aversa et al. (2004), Kielian (2004)
CCL3	In vitro, LPS,	Kielian (2004), Kremlev et al. (2004)
CCL4	In vitro, viral Tat protein	D'Aversa et al. (2004), Kielian (2004)
CCL5	In vitro, LPS	Kielian (2004), Kremlev et al. (2004)
CCL7	In vitro, TLR2 signal	Aravalli et al. (2005)
CCL8	In vitro, TLR2 signal	Aravalli et al. (2005)
CCL9	In vitro, TLR2 signal	Aravalli et al. (2005)
CCL11	In vitro, neuropeptide	Wainwright et al. (2008)
CXCL1	In vitro, TLR2 signal	Aravalli et al. (2005)
CXCL2	In vitro, ATP	Kielian (2004), Shiratori et al. (2010)
CXCL4	In vitro, TLR2 signal	Aravalli et al. (2005)
CXCL5	In vitro, TLR2 signal	Aravalli et al. (2005)
CXCL8	IFNγ/sCD40L, viral Tat protein, *S. aureus*	D'Aversa et al. (2004), D'Aversa et al. (2008), Kielian (2004)
CXCL9	In vitro, TLR2 signal, IFNγ	Aravalli et al. (2005), Ellis et al. (2010)
CXCL10	In vitro, LPS, TLR2 signal, IFNγ, viral Tat protein, *S. aureus*	Aravalli et al. (2005), D'Aversa et al. (2004), Ellis et al. (2010), Kielian (2004), Kremlev et al. (2004)
CXCL16	Ex vivo, glioma	Ludwig et al. (2005)

[a]Selected reports, influenced by activation stimuli, model, age, and species from which microglia were derived

of defense against bacteria and viruses, as well as recognition of endogenous danger signals (Kielian et al. 2005; Ravindran et al. 2010; Zhou et al. 2006). Resultant pro-inflammatory mediator and reactive oxygen species (ROS) production triggered by TLR ligands can be either beneficial or detrimental to the host. For example, TLR9 activation can enhance microglial phagocytosis of amyloid-beta (Aβ) as reported in primary neuron-microglia cocultures (Doi et al. 2009), and TLR2 signaling is pivotal for nitric oxide (NO) production and anti-bacterial responses following *Streptococcus* challenge, leading to neuronal cell death (Lehnardt et al. 2006). Conversely, persistent TLR activation may cause exaggerated immune responses, resulting in damage to neurons and oligodendrocytes (Lehnardt et al. 2002; Lehnardt et al. 2003). Microglial activation is evident following a single systemic LPS injection, which also triggers the expression of numerous pro-inflammatory cytokines and chemokines (Chen et al. 2012; Lehnardt et al. 2002; Rivest 2009). Although microglial activation was originally thought to result from the action of systemically induced cytokines elicited by TLR4 engagement, strong evidence suggests that TLR4 expression within the CNS is essential for eliciting central inflammation either via direct activation of resident microglia or indirectly through cerebral vascular endothelial cells (Chen et al. 2012; Lehnardt et al. 2002). In addition, injection of LPS directly into the brain induces robust and transient expression of numerous inflammatory mediators, including IL-1, TNFα, IL-6, the subunit p40 of IL-12 (IL-12p40), and TGFβ through TLR4 and its adapter, MyD88 (Glezer et al. 2007b). Although astrocytes, endothelial cells, and neurons have been shown to express TLR4 (Bowman et al. 2003; Liu and Kielian 2011; Rolls et al. 2007; Zhou et al. 2006), their role in immunity in response to TLR4 ligands is considered to be limited as compared to microglia (Glezer et al. 2006; Glezer et al. 2007a).

Chronic neurodegeneration also leads to microglial activation (Dheen et al. 2007; Nakajima and Kohsaka 2001) (see Chap. 18 about microglial involvement in neurodegenerative diseases). Aβ, the principal component of Alzheimer's disease (AD)-associated senile plaques, has been shown to induce production of IL-1β, IL-6, TNFα, and IL-18 in white and gray matter microglia cultured from human autopsy tissue (Lue et al. 2001a; Rogers and Lue 2001). IL-1β and IL-18 are produced after a series of signaling events that involve Nod-Like receptors (NLRs), inducing the formation of a multi-cellular complex known as the inflammasome (see below). It was shown that Aβ can activate microglia through the NLRP3 inflammasome pathway (Blasko et al. 2004; Halle et al. 2008) and the subsequent production of IL-1β has been proposed to worsen neurodegeneration (Lindberg et al. 2005). In Parkinson's disease (PD), α-synuclein aggregates are considered central to the pathology, associated with a progressive neuronal loss in the substantia nigra pars compacta (SNpc). In fact, it has been shown that microglial activation is widespread in the SNpc (McGeer et al. 1988), but also in the hippocampus, putamen, and cingulate cortex of PD patient brains (Imamura et al. 2003). A correlation has also been demonstrated between microglial activation and the loss of dopaminergic terminals in the postmortem midbrain of PD patients (Imamura et al. 2003). Indeed, microglia exposed to α-synuclein in vitro express high levels of IL-1β, TNFα, and IL-6 (Su et al. 2008) and a recent study has reported that α-synuclein can also activate the inflammasome (Codolo et al. 2013).

5.1.3 Phagocytosis: Scavenging Functions/Clearance

Phagocytosis is an evolutionarily conserved process serving to remove cellular debris and dying cells or aggregated proteins that can be toxic and trigger exaggerated immune responses. It refers to the recognition, engulfment, and degradation of different types of cargo including dead cells, debris, bacteria, Aβ, dendritic spines, etc. (Sierra et al. 2013). From *Drosophila* (Kurant 2011) to higher eukaryotes, efforts have been directed at understanding the molecular basis underlying the phagocytosis of endogenous debris and pathogens. Understanding antigen uptake and the pathways that lead to phagocytosis and antigen presentation for proper T-cell activation has also been a topic of study for many years. Microglia are considered a first line of defense against invading pathogens and they respond rapidly to inflammatory stimuli. A primary function of microglia is to eliminate harmful exogenous insults as well as endogenous proteins or cellular debris not only during

CNS injury, but also during development and adult neurogenesis (Sierra et al. 2013). There is ample evidence that microglia can phagocytose neurons during neuroinflammatory conditions. Generally, neuronal debris and degenerating axons are readily engulfed by microglia utilizing the scavenging receptor CD68 (Fraser et al. 2010; Tanaka et al. 2009). It is thought that the nucleotide ATP released from dying cells is not only an important trigger for microglial migration to sites of injury, via purinergic receptor and inflammasome activation, but also regulates the phagocytic process itself (Koizumi et al. 2007). Upon axonal degeneration, it was also shown that microglial release of type I interferons (IFNα and IFNβ, which are important for protection against many viral infections) can provide additional inflammatory feedback to maintain phagocytic activity, while creating a more permissive environment for axonal outgrowth in vitro (Hosmane et al. 2012). Interestingly, activated microglia have been shown to phagocytose viable neurons (Fricker et al. 2012), an in vitro observed phenomenon termed "phagoptosis" which is triggered by inflammation and the transient and reversible exposure of membrane phosphatidylserine (PS) on stressed cells. However, the physiological implications of this process remain unclear (Brown and Neher 2012). Other work has shown that microglia play a neuroprotective role during CNS injury by their rapid engulfment of apoptotic polymorphonuclear neutrophils (PMN) (Neumann et al. 2008), in addition to inducing T-cell apoptosis (Magnus et al. 2001)—the latter representing a potential mechanism to subvert autoimmunity. Besides the role of microglia in the phagocytosis of bacterial and viral particles during CNS infections, several studies have highlighted the importance of microglial phagocytosis in response to Aβ deposition in the brain. For example, microglia have been shown to clear Aβ plaques through Fc receptor-mediated phagocytosis and peptide degradation (Bard et al. 2000), although they do not seem to reduce overall the Aβ load in the AD brain. Other surface receptors implicated in fibrillar Aβ engulfment include the scavenger receptor CD36, $\alpha_6\beta_1$ integrin, and the integrin and thrombospondin-1 receptor CD47 (Koenigsknecht and Landreth 2004).

Another interesting area in the process of phagocytosis relates to the pathways activated during physiological versus inflammatory conditions. Initial approaches using opsonized beads (i.e., polystyrene microspheres) in vivo proposed two mechanisms of phagocytosis. Inflammatory microglial phagocytosis is accompanied by the release of TNFα, IL-1β, ROS, and NO. In contrast, homeostatic phagocytosis, which is important to clear apoptotic cells or myelin debris, correlates with the induction of anti-inflammatory factors such as IL-10, TGF-β2, prostaglandin E2, and platelet-activating factor (PAF) (Ryu et al. 2012). Moreover, the uptake of apoptotic cells by microglia can downregulate pro-inflammatory cytokines, including TNFα, IL-12, and IL-1β (Ryu et al. 2012). Of relevance are the differences in lysosomal attributes between various cell types, including microglia, macrophages, monocyte-derived dendritic cells (DCs), and neutrophils. For example, a comparison of primary mouse microglia and J774 macrophages revealed that microglia contain higher levels of many lysosomal proteases than macrophages (Majumdar et al. 2007). However, the microglial lysosomes were less acidic with a pH ~ 6, compared to pH ~ 5 in macrophages, suggesting a decreased lysosomal enzymatic activity.

Interestingly, treatment with macrophage colony-stimulating factor (M-CSF) and IL-6 correlated with lysosomal acidification, enabling microglial cells to degrade Aβ fibrils effectively (Majumdar et al. 2007). Immunohistochemical analyses of primary microglia also showed that the ClC-7 chloride transporter, which acidifies lysosomes via cotransporting protons and chloride ions, is not delivered efficiently to microglial lysosomes, appearing mistargeted and mobilized to the ER for degradation. In this study, M-CSF was also shown to induce ClC-7 trafficking to the lysosomes, revealing an important pathway for the phagocytosis of different types of cargo upon microglial activation through M-CSF (Majumdar et al. 2011). As mentioned above, the relationship between M-CSF microglial activation and lysosome acidification suggests a protective role in AD, by promoting degradation of Aβ fibrils and hence reducing plaques deposition. However, these in vitro findings may need to be interpreted with caution, as conflicting results that microglial phagocytosis is the primary cause of neuronal cell death in the presence of low levels of extracellular Aβ were also obtained in vitro (Neniskyte et al. 2011). In mouse models of AD it was additionally shown that CX3CR1 deletion is associated with an increased phagocytic ability, leading to greater Aβ contents inside of microglial phagolysosomes (Liu et al. 2010b). Therefore, cell–cell interactions between microglia and neurons, notably through CX3CR1-CX3CL1 communication, and the cytokine environment may collectively determine whether microglia are equipped to initiate phagocytosis.

Microglia express a wide range of TLRs, which makes them effective defenders against invading pathogens. For example, peptidoglycan (PGN) recognition is mediated by TLR2 in microglia (Kielian et al. 2005), TLR4 is necessary for *Citrobacter koseri*-mediated microglial activation and response to LPS (Lehnardt et al. 2002; Liu and Kielian 2009; Olson and Miller 2004), while TLR stimulation is important for eliciting the release of inflammatory mediators following *Streptococcus pneumonia* (Ribes et al. 2010) and *Staphylococcus aureus* challenge (Kochan et al. 2012). Studies with cultured macrophages have demonstrated that several TLRs localize to phagosomes (i.e., the vesicles formed around bacteria once engulfed by phagocytosis) following bacterial uptake (Husebye et al. 2010; Underhill et al. 1999), although similar findings have not yet been reported in microglia.

Recognition of dying cells is a key mechanism in the process of cellular digestion. It is clear that the classical "eat me" signal PS, apparently specific to apoptotic cells, is recognized by the phagocytic receptor TIM-4 (T cell immunoglobulin mucin-4) (Kobayashi et al. 2007; Santiago et al. 2007; Savill and Gregory 2007). The phagocytic "synapse" is now considered as a complex structure, which involves PS and the receptors TIM-1 and TIM-4 on phagocytic cells. The seven transmembrane G-protein-coupled receptor BAL1 (B aggressive lymphoma 1) mediates PS binding on the extracellular surface, and their intracytoplasmic portions form complexes with modular proteins engulfment and motility (ELMO)-Dock180-Rac involved in phagocytic signaling (Kobayashi et al. 2007; Santiago et al. 2007). This information brings us closer to understanding the mechanisms by which cellular debris in the form of exosomes regulate immune responses (Sokolowski and Mandell 2011).

5.1.4 Microglia and Inflammasome Activation

Microglia are capable of recognizing a diverse array of infectious agents, including bacteria, fungi, and viruses (Esen and Kielian 2006; Kaushik et al. 2011; Liu and Kielian 2009; Rambach et al. 2010), by expressing various PRRs, namely TLRs and NLRs (Hanamsagar et al. 2011; Jack et al. 2005b; Lee et al. 2013; Regnier-Vigouroux 2003). The recognition of PAMPs by TLRs makes it nearly impossible for a pathogen to go undetected, as it is difficult to evade the immune system by mutating an essential pathogenic motif. Microglia respond to pathogens by producing a wide range of pro-inflammatory mediators, including ROS and nitrogen species, cytokines, and chemokines (Aloisi 2001; Tambuyzer et al. 2009). In a similar manner, microglia are activated in response to endogenous danger signals such as the nucleotide ATP or misfolded proteins, which also culminate into pro-inflammatory cytokine secretion. It is thought that recognition of these DAMPs occurs via the activation of NLRs within the cytoplasm of host cells. NLR inflammasome activation leads to the processing and release of pro-inflammatory cytokines, including IL-1β and IL-18 which have been implicated in the pathophysiology of numerous CNS neurodegenerative diseases, including AD and PD (Benzing et al. 1999; Koprich et al. 2008). Acute brain injuries such as stroke, trauma, and hemorrhage are also characterized by neuroinflammation and IL-1 release, a process which is particularly linked to disease exacerbation (Denes et al. 2011; Lu et al. 2005; Masada et al. 2003). A role for IL-1β and IL-18 has also been described in several CNS infection models, including bacterial meningitis, brain abscess, and human immunodeficiency virus (HIV)-associated dementia (Ghorpade et al. 2003; Iannello et al. 2010; Kielian et al. 2004; Saukkonen et al. 1990; Xiong et al. 2012) (see Chap. 15 for further reading on HIV). IL-1β and its receptor IL-1R play a crucial role in the amplification of cytokine/chemokine networks, and bacterial clearance during acute *S. aureus* infection (Kielian et al. 2004; Xiong et al. 2012). IL-1β and IL-18 also regulate the induction of adaptive immunity, and both cytokines have been shown to influence disease development in experimental autoimmune encephalomyelitis (EAE), an animal model for multiple sclerosis (MS), an inflammatory-mediated demyelinating disease of the human CNS (Gris et al. 2010; Sutton et al. 2006) (Fig. 5.1) (see Chap. 16 for further reading on MS). The clinical course of MS is variable but usually initiates with reversible episodes of neurological disability that transform into a disease of continuous and irreversible neurological decline. Axonal damage and neurodegeneration are a major cause of irreversible neurological disability in MS. Inflammation and demyelination are additional hallmarks of the disease that manifest in clinical symptoms such as numbness, muscle spasms, optic neuritis, neuropathic pain, and paralysis (Dutta and Trapp 2011; Trapp et al. 1999; Trapp and Nave 2008).

In AD, extracellular Aβ triggers activation of the NLRP3 inflammasome in microglia, due to lysosomal damage and cathepsin B release following fibril uptake (Halle et al. 2008). Enhanced caspase-1 expression has been reported in the brains of AD patients, while double transgenic APP/PS1 mice harboring the human amyloid precursor protein (APP) and the presenilin 1 (PS1) mutation for familial AD demonstrated reduced caspase-1, IL-1β, and Aβ burdens when lacking NLRP3 or caspase-1 (Heneka et al. 2013). NLRP3 deficiency has been associated with a skewing of microglial activation towards an M2-like anti-inflammatory state, suggesting an important regulatory role for the inflammasome in microglial-induced inflammation. It is not clear whether other types of NLR inflammasomes could serve as platforms for IL-1β processing in response to Aβ, as their contribution has not yet been tested in AD animal models. However, a recent study that examined NLRP1 gene polymorphisms in a cohort of AD patients revealed an association of four non-synonymous polymorphisms in NLRP1 with AD, suggesting that these mutations may be involved in the predisposition to AD (Pontillo et al. 2012). Nevertheless, a direct cause-and-effect relationship remains to be defined.

The EAE animal model commonly requires injection of *Mycobacterium* along with the myelin peptides, to render the endogenous protein immunogenic. It is thought that the initial detection of danger signals (which remain ill-defined) within the CNS is mediated by microglia through their ability to present antigen and secrete immune molecules that can recruit peripheral immune cells into the CNS. Infiltrating macrophages and T cells can damage neurons, which provides signals to perpetuate microglial activation. If not tightly regulated, this can lead to progressive and irreversible tissue damage (Jack et al. 2005a). Microglia can phagocytose myelin debris, which triggers IL-1β release (Williams et al. 1994); however, additional studies are needed to better define the role of microglial inflammasomes during MS development and/or progression. In general, NLRP3 inflammasome activation has been reported as detrimental to the host in the context of EAE. For example, NLRP3-deficient mice displayed reduced neuroinflammation and delayed myelin loss in a model of demyelination induced by cuprizone, i.e., a dietary chelator commonly used to induce demyelinating lesion in the corpus callosum of rodents as a model of MS (Jha et al. 2010). Similarly, in myelin-oligodendrocyte glycoprotein (MOG)-induced EAE, NLRP3 deficiency reduced disease severity by inhibiting inflammatory cell infiltrates into the spinal cord (Gris et al. 2010). There is also evidence that the inflammasome adaptor protein Apoptosis-associated Speck-like protein containing CARD (ASC) contributes to the progression of EAE, independently from its inflammasome activity; although no association with NLRP3 expression was found in that study (Shaw et al. 2010). Nonetheless, in all cases, the cellular basis of inflammasome activation remains to be determined.

Recently, it has been demonstrated that activated microglia are critical for the maintenance of chronic pain following traumatic spinal cord injury (SCI) (Hains and Waxman 2006). Additionally, microglia are among the first responders to neuronal damage and injury and produce a vast range of pro-inflammatory mediators, including IL-1β, within hours following SCI (Pineau and Lacroix 2007). The NLRP1 inflammasome may be involved in IL-1β and IL-18 processing following SCI through the activation of $P2X_4$ receptor in neurons (de Rivero Vaccari et al. 2008). Currently, it is thought that the majority of IL-1β release that follows SCI originates from neurons (Bernier 2012), and additional studies are needed to determine the role of microglia in inflammasome activation during SCI and chronic pain (see Chaps. 11 and 19 for further reading on these topics).

Although the CNS is usually well-protected from microbial invasion, in many instances are bacterial, fungal, viral, parasitic, and prion infections known to be life-threatening. The role of microglia in the rapid recognition and responses to pathogens has been extensively studied (Cosenza et al. 2002; Hanisch et al. 2001; Kielian et al. 2002; Lee et al. 1995). However, the physiological importance of inflammasome activation in the context of infectious diseases has only recently begun to be investigated. In particular, it was shown that cultured microglia respond to live *S. aureus* by producing IL-1β in a NLRP3- and ASC-dependent manner (Hanamsagar et al. 2011). The same study demonstrated the existence of redundant mechanisms for microglial IL-1β release, including bacterial pore-forming toxins and extracellular ATP, as well as caspase-1 and cathepsin B activity. A recent study with *Legionella pneumophila* demonstrated that flagellin recognition by the NLRC4 inflammasome may be important in restricting bacterial replication within microglia (Jamilloux et al. 2013). Neurotoxin prion infection can also lead to NLRP3 inflammasome activation in microglia (Shi et al. 2012), defining a novel trigger for the NLRP3 inflammasome apart from the traditional stimuli identified to date. Viral infections of the CNS were also shown to activate microglia, and a recent study further suggested that ROS production and subsequent NLRP3 inflammasome activation are critical during West Nile virus infection (Kaushik et al. 2012). Other inflammasomes such as HIN200 or AIM2, which are known to respond directly to viral nucleic acids, may also be involved; however, their role in microglia remains to be investigated.

5.2 Microglia in the Regulation of Adaptive Immunity

As demonstrated in the previous section, microglia, the main immune responsive glial cell population of the CNS, do not only produce a myriad of inflammatory cytokines and chemokines, but also receive inflammatory signals from nearby astrocytes, neurons, endothelial cells, and infiltrating leukocytes. Although chemokines and chemokine receptors are key players in the innate immune response, there are also pathways that directly link chemokine receptor signaling on microglia to the generation of effector T cells. In addition, alternative molecular mechanisms further highlight the role of microglial phagocytosis and antigen processing in effector T-cell responses in the CNS. Therefore, the type of CNS insult, the local inflammatory response, and the molecular program of microglial cells altogether contribute to the contextual regulation of neuroprotective versus neurotoxic pathways.

5.2.1 Antigen Presentation

T cells can recognize antigens only in the form of a peptide bound to MHC on the surface of antigen-presenting cells (APCs), in a process called antigen presentation. Professional APCs (e.g., macrophages, DCs identified by the expression of CD11c, and B lymphocytes) express MHC class II, while non-professional APCs (e.g., fibroblast, thymic epithelial cells, glial cells, pancreatic beta cells, and vascular endothelial cells) express MHC class I. Only professional APCs can activate a helper T cell that has never encountered the antigen before. Therefore, it is well established that adaptive immune responses depend on the function of professional APCs in the CNS. In addition to the peptide-MHC molecule interactions, contacts between the T cell and APCs via costimulatory molecules (such as CD86, intercellular adhesion molecule 1 (ICAM-1), and CD40) are required for full T-cell activation. Without this costimulatory signal, T cells will arrest in a nonactivated state, referred to as anergy. Considering that the immune system can be detrimental or beneficial to brain function, either by causing tissue damage or contributing to repair (Shaked et al. 2004; Wraith and Nicholson 2012), microglial ability to present antigens to T cells through MHC class II could allow these normally quiescent cells to play a yet undescribed role in shaping the outcome of certain neurological diseases where antigen presentation is critically involved such as MS and EAE (O'Keefe et al. 2002). Some studies have demonstrated that treatment of microglia with granulocyte macrophage colony-stimulating factor (GM-CSF) can transform these cells into a DC-like phenotype in vitro (Ponomarev et al. 2005). Usually identified as $CD11c^+$ cells, brain DCs are found in perivascular regions of the CNS and in peripheral blood but might be derived from a microglial precursor as well (Prodinger et al. 2011). However, it remains unresolved whether APCs have the capacity to leave the CNS and transport CNS-derived antigens to the draining lymph nodes. Given the observations that the cerebrospinal fluid (CSF) and CNS parenchyma lack naïve T cells under physiological conditions, it is commonly accepted that adaptive immune responses against CNS antigens are initiated in the periphery, and subsequently propagated to the CNS by circulating memory T cells, which are restimulated by antigens within the CNS (Ransohoff and Engelhardt 2012). The involvement of T cell–APC interactions in EAE has been evaluated using adoptive transfer experiments with activated T cells, either myelin-specific or ovalbumin-specific. In these experiments, the ovalbumin-specific T cells remained in the subarachnoid space (SAS), whereas myelin-specific T cells readily invaded the parenchyma. Ovalbumin-specific T cells were only detected in the brain parenchyma when ovalbumin-pulsed APCs were placed in the SAS (Bartholomaus et al. 2009). Therefore, these data suggest that antigen-specific interactions between T cells and APCs occurring within the SAS are essential for the development of EAE (Ransohoff and Engelhardt 2012). The fact that most studies exploring microglial interactions with T cell activation were performed in vitro prompts careful interpretation, as it was shown that microglia that develop in mixed glial cultures display a more activated phenotype than microglia in pure cultures (Carson et al. 1998). In vivo microglia express low levels of accessory molecules required for efficient antigen presentation and have weak antigen-presenting activity (Ransohoff and Engelhardt 2012). Although adult microglia can express costimulatory molecules under certain conditions (Carson et al. 1998; Garcia et al. 2013; Zhang et al. 2002), they failed to present a peptide antigen to naïve T cells in vitro (Carson et al. 1998). Furthermore, adult microglia were less efficient, when compared to $CD11c^+$ DCs, to restimulate primed T cells and to induce T cell proliferation in vitro (Ford et al. 1995; Garcia et al. 2013). Similarly, during Theiler's murine encephalomyelitis virus-induced demyelinating disease, CNS-infiltrating macrophages were more highly activated based on their MHC class II expression than resident microglia. However, microglia isolated at the onset of disease were as efficient as infiltrating macrophages at inducing Th1 proliferation and IFNγ production (Mack et al. 2003). Although it was reported that EAE disease severity is ameliorated when microglia are selectively depleted before disease induction via systemic administration of ganciclovir (an antiviral nucleotide analogue) to mice engineered to express the thymidine kinase of herpes simplex virus (HSVTK) under the CD11b promoter (Heppner et al. 2005) (see Chap. 16 for further reading about this model), the exact mechanism linking microglia to APC functions in vivo remains unsolved.

Although microglia display enhanced clearance of myelin debris in CNS autoimmunity models (Nielsen et al. 2009), compared with naïve/physiological conditions, it is still unclear how the scavenging mechanism relating to antigen processing and presentation differs between microglia and professional CD11c-positive DCs. Recently it was shown that oligodendrocytes secrete small membrane vesicles containing myelin antigens, called exosomes, which are efficiently taken up by microglia both in vitro and in vivo. However, these exosomes are preferentially internalized by microglia devoid of antigen presenting capacity (Fitzner et al. 2011). Therefore, it was proposed that a decreased exosome internalization by MHC class II-positive microglia might allow myelin antigens to be transported with the interstitial fluid to the meningeal macrophages, which could then present myelin-derived antigen to myelin-specific T cells present in the cerebrospinal fluid (Fitzner et al. 2011; Ransohoff and Engelhardt 2012). It has been demonstrated, however, that IFNγ enables brain CD11c-expressing cells to become effective APCs and to induce antigen-specific T cells expressing the CD4 glycoprotein at their surface (CD4+) to proliferate and secrete cytokines specific to the differentiated Th1/Th17 effector T cell subtypes, which are considered as encephalitogenic (Gottfried-Blackmore et al. 2009). In an entorhinal cortex lesion model, the fact that microglia do not upregulate MHC class II also indicates that they could represent a DC population capable of producing immunological tolerance, i.e., a condition where the immune system does not attack an antigen, under particular circumstances. However, evidence is still lacking to validate a role for microglia expressing the CD11c marker in transporting antigens from the brain to lymphoid organs (Prodinger et al. 2011). Not all microglia express CD11c, only a subpopulation of about 1.1 ± 0.1 % of microglia in naïve controls, and 4 ± 0.8 % at 3 weeks after cuprizone treatment expressed CD11c (Remington et al. 2007). It was shown that CD11c-positive microglial cells in culture are approximately three times better at stimulating T cell activation than CD11c-negative cells, indicative of enhanced antigen presentation capacity associated with this phenotype. Also, resident brain microglia appeared to be key functional players during adaptive responses in the sub-chronic 1-methyl-4-phenyl-1,2,3,6-tetrahydropyridine (MPTP, dopaminergic neurotoxin) mouse model of PD that is associated with T-cell recruitment into the CNS (Depboylu et al. 2012). Using bone marrow chimeric mice to distinguish the resident microglial population from infiltrating leukocytes, in combination with retrograde neuronal tracing, it was shown that resident microglia upregulate MHC class II after MPTP administration. Importantly, microglia also contained neuronal tracer and appeared in close contact with CD4+ T cells in the lesion area. Therefore, resident microglia represent likely candidates for the presentation of antigens to infiltrating activated T lymphocytes during dopaminergic cell death.

More than a decade ago, when the concept of attenuating Aβ accumulation in the brain by eliciting adaptive immune responses emerged, it was the first time that a self-peptide was introduced as a vaccine. Preclinical studies were successful, but the initial clinical trials were halted because of the development of severe inflammatory reactions in the vaccinated AD patients (Nicoll et al. 2003; Schenk 2008). While a number of reports suggested that T cells were activated in the patients and could be found both in the periphery and brain tissues, the question on how those T cells gained access to the brain has been investigated by some groups. In this regard, microglia were proposed to be indirectly involved in the transendothelial migration and activation of T cells. Their involvement with the regulation of adaptive T cells effector responses was evidenced by their high levels of TNFα release in response to Aβ application in vitro, thus resulting in over-expression of the chemokine receptor CXCR2 (mouse equivalent of the human IL-8 receptor) on peripheral T cells derived from AD patients (Liu et al. 2010a). During EAE, disease severity was significantly exacerbated in hypermyelinated PLP-Akt-DD transgenic mice compared to wild-type animals with "normal" myelin content (Jaini et al. 2013). PLP-Akt-DD transgenic animals have been engineered to drive, under the proteolipid PLP promoter, expression of the constitutively active Akt-1 which harbors aspartic acid (D) at positions Thr308 and Ser473, two sites where phosphorylation leads to full activation of Akt. In these mice, expression of Akt-DD dramatically increased myelin production, correlating with increased neurological score and exacerbated neuronal damage. There was also an increased number of resident microglia in the hypermyelinated mice, but their individual capacity to present antigens was not increased, as revealed by the levels of IL-17 (a pro-inflammatory cytokine inducing and mediating pro-inflammatory responses) secreted by T cells isolated from these mice in the presence of microglia (Jaini et al. 2013). This work suggests that the load of self-antigens imposed on the APCs could have an important influence on the disease onset and progression (Fig. 5.1). Therefore, the drastic differences between microglia versus professional DCs, as APCs, support the notion that microglial activity, if well controlled, is a crucial step in determining neuronal survival.

5.2.2 Chemokine Receptors

Microglia express most chemokine receptors in vivo, including CCR3, CCR5, CCR8, CXCR3, CXCR4, and CX3CR1, depending on the CNS microenvironment that shapes their effector phenotype (Biber et al. 2002; Cardona and Ransohoff 2007; Glabinsk and Ransohoff 2001; Prinz and Priller 2010; Ransohoff 2002; Trebst et al. 2001; Trebst et al. 2003), with actions that promote either tissue damage and neurotoxicity or neuroprotection and tissue repair (Cardona et al. 2006; Garcia et al. 2013; Prinz et al. 2011) (also see Chap. 6). Several studies support the concept that microglia are a heterogeneous population based on the selective expression of chemokine receptors. For example, transcriptome analyses in combination with protein validation revealed that nearly all microglial cells express high levels of complement C1q, while only subsets of microglia express the CCR1 ligand, CXCL14, and TREM2 (triggering receptor expressed on myeloid cells 2; which regulates microglial activation and phagocytosis) (Schmid et al. 2009). Adding another layer of complexity, the regulation of microglial function is also influenced by the other CNS resident cells. For example, several studies in PD models suggest a scenario in which astrocyte-microglial communication impacts T cell responses. Astrocytes exposed in vitro to neuron-derived α-synuclein induce expression of transcripts for chemokines such as CCL3-7, CCL12, 19 and 20, and CXCL1,2,4,5,9,10, 11, 12, and 16, and CX3CL1 (Harms et al. 2013). Microglial expression of the chemokine receptors CCR3, CCR5, CXCR3 and CX3CR1 makes them highly responsive to their respective ligands produced by astrocytes (Cowell et al. 2006; Flynn et al. 2003; Gautier et al. 2012). Chemokine receptor-ligand interactions also correlated with morphological changes in microglia and increased expression of MHC class II in vivo (Garcia et al. 2013). Most importantly, microglial responses to aggregated α-synuclein induced antigen processing and presentation, driving T cell proliferation and cytokine release, as well as neurodegeneration in a mouse model of PD in vivo (Harms et al. 2013), suggesting that antigen presentation could promote an inflammatory environment detrimental to neuronal survival. A similar role for microglia/MHC class II was also demonstrated in a facial nerve axotomy model, in which the physical distance between the injury to the facial nerve and the neuronal localization in the brainstem provides a good model to avoid direct CNS trauma and/or disruption of the BBB (Byram et al. 2004). Therefore, these results implicate a central role for microglial MHC class II in the activation of adaptive immune responses, indicating that this process could be viewed as a neuroprotective target.

Microglia represent important components of the CNS immune response in MS and EAE, alongside peripherally derived macrophages and DCs. The mechanism by which these cell types repopulate during acute and chronic brain inflammation have been explored. The chemokine receptor CCR2 that signals in response to several monocyte chemoattractant molecules (known as CCL2, CCL8, and CCL16) is crucial for the accumulation of macrophages and DCs to the sites of inflammation, although its role in mobilizing microglia is still controversial. There is evidence suggesting that $CD11b^+$ $Ly6C^{hi}CCR2^+CD62L^+$ (or adhesion molecule L-selectin+) monocytes are mobilized from the bone marrow into the bloodstream, via a GM-CSF-dependent pathway before EAE relapses (King et al. 2009). However, microglia are recognized by a $CD11b^+$ $Ly6C^-CCR2^-CD62L^+$ phenotype (Carson et al. 1998), and the lack of microglial CCR2 upregulation in CCR2-RFP reporter mice supports a selective role for CCR2 in the trafficking of macrophages (King et al. 2009; Saederup et al. 2010). However, it was reported that over-expression of

CCL2 also leads to morphological activation of microglia and pro-inflammatory cytokine release in the CNS (Selenica et al. 2013), although a direct interaction between exogenous CCL2 and CCR2 on microglia is uncertain. CCR2 also targets the recruitment of $CCR2^{+}CD11c^{+}$ DCs. As expected, the expression of multiple chemokine receptors by myeloid cells may account for the synergistic effects observed when analyzing DC trafficking or microglial mobilization within the CNS.

The pathways that influence chemokine receptor expression are complex and cytokines are the most studied candidates in this process. However, the highly conserved cell–cell communication pathway Notch, initially identified as a pleiotropic mediator of cell fate in invertebrates, has emerged as an important regulator of immune cell development and function. It was recently shown that microglia express the Notch ligand DLL4 and that DLL4 blockade reduces the neurological symptoms of EAE, correlating with a downregulation of CCR2 and CCR6 expression on T cells. Although microglia are not the only cell type expressing DLL4 in the brain (Benedito and Duarte 2005; Shutter et al. 2000), microglia–T cells interactions mediated by DLL4 and Notch receptors could contribute to the regulation of T cells chemokine receptors expression, and their responses to organ-specific chemokine production (Reynolds et al. 2011).

In addition to microglia–T cell interactions, microglia-neuron communication is now recognized as a key mechanism in the regulation of microglial function. Several studies have confirmed the predominant microglial expression of CX3CR1 (Cardona et al. 2006; Gautier et al. 2012; Jung et al. 2000), whose ligand CX3CL1 is expressed in neuronal membranes (Bazan et al. 1997; Mizoue et al. 1999; Rossi et al. 1998). The chemokine module is released by action of proteases including A Disintegrin and metalloproteinase domain-containing protein 10 and 17 (ADAM10 and ADAM10) and cathepsin S (Clark et al. 2009; Garton et al. 2001; Hundhausen et al. 2003). Binding of CX3CL1 to CX3CR1 on microglia has been viewed as neuroprotective because this interaction inhibits microglial activation in selected models of neurodegeneration (Bachstetter et al. 2011; Cardona et al. 2006; Rogers et al. 2011). Although some studies suggest that CX3CR1 deficiency could be neuroprotective and/or anti-inflammatory in various contexts of disease such as AD (Fuhrmann et al. 2010; Lee et al. 2010; Mattison et al. 2013), in EAE, the absence of CX3CR1 correlated with an enhanced neurological disease and increased microglial activation, reflected by their increased MCH class II expression and proliferation. These features of microglia also correlated with an increased T cell proliferation. Furthermore, higher IFNγ and IL-17 levels were detected in cerebellar and spinal cord tissues of CX3CR1-deficient mice (Garcia et al. 2013).

Overall chemokine receptors regulate microglial mobilization within the CNS, but the observation of higher frequencies of IFNγ- and IL-17-producing T cells in the lymphoid tissues of CX3CR1-deficient mice, and the enhanced T cell proliferation induced by CX3CR1-deficient DCs, demonstrate that besides their role in chemoattraction, some chemokine receptors such as CX3CR1 may contribute to the establishment of adaptive immune responses via regulation of antigen presentation function.

5.3 Microglia and the BBB

The BBB is the interface between the CNS and the periphery, fulfilling two main functions, acting as a physical barrier and a selective exchange barrier aimed to maintain the proper CNS microenvironment for optimal neuronal function. This mechanical separation of the CNS is accomplished by the presence of specialized endothelial cells tightly attached one to another via tight junctions and adherent junctions (Hawkins and Davis 2005; Hermann and ElAli 2012). These junctions are formed by the tight junction proteins occludins and claudins, as well as the junctional adhesion molecules E-cadherin, P-cadherin, and N-cadherin, whose ultimate role is to restrict blood-borne molecules and peripheral cells from entering the CNS (Pardridge 2003; Wilson et al. 2010). Extensive work in the last decade has unraveled the presence of a specialized intrinsic innate immune system in the CNS, which also involves the BBB as an active contributor, as opposed to being an immunological neutral and passive barrier (Muldoon et al. 2013). It has been observed not only that peripheral immune cells can cross an intact BBB (Carson et al. 2006), but that the BBB modulates the differentiation of infiltrating CD14+ monocytes into DCs through the influence of BBB-secreted TGFβ and GM-CSF. These data confer a more active role of the BBB in the intrinsic innate immunity of the CNS. Despite the limited infiltration of peripheral immune cells into the CNS under physiological conditions, it is well known that neutrophils, eosinophils, T lymphocytes, monocytes, and other immune cells can infiltrate the CNS parenchyma after injuries, infections, and chronic diseases such as MS (Wilson et al. 2010).

Although it is unclear whether BBB dysfunction during pathological conditions is due to a loss of signals provided by the CNS, or to breakdown signals produced during pathology, recent work has shown that communication between the BBB and microglia is one of the early events involved in EAE. In particular, disruption of the barrier and subsequent leakage of the plasma protein fibrinogen has been identified as a signal triggering microglial motility towards perivascular regions, microglial activation and clustering, as well as axonal damage, using two-photon in vivo imaging. Importantly, these changes in microglia were shown to occur prior to the onset of neurological changes, suggesting their relevance to the disease pathogenesis (Davalos et al. 2012) (also see Chap. 4 for more information on these findings).

Activated microglia and infiltrating macrophages secrete a wide range of proinflammatory signals, including NO and ROS, which have been categorized as a major mechanism leading to axonal damage in MS (Lassmann 2010), through the initiation of mitochondrial dysfunction which, in turn, leads to the production of mitochondrial-derived ROS (Haider et al. 2011). Indeed, increased ROS generation has been observed in close proximity to fibrin-mediated microglial clusters at the peak of EAE, compared to distal areas with normal microglial density, or healthy controls (Davalos et al. 2012). Since fibrin is formed from fibrinogen through the action of the protease thrombin, it represents a molecular marker for BBB disruption and fibrinogen leakage into the CNS parenchyma. Interestingly, in vitro treatment of microglia with fibrinogen induced a 5.1-fold increase of H_2O_2 and a 7.4-fold

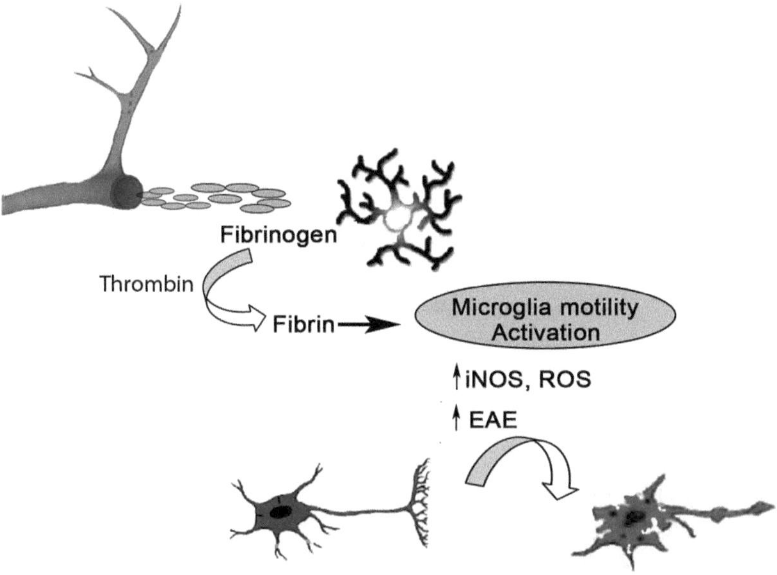

Fig. 5.2 Microglia and the interface to peripheral communication. Fibrinogen appears to be a key modulator of microglial activity as a result of early blood–brain barrier (BBB) damage. Once extravasated, it is lysed into fibrin by thrombin and interacts with parenchymal microglia, leading to an increased expression of inducible nitric oxide synthase (iNOS) and reactive oxygen species (ROS) that accounts for neuronal damage associated with certain neuroinflammatory conditions such as EAE

increase in inducible nitric oxide synthase (iNOS) gene expression (Davalos et al. 2012), strongly suggesting that fibrinogen might contribute to ROS generation during inflammatory demyelinating lesions. These data highlight the importance of fibrin deposition for the development of axonal degeneration within microglial clusters. Postmortem analysis of MS brains has reinforced this finding of microglial clustering within normal-appearing white matter (van Horssen et al. 2012), although microglial association with the disrupted BBB remains a challenge to assess, due to a limited effectiveness of molecular probes to identify microglia at their different states of "activation" in heavily fixed human specimens, which also complicates investigation of their association with BBB leakiness.

This aggregation of microglial cells at perivascular areas before disease onset may constitute an early indicator of new lesion formation, as demyelination and axonal damage are exacerbated at the peak of EAE, coincident with rapid microglial process extension and retraction at the sites of BBB leakage showing fibrin deposition (Fig. 5.2). Despite the strong correlation between perivascular fibrinogen deposition and microglial clusters formation at the early stages of neuroinflammation in EAE, the role of fibrinogen or other blood products within the CNS and their interaction with microglial cells in the context of other CNS pathologies remains to be investigated.

Both in vivo and in vitro studies support the idea that microglia release inflammatory mediators that regulate BBB permeability. Coculture models of microvascular endothelial cells showed that LPS-activated microglia induce BBB leakiness by producing ROS through nicotinamide adenine dinucleotide phosphate (NADPH) oxidase (Matsumoto et al. 2012; Sumi et al. 2010) and TNFα release (Nishioku et al. 2010). Microglial activation is evident in the aged CNS and breakdown of the BBB and blood–retinal barrier in the aging brain suggests that microglial actions at the vasculature interface may contribute not only to neuroinflammatory conditions, but also to the pathogenesis of age-related neurological disorders (see Chap. 13 for additional information on aging). Further understanding of the mechanisms involved in the coupling between the BBB and microglia in innate immunity will require to investigate the role of microglia in the regulation of BBB dynamics during normal physiological conditions, considering that primary microglial cells in culture express a recently discovered tight junction protein, tricellulin (Mariano et al. 2011), whose role in the CNS and immune function remains to be elucidated.

5.4 Summary

It is now well accepted that microglia do not only participate in the genesis of immune responses during CNS infection/injury, but also play key roles in maintaining homeostasis. During CNS development and inflammatory conditions, microglia display the ability to phagocytose debris and apoptotic cells, release cytokines, and upregulate MHC class II to modulate effector T cell responses. These functions of microglia do not occur in isolation but are influenced by neighboring cells, including astrocytes, neurons, and the BBB. Microglia are capable of producing a vast array of cytokines and chemokines and are therefore able to influence the CNS environment in a very effective manner. Microglia can also respond to soluble mediators produced by neurons and astrocytes, and a current challenge is to dissect the signals that enable such an orchestrated network to sustain optimal CNS function. The most recognized function of microglia as phagocytes and scavengers is temporally regulated to allow for appropriate antigen recognition and presentation during adaptive immune responses. As our knowledge of microglial functions during homeostasis and pathology continues to expand, it may be possible to consider harnessing microglial activity to benefit CNS integrity and improve outcomes following neuroinflammatory injury. For many years researchers have been studying the mechanisms that lead to microglial activation and enhance their phagocytic properties. Another interesting venue involves the use of microglia as shuttles of anti-inflammatory molecules. Although the issue of identifying microglial origins is still a moving target, progress has been made to show that microglia are a distinct myeloid subset with unique properties, making them distinct from DCs and macrophages. Balancing neuroprotective versus neurotoxic properties of microglia has been and will continue to be a topic of arduous investigation, with the goal of modulating CNS inflammation due to autoimmunity, infection, or neurodegeneration.

References

Alliot F, Godin I, Pessac B (1999) Microglia derive from progenitors, originating from the yolk sac, and which proliferate in the brain. Brain Res Dev Brain Res 117:145–152

Aloisi F (2001) Immune function of microglia. Glia 36:165–179

Aloisi F, Penna G, Cerase J et al (1997) IL-12 production by central nervous system microglia is inhibited by astrocytes. J Immunol 159:1604–1612

Aravalli RN, Hu S, Rowen TN et al (2005) Cutting edge: TLR2-mediated proinflammatory cytokine and chemokine production by microglial cells in response to herpes simplex virus. J Immunol 175:4189–4193

Bachstetter AD, Morganti JM, Jernberg J et al (2011) Fractalkine and CX 3 CR1 regulate hippocampal neurogenesis in adult and aged rats. Neurobiol Aging 32:2030–2044

Bard F, Cannon C, Barbour R et al (2000) Peripherally administered antibodies against amyloid beta-peptide enter the central nervous system and reduce pathology in a mouse model of Alzheimer disease. Nat Med 6:916–919

Bartholomaus I, Kawakami N, Odoardi F et al (2009) Effector T cell interactions with meningeal vascular structures in nascent autoimmune CNS lesions. Nature 462:94–98

Bazan JF, Bacon KB, Hardiman G et al (1997) A new class of membrane-bound chemokine with a CX3C motif. Nature 385:640–644

Becher B, Dodelet V, Fedorowicz V et al (1996) Soluble tumor necrosis factor receptor inhibits interleukin 12 production by stimulated human adult microglial cells in vitro. J Clin Invest 98:1539–1543

Benedito R, Duarte A (2005) Expression of Dll4 during mouse embryogenesis suggests multiple developmental roles. Gene Expr Patterns 5:750–755

Benzing WC, Wujek JR, Ward EK et al (1999) Evidence for glial-mediated inflammation in aged APP(SW) transgenic mice. Neurobiol Aging 20:581–589

Bernier LP (2012) Purinergic regulation of inflammasome activation after central nervous system injury. J Gen Physiol 140:571–575

Biber K, Dijkstra I, Trebst C et al (2002) Functional expression of CXCR3 in cultured mouse and human astrocytes and microglia. Neuroscience 112:487–497

Blasko I, Stampfer-Kountchev M, Robatscher P et al (2004) How chronic inflammation can affect the brain and support the development of Alzheimer's disease in old age: the role of microglia and astrocytes. Aging Cell 3:169–176

Bowman CC, Rasley A, Tranguch SL et al (2003) Cultured astrocytes express toll-like receptors for bacterial products. Glia 43:281–291

Boya J, Calvo J, Prado A (1979) The origin of microglial cells. J Anat 129:177–186

Brown GC, Neher JJ (2012) Eaten alive! Cell death by primary phagocytosis: 'phagoptosis'. Trends Biochem Sci 37:325–332

Bsibsi M, Ravid R, Gveric D et al (2002) Broad expression of Toll-like receptors in the human central nervous system. J Neuropathol Exp Neurol 61:1013–1021

Byram SC, Carson MJ, DeBoy CA et al (2004) CD4-positive T cell-mediated neuroprotection requires dual compartment antigen presentation. J Neurosci 24:4333–4339

Cardona AE, Ransohoff RM (2007) Chemokine receptors in neuroinflammation. In: Harrison JK and Lukas NW (eds) The chemokine receptors. Humana Press, Totowa, NJ, pp 351–370

Cardona A, Pioro EP, Sasse ME et al (2006) Control of microglial neurotoxicity by the fractalkine receptor. Nat Neurosci 9:917–924

Carson MJ, Reilly CR, Sutcliffe JG et al (1998) Mature microglia resemble immature antigen-presenting cells. Glia 22:72–85

Carson MJ, Doose JM, Melchior B et al (2006) CNS immune privilege: hiding in plain sight. Immunol Rev 213:48–65

Chapman GA, Moores K, Harrison D et al (2000) Fractalkine cleavage from neuronal membranes represents an acute event in the inflammatory response to excitotoxic brain damage. J Neurosci 20:RC87

Chen Z, Jalabi W, Shpargel KB et al (2012) Lipopolysaccharide-induced microglial activation and neuroprotection against experimental brain injury is independent of hematogenous TLR4. J Neurosci 32:11706–11715
Clark AK, Yip PK, Malcangio M (2009) The liberation of fractalkine in the dorsal horn requires microglial cathepsin S. J Neurosci 29:6945–6954
Codolo G, Plotegher N, Pozzobon T et al (2013) Triggering of inflammasome by aggregated alpha-synuclein, an inflammatory response in synucleinopathies. PLoS One 8:e55375
Cosenza MA, Zhao ML, Si Q et al (2002) Human brain parenchymal microglia express CD14 and CD45 and are productively infected by HIV-1 in HIV-1 encephalitis. Brain Pathol 12: 442–455
Cowell RM, Xu H, Parent JM et al (2006) Microglial expression of chemokine receptor CCR5 during rat forebrain development and after perinatal hypoxia-ischemia. J Neuroimmunol 173:155–165
D'Aversa TG, Yu KO, Berman JW (2004) Expression of chemokines by human fetal microglia after treatment with the human immunodeficiency virus type 1 protein Tat. J Neurovirol 10: 86–97
D'Aversa TG, Eugenin EA, Berman JW (2008) CD40-CD40 ligand interactions in human microglia induce CXCL8 (interleukin-8) secretion by a mechanism dependent on activation of ERK1/2 and nuclear translocation of nuclear factor-kappaB (NFkappaB) and activator protein-1 (AP-1). J Neurosci Res 86:630–639
da Cunha A, Jefferson JJ, Tyor WR et al (1997) Transforming growth factor-beta1 in adult human microglia and its stimulated production by interleukin-1. J Interferon Cytokine Res 17:655–664
Davalos D, Grutzendler J, Yang G et al (2005) ATP mediates rapid microglial response to local brain injury in vivo. Nat Neurosci 8:752–758
Davalos D, Ryu JK, Merlini M et al (2012) Fibrinogen-induced perivascular microglial clustering is required for the development of axonal damage in neuroinflammation. Nat Commun 3:1227
de Rivero Vaccari JP, Lotocki G, Marcillo AE et al (2008) A molecular platform in neurons regulates inflammation after spinal cord injury. J Neurosci 28:3404–3414
Denes A, Pinteaux E, Rothwell NJ et al (2011) Interleukin-1 and stroke: biomarker, harbinger of damage, and therapeutic target. Cerebrovasc Dis 32:517–527
Depboylu C, Stricker S, Ghobril JP et al (2012) Brain-resident microglia predominate over infiltrating myeloid cells in activation, phagocytosis and interaction with T-lymphocytes in the MPTP mouse model of Parkinson disease. Exp Neurol 238:183–191
Dheen ST, Kaur C, Ling EA (2007) Microglial activation and its implications in the brain diseases. Curr Med Chem 14:1189–1197
Dinarello CA (2007) Historical insights into cytokines. Eur J Immunol 37(suppl 1):S34–S45
Doi Y, Mizuno T, Maki Y et al (2009) Microglia activated with the toll-like receptor 9 ligand CpG attenuate oligomeric amyloid {beta} neurotoxicity in in vitro and in vivo models of Alzheimer's disease. Am J Pathol 175:2121–2132
Ducharme G, Newell EW, Pinto C et al (2007) Small-conductance Cl- channels contribute to volume regulation and phagocytosis in microglia. Eur J Neurosci 26:2119–2130
Dutta R, Trapp BD (2011) Mechanisms of neuronal dysfunction and degeneration in multiple sclerosis. Prog Neurobiol 93:1–12
Ellis SL, Gysbers V, Manders PM et al (2010) The cell-specific induction of CXC chemokine ligand 9 mediated by IFNgamma in microglia of the central nervous system is determined by the myeloid transcription factor PU.1. J Immunol 185:1864–1877
Esen N, Kielian T (2006) Central role for MyD88 in the responses of microglia to pathogen-associated molecular patterns. J Immunol 176:6802–6811
Fitzner D, Schnaars M, van Rossum D et al (2011) Selective transfer of exosomes from oligodendrocytes to microglia by macropinocytosis. J Cell Sci 124:447–458
Flynn G, Maru S, Loughlin J et al (2003) Regulation of chemokine receptor expression in human microglia and astrocytes. J Neuroimmunol 136:84–93
Fogg DK, Sibon C, Miled C et al (2006) A clonogenic bone marrow progenitor specific for macrophages and dendritic cells. Science 311:83–87

Ford AL, Goodsall AL, Hickey WF et al (1995) Normal adult ramified microglia separated from other central nervous system macrophages by flow cytometric sorting. Phenotypic differences defined and direct ex vivo antigen presentation to myelin basic protein-reactive CD4+ T cells compared. J Immunol 154:4309–4321

Fraser DA, Pisalyaput K, Tenner AJ (2010) C1q enhances microglial clearance of apoptotic neurons and neuronal blebs, and modulates subsequent inflammatory cytokine production. J Neurochem 112:733–743

Fricker M, Neher JJ, Zhao JW et al (2012) MFG-E8 mediates primary phagocytosis of viable neurons during neuroinflammation. J Neurosci 32:2657–2666

Fuhrmann M, Bittner T, Jung CK et al (2010) Microglial Cx3cr1 knockout prevents neuron loss in a mouse model of Alzheimer's disease. Nat Neurosci 13:411–413

Garcia J, Pino PA, Mizutani M et al (2013) Regulation of adaptive immunity by the fractalkine receptor during autoimmune inflammation. J Immunol 191:1063–1072

Garton KJ, Gough PJ, Blobel CP et al (2001) Tumor necrosis factor-alpha-converting enzyme (ADAM17) mediates the cleavage and shedding of fractalkine (CX3CL1). J Biol Chem 276: 37993–38001

Gautier EL, Shay T, Miller J et al (2012) Gene-expression profiles and transcriptional regulatory pathways that underlie the identity and diversity of mouse tissue macrophages. Nat Immunol 13:1118–1128

Ghorpade A, Holter S, Borgmann K et al (2003) HIV-1 and IL-1 beta regulate Fas ligand expression in human astrocytes through the NF-kappa B pathway. J Neuroimmunol 141:141–149

Ginhoux F, Greter M, Leboeuf M et al (2010) Fate mapping analysis reveals that adult microglia derive from primitive macrophages. Science 330:841–845

Glabinsk AR, Ransohoff RM (2001) Targeting the chemokine system for multiple sclerosis treatment. Curr Opin Investig Drugs 2:1712–1719

Glezer I, Lapointe A, Rivest S (2006) Innate immunity triggers oligodendrocyte progenitor reactivity and confines damages to brain injuries. FASEB J 20:750–752

Glezer I, Chernomoretz A, David S et al (2007a) Genes involved in the balance between neuronal survival and death during inflammation. PLoS One 2:e310

Glezer I, Simard AR, Rivest S (2007b) Neuroprotective role of the innate immune system by microglia. Neuroscience 147:867–883

Gottfried-Blackmore A, Kaunzner UW, Idoyaga J et al (2009) Acute in vivo exposure to interferon-gamma enables resident brain dendritic cells to become effective antigen presenting cells. Proc Natl Acad Sci U S A 106:20918–20923

Greenwood J, Heasman SJ, Alvarez JI et al (2011) Review: leucocyte-endothelial cell crosstalk at the blood-brain barrier: a prerequisite for successful immune cell entry to the brain. Neuropathol Appl Neurobiol 37:24–39

Gris D, Ye Z, Iocca HA et al (2010) NLRP3 plays a critical role in the development of experimental autoimmune encephalomyelitis by mediating Th1 and Th17 responses. J Immunol 185:974–981

Hadas S, Spira M, Hanisch UK et al (2012) Complement receptor-3 negatively regulates the phagocytosis of degenerated myelin through tyrosine kinase Syk and cofilin. J Neuroinflammation 9:166

Haider L, Fischer MT, Frischer JM et al (2011) Oxidative damage in multiple sclerosis lesions. Brain 134:1914–1924

Hains BC, Waxman SG (2006) Activated microglia contribute to the maintenance of chronic pain after spinal cord injury. J Neurosci 26:4308–4317

Halle A, Hornung V, Petzold GC et al (2008) The NALP3 inflammasome is involved in the innate immune response to amyloid-beta. Nat Immunol 9:857–865

Hanamsagar R, Torres V, Kielian T (2011) Inflammasome activation and IL-1beta/IL-18 processing are influenced by distinct pathways in microglia. J Neurochem 119:736–748

Hanamsagar R, Hanke ML, Kielian T (2012) Toll-like receptor (TLR) and inflammasome actions in the central nervous system. Trends Immunol 33:333–342

Hanisch UK (2002) Microglia as a source and target of cytokines. Glia 40:140–155

Hanisch UK, Prinz M, Angstwurm K et al (2001) The protein tyrosine kinase inhibitor AG126 prevents the massive microglial cytokine induction by pneumococcal cell walls. Eur J Immunol 31:2104–2115

Harms AS, Cao S, Rowse AL et al (2013) MHCII is required for alpha-synuclein-induced activation of microglia, CD4 T cell proliferation, and dopaminergic neurodegeneration. J Neurosci 33:9592–9600

Harrison JK, Jiang Y, Chen S et al (1998) Role for neuronally derived fractalkine in mediating interactions between neurons and CX3CR1-expressing microglia. Proc Natl Acad Sci U S A 95:10896–10901

Hawkins BT, Davis TP (2005) The blood-brain barrier/neurovascular unit in health and disease. Pharmacol Rev 57:173–185

Heneka MT, Kummer MP, Stutz A et al (2013) NLRP3 is activated in Alzheimer's disease and contributes to pathology in APP/PS1 mice. Nature 493:674–678

Heppner FL, Greter M, Marino D et al (2005) Experimental autoimmune encephalomyelitis repressed by microglial paralysis. Nat Med 11:146–152

Hermann DM, ElAli A (2012) The abluminal endothelial membrane in neurovascular remodeling in health and disease. Sci Signal 5:re4

Hosmane S, Tegenge MA, Rajbhandari L et al (2012) Toll/interleukin-1 receptor domain-containing adapter inducing interferon-beta mediates microglial phagocytosis of degenerating axons. J Neurosci 32:7745–7757

Hundhausen C, Misztela D, Berkhout TA et al (2003) The disintegrin-like metalloproteinase ADAM10 is involved in constitutive cleavage of CX3CL1 (fractalkine) and regulates CX3CL1-mediated cell-cell adhesion. Blood 102:1186–1195

Husebye H, Aune MH, Stenvik J et al (2010) The Rab11a GTPase controls Toll-like receptor 4-induced activation of interferon regulatory factor-3 on phagosomes. Immunity 33:583–596

Iannello A, Samarani S, Debbeche O et al (2010) Potential role of IL-18 in the immunopathogenesis of AIDS, HIV-associated lipodystrophy and related clinical conditions. Curr HIV Res 8:147–164

Imamura K, Hishikawa N, Sawada M et al (2003) Distribution of major histocompatibility complex class II-positive microglia and cytokine profile of Parkinson's disease brains. Acta Neuropathol 106:518–526

Jack C, Ruffini F, Bar-Or A et al (2005a) Microglia and multiple sclerosis. J Neurosci Res 81:363–373

Jack CS, Arbour N, Manusow J et al (2005b) TLR signaling tailors innate immune responses in human microglia and astrocytes. J Immunol 175:4320–4330

Jaini R, Popescu DC, Flask CA et al (2013) Myelin antigen load influences antigen presentation and severity of central nervous system autoimmunity. J Neuroimmunol 259:37–46

Jamilloux Y, Pierini R, Querenet M et al (2013) Inflammasome activation restricts Legionella pneumophila replication in primary microglial cells through flagellin detection. Glia 61: 539–549

Jha S, Srivastava SY, Brickey WJ et al (2010) The inflammasome sensor, NLRP3, regulates CNS inflammation and demyelination via caspase-1 and interleukin-18. J Neurosci 30: 15811–15820

Jung S, Aliberti J, Graemmel P et al (2000) Analysis of fractalkine receptor CX(3)CR1 function by targeted deletion and green fluorescent protein reporter gene insertion. Mol Cell Biol 20:4106–4114

Kaushik DK, Gupta M, Basu A (2011) Microglial response to viral challenges: every silver lining comes with a cloud. Front Biosci 16:2187–2205

Kaushik DK, Gupta M, Kumawat KL et al (2012) NLRP3 inflammasome: key mediator of neuroinflammation in murine Japanese encephalitis. PLoS One 7:e32270

Kielian T (2004) Microglia and chemokines in infectious diseases of the nervous system: views and reviews. Front Biosci 9:732–750

Kielian T, Barry B, Hickey WF (2001) CXC chemokine receptor-2 ligands are required for neutrophil-mediated host defense in experimental brain abscesses. J Immunol 166:4634–4643
Kielian T, Mayes P, Kielian M (2002) Characterization of microglial responses to Staphylococcus aureus: effects on cytokine, costimulatory molecule, and Toll-like receptor expression. J Neuroimmunol 130:86–99
Kielian T, Bearden ED, Baldwin AC et al (2004) IL-1 and TNF-alpha play a pivotal role in the host immune response in a mouse model of Staphylococcus aureus-induced experimental brain abscess. J Neuropathol Exp Neurol 63:381–396
Kielian T, Esen N, Bearden ED (2005) Toll-like receptor 2 (TLR2) is pivotal for recognition of S. aureus peptidoglycan but not intact bacteria by microglia. Glia 49:567–576
King IL, Dickendesher TL, Segal BM (2009) Circulating Ly-6C+ myeloid precursors migrate to the CNS and play a pathogenic role during autoimmune demyelinating disease. Blood 113:3190–3197
Kobayashi N, Karisola P, Pena-Cruz V et al (2007) TIM-1 and TIM-4 glycoproteins bind phosphatidylserine and mediate uptake of apoptotic cells. Immunity 27:927–940
Kochan T, Singla A, Tosi J et al (2012) Toll-like receptor 2 ligand pretreatment attenuates retinal microglial inflammatory response but enhances phagocytic activity toward Staphylococcus aureus. Infect Immun 80:2076–2088
Koenigsknecht J, Landreth G (2004) Microglial phagocytosis of fibrillar beta-amyloid through a beta1 integrin-dependent mechanism. J Neurosci 24:9838–9846
Koizumi S, Shigemoto-Mogami Y, Nasu-Tada K et al (2007) UDP acting at P2Y6 receptors is a mediator of microglial phagocytosis. Nature 446:1091–1095
Koka P, He K, Zack JA et al (1995) Human immunodeficiency virus 1 envelope proteins induce interleukin 1, tumor necrosis factor alpha, and nitric oxide in glial cultures derived from fetal, neonatal, and adult human brain. J Exp Med 182:941–951
Koprich JB, Reske-Nielsen C, Mithal P et al (2008) Neuroinflammation mediated by IL-1beta increases susceptibility of dopamine neurons to degeneration in an animal model of Parkinson's disease. J Neuroinflammation 5:8
Kremlev SG, Roberts RL, Palmer C (2004) Differential expression of chemokines and chemokine receptors during microglial activation and inhibition. J Neuroimmunol 149:1–9
Kurant E (2011) Keeping the CNS clear: glial phagocytic functions in Drosophila. Glia 59: 1304–1311
Lassmann H (2010) Axonal and neuronal pathology in multiple sclerosis: what have we learnt from animal models. Exp Neurol 225:2–8
Ledeboer A, Breve JJ, Wierinckx A et al (2002) Expression and regulation of interleukin-10 and interleukin-10 receptor in rat astroglial and microglial cells. Eur J Neurosci 16:1175–1185
Lee SC, Liu W, Dickson DW et al (1993) Cytokine production by human fetal microglia and astrocytes. Differential induction by lipopolysaccharide and IL-1 beta. J Immunol 150:2659–2667
Lee SC, Kress Y, Dickson DW et al (1995) Human microglia mediate anti-Cryptococcus neoformans activity in the presence of specific antibody. J Neuroimmunol 62:43–52
Lee S, Varvel NH, Konerth ME et al (2010) CX3CR1 deficiency alters microglial activation and reduces beta-amyloid deposition in two Alzheimer's disease mouse models. Am J Pathol 177:2549–2562
Lee HM, Kang J, Lee SJ et al (2013) Microglial activation of the NLRP3 inflammasome by the priming signals derived from macrophages infected with mycobacteria. Glia 61:441–452
Lehnardt S, Lachance C, Patrizi S et al (2002) The toll-like receptor TLR4 is necessary for lipopolysaccharide-induced oligodendrocyte injury in the CNS. J Neurosci 22:2478–2486
Lehnardt S, Massillon L, Follett P et al (2003) Activation of innate immunity in the CNS triggers neurodegeneration through a Toll-like receptor 4-dependent pathway. Proc Natl Acad Sci U S A 100:8514–8519
Lehnardt S, Henneke P, Lien E et al (2006) A mechanism for neurodegeneration induced by group B streptococci through activation of the TLR2/MyD88 pathway in microglia. J Immunol 177:583–592

Li J, Gran B, Zhang GX et al (2003) Differential expression and regulation of IL-23 and IL-12 subunits and receptors in adult mouse microglia. J Neurol Sci 215:95–103
Lindberg C, Chromek M, Ahrengart L et al (2005) Soluble interleukin-1 receptor type II, IL-18 and caspase-1 in mild cognitive impairment and severe Alzheimer's disease. Neurochem Int 46:551–557
Liu S, Kielian T (2009) Microglial activation by Citrobacter koseri is mediated by Citrobacter koseri is mediated by TLR4- and MyD88-dependent pathways. J Immunol 183:5537–5547
Liu S, Kielian T (2011) MyD88 is pivotal for immune recognition of Citrobacter koseri and astrocyte activation during CNS infection. J Neuroinflammation 8:35
Liu YJ, Guo DW, Tian L et al (2010a) Peripheral T cells derived from Alzheimer's disease patients overexpress CXCR2 contributing to its transendothelial migration, which is microglial TNF-alpha-dependent. Neurobiol Aging 31:175–188
Liu Z, Condello C, Schain A et al (2010b) CX3CR1 in microglia regulates brain amyloid deposition through selective protofibrillar amyloid-beta phagocytosis. J Neurosci 30:17091–17101
Lokensgard JR, Hu S, Sheng W et al (2001) Robust expression of TNF-alpha, IL-1beta, RANTES, and IP-10 by human microglial cells during nonproductive infection with herpes simplex virus. J Neurovirol 7:208–219
Lu KT, Wang YW, Yang JT et al (2005) Effect of interleukin-1 on traumatic brain injury-induced damage to hippocampal neurons. J Neurotrauma 22:885–895
Ludwig A, Schulte A, Schnack C et al (2005) Enhanced expression and shedding of the transmembrane chemokine CXCL16 by reactive astrocytes and glioma cells. J Neurochem 93:1293–1303
Lue LF, Rydel R, Brigham EF et al (2001a) Inflammatory repertoire of Alzheimer's disease and nondemented elderly microglia in vitro. Glia 35:72–79
Lue LF, Walker DG, Rogers J (2001b) Modeling microglial activation in Alzheimer's disease with human postmortem microglial cultures. Neurobiol Aging 22:945–956
Mack CL, Vanderlugt-Castaneda CL, Neville KL et al (2003) Microglia are activated to become competent antigen presenting and effector cells in the inflammatory environment of the Theiler's virus model of multiple sclerosis. J Neuroimmunol 144:68–79
Magnus T, Chan A, Grauer O et al (2001) Microglial phagocytosis of apoptotic inflammatory T cells leads to down-regulation of microglial immune activation. J Immunol 167:5004–5010
Majumdar A, Cruz D, Asamoah N et al (2007) Activation of microglia acidifies lysosomes and leads to degradation of Alzheimer amyloid fibrils. Mol Biol Cell 18:1490–1496
Majumdar A, Capetillo-Zarate E, Cruz D et al (2011) Degradation of Alzheimer's amyloid fibrils by microglia requires delivery of ClC-7 to lysosomes. Mol Biol Cell 22:1664–1676
Mariano C, Silva SL, Pereira P et al (2011) Evidence of tricellulin expression by immune cells, particularly microglia. Biochem Biophys Res Commun 409:799–802
Masada T, Hua Y, Xi G et al (2003) Overexpression of interleukin-1 receptor antagonist reduces brain edema induced by intracerebral hemorrhage and thrombin. Acta Neurochir Suppl 86: 463–467
Matsumoto J, Dohgu S, Takata F et al (2012) Lipopolysaccharide-activated microglia lower P-glycoprotein function in brain microvascular endothelial cells. Neurosci Lett 524:45–48
Mattison HA, Nie H, Gao H et al (2013) Suppressed pro-inflammatory response of microglia in CX3CR1 knockout mice. J Neuroimmunol 257:110–115
McGeer PL, Itagaki S, Boyes BE et al (1988) Reactive microglia are positive for HLA-DR in the substantia nigra of Parkinson's and Alzheimer's disease brains. Neurology 38:1285–1291
Mizoue LS, Bazan JF, Johnson EC et al (1999) Solution structure and dynamics of the CX3C chemokine domain of fractalkine and its interaction with an N-terminal fragment of CX3CR1. Biochemistry 38:1402–1414
Mizutani M, Pino PA, Saederup N et al (2012) The fractalkine receptor but not CCR2 is present on microglia from embryonic development throughout adulthood. J Immunol 188:29–36
Muldoon LL, Alvarez JI, Begley DJ et al (2013) Immunologic privilege in the central nervous system and the blood-brain barrier. J Cereb Blood Flow Metab 33:13–21

Nakajima K, Kohsaka S (2001) Microglia activation and their significance in the central nervous system. J Biochem 130:169–175

Nemeth ZH, Deitch EA, Davidson MT et al (2004) Disruption of the actin cytoskeleton results in nuclear factor-kappaB activation and inflammatory mediator production in cultured human intestinal epithelial cells. J Cell Physiol 200:71–81

Neniskyte U, Neher JJ, Brown GC (2011) Neuronal death induced by nanomolar amyloid beta is mediated by primary phagocytosis of neurons by microglia. J Biol Chem 286:39904–39913

Neumann J, Sauerzweig S, Ronicke R et al (2008) Microglia cells protect neurons by direct engulfment of invading neutrophil granulocytes: a new mechanism of CNS immune privilege. J Neurosci 28:5965–5975

Nicoll JA, Wilkinson D, Holmes C et al (2003) Neuropathology of human Alzheimer disease after immunization with amyloid-beta peptide: a case report. Nat Med 9:448–452

Nielsen HH, Ladeby R, Fenger C et al (2009) Enhanced microglial clearance of myelin debris in T cell-infiltrated central nervous system. J Neuropathol Exp Neurol 68:845–856

Nimmerjahn A, Kirchhoff F, Helmchen F (2005) Resting microglial cells are highly dynamic surveillants of brain parenchyma in vivo. Science 308:1314–1318

Nishioku T, Matsumoto J, Dohgu S et al (2010) Tumor necrosis factor-alpha mediates the blood-brain barrier dysfunction induced by activated microglia in mouse brain microvascular endothelial cells. J Pharmacol Sci 112:251–254

O'Keefe GM, Nguyen VT, Benveniste EN (2002) Regulation and function of class II major histocompatibility complex, CD40, and B7 expression in macrophages and microglia: implications in neurological diseases. J Neurovirol 8:496–512

Olson JK, Miller SD (2004) Microglia initiate central nervous system innate and adaptive immune responses through multiple TLRs. J Immunol 173:3916–3924

Pardridge WM (2003) Blood-brain barrier drug targeting: the future of brain drug development. Mol Interv 3(90–105):51

Paloneva J, Kestilä M, Wu J, Salminen A, Böhling T, Ruotsalainen V, Hakola P, Bakker AB, Phillips JH, Pekkarinen P, Lanier LL, Timonen T, Peltonen L (2000) Loss-of-function mutations in TYROBP (DAP12) result in a presenile dementia with bone cysts. Nat Genet 25(3):357–61, PMID: 10888890

Parwaresch MR, Wacker HH (1984) Origin and kinetics of resident tissue macrophages. Parabiosis studies with radiolabelled leucocytes. Cell Tissue Kinet 17:25–39

Pineau I, Lacroix S (2007) Proinflammatory cytokine synthesis in the injured mouse spinal cord: multiphasic expression pattern and identification of the cell types involved. J Comp Neurol 500:267–285

Ponomarev ED, Novikova M, Maresz K et al (2005) Development of a culture system that supports adult microglial cell proliferation and maintenance in the resting state. J Immunol Methods 300:32–46

Pontillo A, Catamo E, Arosio B et al (2012) NALP1/NLRP1 genetic variants are associated with Alzheimer disease. Alzheimer Dis Assoc Disord 26:277–281

Prinz M, Priller J (2010) Tickets to the brain: role of CCR2 and CX(3)CR1 in myeloid cell entry in the CNS. J Neuroimmunol 224:80–84

Prinz M, Priller J, Sisodia SS et al (2011) Heterogeneity of CNS myeloid cells and their roles in neurodegeneration. Nat Neurosci 14:1227–1235

Prodinger C, Bunse J, Kruger M et al (2011) CD11c-expressing cells reside in the juxtavascular parenchyma and extend processes into the glia limitans of the mouse nervous system. Acta Neuropathol 121:445–458

Pulliam L, Moore D, West DC (1995) Human cytomegalovirus induces IL-6 and TNF alpha from macrophages and microglial cells: possible role in neurotoxicity. J Neurovirol 1:219–227

Rambach G, Dum D, Mohsenipour I et al (2010) Secretion of a fungal protease represents a complement evasion mechanism in cerebral aspergillosis. Mol Immunol 47:1438–1449

Ransohoff RM (2002) The chemokine system in neuroinflammation: an update. J Infect Dis 186(suppl 2):S152–S156

Ransohoff RM (2009) Chemokines and chemokine receptors: standing at the crossroads of immunobiology and neurobiology. Immunity 31:711–721

Ransohoff RM, Engelhardt B (2012) The anatomical and cellular basis of immune surveillance in the central nervous system. Nat Rev Immunol 12:623–635

Ravindran C, Cheng YC, Liang SM (2010) CpG-ODNs induces up-regulated expression of chemokine CCL9 in mouse macrophages and microglia. Cell Immunol 260:113–118

Regnier-Vigouroux A (2003) The mannose receptor in the brain. Int Rev Cytol 226:321–342

Remington LT, Babcock AA, Zehntner SP et al (2007) Microglial recruitment, activation, and proliferation in response to primary demyelination. Am J Pathol 170:1713–1724

Renner NA, Sansing HA, Morici LA et al (2012) Microglia activation by SIV-infected macrophages: alterations in morphology and cytokine secretion. J Neurovirol 18:213–221

Reynolds ND, Lukacs NW, Long N et al (2011) Delta-like ligand 4 regulates central nervous system T cell accumulation during experimental autoimmune encephalomyelitis. J Immunol 187:2803–2813

Ribes S, Ebert S, Regen T et al (2010) Toll-like receptor stimulation enhances phagocytosis and intracellular killing of nonencapsulated and encapsulated Streptococcus pneumoniae by murine microglia. Infect Immun 78:865–871

Rivest S (2009) Regulation of innate immune responses in the brain. Nat Rev Immunol 9: 429–439

Rogers J, Lue LF (2001) Microglial chemotaxis, activation, and phagocytosis of amyloid beta-peptide as linked phenomena in Alzheimer's disease. Neurochem Int 39:333–340

Rogers JT, Morganti JM, Bachstetter AD et al (2011) CX3CR1 deficiency leads to impairment of hippocampal cognitive function and synaptic plasticity. J Neurosci 31:16241–16250

Rolls A, Shechter R, London A et al (2007) Toll-like receptors modulate adult hippocampal neurogenesis. Nat Cell Biol 9:1081–1088

Rossi DL, Hardiman G, Copeland NG et al (1998) Cloning and characterization of a new type of mouse chemokine. Genomics 47:163–170

Ryu KY, Cho GS, Piao HZ et al (2012) Role of TGF-beta in survival of phagocytizing microglia: autocrine suppression of TNF-alpha production and oxidative stress. Exp Neurobiol 21: 151–157

Saederup N, Cardona AE, Croft K et al (2010) Selective chemokine receptor usage by central nervous system myeloid cells in CCR2-red fluorescent protein knock-in mice. PLoS One 5:e13693

Santiago C, Ballesteros A, Martinez-Munoz L et al (2007) Structures of T cell immunoglobulin mucin protein 4 show a metal-ion-dependent ligand binding site where phosphatidylserine binds. Immunity 27:941–951

Saukkonen K, Sande S, Cioffe C et al (1990) The role of cytokines in the generation of inflammation and tissue damage in experimental gram-positive meningitis. J Exp Med 171:439–448

Savill J, Gregory C (2007) Apoptotic PS to phagocyte TIM-4: eat me. Immunity 27:830–832

Schenk D (2008) Current challenges for the successful treatment and prevention of Alzheimer's disease: treating the pathologies of the disease to change its clinical course. Alzheimers Dement 4:S119–S121

Schmid CD, Melchior B, Masek K et al (2009) Differential gene expression in LPS/IFNgamma activated microglia and macrophages: in vitro versus in vivo. J Neurochem 109(suppl 1): 117–125

Schwab JM, Schluesener HJ, Seid K et al (2001) IL-16 is differentially expressed in the developing human fetal brain by microglial cells in zones of neuropoesis. Int J Dev Neurosci 19:93–100

Selenica ML, Alvarez JA, Nash KR et al (2013) Diverse activation of microglia by chemokine (C-C motif) ligand 2 overexpression in brain. J Neuroinflammation 10:86

Shaked I, Porat Z, Gersner R et al (2004) Early activation of microglia as antigen-presenting cells correlates with T cell-mediated protection and repair of the injured central nervous system. J Neuroimmunol 146:84–93

Shaw PJ, Lukens JR, Burns S et al (2010) Cutting edge: critical role for PYCARD/ASC in the development of experimental autoimmune encephalomyelitis. J Immunol 184:4610–4614

Shi F, Yang L, Kouadir M et al (2012) The NALP3 inflammasome is involved in neurotoxic prion peptide-induced microglial activation. J Neuroinflammation 9:73

Shiratori M, Tozaki-Saitoh H, Yoshitake M et al (2010) P2X7 receptor activation induces CXCL2 production in microglia through NFAT and PKC/MAPK pathways. J Neurochem 114: 810–819

Shutter JR, Scully S, Fan W et al (2000) Dll4, a novel Notch ligand expressed in arterial endothelium. Genes Dev 14:1313–1318

Sierra A, Gottfried-Blackmore AC, McEwen BS et al (2007) Microglia derived from aging mice exhibit an altered inflammatory profile. Glia 55:412–424

Sierra A, Abiega O, Shahraz A et al (2013) Janus-faced microglia: beneficial and detrimental consequences of microglial phagocytosis. Front Cell Neurosci 7:6

Sokolowski JD, Mandell JW (2011) Phagocytic clearance in neurodegeneration. Am J Pathol 178:1416–1428

Stalder AK, Pagenstecher A, Yu NC et al (1997) Lipopolysaccharide-induced IL-12 expression in the central nervous system and cultured astrocytes and microglia. J Immunol 159:1344–1351

Stevens B, Allen NJ, Vazquez LE et al (2007) The classical complement cascade mediates CNS synapse elimination. Cell 131:1164–1178

Streit WJ (2002) Microglia and the response to brain injury. Ernst Schering Res Found Workshop (39):11–24

Streit WJ, Davis CN, Harrison JK (2005) Role of fractalkine (CX3CL1) in regulating neuron-microglia interactions: development of viral-based CX3CR1 antagonists. Curr Alzheimer Res 2:187–189

Su X, Maguire-Zeiss KA, Giuliano R et al (2008) Synuclein activates microglia in a model of Parkinson's disease. Neurobiol Aging 29:1690–1701

Sumi N, Nishioku T, Takata F et al (2010) Lipopolysaccharide-activated microglia induce dysfunction of the blood-brain barrier in rat microvascular endothelial cells co-cultured with microglia. Cell Mol Neurobiol 30:247–253

Sutton C, Brereton C, Keogh B et al (2006) A crucial role for interleukin (IL)-1 in the induction of IL-17-producing T cells that mediate autoimmune encephalomyelitis. J Exp Med 203: 1685–1691

Tambuyzer BR, Ponsaerts P, Nouwen EJ (2009) Microglia: gatekeepers of central nervous system immunology. J Leukoc Biol 85:352–370

Tanaka T, Ueno M, Yamashita T (2009) Engulfment of axon debris by microglia requires p38 MAPK activity. J Biol Chem 284:21626–21636

Trapp BD, Nave KA (2008) Multiple sclerosis: an immune or neurodegenerative disorder? Annu Rev Neurosci 31:247–269

Trapp BD, Ransohoff R, Rudick R (1999) Axonal pathology in multiple sclerosis: relationship to neurologic disability. Curr Opin Neurol 12:295–302

Trebst C, Sorensen TL, Kivisakk P et al (2001) CCR1+/CCR5+ mononuclear phagocytes accumulate in the central nervous system of patients with multiple sclerosis. Am J Pathol 159:1701–1710

Trebst C, Staugaitis SM, Kivisakk P et al (2003) CC chemokine receptor 8 in the central nervous system is associated with phagocytic macrophages. Am J Pathol 162:427–438

Underhill DM, Ozinsky A, Hajjar AM et al (1999) The Toll-like receptor 2 is recruited to macrophage phagosomes and discriminates between pathogens. Nature 401:811–815

van Horssen J, Singh S, van der Pol S et al (2012) Clusters of activated microglia in normal-appearing white matter show signs of innate immune activation. J Neuroinflammation 9:156

Wainwright DA, Xin J, Sanders VM et al (2008) Differential actions of pituitary adenylyl cyclase-activating polypeptide and interferon gamma on Th2- and Th1-associated chemokine expression in cultured murine microglia. J Neurodegener Regen 1:31–34

Williams K, Ulvestad E, Waage A et al (1994) Activation of adult human derived microglia by myelin phagocytosis in vitro. J Neurosci Res 38:433–443

Williams K, Dooley N, Ulvestad E et al (1996) IL-10 production by adult human derived microglial cells. Neurochem Int 29:55–64

Wilms H, Rosenstiel P, Sievers J et al (2003) Activation of microglia by human neuromelanin is NF-kappaB dependent and involves p38 mitogen-activated protein kinase: implications for Parkinson's disease. FASEB J 17:500–502
Wilson EH, Weninger W, Hunter CA (2010) Trafficking of immune cells in the central nervous system. J Clin Invest 120:1368–1379
Wraith DC, Nicholson LB (2012) The adaptive immune system in diseases of the central nervous system. J Clin Invest 122:1172–1179
Xiang Z, Chen M, Ping J et al (2006) Microglial morphology and its transformation after challenge by extracellular ATP in vitro. J Neurosci Res 83:91–101
Xiong J, Burkovetskaya M, Karpuk N et al (2012) IL-1RI (interleukin-1 receptor type I) signalling is essential for host defence and hemichannel activity during acute central nervous system bacterial infection. ASN Neuro 4:e00085
Zhang GX, Li J, Ventura E et al (2002) Parenchymal microglia of naive adult C57BL/6J mice express high levels of B7.1, B7.2, and MHC class II. Exp Mol Pathol 73:35–45
Zhou H, Lapointe BM, Clark SR et al (2006) A requirement for microglial TLR4 in leukocyte recruitment into brain in response to lipopolysaccharide. J Immunol 177:8103–8110
Zierler S, Frei E, Grissmer S et al (2008) Chloride influx provokes lamellipodium formation in microglial cells. Cell Physiol Biochem 21:55–62

Chapter 6
Neuroprotection Versus Neurotoxicity

Knut Biber and Michelle L. Block

Abstract Several original concepts concerning microglia have changed in the last decade. Ramified microglia are no longer seen to be "resting" cells and it also is very obvious today that microglia responses are by no means stereotypic, but manifold and targeted. Moreover, there is good evidence that microglia are not only important in brain pathology, but that they also play important roles in the healthy brain. One long-standing aspect of microglia biology, however, was never questioned: their involvement in brain disease. Based on morphological changes (retraction of processes and amoeboid shape) that inevitably occur in these cells in case of damage to the central nervous system, microglia in the diseased brain were called "activated". Because "activated" microglia were always found in direct neighbourhood to dead or dying neurons, and since it is known now for more than 20 years that cultured microglia release numerous factors that are able to kill neurons, microglia "activation" was often seen as neurotoxic. From an evolutionary point of view, however, it is difficult to understand why an important, mostly post-mitotic and highly vulnerable organ like the brain would host numerous toxic cells. How microglia can protect the nervous tissue and what might go awry when microglia turn neurotoxic will be discussed in this chapter.

Keywords Microglia • Neurotoxicity • Neuroprotection • Immune stimuli • Neuronal damage • Disease • Mutated proteins • Pesticides • Pollution • Mouse models • Experimental approaches

K. Biber, Ph.D. (✉)
Department of Psychiatry and Psychotherapy, Molecular Psychiatry,
University Hospital Freiburg, Hauptstr. 5, Freiburg 79104, Germany

Department of Neurosciences, University Medical Center Groningen,
Ant. Deusinglaan 1, Groningen 9713AV, The Netherlands
e-mail: knut.biber@uniklinik-freiburg.de

M.L. Block, Ph.D. (✉)
Department of Anatomy and Neurobiology, Virginia Commonwealth University,
Medical Campus, 1101 East Marshall Street Sanger Hall 9-048, Richmond, VA 23298, USA

Department of Anatomy & Cell Biology, Indiana University School of Medicine
and The Stark Neuroscience Research Institute, 635 Barnhill Drive, MS-5022 Indianapolis,
IN 46202, USA
e-mail: mlblock@iupui.edu

M.-È. Tremblay, A. Sierra (eds.), *Microglia in Health and Disease*,
DOI 10.1007/978-1-4939-1429-6_6

Bullet Points

- Microglia are the myeloid cells of the brain and play an important role in all brain diseases.
- Because "activated" microglia are always found in direct neighbourhood to dead or dying neurons, and since microglia in culture are able to kill neurons, microglia "activation" in the brain is often seen as neurotoxic response.
- The inhibition or deletion of microglia, however, often results in a worsening of disease symptoms, indicating that microglia activity can be protective in the brain.
- How microglia can protect the nervous tissue and what might go awry when microglia turn neurotoxic will be discussed in this chapter.

6.1 Microglia Physiology and Function

Microglia research truly started in 1932 by Pío del Río-Hortega, who introduced microglia (and numerous, even today, still valid postulates concerning these cells) in a chapter for "Cytology and Cellular Pathology of the Nervous System", a milestone in neuroscience that was edited by Wilder Penfield (del Rio-Hortega 1932); see for an excellent review about the history of microglia research (Kettenmann et al. 2011) and Chap. 2.

Despite del Río-Hortega's work and the results of many others that continued the study of microglia after him, these cells have been widely neglected by neuroscientists to a point that even in leading textbooks of neuropathology of the early 90s their mere existence was doubted (see for details Graeber 2010). This ignorance most likely was due to the fact that the original method for staining microglia was tedious and highly variable. The 90s of the last century saw technological progress in microglia research. The introduction of microglia cultures, more easy and reliable staining techniques (antibodies, lectins), and the "green microglia" mouse (CX3CR1-EGFP strain introduced by Steffen Jung (2000)) made these cells more accessible for investigators. As a result, microglia research has intensified enormously in the last decade. Many surprising findings about microglia have been published and recent results have begun to provide convincing answers to the most basic questions in microglial biology, namely the origin of these cells or the issue whether or not microglia are replaced by peripheral cells throughout the lifetime of an individual (Ajami et al. 2007; Ginhoux et al. 2010; Kierdorf et al. 2013). Microglia in the healthy brain are characterized by a so-called ramified morphology, small cell bodies, and fine heavily branched processes. Ramified microglia are sessile cells, meaning their cell bodies do not move, which is in contrast to their processes that display the most pronounced motility yet described in the mammalian brain. This process movement is currently discussed to be part of the surveillance function, since in the direct vicinity of microglia other cellular elements are regularly touched by these cells (for further reading, see Chap. 4). Ramified microglia are more or less evenly distributed throughout the brain and claim their own "territory". It is therefore estimated that due to their process motility the brain's parenchyma is "completely investigated" every few hours (Davalos et al. 2005; Nimmerjahn et al. 2005; Raivich 2005). In the last few years it was found that especially active synapses attract microglia process "interactions" and a role for these cells in regulating synaptic function is currently discussed elsewhere (Bechade et al. 2013; Miyamoto et al. 2013) (For further details, see Chap. 9). However, despite the fact that microglia research has enormously intensified and that microglia have been the subject of numerous recent and excellent reviews, we still know very little about their physiology and the function of ramified microglia in the healthy brain. As such, these cells remain among the most mysterious cells of the brain (Hanisch and Kettenmann 2007; Colton 2009; Ransohoff and Perry 2009; Yong and Rivest 2009; Graeber 2010; Parkhurst and Gan 2010; Ransohoff and Cardona 2010; Prinz and Mildner 2011; Tremblay et al. 2011; Aguzzi et al. 2013; Hanisch 2013; Hellwig et al. 2013; Kettenmann et al. 2013; Miyamoto et al. 2013; Sierra et al. 2013).

More is known about the function of microglia in brain disease. However, simplified concepts about their "activation" (see next paragraph) and the experimental difficulty to distinguish microglia from peripheral myeloid cells that often enter the diseased brain (Ransohoff and Cardona 2010; Prinz and Mildner 2011; Hellwig et al. 2013; London et al. 2013) have yet hampered the development of a clear view about whether or not these cells contribute to either initiation, course, or end of a given brain disease. Here we will discuss current knowledge about the role of microglia in brain disease with particular emphasis about microglia in neuroprotection or neurotoxicity.

6.2 Microglial Activation and Neuroinflammation Defined: Surveillance and Response

Almost any type of brain injury or disease inevitably leads to a fast morphological response of microglia. Ramified microglia retract their fine processes and acquire a morphology which resembles that of phagocytic macrophages, a transition that was originally described by del Río-Hortega. This morphological reaction often is correlated with a migratory behaviour and/or proliferation of the cells. Because the morphological complexity of this transition is reduced (basically only one morphological phenotype is the result of this process), it was originally defined as a stereotypic and graded process (Kreutzberg 1996; Streit 2002; van Rossum and Hanisch 2004; Hanisch and Kettenmann 2007). Based on these morphological data it was widely assumed that ramified microglia in the healthy brain would be inactive or resting and that the morphological transition of microglia in response to neuronal injury sets these cells into action, leading to describe this process as "microglia activation". Moreover, because the morphological transition appears to be a stereotypic process, it was more or less believed that "activated" microglia would respond in a stereotypic way with limited response variability. This general and simple concept of microglia "activation" is not valid anymore (see for recent reviews: Hanisch and Kettenmann 2007; Colton 2009; Ransohoff and Perry 2009; Yong and Rivest 2009; Graeber 2010; Parkhurst and Gan 2010; Ransohoff and Cardona 2010; Kettenmann et al. 2011; Prinz and Mildner 2011; Tremblay et al. 2011; Aguzzi et al. 2013; Hanisch 2013; Hellwig et al. 2013; Kierdorf et al. 2013; London et al. 2013; Sierra et al. 2013). Not only is it clear today that ramified microglia are by no means resting cells (see above), it is also apparent now that microglia respond with a variety of different reactions by integrating multifarious inputs (Hanisch and Kettenmann 2007; Colton 2009; Ransohoff and Perry 2009; Yong and Rivest 2009; Graeber 2010; Parkhurst and Gan 2010; Ransohoff and Cardona 2010; Kettenmann et al. 2011, 2013; Prinz and Mildner 2011; Tremblay et al. 2011; Aguzzi et al. 2013; Hanisch 2013; Hellwig et al. 2013; Kierdorf et al. 2013; London et al. 2013; Sierra et al. 2013). In line with this, microglia responses are not inevitably neurotoxic as often believed. Various neuroprotective effects of "activated" microglia have been demonstrated recently in vivo (Boillee et al. 2006; Turrin and Rivest 2006; El Khoury et al. 2007; Lalancette-Hebert et al. 2007; Lambertsen et al. 2009). Conversely, neurotoxicity might occur in case of overshooting an uncontrolled response of microglia (van Rossum and Hanisch 2004; Cardona et al. 2006, and many others) or when microglia function is impaired (Boillee et al. 2006; Streit 2006; Neumann and Takahashi 2007). Moreover, microglia function is not only important in neurodegenerative and neuroinflammatory diseases, but there is accumulating evidence that immune malfunction is also involved in psychiatric diseases like schizophrenia and mood disorders. The control of microglia function has therefore recently become of interest in biological psychiatry (see for review: Muller and Schwarz 2007; Bernstein et al. 2009; Beumer et al. 2012; Blank and Prinz 2013; Stertz et al. 2013). Thus, microglia are currently considered to be involved in a wide range of brain disorders.

As described above, microglia function has long been correlated to morphological criteria. Ramified microglia in the healthy brain were described as "resting", whereas more amoeboid microglia in the diseased brain were thought to be "activated". An overwhelming body of evidence from the last decade has convincingly shown that this simple discrimination does not hold true. Since there are no inactive microglia (the moniker "never resting microglia" is quite popular among microglia-minded scientists at the moment (Karperien et al. 2013)), the terms microglia "activation" or "activated" microglia are misleading. Ramified microglia are clearly surveillance cells that most likely execute numerous functions in the healthy brain. These cells have been referred to as "cops on the beat" (Raivich 2005), "constant gardeners" (Hughes 2012), "industrious housekeepers" (Streit and Xue 2009) or "garbage men" (Kettenmann 2007). However, even these descriptions most likely do not reflect the whole functional spectrum of ramified microglia in the healthy brain. The various functions of microglia are controlled not only by a whole plethora of receptors for exogenous signals (Biber et al. 2007; Hanisch and Kettenmann 2007; Kettenmann 2007; Ransohoff and Perry 2009; Ransohoff and Cardona 2010; Kettenmann et al. 2011; Domercq et al. 2013), but also endogenously by subsets of transcription factors (Kierdorf and Prinz 2013). Microglia action may or may not involve major morphological transitions (Vinet et al. 2012; Karperien et al. 2013) and the roles of amoeboid microglia are manifold (Kettenmann et al. 2011). Microglia merely adjust their current activity in response to changes in their direct vicinity, or as reactions to endogenous changes, such as aging (Kierdorf and Prinz 2013; Wong 2013). As such, the operational definition of microglial activation is a broad concept that only refers to the general process of altering either the physical characteristics, activity, or function of microglia. Given the variety of activation phenotypes possible, the term activation only notes that a cellular change occurred, providing little to no information about cellular function.

Consistent with this premise, simple morphological criteria (ramified versus amoeboid) are also poor indicators of microglial function. For example, the presence of amoeboid microglia near dead (or dying) neurons has often been interpreted as a proof of a neurotoxic microglia response (Hellwig et al. 2013). Based on our current knowledge about microglia, such morphological correlation does not provide any information about causality or function of these cells. It is thus clear that without (i) a detailed analysis of the causal relationship between microglial function and neuronal fate, and (ii) a thorough understanding of all aspects of the microglia response in a given disease (model), it is difficult to fully appreciate what microglia are doing in the brain. New mouse models that allow a more targeted analysis of microglia in vivo may be helpful here (Mizutani et al. 2012; Varvel et al. 2012; Neumann and Wekerle 2013; Wolf et al. 2013). Whether or not a more sophisticated, fractal analysis of microglia morphology will be useful as predictor of microglia function is discussed elsewhere (Karperien et al. 2013).

6.3 Microglia in Central Nervous System Disease and Neuron Damage

Hallmark studies in the 1980s analysing postmortem tissue samples from Parkinson's disease (PD) (McGeer et al. 1988) and Alzheimer's disease (AD) patients (McGeer et al. 1987, 1988) were the first to suggest that microglia positive for human leucocyte antigen (HLA) (a surface receptor complex involved in antigen presentation) were linked to active neurodegeneration (McGeer et al. 1988), which marked the beginning of the field of research dedicated to understanding the involvement of microglia in central nervous system (CNS) diseases. Since then, microglia have been identified as critical for the maintenance of normal, healthy CNS physiology and repair, in addition to being implicated as instigating actors of progressive neuronal damage in neurodegenerative disorders/CNS damage (Perry et al. 2010), including traumatic brain injury (Woodcock and Morganti-Kossmann 2013), AD (Naert and Rivest 2011), Amyotrophic Lateral Sclerosis (ALS) (Phani et al. 2012), Multiple Sclerosis (Doring and Yong 2011), Huntington's disease (Moller 2010), and PD (German et al. 2011).

Most information linking microglia to human neurodegenerative disease/CNS damage is correlative, demonstrating changes in morphology and elevation of markers in the damaged brain regions, where heightened levels of pro-inflammatory molecules are often localized to microglia. PD, a movement disorder that is the second most prevalent neurodegenerative disease (Dorsey et al. 2007), is a classic example of how the microglial response is intricately linked to progressive neuron damage. PD is characterized by the progressive and selective degeneration of dopamine (DA) neurons in the substantia nigra pars compacta (SNpc) (Kish et al. 1992). Importantly, neuroinflammation is present, as pro-inflammatory cytokine levels are elevated in the SNpc of PD patients (Reale et al. 2009; German et al. 2011; Panaro and Cianciulli 2012). Postmortem tissue analysis also reveals elevation of HLA-DR positive microglia in the SNpc of PD patients (McGeer et al. 1988; Imamura et al. 2003; Orr et al. 2005), confirming a phenotypic change in microglia in regions of ongoing neuronal damage. Interestingly, microglial markers are also elevated in other brain regions implicated in PD outside of the SNpc (Imamura et al. 2003; Gerhard et al. 2006), consistent with the fact that despite hallmark DA neuron loss, PD neuropathology extends to many brain regions. For example, PD patients have shown significantly higher numbers of Major Histocompatibility Complex (MHC) class II-positive microglia in the hippocampus (HC), transentorhinal cortex, cingulate cortex, and temporal cortex, regions outside of the SN with reported neurite damage in PD (Imamura et al. 2003). MHC class II is crucial for antigen presentation, its role in PD however remains undetermined. It is known that MHC class II microglia in the SNpc of PD patient tissue are also positive for tumour necrosis factor α (TNFα) and interleukin 6 (IL-6) (Imamura et al. 2003), demonstrating that these cells are at least one source of CNS cytokines. Positron emission tomography (PET) imaging with the radiolabeled ligand ([11C](R)-PK11195) that binds the peripheral benzodiazepine receptor which in the brain is predominantly expressed by "activated" microglia confirms the presence of "activated" microglia in the SNpc of living PD patients (Gerhard et al. 2006), where increased levels of [(11)C](R)-PK11195 binding is associated with the loss of nigrostriatal DA neuron terminal function and disease severity (Ouchi et al. 2005). PET imaging also demonstrates that PK11195 binding is elevated in the SNpc early in the disease process of both PD and Lewy body dementia (PD-related disease) patients (Iannaccone et al. 2013), supporting that microglia are active throughout the disease process. The premise of a pro-inflammatory phenotype in microglial for AD is also supported by analysis of postmortem brains from AD patients (McGeer et al. 1987; Rogers et al. 1988), where the microglial response precedes neutrophil damage (Cagnin et al. 2001), suggesting a potential causal role. Together, evidence points to neuroimmune perturbation and microglial response in PD, AD, and in fact many human CNS diseases (Smith et al. 2012). Yet these important, but descriptive studies provide little functional data about causal relationships.

The use of animal models of CNS disease provides the experimental work implicating microglia as a culpable source of chronic cytokines and reactive oxygen species (ROS) that are toxic to neurons. As innate immune cells, microglia are capable of upregulating an entire battery of compounds identified as toxic to neurons (Phani et al. 2012), inducing TNFα (Harms et al. 2012; Abo-Ouf et al. 2013), inducible nitric oxide synthase (iNOS) (production of nitric oxide) (Brown 2007), NADH oxidase 2 (NOX2) (production of superoxide) (Brown 2007; Sorce and Krause 2009), cyclo oxygenase 2 (COX2) (production of prostaglandin PGE_2) (Jiang et al. 2011), interleukin 1β (IL-1β) (Turola et al. 2012; Ye et al. 2013), and interferon γ (INFγ) (Mount et al. 2007). While not all pro-inflammatory factors are neurotoxic in all disease models and brain regions (Sriram and O'Callaghan 2007; Nadeau et al. 2011), multiple neurodegenerative disease models are benefited by an anti-inflammatory approach (Choi et al. 2013; Kay and Palmer 2013; Ramsey and Tansey 2014; Tabas and Glass 2013). For example, neuronal damage in a PD mouse model is significantly reduced in the SNpc of mutant mice with deficient production of pro-inflammatory factors, such as superoxide (Wu et al. 2003; Zhang et al. 2004), prostaglandins (Feng et al. 2002; Teismann et al. 2003), and TNFα (Sriram et al. 2002). Importantly, several studies have documented that microglia are a source of these potentially neurotoxic molecules in human neuron damage and neurodegenerative disease (Phani et al. 2012; Smith et al. 2012; Woodcock and Morganti-Kossmann 2013).

More recent and convincing evidence employs in vivo studies and myeloid cell-specific gene deletion (which includes microglia, neutrophils, monocytes, other macrophages, and other non-neuronal cells) in CNS disease/damage models to reveal that only certain components of the microglial response can drive neuronal damage. The inhibitor of kappa light polypeptide gene enhancer in B-cells, kinase beta (Ikkβ) is a critical signalling component for nuclear factor kappa B (NF-κB) activation and is also necessary for the production of many pro-inflammatory cytokines. In a mouse model with myeloid cell-specific deletion of Ikkβ an attenuated Kainic acid-induced hippocampal neuronal cell death was reported, suggesting a deleterious role of cytokines in this pathology (Cho et al. 2008). Identifying another microglia/macrophage-specific mechanism, the deletion of the NR1 subunit of the NMDA receptor in myeloid cells has been shown to attenuate neuronal damage in murine models of both traumatic brain injury and excitotoxicity (Kaindl et al. 2012). Further, microglial/macrophage deletion of p38α, a kinase critical to pro-inflammatory signalling, has been shown to attenuate neuroinflammation (Bachstetter et al. 2011), implicating microglia as a source of cytokines in the brain, and is neuroprotective for both lipopolysaccharide (LPS)-induced neuronal damage (Xing et al. 2011) and traumatic brain injury (Bachstetter et al. 2013). Glucocorticoid receptor deletion in macrophages and microglia also demonstrates enhanced neuroinflammation and neuronal damage in a mouse model of PD (Ros-Bernal et al. 2011). Thus, together, in vivo and in vitro studies indicate that while microglia have the potential to actively initiate neuronal damage, and rather than framing microglia as a neurotoxic cell type, evidence supports that it is the dysregulation of only a handful of specific functions that results in microglia-mediated neuronal toxicity. Extensive research continues to explore the mechanisms shifting microglia to a neurotoxic phenotype.

6.4 Beneficial Microglia Responses

Various neuroprotective effects of microglia have been demonstrated in vivo in various experimental stroke paradigms. Microglia were found beneficial in ischemia in neonates (Faustino et al. 2011) and adults (Simard et al. 2006; Lalancette-Hebert et al. 2007; Yanagisawa et al. 2008; Cipriani et al. 2011), as well as in experimental stroke-like conditions in organotypic hippocampal slice cultures (OHSCs) (Neumann et al. 2006; Vinet et al. 2012). Transplantation of exogenous microglia into rodents (intraventricular or intra-arterial) before (Kitamura et al. 2005; Hayashi et al. 2006) or after (Neumann et al. 2006; Imai et al. 2007; Narantuya et al. 2010) experimental stroke reduced neural damage and improved neurological outcome of the animals. The beneficial response of microglia is not only limited to stroke conditions, since these cells were also found to be protective in other brain diseases in which neurons suffer from excitotoxicity such as Huntington's disease or traumatic brain injury (Simard and Rivest 2006; Palazuelos et al. 2009).

The range of potential neuroprotective mediators that are released from microglia in the ischemic brain is wide as there is evidence that these cells express neurotrophic factors such as brain-derived neurotrophic factor (BDNF), glia cell-derived neurotrophic factor (GDNF), vascular endothelial growth factor (VEGF) (Madinier et al. 2009; Narantuya et al. 2010), or release protective adenosine (Cipriani et al. 2011). Others have described that the pro-inflammatory cytokine TNFα in microglia acts as neuroprotective agent in stroke conditions (Sriram et al. 2006; Lambertsen et al. 2009) or provide more general evidence for protective role of microglia-related inflammation (Simard and Rivest 2006; Anrather et al. 2011). However, the role of inflammation in stroke is manifold. TNFα for example was found to be protective in the ischemic hippocampus, but promoted neuronal loss in the striatum in response to stroke (Sriram et al. 2006). Similarly most inflammatory mediators have double edged functions (protective or detrimental) in stroke showing that the overall outcome of their actions depends on a variety of conditions like temporary aspects (at what time point during the disease course) or spatial aspects (where is a cytokine released) (Iadecola and Anrather 2011). To unravel the protective side of microglia activity in excitotoxicity will be a major challenge for neuroscientists in the future, as these cells may be valid drug targets to prevent or treat brain diseases in which neurons suffer from being overexcited.

6.5 Loss of Microglial Function: Neuropathology

The loss of microglia function can experimentally be investigated in mouse models in which microglia are mutated, or in models that allow the specific inhibition or depletion of microglia.

6.5.1 Mutating Microglia Function

It is clear today that microglia in the brain are under constant restraint, particularly because they specifically express receptors for a variety of inhibitory factors that are constitutively expressed in the brain, mostly by neurons (Biber et al. 2007; Ransohoff and Perry 2009). The most prominent ligand-receptor pairs in this respect are CX3-chemokine ligand1-CX3-chemokine receptor1 (CX3CL1-CX3CR1) and OX-2 membrane glycoprotein (also called cluster of differentiation 200) and its receptor (CD200-CD200R), and mutations of these ligand-receptors pairs in mice have revealed much about microglia. Regarding the CX3CR1-CX3CL1 axis, one of the most used mouse model in microglia research is the CX3CR1-EGFP mouse line in which all microglia are GFP-positive (Jung et al. 2000). The consequences of CX3CR1 deletion in microglia largely depends on the mouse model used (see for extensive review: Prinz and Mildner 2011; Ransohoff and Prinz 2013; Wolf et al. 2013). However, the overall idea at the moment is that a lack of CX3CR1 leads to microglia "hyperactivity" in the diseased brain, thereby unleashing potential neurotoxic properties (Wolf et al. 2013). Accordingly, administration of CX3CL1 into the brain causes neuroprotection in experimental stroke and two models of PD (Cipriani et al. 2011; Pabon et al. 2011; Morganti et al. 2012). Similarly, removing the inhibitory input that is normally modulated by CD200 (i.e., as in CD200R knockout mice) reportedly promotes microglial morphological transition even in the healthy brain (Hoek et al. 2000) and leads to an exaggerated disease course both in experimental autoimmune encephalomyelitis (EAE) (an animal model of Multiple Sclerosis) (Broderick et al. 2002) and retinal inflammation (Hoek et al. 2000).

In ALS, mutations within the ubiquitously expressed enzyme superoxide dismutase 1 (SOD1) gene are responsible for about a quarter of the inherited disease cases. Accordingly, mice that express mutant human SOD1 exhibit motor neuron degeneration and a decreased life span (see for review: Lobsiger et al. 2009). The role of microglia in this disease has been investigated in various elegant experiments in which mutated SOD1 was expressed in specific cell types. The conclusion that arose from these experiments is that microglia with mutated SOD1 do not initiate motor neuron degeneration but rather accelerate disease progression (Xiao et al. 2007) (see for review: Lobsiger et al. 2009). The replacement of SOD1 mutated microglia with wild-type cells slowed down disease progression and prolonged the life span of the animals (Beers et al. 2006), which required functional myeloid differentiation primary response 88 (MyD88) signaling in microglia (Kang and Rivest 2007). MyD88 is an intracellular adapter protein that is required for the signaling of various pro-inflammatory receptors. Triggering receptor expressed on myeloid cells 2 (TREM2) is another receptor that is in brain exclusively expressed in microglia (for review see: Linnartz et al. 2010). TREM2 belongs to the family of immunoreceptor tyrosine-based activation motif (ITAM) receptors for which the ligand has yet not been identified. Activation of TREM2 stimulates phagocytic activity in microglia and downregulates TNFα and iNOS expression (Takahashi et al. 2005). TREM2 is thus an anti-inflammatory receptor that at the same time promotes phagocytic activity. TREM2 is intracellularly coupled to the adapter protein DAP12 (for review see: Linnartz et al. 2010), and interestingly, loss of function mutations of either TREM2 or DAP12 lead to a rare chronic neurodegenerative disease known as Nasu-Hakola or polycystic lipomembranous osteodysplasia with sclerosing leukoencephalopathy (PLOSL) (Colonna 2003). From the above description, it can thus be concluded that mutations in microglia mostly lead to or enhance the severity of a given brain disease or model.

6.5.2 Models of Microglial Inhibition or Depletion

6.5.2.1 Microglia Depletion with Clodronate

The bisphosphonate drug clodronate is toxic to cells of the myeloid lineage and can be used to selectively deplete them in vivo and in vitro (Buiting and Van Rooijen 1994). Since microglia are of myeloid origin, clodronate can also be used to deplete microglia in cell culture, OHSC, and in vivo (Kohl et al. 2003; Lauro et al. 2010; Drabek et al. 2012). OHSC was used to address the function of microglia in NMDA-induced neuronal loss by depleting resident microglia and then replenishing them with ectopic microglia (Vinet et al. 2012). It was found that neuronal cell loss was prominently increased in the absence of microglia (Vinet et al. 2012). It was also shown that when microglia-free OHSCs were replenished with microglia, these cells invaded the tissue, distributed themselves evenly across the slice, and acquired an in vivo-like, ramified morphology (Vinet et al. 2012). Most importantly, neurons in the presence of these ectopic microglia were protected from NMDA-induced toxicity to the same extent as in non-depleted control slices (Vinet et al. 2012). These findings convincingly show not only that microglia have a neuroprotective capacity, but also that this property applies to ramified microglia (Vinet et al. 2012). Thus, neurons are protected in the vicinity of ramified microglia, while removing microglia from the local environment renders neurons more vulnerable to excitotoxicity.

6.5.2.2 CD11b HSVTK Mouse Lines

Another way to specifically target microglia is through the use of transgenic mouse strains in which the herpes simplex virus thymidine kinase (HSVTK) is placed under the control of the CD11b promoter, which is active in most cells of myeloid origin (Heppner et al. 2005; Gowing et al. 2006). Application of gancyclovir to CD11b-HSVTK animals is mostly toxic to proliferating CD11b+ cells (Heppner et al. 2005; Gowing et al. 2006). The effects of gancyclovir on microglia in vivo are dependent on the application route of the drug in these animals. If peripheral gancyclovir application (intraperitoneal injection or oral application) is used, transplantation of wild-type bone marrow is required to spare the peripheral myeloid compartment from gancyclovir treatment. In these resulting chimeric animals, gancyclovir application leads to the inhibition of morphological microglia transition to amoeboid cells in the case of EAE (referred to as microglia paralysis (Heppner et al. 2005)) or to the death of microglia undergoing proliferation after experimental stroke (Lalancette-Hebert et al. 2007). Whereas the inhibition of morphological microglia transition (microglia paralysis) was protective in EAE (delayed disease onset and reduced clinical scores) (Heppner et al. 2005), the ablation of microglia proliferation in the stroke model led to a larger stroke lesion area and increased neuronal death (Lalancette-Hebert et al. 2007).

More recent studies using this mouse line have changed the application route of gancyclovir from peripheral to central, which specifically depletes the treated brain tissue of ramified microglia without affecting the peripheral myeloid cells (Gowing et al. 2006; Grathwohl et al. 2009; Mirrione et al. 2010; Varvel et al. 2012). In the corresponding studies it was shown that the depletion of microglia by gancyclovir did not affect the development of amyloid-beta (Aβ) plaques in two different mouse models of AD (Grathwohl et al. 2009) nor did the absence of microglia changed disease progression and motor neuron degeneration in the SOD mouse model of ALS (Gowing et al. 2008). However, in the case of pilocarpine-induced seizures, the depletion of microglia prevented the seizure-reducing effect of LPS pre-conditioning, indicating that the inflammatory capacity of microglia was beneficial in this mouse model (Mirrione et al. 2010). Taken together, it can be concluded that gancyclovir-dependent inhibition of microglia function in CD11b-HSVTK animals is only advantageous in one disease model, which is EAE (Heppner et al. 2005). All other reports either provided evidence for a beneficial role of microglial function in vivo (Lalancette-Hebert et al. 2007; Mirrione et al. 2010) or showed no effect of blunting the microglial response (Gowing et al. 2008). It should be noted here that the latter studies inhibited or depleted microglia for a limited time, and at rather late stages of chronic disease models (Gowing et al. 2008; Grathwohl et al. 2009), which may explain the surprising lack of effect. The inhibition or depletion of microglial function may have been too late or too short to unravel the role of these cells in mouse models of AD and ALS (Gowing et al. 2008; Grathwohl et al. 2009). Thus, inhibition or depletion of microglia for longer time periods may be required for chronic disease models.

Taken together, mutating or depleting microglia is rarely correlated to an improved outcome in various brain disease models. These findings together with the discussed reports about the consequences of mutating microglia in the brain clearly argue for a protective role of the innate immune cells of the brain.

6.6 Regulating Microglia Responses

Current theories on microglia-mediated neuronal damage holds that environmental compounds (Taetzsch and Block 2013), neuronal damage (Gao et al. 2003b; Perry and Teeling 2013), aging (Norden and Godbout 2013; Perry and Teeling 2013), and

CNS diseases (Perry and Teeling 2013) may prime microglia to be overly sensitive to stimuli, thus promoting an exaggerated (Ransohoff and Cardona 2010; Xiong and Kielian 2013) and chronic (Ransohoff and Cardona 2010; Gomez-Nicola et al. 2013) response to exert toxicity.

As described above there have recently been tremendous changes in our understanding of microglia "activation", yet different states (or phenotypes) of microglial responses have not been clearly defined so far. This is different from macrophage biology where the response can be classified as an M1 (classical/pro-inflammatory activation), M2a (alternative activation/anti-inflammatory activation), and M2c (deactivation/wound healing activation) response (Ginhoux et al. 2010; Boche et al. 2013; Jang et al. 2013). M1 activation is characterized by the upregulation of inflammatory mediators (ex. TNFα, IL-1β, COX2, and iNOS) and the production of ROS (ex., H_2O_2, peroxynitrite ($ONOO^-$)) (Block et al. 2007). Critical for the regulation of the immune response, the initial M1 response is typically followed by a secondary M2 activation that is important for wound healing and resolving inflammation, which is marked by the expression of factors such as Arginase1 (AR1, decreased iNOS activity), chitinase 3-like 3 (Ym1, tissue remodeling), and Found in Inflammatory Zone 1 (Fzz1, tissue remodeling) (Jung et al. 2000; Ginhoux et al. 2010; Boche et al. 2013). Microglia, being of myeloid origin, can express M1/M2 markers that might provide some insight into their phenotype. However, microglia are highly plastic cells that may rapidly transit between different states (or phenotypes) and it is at the moment controversially debated whether or not the M1/M2 classification also is valid in the microglia field. Here, we will use the terms M1-like and M2-like as descriptions for marker patterns that resemble the markers expressed by M1/M2 macrophages. However, these terms are not used as descriptors of function; the actual impact of the microglial response on neuronal survival will depend on timing, the degree of the response, the chronic nature of the response, and may very well be model/disease-specific.

6.7 The Excessive and Chronic Microglial Response

Microglial immune responses to pathogens, environmental toxins, and neuronal damage join the long list of beneficial responses that microglia perform to promote CNS health in the case of normal physiology (Varnum and Ikezu 2012). However, accumulating evidence indicates that a deleterious and neurotoxic microglial phenotype occurs when microglia activation is dysregulated to become a polarized M1-like phenotype (Hu et al. 2012; Bechade et al. 2013), which is defined by an exaggerated pro-inflammatory response (M1-like) with impaired resolution (M2-like response) (Aguzzi et al. 2013; Jang et al. 2013). Thus, while microglia expressing M2-like markers are associated with repair such as remyelination (Miron et al. 2013) and desensitization to pro-inflammatory stimuli (Ajmone-Cat et al. 2013), microglia with M1-like markers are implicated in chronic neuroinflammation and progressive neuron damage (Hu et al. 2012; Ardeljan and Chan 2013; Blandini 2013). As such, current research is focusing on mechanisms and techniques capable of shifting microglia to a polarized M2-like phenotype for neuroprotection (Zhang et al. 2013).

6.8 Triggers of the Microglial M1-Like Expression Pattern

At present, there is also considerable interest in identifying the circumstances instigating the pathological microglia response.

6.8.1 Immune Stimuli

There is a wealth of evidence that microglia detect and respond to paracrine and autocrine pro-inflammatory signals implicated in M1 polarization, such as TNFα, ROS, ATP, IL-1β, chemokines, etc. (Hanamsagar et al. 2011; Harms et al. 2012). Interestingly, aging (Michaud et al. 2013) and neurodegenerative diseases such as PD (Reale et al. 2009) and AD (Michaud et al. 2013) have been linked with peripheral immune dysregulation, where pro-inflammatory markers are elevated in the blood. For example, pro-inflammatrory cytokines are elevated in PD patient blood (Chen et al. 2008) and upregulated in the circulating white blood cells, both at basal levels and in response to LPS (Reale et al. 2009), indicating that these peripheral immune cells are biologically altered during the process of CNS pathology. Peripheral inflammation is documented to transfer to the brain in adult animals (Qin et al. 2007) and in utero (Carvey et al. 2003) to activate microglia through circulating cytokines. Importantly, the peripheral immune response has been documented to shift the neuroinflammatory phenotype in the brain, modulating and priming CNS macrophages to be more sensitive to additional pro-inflammatory stimuli (Perry 2004; Puntener et al. 2012). In addition to systemic bacterial infection/cytokines, microglia are also able to detect and respond to other forms of peripheral pathology that elevate circulating cytokines, such as kidney damage (Liu et al. 2008) and intestinal reperfusion injury (Hsieh et al. 2011), posing the interesting premise that microglia may survey and be reprogrammed by peripheral disease/damage.

6.8.2 Neuronal Damage

Critical for wound healing, the microglial response to CNS injury or neuron damage (i.e., reactive microgliosis) was initially perceived as only a transient event (Streit et al. 1999). However, current views hold that this response has the potential to be both long-lived and self-propagating (Gao et al. 2003a; Huh et al. 2003; McGeer et al. 2003). In normal physiology, particularly in the case of apoptosis, neuronal death should not promote M1-like polarization in microglia (Minghetti et al. 2005). However, in disease and pathology, reactive microgliosis is often documented being M1-like and has been implicated as a toxic component of many neurodegenerative diseases (see below for examples) (Eikelenboom et al. 2002; Wenk 2003; Sanchez-Moreno et al. 2004).

One key proposed mechanism through which dying neurons are believed to promote the M1-like response in microglia is the loss of neuron–microglia interactions. Thus, removal of inhibition is a discussed component of initiation of the microglial M1-like response in vivo, where neurons routinely provide many of these anti-inflammatory signals for homeostatic maintenance of the microglial phenotype in normal physiology (Ransohoff and Cardona 2010). More specifically, the disruption of anti-inflammatory cell–cell interactions such as CD200-CD200R-mediated (Hoek et al. 2000) in addition to the depletion of inhibitory-soluble ligands released by damaged neurons, such as CX3CL1 or fractalkine (Suzumura 2013), may set the stage for neurotoxic reactive microgliosis (see below for more detailed information). Furthermore, neuron injury signals released by damaged neurons that can either initiate or amplify the M1-like response have also been identified, including, matrix metalloproteinase-3 (MMP-3) (Kim et al. in review), α synuclein (Zhang et al. in press), μ calpain (Levesque et al. 2010), neuromelanin (Wilms et al. 2003; Zecca et al. 2003), ATP/UTP (Domercq et al. 2013), oxidized mSOD1 protein (Appel et al. 2011), glutamate (Domercq et al. 2013), extracellular nucleotides (Domercq et al. 2013), zinc (Kauppinen et al. 2008), and Heat Shock Protein 60 (HSP60) (Zhang et al. 2012). Together, it is both the removal of inhibitory signals essential in neuron-microglia communication and the release of M1 triggers from damaged neurons that interact to culminate in chronic dysregulation of the microglia in response to neuronal injury.

6.8.3 Endogenous Disease Proteins

Several hallmark proteins from neurodegenerative diseases and CNS pathology also directly activate the microglial M1-like response (see further information in Chap. 18). In the case of AD, the Amyloid Hypothesis postulates that Aβ has a causative role in AD pathology, which may occur through toxicity directly to neurons (Yankner 1989; Yankner et al. 1990) and microglia-mediated neuron damage (Combs et al. 2000; Qin et al. 2002). It is well known that microglia migrate to and cluster around senile plaques containing Aβ and neurofibrillary tangles (McGeer et al. 1987; Sasaki et al. 1997), change morphology (Meda et al. 1995), and produce pro-inflammatory factors, such as nitric oxide (Li et al. 1996), superoxide (Qin et al. 2002; Wilkinson and Landreth 2006), and TNFα (Dheen et al. 2004) in response to Aβ.

α Synuclein is another example of a disease protein that is known to directly interact with microglia to elicit an M1-like expression pattern. α Synuclein is a component of Lewy Bodies, a hallmark feature of PD pathology (Goedert et al. 2013; Trinh and Farrer 2013). Levels of the α Synuclein protein are elevated in the midbrain of sporadic PD patients (Chiba-Falek et al. 2006; Shi et al. 2011) and overexpression of wild-type α Synuclein (SNCA) due to genetic multiplication causes early onset, autosomal dominant-familial PD (Sironi et al. 2010), implicating the protein in PD. Elevation of α Synuclein occurs early in PD progression and, as such, has been proposed as a preclinical marker of PD (Chahine and Stern 2011; Shi et al. 2011). Importantly, the increase in microglial markers in the PD SNpc is linked to the degree of α Synuclein deposition (Croisier et al. 2005). Both the α synuclein monomer (Lee et al. 2010) and the aggregated (Zhang et al. 2005, 2007; Lee et al. 2010) forms have been shown to cause the production of pro-inflammatory cytokines (TNFα and IL-1β), ROS, and microglia-mediated neurotoxicity in vitro.

6.8.4 Pesticides

Several pesticides are linked to an increased incidence to PD and a direct interaction with M1-like polarization of microglia. Paraquat (*N,N′*-dimethyl-4,4′-bipyridinium dichloride) is an extensively used herbicide associated with increased PD risk (Hertzman et al. 1990; Liou et al. 1997; Costello et al. 2009; Wang et al. 2011). Paraquat is believed to cross the blood-brain barrier through the neutral amino acid transporter following systemic exposure (Shimizu et al. 2001; Chanyachukul et al. 2004), thereby causing neuroinflammation (Mitra et al. 2011; Mangano et al. 2012) and selective toxicity to DA neurons of the SNpc (McCormack et al. 2002) in vivo (Fei et al. 2008). In vitro data suggest that while paraquat directly damages neurons at high concentrations (Richardson et al. 2005), lower concentrations are thought to be neurotoxic predominantly through microglial activation (Wu et al. 2005). A single exposure to paraquat in mice causes changes in microglia morphology in the SNpc, thus indicating immune perturbation, where subsequent exposures to either the pesticide paraquat (Purisai et al. 2007), the fungicide maneb (Cicchetti et al. 2005), or LPS (Purisai et al. 2007) results in enhancement of the pro-inflammatory response. In vitro studies in primary cultures have shown that microglia treated with paraquat do not initiate the full M1-like response, but rather, they only produce ROS (Bonneh-Barkay et al. 2005; Wu et al. 2005; Miller et al. 2007). As such, inhibition of iNOS (Yadav et al. 2012) and NOX2 (Taetzsch and Block 2013), as well as co-treatment with a superoxide dismutase mimetic (Peng et al. 2009), decreases the microglial response to paraquat. Exposure to the pesticides rotenone, dieldrin, and lindane is also associated with an increased risk for PD (Dhillon et al. 2008; Tanner et al. 2011; Weisskopf et al. 2010; Mao and Liu 2008), where activated microglia have been shown to produce extracellular superoxide (Mao and Liu 2008; Zhou et al. 2012). Lindane, dieldrin, paraquat, and rotenone also activate a partial M1-like microglial response with ROS production (Taetzsch and Block 2013), but the mechanisms of action are poorly understood.

6.8.5 Air Pollution

While a chronic microglial M1-like response is implicated in pathology, recent reports indicate that microglia are capable of detecting exposure to various forms of air pollution, which may arguably be one of the most common environmental sources of microglial M1-like activation (Block et al. 2012). Air pollution is a complex mixture derived from numerous natural and anthropogenic sources, including particulate matter (PM); ozone, carbon sulfur oxides, nitrogen oxides, methane, and other gases, volatile organic compounds (e.g. benzene, toluene, and xylene), and metals (e.g. lead, manganese, vanadium, iron) (Block et al. 2012). Elevated exposure to air pollution is linked to CNS diseases, behaviour deficits, neuroinflammation, and neuropathology in human and animal studies (Block and Calderon-Garciduenas 2009; Lucchini et al. 2012). Imaging of postmortem human brain tissue has even identified the particle components of urban air pollution in the brain parenchyma (Calderon-Garciduenas et al. 2008). Animal studies have shown that exposure to ambient PM (Campbell et al. 2005, 2009), ozone (Santiago-Lopez et al. 2010), diesel exhaust (Gerlofs-Nijland et al. 2010; Levesque et al. 2011a, b), and manganese (Elder et al. 2006; Antonini et al. 2009) results in elevated cytokine expression and oxidative stress in the brain. Importantly, analysis of postmortem human samples links air pollution to a M1-like microglial response in humans (Calderon-Garciduenas et al. 2008) and animal studies (Levesque et al. 2011b; Morgan et al. 2011; Bolton et al. 2012). Current studies are centred on understanding the particular contexts where this microglial response impacts neuron survival.

6.9 Triggers of the Microglial M2-Like Expression Pattern

Since microglia expressing M2-like patterns are currently seen as beneficial elements, it would be desirable to gain more information about the signals that initiate M2-like patterns in microglia. Little is known about the triggers that lead to microglial M2-like expression patterns. In macrophages, the cytokine Interleukin-4 (IL-4) is often used as a stimulus for M2 polarization, and it was shown some years ago that IL-4 induces cultured microglia to promote oligodendrogenesis (Butovsky et al. 2006). A year later it was described from the same group that IL-4-treated microglia express some dendritic cell-like markers (Butovsky et al. 2007), but whether or not IL-4 causes a M2-like microglia pattern was not investigated. Data in favour of this assumption were published by Ponomarev and colleagues. Using knockout animals and bone-marrow chimeras they showed that IL-4 in the CNS but not in the periphery is required to cause a M2-like microglia response, correlated in that case to diminished EAE disease course (Ponomarev et al. 2007). In aged mice it was found that ex vivo microglia are less sensitive to IL-4 treatment compared to microglia derived from young adult mice (Fenn et al. 2012), suggesting that aging reduces the protective properties of this cytokine. In this regard it is interesting to note that aging is associated with an M1-like pro-inflammatory response in microglia (for further details see Chap. 13). Recently, it was found that the microglia cell line BV-2, when treated with IL-4, express some major M2 markers and are found to be neuroprotective in OHSCs subjected to oxygen-glucose deprivation (Girard et al. 2013). It should be noted here, however, that untreated BV2 cells in the same paradigm display more pronounced neuroprotective properties than IL-4 treated cells (Girard et al. 2013) in agreement with earlier findings (Neumann et al. 2006).

Interestingly, various lines of evidence suggest that chronic treatment with low doses of LPS causes an M2-like pattern in brain microglia, which might be the potential mechanism for the known LPS-dependent preconditioning (Chen et al. 2012; Ajmone-Cat et al. 2013). The potential therapeutic value of LPS-dependent preconditioning, however, remains to be established. In conclusion, while microglia express many M2 markers during different functions, it is unlikely that microglia follow a strict macrophage activation pattern typical of peripheral myeloid cells. Significant study is needed to resolve the role of M2-like activation in microglial function and phenotype.

6.10 Implications

Microglia are the myeloid cells of the CNS. They are derived from a specific embryonic myeloid cell population and these cells invade the developing brain very early. They become true brain cells in a sense as they do not leave the brain and are a self-renewable population that is not replaced by peripheral myeloid cells. Microglia are protective elements in the brain, because mutating or deleting them is seldom associated with a beneficial outcome in an acute injury situation. This however might change in chronic disease or the aged brain. The chronic exposure to a variety of stimuli may lead to the development of microglia with a M1-like expression pattern; aging may cause similar processes. These often called "primed" microglia may initiate or participate in neurodegenerative diseases, thus turning the original beneficial phenotype of microglia into a potentially neurotoxic one. It will be a major challenge for future research to understand the molecular mechanisms that trigger the development of neurotoxic microglia.

References

Abo-Ouf H, Hooper AW, White EJ, van Rensburg HJ, Trigatti BL, Igdoura SA (2013) Deletion of tumor necrosis factor-alpha ameliorates neurodegeneration in Sandhoff disease mice. Hum Mol Genet 22:3960–3975

Aguzzi A, Barres BA, Bennett ML (2013) Microglia: scapegoat, saboteur, or something else? Science 339:156–161

Ajami B, Bennett JL, Krieger C, Tetzlaff W, Rossi FM (2007) Local self-renewal can sustain CNS microglia maintenance and function throughout adult life. Nat Neurosci 10:1538–1543

Anrather J, Gallo EF, Kawano T, Orio M, Abe T, Gooden C et al (2011) Purinergic signaling induces cyclooxygenase-1-dependent prostanoid synthesis in microglia: roles in the outcome of excitotoxic brain injury. PLoS One 6:e25916

Ajmone-Cat MA, Mancini M, De Simone R, Cilli P, Minghetti L (2013) Microglial polarization and plasticity: evidence from organotypic hippocampal slice cultures. Glia 61:1698–1711

Antonini JM, Sriram K, Benkovic SA, Roberts JR, Stone S, Chen BT et al (2009) Mild steel welding fume causes manganese accumulation and subtle neuroinflammatory changes but not overt neuronal damage in discrete brain regions of rats after short-term inhalation exposure. Neurotoxicology 30:915–925

Appel SH, Zhao W, Beers DR, Henkel JS (2011) The microglial-motoneuron dialogue in ALS. Acta Myol 30:4–8

Ardeljan D, Chan CC (2013) Aging is not a disease: distinguishing age-related macular degeneration from aging. Prog Retin Eye Res 37:68–89

Bachstetter AD, Rowe RK, Kaneko M, Goulding D, Lifshitz J, Van Eldik LJ (2013) The p38alpha MAPK regulates microglial responsiveness to diffuse traumatic brain injury. J Neurosci 33:6143–6153

Bachstetter AD, Xing B, de Almeida L, Dimayuga ER, Watterson DM, Van Eldik LJ (2011) Microglial p38alpha MAPK is a key regulator of proinflammatory cytokine up-regulation induced by toll-like receptor (TLR) ligands or beta-amyloid (Abeta). J Neuroinflammation 8:79

Bechade C, Cantaut-Belarif Y, Bessis A (2013) Microglial control of neuronal activity. Front Cell Neurosci 7:32

Beers DR, Henkel JS, Xiao Q, Zhao W, Wang J, Yen AA et al (2006) Wild-type microglia extend survival in PU.1 knockout mice with familial amyotrophic lateral sclerosis. Proc Natl Acad Sci U S A 103:16021–16026

Bernstein HG, Steiner J, Bogerts B (2009) Glial cells in schizophrenia: pathophysiological significance and possible consequences for therapy. Expert Rev Neurother 9:1059–1071

Beumer W, Gibney SM, Drexhage RC, Pont-Lezica L, Doorduin J, Klein HC et al (2012) The immune theory of psychiatric diseases: a key role for activated microglia and circulating monocytes. J Leukoc Biol 92:959–975

Biber K, Neumann H, Inoue K, Boddeke HW (2007) Neuronal 'On' and 'Off' signals control microglia. Trends Neurosci 30:596–602

Blandini F (2013) Neural and immune mechanisms in the pathogenesis of Parkinson's disease. J Neuroimmune Pharmacol 8:189–201

Blank T, Prinz M (2013) Microglia as modulators of cognition and neuropsychiatric disorders. Glia 61:62–70

Block ML, Calderon-Garciduenas L (2009) Air pollution: mechanisms of neuroinflammation and CNS disease. Trends Neurosci 32:506–516

Block ML, Elder A, Auten RL, Bilbo SD, Chen H, Chen JC et al (2012) The outdoor air pollution and brain health workshop. Neurotoxicology 33:972–984

Block ML, Zecca L, Hong JS (2007) Microglia-mediated neurotoxicity: uncovering the molecular mechanisms. Nat Rev Neurosci 8:57–69

Boche D, Perry VH, Nicoll JA (2013) Review: activation patterns of microglia and their identification in the human brain. Neuropathol Appl Neurobiol 39:3–18

Boillee S, Yamanaka K, Lobsiger CS, Copeland NG, Jenkins NA, Kassiotis G et al (2006) Onset and progression in inherited ALS determined by motor neurons and microglia. Science 312:1389–1392

Bolton JL, Smith SH, Huff NC, Gilmour MI, Foster WM, Auten RL et al (2012) Prenatal air pollution exposure induces neuroinflammation and predisposes offspring to weight gain in adulthood in a sex-specific manner. FASEB J 26:4743–4754

Bonneh-Barkay D, Reaney SH, Langston WJ, Di Monte DA (2005) Redox cycling of the herbicide paraquat in microglial cultures. Brain Res Mol Brain Res 134:52–56

Broderick C, Hoek RM, Forrester JV, Liversidge J, Sedgwick JD, Dick AD (2002) Constitutive retinal CD200 expression regulates resident microglia and activation state of inflammatory cells during experimental autoimmune uveoretinitis. Am J Pathol 161:1669–1677

Brown GC (2007) Mechanisms of inflammatory neurodegeneration: iNOS and NADPH oxidase. Biochem Soc Trans 35:1119–1121

Butovsky O, Landa G, Kunis G, Ziv Y, Avidan H, Greenberg N, Schwartz A, Smirnov I, Pollack A, Jung S, Schwartz M (2006) Induction and blockage of oligodendrogenesis by differently activated microglia in an animal model of multiple sclerosis. J Clin Invest 116(4):905–915

Butovsky O, Bukshpan S, Kunis G, Jung S, Schwartz M (2007) Microglia can be induced by IFN-gamma or IL-4 to express neural or dendritic-like markers. Mol Cell Neurosci 35(3):490–500

Buiting AM, Van Rooijen N (1994) Liposome mediated depletion of macrophages: an approach for fundamental studies. J Drug Target 2:357–362

Cagnin A, Brooks DJ, Kennedy AM, Gunn RN, Myers R, Turkheimer FE et al (2001) In-vivo measurement of activated microglia in dementia. Lancet 358:461–467

Calderon-Garciduenas L, Solt AC, Henriquez-Roldan C, Torres-Jardon R, Nuse B, Herritt L et al (2008) Long-term air pollution exposure is associated with neuroinflammation, an altered innate immune response, disruption of the blood-brain barrier, ultrafine particulate deposition, and accumulation of amyloid beta-42 and alpha-synuclein in children and young adults. Toxicol Pathol 36:289–310

Campbell A, Araujo JA, Li H, Sioutas C, Kleinman M (2009) Particulate matter induced enhancement of inflammatory markers in the brains of apolipoprotein E knockout mice. J Nanosci Nanotechnol 9:5099–5104

Campbell A, Oldham M, Becaria A, Bondy SC, Meacher D, Sioutas C et al (2005) Particulate matter in polluted air may increase biomarkers of inflammation in mouse brain. Neurotoxicology 26:133–140

Cardona AE, Pioro EP, Sasse ME, Kostenko V, Cardona SM, Dijkstra IM et al (2006) Control of microglial neurotoxicity by the fractalkine receptor. Nat Neurosci 9:917–924

Carvey PM, Chang Q, Lipton JW, Ling Z (2003) Prenatal exposure to the bacteriotoxin lipopolysaccharide leads to long-term losses of dopamine neurons in offspring: a potential, new model of Parkinson's disease. Front Biosci 8:s826–s837

Chahine LM, Stern MB (2011) Diagnostic markers for Parkinson's disease. Curr Opin Neurol 24:309–317

Chanyachukul T, Yoovathaworn K, Thongsaard W, Chongthammakun S, Navasumrit P, Satayavivad J (2004) Attenuation of paraquat-induced motor behavior and neurochemical disturbances by L-valine in vivo. Toxicol Lett 150:259–269

Chen H, O'Reilly EJ, Schwarzschild MA, Ascherio A (2008) Peripheral inflammatory biomarkers and risk of Parkinson's disease. Am J Epidemiol 167:90–95

Chiba-Falek O, Lopez GJ, Nussbaum RL (2006) Levels of alpha-synuclein mRNA in sporadic Parkinson disease patients. Mov Disord 21:1703–1708

Cho IH, Hong J, Suh EC, Kim JH, Lee H, Lee JE et al (2008) Role of microglial IKKbeta in kainic acid-induced hippocampal neuronal cell death. Brain 131:3019–3033

Choi SH, Aid S, Caracciolo L, Minami SS, Niikura T, Matsuoka Y et al (2013) Cyclooxygenase-1 inhibition reduces amyloid pathology and improves memory deficits in a mouse model of Alzheimer's disease. J Neurochem 124:59–68

Cicchetti F, Lapointe N, Roberge-Tremblay A, Saint-Pierre M, Jimenez L, Ficke BW et al (2005) Systemic exposure to paraquat and maneb models early Parkinson's disease in young adult rats. Neurobiol Dis 20:360–371

Cipriani R, Villa P, Chece G, Lauro C, Paladini A, Micotti E et al (2011) CX3CL1 is neuroprotective in permanent focal cerebral ischemia in rodents. J Neurosci 31:16327–16335

Colonna M (2003) TREMs in the immune system and beyond. Nat Rev Immunol 3:445–453

Colton CA (2009) Heterogeneity of microglial activation in the innate immune response in the brain. J Neuroimmune Pharmacol 4:399–418

Combs CK, Johnson DE, Karlo JC, Cannady SB, Landreth GE (2000) Inflammatory mechanisms in Alzheimer's disease: inhibition of beta-amyloid-stimulated proinflammatory responses and neurotoxicity by PPARgamma agonists. J Neurosci 20:558–567

Costello S, Cockburn M, Bronstein J, Zhang X, Ritz B (2009) Parkinson's disease and residential exposure to maneb and paraquat from agricultural applications in the central valley of California. Am J Epidemiol 169:919–926

Croisier E, Moran LB, Dexter DT, Pearce RK, Graeber MB (2005) Microglial inflammation in the parkinsonian substantia nigra: relationship to alpha-synuclein deposition. J Neuroinflammation 2:14

Davalos D, Grutzendler J, Yang G, Kim JV, Zuo Y, Jung S et al (2005) ATP mediates rapid microglial response to local brain injury in vivo. Nat Neurosci 8:752–758

del Rio-Hortega P (1932) Cytology and cellular pathology of the nervous system. Can Med Assoc J 27(5):576

Dheen ST, Jun Y, Yan Z, Tay SS, Ang Ling E (2004) Retinoic acid inhibits expression of TNF-alpha and iNOS in activated rat microglia. Glia 50:21–31

Dhillon AS, Tarbutton GL, Levin JL, Plotkin GM, Lowry LK, Nalbone JT et al (2008) Pesticide/environmental exposures and Parkinson's disease in East Texas. J Agromedicine 13:37–48

Domercq M, Vazquez-Villoldo N, Matute C (2013) Neurotransmitter signaling in the pathophysiology of microglia. Front Cell Neurosci 7:49

Doring A, Yong VW (2011) The good, the bad and the ugly. Macrophages/microglia with a focus on myelin repair. Front Biosci (Schol Ed) 3:846–856

Dorsey ER, Constantinescu R, Thompson JP, Biglan KM, Holloway RG, Kieburtz K et al (2007) Projected number of people with Parkinson disease in the most populous nations, 2005 through 2030. Neurology 68:384–386

Drabek T, Janata A, Jackson EK, End B, Stezoski J, Vagni VA et al (2012) Microglial depletion using intrahippocampal injection of liposome-encapsulated clodronate in prolonged hypothermic cardiac arrest in rats. Resuscitation 83:517–526

Eikelenboom P, Bate C, Van Gool WA, Hoozemans JJ, Rozemuller JM, Veerhuis R et al (2002) Neuroinflammation in Alzheimer's disease and prion disease. Glia 40:232–239

El Khoury J, Toft M, Hickman SE, Means TK, Terada K, Geula C et al (2007) Ccr2 deficiency impairs microglial accumulation and accelerates progression of Alzheimer-like disease. Nat Med 13:432–438

Elder A, Gelein R, Silva V, Feikert T, Opanashuk L, Carter J et al (2006) Translocation of inhaled ultrafine manganese oxide particles to the central nervous system. Environ Health Perspect 114:1172–1178

Faustino JV, Wang X, Johnson CE, Klibanov A, Derugin N, Wendland MF et al (2011) Microglial cells contribute to endogenous brain defenses after acute neonatal focal stroke. J Neurosci 31:12992–13001

Fei Q, McCormack AL, Di Monte DA, Ethell DW (2008) Paraquat neurotoxicity is mediated by a Bak-dependent mechanism. J Biol Chem 283:3357–3364

Fenn AM, Henry CJ, Huang Y, Dugan A, Godbout JP (2012) Lipopolysaccharide-induced interleukin (IL)-4 receptor-α expression and corresponding sensitivity to the M2 promoting effects of IL-4 are impaired in microglia of aged mice. Brain Behav Immun 26(5):766–777

Feng ZH, Wang TG, Li DD, Fung P, Wilson BC, Liu B et al (2002) Cyclooxygenase-2-deficient mice are resistant to 1-methyl-4-phenyl1, 2, 3, 6-tetrahydropyridine-induced damage of dopaminergic neurons in the substantia nigra. Neurosci Lett 329:354–358

Gao HM, Liu B, Zhang W, Hong JS (2003a) Critical role of microglial NADPH oxidase-derived free radicals in the in vitro MPTP model of Parkinson's disease. FASEB J 17:1954–1956

Gao HM, Liu B, Zhang W, Hong JS (2003b) Synergistic dopaminergic neurotoxicity of MPTP and inflammogen lipopolysaccharide: relevance to the etiology of Parkinson's disease. FASEB J 17:1957–1959

Gerhard A, Pavese N, Hotton G, Turkheimer F, Es M, Hammers A et al (2006) In vivo imaging of microglial activation with [11C](R)-PK11195 PET in idiopathic Parkinson's disease. Neurobiol Dis 21:404–412

Gerlofs-Nijland ME, van Berlo D, Cassee FR, Schins RP, Wang K, Campbell A (2010) Effect of prolonged exposure to diesel engine exhaust on proinflammatory markers in different regions of the rat brain. Part Fibre Toxicol 7:12

German DC, Eagar T, Sonsalla PK (2011) Parkinson's disease: a role for the immune system. Curr Mol Pharmacol

Ginhoux F, Greter M, Leboeuf M, Nandi S, See P, Gokhan S et al (2010) Fate mapping analysis reveals that adult microglia derive from primitive macrophages. Science 330:841–845

Girard S, Brough D, Lopez-Castejon G, Giles J, Rothwell NJ, Allan SM (2013) Microglia and macrophages differentially modulate cell death after brain injury caused by oxygen-glucose deprivation in organotypic brain slices. Glia 61(5):813–824

Goedert M, Spillantini MG, Del Tredici K, Braak H (2013) 100 years of Lewy pathology. Nat Rev Neurol 9:13–24

Gomez-Nicola D, Fransen NL, Suzzi S, Perry VH (2013) Regulation of microglial proliferation during chronic neurodegeneration. J Neurosci 33:2481–2493

Gowing G, Philips T, Van Wijmeersch B, Audet JN, Dewil M, Van Den Bosch L et al (2008) Ablation of proliferating microglia does not affect motor neuron degeneration in amyotrophic lateral sclerosis caused by mutant superoxide dismutase. J Neurosci 28:10234–10244

Gowing G, Vallieres L, Julien JP (2006) Mouse model for ablation of proliferating microglia in acute CNS injuries. Glia 53:331–337

Graeber MB (2010) Changing face of microglia. Science 330:783–788

Grathwohl SA, Kalin RE, Bolmont T, Prokop S, Winkelmann G, Kaeser SA et al (2009) Formation and maintenance of Alzheimer's disease beta-amyloid plaques in the absence of microglia. Nat Neurosci 12:1361–1363

Hanamsagar R, Torres V, Kielian T (2011) Inflammasome activation and IL-1beta/IL-18 processing are influenced by distinct pathways in microglia. J Neurochem 119:736–748

Hanisch UK (2013) Functional diversity of microglia—how heterogeneous are they to begin with? Front Cell Neurosci 7:65

Hanisch UK, Kettenmann H (2007) Microglia: active sensor and versatile effector cells in the normal and pathologic brain. Nat Neurosci 10:1387–1394

Harms AS, Lee JK, Nguyen TA, Chang J, Ruhn KM, Trevino I et al (2012) Regulation of microglia effector functions by tumor necrosis factor signaling. Glia 60:189–202

Hayashi Y, Tomimatsu Y, Suzuki H, Yamada J, Wu Z, Yao H et al (2006) The intra-arterial injection of microglia protects hippocampal CA1 neurons against global ischemia-induced functional deficits in rats. Neuroscience 142:87–96

Hellwig S, Heinrich A, Biber K (2013) The brain's best friend: microglial neurotoxicity revisited. Front Cell Neurosci 7:71

Heppner FL, Greter M, Marino D, Falsig J, Raivich G, Hovelmeyer N et al (2005) Experimental autoimmune encephalomyelitis repressed by microglial paralysis. Nat Med 11:146–152

Hertzman C, Wiens M, Bowering D, Snow B, Calne D (1990) Parkinson's disease: a case-control study of occupational and environmental risk factors. Am J Ind Med 17:349–355

Hoek RM, Ruuls SR, Murphy CA, Wright GJ, Goddard R, Zurawski SM et al (2000) Down-regulation of the macrophage lineage through interaction with OX2 (CD200). Science 290:1768–1771

Hsieh YH, McCartney K, Moore TA, Thundyil J, Gelderblom M, Manzanero S et al (2011) Intestinal ischemia-reperfusion injury leads to inflammatory changes in the brain. Shock 36:424–430

Hu X, Li P, Guo Y, Wang H, Leak RK, Chen S et al (2012) Microglia/macrophage polarization dynamics reveal novel mechanism of injury expansion after focal cerebral ischemia. Stroke 43:3063–3070

Hughes V (2012) Microglia: the constant gardeners. Nature 485:570–572

Huh Y, Jung JW, Park C, Ryu JR, Shin CY, Kim WK et al (2003) Microglial activation and tyrosine hydroxylase immunoreactivity in the substantia nigral region following transient focal ischemia in rats. Neurosci Lett 349:63–67

Iadecola C, Anrather J (2011) Stroke research at a crossroad: asking the brain for directions. Nat Neurosci 14:1363–1368

Iannaccone S, Cerami C, Alessio M, Garibotto V, Panzacchi A, Olivieri S et al (2013) In vivo microglia activation in very early dementia with Lewy bodies, comparison with Parkinson's disease. Parkinsonism Relat Disord 19:47–52

Imai F, Suzuki H, Oda J, Ninomiya T, Ono K, Sano H et al (2007) Neuroprotective effect of exogenous microglia in global brain ischemia. J Cereb Blood Flow Metab 27:488–500

Imamura K, Hishikawa N, Sawada M, Nagatsu T, Yoshida M, Hashizume Y (2003) Distribution of major histocompatibility complex class II-positive microglia and cytokine profile of Parkinson's disease brains. Acta Neuropathol 106:518–526

Jang E, Lee S, Kim JH, Seo JW, Lee WH, Mori K et al (2013) Secreted protein lipocalin-2 promotes microglial M1 polarization. FASEB J 27:1176–1190

Jiang C, Cui K, Wang J, He Y (2011) Microglia and cyclooxygenase-2: possible therapeutic targets of progesterone for stroke. Int Immunopharmacol 11:1925–1931

Jung S, Aliberti J, Graemmel P, Sunshine MJ, Kreutzberg GW, Sher A et al (2000) Analysis of fractalkine receptor CX(3)CR1 function by targeted deletion and green fluorescent protein reporter gene insertion. Mol Cell Biol 20:4106–4114

Kaindl AM, Degos V, Peineau S, Gouadon E, Chhor V, Loron G et al (2012) Activation of microglial N-methyl-D-aspartate receptors triggers inflammation and neuronal cell death in the developing and mature brain. Ann Neurol 72:536–549

Kang J, Rivest S (2007) MyD88-deficient bone marrow cells accelerate onset and reduce survival in a mouse model of amyotrophic lateral sclerosis. J Cell Biol 179:1219–1230

Karperien A, Ahammer H, Jelinek HF (2013) Quantitating the subtleties of microglial morphology with fractal analysis. Front Cell Neurosci 7:3

Kauppinen TM, Higashi Y, Suh SW, Escartin C, Nagasawa K, Swanson RA (2008) Zinc triggers microglial activation. J Neurosci 28:5827–5835

Kay GW, Palmer DN (2013) Chronic oral administration of minocycline to sheep with ovine CLN6 neuronal ceroid lipofuscinosis maintains pharmacological concentrations in the brain but does not suppress neuroinflammation or disease progression. J Neuroinflammation 10:97

Kettenmann H (2007) Neuroscience: the brain's garbage men. Nature 446:987–989

Kettenmann H, Hanisch UK, Noda M, Verkhratsky A (2011) Physiology of microglia. Physiol Rev 91:461–553

Kettenmann H, Kirchhoff F, Verkhratsky A (2013) Microglia: new roles for the synaptic stripper. Neuron 77:10–18

Kierdorf K, Erny D, Goldmann T, Sander V, Schulz C, Perdiguero EG et al (2013) Microglia emerge from erythromyeloid precursors via Pu.1- and Irf8-dependent pathways. Nat Neurosci 16:273–280

Kierdorf K, Prinz M (2013) Factors regulating microglia activation. Front Cell Neurosci 7:44

Kim YS, Choi DH, Block ML, Yang L, Lorenz S, Shin DH et al (in review) Matrix metalloproteinase-3 inhibition prevents dopamine neuronal degeneration. Neuroscience

Kish SJ, Shannak K, Rajput A, Deck JH, Hornykiewicz O (1992) Aging produces a specific pattern of striatal dopamine loss: implications for the etiology of idiopathic Parkinson's disease. J Neurochem 58:642–648

Kitamura Y, Yanagisawa D, Inden M, Takata K, Tsuchiya D, Kawasaki T et al (2005) Recovery of focal brain ischemia-induced behavioral dysfunction by intracerebroventricular injection of microglia. J Pharmacol Sci 97:289–293

Kohl A, Dehghani F, Korf HW, Hailer NP (2003) The bisphosphonate clodronate depletes microglial cells in excitotoxically injured organotypic hippocampal slice cultures. Exp Neurol 181:1–11

Kreutzberg GW (1996) Microglia: a sensor for pathological events in the CNS. Trends Neurosci 19:312–318

Lalancette-Hebert M, Gowing G, Simard A, Weng YC, Kriz J (2007) Selective ablation of proliferating microglial cells exacerbates ischemic injury in the brain. J Neurosci 27:2596–2605

Lambertsen KL, Clausen BH, Babcock AA, Gregersen R, Fenger C, Nielsen HH et al (2009) Microglia protect neurons against ischemia by synthesis of tumor necrosis factor. J Neurosci 29:1319–1330

Lauro C, Cipriani R, Catalano M, Trettel F, Chece G, Brusadin V et al (2010) Adenosine A1 receptors and microglial cells mediate CX3CL1-induced protection of hippocampal neurons against Glu-induced death. Neuropsychopharmacology 35:1550–1559
Lee EJ, Woo MS, Moon PG, Baek MC, Choi IY, Kim WK et al (2010) Alpha-synuclein activates microglia by inducing the expressions of matrix metalloproteinases and the subsequent activation of protease-activated receptor-1. J Immunol 185:615–623
Levesque S, Surace MJ, McDonald J, Block ML (2011a) Air pollution & the brain: subchronic diesel exhaust exposure causes neuroinflammation and elevates early markers of neurodegenerative disease. J Neuroinflammation 8:105
Levesque S, Taetzsch T, Lull ME, Kodavanti U, Stadler K, Wagner A et al (2011b) Diesel exhaust activates and primes microglia: air pollution, neuroinflammation, and regulation of dopaminergic neurotoxicity. Environ Health Perspect 119:1149–1155
Levesque S, Wilson B, Gregoria V, Thorpe LB, Dallas S, Polikov VS et al (2010) Reactive microgliosis: extracellular micro-calpain and microglia-mediated dopaminergic neurotoxicity. Brain 133:808–821
Li MSM, Ohnishi K, Ichimori Y (1996) beta-Amyloid protein-dependent nitric oxide production from microglial cells and neurotoxicity. Brain Res 720:93–100
Linnartz B, Wang Y, Neumann H (2010) Microglial immunoreceptor tyrosine-based activation and inhibition motif signaling in neuroinflammation. Int J Alzheimers Dis 2010 pii: 587463. doi: 10.4061/2010/587463
Liou HH, Tsai MC, Chen CJ, Jeng JS, Chang YC, Chen SY et al (1997) Environmental risk factors and Parkinson's disease: a case-control study in Taiwan. Neurology 48:1583–1588
Liu M, Liang Y, Chigurupati S, Lathia JD, Pletnikov M, Sun Z et al (2008) Acute kidney injury leads to inflammation and functional changes in the brain. J Am Soc Nephrol 19:1360–1370
Lobsiger CS, Boillee S, McAlonis-Downes M, Khan AM, Feltri ML, Yamanaka K et al (2009) Schwann cells expressing dismutase active mutant SOD1 unexpectedly slow disease progression in ALS mice. Proc Natl Acad Sci U S A 106:4465–4470
London A, Cohen M, Schwartz M (2013) Microglia and monocyte-derived macrophages: functionally distinct populations that act in concert in CNS plasticity and repair. Front Cell Neurosci 7:34
Lucchini RG, Dorman DC, Elder A, Veronesi B (2012) Neurological impacts from inhalation of pollutants and the nose-brain connection. Neurotoxicology 33:838–841
Madinier A, Bertrand N, Mossiat C, Prigent-Tessier A, Beley A, Marie C et al (2009) Microglial involvement in neuroplastic changes following focal brain ischemia in rats. PLoS One 4:e8101
Mangano EN, Litteljohn D, So R, Nelson E, Peters S, Bethune C et al (2012) Interferon-gamma plays a role in paraquat-induced neurodegeneration involving oxidative and proinflammatory pathways. Neurobiol Aging 33:1411–1426
Mao H, Liu B (2008) Synergistic microglial reactive oxygen species generation induced by pesticides lindane and dieldrin. Neuroreport 19:1317–1320
McCormack AL, Thiruchelvam M, Manning-Bog AB, Thiffault C, Langston JW, Cory-Slechta DA et al (2002) Environmental risk factors and Parkinson's disease: selective degeneration of nigral dopaminergic neurons caused by the herbicide paraquat. Neurobiol Dis 10:119–127
McGeer PL, Itagaki S, Boyes BE, McGeer EG (1988) Reactive microglia are positive for HLA-DR in the substantia nigra of Parkinson's and Alzheimer's disease brains. Neurology 38: 1285–1291
McGeer PL, Itagaki S, Tago H, McGeer EG (1987) Reactive microglia in patients with senile dementia of the Alzheimer type are positive for the histocompatibility glycoprotein HLA-DR. Neurosci Lett 79:195–200
McGeer PL, Schwab C, Parent A, Doudet D (2003) Presence of reactive microglia in monkey substantia nigra years after 1-methyl-4-phenyl-1,2,3,6-tetrahydropyridine administration. Ann Neurol 54:599–604
Meda L, Cassatella MA, Szendrei GI, Otvos L Jr, Baron P, Villalba M et al (1995) Activation of microglial cells by beta-amyloid protein and interferon-gamma. Nature 374:647–650
Michaud M, Balardy L, Moulis G, Gaudin C, Peyrot C, Vellas B et al (2013) Proinflammatory cytokines, aging, and age-related diseases. J Am Med Dir Assoc 14:877–882

Miller RL, Sun GY, Sun AY (2007) Cytotoxicity of paraquat in microglial cells: Involvement of PKCdelta- and ERK1/2-dependent NADPH oxidase. Brain Res 1167:129–139
Minghetti L, Ajmone-Cat MA, De Berardinis MA, De Simone R (2005) Microglial activation in chronic neurodegenerative diseases: roles of apoptotic neurons and chronic stimulation. Brain Res Brain Res Rev 48:251–256
Miron VE, Boyd A, Zhao JW, Yuen TJ, Ruckh JM, Shadrach JL van Wijngaarden P, Wagers AJ, Williams A, Franklin RJ, ffrench-Constant C 2013) M2 microglia and macrophages drive oligodendrocyte differentiation during CNS remyelination. Nat Neurosci 16:1211–1218
Mirrione MM, Konomos DK, Gravanis I, Dewey SL, Aguzzi A, Heppner FL et al (2010) Microglial ablation and lipopolysaccharide preconditioning affects pilocarpine-induced seizures in mice. Neurobiol Dis 39:85–97
Mitra S, Chakrabarti N, Bhattacharyya A (2011) Differential regional expression patterns of alpha-synuclein, TNF-alpha, and IL-1beta; and variable status of dopaminergic neurotoxicity in mouse brain after Paraquat treatment. J Neuroinflammation 8:163
Miyamoto A, Wake H, Moorhouse AJ, Nabekura J (2013) Microglia and synapse interactions: fine tuning neural circuits and candidate molecules. Front Cell Neurosci 7:70
Mizutani M, Pino PA, Saederup N, Charo IF, Ransohoff RM, Cardona AE (2012) The fractalkine receptor but not CCR2 is present on microglia from embryonic development throughout adulthood. J Immunol 188:29–36
Moller T (2010) Neuroinflammation in Huntington's disease. J Neural Transm 117:1001–1008
Morgan TE, Davis DA, Iwata N, Tanner JA, Snyder D, Ning Z et al (2011) Glutamatergic neurons in rodent models respond to nanoscale particulate urban air pollutants in vivo and in vitro. Environ Health Perspect 119:1003–1009
Morganti JM, Nash KR, Grimmig BA, Ranjit S, Small B, Bickford PC et al (2012) The soluble isoform of CX3CL1 is necessary for neuroprotection in a mouse model of Parkinson's disease. J Neurosci 32:14592–14601
Mount MP, Lira A, Grimes D, Smith PD, Faucher S, Slack R et al (2007) Involvement of interferon-gamma in microglial-mediated loss of dopaminergic neurons. J Neurosci 27:3328–3337
Muller N, Schwarz MJ (2007) The immune-mediated alteration of serotonin and glutamate: towards an integrated view of depression. Mol Psychiatry 12:988–1000
Nadeau S, Filali M, Zhang J, Kerr BJ, Rivest S, Soulet D et al (2011) Functional recovery after peripheral nerve injury is dependent on the pro-inflammatory cytokines IL-1beta and TNF: implications for neuropathic pain. J Neurosci 31:12533–12542
Naert G, Rivest S (2011) The role of microglial cell subsets in Alzheimer's disease. Curr Alzheimer Res 8:151–155
Narantuya D, Nagai A, Sheikh AM, Masuda J, Kobayashi S, Yamaguchi S et al (2010) Human microglia transplanted in rat focal ischemia brain induce neuroprotection and behavioral improvement. PLoS One 5:e11746
Neumann H, Takahashi K (2007) Essential role of the microglial triggering receptor expressed on myeloid cells-2 (TREM2) for central nervous tissue immune homeostasis. J Neuroimmunol 184:92–99
Neumann H, Wekerle H (2013) Brain microglia: watchdogs with pedigree. Nat Neurosci 16: 253–255
Neumann J, Gunzer M, Gutzeit HO, Ullrich O, Reymann KG, Dinkel K (2006) Microglia provide neuroprotection after ischemia. FASEB J 20(6):714–716
Nimmerjahn A, Kirchhoff F, Helmchen F (2005) Resting microglial cells are highly dynamic surveillants of brain parenchyma in vivo. Science 308:1314–1318
Norden DM, Godbout JP (2013) Review: microglia of the aged brain: primed to be activated and resistant to regulation. Neuropathol Appl Neurobiol 39:19–34
Orr CF, Rowe DB, Mizuno Y, Mori H, Halliday GM (2005) A possible role for humoral immunity in the pathogenesis of Parkinson's disease. Brain 128:2665–2674
Ouchi Y, Yoshikawa E, Sekine Y, Futatsubashi M, Kanno T, Ogusu T et al (2005) Microglial activation and dopamine terminal loss in early Parkinson's disease. Ann Neurol 57:168–175

Pabon MM, Bachstetter AD, Hudson CE, Gemma C, Bickford PC (2011) CX3CL1 reduces neurotoxicity and microglial activation in a rat model of Parkinson's disease. J Neuroinflammation 8:9
Palazuelos J, Aguado T, Pazos MR, Julien B, Carrasco C, Resel E et al (2009) Microglial CB2 cannabinoid receptors are neuroprotective in Huntington's disease excitotoxicity. Brain 132:3152–3164
Panaro MA, Cianciulli A (2012) Current opinions and perspectives on the role of immune system in the pathogenesis of Parkinson's disease. Curr Pharm Des 18:200–208
Parkhurst CN, Gan WB (2010) Microglia dynamics and function in the CNS. Curr Opin Neurobiol 20:595–600
Peng J, Stevenson FF, Oo ML, Andersen JK (2009) Iron-enhanced paraquat-mediated dopaminergic cell death due to increased oxidative stress as a consequence of microglial activation. Free Radic Biol Med 46:312–320
Perry VH (2004) The influence of systemic inflammation on inflammation in the brain: implications for chronic neurodegenerative disease. Brain Behav Immun 18:407–413
Perry VH, Nicoll JA, Holmes C (2010) Microglia in neurodegenerative disease. Nat Rev Neurol 6:193–201
Perry VH, Teeling J (2013) Microglia and macrophages of the central nervous system: the contribution of microglia priming and systemic inflammation to chronic neurodegeneration. Semin Immunopathol 35:601–612
Phani S, Re DB, Przedborski S (2012) The role of the innate immune system in ALS. Front Pharmacol 3:150
Prinz M, Mildner A (2011) Microglia in the CNS: immigrants from another world. Glia 59:177–187
Ponomarev ED, Maresz K, Tan Y, Dittel BN (2007) CNS-derived interleukin-4 is essential for the regulation of autoimmune inflammation and induces a state of alternative activation in microglial cells. J Neurosci 27(40):10714–10721
Puntener U, Booth SG, Perry VH, Teeling JL (2012) Long-term impact of systemic bacterial infection on the cerebral vasculature and microglia. J Neuroinflammation 9:146
Purisai MG, McCormack AL, Cumine S, Li J, Isla MZ, Di Monte DA (2007) Microglial activation as a priming event leading to paraquat-induced dopaminergic cell degeneration. Neurobiol Dis 25:392–400
Qin L, Liu Y, Cooper C, Liu B, Wilson B, Hong JS (2002) Microglia enhance beta-amyloid peptide-induced toxicity in cortical and mesencephalic neurons by producing reactive oxygen species. J Neurochem 83:973–983
Qin L, Wu X, Block ML, Liu Y, Breese GR, Hong JS et al (2007) Systemic LPS causes chronic neuroinflammation and progressive neurodegeneration. Glia 55:453–462
Raivich G (2005) Like cops on the beat: the active role of resting microglia. Trends Neurosci 28:571–573
Ramsey CP, Tansey MG (2014) A survey from 2012 of evidence for the role of neuroinflammation in neurotoxin animal models of Parkinson's disease and potential molecular targets. Exp Neurol 256:126–132
Ransohoff RM, Cardona AE (2010) The myeloid cells of the central nervous system parenchyma. Nature 468:253–262
Ransohoff RM, Perry VH (2009) Microglial physiology: unique stimuli, specialized responses. Annu Rev Immunol 27:119–145
Ransohoff RM, Prinz M (2013) Editors' preface: microglia–a new era dawns. Glia 61:1–2
Reale M, Iarlori C, Thomas A, Gambi D, Perfetti B, Di Nicola M et al (2009) Peripheral cytokines profile in Parkinson's disease. Brain Behav Immun 23:55–63
Richardson JR, Quan Y, Sherer TB, Greenamyre JT, Miller GW (2005) Paraquat neurotoxicity is distinct from that of MPTP and rotenone. Toxicol Sci 88:193–201
Rogers J, Luber-Narod J, Styren SD, Civin WH (1988) Expression of immune system-associated antigens by cells of the human central nervous system: relationship to the pathology of Alzheimer's disease. Neurobiol Aging 9:339–349

Ros-Bernal F, Hunot S, Herrero MT, Parnadeau S, Corvol JC, Lu L et al (2011) Microglial glucocorticoid receptors play a pivotal role in regulating dopaminergic neurodegeneration in parkinsonism. Proc Natl Acad Sci U S A 108:6632–6637

Sanchez-Moreno C, Dashe JF, Scott T, Thaler D, Folstein MF, Martin A (2004) Decreased levels of plasma vitamin C and increased concentrations of inflammatory and oxidative stress markers after stroke. Stroke 35:163–168

Santiago-Lopez D, Bautista-Martinez JA, Reyes-Hernandez CI, Aguilar-Martinez M, Rivas-Arancibia S (2010) Oxidative stress, progressive damage in the substantia nigra and plasma dopamine oxidation, in rats chronically exposed to ozone. Toxicol Lett 197:193–200

Sasaki A, Yamaguchi H, Ogawa A, Sugihara S, Nakazato Y (1997) Microglial activation in early stages of amyloid beta protein deposition. Acta Neuropathol (Berl) 94:316–322

Shi M, Bradner J, Hancock AM, Chung KA, Quinn JF, Peskind ER et al (2011) Cerebrospinal fluid biomarkers for Parkinson disease diagnosis and progression. Ann Neurol 69:570–580

Shimizu K, Ohtaki K, Matsubara K, Aoyama K, Uezono T, Saito O et al (2001) Carrier-mediated processes in blood–brain barrier penetration and neural uptake of paraquat. Brain Res 906: 135–142

Sierra A, Abiega O, Shahraz A, Neumann H (2013) Janus-faced microglia: beneficial and detrimental consequences of microglial phagocytosis. Front Cell Neurosci 7:6

Simard AR, Rivest S (2006) Neuroprotective properties of the innate immune system and bone marrow stem cells in Alzheimer's disease. Mol Psychiatry 11:327–335

Simard AR, Soulet D, Gowing G, Julien JP, Rivest S (2006) Bone marrow-derived microglia play a critical role in restricting senile plaque formation in Alzheimer's disease. Neuron 49: 489–502

Sironi F, Trotta L, Antonini A, Zini M, Ciccone R, Della Mina E et al (2010) alpha-Synuclein multiplication analysis in Italian familial Parkinson disease. Parkinsonism Relat Disord 16:228–231

Smith JA, Das A, Ray SK, Banik NL (2012) Role of pro-inflammatory cytokines released from microglia in neurodegenerative diseases. Brain Res Bull 87:10–20

Sorce S, Krause KH (2009) NOX enzymes in the central nervous system: from signaling to disease. Antioxid Redox Signal 11:2481–2504

Sriram K, Matheson JM, Benkovic SA, Miller DB, Luster MI, O'Callaghan JP (2006) Deficiency of TNF receptors suppresses microglial activation and alters the susceptibility of brain regions to MPTP-induced neurotoxicity: role of TNF-alpha. FASEB J 20:670–682

Sriram K, Matheson JM, Benkovic SA, Miller DB, Luster MI, O'Callaghan JP (2002) Mice deficient in TNF receptors are protected against dopaminergic neurotoxicity: implications for Parkinson's disease. FASEB J 16:1474–1476

Sriram K, O'Callaghan JP (2007) Divergent roles for tumor necrosis factor-alpha in the brain. J Neuroimmune Pharmacol 2:140–153

Stertz L, Magalhaes PV, Kapczinski F (2013) Is bipolar disorder an inflammatory condition? The relevance of microglial activation. Curr Opin Psychiatry 26:19–26

Streit WJ (2002) Microglia and the response to brain injury. Ernst Schering Res Found Workshop (39):11–24

Streit WJ (2006) Microglial senescence: does the brain's immune system have an expiration date? Trends Neurosci 29:506–510

Streit WJ, Walter SA, Pennell NA (1999) Reactive microgliosis. Prog Neurobiol 57:563–581

Streit WJ, Xue QS (2009) Life and death of microglia. J Neuroimmune Pharmacol 4:371–379

Suzumura A (2013) Neuron-microglia interaction in neuroinflammation. Curr Protein Pept Sci 14:16–20

Tabas I, Glass CK (2013) Anti-inflammatory therapy in chronic disease: challenges and opportunities. Science 339:166–172

Taetzsch T, Block ML (2013) Pesticides, microglial NOX2, and Parkinson's disease. J Biochem Mol Toxicol 27:137–149

Takahashi K, Rochford CD, Neumann H (2005) Clearance of apoptotic neurons without inflammation by microglial triggering receptor expressed on myeloid cells-2. J Exp Med 201:647–657

Tanner CM, Kamel F, Ross GW, Hoppin JA, Goldman SM, Korell M et al (2011) Rotenone, paraquat, and Parkinson's disease. Environ Health Perspect 119:866–872
Teismann P, Vila M, Choi DK, Tieu K, Wu DC, Jackson-Lewis V et al (2003) COX-2 and neurodegeneration in Parkinson's disease. Ann N Y Acad Sci 991:272–277
Tremblay ME, Stevens B, Sierra A, Wake H, Bessis A, Nimmerjahn A (2011) The role of microglia in the healthy brain. J Neurosci 31:16064–16069
Trinh J, Farrer M (2013) Advances in the genetics of Parkinson disease. Nat Rev Neurol 9: 445–454
Turola E, Furlan R, Bianco F, Matteoli M, Verderio C (2012) Microglial microvesicle secretion and intercellular signaling. Front Physiol 3:149
Turrin NP, Rivest S (2006) Molecular and cellular immune mediators of neuroprotection. Mol Neurobiol 34:221–242
van Rossum D, Hanisch UK (2004) Microglia. Metab Brain Dis 19:393–411
Varnum MM, Ikezu T (2012) The classification of microglial activation phenotypes on neurodegeneration and regeneration in Alzheimer's disease brain. Arch Immunol Ther Exp (Warsz) 60:251–266
Varvel NH, Grathwohl SA, Baumann F, Liebig C, Bosch A, Brawek B et al (2012) Microglial repopulation model reveals a robust homeostatic process for replacing CNS myeloid cells. Proc Natl Acad Sci U S A 109:18150–18155
Vinet J, Weering HR, Heinrich A, Kalin RE, Wegner A, Brouwer N et al (2012) Neuroprotective function for ramified microglia in hippocampal excitotoxicity. J Neuroinflammation 9:27
Wang A, Costello S, Cockburn M, Zhang X, Bronstein J, Ritz B (2011) Parkinson's disease risk from ambient exposure to pesticides. Eur J Epidemiol 26:547–555
Weisskopf MG, Knekt P, O'Reilly EJ, Lyytinen J, Reunanen A, Laden F et al (2010) Persistent organochlorine pesticides in serum and risk of Parkinson disease. Neurology 74:1055–1061
Wenk GL (2003) Neuropathologic changes in Alzheimer's disease. J Clin Psychiatry 64(suppl 9):7–10
Wilkinson BL, Landreth GE (2006) The microglial NADPH oxidase complex as a source of oxidative stress in Alzheimer's disease. J Neuroinflammation 3:30
Wilms H, Rosenstiel P, Sievers J, Deuschl G, Zecca L, Lucius R (2003) Activation of microglia by human neuromelanin is NF-kappaB dependent and involves p38 mitogen-activated protein kinase: implications for Parkinson's disease. FASEB J 17:500–502
Wolf Y, Yona S, Kim KW, Jung S (2013) Microglia, seen from the CX3CR1 angle. Front Cell Neurosci 7:26
Wong WT (2013) Microglial aging in the healthy CNS: phenotypes, drivers, and rejuvenation. Front Cell Neurosci 7:22
Woodcock T, Morganti-Kossmann MC (2013) The role of markers of inflammation in traumatic brain injury. Front Neurol 4:18
Wu DC, Teismann P, Tieu K, Vila M, Jackson-Lewis V, Ischiropoulos H et al (2003) NADPH oxidase mediates oxidative stress in the 1-methyl-4-phenyl-1,2,3,6-tetrahydropyridine model of Parkinson's disease. Proc Natl Acad Sci U S A 100:6145–6150
Wu XF, Block ML, Zhang W, Qin L, Wilson B, Zhang WQ et al (2005) The role of microglia in paraquat-induced dopaminergic neurotoxicity. Antioxid Redox Signal 7:654–661
Xiao Q, Zhao W, Beers DR, Yen AA, Xie W, Henkel JS et al (2007) Mutant SOD1(G93A) microglia are more neurotoxic relative to wild-type microglia. J Neurochem 102:2008–2019
Xing B, Bachstetter AD, Van Eldik LJ (2011) Microglial p38alpha MAPK is critical for LPS-induced neuron degeneration, through a mechanism involving TNFalpha. Mol Neurodegener 6:84
Xiong J, Kielian T (2013) Microglia in Juvenile Neuronal Ceroid Lipofuscinosis (JNCL) are primed towards a proinflammatory phenotype. J Neurochem 127:245–258
Yadav S, Gupta SP, Srivastava G, Srivastava PK, Singh MP (2012) Role of secondary mediators in caffeine-mediated neuroprotection in maneb- and paraquat-induced Parkinson's disease phenotype in the mouse. Neurochem Res 37:875–884

Yanagisawa D, Kitamura Y, Takata K, Hide I, Nakata Y, Taniguchi T (2008) Possible involvement of P2X7 receptor activation in microglial neuroprotection against focal cerebral ischemia in rats. Biol Pharm Bull 31:1121–1130

Yankner BA (1989) Amyloid and Alzheimer's disease–cause or effect? Neurobiol Aging 10: 470–471, discussion 477–478

Yankner BA, Duffy LK, Kirschner DA (1990) Neurotrophic and neurotoxic effects of amyloid beta protein: reversal by tachykinin neuropeptides. Science 250:279–282

Ye L, Huang Y, Zhao L, Li Y, Sun L, Zhou Y et al (2013) IL-1beta and TNF-alpha induce neurotoxicity through glutamate production: a potential role for neuronal glutaminase. J Neurochem 125:897–908

Yong VW, Rivest S (2009) Taking advantage of the systemic immune system to cure brain diseases. Neuron 64:55–60

Zecca L, Zucca FA, Wilms H, Sulzer D (2003) Neuromelanin of the substantia nigra: a neuronal black hole with protective and toxic characteristics. Trends Neurosci 26:578–580

Zhang D, Sun L, Zhu H, Wang L, Wu W, Xie J et al (2012) Microglial LOX-1 reacts with extracellular HSP60 to bridge neuroinflammation and neurotoxicity. Neurochem Int 61:1021–1035

Zhang H, Li Y, Yu J, Guo M, Meng J, Liu C et al (2013) Rho kinase inhibitor Fasudil regulates microglia polarization and function. Neuroimmunomodulation 20:313–322

Zhang W, Dallas S, Zhang D, Guo JP, Pang H, Wilson B et al (2007) Microglial PHOX and Mac-1 are essential to the enhanced dopaminergic neurodegeneration elicited by A30P and A53T mutant alpha-synuclein. Glia 55:1178–1188

Zhang W, Wang T, Pei Z, Miller DS, Wu X, Block ML et al (2005) Aggregated alpha-synuclein activates microglia: a process leading to disease progression in Parkinson's disease. FASEB J 19:533–542

Zhang W, Wang T, Pei Z, Miller DS, Wu X, Yang M et al. (in press) a-Synuclein activates microglia: a process leading to disease progression in Parkinson's disease? J Neurosci

Zhang W, Wang T, Qin L, Gao HM, Wilson B, Ali SF et al (2004) Neuroprotective effect of dextromethorphan in the MPTP Parkinson's disease model: role of NADPH oxidase. FASEB J 18:589–591

Zhou H, Zhang F, Chen SH, Zhang D, Wilson B, Hong JS et al (2012) Rotenone activates phagocyte NADPH oxidase by binding to its membrane subunit gp91phox. Free Radic Biol Med 52:303–313

Part II
What Is on Stage
The New Roles

Chapter 7
Developmental Neuronal Elimination

José L. Marín-Teva, Julio Navascués, Ana Sierra, and Michel Mallat

Abstract Microglia, the brain's innate immune cell type, are cells of mesodermal origin that populate the central nervous system (CNS) during early development. Their functions which are best characterized in the developing CNS are related to programmed cell death (PCD), a physiological process that massively affects neural cell lineages and contributes to brain morphogenesis and neuronal network maturation. Although relatively scarce before advanced developmental stages, microglia can remove dead cells in an effective manner due to their migratory and phagocytic behavior. Recent studies indicate that microglia do not only scavenge cell corpses, but also eliminate neural progenitors cells and trigger or induce PCD in different types of developing neurons. Conversely, microglia were also found to promote the neuronal survival by their release of trophic factors. In this chapter we shall discuss the functional involvement of microglia in the loss of neural cells during normal development and review the mechanisms and cell signalling that underlie microglial regulation of PCD and elimination of dead cells.

Keywords Microglia • Neural progenitors • Apoptotic neural cells • Programmed cell death • Survival • Physiology • Development • Phagocytosis • Trophic factors

Bullet Points

- Microglial cells remove apoptotic neural cells by phagocytosis in different central nervous system (CNS) regions from early to advanced development stages.
- The contribution of microglia to the elimination of apoptotic neural cells varies depending on the species, developmental stage, and CNS region.

J.L. Marín-Teva, Ph.D. (✉) • J. Navascués, Ph.D. • A. Sierra
Facultad de Ciencias, Departamento de Biología Celular, Universidad de Granada,
Avda. Fuentenueva s/n, Granada E-18071, Spain
e-mail: jlmarin@ugr.es; navascue@ugr.es; asierra@ugr.es

M. Mallat, M.D. Ph.D.
Centre de Recherche de l'Institut du Cerveau et de la Moelle Epinière,
UPMC/InsermUMR_S1127, CNRS UMR 7225, Hôpital de la Salpêtrière,
47 boulevard de l'Hôpital, Paris 75013, France
e-mail: michel.mallat@upmc.fr

M.-È. Tremblay, A. Sierra (eds.), *Microglia in Health and Disease*,
DOI 10.1007/978-1-4939-1429-6_7

- Secreted "find-me" and exposed "eat-me" signals comprise the central elements of apoptotic cell removal by microglial cells.
- Microglia not only scavenge cell corpses, but also eliminate neural progenitors and trigger or induce programmed cell death in different types of developing neurons.
- Microglia can also promote the survival of some neurons during development through the release of trophic factors.

7.1 Introduction

Microglia are the resident macrophages of the central nervous system (CNS). Since the first studies by del Rio-Hortega, it has generally been accepted that ramified microglia are differentiated cells in the CNS that arise from precursor cells of mesodermal origin with ameboid morphology, which are called ameboid microglia and are present in the developing CNS (Cuadros et al. 1992; del Rio-Hortega 1932; Cuadros and Navascués 1998; Chan et al. 2007; Prinz and Mildner 2011; Prinz et al. 2011). It is now established that most if not all microglia originate from yolk sac-derived primitive myeloid progenitors that seed the CNS during early developmental stages (Alliot et al. 1999; Herbomel et al. 2001; Chan et al. 2007; Ginhoux et al. 2010, 2013; Mizutani et al. 2012; Prinz et al. 2011; Schulz et al. 2012; Kierdorf et al. 2013). During development, microglia infiltrate all CNS regions through extensive cell migration and proliferation. They progressively differentiate into mature ramified cells and become spread throughout the adult CNS. Ramified microglia appear to continuously scan the extracellular environment by extending and retracting their cell processes. Contact inhibition between microglial cells leads to a characteristic spatial arrangement in which microglial cells occupy non-overlapping territories (Davalos et al. 2005; Nimmerjahn et al. 2005). Microglia are highly plastic cells that strongly react to any lesion or pathologies affecting the CNS, and they play key roles in neuroinflammatory reactions (Hanisch and Kettenmann 2007; Saijo and Glass 2011).

Since the discovery of microglial cells, their biology has been widely considered with relation to programmed cell death (PCD) taking place in the CNS during normal development. Histological studies have addressed temporal and spatial relationships between the spread of microglial cells and PCD in the CNS of developing vertebrates (Perry et al. 1985; Ferrer et al. 1990; Ashwell 1990, 1991; Perry and Gordon 1991; Marín-Teva et al. 1999; Peri and Nusslein-Volhard 2008; Calderó et al. 2009; Rigato et al. 2011). Microglial phagocytosis of dying cells has been described in various regions of the developing CNS (Ferrer et al. 1990; Marín-Teva et al. 1999; Dalmau et al. 2003; Peri and Nusslein-Volhard 2008). Besides their debris clearance function, microglial cells actively contribute to cell–cell interactions that trigger or ensure the execution of PCD (Mallat et al. 2005; Bessis et al. 2007; Marín-Teva et al. 2011). Lately, microglial cells were also shown to promote neuronal survival in the developing CNS (Ueno et al. 2013).

In this chapter, we shall review current knowledge on the role of microglial cells in neuronal elimination/survival during normal development.

7.2 Programmed Cell Death of Neurons in the Developing CNS: Where, When, and Which

The large-scale loss of neural cells occurs as a normal and essential stage in CNS maturation during the embryonic and postnatal development of many vertebrate species (Oppenheim 1991; Kuan et al. 2000; Yuan and Yankner 2000; Buss et al. 2006). Cell death affects neuronal and glial cell lineages at different stages of development. The loss of neural cells contributes to the morphological sculpting of the developing CNS and is one of the regressive events involved in the remodelling and functional adaptation of neuronal networks (Oppenheim 1991; Roth and D'Sa 2001). Although a variety of forms of cell death have been reported depending on morphological and biochemical criteria (Edinger and Thompson 2004; Lockshin and Zakeri 2002), this physiological PCD appears to occur by two major pathways: apoptosis, recognized by cell rounding, DNA fragmentation, externalization of phosphatidyl serine (PS), caspase activation, and the absence of inflammatory reaction; and autophagy, characterized by the presence of large vacuoles and the fact that cells can be rescued from death until very late in the process. The boundary between apoptosis and autophagy is not sharply defined, and a complex interplay between the two forms of cell death has been described (reviewed in Booth et al. 2013). However, apoptosis is the dominant type of PCD during normal CNS development (Clarke 1990). The major effect of PCD on projecting neurons during development is well-documented. Thus, during the period in which neuronal connections are established, up to 50 % of numerous types of differentiated neurons undergo cell death (Oppenheim 1991; Raff et al. 1993). This neuronal death is thought to be in part triggered by the shortage of neurotrophic factors (neural growth factor, NGF; brain-derived neurotrophic factor, BDNF; etc.) released by the target cells innervated by these neurons (Barde 1989; Oppenheim 1989; Raff et al. 1993; Snider 1994). Even though less documented, PCD is also observed earlier in development, before synaptogenesis, in populations of undifferentiated cells such as the proliferating neuroepithelial cells and newly postmitotic neuroblasts, where it participates to specific functions (e.g., morphogenic sculpting of the early CNS) and involves various regulatory mechanisms (Kuan et al. 2000; de la Rosa and de Pablo 2000; Yeo and Gautier 2004; Valenciano et al. 2009). It is generally considered that the PCD of neural cells during development requires relatively conserved molecular pathways. These include "proapoptotic" genes of the Bcl-2 family, the apoptosome (cytochrome c, Apaf-1, caspase-9) and downstream caspases (e.g., caspase-3), which lead to the formation of apoptotic bodies that are rapidly phagocytosed and digested by different types of "professional" and "non-professional" phagocytes. Some studies on the in vivo role of caspases in the normal PCD of developing neurons have shown that caspase activation is involved in most cases of neuronal PCD, but is only necessary for the PCD of immature neurons or neuronal precursors at early developmental stages, when neurogenesis is ongoing (Kuan et al. 2000; Oppenheim et al. 2008). Genetic deletion or pharmacological inhibition of caspases was shown to prevent this early type of PCD in the CNS, while being ineffective at preventing the normal PCD of postmitotic neurons at later developmental stages during the establishment of synaptic connections (Oppenheim et al. 2001, 2008; Yaginuma et al. 2001; Boya et al. 2008). In the absence of caspases, these postmitotic neurons undergo quantitatively normal amounts of PCD by a different caspase-independent pathway that exhibits signs of autophagy (Oppenheim et al. 2008). Recent studies have also reported that the authophagic machinery provides the energy required for proper cell corpse removal and further degradation of apoptotic cells during the neurogenesis period (Mellén et al. 2008; Boya et al. 2008; Aburto et al. 2012).

7.3 Scavenger Role of Microglia in Different CNS Regions from Early to Advanced Development

Cell apoptosis is marked by DNA fragmentation and caspase-triggered cleavage of the cellular proteome (Nicholson 1999), which lead to cell shrinkage but spare the integrity of the plasma membrane up to an advanced stage of the death process. Swift elimination of apoptotic cells by tissue phagocytes is important to prevent secondary cell necrosis involving plasma membrane disruption and leakage of intracellular compounds in the extracellular space, which may be responsible for inflammatory reactions or autoimmune diseases (Nagata et al. 2010). Microglial cells remove apoptotic neural cells by phagocytosis in different CNS regions from early (Sorokin et al. 1992; Cuadros et al. 1991, 1993; Herbomel et al. 1999, 2001; Lichanska and Hume 2000) to more advanced development stages (Hume et al. 1983; Ashwell et al. 1989; Ashwell 1990, 1991; Pearson et al. 1993; Perry et al. 1985; Ferrer et al. 1990; Perry and Gordon 1991; Thanos 1991; Moujahid et al. 1996; Egensperger et al. 1996; Marín-Teva et al. 1999; Upender and Naegele 1999; Rakic and Zecevic 2000; Dalmau et al. 2003; Peri and Nusslein-Volhard 2008; Calderó et al. 2009). In the adult mouse CNS, phagocytic microglia ensure the elimination of dying neuroblasts that derive from neural stem cells in the hippocampal neurogenic niche (Sierra et al. 2010) (see Chap. 10).

The contribution of microglia to the elimination of apoptotic neural cells appears to vary depending on the species, developmental stage, and CNS region studied (Dalmau et al. 2003; Calderó et al. 2009; Sierra et al. 2013). For example, in the cerebral cortex, subcortical white matter, and hippocampus of the in vivo perinatal rat brain, microglial cells engulf virtually all cells undergoing PCD (Dalmau et al. 2003). However, low microglial cell density may be limiting for scavenging activity, especially at the earliest stages of CNS development, when microglial cells remain scarce. During early development of the CNS, in which the phagocytic capacities of microglia appear to be overwhelmed by the number of dying cells, other non-professional phagocytes, such as neighboring neuroepithelial cells, neuroblasts, retinal Müller cells, cerebellar Bergmann glia, and spinal cord astrocytes also contribute to the elimination of apoptotic cells or bodies (O'Connor and Wyttenbach 1974; García-Porrero and Ojeda 1979; Kálmán 1989; Cuadros et al. 1991; Egensperger et al. 1996; Marín-Teva et al. 1999; Parnaik et al. 2000; Mellén et al. 2008).

7.4 How Do Microglia Find and Selectively Engulf Dead Cells in the Developing CNS?

The molecular signalling that triggers microglial phagocytosis during normal development remains a key question to be solved, but this issue clearly benefits from the recent discovery of signals that are generated by apoptotic cells and sensed by phagocytes.

Secreted "find-me," exposed "eat-me," and disabled "don't-eat-me" signals comprise the central elements of apoptotic cell removal by professional phagocytes (Savill and Fadok 2000; Lauber et al. 2004; Ravichandran 2010). Soluble forms of lysophosphatidylcholine (LPC), as well as the chemokine and adhesion molecule CX3CL1 (also known as fractalkine) and the nucleotides ATP and UTP, are known to act as "find-me" signals that are released by apoptotic cells and attract phagocytes (Lauber et al. 2004; Ravichandran 2010). LPC has an attractive chemotactic effect on human monocytic cells and macrophages (Lauber et al. 2003). Moreover, it is released from apoptotic cells through the caspase-3-mediated activation of calcium-independent phospholipase A_2 and is recognized by the phagocyte G-protein-coupled receptor G2A (Lauber et al. 2003; Peter et al. 2008). On the other hand, fractalkine was found to be released before the loss of plasma membrane integrity by apoptotic lymphocytes via a caspase-regulated mechanism and to have an attractive effect on macrophages expressing the fractalkine receptor CX3CR1, as inferred from in vitro and in vivo studies carried out in humans and mice (Truman et al. 2008). Microglial cells are known to be the only CNS cells that express CX3CR1 (Jung et al. 2000). Interestingly, fractalkine released from cultured neurons damaged by glutamatergic excitotoxicity promotes the phagocytosis of cell debris by microglial cells (Noda et al. 2011). However, the actual role of G2A or CX3CR1 in the elimination of neural cell corpses during development has not yet been established. Finally, it was reported that the caspase-dependent release of ATP and UTP during the early stages of apoptosis in thymocytes acts as a "find-me" signal to promote phagocytic clearance by human monocytes (Elliott et al. 2009). Nucleotides could possibly act as "find-me" signals in the normal developing CNS, but this hypothesis needs to be experimentally tested. In particular, extracellular ATP/ADP has chemotactic effects on microglial cells by binding to $P2Y_{12}$ purinergic receptors (Davalos et al. 2005; Nimmerjahn et al. 2005; Haynes et al. 2006; Ohsawa and Kohsaka 2011). Costimulation of microglial P2Y receptors and adenosine (A1 type) receptors appears to be required for microglial chemotaxis towards ATP. This cell response is prevented in purified microglia, obtained by shaking mixed glial cell cultures, which are derived from the brain of newborn mice deficient in the expression of CD39, an ectonucleotidase that degrades nucleotides to nucleosides (Farber et al. 2008).

Once recruited in the vicinity of apoptotic cells, phagocytes can selectively recognize the dead cells as prey to be engulfed due to their expression of "eat-me" signals. In mammals, the best-characterized "eat-me" signal is PS, a cell membrane component that translocates from the inner to outer leaflet of the plasma membrane during the apoptotic process (Fadok et al. 2000). Externalized PS can stimulate phagocytosis of apoptotic cells through direct binding to phagocyte receptors such as T cell immunoglobulin- and mucin-domain containing molecule 4 (Tim-4) (Miyanishi et al. 2007), brain-specific angiogenesis inhibitor-1 (BAI1) (Park et al. 2007) or stabilin-2 (Park et al. 2008). PS recognition by phagocytes can also involve several "bridging" molecules, binding to both PS and phagocyte receptors, which are released in extracellular fluid (Erwig and Henson 2008). Among these, milk fat globule epidermal growth factor-8 (MFG-E8) and C3bi can bind to PS and are then recognized by integrins expressed by macrophages/microglia such as vitronectin receptor (αvβ3) and complement-receptor-3 (CR3/CD11b), respectively (Hanayama et al. 2002; Mevorach et al. 1998; Savill and Fadok 2000). The bridging molecules and receptors responsible for PS-mediated clearance of dead cells (see Sierra et al. 2013 for a review) are not yet clearly defined in the developing CNS, but the capacity of purified microglial cells to engulf PS-coated cells or particles is well-documented (Witting et al. 2000; Konduru et al. 2009; Neher et al. 2011; Liu et al. 2013).

Another possible "eat-me" signal is the endogenous cellular ligand for the triggering receptor expressed on myeloid cells 2 (TREM2). TREM2 is specifically expressed by microglial cells, as demonstrated by in vivo immunocytochemical studies in different regions of the mouse brain (Hsieh et al. 2009). Its ligand is up-regulated on apoptotic neurons, mediates signal transduction by association with DNAX adaptor protein-12 (DAP12) on microglia, and promotes the phagocytosis of dying neurons in cell cultures (Takahashi et al. 2005; Hsieh et al. 2009; Neumann et al. 2009). Finally, another in vivo and in vitro study demonstrated that activation of $P2Y_6$ receptors in microglial cells by UTP/UDP released from damaged neurons triggers the clearance of dying cells by phagocytosis (Koizumi et al. 2007). Hence, in addition to PS, the ligands for TREM2 and UTP/UDP may be considered as putative "eat-me" signals promoting microglial phagocytosis during development.

7.5 PCD and Microglial Cell Distribution in the Developing CNS

Besides its role in the elimination of dead cells, apoptotic cell-to-phagocyte signalling clearly fulfills the expected criteria for a mechanism that underlies microglia distribution in the developing CNS. Consistent with the recent demonstration that dying cells release chemotactic "find-me" signals, the codetection of microglia and dying cells provided evidence that local PCD may account for the entry, local spread, or transient clustering of macrophages/microglia in different CNS regions of embryonic vertebrates, including the retina (Santos et al. 2008), spinal cord (Rigato et al. 2011), choroid plexus (Swinnen et al. 2013), cerebellum, or cerebral cortex (Ashwell 1990, 1991). However, microglia infiltration in cell layers of the developing CNS is not always correlated with the occurrence of PCD (Marín-Teva et al. 1999), and it involves different molecular cues unrelated to the signalling generated by dying cells. In particular, developing neural cells can produce chemoattractants that target macrophages/microglia, such as chemokines (Ransohoff 2009), macrophage colony-stimulating factor 1 (M-CSF), or vascular endothelial growth factor A (VEGF-A) (Breier et al. 1992; Nandi et al. 2012; Lelli et al. 2013). In zebra fish embryos, M-CSF receptor expressed by mononuclear phagocytes is required for the early infiltration of yolk-sac-derived phagocytes in the CNS (Herbomel et al. 2001), whereas VEGF receptor 1 signalling promotes early postnatal infiltration of microglia in deep layers of the mouse cerebral cortex (Lelli et al. 2013). Furthermore, microglia appear to use radially oriented processes of glial or neural progenitors cells as a substrate for cell migration in the retina (Sánchez-López et al. 2004), spinal cord (Rigato et al. 2011), and cerebral cortex (Swinnen et al. 2013). Strikingly, ameboid microglial cells migrate in embryonic quail retina following well-defined routes that are not altered when they pass close to regions of abundant cell death, and they only engulf apoptotic bodies encountered along these routes (Marín-Teva et al. 1999). This observation suggests that cell guiding cues unrelated to PCD may outcompete "find-me" signals emitted by dead cells. Alternatively, the capacity of microglial cells to sense "find-me" (and/or "eat-me") signals may vary according to their localization in the developing CNS. In connection with this possibility, the heterogeneity of microglial cell phenotypes during normal development is well-documented (Hristova et al. 2010; Verney et al. 2010; Scheffel et al. 2012; Arnoux et al. 2013). Moreover, recent studies have emphasized that the functional implications of microglial recruitment are not limited to the scavenging of dead cells. Microglia fulfill other roles during development, including the remodelling of developing synapses (Hoshiko et al. 2012; Tremblay et al. 2010; Paolicelli et al. 2011; Schafer et al. 2012), homeostatic regulation of neuronal firing (Ji et al. 2013) (Chap. 9), stimulation of angiogenesis (Checchin et al. 2006; Fantin et al. 2010) (Chap. 8), and the induction or prevention of developmental cell death, which is discussed below.

7.6 Switching Programmed Cell Death On and Off in the Developing CNS: A Dual Role for Microglia

Various studies have demonstrated that microglial cells not only have a scavenger role during development, but can also trigger or promote PCD in different types of developing neurons. The mechanisms involved in the induction of PCD by microglial cells are diverse and can be classified into two groups: phagocytosis-unrelated PCD induction; and PCD triggering/execution involving phagocytosis, designated as "engulfment-promoted PCD."

The former group relies in part on the capacity of microglia to produce neuronal growth factors such as neurotrophins (Mallat et al. 1989; Elkabes et al. 1996). Frade and Barde (1998) showed that nerve growth factor (NGF) is produced in vivo during early chick embryo development by primitive macrophages and microglial cells, which are present in the vitreous body and scattered within the retinal neuroepithelium between embryonic day 3 (E3) and E6. In addition, in vitro experiments in cultured eye cups demonstrated that macrophage/microglial production of NGF triggers PCD in retinal progenitor cells expressing neurotrophin receptor p75 at E3 (Fig. 7.1a). Another type of phagocytosis-independent involvement of macrophages in PCD was observed in the embryonic rat spinal cord, in which motoneurons undergo PCD from E15. Experiments in cultured explants of ventral horns of rat embryo spinal cord showed that embryonic motoneurons acquire the competence to undergo PCD between E12 and E13, 2 days before the onset of cell death (Sedel et al. 2004). This neuronal commitment to a death fate appears to be driven by tumor necrosis factor α (TNF-α) released from primitive macrophages that invade the surrounding somitic mesenchyme at these developmental stages (Fig. 7.1b). Macrophage-derived TNF-α signals through the TNF-α receptor 1 expressed by embryonic motoneurons (Sedel et al. 2004). Subsequent in vivo studies provided evidence that these somitic macrophages eventually infiltrate the spinal cord and therefore contribute to microglial development (Rigato et al. 2011).

Engulfment-promoted PCD was observed in postmitotic neurons in the cerebellum of 3 day-old (P3) mice (Marín-Teva et al. 2004); at this developmental stage, Purkinje cells initiate a death program marked by caspase-3 activation. In vivo observations in the P3 mouse cerebellum and in vitro experimental studies in P3 cerebellar slices showed that the majority of these neurons are engulfed by microglial cells that actively prevent abortion of the death program in the engulfed cells (Marín-Teva et al. 2004). In the developing mouse CNS, microglia express all of the genes encoding multimeric Nox1- and Nox2-dependent NADPH oxidases that catalyze the formation of superoxide ions (Bedard and Krause 2007; Chéret et al. 2008). Activation of microglial NADPH oxidases is detectable at early postnatal stages and is required for microglia-promoted Purkinje cell death (Fig. 7.1c_1) (Marín-Teva et al. 2004; Lelli et al. 2013). This mechanism is reminiscent of the innate immune response in which bacteria or fungi ingested by neutrophils or macrophages are neutralized by reactive oxygen species derived from the activation of phagocyte Nox2-dependent NADPH-oxidase (Bedard and Krause 2007; Lam et al. 2010). It is noteworthy that the involvement of engulfing cells in the execution of developmental cell death is not limited to vertebrates (Mallat et al. 2005). It is well documented

Fig. 7.1 (continued) MiCs promote neuron death by the engulfment of differentiated neurons (DNs) undergoing early reversible steps of cell death (**c_1**) and healthy neural progenitor cells (**c_2**). (**c_1**). In the developing cerebellum (Marín-Teva et al. 2004) and hippocampus (Wakselman et al. 2008), DNs in early stages of cell death express activated caspase-3 and show putative "eat-me" signals on their surface. These signals are recognized by the β2-integrin CD11b and the immunoreceptor DAP12 on the MiC surface, thereby triggering neuron engulfment and the progression of late stages of cell death, which are promoted by microglial-derived apoptotic effectors such as superoxide ions ($O_2^{\bullet-}$). (**c_2**). MiCs colonize proliferative zones in the developing cerebral cortex and phagocytose postmitotic neural progenitors (NPs) that show no sign of cell death, thereby contributing to regulate the size of the NP pool (Cunningham et al. 2013). (**d**) In the developing cerebral cortex, MiCs recruited in the subcortical white matter limit the extent of PCD in DNs of the cortical layer V through production of insulin-like growth factor 1 (IGF-1) (Ueno et al. 2013)

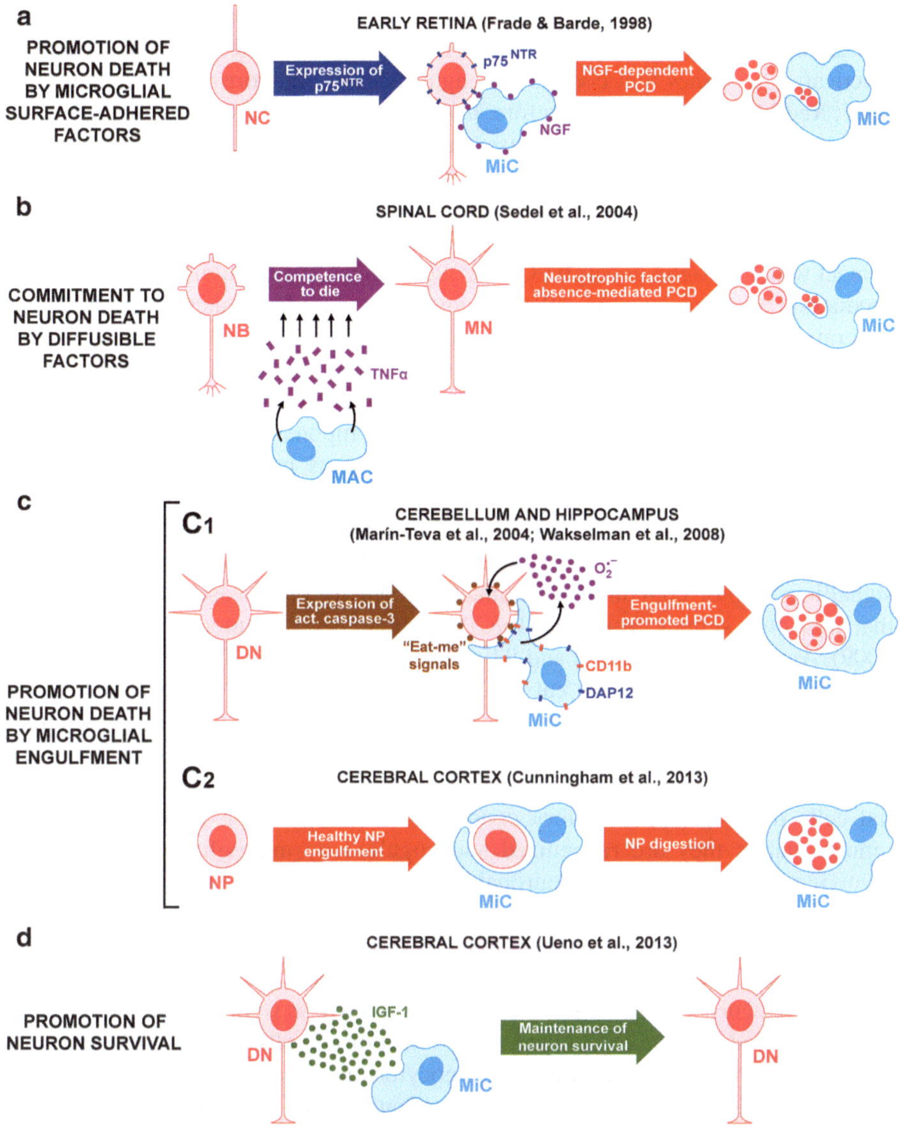

Fig. 7.1 Dual role of microglial cells in promoting either the death or survival of neurons and neural progenitor cells, as demonstrated by studies performed in different parts of the developing central nervous system (CNS). (**a**) In the early retina, ameboid microglial cells (MiCs) are involved in the death of some neuroepithelial cells (NCs) expressing the p75 neurotrophin receptor ($p75^{NTR}$) by producing nerve growth factor (NGF). The NGF secreted by MiCs remains bound to the cell surface and promotes NC death after $p75^{NTR}$ stimulation (Frade and Barde 1998). (**b**) Primitive macrophages (MACs) have a role in committing differentiating motoneurons (MNs) in the developing spinal cord to death (Sedel et al. 2004). These MNs acquire competence to die through signalling via the tumor necrosis factor α (TNF-α) released from MACs in adjacent somites. Two days after TNF-α signalling, committed MNs undergo neurotrophic PCD, and cell debris is phagocytosed by ameboid MiCs invading the spinal cord. (**c**) In other developing CNS regions,

in the nematode *Caenorhabditis elegans*, in which "death-committed" cells expressing activated CED-3 (homologue to vertebrate caspase-3) can be rescued by inactivation of genes that control the engulfment of cell corpses (Hoeppner et al. 2001; Reddien et al. 2001). Therefore, it was suggested that live Purkinje cells committed to a death fate readily express "eat me" signals, which stimulate engulfing behavior and superoxide production in microglia (Marín-Teva et al. 2004).

Further insights into this issue came from an in vivo study in newborn mice, which showed that, similar to developing Purkinje cells, hippocampal neurons in contact with microglia undergo PCD through a mechanism requiring the microglial production of superoxide ions (Wakselman et al. 2008). Both microglial superoxide generation and neuronal death were found to be reduced in newborn mice bearing a mutation preventing expression of the CD11b integrin subunit or activity of the DAP12 signalling protein. CD11b/CD18 (αMβ2) functions as an engulfment receptor of prey tagged with the C3bi complement component (Bohana-Kashtan et al. 2004). The involvement of complement components in developmental neuronal death has not yet been established, whereas a recent investigation shows that C3bi-CD11b signalling contributes to microglia-mediated synapse elimination during normal development (Schafer et al. 2012). DAP12 is a transmembrane-anchored signalling adaptor containing an immunoreceptor tyrosine-based activation motif that transmits signals from CD11b/CD18 and other immunoreceptors (Ivashkiv 2009). Among these, TREM2 and β3 integrins are expressed by microglia and can directly or indirectly bind to "eat-me" signals, including PS. Whether these receptors contribute to the DAP12-dependent loss of hippocampal neurons (Wakselman et al. 2008) has not been addressed. However, recent in vitro studies show that a transient PS externalization can occur in neurons cultured in the presence of microglial cells under proinflammatory conditions. Although these PS-tagged neurons remain viable, they are recognized and engulfed by microglia via mechanisms involving the cross-linking of neuronal PS and microglial αvβ3 integrin (vitronectin receptor) by the MFG-E8 bridging protein, and in fact, neuronal death is executed by microglial phagocytosis (Neher et al. 2011; Fricker et al. 2012).

In the developing CNS, the engulfment-mediated loss of cells is not restricted to postmitotic neurons. Thus, a study carried out in vivo, as well as in in vitro cultured slices, revealed that microglia limit the production of cells in the developing cerebral cortex of macaques and rats by phagocytosing neural progenitor cells, mostly during late stages of cortical neurogenesis (Cunningham et al. 2013). Interestingly, this study found that most neural progenitor cells contacted or engulfed by microglial cells in the cortical proliferative zones show no signs of cell death or apoptosis as defined by the expression of cleaved caspase-3, TUNEL-labelling of fragmented DNA, PS exposure, or nuclear breakdown. The mechanisms by which microglia recognize these healthy progenitor cells as a prey to be engulfed have not been determined. Nevertheless, these findings reveal a new type of engulfment-promoted cell elimination during development, with microglial cells eliminating viable neural progenitor cells through a process unrelated to apoptosis (Fig. 7.1c_2).

Besides engulfing behavior and the generation of prodeath signals, microglia can produce various neuronal growth factors that may directly support the survival of neurons or the growth of their processes. These functional capacities were documented in culture studies more than 20 years ago (Mallat and Chamak 1994), and their relevance to normal brain development was recently demonstrated in an in vivo and in vitro study of the mouse cerebral cortex (Fig. 7.1d) (Ueno et al. 2013). It was observed that, during early postnatal stages (P3–P7), microglial cells recruited in the subcortical white matter limit the extent of PCD in corticospinal and callosal neurons, the cell bodies of which are localized in the cortical cell layer V. This neurotrophic influence is mediated by the microglial production of insulin-like growth factor 1 (IGF-1), as shown in vitro in cocultures of cortical neurons and microglia (Ueno et al. 2013). The apparently contradictory roles of microglia (elimination of viable cells versus trophic effect on neuron survival) revealed by the two studies performed in the rat and mouse cerebral cortex (Cunningham et al. 2013; Ueno et al. 2013, Fig. 7.1c_2, d) are in agreement with the current view that microglia have a dual role on cell death depending on the microenvironment and interactions with other cell types (Hanisch and Kettenmann 2007; Mallat and Chamak 1994; Czeh et al. 2011). Microglia analyzed in the above studies act in different microenvironments and at different developmental periods. Thus, the study by Cunningham et al. (2013) reported that microglial phagocytosis of neural precursor cells occurs in the proliferative ventricular zone of E19-P2 rat embryo cerebral cortex, whereas Ueno et al. (2013) observed microglia-promoted survival of differentiated corticospinal neurons, through IGF-1 signaling, in the cortical layer V of the P3–P5 mouse brain. These differences may reflect functional heterogeneity of microglia driven by diverse environmental cues. They may also arise from the selective capacities of target cells to express compounds required for microglial phagocytic or neurotrophic activities, such as "eat-me" signals, the IGF-1 receptor or other signaling components.

7.7 Concluding Remarks

Taken together, current evidence indicates that microglia not only eliminate dead cells, but also play an important role in cell–cell interactions that regulate PCD in the developing CNS. Notably, the signalling by which microglia can promote PCD is diversified. Studies of the developing cerebral cortex have revealed that microglia exert a dual influence on local cell production, promoting the elimination of progenitor cells but preventing the loss of differentiated neurons. It is now clear that the role of microglial phagocytosis extends beyond the elimination of dead cells, as it plays a part in the mechanisms that determine the death fate of neural cells during brain development. The engulfment of apoptotic cells was also shown to impact on the ability of the engulfing cells to produce cytokines or lipid mediators (Elliott and Ravichandran 2010). Therefore, microglial phagocytosis could additionally modulate microglial capacity to influence neural cell survival, growth, or differentiation during normal development. This issue warrants further investigation. The variety of mechanisms by which microglia regulate developmental neural cell death, as reviewed here, is consistent with the high functional plasticity of microglia, whose behavioral repertoire appears to be location- and time-specific. The physiological effects of microglia on neural cell survival is most likely tuned by the local cell environment or by specific interactions between microglia and target cells, which may change according to the CNS region and developmental time.

Acknowledgments This study was supported by grants from the Spanish Ministry of Science and Innovation (BFU2010-19981) and from Agence Nationale de la Recherche (ANR-06-NEURO-028). The authors thank Richard Davies for improving the English.

Statement of Interest: None.

References

Aburto MR, Sánchez-Calderón H, Hurlé JM, Varela-Nieto I, Magariños M (2012) Early otic development depends on autophagy for apoptotic cell clearance and neural differentiation. Cell Death Dis 3:e394

Alliot F, Godin I, Pessac B (1999) Microglia derive from progenitors, originating from the yolk sac, and which proliferate in the brain. Brain Res Dev Brain Res 117:145–152

Arnoux I, Hoshiko M, Mandavy L, Avignone E, Yamamoto N, Audinat E (2013) Adaptive phenotype of microglial cells during the normal postnatal development of the somatosensory "Barrel" cortex. Glia 61:1582–1594

Ashwell K (1990) Microglia and cell death in the developing mouse cerebellum. Brain Res Dev Brain Res 55:219–230

Ashwell K (1991) The distribution of microglia and cell death in the fetal rat forebrain. Brain Res Dev Brain Res 58:1–12

Ashwell KW, Holländer H, Streit W, Stone J (1989) The appearance and distribution of microglia in the developing retina of the rat. Vis Neurosci 2:437–448

Barde YA (1989) Trophic factors and neuronal survival. Neuron 2:1525–1534

Bedard K, Krause KH (2007) The NOX family of ROS-generating NADPH oxidases: physiology and pathophysiology. Physiol Rev 87:245–313

Bessis A, Bechade C, Bernard D, Roumier A (2007) Microglial control of neuronal death and synaptic properties. Glia 55:233–238

Bohana-Kashtan O, Ziporen L, Donin N, Kraus S, Fishelson Z (2004) Cell signals transduced by complement. Mol Immunol 41:583–597

Booth LA, Tavallai S, Hamed HA, Cruickshanks N, Dent P (2013) The role of cell signalling in the crosstalk between autophagy and apoptosis. Cell Signal 26:549–555

Boya P, Mellén MA, de la Rosa EJ (2008) How autophagy is related to programmed cell death during the development of the nervous system. Biochem Soc Trans 36:813–817

Breier G, Albrecht U, Sterrer S, Risau W (1992) Expression of vascular endothelial growth factor during embryonic angiogenesis and endothelial cell differentiation. Development 114:521–532

Buss RR, Sun W, Oppenheim RW (2006) Adaptive roles of programmed cell death during nervous system development. Annu Rev Neurosci 29:1–35

Calderó J, Brunet N, Ciutat D, Hereu M, Esquerda JE (2009) Development of microglia in the chick embryo spinal cord: implications in the regulation of motoneuronal survival and death. J Neurosci Res 87:2447–2466

Chan WY, Kohsaka S, Rezaie P (2007) The origin and cell lineage of microglia: new concepts. Brain Res Rev 53:344–354

Checchin D, Sennlaub F, Levavasseur E, Leduc M, Chemtob S (2006) Potential role of microglia in retinal blood vessel formation. Invest Ophthalmol Vis Sci 47:3595–3602

Chéret C, Gervais A, Lelli A, Colin C, Amar L, Ravassard P, Mallet J, Cumano A, Krause KH, Mallat M (2008) Neurotoxic activation of microglia is promoted by a Nox1-dependent NADPH oxidase. J Neurosci 28:12039–12051
Clarke PG (1990) Developmental cell death: morphological diversity and multiple mechanisms. Anat Embryol (Berl) 181:195–213
Cuadros MA, Navascués J (1998) The origin and differentiation of microglial cells during development. Prog Neurobiol 56:173–189
Cuadros MA, García-Martín M, Martin C, Ríos A (1991) Haemopoietic phagocytes in the early differentiating avian retina. J Anat 177:145–158
Cuadros MA, Coltey P, Carmen Nieto M, Martin C (1992) Demonstration of a phagocytic cell system belonging to the hemopoietic lineage and originating from the yolk sac in the early avian embryo. Development 115:157–168
Cuadros MA, Martin C, Coltey P, Almendros A, Navascués J (1993) First appearance, distribution, and origin of macrophages in the early development of the avian central nervous system. J Comp Neurol 330:113–129
Cunningham CL, Martínez-Cerdeño V, Noctor SC (2013) Microglia regulate the number of neural precursor cells in the developing cerebral cortex. J Neurosci 33:4216–4233
Czeh M, Gressens P, Kaindl AM (2011) The yin and yang of microglia. Dev Neurosci 33: 199–209
Dalmau I, Vela JM, González B, Finsen B, Castellano B (2003) Dynamics of microglia in the developing rat brain. J Comp Neurol 458:144–157
Davalos D, Grutzendler J, Yang G, Kim JV, Zuo Y, Jung S, Littman DR, Dustin ML, Gan WB (2005) ATP mediates rapid microglial response to local brain injury in vivo. Nat Neurosci 8: 752–758
de la Rosa EJ, de Pablo F (2000) Cell death in early neural development: beyond the neurotrophic theory. Trends Neurosci 23:454–458
del Rio-Hortega P (1932) Microglia. In: Penfield W (ed) Cytology and cellular pathology of the nervous system, vol 2. Hafner, New York, pp 483–534
Edinger AL, Thompson CB (2004) Death by design: apoptosis, necrosis and autophagy. Curr Opin Cell Biol 16:663–669
Egensperger R, Maslim J, Bisti S, Holländer H, Stone J (1996) Fate of DNA from retinal cells dying during development: uptake by microglia and macroglia (Müller cells). Brain Rev Dev Brain Res 97:1–8
Elkabes S, DiCicco-Bloom EM, Black IB (1996) Brain microglia/macrophages express neurotrophins that selectively regulate microglial proliferation and function. J Neurosci 16: 2508–2521
Elliott MR, Ravichandran KS (2010) Clearance of apoptotic cells: implications in health and disease. J Cell Biol 189:1059–1070
Elliott MR, Chekeni FB, Trampont PC, Lazarowski ER, Kadl A, Walk SF, Park D, Woodson RI, Ostankovich M, Sharma P, Lysiak JJ, Harden TK, Leitinger N, Ravichandran KS (2009) Nucleotides released by apoptotic cells act as a find-me signal to promote phagocytic clearance. Nature 461:282–286
Erwig LP, Henson PM (2008) Clearance of apoptotic cells by phagocytes. Cell Death Differ 15: 243–250
Fadok VA, Bratton DL, Rose DM, Pearson A, Ezekewitz RA, Henson PM (2000) A receptor for phosphatidylserine-specific clearance of apoptotic cells. Nature 405:85–90
Fantin A, Vieira JM, Gestri G, Denti L, Schwarz Q, Prykhozhij S, Peri F, Wilson SW, Ruhrberg C (2010) Tissue macrophages act as cellular chaperones for vascular anastomosis downstream of VEGF-mediated endothelial tip cell induction. Blood 116:829–840
Farber K, Markworth S, Pannasch U, Nolte C, Prinz V, Kronenberg G, Gertz K, Endres M, Bechmann I, Enjyoji K, Robson SC, Kettenmann H (2008) The ectonucleotidase cd39/ENTPDase1 modulates purinergic-mediated microglial migration. Glia 56:331–341
Ferrer I, Bernet E, Soriano E, del Rio T, Fonseca M (1990) Naturally occurring cell death in the cerebral cortex of the rat and removal of dead cells by transitory phagocytes. Neuroscience 39: 451–458

Frade JM, Barde YA (1998) Microglia-derived nerve growth factor causes cell death in the developing retina. Neuron 20:35–41
Fricker M, Neher JJ, Zhao JW, Thery C, Tolkovsky AM, Brown GC (2012) MFG-E8 mediates primary phagocytosis of viable neurons during neuroinflammation. J Neurosci 32:2657–2666
García-Porrero JA, Ojeda JL (1979) Cell death and phagocytosis in the neuroepithelium of the developing retina. A TEM and SEM study. Experientia 35:375–376
Ginhoux F, Greter M, Leboeuf M, Nandi S, See P, Gokhan S, Mehler MF, Conway SJ, Ng LG, Stanley ER, Samokhvalov IM, Merad M (2010) Fate mapping analysis reveals that adult microglia derive from primitive macrophages. Science 330:841–845
Ginhoux F, Lim S, Hoeffel G, Low D, Huber T (2013) Origin and differentiation of microglia. Front Cell Neurosci 7:45
Hanayama R, Tanaka M, Miwa K, Shinohara A, Iwamatsu A, Nagata S (2002) Identification of a factor that links apoptotic cells to phagocytes. Nature 417:182–187
Hanisch UK, Kettenmann H (2007) Microglia: active sensor and versatile effector cells in the normal and pathologic brain. Nat Neurosci 10:1387–1394
Haynes SE, Hollopeter G, Yang G, Kurpius D, Dailey ME, Gan WB, Julius D (2006) The P2Y12 receptor regulates microglial activation by extracellular nucleotides. Nat Neurosci 9: 1512–1519
Herbomel P, Thisse B, Thisse C (1999) Ontogeny and behaviour of early macrophages in the zebrafish embryo. Development 126:3735–3745
Herbomel P, Thisse B, Thisse C (2001) Zebrafish early macrophages colonize cephalic mesenchyme and developing brain, retina, and epidermis through a M-CSF receptor-dependent invasive process. Dev Biol 238:274–288
Hoeppner DJ, Hengartner MO, Schnabel R (2001) Engulfment genes cooperate with ced-3 to promote cell death in Caenorhabditis elegans. Nature 412:202–206
Hoshiko M, Arnoux I, Avignone E, Yamamoto N, Audinat E (2012) Deficiency of the microglial receptor CX3CR1 impairs postnatal functional development of thalamocortical synapses in the barrel cortex. J Neurosci 32:15106–15111
Hristova M, Cuthill D, Zbarsky V, Acosta-Saltos A, Wallace A, Blight K, Buckley SM, Peebles D, Heuer H, Waddington SN, Raivich G (2010) Activation and deactivation of periventricular white matter phagocytes during postnatal mouse development. Glia 58:11–28
Hsieh CL, Koike M, Spusta SC, Niemi EC, Yenari M, Nakamura MC, Seaman WE (2009) A role for TREM2 ligands in the phagocytosis of apoptotic neuronal cells by microglia. J Neurochem 109:1144–1156
Hume DA, Perry VH, Gordon S (1983) Immunohistochemical localization of a macrophage-specific antigen in developing mouse retina: phagocytosis of dying neurons and differentiation of microglial cells to form a regular array in the plexiform layers. J Cell Biol 97:253–257
Ivashkiv LB (2009) Cross-regulation of signaling by ITAM-associated receptors. Nat Immunol 10:340–347
Ji K, Akgul G, Wollmuth LP, Tsirka SE (2013) Microglia actively regulate the number of functional synapses. PLoS One 8:e56293
Jung S, Aliberti J, Graemmel P, Sunshine MJ, Kreutzberg GW, Sher A, Littman DR (2000) Analysis of fractalkine receptor CX(3)CR1 function by targeted deletion and green fluorescent protein reporter gene insertion. Mol Cell Biol 20:4106–4114
Kálmán M (1989) Dead cells can be phagocytosed by any neighboring cell in early developing rat brain. Int J Neurosci 46:139–145
Kierdorf K, Erny D, Goldmann T, Sander V, Schulz C, Perdiguero EG, Wieghofer P, Heinrich A, Riemke P, Hölscher C, Müller DN, Luckow B, Brocker T, Debowski K, Fritz G, Opdenakker G, Diefenbach A, Biber K, Heikenwalder M, Geissmann F, Rosenbauer F, Prinz M (2013) Microglia emerge from erythromyeloid precursors via Pu.1- and Irf8-dependent pathways. Nat Neurosci 16:273–280
Koizumi S, Shigemoto-Mogami Y, Nasu-Tada K, Shinozaki Y, Ohsawa K, Tsuda M, Joshi BV, Jacobson KA, Kohsaka S, Inoue K (2007) UDP acting at P2Y6 receptors is a mediator of microglial phagocytosis. Nature 446:1091–1095

Konduru NV, Tyurina YY, Feng W, Basova LV, Belikova NA, Bayir H, Clark K, Rubin M, Stolz D, Vallhov H, Scheynius A, Witasp E, Fadeel B, Kichambare PD, Star A, Kisin ER, Murray AR, Shvedova AA, Kagan VE (2009) Phosphatidylserine targets single-walled carbon nanotubes to professional phagocytes in vitro and in vivo. PLoS One 4:e4398
Kuan CY, Roth KA, Flavell RA, Rakic P (2000) Mechanisms of programmed cell death in the developing brain. Trends Neurosci 23:291–297
Lam GY, Huang J, Brumell JH (2010) The many roles of NOX2 NADPH oxidase-derived ROS in immunity. Semin Immunopathol 32:415–430
Lauber K, Bohn E, Kröber SM, Xiao YJ, Blumenthal SG, Lindemann RK, Marini P, Wiedig C, Zobywalski A, Baksh S, Xu Y, Autenrieth IB, Schulze-Osthoff K, Belka C, Stuhler G, Wesselborg S (2003) Apoptotic cells induce migration of phagocytes via caspase-3-mediated release of a lipid attraction signal. Cell 113:717–730
Lauber K, Blumenthal SG, Waibel M, Wesselborg S (2004) Clearance of apoptotic cells: getting rid of the corpses. Mol Cell 14:277–287
Lelli A, Gervais A, Colin C, Chéret C, Ruiz de Almodovar C, Carmeliet P, Krause KH, Boillée S, Mallat M (2013) The NADPH oxidase Nox2 regulates VEGFR1/CSF-1R-mediated microglial chemotaxis and promotes early postnatal infiltration of phagocytes in the subventricular zone of the mouse cerebral cortex. Glia 61:1542–1555
Lichanska AM, Hume DA (2000) Origins and functions of phagocytes in the embryo. Exp Hematol 28:601–611
Liu Y, Yang X, Guo C, Nie P, Ma J (2013) Essential role of MFG-E8 for phagocytic properties of microglial cells. PLoS One 8:e55754
Lockshin RA, Zakeri Z (2002) Caspase-independent cell deaths. Curr Opin Cell Biol 14:727–733
Mallat M, Chamak B (1994) Brain macrophages: neurotoxic or neurotrophic effector cells? J Leukoc Biol 56:416–422
Mallat M, Houlgatte R, Brachet P, Prochiantz A (1989) Lipopolysaccharide-stimulated rat brain macrophages release NGF in vitro. Dev Biol 133:309–311
Mallat M, Marín-Teva JL, Chéret C (2005) Phagocytosis in the developing CNS: more than clearing the corpses. Curr Opin Neurobiol 15:101–107
Marín-Teva JL, Cuadros MA, Calvente R, Almendros A, Navascués J (1999) Naturally occurring cell death and migration of microglial precursors in the quail retina during normal development. J Comp Neurol 412:255–275
Marín-Teva JL, Dusart I, Colin C, Gervais A, van Rooijen N, Mallat M (2004) Microglia promote the death of developing Purkinje cells. Neuron 41:535–547
Marín-Teva JL, Cuadros MA, Martín-Oliva D, Navascués J (2011) Microglia and neuronal cell death. Neuron Glia Biol 7:25–40
Mellén MA, de la Rosa EJ, Boya P (2008) The autophagic machinery is necessary for removal of cell corpses from the developing retinal neuroepithelium. Cell Death Differ 15:1279–1290
Mevorach D, Mascarenhas JO, Gershov D, Elkon KB (1998) Complement-dependent clearance of apoptotic cells by human macrophages. J Exp Med 188:2313–2320
Miyanishi M, Tada K, Koike M, Uchiyama Y, Kitamura T, Nagata S (2007) Identification of Tim4 as a phosphatidylserine receptor. Nature 450:435–439
Mizutani M, Pino PA, Saederup N, Charo IF, Ransohoff RM, Cardona AE (2012) The fractalkine receptor but not CCR2 is present on microglia from embryonic development throughout adulthood. J Immunol 188:29–36
Moujahid A, Navascués J, Marín-Teva JL, Cuadros MA (1996) Macrophages during avian optic nerve development: relationship to cell death and differentiation into microglia. Anat Embryol (Berl) 193:131–144
Nagata S, Hanayama R, Kawane K (2010) Autoimmunity and the clearance of dead cells. Cell 140:619–630
Nandi S, Gokhan S, Dai XM, Wei S, Enikolopov G, Lin H, Mehler MF, Stanley ER (2012) The CSF-1 receptor ligands IL-34 and CSF-1 exhibit distinct developmental brain expression patterns and regulate neural progenitor cell maintenance and maturation. Dev Biol 367:100–113

Neher JJ, Neniskyte U, Zhao JW, Bal-Price A, Tolkovsky AM, Brown GC (2011) Inhibition of microglial phagocytosis is sufficient to prevent inflammatory neuronal death. J Immunol 186:4973–4983

Neumann H, Kotter MR, Franklin RJ (2009) Debris clearance by microglia: an essential link between degeneration and regeneration. Brain 132:288–295

Nicholson DW (1999) Caspase structure, proteolytic substrates, and function during apoptotic cell death. Cell Death Differ 6:1028–1042

Nimmerjahn A, Kirchhoff F, Helmchen F (2005) Resting microglial cells are highly dynamic surveillants of brain parenchyma in vivo. Science 308:1314–1318

Noda M, Doi Y, Liang J, Kawanokuchi J, Sonobe Y, Takeuchi H, Mizuno T, Suzumura A (2011) Fractalkine attenuates excito-neurotoxicity via microglial clearance of damaged neurons and antioxidant enzyme heme oxygenase-1 expression. J Biol Chem 286:2308–2319

O'Connor TM, Wyttenbach CR (1974) Cell death in the embryonic chick spinal cord. J Cell Biol 60:448–459

Ohsawa K, Kohsaka S (2011) Dynamic motility of microglia: purinergic modulation of microglial movement in the normal and pathological brain. Glia 59:1793–1799

Oppenheim RW (1989) The neurotrophic theory and naturally occurring motoneuron death. Trends Neurosci 12:252–255

Oppenheim RW (1991) Cell death during development of the nervous system. Annu Rev Neurosci 14:453–501

Oppenheim RW, Flavell RA, Vinsant S, Prevette D, Kuan CY, Rakic P (2001) Programmed cell death of developing mammalian neurons after genetic deletion of caspases. J Neurosci 21:4752–4760

Oppenheim RW, Blomgren K, Ethell DW, Koike M, Komatsu M, Prevette D, Roth KA, Uchiyama Y, Vinsant S, Zhu C (2008) Developing postmitotic mammalian neurons in vivo lacking Apaf-1 undergo programmed cell death by a caspase-independent, nonapoptotic pathway involving autophagy. J Neurosci 28:1490–1497

Paolicelli RC, Bolasco G, Pagani F, Maggi L, Scianni M, Panzanelli P, Giustetto M, Ferreira TA, Guiducci E, Dumas L, Ragozzino D, Gross CT (2011) Synaptic pruning by microglia is necessary for normal brain development. Science 333:1456–1458

Park D, Tosello-Trampont AC, Elliott MR, Lu M, Haney LB, Ma Z, Klibanov AL, Mandell JW, Ravichandran KS (2007) BAI1 is an engulfment receptor for apoptotic cells upstream of the ELMO/Dock180/Rac module. Nature 450:430–434

Park SY, Jung MY, Kim HJ, Lee SJ, Kim SY, Lee BH, Kwon TH, Park RW, Kim IS (2008) Rapid cell corpse clearance by stabilin-2, a membrane phosphatidylserine receptor. Cell Death Differ 15:192–201

Parnaik R, Raff MC, Scholes J (2000) Differences between the clearance of apoptotic cells by professional and non-professional phagocytes. Curr Biol 10:857–860

Pearson HE, Payne BR, Cunningham TJ (1993) Microglial invasion and activation in response to naturally occurring neuronal degeneration in the ganglion cell layer of the postnatal cat retina. Brain Res Dev Brain Res 76:249–255

Peri F, Nusslein-Volhard C (2008) Live imaging of neuronal degradation by microglia reveals a role for v0-ATPase a1 in phagosomal fusion in vivo. Cell 133:916–927

Perry VH, Gordon S (1991) Macrophages and the nervous system. Int Rev Cytol 125:203–244

Perry VH, Hume DA, Gordon S (1985) Immunohistochemical localization of macrophages and microglia in the adult and developing mouse brain. Neuroscience 15:313–326

Peter C, Waibel M, Radu CG, Yang LV, Witte ON, Schulze-Osthoff K, Wesselborg S, Lauber K (2008) Migration to apoptotic "find-me" signals is mediated via the phagocyte receptor G2A. J Biol Chem 283:5296–5305

Prinz M, Mildner A (2011) Microglia in the CNS: immigrants from another world. Glia 59: 177–187

Prinz M, Priller J, Sisodia SS, Ransohoff RM (2011) Heterogeneity of CNS myeloid cells and their roles in neurodegeneration. Nat Neurosci 14:1227–1235

Raff MC, Barres BA, Burne JF, Coles HS, Ishizaki Y, Jacobson MD (1993) Programmed cell death and the control of cell survival: lessons from the nervous system. Science 262:695–700
Rakic S, Zecevic N (2000) Programmed cell death in the developing human telencephalon. Eur J Neurosci 12:2721–2734
Ransohoff RM (2009) Chemokines and chemokine receptors: standing at the crossroads of immunobiology and neurobiology. Immunity 31:711–721
Ravichandran KS (2010) Find-me and eat-me signals in apoptotic cell clearance: progress and conundrums. J Exp Med 207:1807–1817
Reddien PW, Cameron S, Horvitz HR (2001) Phagocytosis promotes programmed cell death in C. elegans. Nature 412:198–202
Rigato C, Buckinx R, Le-Corronc H, Rigo JM, Legendre P (2011) Pattern of invasion of the embryonic mouse spinal cord by microglial cells at the time of the onset of functional neuronal networks. Glia 59:675–695
Roth KA, D'Sa C (2001) Apoptosis and brain development. Ment Retard Dev Disabil Res Rev 7: 261–266
Saijo K, Glass CK (2011) Microglial cell origin and phenotypes in health and disease. Nat Rev Immunol 11:775–787
Sánchez-López A, Cuadros MA, Calvente R, Tassi M, Marín-Teva JL, Navascués J (2004) Radial migration of developing microglial cells in quail retina: a confocal microscopy study. Glia 46: 261–273
Santos AM, Calvente R, Tassi M, Carrasco MC, Martín-Oliva D, Marín-Teva JL, Navascués J, Cuadros MA (2008) Embryonic and postnatal development of microglial cells in the mouse retina. J Comp Neurol 506:224–239
Savill J, Fadok V (2000) Corpse clearance defines the meaning of cell death. Nature 407: 784–788
Schafer DP, Lehrman EK, Kautzman AG, Koyama R, Mardinly AR, Yamasaki R, Ransohoff RM, Greenberg ME, Barres BA, Stevens B (2012) Microglia sculpt postnatal neural circuits in an activity and complement-dependent manner. Neuron 74:691–705
Scheffel J, Regen T, Van Rossum D, Seifert S, Ribes S, Nau R, Parsa R, Harris RA, Boddeke HW, Chuang HN, Pukrop T, Wessels JT, Jurgens T, Merkler D, Bruck W, Schnaars M, Simons M, Kettenmann H, Hanisch UK (2012) Toll-like receptor activation reveals developmental reorganization and unmasks responder subsets of microglia. Glia 60:1930–1943
Schulz C, Gomez Perdiguero E, Chorro L, Szabo-Rogers H, Cagnard N, Kierdorf K, Prinz M, Wu B, Jacobsen SE, Pollard JW, Frampton J, Liu KJ, Geissmann F (2012) A lineage of myeloid cells independent of Myb and hematopoietic stem cells. Science 336:86–90
Sedel F, Béchade C, Vyas S, Triller A (2004) Macrophage-derived tumor necrosis factor alpha, an early developmental signal for motoneuron death. J Neurosci 24:2236–2246
Sierra A, Encinas JM, Deudero JJ, Chancey JH, Enikolopov G, Overstreet-Wadiche LS, Tsirka SE, Maletic-Savatic M (2010) Microglia shape adult hippocampal neurogenesis through apoptosis-coupled phagocytosis. Cell Stem Cell 7:483–495
Sierra A, Abiega O, Shahraz A, Neumann H (2013) Janus-faced microglia: beneficial and detrimental consequences of microglial phagocytosis. Front Cell Neurosci 7:6
Snider WD (1994) Functions of the neurotrophins during nervous system development: what the knockouts are teaching us. Cell 77:627–638
Sorokin SP, Hoyt RF Jr, Blunt DG, McNelly NA (1992) Macrophage development: II. Early ontogeny of macrophage populations in brain, liver, and lungs of rat embryos as revealed by a lectin marker. Anat Rec 232:527–550
Swinnen N, Smolders S, Avila A, Notelaers K, Paesen R, Ameloot M, Brône B, Legendre P, Rigo JM (2013) Complex invasion pattern of the cerebral cortex by microglial cells during development of the mouse embryo. Glia 61:150–163
Takahashi K, Rochford CD, Neumann H (2005) Clearance of apoptotic neurons without inflammation by microglial triggering receptor expressed on myeloid cells-2. J Exp Med 201:647–657
Thanos S (1991) The relationship of microglial cells to dying neurons during natural neuronal cell death and axotomy-induced degeneration of the rat retina. Eur J Neurosci 3:1189–1207

Tremblay ME, Lowery RL, Majewska AK (2010) Microglial interactions with synapses are modulated by visual experience. PLoS Biol 8:e1000527

Truman LA, Ford CA, Pasikowska M, Pound JD, Wilkinson SJ, Dumitriu IE, Melville L, Melrose LA, Ogden CA, Nibbs R, Graham G, Combadiere C, Gregory CD (2008) CX3CL1/fractalkine is released from apoptotic lymphocytes to stimulate macrophage chemotaxis. Blood 112: 5026–5036

Ueno M, Fujita Y, Tanaka T, Nakamura Y, Kikuta J, Ishii M, Yamashita T (2013) Layer V cortical neurons require microglial support for survival during postnatal development. Nat Neurosci 16:543–551

Upender MB, Naegele JR (1999) Activation of microglia during developmentally regulated cell death in the cerebral cortex. Dev Neurosci 21:491–505

Valenciano AI, Boya P, de la Rosa EJ (2009) Early neural cell death: numbers and cues from the developing neuroretina. Int J Dev Biol 53:1515–1528

Verney C, Monier A, Fallet-Bianco C, Gressens P (2010) Early microglial colonization of the human forebrain and possible involvement in periventricular white-matter injury of preterm infants. J Anat 217:436–448

Wakselman S, Béchade C, Roumier A, Bernard D, Triller A, Bessis A (2008) Developmental neuronal death in hippocampus requires the microglial CD11b integrin and DAP12 immunoreceptor. J Neurosci 28:8138–8143

Witting A, Muller P, Herrmann A, Kettenmann H, Nolte C (2000) Phagocytic clearance of apoptotic neurons by microglia/brain macrophages in vitro: involvement of lectin-, integrin-, and phosphatidylserine-mediated recognition. J Neurochem 75:1060–1070

Yaginuma H, Shiraiwa N, Shimada T, Nishiyama K, Hong J, Wang S, Momoi T, Uchiyama Y, Oppenheim RW (2001) Caspase activity is involved in, but is dispensable for, early motoneuron death in the chick embryo cervical spinal cord. Mol Cell Neurosci 18:168–182

Yeo W, Gautier J (2004) Early neural cell death: dying to become neurons. Dev Biol 274: 233–244

Yuan J, Yankner BA (2000) Apoptosis in the nervous system. Nature 407:802–809

Chapter 8
Developmental Vascularization, Neurogenesis, Myelination, and Astrogliogenesis

G. Jean Harry and Lorena Pont-Lezica

Abstract The temporal and spatial pattern of microglia colonization of the nervous system implies a role in neural cell proliferation and differentiation, as well as neurovascularization and postnatal myelination. Microglia established within the developing nervous system assume a neural-specific identity and contribute to key developmental events. Their association around blood vessels implicates them in development of the vascular system. A similar association has been reported for neural cell proliferation and associated phenotypic shifts, and also for cell fate differentiation to neuronal or glial phenotypes. These processes are accomplished by phagocytic activities, cell–cell contact relationships, and secretion of various factors. This chapter will provide an assessment of data currently available from in vivo and in vitro studies evaluating the dynamic, interactive nature of these processes throughout the progression of nervous system development.

Keywords Microglia • Vascularization • Neurogenesis • Astrogliogenesis • Oligodendrocyte development • Myelination • Radial glia

Bullet Points

- Microglia display a spatial and temporal relationship with neural cell types during critical stages of nervous system development including neurogenesis, astrogliogenesis, neural cell specification and differentiation, myelination, and vascularization.

G.J. Harry, Ph.D. (✉)
National Toxicology Program Laboratory, National Institute of Environmental Health Sciences/National Institutes of Health, MD C1-04, 111 T.W. Alexander Drive, Research Triangle Park, NC 27709, USA
e-mail: harry@niehs.nih.gov

L. Pont-Lezica, Ph.D.
Cellular Biology of the Synapse Laboratory, Institut de Biologie de l'Ecole Normale Supérieure, Neuronal Functions and Microglia group, Inserm U1024/CNRS UMR8197, Paris, France
e-mail: lorena_pl@yahoo.com

M.-È. Tremblay, A. Sierra (eds.), *Microglia in Health and Disease*,
DOI 10.1007/978-1-4939-1429-6_8

- Microglia can produce a variety of molecular factors important for normal CNS development, but their in vivo contribution to most critical stages remains to be determined.
- Microglia are linked with neurovascularization but may not be the leading influence on this process or directly follow the vasculature for brain entry.

8.1 Introduction

Development of the nervous system follows a temporal and spatial sequence of critical events. The nervous system begins to emerge with a thickening of the dorsal ectoderm forming the neural plate along the dorsal side of the embryo. The neural plate matures via neurulation into the neural tube, which gives rise to the central nervous system (CNS), and the neural crest, which gives rise to the peripheral nervous system (PNS). At this stage, the parenchymal wall is composed of the columnar cells of the neuroepithelium that undergo expansion, with some cells dividing to produce the first neurons (reviewed in (Kriegstein and Alvarez-Buylla 2009; Bergstrom and Forsberg-Nilsson 2012)). Neuroepithelial cells transition to radial glia (RG) cells, which divide symmetrically to maintain the progenitor population, or asymmetrically to produce a self-renewing RG cell and an intermediate neuronal or glial progenitor (Noctor et al. 2001; Malatesta et al. 2003; Gotz and Huttner 2005). Neurons are usually produced first, followed closely by oligodendrocytes and astrocytes. As previously described in Chaps. 1 and 2, microglial progenitors migrate from the yolk sac and are first found within the brain parenchyma of the mouse from embryonic day (ED)9–10 (Alliot et al. 1999; Ginhoux et al. 2010), a time point which corresponds to the initiation of neurogenesis. The co-localization of microglia with several components of the developing CNS suggests not only a timing association but also the potential for cell–cell interactions that regulate progenitor proliferation, precursor migration, and neuronal/glia differentiation, as well as establish the vascular network of the brain. This chapter will summarize the existing literature supporting this assumption of interactions between microglia and other neural cells during normal CNS development.

8.2 Microglia During Embryogenesis

Microglia are present in the developing brain from the very early stages of neurogenesis when neuroepithelial cells transform into RG, the cortical progenitor cells (Fig. 8.1). A role for microglia in maintaining the RG population and/or influencing progenitor specification and differentiation is implied by their localization along the embryonic proliferative ventricular zone (VZ), although direct evidence is still missing. During the second trimester of human gestation, microglia can be

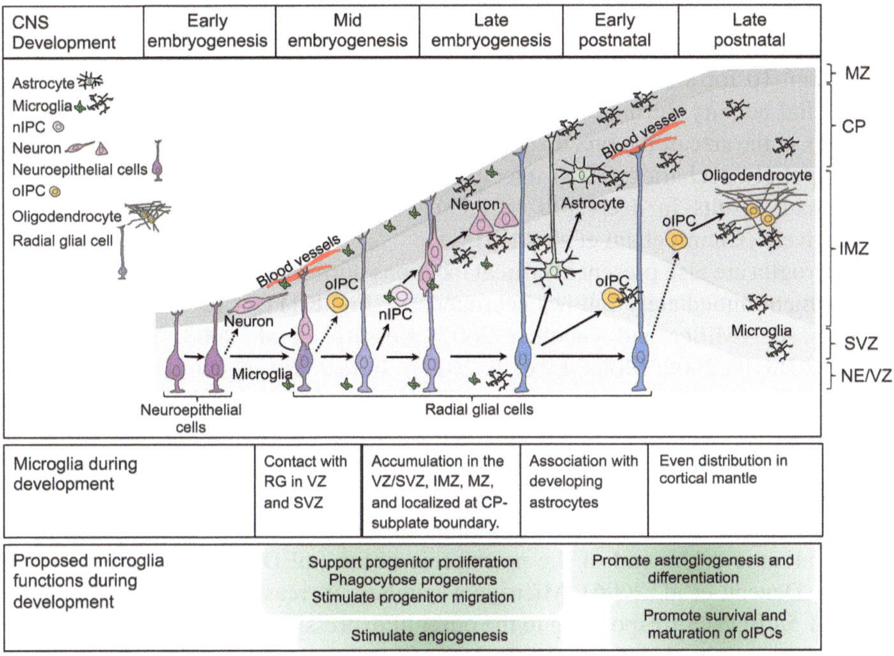

Fig. 8.1 Development of the cortical neural cell types. Embryogenesis of the CNS. Early in development, neuroepithelial cells expand by symmetrical division. Some neuroepithelial cells likely generate early neurons. Neuroepithelial cells later elongate and convert into radial glial (RG) cells. RG divide asymmetrically to generate one RG cell and one intermediate progenitor cell (IPCs). IPCs in the SVZ divide to give rise to neurons (nIPCs) or oligodendrocytes (oIPCs). The progeny from RG and IPCs move into the intermediate zone for differentiation. At the end of embryonic development, glial-committed RG detach from the apical side and convert into astrocytes. oIPC production continues. *Solid arrows* are supported by experimental evidence; *dashed arrows* are hypothetical. Microglial distribution changes across developmental time. Potential stage-specific functions of microglia during development of neural cell types: stimulation of progenitor proliferation and migration, phagocytosis of progenitors and apoptotic cells, stimulation of angiogenesis, promotion of astrogliogenesis and astrocyte differentiation, promotion of survival and maturation of OPCs. Adapted from Kriegstein and Alvarez-Buylla (2009). *CP* cortical plate, *IPC* intermediate progenitor cell, *IMZ* intermediate zone, *MZ* marginal zone, *NE* neuroepithelium, *nIPC* neurogenic progenitor cell, *oIPC* oligodendrocytic progenitor cell, *RG* radial glia, *SVZ* subventricular zone, *VZ* ventricular zone

visualized in the VZ and subventricular zone (SVZ) (Verney et al. 2010). When examining this localization and timing in the macaque monkey, a monolayer of microglia is observed between the VZ and SVZ at ED50–65 with a massive accumulation in the SVZ at ED80 (Cunningham et al. 2013). In a comparable time frame of brain development, murine microglia are present as early as ED8 (Alliot et al. 1999), and they accumulate within the VZ/SVZ from ED12, where they are observed to contact immature RG (Antony et al. 2011; Swinnen et al. 2012). The exact function of microglia at this time with regards to RG has not been determined but in both the macaque and rat, these cells have been shown to engulf neural progenitors

((Cunningham et al. 2013); Chaps. 7 and 10), thereby regulating the number of embryonic neural precursors through phagocytosis (Cunningham et al. 2013) (also see Chap. 10 for a similar role during adult hippocampal neurogenesis). Altering microglial activity during embryonic neurogenesis in vivo also leads to dramatic changes in the size of the precursor pool, and in this context, inducing inflammation with lipopolysaccharide (LPS) injection or eliminating microglia with liposomal clodronate results in decreased or increased numbers of neuronal progenitors, respectively (Cunningham et al. 2013).

Microglia are also present at critical locations during the process of astrogliogenesis, which immediately follows neurogenesis (Fig. 8.1) (for reviews on astrogliogenesis see (Miller and Gauthier 2007; Kessaris et al. 2008; Kriegstein and Alvarez-Buylla 2009; Molofsky et al. 2012)). In fact, the sequential appearance of microglia and astrocytes in the same regions during CNS development has led to the hypothesis that microglia could play a role in RG differentiation into astrocytes (Voigt 1989; Aquino et al. 1996; Kalman and Ajtai 2001; Rezaie et al. 2003; Fox et al. 2004; Monier et al. 2006). This phenomenon is particularly well-documented and well-described in the developing spinal cord where microglia first appear in the marginal zone at the end of the neurogenic phase (mED10.5 in mouse, cED3 in chick) (Deneen et al. 2006). Microglial numbers increase in the VZ and peak at mED11.5/cED8, corresponding to the onset of expression of astrocyte specification markers such as nuclear factor 1A/B (NFIA/IB). Subsequently, microglia are redistributed to the surrounding gray matter (mED13.5, cED10) (Caldero et al. 2009; Rigato et al. 2011), coinciding with the migration of astrocyte progenitors (mED13.5–15.5, cED6–7), followed by their differentiation at mED17.5–18.5, cED10 (Deneen et al. 2006).

In addition, the physical proximity of microglia with glial fibrillary acidic protein (GFAP)-positive cells suggests a potential cell–cell interaction during astrocyte development. Verney and colleagues (2010) consistently found ionized calcium binding adaptor molecule 1 (Iba1)-positive microglia juxtaposed with GFAP+ cells within the intermediate zone (IMZ) of humans between 19 and 30 gestation weeks (GW). Microglia density is higher during astrocyte differentiation, at 25GW, than later, once differentiation is complete (32GW) (Rezaie et al. 2003; Verney et al. 2010). In the rat, microglia are seen to associate with vimentin+ radial glia from ED19 in the hippocampus and then with vimentin+/GFAP+ cells at ED21, corresponding to the beginning of astrocyte differentiation (Dalmau et al. 1997). In the developing human spinal cord, microglia are associated with vimentin+/GFAP+RG in the ependymal layer at 9GW and in the marginal zone at 16GW (Rezaie et al. 1999). In the developing mouse spinal cord at E13.5–14.5, 50 % of microglia contact nestin+ RG (Rigato et al. 2011). While the physical proximity of microglia and RG within the defined developmental time periods of neurogenesis and astrogliogenesis has been observed, the actual functional relationship between these cells has not been determined.

8.2.1 In Vitro Experimental Evidence

In light of these in vivo observations, microglia could have a role in maintaining the progenitor cell pool and regulating progenitor cell migration, or in influencing the timing of neuronal or glial differentiation. If early RG are already committed to a neuronal or glial fate as proposed by Malatesta and colleagues, microglia may assist in maintaining their phenotype as well (Malatesta et al. 2000). The few in vitro studies examining the role of microglia in the development of other neural cell types using embryonic neural progenitor cells (eNPCs) also generated fairly consistent results suggesting an enhancing effect of microglia on cell proliferation. eNPCs are typically isolated from embryonic VZ and SVZ and, when cultured under appropriate conditions, they acquire a RG-like phenotype and express RG markers such as nestin, radial glial cell marker-2 (RC2), and brain lipid binding protein (Malatesta et al. 2008; Miyata et al. 2010; Bergstrom and Forsberg-Nilsson 2012). eNPCs are multipotent cells which can be directed to differentiate into neurons, astrocytes, or oligodendrocytes (Malatesta et al. 2008; Bergstrom and Forsberg-Nilsson 2012). In vitro, the presence of microglia or microglia-conditioned medium increases eNPC proliferation (Cacci et al. 2008; Antony et al. 2011). In particular, Antony and colleagues (2011) looked at the properties of cortical progenitors isolated from mice deficient in the transcription factor, Pu.1 (Pu.1 KO), expressed solely in the hematopoietic system and essential for development of the monocytic lineage. Comparison of early embryonic (ED12–13) cortical precursor cultures from wild type and microglia-depleted Pu.1 KO mice showed a 25 % decrease in the number of Ki67+ proliferating cells in the absence of microglia. The addition of exogenous microglia was sufficient to rescue this phenotype and return the level of proliferation to within normal range. In addition, it was observed that the proportion of astrocytes in Pu.1 KO cultures was diminished relative to wild type, and this was also rescued by exogenous microglia (Antony et al. 2011). Along this continuum, the addition of an excess of exogenous microglia increased eNPC differentiation into astrocytes (Antony et al. 2011). This role in promoting differentiation of NPC to an astrocyte lineage has been reported in additional studies (Nakanishi et al. 2007; Cacci et al. 2008; Zhu et al. 2008; Ajmone-Cat et al. 2010; Gu et al. 2011). Zhu and colleagues (2008) reported that a 1-day exposure of neural stem cells (NSCs) to conditioned media obtained from unstimulated microglia increased their expression of the pro-gliogenic genes Notch1-3, Hes5, Id3, and Sox9, while decreasing their expression of the pro-neurogenic genes Mash1, NeuroD1, NeuroD2 in eNPCs (Zhu et al. 2008). Interestingly, this pro-gliogenic effect was augmented when microglia were stimulated with LPS (Cacci et al. 2008; Ajmone-Cat et al. 2010). It remains to be determined if microglia specifically stimulate the proliferation of pre-committed astrocyte precursors, or whether they have two independent functions, supporting precursor proliferation and promoting the astrocytic cell fate.

8.2.2 Microglial Factors and CNS Development

Although the data relative to microglial involvement in developmental neurogenesis is limited, besides their roles in regulating the pool of embryonic neural precursors, a number of factors known to be released by microglia (Kim and de Vellis 2005; Garden and Moller 2006; Tambuyzer et al. 2009; Kettenmann et al. 2011) have been demonstrated to influence brain development (Tran and Miller 2003; Adler et al. 2005; Deverman and Patterson 2009) (see Table 8.1). For example, microglia can release trophic factors essential for the proliferation and maintenance of cortical progenitors such as insulin-like growth factor 1 (IGF-1), basic fibroblast growth factor (bFGF/FGF2), and leukemia inhibitory factor (LIF). They are also known to produce chemokines, such as macrophage inflammatory protein-1α (MIP-1α/CCL-3) and monocyte chemotactic protein-1 (MCP-1/CCL-2), which can regulate adult progenitor migration in vitro. These chemokines are expressed in the developing CNS at the time of microglia-NPC association. The influence of pro-inflammatory cytokines on NSC in culture suggests a potential role for microglia in determining the survival and differentiation of the cells into a neuronal phenotype. The addition of factors such as interferon γ (IFNγ), transforming growth factor β (TGFβ), glial cell line-derived neurotrophic factor (GDNF), and interleukin 1β (IL-1β) has been reported to contribute to NSC proliferation, survival, and differentiation to a neuronal phenotype (Nakajima and Kohsaka 2004; Carpentier and Palmer 2009; Gonzalez-Perez et al. 2012). Some of these factors can contribute to neuronal survival and maturation in vivo. For example, GDNF is important for motor neuron survival in vivo (Yan et al. 1995) (for review see (Nakajima and Kohsaka 2004)) and IGF-1 for neurite growth and synaptogenesis as evaluated in vitro and in vivo (reviewed (D'Ercole and Ye 2008)).

The molecular mechanisms responsible for initiation of astrogliogenesis include a Notch-dependent epigenetic switch relieving the silencing of astrocytic genes, the modulation of activator/repressor transcriptional complexes, and the expression of pro-gliogenic cytokines from the IL-6 family (for reviews see (Miller and Gauthier 2007; Chenn 2009; Freeman 2010; Rowitch and Kriegstein 2010; Spooren et al. 2011)). Additional factors such as bone morphogenic proteins (BMPs), epidermal growth factor (EGF), and FGF2 influence NPC commitment to the astrocyte lineage (Wakamatsu 2004; Kessaris et al. 2008; Deverman and Patterson 2009). Newborn neurons are considered a primary source of pro-gliogenic factors (Barnabe-Heider et al. 2005; Malatesta et al. 2008; Miyata et al. 2010); however, microglia are known to express critical stimulatory factors such as the Notch1 ligands, jagged-1, and Delta-1 (Grandbarbe et al. 2007; Cao et al. 2008; Jurynczyk and Selmaj 2010), IL-6, LIF (Nakanishi et al. 2007; Zhu et al. 2008), MCP-1 (Gordon et al. 2012), BMP2, and FGF2 (Zhu et al. 2008). This capacity for production and release of critical astrogliogenic factors suggests that microglia have the potential to influence cell fate (Rezaie et al. 2002a; Gordon et al. 2012). With regards to a temporal association, the in vivo expression of IL-1α and IL-1β peaks at the onset of astrogliogenesis,

Table 8.1 Function of selected cytokines during CNS development and evidence for their expression in microglia

Protein	Developmental function(s)	References	In vitro and in vivo expression by microglia	References
IGF-1	Embryonic rodent NPC proliferation and survival in vitro	Torres-Aleman et al. (1990), Drago et al. (1991), Arsenijevic et al. (2001), Mairet-Coello et al. (2009)	Cultured newborn mouse microglia.	Butovsky et al. (2006a)
	Adult rodent NPC proliferation and survival in vitro	Aberg et al. (2003)	Ameboid microglia between P0 and P7, and adult rodent brain	Kaur et al. (2006), Hristova et al. (2010), Ueno et al. (2013)
	NSC pool and brain size	Beck et al. (1995)		
	Embryonic NPC proliferation in vivo	Hodge et al. (2004), Popken et al. (2004), Mairet-Coello et al. (2009)		
	Oligodendrocyte survival and myelination	D'Ercole and Ye (2008), Kaur and Ling (2009)		
	Role in adult neurogenesis	Ziv and Schwartz (2008)		
FGF2/bFGF	Amplification and self-renewal of NSCs, negative regulation of neurogenesis	Vaccarino et al. (1999), Raballo et al. (2000)	Cultured newborn rat and human fetal microglia	Shimojo et al. (1991), Araujo and Cotman (1992), Presta et al. (1995)
	Astrogliogenesis	Morrow et al. (2001), Miller and Gauthier (2007)		
	Oligodendrogenesis	Kessaris et al. (2008)		

(continued)

Table 8.1 (continued)

Protein	Developmental function(s)	References	In vitro and in vivo expression by microglia	References
LIF	Amplification and self-renewal of NSCs	Chang et al. (2004), Hellstrom et al. (2009)	Cultured newborn rat microglia	Nakanishi et al. (2007), Zhu et al. (2008)
	Astrogliogenesis	Miller and Gauthier (2007), Rowitch and Kriegstein (2010), Spooren et al. (2011)	Upregulated in SVZ microglia in response to insult	Covey and Levison (2007), Hellstrom et al. (2009)
TGFβ	Negative regulation of NSC expansion in vitro and in vivo	Miller and Luo (2002), Falk et al. (2008)	Cultured newborn rat microglia	Smith and Hale (1997), Nakajima et al. (2007), Cacci et al. (2008)
	Survival and differentiation of embryonic dopaminergic neurons in vitro	Krieglstein et al. (1995)	Upregulated in vivo in adult microglia in response to injury	Battista et al. (2006)
	Survival and differentiation of adult hippocampal NSCs in vitro	Battista et al. (2006)		
GDNF	Survival and differentiation of embryonic dopaminergic neurons in vitro	Lin et al. (1993), Krieglstein et al. (1995), Ling et al. (1998)	Cultured newborn rat microglia	Nakajima et al. (2007)
	Motor neuron survival in vivo	Yan et al. (1995)		

(continued)

Table 8.1 (continued)

Protein	Developmental function(s)	References	In vitro and in vivo expression by microglia	References
MIP-1α/ CCL-3	Proliferation and survival of adult precursors in vitro	Gordon et al. (2012)	Cultured human fetal microglia	McManus et al. (1998), Hua and Lee (2000), Rezaie et al. (2002a, b)
	Adult NSC migration in vitro, and in vivo after insult	Widera et al. (2004), Gordon et al. (2009)	Microglia in IMZ and choroid plexus in human cerebral cortex at 16–22GW	Rezaie and Male (1999)
	Differentiation of adult precursors towards neuronal and oligodendrocytic fates in vitro	Gordon et al. (2012)	Microglia in human spinal cord at 12GW	Rezaie and Male (1999), Rezaie et al. (1999)
			Microglia of rat cerebellum at PND7 and PND14	Cowell and Silverstein (2003)
MCP-1/ CCL-2	Adult NSC migration in vitro, and in vivo after insult	Widera et al. (2004), Tran et al. (2007), Gordon et al. (2009)	Cultured human fetal microglia	McManus et al. (1998), Hua and Lee (2000)
	Differentiation of adult precursors towards neuronal and oligodendrocytic fates in vitro	Gordon et al. (2012)	Microglia in IMZ and choroid plexus in human cerebral cortex at 16–22GW	Rezaie and Male (1999)
IL-1α	Pro-gliogenic in vitro	Giulian et al. (1988), Johansson et al. (2008), Ajmone-Cat et al. (2010)	Peak expression between ED18 and birth in the rat	Giulian et al. (1988)
	Embryonic spinal cord neuron survival in vitro	Brenneman et al. (1992)	Rodent newborn microglia following acute activation	Giulian et al. (1988), Cacci et al. (2008)

(continued)

Table 8.1 (continued)

Protein	Developmental function(s)	References	In vitro and in vivo expression by microglia	References
IL-1β	Pro-gliogenic in vitro	Vela et al. (1995), Ajmone-Cat et al. (2010), Green et al. (2012)	Cultured human microglia	Liu et al. (1998)
	Effects on proliferation and survival of rodent eNPCs in vitro highly debated.	Akaneya et al. (1995), de la Mano et al. (2007), Wang et al. (2007), Koo and Duman (2008), Ajmone-Cat et al. (2010), Green et al. (2012)	Expression throughout prenatal and postnatal CNS: human forebrain from 5GW, sheep neocortex from ED35–40; phagocytic microglia prior to ED80, mouse from ED14, rat and chick spinal cords from ED12 and stage 17HH.	Giulian et al. (1988), Pillay et al. (1993), Mousa et al. (1999), Dziegielewska et al. (2000), de la Mano et al. (2007)
	Discrepancies could be due to age of NPCs, region, and dosage.		Rodent newborn microglia following acute activation	Giulian et al. (1988), Cacci et al. (2008)
	Differentiation of human and rodent eNPCs in vitro	Ling et al. (1998), Storch et al. (2001), Johansson and Stromberg (2002), Riaz et al. (2004)		
	Inhibition in vivo during development results in memory and learning deficits in the adult	Goshen et al. (2007)		

(continued)

Table 8.1 (continued)

Protein	Developmental function(s)	References	In vitro and in vivo expression by microglia	References
INFγ	Decreased proliferation of rodent postnatal NSCs in vitro	Ben-Hur et al. (2003), Wong et al. (2004), Li et al. (2008)	Cultured newborn rat microglia	De Simone et al. (1998)
	Neuronal differentiation of rodent postnatal NSCs in vitro	Wong et al. (2004), Li et al. (2008)		
	Increased neurosphere formation by eNPCs in vitro	Li et al. (2008)		
Notch ligands: jagged-1, Delta-1	Proliferation and maintenance of the progenitor pool	Zhou et al. (2010)	Cultured newborn rat microglia, upregulation upon inflammatory activation	Grandbarbe et al. (2007), Cao et al. (2008)
	Astrogliogenesis in vitro	Tanigaki et al. (2001)	SVZ and corpus callosum of postnatal rat	Cao et al. (2008)

IGF-1 Insulin-like growth factor 1, *FGF2/bFGF2* fibroblast growth factor 2/basic fibroblast growth factor, *LIF* leukemia inhibitory factor, *TGFβ* transforming growth factor β, *GDNF* glial cell line-derived neurotrophic factor, *MIP-1α/CCL-3* macrophage inflammatory protein-1α/chemokine (C-C) motif ligand-3, *MCP-1/CCL-2* macrophage chemotactic protein-1/chemokine (C-C) motif ligand-2, *IL-1α* interleukin 1α, *IL-1β* interleukin 1β, *INFγ* interferon γ

thus providing further support for a pro-gliogenic implication (Giulian et al. 1988; Green et al. 2012).

It has been proposed that the trophic contributions of microglia to neuro- and gliogenesis following injury might recapitulate similar functions of these myeloid cells during CNS development. In fact, many of the microglia-derived factors identified above have known trophic functions in the adult, or become upregulated in response to injury. For example, a role for IGF-1 in adult neurogenesis (reviewed in (Ziv and Schwartz 2008)) and neuronal survival in response to injury (Guan et al. 2003; Fushimi and Shirabe 2004; Choi et al. 2008; Wine et al. 2009) has been demonstrated. In response to widespread insults such as hypoxia, ischemia or ionizing radiation, microglial production of LIF is increased in the perinatal and adult SVZ (Covey and Levison 2007; Hellstrom et al. 2009), and in this context, TGFβ and GDNF are upregulated locally at the sites of injury (Kearns and Gash 1995; Kreutzberg 1996; Batchelor et al. 1999; Suzuki et al. 2001; Nakajima and Kohsaka 2004). In the adult, an injury-induced increase in astrogliogenesis has been reported in the striatum, at the site of neurotoxin 1-methyl-4-phenyl-1,2,3,6-tetrahydropyridine (MPTP) lesion, a model of Parkinson's disease (Mao et al. 2001). The authors speculated that astrogliogenesis could be occurring locally, based on the clustered distribution of newborn astrocytes in the striatum and their progressive increase in expression of the astrocytic marker S100 calcium binding protein β (S100β) between 3 days and 1 month post-lesion (Mao et al. 2001). In the neonatal SVZ, Bain and colleagues also examined the identity of newborn cells generated in response to hypoxia (Bain et al. 2010). Similarly to the cell characterization in the focal lesion, the majority of newly generated cells expressed markers of astrocytes, beginning with the marker of immature astrocytes, vimentin. Over 3 weeks, these newborn astrocytes matured and migrated out of the SVZ into the subcortical white matter, where this increase in astrocytes number was accompanied by a concurrent decrease in oligodendrocytes number (Bain et al. 2010). The authors suggested an association between the localized proliferation of astrocytes and the induction of IL-1α, IL-1β, IL-6, and LIF cytokines (Kim and de Vellis 2005; Deverman and Patterson 2009; Bain et al. 2010; Robel et al. 2011; Gonzalez-Perez et al. 2012). Such studies of injury-induced cell proliferation lead to the speculation that cytokine release may also influence cell proliferation and fate during development.

8.3 Microglia and Oligodendrocyte Development

In vivo characterization of the oligodendrocyte—the myelin-producing cell of the CNS—developmental lineage has been previously limited due to the lack of cell-specific markers to uniquely identify oligodendrocyte progenitor cells (OPCs). From the available data, it is evident that oligodendrocytes originate as progenitor cells from the ventral neuroepithelium during mid-neurogenesis. Neural precursors giving rise to oligodendrocytes were first identified in the rodent at ED12-14 by their expression of platelet-derived growth factor alpha receptor (PDGFαR; (Pringle et al. 1992; Pringle and Richardson 1993; Ellison and de Vellis 1994)). These cells localized to the ventral-half of the ventricular zones of the spinal cord, diencephalon, and telencephalon (Pringle and Richardson 1993; Woodruff et al. 2001; Tripathi et al. 2011). Early work using staining for the oligodendrocytic marker *Griffonia simplicifolia (GSL) II* detected a subpopulation of RG cells in the embryonic rat brain. These cells differed from the nestin+ RG cells, implicating them as possible precursors for oligodendrocytes (Hurley and Streit 1995). The authors concluded that lectin binding with *GSL II* could serve as a differentiation marker for cells committed to the oligodendroglial lineage and suggested that RG precursors could provide a source of oligodendroglia. In the embryonic telencephalon, oligodendrocytes progenitors (OLPs) in the ventral forebrain, anterior hypothalamic region, and the medial ganglionic eminence were demonstrated to migrate tangentially into the dorsal telencephalon (Kessaris et al. 2006). Once the first wave of OLPs populates the entire embryonic telencephalon they are joined by a second wave of OLPs from the lateral/caudal ganglionic eminence. This is followed by a third wave of OLPs within the postnatal cortex (Kessaris et al. 2006). Initially, the primitive precursor cells differentiated into proliferative, migratory, bipolar O2A progenitor cells that are bipotential, being capable of differentiating into either astrocytes or oligodendrocytes. At ED15, O2A glial-lineage cells co-expressing PDGFαR and neuron-glial antigen 2 (NG2) proteoglycan were reported to be arranged in small clusters adjacent to the central canal in the ventral spinal cord. By ED17, small process-bearing PDGFαR+/NG2+ cells are observed in many regions of the CNS outside of the ventricular zone (Nishiyama et al. 1996). Between postnatal day (PND)0 and PND7, PDGFαR+ cells in the rat brain increase in number, followed by a decline in number over the lifespan of the animal (He et al. 2009).

The oligodendrocyte lineage progresses through several stages. As PDGFαR+/NG2+ cells mature into oligodendrocytes they enter an intermediate pro-oligodendrocyte state that is recognized by a loss of these antigens and the concomitant expression of the oligodendrocyte marker, O4 (Pfeiffer et al. 1993). A wave of differentiation beginning at PND4 spreads outwards from the corpus callosum to the pial surface (Reynolds and Hardy 1997); arriving at the mature post-mitotic myelin-forming oligodendrocyte stage around PND20. In this final stage, high levels of myelin basic protein, proteolipid protein, and other myelin-related proteins are expressed. Several growth factors produced by glia (astrocytes and microglia) such as, IGF-1 (McMorris et al. 1990; Ishibashi et al. 2009), platelet-derived growth factor AA (PDGF-AA) (Raff et al. 1988), LIF (Mayer et al. 1994), FGF2 (Mayer et al. 1994), and osteopontin (Selvaraju et al. 2004) control proliferation and survival of oligodendrocytes and their precursors (Park et al. 2001). In the immature OPC stage, the cells display a greater level of vulnerability to various types of insults as compared to mature oligodendrocytes (Oka et al. 1993; Baerwald and Popko 1998).

In the later stages of oligodendrocyte maturation, the cells initiate the process of myelinating axons. Myelination is a major metabolic and structural event occurring in the normal progression of CNS development. The prominent phase of myelination proceeds over an extended postnatal period continuing into young adulthood. In the rat, extensive cellular proliferation of OPCs takes place around PND10 with the initiation of myelination occurring in the cerebrum between PND10 and PND12. The peak period of myelin accumulation around the time of weaning at PND21 involves a complex interaction between oligodendrocytes and axons. This peak is followed by further maturation and compaction of the myelin sheath. The enormous amount of myelin formed and maintained at considerable distance from the supporting glial cell body places an excessive burden on the myelinating oligodendrocyte to generate and maintain the myelin sheath. During this period the cells are maximally stressed in terms of metabolic and synthetic capacity, which may account for their elevated vulnerability to changes in their environment (Morell et al. 1994). During early postnatal development, microglia are present within the myelin tracts, evident in the subcortical white matter during the first week after birth in rodents (Hristova et al. 2010). While a higher level of microglia proliferation was not detected in the white matter during the postnatal period, a higher level of microglial cell death was found to occur in the periventricular white matter. Over this period, mRNA levels of IGF-1 and macrophage colony stimulating factor (M-CSF) were elevated in periventricular white matter phagocytes (microglia), possibly influencing microglia cell number and oligodendrocyte maturation (Hristova et al. 2010).

Both inflammation and changes in white matter microglia have been associated with alterations in oligodendroglia progenitor cells or myelination. For example, the induction of a pro-inflammatory M1-like state of cultured microglia by INFγ preferentially induced neurogenesis while, induction of an anti-inflammatory/repair M2-like activation state by IL-4 induced differentiation of adult NPCs to oligodendrocytes (Butovsky et al. 2006b) (see Chap. 6 for further details on the M1 and M2 macrophage phenotypes). This observation suggested that the M1/M2 inflammatory environment of neural progenitor cells significantly impacted cell fate during brain development. Clinically, the condition of pediatric white matter brain injury known as periventricular leukomalacia is considered a human example of developmental damage to OPC and hypomyelination. An association between periventricular leukomalacia and maternal infection/inflammation or fetal inflammatory (Leviton et al. 1999) has led to speculation that inflammation and microglia responses can be detrimental to oligodendrocytes and myelination (Volpe 2001). Work from Pang and colleagues, demonstrated that astrocyte and microglia-mediated effects of LPS on OPCs are sufficient to impede oligodendroglia lineage progression and survival in vitro (Pang et al. 2000). LPS diminishes the production of IGF-1 and ciliary neurotrophic factor (CNTF) by microglia, potentially exacerbating the negative effects of nitric oxide and pro-inflammatory cytokines on oligodendrocyte survival (Pang et al. 2010). The significance of this loss in trophic support is evidenced by the ability of exogenous IGF-1 and CNTF to provide a level of protection from LPS induced cell death (Pang et al. 2010).

Homozygous loss-of-function mutations in TREM2, encoding the triggering receptor expressed on myeloid cells 2 protein, are associated with an autosomal recessive form of early onset dementia and altered myelination (Klunemann et al. 2005). TREM2 plays a role in limiting the pro-inflammatory phenotype of macrophages and promoting the expression of molecules associated with antigen presentation. TREM2 is distributed intracellularly in two pools. One deposit is in the Golgi complex, the second is in a population of exocytic vesicles that are distinct from endosomes and lysosomes. This second population is continuously translocated to the cell surface and recycled back from the cell surface (Prada et al. 2006). From this work, the TREM2 surface density and likely its response to activation can be increased as a function of cell activation. $TREM2^+/DAP12^+$ microglia are found adjacent to oligodendrocytes in PND1 mice prior to the initiation of myelination (Thrash et al. 2009). Given that TREM2 inhibits microglial and macrophage cytokine expression and promotes microglial phagocytosis of cellular debris (Takahashi et al. 2007), one could make the assumption that the close apposition of $TREM2^+/DAP12^+$ microglia to oligodendrocytes prior to myelination represented a regulatory cell–cell interaction required for normal brain development.

In early studies by Nicholas and colleagues (2001), non-stimulated microglia were reported to significantly contribute to the survival and maturation of OPCs in culture.

This was demonstrated by the inhibition of OPC apoptosis and promotion of oligodendrocyte maturation as determined by staining for galactocerebroside, a marker of oligodendrocytes. This observation was supported by the work of Miller and colleagues reporting a reduction in caspase activation in oligodendrocytes when cells were cultured in the presence of microglia (Miller et al. 2007). Subsequent studies provided additional evidence that factors released normally by non-stimulated microglia into the cell culture media attenuated OPC proliferation and did not diminish cell survival (Taylor et al. 2010). However, conditioned media from LPS-stimulated cultured microglia was sufficient to induce OPC proliferation. This effect was attributed to the release of tumor necrosis factor alpha (TNFα) and IL-6 by the stimulated microglia (Taylor et al. 2010). Work by Filipovic and Zecevic (2005) suggested that the OPC proliferative effect of LPS-stimulated microglia-conditioned media involved Golli proteins, products of the myelin basic protein gene, as mediators of OPC proliferation and differentiation (Paez et al. 2012).

8.4 Microglia and the Vasculature

In the mid 1930s, three investigations of the relationship between microglia and blood vessels were reported in the literature. Initially, Von Sántha described the presence of microglia and their intimate association with blood vessels prior to mid-gestation in numerous non-primate species (von Sántha 1932). In the rat embryo, the earliest appearance of microglia in the rhombencephalon and diencephalon coincided with the onset of vascularization (von Sántha and Juba 1933). This association between microglia and the vascular system can also be observed in the developing human brain with a temporal association of microglia progenitor influx occurring with vascularization (Kershman 1939). It was not until the 1970s that research extended the associations reported in earlier studies, finding precursors of ramified brain-resident microglia expressing the adult myeloid cell marker, F4/80, in the mature rat brain. These precursor cells were found in the vicinity of brain capillaries and characterized by a round or irregular-shaped cell body, with or without pseudopodia (Imamoto and Leblond 1978). When the fetal human brain was examined for precursor or immature cell location between 16 and 22 weeks of gestation, microglia were located at highly vascularized sites (Rezaie et al. 1997). A contact relationship was implied in the human brain with the expression of intracellular adhesion molecule (ICAM)-2 on cerebral endothelium (Rezaie et al. 1997). These human microglial progenitors showed a close association with the parenchymal wall of penetrating radial blood vessels (Rezaie et al. 2005). While these studies demonstrated a temporal and spatial association between microglia and developing vasculature the functional implications of co-localization has not been clearly defined to date (for review see (Arnold and Betsholtz 2013)).

8.4.1 Ontogeny of Vascularization and Microglia Interactions

Vascularization of the CNS is initiated exclusively by angiogenesis. Tip cells in nascent capillaries anastomose with each other during vascular sprouting. This creates vascular loops moving toward the center of the neural tube establishing a temporary plexus (perineural vascular plexus (PNVP)) around the ventricular spaces and the spinal cord central canal. In rodents, this starts at ED9 and continues until ED15 (Risau 1997). The process follows a caudal-cephalic gradient commencing at the myelencephalon and ascending through the metencephalon, mesencephalon, diencephalon, and telencephalon (Marin-Padilla 1985). Capillary sprouts perforate the CNS and cell–cell communication directs microglia migration towards developing vessels (Fig. 8.2). When examined in cultured tissue slices, the addition of microglia in the aortic ring model was sufficient to stimulate vessel sprouting. This effect could be partially reproduced with the addition of microglial conditioned media suggesting a potential role for cell–cell communication via contact and

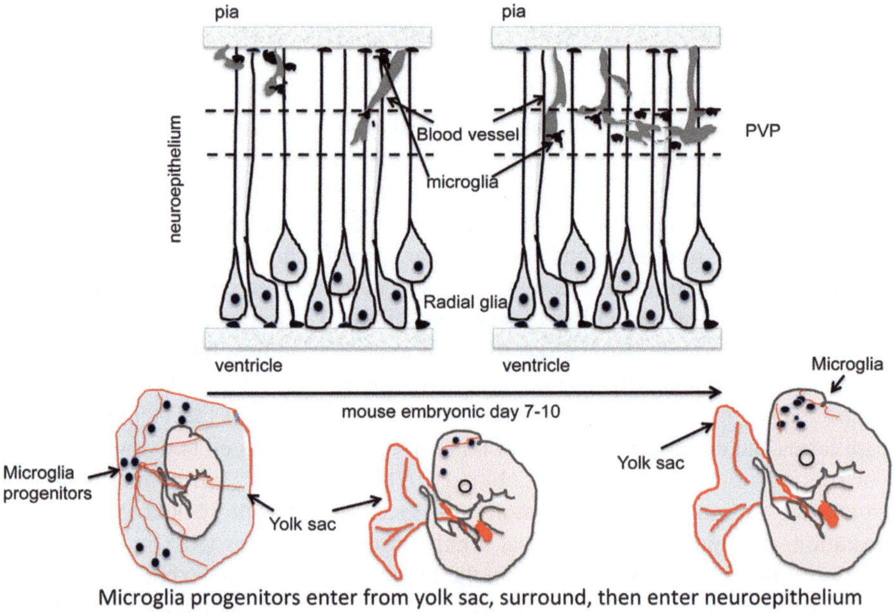

Fig. 8.2 Microglial progenitors migrate from the yolk sac to the embryo. Microglia in the neuroepithelium are evident in the mouse by embryonic day (ED)10. The new microglia associated with radial glia transgressing from the ventricle to the pial and with blood vessels engrossing into the brain from the pial surface. Initially the association with blood vessels is observed at the pial surface with microglia displaying a rounded or stunted process-bearing phenotype (noted as small dark cells located along the pial surface or within the periventricular vascular plexus (PVP) in contact with blood vessels). As the blood vessels further develop the association of microglia is demonstrated in the PVP where they are implicated in promoting fusion of vascular tip cells and maturation of the neurovasculature

soluble microglial-derived factors to stimulate angiogenic activity (Rymo et al. 2011). It has been suggested that microglia serve an angiogenic role similar to that of macrophages, including release of factors such as TGFβ, FGF2, matrix metalloproteinases, and cytokines to break down the basement membrane directly or via stimulation of other cells (Polverini et al. 1977; Sunderkotter et al. 1991; Lingen 2001) in order to regulate angiogenesis of the developing brain (Kurz 2000; Checchin et al. 2006).

Once the neurovasculature is established, end-feet processes of juxtavascular microglia physically contact the basal lamina of microvessels in the rat cerebrum and spinal cord (Lassmann et al. 1991). Contact was observed with vessels of various sizes and types, including arteries, veins, and capillaries. While residing in the parenchyma, this specialized population of microglia remains distinct from other parenchyma microglia by the nature of their contact with blood vessels. Interaction with blood vessels is maintained and can be stimulated upon vessel injury. Using time-lapsed microscopy of brain slices from immature postnatal rodents, Grossmann and colleagues demonstrated that, following traumatic stimulation, juxtavascular microglia easily migrate along the parenchymal surface of the vessel (Grossmann et al. 2002). In contrast, any stimulation of movement along blood vessels for parenchyma microglia required a distinct retraction of microglia processes allowing for greater mobility. In general, when microglia were tracked along vessels they displayed a flattened morphology over the vessel surface, extending a leading process parallel to the vessel. A heterogeneous response was also observed with some cells showing a transient response to stimulation while other cells maintain migration for hours and cover extended distances. In addition to these migrating cells, a second group of blood vessel-associated microglia was observed in the injured tissue. These cells remained stationary and apparently anchored to the blood vessel, yet maintaining motility by actively extending processes into the surrounding parenchymal tissue (Grossmann et al. 2002). Thus, with physical injury, some microglia were observed to migrate along the blood vessels while other activated microglia allowed for additional contact into the parenchyma by extension of processes. While it has not been examined, it is likely, given the postnatal age at which microglia were examined, that this distinction remains in the adult brain.

Influence of the brain vasculature on microglia can be observed in morphological phenotype and development. During the normal maturation of microglia, cells shift from a round non-process-bearing cells to cells of a fully ramified morphology. A delay in maturation is observed under conditions of decreased capillary flow, suggesting a direct influence of capillary flow on normal developmental process (Masuda et al. 2011). An influence of the vascular system on microglia has been implied in the adult brain. Microglia located in brain regions lacking a blood–brain barrier (BBB), including the circumventricular organs and the choroid plexus, display a morphological phenotype characterized by shorter thicker processes as compared to parenchymal microglia (Perry and Gordon 1987). In this case, differences observed in the adult brain have been interpreted as indicative of the cells being maintained in a higher activation state. It has also been speculated that specific characteristics of these local microglia may reflect an influence of serum proteins made available due to the diminished BBB (Chamak and Mallat 1991; Fujita et al. 1996). However, these differences observed in development and in the adult brain did not appear associated only with blood vessel presence. Microglia located in brain areas receiving the best innervation of blood vessels display a more complex process ramification pattern, while in areas lacking a BBB and often diminished in vascularization, microglia display a stunted morphology (Rowan and Maxwell 1981; Vela et al. 1995). This may be related to the lower level of innervation or alternatively due to the leakage of blood-borne factors into the area stimulating microglia.

8.4.2 Association of Vascularization and Microglia Colonization

While alternative routes have been considered, the association of microglia around blood vessels continues to implicate the vascular system as an avenue for microglial entry into the brain. However, there remains little hard data to demonstrate that microglia use a vascular route of entry. When examined in the mouse spinal cord, microglia were found to reach the area surrounding the embryonic mouse spinal cord through the developing vasculature. Using a green fluorescent protein (eGFP) marker, microglia were identified as early as ED11.5 inside the primitive arterial tract. Between ED12.5 and 14.5, these cells proliferated and accumulated in the dorsal part of the PNVP (Rigato et al. 2011). Within the external dorsolateral region of the spinal cord, amoeboid microglia clustered close to terminals of dying dorsal root ganglia neurons and within the lateral motor columns at the onset of the developmental motor neuron death. This occurred between ED12.5 and 13.5, and in the first 24 h, ramified microglia within the parenchyma interacted with growing capillaries. Around ED14.5, approximately 30 % of the total microglia within the gray matter initiated contact with growing capillaries. While microglia were evident in the surround area, further examination of the regions by von Willenbrand factor immunoreactivity of the blood vessel and eGFP cell bodies in the gray matter led Rigato and colleagues to conclude that, while microglia make contact with developing blood capillaries in the spinal cord, there is little evidence that they actually utilize this route for colonization (Rigato et al. 2011).

In the avian embryo, the use of chick-quail transplantation and parabiosis chimeras suggested that vascularization is not required for microglia colonization of the CNS. Cells were observed to invade the CNS through the pial basal lamina before vascularization occurred (Cuadros et al. 1993; Kurz and Christ 1998). In the zebrafish embryo, live recordings of cell movement showed migration of microglia through the cephalic mesenchyme toward the brain pial surface and roof of the fourth ventricle with colonization of the brain and retina occurring through a M-CSF receptor-dependent process (Herbomel et al. 2001).

In the retina, heterogeneous populations of microglia are believed to be of hemangioblastic mesodermal origin (Streit 2001). In fetal and adult human retina, microglia are restricted to the inner vascularized regions. These include the nerve fiber/ganglion cell layer and the inner and outer plexiform layers (Diaz-Araya et al. 1995b; Provis et al. 1995). During human retinal development, microglia lacking macrophage markers are present by 10GW. This is prior to astrocyte invasion and the onset of vasculogenesis (Diaz-Araya et al. 1995a). At approximately 14GW, macrophage antigen expressing microglia were detected, coinciding with the beginning of vascularization (Diaz-Araya et al. 1995a). These cells may represent vessel-associated perivascular microglia detected in the adult retina (Provis et al. 1995, 1996). With further maturation, microglia extend processes to orientate toward and contact blood vessels (Ling 1982; Ashwell et al. 1989; Diaz-Araya et al. 1995b; Pennell and Streit 1997). Using clodronate liposomes to selectively deplete circulating monocytes or retinal microglia, Checchin and colleagues (2006) reported that selectively reducing retinal microglia number diminished vascular density. However, depletion of microglia by downregulation of M-CSF diminished branching anastomosis during retinal development without altering the number of endothelial tip cells and filopodia (Kubota et al. 2009). A diminished number of vascular branch points was also observed in the hindbrain after M-CSF-dependent depletion of microglia (Fantin et al. 2010). Not only were branch points decreased in the absence of microglia but also the angle of filopodia extending from tip cells (Rymo et al. 2011). Further work by Unoki and colleagues found that clodronate liposome depletion of microglia in ex vivo retinas significantly reduced the potency of vascular endothelial growth factor (VEGF)-induced neovascular sprouting (Unoki et al. 2010).

Cell–cell communication between microglia and the vasculature in the retina has been demonstrated. Notch-activated retinal microglia were found in close association with Notch ligand, Delta-like ligand 4 (Dll4)-expressing endothelial tip cells (Outtz et al. 2011). This proximity was supported by the observation that a genetic deletion of Notch1 in retinal microglia decreased the number of microglia localized to the vasculature. In vitro, the presence of aortic rings serves as an attractant for microglia, and both the physical presence of microglia and their released factors stimulate vascular branching from the ring (Rymo et al. 2011). In the communication between microglia and aortic rings, neutralizing VEGF receptor activation did not alter the induction of branching (Rymo et al. 2011). Stefater and colleagues reported that retinal myeloid cells produce Wnt ligands that suppress angiogenic branching through a non-canonical, Wnt-Flt1 pathway (Stefater et al. 2011). Haploinsufficiency of Wnt5a, Wnt11, the common Wnt-ligand transporter, Wls, or the VEGF inhibitory protein, sFlt1, resulted in increased vascular branching in the deep vascular plexus of the retina. However, Fantin and colleagues reported that the lack/absence of microglia had no effect on branch density in this region (Fantin et al. 2010). This apparent difference could be due to a number of factors. First, Stefater and colleagues examined myeloid cells and not microglial cells specifically (Stefater et al. 2011). Second, it is possible that the diminished number of microglia in the M-CSF osteopetrotic (op/op) mice used by Fantin and colleagues did not result in a significant depletion of released factors (Fantin et al. 2010). Rather, a lower threshold for activation or a higher level of soluble factor release may have compensated for the diminished number of microglia. Such a compensatory response has been demonstrated in op/op mice following brain injury (Bruccoleri and Harry 2000). Thus, with further investigation the findings of Rymo and colleagues (2011) and Fantin and colleagues (2010) may not be as discordant as they appear.

8.5 Conclusions

While substantial data exists that supports an essential role for microglia during the formation and maturation of the CNS, the exact nature of such a role remains to be determined. The timing and location of microglia with the brain vasculature, radial glia, and neural, astroglial and oligodendroglial progenitor cells imply a close and functional association. The experimental data available thus far, suggests a role for microglia in the regulation of angiogenesis and in the maintenance of the progenitor pool via phagocytosis and release of factors that promote progenitor proliferation. Additional evidence exists supporting a microglia-influence on cell fate via the secretion of pro-neurogenic and pro-gliogenic molecules. However, microglia are not the only source of proteins necessary for normal development, and it remains to be determined whether these myeloid cells secrete any critical components at the pertinent developmental time points, and thus, whether they are recruited as essential players.

Acknowledgments This research was supported by the National Toxicology Program Division, the National Institute of Environmental Health Sciences, the National Institutes of Health, and the Department of Health and Human Services #1Z01ES101623 and ES021164. The views expressed in this article are those of the author and they do not necessarily represent the views or policies of the National Toxicology Program.

L. Pont-Lezica was supported by the seventh framework program Moodinflame (222963).

References

Aberg MA, Aberg ND, Palmer TD, Alborn AM, Carlsson-Skwirut C, Bang P et al (2003) IGF-I has a direct proliferative effect in adult hippocampal progenitor cells. Mol Cell Neurosci 24:23–40

Adler MW, Geller EB, Chen X, Rogers TJ (2005) Viewing chemokines as a third major system of communication in the brain. AAPS J 7:E865–E870

Ajmone-Cat MA, Cacci E, Ragazzoni Y, Minghetti L, Biagioni S (2010) Pro-gliogenic effect of IL-1alpha in the differentiation of embryonic neural precursor cells in vitro. J Neurochem 113:1060–1072

Akaneya Y, Takahashi M, Hatanaka H (1995) Interleukin-1 beta enhances survival and interleukin-6 protects against MPP+neurotoxicity in cultures of fetal rat dopaminergic neurons. Exp Neurol 136:44–52

Alliot F, Godin I, Pessac B (1999) Microglia derive from progenitors, originating from the yolk sac, and which proliferate in the brain. Brain Res Dev Brain Res 117:145–152

Antony JM, Paquin A, Nutt SL, Kaplan DR, Miller FD (2011) Endogenous microglia regulate development of embryonic cortical precursor cells. J Neurosci Res 89:286–298

Aquino DA, Padin C, Perez JM, Peng D, Lyman WD, Chiu FC (1996) Analysis of glial fibrillary acidic protein, neurofilament protein, actin and heat shock proteins in human fetal brain during the second trimester. Brain Res Dev Brain Res 91:1–10

Araujo DM, Cotman CW (1992) Basic FGF in astroglial, microglial, and neuronal cultures: characterization of binding sites and modulation of release by lymphokines and trophic factors. J Neurosci 12:1668–1678

Arnold T, Betsholtz C (2013) The importance of microglia in the development of the vasculature in the central nervous system. Vasc Cell 5:4

Arsenijevic Y, Weiss S, Schneider B, Aebischer P (2001) Insulin-like growth factor-I is necessary for neural stem cell proliferation and demonstrates distinct actions of epidermal growth factor and fibroblast growth factor-2. J Neurosci 21:7194–7202

Ashwell KW, Hollander H, Streit W, Stone J (1989) The appearance and distribution of microglia in the developing retina of the rat. Vis Neurosci 2:437–448

Baerwald KD, Popko B (1998) Developing and mature oligodendrocytes respond differently to the immune cytokine interferon-gamma. J Neurosci Res 52:230–239

Bain JM, Ziegler A, Yang Z, Levison SW, Sen E (2010) TGFbeta1 stimulates the over-production of white matter astrocytes from precursors of the "brain marrow" in a rodent model of neonatal encephalopathy. PLoS One 5:e9567

Barnabe-Heider F, Wasylnka JA, Fernandes KJ, Porsche C, Sendtner M, Kaplan DR et al (2005) Evidence that embryonic neurons regulate the onset of cortical gliogenesis via cardiotrophin-1. Neuron 48:253–265

Batchelor PE, Liberatore GT, Wong JY, Porritt MJ, Frerichs F, Donnan GA et al (1999) Activated macrophages and microglia induce dopaminergic sprouting in the injured striatum and express brain-derived neurotrophic factor and glial cell line-derived neurotrophic factor. J Neurosci 19:1708–1716

Battista D, Ferrari CC, Gage FH, Pitossi FJ (2006) Neurogenic niche modulation by activated microglia: transforming growth factor beta increases neurogenesis in the adult dentate gyrus. Eur J Neurosci 23:83–93

Beck KD, Powell-Braxton L, Widmer HR, Valverde J, Hefti F (1995) Igf1 gene disruption results in reduced brain size, CNS hypomyelination, and loss of hippocampal granule and striatal parvalbumin-containing neurons. Neuron 14:717–730

Ben-Hur T, Ben-Menachem O, Furer V, Einstein O, Mizrachi-Kol R, Grigoriadis N (2003) Effects of proinflammatory cytokines on the growth, fate, and motility of multipotential neural precursor cells. Mol Cell Neurosci 24:623–631

Bergstrom T, Forsberg-Nilsson K (2012) Neural stem cells: brain building blocks and beyond. Ups J Med Sci 117:132–142

Brenneman DE, Schultzberg M, Bartfai T, Gozes I (1992) Cytokine regulation of neuronal survival. J Neurochem 58:454–460

Bruccoleri A, Harry GJ (2000) Chemical-induced hippocampal neurodegeneration and elevations in TNFalpha, TNFbeta, IL-1alpha, IP-10, and MCP-1 mRNA in osteopetrotic (op/op) mice. J Neurosci Res 62:146–155

Butovsky O, Landa G, Kunis G, Ziv Y, Avidan H, Greenberg N et al (2006a) Induction and blockage of oligodendrogenesis by differently activated microglia in an animal model of multiple sclerosis. J Clin Invest 116:905–915

Butovsky O, Ziv Y, Schwartz A, Landa G, Talpalar AE, Pluchino S et al (2006b) Microglia activated by IL-4 or IFN-gamma differentially induce neurogenesis and oligodendrogenesis from adult stem/progenitor cells. Mol Cell Neurosci 31:149–160

Cacci E, Ajmone-Cat MA, Anelli T, Biagioni S, Minghetti L (2008) In vitro neuronal and glial differentiation from embryonic or adult neural precursor cells are differently affected by chronic or acute activation of microglia. Glia 56:412–425

Caldero J, Brunet N, Ciutat D, Hereu M, Esquerda JE (2009) Development of microglia in the chick embryo spinal cord: implications in the regulation of motoneuronal survival and death. J Neurosci Res 87:2447–2466

Cao Q, Lu J, Kaur C, Sivakumar V, Li F, Cheah PS et al (2008) Expression of Notch-1 receptor and its ligands Jagged-1 and Delta-1 in amoeboid microglia in postnatal rat brain and murine BV-2 cells. Glia 56:1224–1237

Carpentier PA, Palmer TD (2009) Immune influence on adult neural stem cell regulation and function. Neuron 64:79–92

Chamak B, Mallat M (1991) Fibronectin and laminin regulate the in vitro differentiation of microglial cells. Neuroscience 45:513–527

Chang MY, Park CH, Son H, Lee YS, Lee SH (2004) Developmental stage-dependent self-regulation of embryonic cortical precursor cell survival and differentiation by leukemia inhibitory factor. Cell Death Differ 11:985–996

Checchin D, Sennlaub F, Levavasseur E, Leduc M, Chemtob S (2006) Potential role of microglia in retinal blood vessel formation. Invest Ophthalmol Vis Sci 47:3595–3602

Chenn A (2009) A top-NOTCH way to make astrocytes. Dev Cell 16:158–159

Choi YS, Cho HY, Hoyt KR, Naegele JR, Obrietan K (2008) IGF-1 receptor-mediated ERK/MAPK signaling couples status epilepticus to progenitor cell proliferation in the subgranular layer of the dentate gyrus. Glia 56:791–800

Covey MV, Levison SW (2007) Leukemia inhibitory factor participates in the expansion of neural stem/progenitors after perinatal hypoxia/ischemia. Neuroscience 148:501–509

Cowell RM, Silverstein FS (2003) Developmental changes in the expression of chemokine receptor CCR1 in the rat cerebellum. J Comp Neurol 457:7–23

Cuadros MA, Martin C, Coltey P, Almendros A, Navascues J (1993) First appearance, distribution, and origin of macrophages in the early development of the avian central nervous system. J Comp Neurol 330:113–129

Cunningham CL, Martinez-Cerdeno V, Noctor SC (2013) Microglia regulate the number of neural precursor cells in the developing cerebral cortex. J Neurosci 33:4216–4233

D'Ercole AJ, Ye P (2008) Minireview: expanding the mind: insulin-like growth factor I and brain development. Endocrinology 149:5958–5962

Dalmau I, Finsen B, Tonder N, Zimmer J, Gonzalez B, Castellano B (1997) Development of microglia in the prenatal rat hippocampus. J Comp Neurol 377:70–84

de la Mano A, Gato A, Alonso MI, Carnicero E, Martin C, Moro JA (2007) Role of interleukin-1beta in the control of neuroepithelial proliferation and differentiation of the spinal cord during development. Cytokine 37:128–137

De Simone R, Levi G, Aloisi F (1998) Interferon gamma gene expression in rat central nervous system glial cells. Cytokine 10:418–422

Deneen B, Ho R, Lukaszewicz A, Hochstim CJ, Gronostajski RM, Anderson DJ (2006) The transcription factor NFIA controls the onset of gliogenesis in the developing spinal cord. Neuron 52:953–968

Deverman BE, Patterson PH (2009) Cytokines and CNS development. Neuron 64:61–78

Diaz-Araya CM, Provis JM, Penfold PL (1995a) Ontogeny and cellular expression of MHC and leucocyte antigens in human retina. Glia 15:458–470

Diaz-Araya CM, Provis JM, Penfold PL, Billson FA (1995b) Development of microglial topography in human retina. J Comp Neurol 363:53–68

Drago J, Murphy M, Carroll SM, Harvey RP, Bartlett PF (1991) Fibroblast growth factor-mediated proliferation of central nervous system precursors depends on endogenous production of insulin-like growth factor I. Proc Natl Acad Sci U S A 88:2199–2203

Dziegielewska KM, Moller JE, Potter AM, Ek J, Lane MA, Saunders NR (2000) Acute-phase cytokines IL-1beta and TNF-alpha in brain development. Cell Tissue Res 299:335–345

Ellison JA, de Vellis J (1994) Platelet-derived growth factor receptor is expressed by cells in the early oligodendrocyte lineage. J Neurosci Res 37:116–128

Falk S, Wurdak H, Ittner LM, Ille F, Sumara G, Schmid MT et al (2008) Brain area-specific effect of TGF-beta signaling on Wnt-dependent neural stem cell expansion. Cell Stem Cell 2:472–483

Fantin A, Vieira JM, Gestri G, Denti L, Schwarz Q, Prykhozhij S et al (2010) Tissue macrophages act as cellular chaperones for vascular anastomosis downstream of VEGF-mediated endothelial tip cell induction. Blood 116:829–840

Filipovic R, Zecevic N (2005) Interaction between microglia and oligodendrocyte cell progenitors involves Golli proteins. Ann N Y Acad Sci 1048:166–174

Fox IJ, Paucar AA, Nakano I, Mottahedeh J, Dougherty JD, Kornblum HI (2004) Developmental expression of glial fibrillary acidic protein mRNA in mouse forebrain germinal zones—implications for stem cell biology. Brain Res Dev Brain Res 153:121–125

Freeman MR (2010) Specification and morphogenesis of astrocytes. Science 330:774–778

Fujita H, Tanaka J, Toku K, Tateishi N, Suzuki Y, Matsuda S et al (1996) Effects of GM-CSF and ordinary supplements on the ramification of microglia in culture: a morphometrical study. Glia 18:269–281

Fushimi S, Shirabe T (2004) Expression of insulin-like growth factors in remyelination following ethidium bromide-induced demyelination in the mouse spinal cord. Neuropathology 24: 208–218

Garden GA, Moller T (2006) Microglia biology in health and disease. J Neuroimmune Pharmacol 1:127–137

Ginhoux F, Greter M, Leboeuf M, Nandi S, See P, Gokhan S et al (2010) Fate mapping analysis reveals that adult microglia derive from primitive macrophages. Science 330:841–845

Giulian D, Young DG, Woodward J, Brown DC, Lachman LB (1988) Interleukin-1 is an astroglial growth factor in the developing brain. J Neurosci 8:709–714

Gonzalez-Perez O, Gutierrez-Fernandez F, Lopez-Virgen V, Collas-Aguilar J, Quinones-Hinojosa A, Garcia-Verdugo JM (2012) Immunological regulation of neurogenic niches in the adult brain. Neuroscience 226:270–281

Gordon RJ, McGregor AL, Connor B (2009) Chemokines direct neural progenitor cell migration following striatal cell loss. Mol Cell Neurosci 41:219–232

Gordon RJ, Mehrabi NF, Maucksch C, Connor B (2012) Chemokines influence the migration and fate of neural precursor cells from the young adult and middle-aged rat subventricular zone. Exp Neurol 233:587–594

Goshen I, Kreisel T, Ounallah-Saad H, Renbaum P, Zalzstein Y, Ben-Hur T et al (2007) A dual role for interleukin-1 in hippocampal-dependent memory processes. Psychoneuroendocrinology 32:1106–1115

Gotz M, Huttner WB (2005) The cell biology of neurogenesis. Nat Rev Mol Cell Biol 6:777–788

Grandbarbe L, Michelucci A, Heurtaux T, Hemmer K, Morga E, Heuschling P (2007) Notch signaling modulates the activation of microglial cells. Glia 55:1519–1530

Green HF, Treacy E, Keohane AK, Sullivan AM, O'Keeffe GW, Nolan YM (2012) A role for interleukin-1beta in determining the lineage fate of embryonic rat hippocampal neural precursor cells. Mol Cell Neurosci 49:311–321

Grossmann R, Stence N, Carr J, Fuller L, Waite M, Dailey ME (2002) Juxtavascular microglia migrate along brain microvessels following activation during early postnatal development. Glia 37:229–240

Gu F, Wang J, Fu L, Ma YJ (2011) Co-culture with microglia promotes neural stem cells differentiation into astrocytes. Chin Med J (Engl) 124:3394–3398

Guan J, Bennet L, Gluckman PD, Gunn AJ (2003) Insulin-like growth factor-1 and post-ischemic brain injury. Prog Neurobiol 70:443–462

He Y, Cai W, Wang L, Chen P (2009) A developmental study on the expression of PDGFalphaR immunoreactive cells in the brain of postnatal rats. Neurosci Res 65:272–279

Hellstrom NA, Bjork-Eriksson T, Blomgren K, Kuhn HG (2009) Differential recovery of neural stem cells in the subventricular zone and dentate gyrus after ionizing radiation. Stem Cells 27: 634–641

Herbomel P, Thisse B, Thisse C (2001) Zebrafish early macrophages colonize cephalic mesenchyme and developing brain, retina, and epidermis through a M-CSF receptor-dependent invasive process. Dev Biol 238:274–288

Hodge RD, D'Ercole AJ, O'Kusky JR (2004) Insulin-like growth factor-I accelerates the cell cycle by decreasing G1 phase length and increases cell cycle reentry in the embryonic cerebral cortex. J Neurosci 24:10201–10210

Hristova M, Cuthill D, Zbarsky V, Acosta-Saltos A, Wallace A, Blight K et al (2010) Activation and deactivation of periventricular white matter phagocytes during postnatal mouse development. Glia 58:11–28
Hua LL, Lee SC (2000) Distinct patterns of stimulus-inducible chemokine mRNA accumulation in human fetal astrocytes and microglia. Glia 30:74–81
Hurley SD, Streit WJ (1995) Griffonia simplicifolia II lectin labels a population of radial glial cells in the embryonic rat brain. Dev Neurosci 17:324–334
Imamoto K, Leblond CP (1978) Radioautographic investigation of gliogenesis in the corpus callosum of young rats. II. Origin of microglial cells. J Comp Neurol 180:139–163
Ishibashi T, Lee PR, Baba H, Fields RD (2009) Leukemia inhibitory factor regulates the timing of oligodendrocyte development and myelination in the postnatal optic nerve. J Neurosci Res 87:3343–3355
Johansson S, Stromberg I (2002) Guidance of dopaminergic neuritic growth by immature astrocytes in organotypic cultures of rat fetal ventral mesencephalon. J Comp Neurol 443:237–249
Johansson S, Price J, Modo M (2008) Effect of inflammatory cytokines on major histocompatibility complex expression and differentiation of human neural stem/progenitor cells. Stem Cells 26:2444–2454
Jurynczyk M, Selmaj K (2010) Notch: a new player in MS mechanisms. J Neuroimmunol 218: 3–11
Kalman M, Ajtai BM (2001) A comparison of intermediate filament markers for presumptive astroglia in the developing rat neocortex: immunostaining against nestin reveals more detail, than GFAP or vimentin. Int J Dev Neurosci 19:101–108
Kaur C, Ling EA (2009) Periventricular white matter damage in the hypoxic neonatal brain: role of microglial cells. Prog Neurobiol 87:264–280
Kaur C, Sivakumar V, Dheen ST, Ling EA (2006) Insulin-like growth factor I and II expression and modulation in amoeboid microglial cells by lipopolysaccharide and retinoic acid. Neuroscience 138:1233–1244
Kearns CM, Gash DM (1995) GDNF protects nigral dopamine neurons against 6-hydroxydopamine in vivo. Brain Res 672:104–111
Kershman J (1939) Genesis of microglia in the human brain. Arch Neurol Psychiatry 41:24–50
Kessaris N, Fogarty M, Iannarelli P, Grist M, Wegner M, Richardson WD (2006) Competing waves of oligodendrocytes in the forebrain and postnatal elimination of an embryonic lineage. Nat Neurosci 9:173–179
Kessaris N, Pringle N, Richardson WD (2008) Specification of CNS glia from neural stem cells in the embryonic neuroepithelium. Philos Trans R Soc Lond B Biol Sci 363:71–85
Kettenmann H, Hanisch UK, Noda M, Verkhratsky A (2011) Physiology of microglia. Physiol Rev 91:461–553
Kim SU, de Vellis J (2005) Microglia in health and disease. J Neurosci Res 81:302–313
Klunemann HH, Ridha BH, Magy L, Wherrett JR, Hemelsoet DM, Keen RW et al (2005) The genetic causes of basal ganglia calcification, dementia, and bone cysts: DAP12 and TREM2. Neurology 64:1502–1507
Koo JW, Duman RS (2008) IL-1beta is an essential mediator of the antineurogenic and anhedonic effects of stress. Proc Natl Acad Sci U S A 105:751–756
Kreutzberg GW (1996) Microglia: a sensor for pathological events in the CNS. Trends Neurosci 19:312–318
Krieglstein K, Suter-Crazzolara C, Fischer WH, Unsicker K (1995) TGF-beta superfamily members promote survival of midbrain dopaminergic neurons and protect them against MPP+ toxicity. EMBO J 14:736–742
Kriegstein A, Alvarez-Buylla A (2009) The glial nature of embryonic and adult neural stem cells. Annu Rev Neurosci 32:149–184
Kubota Y, Takubo K, Shimizu T, Ohno H, Kishi K, Shibuya M et al (2009) M-CSF inhibition selectively targets pathological angiogenesis and lymphangiogenesis. J Exp Med 206: 1089–1102

Kurz H (2000) Physiology of angiogenesis. J Neurooncol 50:17–35
Kurz H, Christ B (1998) Embryonic CNS macrophages and microglia do not stem from circulating, but from extravascular precursors. Glia 22:98–102
Lassmann H, Zimprich F, Vass K, Hickey WF (1991) Microglial cells are a component of the perivascular glia limitans. J Neurosci Res 28:236–243
Leviton A, Paneth N, Reuss ML, Susser M, Allred EN, Dammann O et al (1999) Maternal infection, fetal inflammatory response, and brain damage in very low birth weight infants. Developmental Epidemiology Network Investigators. Pediatr Res 46:566–575
Li L, Walker TL, Zhang Y, Mackay EW, Bartlett PF (2008) Endogenous interferon gamma directly regulates neural precursors in the non-inflammatory brain. J Neurosci 30:9038–9050
Lin LF, Doherty DH, Lile JD, Bektesh S, Collins F (1993) GDNF: a glial cell line-derived neurotrophic factor for midbrain dopaminergic neurons. Science 260:1130–1132
Ling EA (1982) A light microscopic demonstration of amoeboid microglia and microglial cells in the retina of rats of various ages. Arch Histol Jpn 45:37–44
Ling ZD, Potter ED, Lipton JW, Carvey PM (1998) Differentiation of mesencephalic progenitor cells into dopaminergic neurons by cytokines. Exp Neurol 149:411–423
Lingen MW (2001) Role of leukocytes and endothelial cells in the development of angiogenesis in inflammation and wound healing. Arch Pathol Lab Med 125:67–71
Liu JS, Amaral TD, Brosnan CF, Lee SC (1998) IFNs are critical regulators of IL-1 receptor antagonist and IL-1 expression in human microglia. J Immunol 161:1989–1996
Mairet-Coello G, Tury A, DiCicco-Bloom E (2009) Insulin-like growth factor-1 promotes G(1)/S cell cycle progression through bidirectional regulation of cyclins and cyclin-dependent kinase inhibitors via the phosphatidylinositol 3-kinase/Akt pathway in developing rat cerebral cortex. J Neurosci 29:775–788
Malatesta P, Hartfuss E, Gotz M (2000) Isolation of radial glial cells by fluorescent-activated cell sorting reveals a neuronal lineage. Development 127:5253–5263
Malatesta P, Hack MA, Hartfuss E, Kettenmann H, Klinkert W, Kirchhoff F et al (2003) Neuronal or glial progeny: regional differences in radial glia fate. Neuron 37:751–764
Malatesta P, Appolloni I, Calzolari F (2008) Radial glia and neural stem cells. Cell Tissue Res 331:165–178
Mao L, Lau YS, Petroske E, Wang JQ (2001) Profound astrogenesis in the striatum of adult mice following nigrostriatal dopaminergic lesion by repeated MPTP administration. Brain Res Dev Brain Res 131:57–65
Marin-Padilla M (1985) Early vascularization of the embryonic cerebral cortex: Golgi and electron microscopic studies. J Comp Neurol 241:237–249
Masuda T, Croom D, Hida H, Kirov SA (2011) Capillary blood flow around microglial somata determines dynamics of microglial processes in ischemic conditions. Glia 59:1744–1753
Mayer M, Bhakoo K, Noble M (1994) Ciliary neurotrophic factor and leukemia inhibitory factor promote the generation, maturation and survival of oligodendrocytes in vitro. Development 120:143–153
McManus C, Berman JW, Brett FM, Staunton H, Farrell M, Brosnan CF (1998) MCP-1, MCP-2 and MCP-3 expression in multiple sclerosis lesions: an immunohistochemical and in situ hybridization study. J Neuroimmunol 86:20–29
McMorris FA, Furlanetto RW, Mozell RL, Carson MJ, Raible DW (1990) Regulation of oligodendrocyte development by insulin-like growth factors and cyclic nucleotides. Ann N Y Acad Sci 605:101–109
Miller FD, Gauthier AS (2007) Timing is everything: making neurons versus glia in the developing cortex. Neuron 54:357–369
Miller MW, Luo J (2002) Effects of ethanol and basic fibroblast growth factor on the transforming growth factor beta1 regulated proliferation of cortical astrocytes and C6 astrocytoma cells. Alcohol Clin Exp Res 26:671–676
Miller BA, Crum JM, Tovar CA, Ferguson AR, Bresnahan JC, Beattie MS (2007) Developmental stage of oligodendrocytes determines their response to activated microglia in vitro. J Neuroinflammation 4:28

Miyata T, Kawaguchi D, Kawaguchi A, Gotoh Y (2010) Mechanisms that regulate the number of neurons during mouse neocortical development. Curr Opin Neurobiol 20:22–28

Molofsky AV, Krencik R, Ullian EM, Tsai HH, Deneen B, Richardson WD et al (2012) Astrocytes and disease: a neurodevelopmental perspective. Genes Dev 26:891–907

Monier A, Evrard P, Gressens P, Verney C (2006) Distribution and differentiation of microglia in the human encephalon during the first two trimesters of gestation. J Comp Neurol 499: 565–582

Morell P, Quarles RH, Norton WT (1994) Myelin formation, structure, and biochemistry. In: Albers RW, Siegel GW, Molinoff P, Agranoff B (eds) Basic neurochemistry, 5th edn. Raven, New York, pp 117–143

Morrow T, Song MR, Ghosh A (2001) Sequential specification of neurons and glia by developmentally regulated extracellular factors. Development 128:3585–3594

Mousa A, Seiger A, Kjaeldgaard A, Bakhiet M (1999) Human first trimester forebrain cells express genes for inflammatory and anti-inflammatory cytokines. Cytokine 11:55–60

Nakajima K, Kohsaka S (2004) Microglia: neuroprotective and neurotrophic cells in the central nervous system. Curr Drug Targets Cardiovasc Haematol Disord 4:65–84

Nakajima K, Tohyama Y, Maeda S, Kohsaka S, Kurihara T (2007) Neuronal regulation by which microglia enhance the production of neurotrophic factors for GABAergic, catecholaminergic, and cholinergic neurons. Neurochem Int 50:807–820

Nakanishi M, Niidome T, Matsuda S, Akaike A, Kihara T, Sugimoto H (2007) Microglia-derived interleukin-6 and leukaemia inhibitory factor promote astrocytic differentiation of neural stem/ progenitor cells. Eur J Neurosci 25:649–658

Nicholas RS, Wing MG, Compston A (2001) Nonactivated microglia promote oligodendrocyte precursor survival and maturation through the transcription factor NF-kappa B. Eur J Neurosci 13:959–967

Nishiyama A, Lin XH, Giese N, Heldin CH, Stallcup WB (1996) Co-localization of NG2 proteoglycan and PDGF alpha-receptor on O2A progenitor cells in the developing rat brain. J Neurosci Res 43:299–314

Noctor SC, Flint AC, Weissman TA, Dammerman RS, Kriegstein AR (2001) Neurons derived from radial glial cells establish radial units in neocortex. Nature 409:714–720

Oka A, Belliveau MJ, Rosenberg PA, Volpe JJ (1993) Vulnerability of oligodendroglia to glutamate: pharmacology, mechanisms, and prevention. J Neurosci 13:1441–1453

Outtz HH, Tattersall IW, Kofler NM, Steinbach N, Kitajewski J (2011) Notch1 controls macrophage recruitment and Notch signaling is activated at sites of endothelial cell anastomosis during retinal angiogenesis in mice. Blood 118:3436–3439

Paez PM, Cheli VT, Ghiani CA, Spreuer V, Handley VW, Campagnoni AT (2012) Golli myelin basic proteins stimulate oligodendrocyte progenitor cell proliferation and differentiation in remyelinating adult mouse brain. Glia 60:1078–1093

Pang Y, Cai Z, Rhodes PG (2000) Effects of lipopolysaccharide on oligodendrocyte progenitor cells are mediated by astrocytes and microglia. J Neurosci Res 62:510–520

Pang Y, Campbell L, Zheng B, Fan L, Cai Z, Rhodes P (2010) Lipopolysaccharide-activated microglia induce death of oligodendrocyte progenitor cells and impede their development. Neuroscience 166:464–475

Park SK, Solomon D, Vartanian T (2001) Growth factor control of CNS myelination. Dev Neurosci 23:327–337

Pennell NA, Streit WJ (1997) Colonization of neural allografts by host microglial cells: relationship to graft neovascularization. Cell Transplant 6:221–230

Perry VH, Gordon S (1987) Modulation of CD4 antigen on macrophages and microglia in rat brain. J Exp Med 166:1138–1143

Pfeiffer SE, Warrington AE, Bansal R (1993) The oligodendrocyte and its many cellular processes. Trends Cell Biol 3:191–197

Pillay V, Savage N, Laburn H (1993) Interleukin-1 receptor antagonist in newborn babies and pregnant women. Pflugers Arch 424:549–551

Polverini PJ, Cotran PS, Gimbrone MA Jr, Unanue ER (1977) Activated macrophages induce vascular proliferation. Nature 269:804–806
Popken GJ, Hodge RD, Ye P, Zhang J, Ng W, O'Kusky JR et al (2004) In vivo effects of insulin-like growth factor-I (IGF-I) on prenatal and early postnatal development of the central nervous system. Eur J Neurosci 19:2056–2068
Prada I, Ongania GN, Buonsanti C, Panina-Bordignon P, Meldolesi J (2006) Triggering receptor expressed in myeloid cells 2 (TREM2) trafficking in microglial cells: continuous shuttling to and from the plasma membrane regulated by cell stimulation. Neuroscience 140:1139–1148
Presta M, Urbinati C, Dell'era P, Lauro GM, Sogos V, Balaci L et al (1995) Expression of basic fibroblast growth factor and its receptors in human fetal microglia cells. Int J Dev Neurosci 13:29–39
Pringle NP, Richardson WD (1993) A singularity of PDGF alpha-receptor expression in the dorso-ventral axis of the neural tube may define the origin of the oligodendrocyte lineage. Development 117:525–533
Pringle NP, Mudhar HS, Collarini EJ, Richardson WD (1992) PDGF receptors in the rat CNS: during late neurogenesis, PDGF alpha-receptor expression appears to be restricted to glial cells of the oligodendrocyte lineage. Development 115:535–551
Provis JM, Penfold PL, Edwards AJ, van Driel D (1995) Human retinal microglia: expression of immune markers and relationship to the glia limitans. Glia 14:243–256
Provis JM, Diaz CM, Penfold PL (1996) Microglia in human retina: a heterogeneous population with distinct ontogenies. Perspect Dev Neurobiol 3:213–222
Raballo R, Rhee J, Lyn-Cook R, Leckman JF, Schwartz ML, Vaccarino FM (2000) Basic fibroblast growth factor (Fgf2) is necessary for cell proliferation and neurogenesis in the developing cerebral cortex. J Neurosci 20:5012–5023
Raff MC, Lillien LE, Richardson WD, Burne JF, Noble MD (1988) Platelet-derived growth factor from astrocytes drives the clock that times oligodendrocyte development in culture. Nature 333:562–565
Reynolds R, Hardy R (1997) Oligodendroglial progenitors labeled with the O4 antibody persist in the adult rat cerebral cortex in vivo. J Neurosci Res 47:455–470
Rezaie P, Male D (1999) Colonisation of the developing human brain and spinal cord by microglia: a review. Microsc Res Tech 45:359–382
Rezaie P, Cairns NJ, Male DK (1997) Expression of adhesion molecules on human fetal cerebral vessels: relationship to microglial colonisation during development. Brain Res Dev Brain Res 104:175–189
Rezaie P, Patel K, Male DK (1999) Microglia in the human fetal spinal cord—patterns of distribution, morphology and phenotype. Brain Res Dev Brain Res 115:71–81
Rezaie P, Trillo-Pazos G, Everall IP, Male DK (2002a) Expression of beta-chemokines and chemokine receptors in human fetal astrocyte and microglial co-cultures: potential role of chemokines in the developing CNS. Glia 37:64–75
Rezaie P, Trillo-Pazos G, Greenwood J, Everall IP, Male DK (2002b) Motility and ramification of human fetal microglia in culture: an investigation using time-lapse video microscopy and image analysis. Exp Cell Res 274:68–82
Rezaie P, Ulfig N, male D (2003) Distribution and morphology of GFAP-positive astrocytes in the human fetal brain at second trimester. Neuroembryology 2:50–63
Rezaie P, Dean A, Male D, Ulfig N (2005) Microglia in the cerebral wall of the human telencephalon at second trimester. Cereb Cortex 15:938–949
Riaz SS, Theofilopoulos S, Jauniaux E, Stern GM, Bradford HF (2004) The differentiation potential of human foetal neuronal progenitor cells in vitro. Brain Res Dev Brain Res 153:39–51
Rigato C, Buckinx R, Le-Corronc H, Rigo JM, Legendre P (2011) Pattern of invasion of the embryonic mouse spinal cord by microglial cells at the time of the onset of functional neuronal networks. Glia 59:675–695
Risau W (1997) Mechanisms of angiogenesis. Nature 386:671–674
Robel S, Berninger B, Gotz M (2011) The stem cell potential of glia: lessons from reactive gliosis. Nat Rev Neurosci 12:88–104

Rowan RA, Maxwell DS (1981) Patterns of vascular sprouting in the postnatal development of the cerebral cortex of the rat. Am J Anat 160:247–255
Rowitch DH, Kriegstein AR (2010) Developmental genetics of vertebrate glial-cell specification. Nature 468:214–222
Rymo SF, Gerhardt H, Wolfhagen Sand F, Lang R, Uv A, Betsholtz C (2011) A two-way communication between microglial cells and angiogenic sprouts regulates angiogenesis in aortic ring cultures. PLoS One 6:e15846
Selvaraju R, Bernasconi L, Losberger C, Graber P, Kadi L, Avellana-Adalid V et al (2004) Osteopontin is upregulated during in vivo demyelination and remyelination and enhances myelin formation in vitro. Mol Cell Neurosci 25:707–721
Shimojo M, Nakajima K, Takei N, Hamanoue M, Kohsaka S (1991) Production of basic fibroblast growth factor in cultured rat brain microglia. Neurosci Lett 123:229–231
Smith GM, Hale JH (1997) Macrophage/microglia regulation of astrocytic tenascin: synergistic action of transforming growth factor-beta and basic fibroblast growth factor. J Neurosci 17: 9624–9633
Spooren A, Kolmus K, Laureys G, Clinckers R, De Keyser J, Haegeman G et al (2011) Interleukin-6, a mental cytokine. Brain Res Rev 67:157–183
Stefater JA 3rd, Lewkowich I, Rao S, Mariggi G, Carpenter AC, Burr AR et al (2011) Regulation of angiogenesis by a non-canonical Wnt-Flt1 pathway in myeloid cells. Nature 474:511–515
Storch A, Paul G, Csete M, Boehm BO, Carvey PM, Kupsch A et al (2001) Long-term proliferation and dopaminergic differentiation of human mesencephalic neural precursor cells. Exp Neurol 170:317–325
Streit WJ (2001) Microglia and macrophages in the developing CNS. Neurotoxicology 22: 619–624
Sunderkotter C, Goebeler M, Schulze-Osthoff K, Bhardwaj R, Sorg C (1991) Macrophage-derived angiogenesis factors. Pharmacol Ther 51:195–216
Suzuki H, Imai F, Kanno T, Sawada M (2001) Preservation of neurotrophin expression in microglia that migrate into the gerbil's brain across the blood-brain barrier. Neurosci Lett 312:95–98
Swinnen N, Smolders S, Avila A, Notelaers K, Paesen R, Ameloot M et al (2012) Complex invasion pattern of the cerebral cortex bymicroglial cells during development of the mouse embryo. Glia 61(2):150–63
Takahashi K, Prinz M, Stagi M, Chechneva O, Neumann H (2007) TREM2-transduced myeloid precursors mediate nervous tissue debris clearance and facilitate recovery in an animal model of multiple sclerosis. PLoS Med 4:e124
Tambuyzer BR, Ponsaerts P, Nouwen EJ (2009) Microglia: gatekeepers of central nervous system immunology. J Leukoc Biol 85:352–370
Tanigaki K, Nogaki F, Takahashi J, Tashiro K, Kurooka H, Honjo T (2001) Notch1 and Notch3 instructively restrict bFGF-responsive multipotent neural progenitor cells to an astroglial fate. Neuron 29:45–55
Taylor DL, Pirianov G, Holland S, McGinnity CJ, Norman AL, Reali C et al (2010) Attenuation of proliferation in oligodendrocyte precursor cells by activated microglia. J Neurosci Res 88: 1632–1644
Thrash JC, Torbett BE, Carson MJ (2009) Developmental regulation of TREM2 and DAP12 expression in the murine CNS: implications for Nasu-Hakola disease. Neurochem Res 34: 38–45
Torres-Aleman I, Naftolin F, Robbins RJ (1990) Trophic effects of insulin-like growth factor-I on fetal rat hypothalamic cells in culture. Neuroscience 35:601–608
Tran PB, Miller RJ (2003) Chemokine receptors in the brain: a developing story. J Comp Neurol 457:1–6
Tran PB, Banisadr G, Ren D, Chenn A, Miller RJ (2007) Chemokine receptor expression by neural progenitor cells in neurogenic regions of mouse brain. J Comp Neurol 500:1007–1033
Tripathi RB, Clarke LE, Burzomato V, Kessaris N, Anderson PN, Attwell D et al (2011) Dorsally and ventrally derived oligodendrocytes have similar electrical properties but myelinate preferred tracts. J Neurosci 31:6809–6819

Ueno M, Fujita Y, Tanaka T, Nakamura Y, Kikuta J, Ishii M et al (2013) Layer V cortical neurons require microglial support for survival during postnatal development. Nat Neurosci 16: 543–551
Unoki N, Murakami T, Nishijima K, Ogino K, van Rooijen N, Yoshimura N (2010) SDF-1/CXCR4 contributes to the activation of tip cells and microglia in retinal angiogenesis. Invest Ophthalmol Vis Sci 51:3362–3371
Vaccarino FM, Schwartz ML, Raballo R, Nilsen J, Rhee J, Zhou M et al (1999) Changes in cerebral cortex size are governed by fibroblast growth factor during embryogenesis. Nat Neurosci 2:246–253
Vela JM, Dalmau I, Gonzalez B, Castellano B (1995) Morphology and distribution of microglial cells in the young and adult mouse cerebellum. J Comp Neurol 361:602–616
Verney C, Monier A, Fallet-Bianco C, Gressens P (2010) Early microglial colonization of the human forebrain and possible involvement in periventricular white-matter injury of preterm infants. J Anat 217:436–448
Voigt T (1989) Development of glial cells in the cerebral wall of ferrets: direct tracing of their transformation from radial glia into astrocytes. J Comp Neurol 289:74–88
Volpe JJ (2001) Neurobiology of periventricular leukomalacia in the premature infant. Pediatr Res 50:553–562
von Sántha K (1932) Untersuchungen uber die Entwicklung der Hortegaschen Mikroglia. Arch fur Psychiatry 96:36–37
von Sántha K, Juba A (1933) Weitere Untersuchungen uber die Entwicklung der Hortegaschen Mikroglia. Arch fur Psychiat 98:598–613
Wakamatsu Y (2004) Understanding glial differentiation in vertebrate nervous system development. Tohoku J Exp Med 203:233–240
Wang X, Fu S, Wang Y, Yu P, Hu J, Gu W et al (2007) Interleukin-1beta mediates proliferation and differentiation of multipotent neural precursor cells through the activation of SAPK/JNK pathway. Mol Cell Neurosci 36:343–354
Widera D, Holtkamp W, Entschladen F, Niggemann B, Zanker K, Kaltschmidt B et al (2004) MCP-1 induces migration of adult neural stem cells. Eur J Cell Biol 83:381–387
Wine RN, McPherson CA, Harry GJ (2009) IGF-1 and pAKT signaling promote hippocampal CA1 neuronal survival following injury to dentate granule cells. Neurotox Res 16:280–292
Wong G, Goldshmit Y, Turnley AM (2004) Interferon-gamma but not TNF alpha promotes neuronal differentiation and neurite outgrowth of murine adult neural stem cells. Exp Neurol 187:171–177
Woodruff RH, Tekki-Kessaris N, Stiles CD, Rowitch DH, Richardson WD (2001) Oligodendrocyte development in the spinal cord and telencephalon: common themes and new perspectives. Int J Dev Neurosci 19:379–385
Yan Q, Matheson C, Lopez OT (1995) In vivo neurotrophic effects of GDNF on neonatal and adult facial motor neurons. Nature 373:341–344
Zhou ZD, Kumari U, Xiao ZC, Tan EK (2010) Notch as a molecular switch in neural stem cells. IUBMB Life 62:618–623
Zhu P, Hata R, Cao F, Gu F, Hanakawa Y, Hashimoto K et al (2008) Ramified microglial cells promote astrogliogenesis and maintenance of neural stem cells through activation of Stat3 function. FASEB J 22:3866–3877
Ziv Y, Schwartz M (2008) Immune-based regulation of adult neurogenesis: implications for learning and memory. Brain Behav Immun 22:167–176

Chapter 9
Developing and Mature Synapses

Marie-Ève Tremblay, Rosa C. Paolicelli, Beth Stevens, Hiroaki Wake, and Alain Bessis

Abstract Microglia are the resident immune cells of the brain. As such, they rapidly detect changes in normal brain homeostasis and accurately respond by fine-tuning in a tightly regulated manner their morphology, gene expression, and functional behaviour. Depending on the nature of these changes, microglia can thicken and retract their processes, proliferate and migrate, release numerous signalling factors and compounds influencing neuronal physiology (e.g., cytokines and trophic factors), in addition to secreting proteases able to transform the extracellular matrix, and phagocytosing various types of cellular debris, etc. Because microglia also transform rapidly (on a time scale of minutes) during experimental procedures, studying these very special cells requires methods that are specifically non-invasive. The development of such methods has provided unprecedented insights, these past few years, into the roles of microglia during normal physiological conditions. In particular, transcranial two-photon in vivo imaging revealed that presumably

M.-È. Tremblay, Ph.D. (✉)
Axe Neurosciences, Centre de Recherche du CHU de Québec, and Department of Molecular Medicine, Université Laval, Québec, QC, Canada
e-mail: tremblay.marie-eve@crchudequebec.ulaval.ca

R.C. Paolicelli, Ph.D.
Division of Psychiatry Research, University of Zurich, Wagistrasse 12, Schlieren 8952, Switzerland
e-mail: rosa.paolicelli@gmail.com

B. Stevens, Ph.D.
Department of Neurology, Harvard Medical School, Center for Life Science, Boston Children's Hospital, F.M. Kirby Neurobiology Center, Room 12-257, 3 Blackfan Circle, Boston, MA 02115, USA
e-mail: Beth.Stevens@childrens.harvard.edu

H. Wake, M.D., Ph.D.
Division of Brain Circuits, National Institute for Basic Biology, 38 Nishigonaka, Myodaiji-cho, Okazaki, Japan
e-mail: hirowake@nips.ac.jp

A. Bessis, Ph.D. (✉)
Institut de Biologie de l'École Normale Supérieure; Institut National de la Santé et de la Recherche Médicale, Centre National de la Recherche Scientifique, Unité Mixte de Recherche, 46 rue d'Ulm, Paris, France
e-mail: alain.bessis@ens.fr

M.-È. Tremblay, A. Sierra (eds.), *Microglia in Health and Disease*,
DOI 10.1007/978-1-4939-1429-6_9

"resting" microglia continuously survey the brain parenchyma with their highly motile processes, in addition to modulating their structural and functional interactions with neuronal circuits along the changes in neuronal activity and behavioural experience occurring throughout the lifespan. In this chapter, we will describe how surveillant microglia interact with synaptic elements, and modulate the number, maturation, function, and plasticity of synapses in the healthy developing, mature, and aging brain, with consequences on neuronal activity, learning and memory, and the behavioural outcome.

Keywords Microglia • Synapses • Physiology • Neuronal circuit remodeling • Maturation • Function • Plasticity • Phagocytosis • Fractalkine • Complement • Brain-derived neurotrophic factor • Purinergic signalling

Bullet points

- Surveillant microglia dynamically interact with neuronal circuits at synapses.
- Axon terminals, dendritic spines, and peri-synaptic astrocytic processes are contacted in the healthy brain, during postnatal development, adolescence, adulthood, and aging.
- Microglia–synapse interactions are modulated by changes in neuronal activity and experience.
- Conversely, microglia contribute to regulating the number, maturation, function, and plasticity of excitatory synapses in normal physiological conditions.
- These new roles of microglia at synapses are mediated via phagocytosis, intercellular contacts, and release of soluble factors.
- The underlying molecular mechanisms discovered so far include fractalkine signalling, the classical complement pathway, brain-derived neurotrophic factor, DAP12, CD200R, and purinergic signalling.

9.1 Introduction

The capacity of the central nervous system (CNS) to adapt to the environment depends on the maintenance and refinement of neuronal circuits. This refinement is mainly achieved through the formation, elimination, or functional modulation of synapses, the elementary cellular components of neurotransmission. Because they have to transmit accurately the message from one neuron to another under all circumstances of life, synapses display strong plasticity, a phenomenon describing their ability to permanently adapt their structure and function to the demands of the circuitry. Synaptic plasticity comprises a vast repertoire of structural (number and shape of pre-synaptic axon terminals, post-synaptic dendritic spines, etc.) and functional changes (pre-synaptic release of neurotransmitters, modification of post-synaptic receptors behaviour, second messenger signalling, etc.) (Bhatt et al. 2009; Holtmaat and Svoboda 2009; Bourne and Harris 2012; Choquet and Triller 2013).

Such precisely coordinated processes do not only require a synergistic choreography at the molecular, cellular, and behavioural levels, but also engage all the various cell types making up the CNS, including the supporting glial cells. The involvement of astrocytes with synapses has long been investigated (Nedergaard 1994; Parpura et al. 1994), and a large number of studies have revealed several aspects of their roles in the regulation of synapse formation, maturation, function, and plasticity (Chung et al. 2013; Clarke and Barres 2013; Pannasch and Rouach 2013). However, microglia are still an understated type of cell in normal CNS physiology, mostly studied in the context of immune and inflammatory responses since their first description by Pio del Rio-Hortega (Prinz and Priller 2014) (see Chap. 2 for further reading.)

As the primary immune effectors cells of the CNS, microglia are accurately equipped to sense and respond to changes in their local environment, associated with neuronal activity, experience, injury, or disease. Microglia express a vast repertoire of membrane receptors for various informative molecules such as neurotransmitters and neuromodulators, pro- and anti-inflammatory cytokines, growth factors, or extracellular matrix (ECM) components. (see Chaps. 3 and 4 for further reading.) These signalling molecules induce a constellation of intracellular responses in microglia, leading to concerted changes in morphology, gene expression, and functional dynamics (Bessis et al. 2007; Biber et al. 2007; Pocock and Kettenmann 2007; Kettenmann et al. 2011). As macrophages or phagocytes in residence into the CNS, microglia also display the ability to migrate and scavenge unwanted cells, cellular components, and debris by active phagocytosis (Gomez Perdiguero et al. 2013; Sierra et al. 2013). This ability is further refined by the extraordinary dynamics of microglial processes frequently interacting with the neuronal circuitry, particularly at synapses. Recently, microglial capacity to eliminate synaptic elements in an activity- and experience-dependent manner, and to promote the formation of dendritic spines, as well as modulate the maturation, function, and plasticity of excitatory synapses was additionally revealed in the developing and mature healthy brain (reviewed in Tremblay 2011; Bechade et al. 2013; Kettenmann et al. 2013; Schafer et al. 2013; Wake et al. 2013).

The combined evidence further advances the view that complex brain functions such as learning and memory emerge from the intercellular relationships between all types of cells in the CNS, acting in concert to achieve the most effective information processing and decision making. This chapter will describe the structural and functional interactions between microglia and synapses in the healthy brain, bridging the cellular/dynamic, ultrastructural and molecular levels, and summarize the key points that emerge from our current understanding of these roles, as well as the most promising directions to pursue in this nascent field of investigation.

9.2 Microglia–Synapse Interactions

Immunocytochemical electron microscopy (immune-EM; Tremblay et al. 2010) surprisingly revealed that almost all microglial processes (94 %) directly juxtapose synaptic elements in the visual cortex of adolescent mice, under physiological conditions. In particular, axon terminals, dendritic spines, peri-synaptic astrocytic processes, and synaptic clefts are contacted, in decreasing order of frequency (Tremblay et al. 2010). More recently, similar interactions were described in the visual and auditory cortex throughout adulthood and normal aging (Tremblay et al. 2012). Conversely, the proportion of each of those synaptic elements contacted by microglial processes could not be determined, since only a subset were stained for the microglial marker "ionized calcium-binding adapter molecule 1" (Iba1) and thus positively identified under these stringent immunocytochemical conditions. In the original characterization, serial section EM with 3D reconstruction showed that a single microglial process contacts multiple synaptic elements at multiple synapses simultaneously, sometimes with morphological specializations such as finger-like protrusions wrapping around dendritic spines and axon terminals (Fig. 9.1a,b). EM also revealed that microglial cell bodies and processes are distinctively surrounded by pockets of extracellular space, varying in volume by two orders of magnitude, in the cerebral cortex of adolescent and adult mice (Tremblay et al. 2010, 2012) (Fig. 9.1a,b). These pockets could result from the release of ECM proteases remodeling the extracellular space, such as metalloproteinases, plasminogens, and cathepsins, which might facilitate microglial dynamics, but also the motility, formation, and elimination of dendritic spines during normal physiological conditions (Tremblay 2011). Clathrin-coated pits were lastly encountered inside microglial processes and synaptic elements, specifically at their sites of contact (Tremblay et al. 2010), thus suggesting reciprocal exchange of molecular signals through clathrin-mediated endocytosis of membrane-bound receptors and their ligands (see Le Roy and Wrana 2005 for details on clathrin-mediated endocytosis). Therefore, these observations suggested that microglia can interact functionally with synaptic structures in the healthy brain (Tremblay and Majewska 2011).

Microglial contacts with subsets of axon terminals and dendritic spines were also encountered in vivo using non-invasive transcranial two-photon imaging. (see Chap. 4 for further reading on the technique.) To this end, Iba1$^{GFP/GFP}$ mice, in which GFP expression is driven by the Iba1 promoter (Hirasawa et al. 2005), were first crossed with Thy1-GFP mice (Feng et al. 2000) to visualize microglia and a subset of neuronal elements in the same color (green), including distal processes, dendritic spines, and axon terminals (Wake et al. 2009). Subsequently, the CX3CR1$^{GFP/-}$ mice in which the fractalkine receptor CX3CR1 only expressed by microglia in the healthy brain is replaced by the GFP reporter (Jung et al. 2000) were crossed with Thy1-YFP mice (Feng et al. 2000), thus providing an exceptional visualization of microglia and a subset of neuronal elements in two different colors (Tremblay et al. 2010). The CX3CR1 $^{GFP/-}$ heterozygous mice are partially deficient in fractalkine signalling (see discussion below), but microglial morphology, dynamic behaviour (Davalos et al. 2005; Nimmerjahn et al. 2005; Wake et al. 2009), dendritic spine turnover (Parkhurst et al. 2013), and microglial interactions with synaptic elements

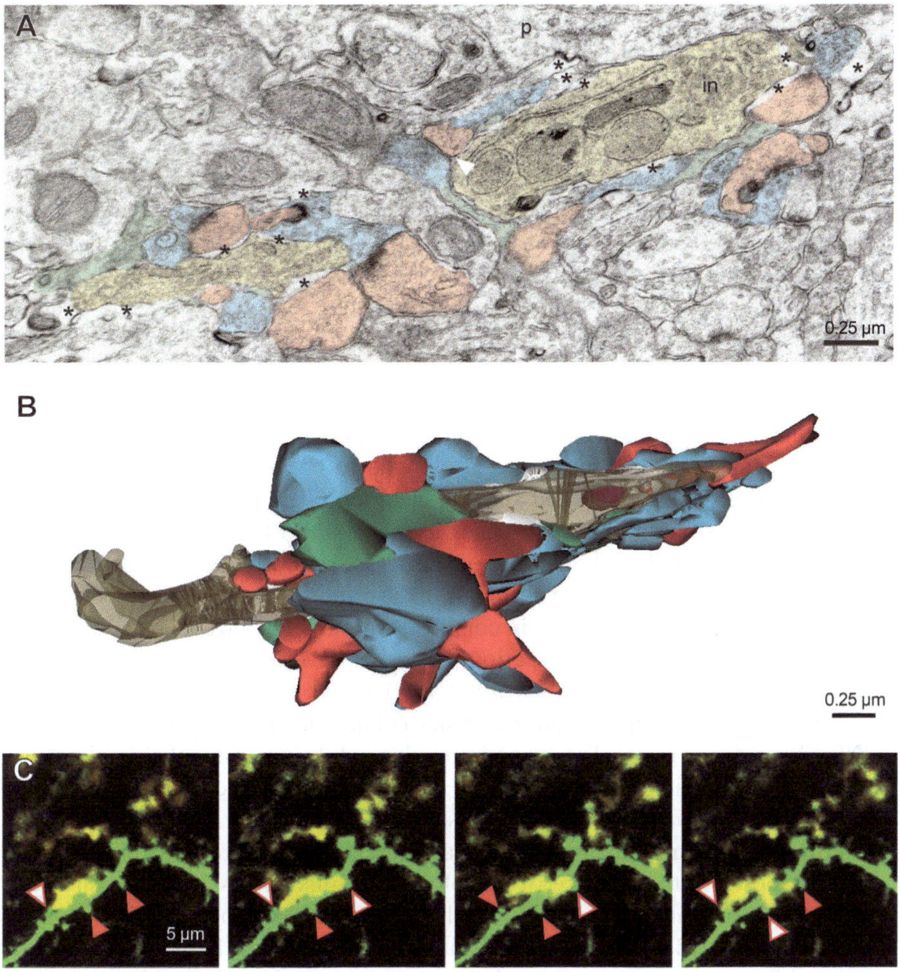

Fig. 9.1 *Microglia–synapse interactions in the healthy brain.* (**a**) Electron micrograph from a series of sections showing a proximal microglial process (cut in transverse, coloured in *yellow*) with a distal protrusion (cut longitudinally) contiguous to a neuronal perikaryon (p), and making direct contacts with dendritic spines (*pink*), axon terminals (*blue*), synaptic cleft (*arrowhead*), and peri-synaptic astrocytic processes (*green*) at postnatal day (P)28. Note the extracellular space pockets surrounding the microglia (*asterisks*). (**b**) Serial section 3D reconstruction of the microglial process and protrusion shown in **a**, uncovering simultaneous contacts with axon terminals (*blue*), dendritic spines (*red*), and peri-synaptic astrocytic processes (*green*) at multiple excitatory synapses. Extracellular space pockets are displayed in *white*, and a phagocytic inclusion within the proximal microglial process is in *purple*. (**c**) Two-photon time-lapse micrographs showing a dynamic microglial process (*yellow*) transiently interacting with dendritic spines (*green*) over the course of 20 min in the visual cortex of a CX3CR1$^{GFP/-}$/Thy1-YFP adolescent mouse. Each frame was captured 5 min apart. *Red arrowheads* indicate non-targeted dendritic spines, and *white arrowheads* targeted ones. Reproduced from Tremblay et al. (2010)

(Wake et al. 2009; Tremblay et al. 2010) were comparable in vivo between these CX3CR1$^{GFP/-}$ and the Iba1-GFP mice in which the CX3CR1 locus is intact. Using these mice, microglial processes were shown to frequently interact with axon terminals and dendritic spines, from the labeled subset of layer V pyramidal neurons, for durations varying between 5 and 30 min in the somatosensory and visual cortices during adolescence and adulthood (Wake et al. 2009; Tremblay et al. 2010) (Fig. 9.1c). Among these synaptic structures, small dendritic spines which are the most structurally dynamic and transient synaptic structures in mouse visual, somatosensory, motor, and frontal cortices in vivo (Trachtenberg et al. 2002; Zuo et al. 2005; Majewska and Sur 2006; Alvarez and Sabatini 2007; Holtmaat and Svoboda 2009) were preferentially targeted by microglial processes, and more frequently eliminated over 2 days than the non-contacted spines as revealed by chronic imaging of the same dendrites (Tremblay et al. 2010), raising the previously unsuspected possibility that microglia could be involved in their elimination. Both in vivo and in situ, microglia–synapse interactions were also found to be modulated by the ongoing changes in neuronal activity and sensory experience. Enucleation of the eyes, or injection of the sodium channel blocker tetrodotoxin (TTX) into the eyes, which silences excitatory and inhibitory neuronal activity, similarly reduced the frequency of microglial contacts with axon terminals in the primary visual cortex in vivo (Wake et al. 2009). Housing the animals in complete darkness for 1 week, from the beginning to the peak of the critical period for visual development, followed by their reexposure to normal circadian daylight for 2 days, a paradigm which increases the motility and elimination of dendritic spines (Majewska and Sur 2003; Keck et al. 2008), also reduced microglial dynamics in vivo, while increasing their perimeter of contact with synaptic elements and their association with the extracellular space in situ in the primary visual cortex of adolescent mice (Tremblay et al. 2010).

Further insight into this hypothesis that microglia may be responsible for eliminating synapses in the absence of pathological insults comes from ultrastructural observations. Immune-EM indeed revealed the presence of phagocytic inclusions positively identified as axon terminals and dendritic spines, based on their ultrastructural features (i.e., synaptic vesicles, post-synaptic densities) or immunostaining for specific markers, inside the cytoplasm and lysosomal compartments of microglial cell bodies and processes, during postnatal development, adolescence, adulthood, and normal aging, in mouse hippocampus, visual cortex and thalamus, and auditory cortex (Tremblay et al. 2010; Paolicelli et al. 2011; Schafer et al. 2012; Tremblay et al. 2012) (see Fig. 9.2 for examples and discussion below).

Fig. 9.2 (continued) t, axon terminal. Scale bars = 250 nm. Reproduced from Tremblay et al. (2010). (**c**) Example of microglial phagocytosis of synaptic elements during age-associated loss of sensory function. The immunostained microglial cell body contains a single cellular inclusion resembling an axon terminal "t" with clearly visible synaptic vesicles (inset) in the visual cortex of a 20-month-old mouse impaired in vision. Annotations as in **a,b**. Scale bars: 1 μm. Reproduced from Tremblay et al. (2012). (**d**) Microglial cell bodies and processes showing reduced engulfment of retinogeniculate projections during early postnatal development, in mice deficient in the complement opsonin C1q or retinal transforming growth factor β (TGF-β) signalling. Volume of individual cells and engulfed cholera toxin-β conjugated to Alexa 488 or Alexa 594 (CTB-488 or CTB-594) injected into the left and right eyes, respectively, to label retinal projections at P10. Reproduced from Bialas and Stevens (2013)

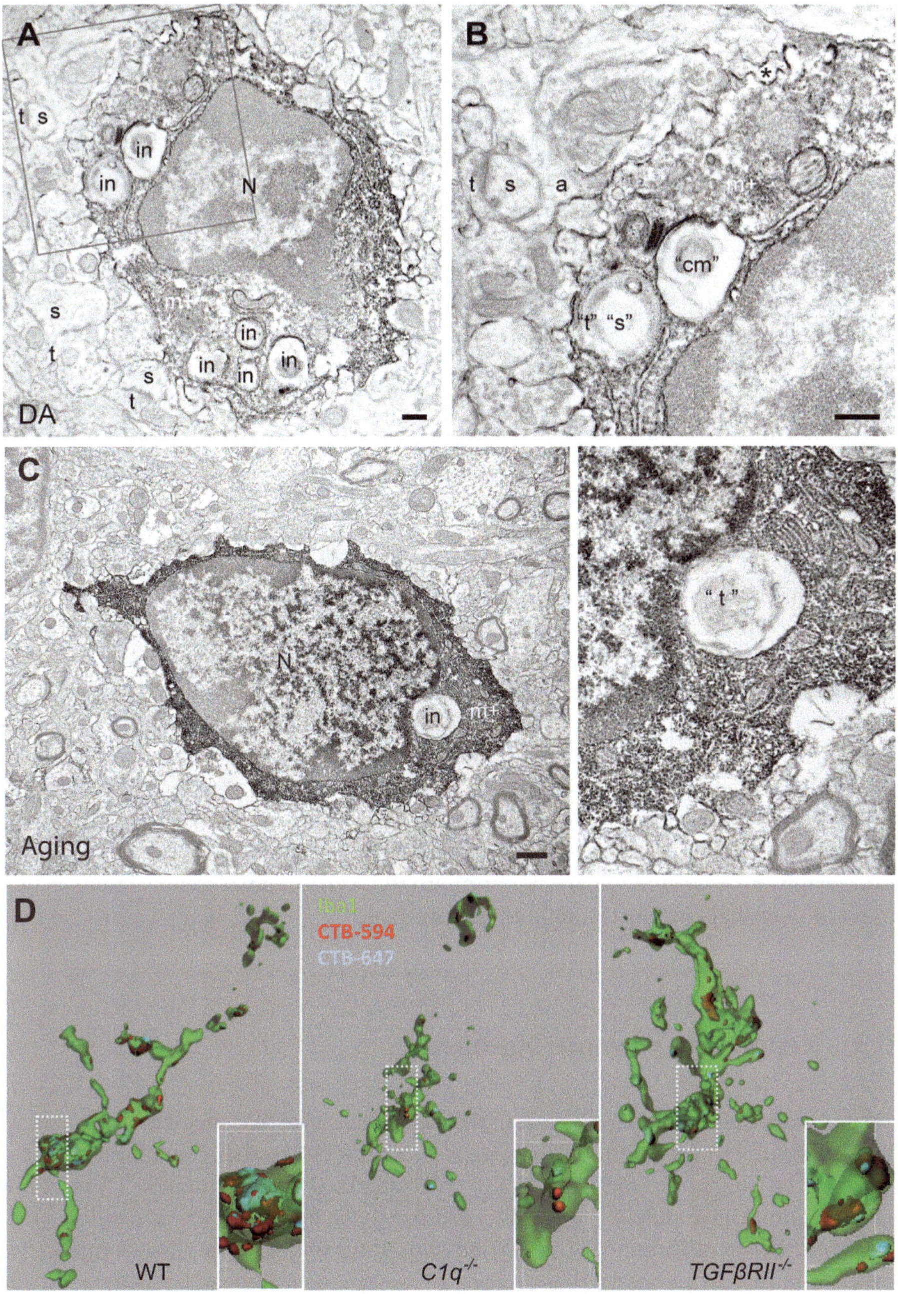

Fig. 9.2 *Microglial engulfment of synaptic elements during normal physiological conditions.* (**a**,**b**) Examples during dark-adaptation (DA), a form of visual plasticity that is induced by housing the animals in complete darkness for 1 week, from the beginning to the peak of the critical period in the visual system, i.e., postnatal day (P)21–28. In A, an Iba1-immunopositive microglial cell body (m+) observed in the visual cortex displays multiple cellular inclusions (in) at P28. (**b**) shows a magnified view of the boxed region in (**a**). One of these inclusions resembles a dendritic spine (s) receiving a synapse from an axon terminal (t), while the other inclusion contains an accumulation of cellular membranes probably undergoing digestion (cm). a, peri-synaptic astrocyte; *N*, nucleus; s, dendritic spine;

The phagocytosed elements displayed ultrastructural features of healthy cells such as an electron-lucent cytoplasm, contrarily to the apoptotic elements which are phagocytosed during injury or disease (Schmechel 1999), and the apoptotic newborn cells which are phagocytosed during adult neurogenesis (Sierra et al. 2010). Microglial processes with phagocytic specializations were also observed in vivo, sometimes encircling neuronal elements, but since these structures were stable throughout the 30–120 min imaging sessions, episodes of microglial engulfment followed by a full disappearance of the synaptic elements were not completely visualized. Both in situ and in vivo, microglial phagocytic structures became more prevalent during light deprivation, and still persisted after reexposure to light (Tremblay et al. 2010) (Fig. 9.2a,b), suggesting a role in the refinement of neuronal circuits during adaptation to a novel environment. More recently, phagocytic inclusions within microglial cell bodies and processes were also observed during adulthood and normal aging. Immune-EM revealed that microglial cell bodies and processes accumulate large amounts of phagocytic inclusions, sometimes with ultrastructural features of axon terminals and dendritic spines, with or without an electron-lucent cytoplasm, but also lysosomal lipopigments, large vesicles, vacuoles, and lipid droplets, in the visual and auditory cortices of two stains of mice normally undergoing complementary age-related loss of vision (CBA/CaJ mice) or hearing (C57Bl/6 J mice) during their second year of life (Fig. 9.2b). This accumulation was exacerbated by the age-related loss of visual or auditory function, with nearly all microglia containing phagocytic inclusions and ~20 % of microglial cells becoming almost completely filled by the inclusions (Tremblay et al. 2012). Together, these observations during adolescence, adulthood, and normal aging suggested that microglial phagocytosis is regulated in an experience-dependent manner in the mature CNS, thus proposing a physiological role for microglia–synapse interactions in the refinement of neuronal circuits throughout the lifespan.

9.3 Regulating Synapse Number

At this early stage of investigation in the field, the molecular mechanisms regulating microglial interactions with synapses, as much as their functional consequences on the brain and behaviour, still remained undetermined. Three main investigations have recently started to address these questions, in situ and in vivo, showing the involvement of fractalkine signalling and the classical complement cascade in regulating microglial pruning of synapses during early postnatal development, and the role of brain-derived neurotrophic factor (BDNF) in promoting the formation of dendritic spines during adolescence and adulthood, as discussed below.

9.3.1 Fractalkine Signalling

The first evidence that microglia–synapse interactions have functional consequences comes from early postnatal development, a period of remarkable plasticity where the remodeling of neuronal circuits is particularly exacerbated. The maturing neurons actively extend supernumerary connections that are progressively removed while others are maintained and strengthened. The term "pruning" refers to a selective developmental process regulated by neuronal activity that involves both large-scale elimination of axons and dendritic arbours, as well as the local removal of axon terminals or dendritic spines without the death of parent neurons (Herrmann and Shatz 1995; Tessier-Lavigne and Goodman 1996; Hua and Smith 2004). The large-scale removal of neuronal processes was initially proposed to be mediated by degeneration, cytoskeletal breakdown, fragmentation, and/or autophagy, while the pruning of synapses was considered to result from cytoskeletal collapse leading to retraction of the synaptic structure (Luo and O'Leary 2005), through coordinated interactions with the ECM and peri-synaptic astrocytic processes (Ethell and Pasquale 2005; Haber et al. 2006; Majewska and Sur 2006; Hotulainen and Hoogenraad 2010). Nevertheless, the cellular and molecular mechanisms that determine whether and how particular subsets of synapses are specifically eliminated remain an important question in the field, and the involvement of microglia in this process has just begun to be unravelled.

Analyses of mice bearing loss-of-function mutations in microglial genes encoding signalling molecules are useful for hinting at possible links between microglial function and neuronal activity. One example of such molecules involved in neuron-microglia communication is fractalkine, a member of the ∂-chemokine subfamily also known as neurotactin or CX3CL1. Fractalkine is expressed almost exclusively by neurons (Tarozzo et al. 2003; Ransohoff 2009) and can be soluble or membrane-bound (Garton et al. 2001; Tsou et al. 2001), but the individual activities of these two forms still need to be clarified, especially in the brain. Most importantly, fractalkine has a unique receptor, CX3CR1, that is only expressed by microglial cells in the healthy brain (Combadiere et al. 1998; Cardona et al. 2006). The involvement of fractalkine signalling in mediating the elimination of synapses during normal development was particularly addressed by examining CX_3CR1-deficient mice in the CA1 region of the hippocampus during the first postnatal weeks, a period of intense synaptic remodeling. In this manner, Paolicelli and colleagues revealed a slower increase of microglial density in the $CX3CR1^{GFP/GFP}$ mice (or $CX3CR1^{KO/KO}$) versus $CX3CR1^{GFP/-}$ littermates between postnatal day (P)8 and 28. This delayed colonization of the maturing hippocampus (see Chaps. 7 and 8 for further reading on the roles of microglia during normal brain development) was found to be accompanied by a transient increase in dendritic spine density on the pyramidal neurons apical dendrites of $CX3CR1^{KO/KO}$ mice compared with wild-type littermates, over the same developmental period (Paolicelli et al. 2011). However, since microglial engulfment of dendritic spines was observed but not quantified, it remains to be elucidated whether fractalkine signalling instructs microglial recruitment to the proximity of maturing dendrites or contributes more directly to regulating microglial phagocytosis of dendritic spines in the developing hippocampus.

In a follow-up study by the same group, it was recently shown that the CX3CR1$^{KO/KO}$ mice additionally display a lasting impairment of hippocampal synaptic connectivity into adolescence. In particular, quantifying at the ultrastructural level the density of multi-synaptic boutons, i.e., axon terminals contacting two dendritic spines, revealed a significant reduction in the knockout mice compared to wild-type littermates at P40 (Zhan et al. 2014) despite a normalized density of dendritic spines observed in adult animals (Paolicelli et al. 2011). Local field potentials (LFPs), which inform about the sum of synaptic activity within a particular volume, were also investigated at P40. As a measure of long-range connectivity, coherence spectra of the LFPs were calculated, under the premise that high coherence values would reflect strongly connected structures. By implanting electrodes in vivo within different brain regions simultaneously, a significant decrease in the coherence between the hippocampus and prefrontal cortex was found in adult CX3CR1 knockout mice, compared to wild-type littermates, thus reflecting a decrease in functional connectivity between these two areas. Supporting these results, the global connectivity assessed by functional magnetic resonance imaging (fMRI) was also significantly reduced across brain regions in the CX3CR1$^{KO/KO}$ mice, with the difference becoming particularly evident for distant regions (Zhan et al. 2014). These changes in neuronal circuits also seem to have behavioural repercussion, since social interactions were found to be altered in the CX3CR1$^{KO/KO}$ mice. In the same study by Zhan and colleagues, impaired social behaviour was particularly observed in juvenile mice, displaying no preference for their own mother over an inert stimulus, as assessed by the homing testing of motivation towards a relevant stimulus, i.e., the odour of the nest. However, no impairment in performing the novel object recognition test, which relies on prefrontal cortex function, was observed in the same mice. Similarly, adult CX3CR1$^{KO/KO}$ mice tested in a standard three-chamber apparatus failed to display significant interest towards a sex-matched social stimulus, compared to wild-type controls (Zhan et al. 2014). No deficit in responding to social olfactory cues was reported in these mice, suggesting that the impairment observed in social behaviour was due to reduced social motivation, rather than difficulties with the discrimination of social cues. Increased grooming behaviour was also reported in adult CX3CR1 knockout mice, when tested in a novel cage for 10 min, suggesting a propensity for increased repetitive behaviour, particularly triggered under stressful conditions (Zhan et al. 2014). Additionally, Rogers and colleagues have reported deficits in different forms of learning and memory in the adult CX3CR1$^{KO/KO}$ mice. Motor learning was found to be compromised in CX3CR1$^{KO/KO}$ mice versus wild-type littermates, using the rotarod test for balance, coordination, physical condition, and motor planning (Rogers et al. 2011). Locomotor and exploratory activity was, however, similar between genotypes, as assessed in the open field test, and no difference was observed in anxiety behaviour, measured by the elevated plus maze. Associative learning and memory was also found to be altered in a standard fear-conditioning paradigm, as the knockout mice failed to display reduced freezing (Rogers et al. 2011). The CX3CR1$^{KO/KO}$ were lastly impaired in the water maze, further supporting a role for fractalkine signalling in modulating hippocampal-dependent learning and memory. These effects could be mediated by microglial release of the pro-inflammatory cytokine interleukine-1β (IL-1β) since intrahippocampal infusion of its antagonist IL-1ra significantly reversed the deficits in cognitive function measured in the knockouts (Rogers et al. 2011). Providing additional insights, Maggi and colleagues also revealed that the $CX3CR1^{KO/KO}$ mice failed to perform better in the water maze following prolonged exposure to an enriched environment, compared with standard housing conditions, even though they were not significantly different from aged-matched wild-type controls in their ability to learn the water maze task in this study (Maggi et al. 2011), an apparent discrepancy which warrants further investigation.

These recent findings are complementing one another in proposing a role for CX3CL1-CX3CR1 interactions in mediating the refinement of neuronal circuits, learning and memory, and the behaviour, as well as the maturation and plasticity of synapses as described in the following section. Since microglia are the only cells expressing CX3CR1 in the healthy brain, they might be crucially involved in all of these processes shown to be influenced by fractalkine signalling during normal physiological conditions (also see Paolicelli et al. 2014). Nevertheless, the microglial effector functions which are precisely recruited, the molecular mechanisms acting downstream of fractalkine signalling, and the respective contributions of soluble versus membrane-bound fractalkine remain to be elucidated.

9.3.2 Complement Proteins

Pruning has been classically described in the mouse retinogeniculate system (Huberman 2007, 2008; Hong and Chen 2011; Schafer et al. 2012). This system is comprised of retinal ganglion cells (RGCs) that project axons to eye-specific territories in the dorsal lateral geniculate nucleus (dLGN) of the thalamus. During early postnatal development, retinogeniculate synapses undergo activity-dependent pruning to achieve the precise connectivity characteristic of the adult system, including proper eye-specific segregation, as required for the development of a normal binocular vision (Chen and Regehr 2000; Hooks and Chen 2006; Huberman 2007, 2008).

In this system, ramified microglia were also revealed to be intimately associated with retinogeniculate synapses during active synaptic remodeling, which peaks around P5 in mice, thereby suggesting microglia as a candidate cellular mediator of synaptic pruning. To test this hypothesis, Schafer and colleagues developed an in vivo engulfment assay in order to determine whether microglia phagocytose retinogeniculate synapses (Schafer et al. 2012). Using the $CXCR1^{GFP/-}$ mice, eye-specific RGC axonal inputs were labelled by injecting Alexa-conjugated anterograde tracers (cholera toxin B) into the eyes of postnatal mice. High-resolution confocal microscopy revealed an internalization of the RGC axonal terminals within microglia throughout the dLGN (see Fig. 9.2c), which was particularly exacerbated at P5, suggesting its developmental regulation. Moreover, immune-EM analyses revealed internalization of pre-synaptic elements contained within the microglial cytoplasm and lysosomal compartments. These observations proposed that microglia-mediated pruning could be actively involved during the peak of this early phase of synaptic remodelling in the retinogeniculate system, i.e., around P5, guiding the search for candidate molecules controlling this form of microglia–synapse interactions.

Complement proteins are a group of innate immune proteins associated with the rapid removal of apoptotic cells and pathogens in the periphery. C1q is the initiating protein of the classical complement cascade. When C1q binds to and coats (or opsonizes) dead cells, pathogens, or cellular debris, it triggers a protease cascade leading to the deposition of the downstream complement protein C3. Opsonization with the activated C3 fragments C3b can directly activate C3 receptors on macrophages and microglia, thereby triggering elimination by phagocytosis, or trigger the terminal activation of the complement cascade, leading to cellular lysis through the formation of a lytic membrane attack complex (Gasque 2004; van Lookeren Campagne et al. 2007). Recent studies using array tomography on 70 nm-thick brain sections revealed that the complement proteins C1q and C3 are widely expressed in the healthy developing LGN where they partially localize to excitatory synapses (Stevens et al. 2007). Mice deficient in C1q and C3 also have sustained defects in structural and functional elimination of synapses in the mouse developing visual system, raising questions about how complement-targeted synapses could be eliminated. Since microglia are the only resident brain cells to express CR3 in the healthy brain (Ransohoff and Perry 2009), these observations suggest C3-CR3-mediated phagocytosis as a potential mechanism underlying microglia-mediated pruning in the developing brain. Consistent with this hypothesis, mice deficient in C3 or CR3 resulted in a ≈50 % deficit in the ability of microglia to engulf RGC axons (Schafer et al. 2012). Moreover, the C3 and CR3 knockout mice have significantly more excitatory synapses, accompanied by defects in eye-specific segregation in the LGN at P32–P35, indicating that altered complement signalling early during development results in a sustained defect of synaptic connectivity into late adolescence (Schafer et al. 2012). In another recent study, the density of axon terminals on layer V pyramidal neurons was additionally found to be significantly increased in cortical slices derived from C1q knockout mice (P27–P31) and accompanied by epileptic seizures at the behavioural level, thus suggesting a failure to prune excitatory synapses during development (Chu et al. 2010).

What controls the timing and location of microglia-mediated synaptic refinement in the brain? Previous findings suggested that an astrocyte-derived factor triggers neuronal expression of C1q mRNA in purified postnatal retinal ganglion neurons in culture (Stevens et al. 2007). As the initiator of the classical complement cascade, C1q is a critical point of regulation in this pathway. C1q is developmentally expressed in postnatal RGCs, especially at P5, the peak of microglia-mediated pruning (Stevens et al. 2007). In the immune system, C1q can be modulated by rapid cytokine signalling pathways. A recent study identified transforming growth factor β (TGFβ) as a factor secreted by astrocytes that regulates C1q expression in RGCs (Bialas and Stevens 2013). This study showed that TGFβ is mostly expressed by astrocytes in the retina during the refinement period and TGFβ receptors (TGFβRII) are developmentally expressed in the postnatal RGCs. Blocking TGFβ signalling in retinal neurons resulted in a significant reduction of C1q expression in postnatal RGCs as well as reduced synaptic localization of complement in the dLGN in situ. Moreover, specific disruption of TGFβRII in retinal neurons, using retina-specific TGFβRII knockout mice generated using a Cre recombinase approach under the Chx10 promoter, which is active at embryonic day 13.5 in the outer neuroblastic layer of the retina only, also inhibited complement and microglia-mediated synaptic pruning in the dLGN (Fig. 9.2c) (Bialas and Stevens 2013). This data suggested that TGFβ-dependent regulation of neuronal C1q in the retina regulates downstream complement-dependent synapse elimination, a novel role for the TGFβ cytokine-signalling pathway in the regulation of microglia-mediated synaptic refinement in the postnatal brain.

Another important regulator of microglia-mediated pruning is neuronal activity. Synapse elimination is thought to result from competition between neighboring axons for territory on a post-synaptic neuron based on differences in patterns or levels of neuronal activity (Shatz 1990; Arnold 1999; Sanes and Lichtman 1999; Huberman et al. 2008). In the retina, spontaneous neuronal activity from both eyes is thought to drive eye-specific segregation and retinogeniculate pruning (Stellwagen and Shatz 2002; Torborg and Feller 2005; Huberman 2007). While the specific properties of retinal activity that guide this process remain elusive, these findings are consistent with a model in which left and right eye retinal axons compete for territory on post-synaptic dLGN relay neurons. Consistent with this idea, recent studies reveal that microglia-mediated pruning of RGC inputs in vivo is an activity-dependent process (Schafer et al. 2012). When competition between RGC inputs originating from the two eyes was enhanced by monocular injection of TTX or forskolin, an activator of adenylate cyclase which raises the level of cyclic AMP thereby increasing spontaneous neuronal activity, microglia preferentially engulfed inputs from the eye with reduced neuronal activity relative to the other eye, leading to an impairment of eye-specific segregation as observed in the C3- and C3R-deficient mice (Schafer et al. 2012). These data are consistent with previous work demonstrating a decreased synaptic territory of the "weaker" inputs and increased territory of "stronger" inputs within the dLGN (Penn et al. 1998; Cook et al. 1999; Stellwagen and Shatz 2002; Del Rio and Feller 2006; Hooks and Chen 2006; Huberman et al. 2008). Although it is not yet known how microglia specifically target the "weaker" synapses, complement is indeed partially required for microglial engulfment, suggesting a possible link between neuronal activity and the complement cascade. Together, these findings suggest a model in which context-dependent molecular cues (i.e., TGFβ) and neuronal activity cooperatively regulate the timing and degree of microglia-mediated pruning during development in the retinogeniculate system. Whether these mechanisms apply to other developmental or adult pruning conditions in health or disease remains to be determined.

9.3.3 Brain-Derived Neurotrophic Factor

Beyond early postnatal development, the underlying molecular mechanisms and functional consequences of microglia–synapse interactions were recently investigated during adolescence and early adulthood. The role of microglial BDNF, a potent regulator of synaptic development and plasticity that is expressed by various cell types including microglia (Chao 2003; Chakravarthy et al. 2006), was particularly examined. Microglial release of BDNF has also been shown to modulate neuronal plasticity in a mouse model of neuropathic pain (see Chap. 11 for further reading.) (Coull et al. 2005). In this study, the BDNF locus was deleted in microglia specifically, following early postnatal development, using CX3CR1-CreER crossed with BDNF-floxed mice treated with tamoxifen to induce BDNF deletion during adolescence. The mice were further crossed with Thy1-YFP mice to visualise neurons and analyse the turnover of dendritic spines (i.e., rates of formation and elimination) using chronic two photon in vivo imaging over 2 days (Parkhurst et al. 2013). This analysis revealed that removing microglial BDNF induces a significant decrease in the formation rate of dendritic spines (i.e., proportion of dendritic spines only observed in the second imaging session) in the motor cortex during learning of the rotarod, leaving unchanged the elimination of dendritic spines (i.e., proportion of dendritic spines only observed in the first imaging session) (Parkhurst et al. 2013). These observations thus revealed an unexpected role for microglia in spinogenesis and possibly in the formation of synapses in the mature healthy brain.

Similar findings were obtained in a microglial depletion model achieved by crossing the CX3CR1-CreER with diphteria toxin receptor (DTR)-floxed mice. These mice were treated with tamoxifen shortly after birth to induce DTR expression, followed by diphtheria toxin (DT) administered 7 days before imaging performed at P19 or P30. This time frame enabled to deplete microglial populations while preventing the brain replenishment by peripheral immune cells. In this manner, microglial depletion caused a $\approx$50 % decrease in the elimination rate of spines, measured over 4 days, at P19, and a $\approx$20 % decrease at P30. Even though microglia–synapse interactions were not visualised, these observations confirm microglial contribution to synaptic pruning in the healthy mature brain. In these microglia-depleted mice, a decreased formation and elimination of dendritic spines were also observed during motor learning on the rotarod. The difference between the microglia-depleted mice and non-depleted controls was significant for both the formation and elimination of dendritic spines at P30, and only for the formation of dendritic spines at P60 (Parkhurst et al. 2013). These observations suggest that the developmental regulation of these processes could follow a different time course, or that different mechanisms could be recruited depending on the stage of the lifespan. Providing additional mechanistic insights, in vitro experiments were additionally performed, showing that media from cultures of purified microglia (microglia-conditioned media) phosphorylates the BDNF tyrosine kinase receptor TrkB in cultures of purified cortical neurons (Parkhurst et al. 2013). Since the autophosphorylation of TrkB is required for its involvement in synaptic plasticity (Chao 2003), these observations suggest that microglial BDNF induces functional signalling in the neurons to induce dendritic spine formation.

At the behavioural level, microglial BDNF removal and microglial depletion were both accompanied by an impairment of motor learning on the rotarod. Auditory fear conditioning mainly relying on hippocampal, amygdala, and auditory cortex function was also impaired in both types of mice compared to wild-type mice, but there were no significant deficits of novel object recognition, a paradigm which depends on the prefrontal cortex, in the microglial BDNF-deficient mice contrarily to the microglia-depleted mice (Parkhurst et al. 2013). These findings suggest that different cellular and molecular mechanisms could be recruited depending on the particular form of memory formation and learning that is involved. The overall findings also underscore an important role of microglia during learning within multiple brain regions, particularly including the motor cortex, auditory cortex, the hippocampus, and amygdala which are together part of the limbic system, and the prefrontal cortex. However, more recent data found that dietary administration of a selective inhibitor of the colony-stimulating factor 1 receptor (CSF1R), which is exclusively expressed by microglia in the healthy brain, also depletes microglia during adulthood, possibly through a reduction of their survival. Noteworthy, depleting microglia with this pharmacological approach does not induce deficit in various forms of learning and memory tasks, including the rotarod (Elmore et al. 2014), suggesting that the method of microglial elimination could bias the interpretation of data. Future work will certainly elucidate this apparent paradox.

From the combined observations, a "two-step model" could be hypothesized, according to which soluble fractalkine would first mediate microglial recruitment, promoting their migration and specific interactions with the relevant neuronal circuits requiring function intervention, where fractalkine is up-regulated to "call" for microglial support during development or learning. According to this model, microglia would first address their terminal processes expressing CX3CR1 to the proximity of membrane-bound fractalkine, to ensure proper contact with neuronal branches and synaptic extremities. Once properly positioned in these hot-calling-spots, microglia would then respond to specific molecular cues such as complement molecules (e.g., C3) selectively targeting the weak and inappropriate synaptic terminals for pruning, or additional and yet unidentified cues specifically dictating the release of BDNF to promote dendritic spine formation.

9.3.4 Overview, Open Questions, and Further Insights

Several studies have now demonstrated that microglia play an active role in mediating both the formation and elimination of synaptic structures (see Fig. 9.3 for a schematic overview of the cellular and molecular mechanisms recruited). These studies have thereby opened many new questions that will certainly be tackled in the future. For instance, do microglia interact with inhibitory synapses during normal physiological conditions? Up to now, microglial-related pruning has only been

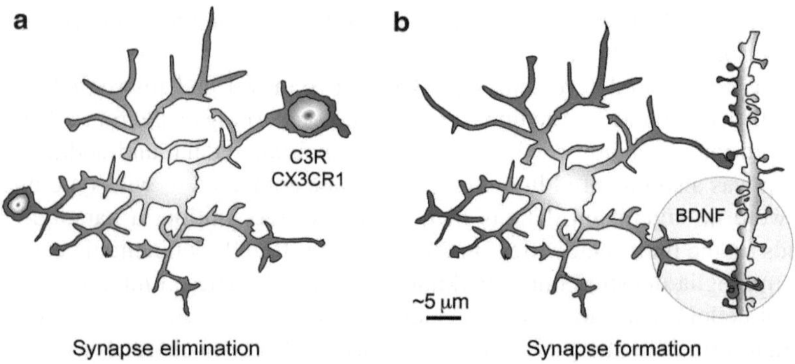

Fig. 9.3 *Microglial regulation of synapse number: an overview of the proposed mechanisms.* (**a**) Microglial pruning of synaptic elements is mediated by the classical complement pathway in the developing retinogeniculate system. In particular, mice deficient in the opsonin C3 or microglial C3 receptor (C3R) displayed a similar impairment in the eye-specific segregation of retinal inputs to the visual thalamus, accompanied by a reduced microglial engulfment of their projections. Additionally, fractalkine signalling could also be involved in mediating microglial pruning of synapses in the hippocampus during early postnatal development, since a reduced microglial colonisation of the hippocampus accompanied by an increased density of dendritic spines has been observed in mice deficient in the fractalkine receptor, CX3CR1, only expressed by microglia in the healthy brain, compared with wild-type littermates. Evidence of reduced synaptic multiplicity (i.e., density of axon terminals contacting two dendritic spines), impairment of synaptic maturation, function and plasticity, learning and memory, and social behaviour were also reported in the CX3CR1 knockouts. (**b**) Microglial release of brain-derived neurotrophic factor (BDNF) was also recently revealed to promote dendritic spine formation during motor learning, in the primary motor cortex

demonstrated for excitatory synapses. Yet, the frequency of miniature inhibitory post-synaptic currents (mIPSCs) has been shown to be reduced in CX3CR1$^{KO/KO}$ mouse hippocampus at P15 (Zhan et al. 2014), indicating a possible functional relationship between microglia and inhibitory synapses. These mIPSCs were recorded from interneurons in the presence of the sodium channel blocker TTX which prevents action potential-induced currents, thus enabling to measure the post-synaptic currents induced by the random release of synaptic vesicles. However, further studies are needed to elucidate whether inhibitory synapses, for example, can serve as a substrate for microglial-mediated pruning. What are the regional and sex-specific differences in microglial involvement with synapses? Complement-deficient animals display more synapses in the visual system (Schafer et al. 2012), but not in the spinal cord (Lobsiger et al. 2013), suggesting tissue-specific action of the complement-dependent pruning. Recently, intracerebroventricular injection of the anti-inflammatory antibiotic minocycline, which is commonly used for preventing microglial activation but also acts on astrocytes and other inflammatory cells, was shown to prevent the occurrence of sex differences in microglia (density, morphology), and the masculinization of dendritic spine density in the preoptic area of the hypothalamus during the critical period for sexual differentiation, i.e., between P0 and P3, with consequences on the adult copulatory behaviour. In this context, microglial regulation of estradiol (aromatized locally from testis-derived testosterone) and prostaglandin levels was proposed as an underlying mechanism (Lenz et al. 2013).

What are the molecular cues which determine microglial recognition and elimination of synaptic elements specifically, i.e., versus other neuronal compartments such as axons and dendrites? Microglial processes sometimes contact dendrites and axons in vivo and in situ, but their consequences still remain undetermined. What are the molecular mechanisms responsible for microglial recognition and elimination of particular subsets of synaptic elements at particular excitatory synapses? Linnartz and colleagues have recently revealed that changes in the neuronal glycocalyx induce activation of the classical complement cascade in vitro. In this study, desialylated (i.e., lacking the terminal sugar residue sialic acid) neurites from hippocampal neurons were found to be preferentially phagocytosed by hippocampal microglial cells in primary coculture via a CR3-dependent mechanism (Linnartz et al. 2012). Do classical mechanisms such as cytoskeletal collapse accompanied by dendritic spine retraction compete or cooperate with microglial involvement in the formation and elimination of synaptic elements? How do microglia cooperate with astrocytes and other types of glial cells in mediating these structural changes at synapses? Future works might unravel novel molecular pathways regulating the refinement of neuronal circuits in the healthy brain and will also determine the relationships between fractalkine, the complement pathway, and BDNF and elucidate the functional relevance and contextual specificities of each of these pathways.

9.4 Regulating Synapse Maturation, Function, and Plasticity

As cells of the immune system, microglia are able to sense and react to modifications of their environment and to physiologically interact with their neighboring cells. Recent studies revealed that microglia functionally communicate with neurons and have the capacity to regulate the maturation, function, and plasticity of synapses, as will be discussed in this section.

9.4.1 Maturation of Synapses

In addition to impairing the elimination of synapses, the brain functional connectivity, learning and memory, and different types of behaviour, removing microglial CX3CR1 has been shown to delay the maturation of excitatory synapses in hippocampal slices. During early postnatal development, the ratio between the amplitude of spontaneous excitatory post-synaptic currents (sEPSC) and the amplitude of miniature (mEPSCs) recorded from hippocampal pyramidal cells progressively increases as neuronal circuits mature (Hsia et al. 1998). Since mEPSCs are recorded in the presence of the sodium channel blocker TTX, they enable to measure the post-synaptic currents induced by the random release of synaptic vesicles. On the other hand, sEPSCs reflect currents which are triggered by an action potential, thus measuring synaptic efficiency. The amplitude of sEPSC normally increases over the course of postnatal development, as functional synapses increase in number and neuronal networks become properly interconnected (Hsia et al. 1998). On the contrary, the amplitude of mEPSCs remains relatively constant across the same developmental period. In hippocampal slices of CX3CR1$^{KO/KO}$ mice compared to wild-type littermates, this sEPSC/mEPSC ratio was shown to be reduced at P15 as well as at P40, thus suggesting a deficit in the maturation of synaptic connectivity (Paolicelli et al. 2011; Zhan et al. 2014). This observation raised the provocative idea that microglia could regulate the maturation of synapses.

Such a role in synaptic maturation is supported by the recent observations of Hoshiko and colleagues, revealing that microglial recruitment to the sites of maturating synapses begins at P5 in the somatosensory cortex and is mediated by fractalkine signalling (Hoshiko et al. 2012). In this study, a reduced density of microglial cells was observed in the thalamocortical clusters, at P6 and P7, but not at P9, in the CX3CR1$^{KO/KO}$ mice versus wild-type littermates. This reduced microglial density inside the clusters was accompanied by an increased density outside the clusters, resulting in an unchanged overall density within the somatosensory cortex. Interestingly, this delayed migration of microglial cells was also paralleled by a transient impairment in the maturation of synaptic properties (Hoshiko et al. 2012). At thalamocortical synapses, the composition of NMDA receptors (NMDARs; named after their agonist *N*-methyl-D-aspartate), which are the main glutamate receptors in the CNS, is known to switch from GluN2B to GluN2A subunits between the first and second postnatal weeks. In this study by Hoshiko and colleagues, a higher proportion of GluN2B-containing NMDARs was reported in the CX3CR1 knockout mice, associated with slower kinetics of the NMDA-mediated synaptic currents. Indeed, the decay time of the NMDAR-mediated responses was found to be significantly reduced at P9–P10 using single-cell recordings in the CX3CR$^{KO/KO}$ mice compared to wild-type littermates. However, the defect was transient, without any difference in the decay time of NMDAR-mediated responses remaining at P27–P33 (Hoshiko et al. 2012). To mediate this delayed maturation of thalamocortical synapses, microglia have been proposed to influence the function of glutamate receptors by their release of signalling factors which remain to be identified and might include IL-1β, tumour necrosis factor α (TNFα), and BDNF, which are known for modulating the expression and function of glutamate receptors in vitro (Beattie et al. 2002; Chao 2003; Zhang et al. 2010; Zhong et al. 2010).

9.4.2 Function and Plasticity of Synapses

Long-term potentiation (LTP) is a long-lasting enhancement of synaptic transmission resulting from strong stimulation of the same synapses. This is the most widely used cellular paradigm of synaptic plasticity. Synaptic alterations including enhanced hippocampal LTP in adolescent brain slices (Roumier et al. 2004, 2008) and impairment of higher brain functions such as loss of memory, aphasia, agraphia, acalculia, and apraxia (Paloneva et al. 2000) have also been demonstrated upon the loss-of-function mutation of DAP12, a transmembrane signalling protein that is exclusively expressed by microglia in the healthy brain (Roumier et al. 2004). Another microglial protein which was shown to be involved in synaptic regulation is CD200R, a membrane protein exclusively expressed by microglia in the CNS during normal physiological conditions that interacts with CD200 expressed by neurons, oligodendrocytes, and astrocytes (Costello et al. 2011). It has been demonstrated that LTP is inhibited in hippocampal slices from adult CD200-deficient mice, further supporting the notion that the integrity of microglial signalling is important for neurotransmission homeostasis (Costello et al. 2011). Finally, hippocampal LTP in the adult was shown to be reduced (Maggi et al. 2009; Rogers et al. 2011) or increased (Maggi et al. 2011) in $CX3CR1^{KO/KO}$ mice depending on the study. Despite the disagreement warranting further investigation, this observation that LTP is modified in CX3CR1-deficient mice suggests that microglia could be involved in regulating synaptic function and plasticity. Paolicelli and colleagues lastly examined the effects of fractalkine signalling on the induction of long-term depression (LTD), another form of synaptic plasticity reducing synaptic transmission on the opposite of LTP, in hippocampal slices from the CX3CR1 knockout mice and their wild-type littermates, reporting no difference between genotypes at P40, despite a significant enhancement at P13 (Paolicelli et al. 2011), thus suggesting that in the mature healthy brain fractalkine signalling selectively modulates LTP. These examples only show correlations between the alterations of microglial signalling and synaptic dysfunction and indirect developmental or systemic issues cannot be excluded in gene knockout animals. However, the recent work of Parkhurst and colleagues demonstrated that genetic depletion of microglia in adult mice induces a robust decrease in the frequency of mEPSCs mediated both by NMDARs and AMPA receptors (AMPARs, named after their agonist α-amino-3-hydroxy-5-methyl-4-isoxazolepropionic acid) (Parkhurst et al. 2013), probably revealing a direct regulation of glutamatergic transmission by microglia.

Several mechanisms have now been identified showing that microglia actively influence the synaptic properties in hippocampal slices. For instance, upon stimulation with the nucleotide ATP, cultured microglia can rapidly shed micro-vesicles, most probably by a mechanism depending on the P2X7 purinergic receptor (Bianco et al. 2005). When these vesicles were applied to cultured hippocampal neurons, they increased the frequency of mEPSCs (Antonucci et al. 2012), raising the hypothesis that neurotransmission could be regulated by physical contacts. Microglia produce a broad spectrum of signalling molecules, from cytokines to neurotransmitters and ECM proteins known to regulate the synaptic function (reviewed in Bechade et al. 2013). In line with such production, several studies have demonstrated a regulation of synaptic properties by microglial signalling molecules. Stimulation of microglia by fractalkine in neuronal cultures was shown to induce a strong and rapid modulation of AMPARs-mediated calcium currents in neurons (Meucci et al. 1998). This modulation has also been confirmed in acute hippocampal slices, in

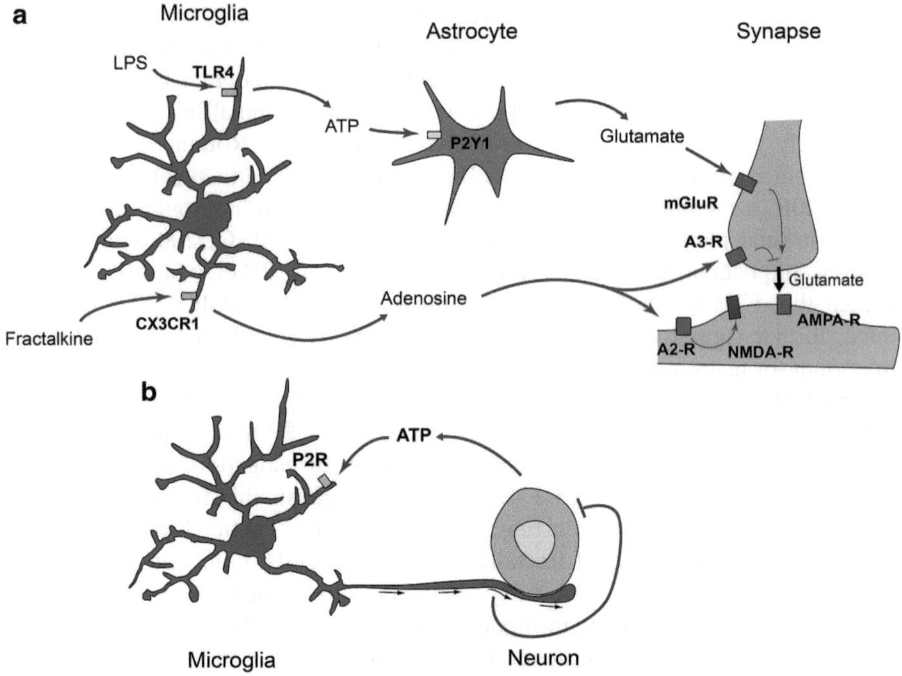

Fig. 9.4 *Microglial regulation of synapse function and plasticity: an overview of the proposed mechanisms*. (**a**) Stimulation of microglia with bacterial lipopolysaccharide (LPS) induces a rapid release of the nucleotide ATP, which binds to P2Y1 on astrocytes. Upon purinergic stimulation, astrocytes release glutamate, inducing an mGluR-dependent release of pre-synaptic glutamate. Stimulation of microglia by fractalkine induces the release of adenosine, which can decrease neuronal activity through A3R and promote NMDAR activity through A2R. (**b**) In zebrafish larva, active neurons also release ATP that attracts microglial bulbous processes. These processes decrease neuronal activity by an as yet unknown mechanism

which stimulation of microglia by fractalkine induces a significant and transient reduction of the amplitude of evoked EPSCs in CA1 pyramidal neurons (Ragozzino et al. 2006). It was further demonstrated that this reduction involves adenosine, supposedly acting on neuronal adenosine receptors 3 (A3R) (Piccinin et al. 2010). A probable mechanism of regulation is fractalkine-induced microglial release of adenosine inhibiting the pre-synaptic release of glutamate (Fig. 9.4a). Alternatively, microglia could produce ATP that is rapidly degraded into adenosine by ectonucleotidases. More recently, the same group showed that fractalkine-induced release of adenosine could also increase the hippocampal NMDA response (Scianni et al. 2013). In this study, the effects of fractalkine on the fEPSPs were abolished by a specific blocker of adenosine receptors subtype 2 (AR2), and in acute slices from AR2 knockout mice, thus proposing a role for A2R in mediating these effects. Fractalkine stimulation of microglia and astrocyte primary culture induced a modest but significant increase in the concentration of D-serine, a co-agonist of NDMA receptors. It was therefore proposed that microglial or astrocytic release of D-serine might potentiate NMDA receptors function downstream of fractalkine signalling (Scianni et al. 2013).

A rapid regulation of neuronal activity by microglia has also been established upon stimulation of microglia by lipopolysaccharides (LPS) (Pascual et al. 2012). LPS is a specific ligand of toll-like receptor 4 (TLR4), which in the brain is exclusively expressed by microglia. Therefore, the analysis of neuronal outcomes following acute application of LPS onto brain tissue can be used to reveal relevant physiological microglia-neuron interacting pathways. Indeed, LPS stimulation of microglia in acute hippocampal slices induces a rapid and transient increase in the frequency of spontaneous synaptic AMPAergic post-synaptic currents in CA1 neurons. This effect does not occur in slices prepared from mice deficient for the myeloid-cell specific transcription factor Pu.1, which lack microglia, showing that the effect of LPS requires microglia. It was then demonstrated that upon LPS stimulation, microglia rapidly produce ATP, which recruits astrocytes (Fig. 9.4a). Astrocytes subsequently release glutamate, and this leads to increased excitatory transmission via a metabotropic glutamate receptor-dependent mechanism (Pascual et al. 2012).

The above-described studies show that stimulation of microglia modulates neuronal activity in vitro and ex vivo. The occurrence of regulating, bidirectional interactions between microglia and neuronal activity during normal physiological conditions has recently been demonstrated in vivo using confocal and two-photon imaging in the zebrafish larva (Li et al. 2012). In this system, microglia were shown to monitor spontaneous or visually evoked neuronal activity and send processes toward the active neurons, detected by their production of ATP (Fig. 9.4b). These contacts between microglial endings and active neurons further induced a rapid decrease in both frequency and amplitude of neuronal activity measured by tracking the changes in calcium events (Li et al. 2012). This study confirms and extends the data obtained upon stimulation of microglia and further demonstrates that microglia could be genuine partners of neuronal activity in the healthy brain. This role of microglia suggests that at least part of the synaptic dysfunction, observed during brain diseases, could be due to the microglial activation or dysfunction. Future work will help to elucidate the mechanisms of synaptic transmission which are specifically influenced by microglia, determine whether their structural interactions with peri-synaptic astrocytic processes are somehow involved in this regulation, and shed light on the respective roles of their interactions with pre-synaptic elements, post-synaptic elements, and peri-synaptic astrocytic processes, compared with their release of soluble mediators, in various contexts of health in the developing and mature brain in vivo.

9.5 Conclusion

Together, these combined findings from imaging, electrophysiology, and behaviour indicate that microglia in the healthy brain frequently interact with synapses, structurally and functionally, in relation with the homeostatic variations in neuronal activity and experience. Their habitual forms of interactions include the phagocytic elimination of synaptic structures and release of soluble factors which can influence the formation, maturation, function, and plasticity of synapses, directly or indirectly. Beyond their decisive role in pathological conditions, microglia are thus emerging as important contributors to normal brain physiology. Future works will elucidate the cellular and molecular mechanisms governing their interactions with neuronal circuits, in cooperation with astrocytes and other types of glial cells, and thereby help unravel their relevance to various contexts of injury and disease.

Acknowledgement This work was supported by grants from the Banting Research Foundation, Scottish Rite Charitable Foundation of Canada, and Natural Sciences and Engineering Research Council of Canada, as well as a start-up fund from the Université Laval and Centre de recherche du CHU de Québec to Marie-Ève Tremblay.

References

Alvarez VA, Sabatini BL (2007) Anatomical and physiological plasticity of dendritic spines. Annu Rev Neurosci 30:79–97

Antonucci F, Turola E, Riganti L, Caleo M, Gabrielli M, Perrotta C, Novellino L, Clementi E, Giussani P, Viani P, Matteoli M, Verderio C (2012) Micro vesicles released from microglia stimulate synaptic activity via enhanced sphingolipid metabolism. EMBO J 31:1231–1240

Arnold SE (1999) Neuro developmental abnormalities in schizophrenia: insights from neuropathology. Dev Psychopathol 11:439–456

Beattie EC, Stellwagen D, Morishita W, Bresnahan JC, Ha BK, Von Zastrow M, Beattie MS, Malenka RC (2002) Control of synaptic strength by glial TNF-alpha. Science 295:2282–2285

Bechade C, Cantaut-Belarif Y, Bessis A (2013) Microglial control of neuronal activity. Front Cell Neurosci 7:32

Bessis A, Bechade C, Bernard D, Roumier A (2007) Microglial control of neuronal death and synaptic properties. Glia 55:233–238

Bhatt DH, Zhang S, Gan WB (2009) Dendritic spine dynamics. Annu Rev Physiol 71:261–282

Bialas AR, Stevens B (2013) TGF-beta signaling regulates neuronal C1q expression and developmental synaptic refinement. Nat Neurosci 16:1773–1782

Bianco F, Pravettoni E, Colombo A, Schenk U, Moller T, Matteoli M, Verderio C (2005) Astrocyte-derived ATP induces vesicle shedding and IL-1 beta release from microglia. J Immunol 174: 7268–7277

Biber K, Neumann H, Inoue K, Boddeke HW (2007) Neuronal 'On' and 'Off' signals control microglia. Trends Neurosci 30:596–602

Bourne JN, Harris KM (2012) Nanoscale analysis of structural synaptic plasticity. Curr Opin Neurobiol 22:372–382

Cardona AE, Pioro EP, Sasse ME, Kostenko V, Cardona SM, Dijkstra IM, Huang D, Kidd G, Dombrowski S, Dutta R, Lee JC, Cook DN, Jung S, Lira SA, Littman DR, Ransohoff RM (2006) Control of microglial neurotoxicity by the fractalkine receptor. Nat Neurosci 9:917–924

Chakravarthy S, Saiepour MH, Bence M, Perry S, Hartman R, Couey JJ, Mansvelder HD, Levelt CN (2006) Postsynaptic TrkB signaling has distinct roles in spine maintenance in adult visual cortex and hippocampus. Proc Natl Acad Sci U S A 103:1071–1076

Chao MV (2003) Neurotrophins and their receptors: a convergence point for many signalling pathways. Nat Rev Neurosci 4:299–309

Chen C, Regehr WG (2000) Developmental remodeling of the retinogeniculate synapse. Neuron 28:955–966

Choquet D, Triller A (2013) The dynamic synapse. Neuron 80:691–703
Chu Y, Jin X, Parada I, Pesic A, Stevens B, Barres B, Prince DA (2010) Enhanced synaptic connectivity and epilepsy in C1q knockout mice. Proc Natl Acad Sci U S A 107:7975–7980
Chung WS, Clarke LE, Wang GX, Stafford BK, Sher A, Chakraborty C, Joung J, Foo LC, Thompson A, Chen C, Smith SJ, Barres BA (2013) Astrocytes mediate synapse elimination through MEGF10 and MERTK pathways. Nature 504:394–400
Clarke LE, Barres BA (2013) Emerging roles of astrocytes in neural circuit development. Nat Rev Neurosci 14:311–321
Combadiere C, Salzwedel K, Smith ED, Tiffany HL, Berger EA, Murphy PM (1998) Identification of CX3CR1. A chemotactic receptor for the human CX3C chemokine fractalkine and a fusion coreceptor for HIV-1. J Biol Chem 273:23799–23804
Cook PM, Prusky G, Ramoa AS (1999) The role of spontaneous retinal activity before eye opening in the maturation of form and function in the retinogeniculate pathway of the ferret. Vis Neurosci 16:491–501
Costello DA, Lyons A, Denieffe S, Browne TC, Cox FF, Lynch MA (2011) Long term potentiation is impaired in membrane glycoprotein CD200-deficient mice: a role for Toll-like receptor activation. J Biol Chem 286:34722–34732
Coull JA, Beggs S, Boudreau D, Boivin D, Tsuda M, Inoue K, Gravel C, Salter MW, De Koninck Y (2005) BDNF from microglia causes the shift in neuronal anion gradient underlying neuropathic pain. Nature 438:1017–1021
Davalos D, Grutzendler J, Yang G, Kim JV, Zuo Y, Jung S, Littman DR, Dustin ML, Gan WB (2005) ATP mediates rapid microglial response to local brain injury in vivo. Nat Neurosci 8:752–758
Del Rio T, Feller MB (2006) Early retinal activity and visual circuit development. Neuron 52: 221–222
Elmore MR, Najafi AR, Koike MA, Dagher NN, Spangenberg EE, Rice RA, Kitazawa M, Matusow B, Nguyen H, West BL, Green KN (2014) Colony-stimulating factor 1 receptor signaling is necessary for microglia viability, unmasking a microglia progenitor cell in the adult brain. Neuron 82:380–397
Ethell IM, Pasquale EB (2005) Molecular mechanisms of dendritic spine development and remodeling. Prog Neurobiol 75:161–205
Feng G, Mellor RH, Bernstein M, Keller-Peck C, Nguyen QT, Wallace M, Nerbonne JM, Lichtman JW, Sanes JR (2000) Imaging neuronal subsets in transgenic mice expressing multiple spectral variants of GFP. Neuron 28:41–51
Garton KJ, Gough PJ, Blobel CP, Murphy G, Greaves DR, Dempsey PJ, Raines EW (2001) Tumor necrosis factor-alpha-converting enzyme (ADAM17) mediates the cleavage and shedding of fractalkine (CX3CL1). J Biol Chem 276:37993–38001
Gasque P (2004) Complement: a unique innate immune sensor for danger signals. Mol Immunol 41:1089–1098
Gomez Perdiguero E, Schulz C, Geissmann F (2013) Development and homeostasis of "resident" myeloid cells: the case of the microglia. Glia 61:112–120
Haber M, Zhou L, Murai KK (2006) Cooperative astrocyte and dendritic spine dynamics at hippocampal excitatory synapses. J Neurosci 26:8881–8891
Herrmann K, Shatz CJ (1995) Blockade of action potential activity alters initial arborization of thalamic axons within cortical layer 4. Proc Natl Acad Sci U S A 92:11244–11248
Hirasawa T, Ohsawa K, Imai Y, Ondo Y, Akazawa C, Uchino S, Kohsaka S (2005) Visualization of microglia in living tissues using Iba1-EGFP transgenic mice. J Neurosci Res 81:357–362
Holtmaat A, Svoboda K (2009) Experience-dependent structural synaptic plasticity in the mammalian brain. Nat Rev Neurosci 10:647–658
Hong YK, Chen C (2011) Wiring and rewiring of the retinogeniculate synapse. Curr Opin Neurobiol 21:228–237
Hooks BM, Chen C (2006) Distinct roles for spontaneous and visual activity in remodeling of the retinogeniculate synapse. Neuron 52:281–291

Hoshiko M, Arnoux I, Avignone E, Yamamoto N, Audinat E (2012) Deficiency of the microglial receptor CX3CR1 impairs postnatal functional development of thalamocortical synapses in the barrel cortex. J Neurosci 32:15106–15111

Hotulainen P, Hoogenraad CC (2010) Actin in dendritic spines: connecting dynamics to function. J Cell Biol 189:619–629

Hsia AY, Malenka RC, Nicoll RA (1998) Development of excitatory circuitry in the hippocampus. J Neurophysiol 79:2013–2024

Hua JY, Smith SJ (2004) Neural activity and the dynamics of central nervous system development. Nat Neurosci 7:327–332

Huberman AD (2007) Mechanisms of eye-specific visual circuit development. Curr Opin Neurobiol 17:73–80

Huberman AD, Feller MB, Chapman B (2008) Mechanisms underlying development of visual maps and receptive fields. Annu Rev Neurosci 31:479–509

Jung S, Aliberti J, Graemmel P, Sunshine MJ, Kreutzberg GW, Sher A, Littman DR (2000) Analysis of fractalkine receptor CX(3)CR1 function by targeted deletion and green fluorescent protein reporter gene insertion. Mol Cell Biol 20:4106–4114

Keck T, Mrsic-Flogel TD, Vaz Afonso M, Eysel UT, Bonhoeffer T, Hubener M (2008) Massive restructuring of neuronal circuits during functional reorganization of adult visual cortex. Nat Neurosci 11:1162–1167

Kettenmann H, Hanisch UK, Noda M, Verkhratsky A (2011) Physiology of microglia. Physiol Rev 91:461–553

Kettenmann H, Kirchhoff F, Verkhratsky A (2013) Microglia: new roles for the synaptic stripper. Neuron 77:10–18

Le Roy C, Wrana JL (2005) Clathrin- and non-clathrin-mediated endocytic regulation of cell signalling. Nat Rev Mol Cell Biol 6:112–126

Lenz KM, Nugent BM, Haliyur R, McCarthy MM (2013) Microglia are essential to masculinization of brain and behavior. J Neurosci 33:2761–2772

Li Y, Du XF, Liu CS, Wen ZL, Du JL (2012) Reciprocal regulation between resting microglial dynamics and neuronal activity in vivo. Dev Cell 23:1189–1202

Linnartz B, Kopatz J, Tenner AJ, Neumann H (2012) Sialic acid on the neuronal glycocalyx prevents complement C1 binding and complement receptor-3-mediated removal by microglia. J Neurosci 32:946–952

Lobsiger CS, Boillee S, Pozniak C, Khan AM, McAlonis-Downes M, Lewcock JW, Cleveland DW (2013) C1q induction and global complement pathway activation do not contribute to ALS toxicity in mutant SOD1 mice. Proc Natl Acad Sci U S A 110:E4385–E4392

Luo L, O'Leary DD (2005) Axon retraction and degeneration in development and disease. Annu Rev Neurosci 28:127–156

Maggi L, Scianni M, Branchi I, D'Andrea I, Lauro C, Limatola C (2011) CX(3)CR1 deficiency alters hippocampal-dependent plasticity phenomena blunting the effects of enriched environment. Front Cell Neurosci 5:22

Maggi L, Trettel F, Scianni M, Bertollini C, Eusebi F, Fredholm BB, Limatola C (2009) LTP impairment by fractalkine/CX3CL1 in mouse hippocampus is mediated through the activity of adenosine receptor type 3 (A3R). J Neuroimmunol 215:36–42

Majewska A, Sur M (2003) Motility of dendritic spines in visual cortex in vivo: changes during the critical period and effects of visual deprivation. Proc Natl Acad Sci U S A 100:16024–16029

Majewska AK, Sur M (2006) Plasticity and specificity of cortical processing networks. Trends Neurosci 29:323–329

Meucci O, Fatatis A, Simen AA, Bushell TJ, Gray PW, Miller RJ (1998) Chemokines regulate hippocampal neuronal signaling and gp120 neurotoxicity. Proc Natl Acad Sci U S A 95:14500–14505

Nedergaard M (1994) Direct signaling from astrocytes to neurons in cultures of mammalian brain cells. Science 263:1768–1771

Nimmerjahn A, Kirchhoff F, Helmchen F (2005) Resting microglial cells are highly dynamic surveillants of brain parenchyma in vivo. Science 308:1314–1318

Pannasch U, Rouach N (2013) Emerging role for astroglial networks in information processing: from synapse to behavior. Trends Neurosci 36:405–417

Paolicelli RC, Bisht K, Tremblay ME (2014) Fractalkine regulation of microglial physiology and consequences on the brain and behaviour. Front Cell Neurosci 8:129

Paolicelli RC, Bolasco G, Pagani F, Maggi L, Scianni M, Panzanelli P, Giustetto M, Ferreira TA, Guiducci E, Dumas L, Ragozzino D, Gross CT (2011) Synaptic pruning by microglia is necessary for normal brain development. Science 333:1456–1458

Parkhurst CN, Yang G, Ninan I, Savas JN, Yates JR 3rd, Lafaille JJ, Hempstead BL, Littman DR, Gan WB (2013) Microglia promote learning-dependent synapse formation through brain-derived neurotrophic factor. Cell 155:1596–1609

Parpura V, Basarsky TA, Liu F, Jeftinija K, Jeftinija S, Haydon PG (1994) Glutamate-mediated astrocyte-neuron signalling. Nature 369:744–747

Pascual O, Ben Achour S, Rostaing P, Triller A, Bessis A (2012) Microglia activation triggers astrocyte-mediated modulation of excitatory neurotransmission. Proc Natl Acad Sci U S A 109:E197–E205

Penn AA, Riquelme PA, Feller MB, Shatz CJ (1998) Competition in retinogeniculate patterning driven by spontaneous activity. Science 279:2108–2112

Piccinin S, Di Angelantonio S, Piccioni A, Volpini R, Cristalli G, Fredholm BB, Limatola C, Eusebi F, Ragozzino D (2010) CX3CL1-induced modulation at CA1 synapses reveals multiple mechanisms of EPSC modulation involving adenosine receptor subtypes. J Neuroimmunol 224:85–92

Pocock JM, Kettenmann H (2007) Neurotransmitter receptors on microglia. Trends Neurosci 30:527–535

Prinz M, Priller J (2014) Microglia and brain macrophages in the molecular age: from origin to neuropsychiatric disease. Nat Rev Neurosci 15:300–312

Ragozzino D, Di Angelantonio S, Trettel F, Bertollini C, Maggi L, Gross C, Charo IF, Limatola C, Eusebi F (2006) Chemokine fractalkine/CX3CL1 negatively modulates active glutamatergic synapses in rat hippocampal neurons. J Neurosci 26:10488–10498

Ransohoff RM (2009) Chemokines and chemokine receptors: standing at the crossroads of immunobiology and neurobiology. Immunity 31:711–721

Ransohoff RM, Perry VH (2009) Microglial physiology: unique stimuli, specialized responses. Annu Rev Immunol 27:119–145

Rogers JT, Morganti JM, Bachstetter AD, Hudson CE, Peters MM, Grimmig BA, Weeber EJ, Bickford PC, Gemma C (2011) CX3CR1 deficiency leads to impairment of hippocampal cognitive function and synaptic plasticity. J Neurosci 31:16241–16250

Roumier A, Bechade C, Poncer JC, Smalla KH, Tomasello E, Vivier E, Gundelfinger ED, Triller A, Bessis A (2004) Impaired synaptic function in the microglial KARAP/DAP12-deficient mouse. J Neurosci 24:11421–11428

Roumier A, Pascual O, Bechade C, Wakselman S, Poncer JC, Real E, Triller A, Bessis A (2008) Prenatal activation of microglia induces delayed impairment of glutamatergic synaptic function. PLoS One 3:e2595

Sanes JR, Lichtman JW (1999) Development of the vertebrate neuromuscular junction. Annu Rev Neurosci 22:389–442

Schafer DP, Lehrman EK, Kautzman AG, Koyama R, Mardinly AR, Yamasaki R, Ransohoff RM, Greenberg ME, Barres BA, Stevens B (2012) Microglia sculpt postnatal neural circuits in an activity and complement-dependent manner. Neuron 74:691–705

Schafer DP, Lehrman EK, Stevens B (2013) The "quad-partite" synapse: microglia–synapse interactions in the developing and mature CNS. Glia 61:24–36

Schmechel DE (1999) Assessment of ultrastructural changes associated with apoptosis. In: Hannun YA, Boustany R-M (eds) Apoptosis in neurobiology. CRC, Boca Raton, pp 161–163

Scianni M, Antonilli L, Chece G, Cristalli G, Di Castro MA, Limatola C, Maggi L (2013) Fractalkine (CX3CL1) enhances hippocampal *N*-methyl-D-aspartate receptor (NMDAR) function via D-serine and adenosine receptor type A2 (A2AR) activity. J Neuroinflammation 10:108

Shatz CJ (1990) Competitive interactions between retinal ganglion cells during prenatal development. J Neurobiol 21:197–211

Sierra A, Abiega O, Shahraz A, Neumann H (2013) Janus-faced microglia: beneficial and detrimental consequences of microglial phagocytosis. Front Cell Neurosci 7:6

Sierra A, Encinas JM, Deudero JJ, Chancey JH, Enikolopov G, Overstreet-Wadiche LS, Tsirka SE, Maletic-Savatic M (2010) Microglia shape adult hippocampal neurogenesis through apoptosis-coupled phagocytosis. Cell Stem Cell 7:483–495

Stellwagen D, Shatz CJ (2002) An instructive role for retinal waves in the development of retinogeniculate connectivity. Neuron 33:357–367

Stevens B, Allen NJ, Vazquez LE, Howell GR, Christopherson KS, Nouri N, Micheva KD, Mehalow AK, Huberman AD, Stafford B, Sher A, Litke AM, Lambris JD, Smith SJ, John SW, Barres BA (2007) The classical complement cascade mediates CNS synapse elimination. Cell 131:1164–1178

Tarozzo G, Bortolazzi S, Crochemore C, Chen SC, Lira AS, Abrams JS, Beltramo M (2003) Fractalkine protein localization and gene expression in mouse brain. J Neurosci Res 73:81–88

Tessier-Lavigne M, Goodman CS (1996) The molecular biology of axon guidance. Science 274:1123–1133

Torborg CL, Feller MB (2005) Spontaneous patterned retinal activity and the refinement of retinal projections. Prog Neurobiol 76:213–235

Trachtenberg JT, Chen BE, Knott GW, Feng G, Sanes JR, Welker E, Svoboda K (2002) Long-term in vivo imaging of experience-dependent synaptic plasticity in adult cortex. Nature 420: 788–794

Tremblay ME (2011) The role of microglia at synapses in the healthy CNS: novel insights from recent imaging studies. Neuron Glia Biol 7:67–76

Tremblay ME, Lowery RL, Majewska AK (2010) Microglial interactions with synapses are modulated by visual experience. PLoS Biol 8:e1000527

Tremblay ME, Majewska AK (2011) A role for microglia in synaptic plasticity? Commun Integr Biol 4:220–222

Tremblay ME, Zettel ML, Ison JR, Allen PD, Majewska AK (2012) Effects of aging and sensory loss on glial cells in mouse visual and auditory cortices. Glia 60:541–558

Tsou CL, Haskell CA, Charo IF (2001) Tumor necrosis factor-alpha-converting enzyme mediates the inducible cleavage of fractalkine. J Biol Chem 276:44622–44626

van Lookeren Campagne M, Wiesmann C, Brown EJ (2007) Macrophage complement receptors and pathogen clearance. Cell Microbiol 9:2095–2102

Wake H, Moorhouse AJ, Jinno S, Kohsaka S, Nabekura J (2009) Resting microglia directly monitor the functional state of synapses in vivo and determine the fate of ischemic terminals. J Neurosci 29:3974–3980

Wake H, Moorhouse AJ, Miyamoto A, Nabekura J (2013) Microglia: actively surveying and shaping neuronal circuit structure and function. Trends Neurosci 36:209–217

Zhan Y, Paolicelli RC, Sforazzini F, Weinhard L, Bolasco G, Pagani F, Vyssotski AL, Bifone A, Gozzi A, Ragozzino D, Gross CT (2014) Deficient neuron-microglia signaling results in impaired functional brain connectivity and social behavior. Nat Neurosci 17:400–406

Zhang H, Nei H, Dougherty PM (2010) A p38 mitogen-activated protein kinase-dependent mechanism of disinhibition in spinal synaptic transmission induced by tumor necrosis factor-alpha. J Neurosci 30:12844–12855

Zhong Y, Zhou LJ, Ren WJ, Xin WJ, Li YY, Zhang T, Liu XG (2010) The direction of synaptic plasticity mediated by C-fibers in spinal dorsal horn is decided by Src-family kinases in microglia: the role of tumor necrosis factor-alpha. Brain Behav Immun 24:874–880

Zuo Y, Lin A, Chang P, Gan WB (2005) Development of long-term dendritic spine stability in diverse regions of cerebral cortex. Neuron 46:181–189

Chapter 10
Adult Neurogenesis, Learning and Memory

Amanda Sierra and Marie-Ève Tremblay

Abstract An emerging view in Neuroscience is that cognitive functions such as learning and memory engage all the various cell types making-up the brain, including surveillant microglia, the resident immune cells. These complex functions prominently depend on the structural remodeling of neuronal circuits, rooted in the formation, strengthening, and elimination of synaptic structures, but also on the continuous integration of newborn neurons into the mature circuitry. In this chapter, we will focus our attention on the emerging roles of microglia in adult neurogenesis within its two consensus regions, the hippocampus and the olfactory bulb. We will discuss the underlying cellular and molecular mechanisms based on evidence from both inflammatory and non-inflammatory conditions. Doing so, we will also examine their implications for learning and memory, and the effects of environmental enrichment, running paradigms, normal aging, and neurodegenerative diseases on microglia, adult neurogenesis, and the behavioural outcome. Lastly, we will summarize the key points that emerge from our current understanding of these new roles of microglia in adult neurogenesis, learning and memory, and the most promising directions to pursue in this recent field of investigation.

Keywords Microglia • Adult neurogenesis • Hippocampus • Olfactory bulb • Physiology • Environmental enrichment • Aging • Neurodegenerative diseases • Cytokines • Chemokines • Phagocytosis

A. Sierra, Ph.D. (✉)
Achucarro Basque Center for Neuroscience, University of the Basque Country, Ikerbasque Foundation, Bizkaia Science and Technology Park, Building 205, Zamudio, Bizkaia 48116, Spain
e-mail: a.sierra@ikerbasque.org

M.-È. Tremblay, Ph.D. (✉)
Axe Neurosciences, Centre de Recherche du CHU de Québec, Department of Molecular Medicine, Université Laval, Québec, QC G1V 4G2, Canada
e-mail: tremblay.marie-eve@crchudequebec.ulaval.ca

M.-È. Tremblay, A. Sierra (eds.), *Microglia in Health and Disease*,
DOI 10.1007/978-1-4939-1429-6_10

Bullet points

- Adult neurogenesis is a normal physiological process that is required for learning and memory in the healthy brain.
- Microglia are an essential component of adult neurogenic niches, where they intermingle with the neuroprogenitors and newborn neurons and phagocytose the excess of newborn neurons, but little is known about their role in physiological conditions, including enriched environment and running paradigms.
- In inflammatory conditions, as well as normal aging and neurodegenerative diseases, microglial-derived pro-inflammatory cytokines such as interleukin 1β (IL-1β) and interleukin 6 (IL-6) are considered detrimental for adult neurogenesis.
- The relative contributions of microglia versus other brain resident (astrocytes, endothelial cells, meningeal, and perivascular macrophages) and infiltrating (monocytes, lymphocytes) inflammatory cells to adult neurogenesis remains largely unknown.
- Further research is necessary to determine the impact of microglial activities in neurogenesis-dependent learning and memory, in health and disease.

10.1 Introduction

Neurogenesis is the process by which new neurons are generated. While it is mostly restricted to the embryonic development, it continues throughout adulthood in well-defined areas of the human and rodent brain where the existence of neural stem cells persists. The two consensus areas for adult neurogenesis are the subgranular zone (SGZ) of the hippocampus, in which newborn cells integrate locally as mature granule cells; and the subventricular zone (SVZ) situated throughout the lateral walls of the lateral ventricles, which generates interneurons populating the olfactory bulb (OB). Importantly, neurogenesis has also been reported in the adult human hippocampus (Eriksson et al. 1998; Spalding et al. 2013). In contrast, the occurrence of SVZ neurogenesis in the adult human brain is still under debate (Curtis et al. 2007; Sanai et al. 2007; Bergmann et al. 2012). In addition to holding the potential for regenerating the damaged brain, newborn neurons have been involved in neuronal circuit remodeling, alongside complementary mechanisms such as synaptic plasticity (see Chap. 9), as well as in learning and memory formation processes. In this chapter, we will describe the anatomical niche of SGZ and SVZ neurogenesis, their implication in hippocampal and OB learning and memory, and finally, their regulation by microglia in different contexts of health and disease, including environmental enrichment (EE), normal aging, and neurodegenerative diseases.

10.2 Adult Neurogenic Niches

The neuroprogenitors of the adult hippocampus reside in the SGZ. Morphologically and functionally, they are similar to the radial glial cells which give rise to neurons and astrocytes during embryonic development (Fig. 10.1). They constitute a quiescent population (type 1 cells or QNPs, for quiescent neuroprogenitors) of which only a small subset is recruited to enter the pool of dividing cells, giving rise to a transient population of proliferating cells (type 2 cells, or ANPs, for amplifying neuroprogenitors). After 3–4 days, these cells become postmitotic neuroblasts (or type 3 cells) committed to a neuronal lineage. Within 4 weeks, the newborn cells are integrated as mature neurons into the granule cell layer, which mainly receive inputs in the dentate gyrus molecular and polymorphic layers and project to the stratum lucidum of the hippocampus CA3 region. In the CA3, their axon terminals which are named the mossy fibres make synapses on the apical dendrites of the CA3 pyramidal cells (reviewed in Kempermann et al. (2004), Encinas and Enikolopov (2008)). Similarly, SVZ neuroprogenitors are modified astrocytes (type B cells) located underneath the ependymal cells lining the lateral ventricles (Fig. 10.1). Once they have proliferated, they give rise to a population of transient amplifying cells (type C cells) which in turn differentiate into neuroblasts (type A cells) to start the neuronal differentiation program. As they migrate as chains of cells through the rostral migratory stream (RMS) from the walls of the lateral ventricles to the OB, the type A cells continue to proliferate. Once they reach the OB, they differentiate into dopaminergic interneurons and integrate into the OB circuitry, where they constitute a population of anaxonic granule cells forming reciprocal dendrodendritic synapses within the external plexiform layer (reviewed in Lledo et al. (2008)).

Thus, neurogenesis encompasses a series of processes designated as the neurogenic cascade, comprising the proliferation of neuroprogenitors, survival of newborn cells, differentiation of neuroblasts, and integration of mature neurons in the preexisting circuitry, as well as migration in the case of SVZ neuroblasts. Stringent definitions and methods must therefore be used to analyze each of these processes individually. The most widely used method is the administration of the thymidine analog 5-bromo-2′-deoxyuridine (BrdU), which incorporates into the DNA of dividing cells during the synthesis phase (or S phase) and can be detected using specific antibodies. BrdU is commonly used in pulse-and-chase experiments to follow the progeny of the dividing cells. For instance, the number of BrdU-positive cells observed 2–24 h after administration reflects the proliferation of neuroprogenitors, whereas the number of BrdU cells co-labeled with neuronal markers such as NeuN observed 4 weeks after administration reflects actual neurogenesis. Other methods to study different aspects of the neurogenic cascade include the use of selective antibodies for visualising the neuroprogenitors or neuroblasts, an alternative approach to BrdU administration which does not require in vivo manipulation. Some of the most popular markers are directed against the intermediate filament nestin, to specifically visualise the neuroprogenitors, or the microtubule-associated protein doublecortin, to target the neuroblasts. These markers thus enable to analyse

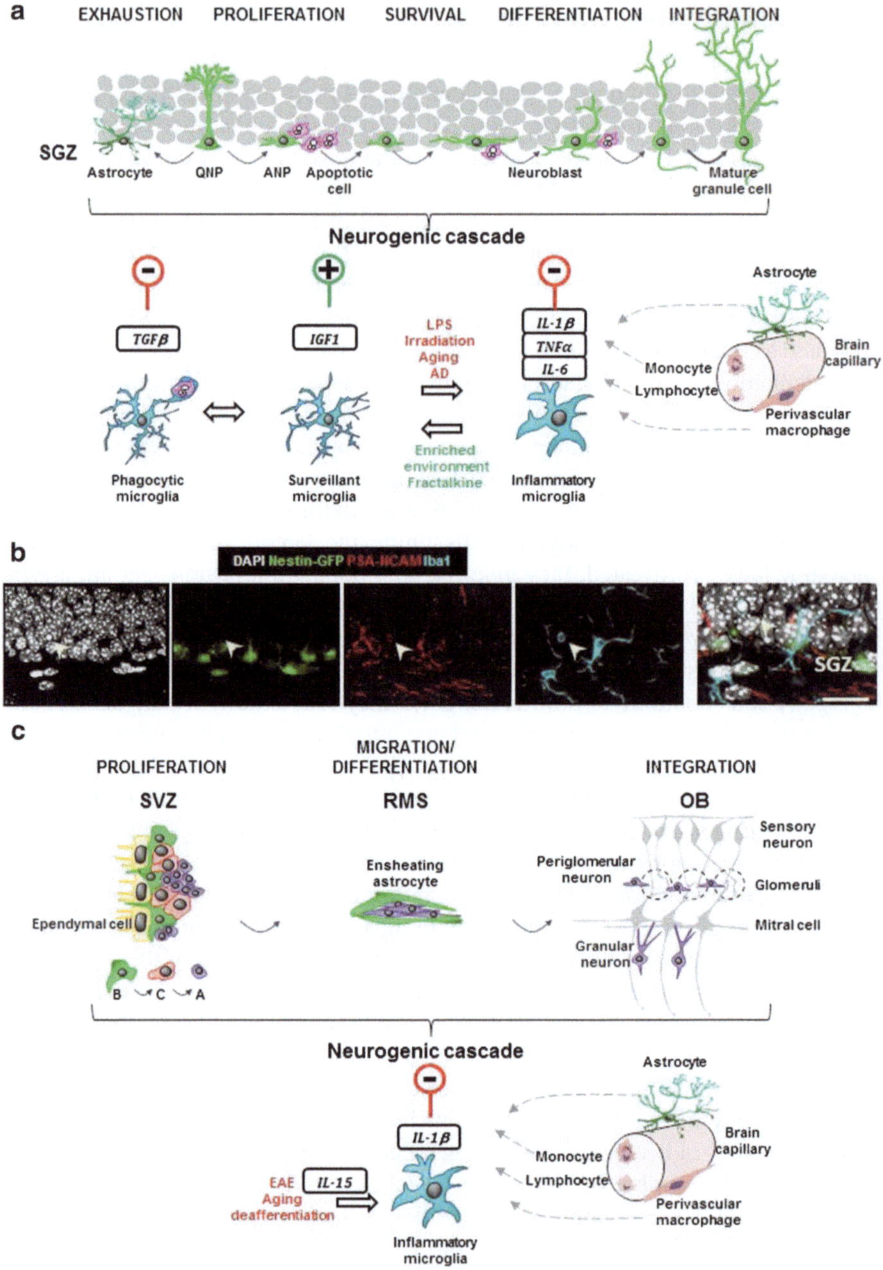

Fig. 10.1 (**a**) Schematic drawing of the adult hippocampal neurogenic cascade, representing the main cell types and stages in the SGZ (see main text for a thorough description of the cascade). In this hippocampus, microglia are physically part of the niche and intermingle with all cell types, and phagocytose the excess of newborn cells that undergo apoptosis in physiological conditions. In vitro, phagocytic microglia is known to release TGFβ, which is anti-neurogenic.

neurogenesis in situ, in postmortem samples. Nonetheless, antibody-based methods can only identify putative neuroprogenitors or neuroblasts, and particular care must be exercised to disregard non-specific or aberrant labeling, particularly in postmortem samples. To date, pulse-and-chase experiments remain the only available method to certify that the newborn cells actually become adult neurons. Additionally, the use of transgenic mice expressing fluorescent reporters in a particular cell type is also extremely useful for visualising how the neuroprogenitors or neuroblasts progress along the neurogenic cascade, from their formation to their integration into the circuitry. For example, the nestin-GFP mice where all neuroprogenitors are labeled with the green fluorescent protein (GFP) can be used in both constitutive or inducible forms (nestin-CreER), in which tamoxifen-induced Cre-mediated recombination leads to the expression of the reporter in the progeny of the neuroprogenitors only. For a thorough description of the methods used to analyze rodent and human neurogenesis, see (Breunig et al. 2007; Sierra et al. 2011).

10.3 The Role of Adult Neurogenesis in Learning and Memory

Adult neurogenesis was initially described by Robert Altman in the 1960s, although it was not until its rediscovery in the early 1990s (Reynolds and Weiss 1992; Richards et al. 1992; Gage et al. 1995; Palmer et al. 1995) that its potential plasticity

Fig. 10.1 (continued) Surveillant microglia may produce proneurogenic factors, such as IGF1, triggered by enriched environment. Fractalkine signaling maintains microglia in an anti-inflammatory state, whereas LPS, irradiation, aging, and AD push microglia towards the production of pro-inflammatory cytokines such as IL-1β, TNFα, and IL-6, with detrimental consequences for neurogenesis. Overall, the role of microglia compared to other inflammatory cells (such as astrocytes, perivascular macrophages, and invading monocytes and lymphocytes) is poorly understood. (**b**) Iba1-labeled microglia (cyan) phagocytosing an apoptotic (pyknotic, visualized with the DNA dye DAPI, white) neuroblast (labeled with polysialated neural cell adhesion molecule, PSA-NCAM, *red*) in the SGZ niche, surrounded by Nestin-GFP-labeled neuroprogenitors (QNPs and ANPs, *green*). Scale bar, 50 μm. (**c**) Schematic drawing of the adult SVZ niche, the RMS, and the OB circuitry, representing the main cell types and stages (see main text for a thorough description). Little is known about the location and role of microglia in physiological conditions in the SVZ niche. In contrast, conditions such as EAE (possibly via IL-15), aging, and differentiation lead to an inflammatory response of microglia and other inflammatory cells (as above), which induce the production of IL-1β and inhibit SVZ neurogenesis. Abbreviations: *AD* Alzheimer's disease, *ANP* amplifying neuroprogenitor, *EAE* experimental acute encephalomyelitis, *IGF-1* insulin-like growth factor 1, *IL-15* interleukin 15, *IL-1β* interleukin 1 beta, *IL-6* interleukin 6, *LPS* bacterial lipopolysaccharides, *OB* olfactory bulb, *QNP* quiescent neuroprogenitor, *RMS* rostral migratory stream, *SGZ* subgranular zone, *SVZ* subventricular zone, *TGFβ* transforming growth factor beta. (**b**) is reprinted from Sierra A, Encinas JM, Deudero JP, Chancey JH, Enikolopov G, Overstreet-Wadiche LS, Tsirka SE, and Maletic-Savatic M. Microglia Shape Adult Hippocampal Neurogenesis through Apoptosis-Coupled Phagocytosis. Cell Stem Cell 7: 483–495, 2010, with permission from Elsevier

strongly suggested its possible involvement in normal brain functions requiring extensive structural remodeling such as learning and memory (Gage 2002). The relationship seems to be bidirectional and at least some forms of learning and/or memory have been demonstrated to both enhance and rely on adult neurogenesis.

The initial reports on neurogenesis, learning, and memory showed that hippocampal-dependent learning, such as spatial navigation in a water maze and conditioning of the eye blink response using a trace protocol, similarly enhanced the number of 8-day-old cells labeled with BrdU prior to the training, probably reflecting an increased survival of the neuroblasts (Gould et al. 1999). In particular, the water maze takes advantage of the animal's ability to swim, forcing the animals to locate a hidden platform to escape from a pool of opaque water. The location of the platform remains constant throughout the training and test sessions, therefore requiring the animals to use spatial mapping from different visual cues located around the pool in order to navigate from a random starting location to the platform. In this context, a decreased latency measured by video tracking analysis is considered to be primarily related to hippocampus-dependent memory. Contextual fear conditioning is another commonly used cognitive task that mainly involves the hippocampus, and to a lesser degree the amygdala, which is also part of the limbic system. This type of classical conditioning requires the animals to learn the association between two stimuli, an arbitrary signal and a significant event such as a danger or a punishment. In a typical fear conditioning task, a brief exposure to a light or a tone is timed with a mild electrical footshock, allowing the animals to form an association not only between the light/tone and the shock, but also between the environmental context and the shock itself. The animal's freezing response is assessed in the conditioning chamber without light/tone presentation in the case of "contextual" conditioning, or by re-exposure to the light/tone in a novel test environment for "classical" conditioning. Alternatively, a mild puff of air administered to the eyes can be used for eliciting an eye blink response. In both cases, the delayed conditioning protocol can be modified by introducing a temporal gap between the light/tone presentation and the shock, a variant generally referred to as "trace" fear conditioning (Sousa et al. 2006; Crawley 2008). For a thorough description of these and other cognitive tests used to analyze learning and memory in rodents and other species, see Sousa et al. (2006); Crawley (2008).

In the hippocampus, the subsequent studies characterizing the effects of learning on adult neurogenesis did however reveal a more complex relationship. For instance, an increased survival of the young newborn cells (1–4 days-old) has been observed in the late phase, but not in the early phase of learning in a mater maze, when over 80 % of the performance improvement has already occurred, thus suggesting that neurogenesis is involved in the consolidation but not in the acquisition of learning (Dobrossy et al. 2003). Surprisingly, the late phase also resulted in decreased survival of the cells born during the early phase (4–8 days-old), potentially due to different levels of growth factors released within the neurogenic niche in the early versus late phases (Dobrossy et al. 2003), although this hypothesis remains to be experimentally addressed. Further research showed that the young (1.5–2 months-old) hippocampal newborn neurons are preferentially activated during memory recall in a water maze, compared to older neurons, as determined by co-labeling of BrdU and immediate-early genes, such as c-Fos and Arc, whose expression correlates with neuronal firing (Kee et al. 2007), and which have been proposed to be necessary for the consolidation of synaptic plasticity and memory (Plath et al. 2006). These young neurons are also more plastic than their older counterparts, and they show enhanced amplitude response after long-term potentiation (LTP, a cellular correlate of learning and memory; see Chap. 9 for further reading) induced with a theta-burst stimulation paradigm (Ge et al. 2007). However, the evidence that neurogenesis is essential for learning and/or memory using ablation protocols is somewhat conflictive, possibly due to limitations of the protocols and/or behavioural paradigms, as well as differences in their application between studies (reviewed by Marin-Burgin and Schinder (2012)). Chemical depletion of proliferating cells with the neurotoxin methylazoxymethanol acetate (MAM), which reduces DNA synthesis, has been shown to impair learning in trace conditioning protocols (Shors et al. 2001). In contrast, depletion of neuroprogenitors in mice expressing a viral thymidine kinase under the nestin promoter (nestin-tk mice) and treated with ganciclovir did not affect learning but induced impairment in retaining the location of a hidden platform in a water maze (Deng et al. 2009). In these mice, the antiviral ganciclovir acts as a pro-drug, not toxic in itself, but converted to toxic drugs by phosphorylation from the viral thymidine kinase. The neuroprogenitors infected with the virus therefore produce highly toxic triphosphates leading to cell death. Similar results were obtained in mice expressing the diphtheria toxin under the control of the nestin-CreER system. By administering the toxin 7 weeks after the induction of the Cre recombinase, the authors were able to ablate young neurons, without preventing learning in a contextual fear task or the water maze. However, post-training ablation of these 7-week-old cells impaired retention of both types of acquired hippocampal-dependent memories (Arruda-Carvalho et al. 2011). Importantly, a gain-of-function approach ablating the pro-apoptotic gene Bax selectively in the neuroprogenitors with the nestin-CreER system, to increase their survival, had opposite effects, namely, the enhancement of hippocampal synaptic plasticity, context discrimination responses using a contextual-fear conditioning test, and encoding of overlapping input patterns (i.e., pattern separation) (Sahay et al. 2011), a necessary property of associative memory networks in which the dentate gyrus and hippocampus CA3 region were shown to play synergistic roles (Leutgeb et al. 2007).

Similarly, SVZ neurogenesis has also been recently involved in olfactory learning (reviewed in Lazarini and Lledo (2011)). In the OB, the survival of young (2–4 weeks-old) newborn neurons labeled with BrdU increases during learning of an olfactory discrimination test (Alonso et al. 2006), which exploits the rodent's tendency to use olfactory cues to forage and is further decreased following 2 weeks of olfactory input deprivation (Yamaguchi and Mori 2005). In a main variant of this task, the animals learn to associate a first odour with a sugar reward and a second odour with the absence of reward. During the test phase, both odours are presented simultaneously in the absence of sugar, and the animals demonstrate that they have learned the association between the first odour and the availability for sugar by digging in that particular odour (Schellinck et al. 2001). The survival of young newborn cells (3–4 weeks) labeled with BrdU was also found to increase during a similar type of operant conditioning in which water was used as a reward, while, on the opposite, the survival of older neurons (5–7 weeks-old) was decreased (Mouret et al. 2008), suggesting that olfactory training could differently affect the number of surviving newborn neurons depending on the timing between learning and their genesis. The data is however conflictive and other groups have reported no effects on survival, but rather an increased functional activation for 5–9 weeks-old BrdU cells, as determined by c-Fos expression during odour-discrimination learning (Belnoue et al. 2011). More recently, the consequences of adult neurogenesis on a variety of olfactory functions were also directly examined using Nestin-Cre;neuron-specific enolase- diphtheria toxin fragment A (NSE-DTA) mice treated with tamoxifen at 2 months of age, where neurons born in the SVZ and SGZ are both ablated (Imayoshi et al. 2008). In these mice, tamoxifen-inducible Cre is expressed under the control of the nestin promoter, whereas the loxP-STOP-loxP-IRES-diphtheria toxin fragment A(DTA) gene cassette is knocked into the 3′-noncoding region of the neuron-specific enolase (NSE, Eno2) gene. Tamoxifen administration induces Cre-mediated recombination in the neuroprogenitors, but DTA is not expressed in these cells because the NSE promoter is inactive. When the neuroprogenitors begin their neuronal differentiation, however, the NSE promoter becomes active and induces the expression of DTA killing the cells. These mutant mice in which the OB structure is severely disrupted (Imayoshi et al. 2008) were able to discriminate odours as competently as control mice (Imayoshi et al. 2008; Sakamoto et al. 2011), but they showed impairment in predator avoidance (i.e., approach towards a fox scent associated with a reward) and innate sex-specific responses which depend on olfaction, including deficits in male-male aggression, male sexual behaviour towards females, and female deficits in fertility and nurturing (Sakamoto et al. 2011).

Altogether the data strongly suggest that SGZ and SVZ neurogenesis are indeed involved in particular forms of hippocampal and OB-dependent learning and memory, respectively, but, nonetheless, more precise experimental manipulations are still needed to precisely delineate the cellular and molecular mechanisms by which the newborn neurons participate in these processes.

10.4 Microglial Involvement in Adult Neurogenesis

Neurogenesis is modulated by a long list of physiological and pathological conditions, acting through a variety of cellular and molecular mechanisms (reviewed in Marin-Burgin and Schinder (2012)). Among these factors, microglial cells are relatively a newcomer. Belonging to an entirely different lineage from the ectodermic-derived neurons, astrocytes, and oligodendrocytes, and historically implicated mostly in orchestrating the immune and inflammatory response to pathological conditions (Chap. 5), it has been only in recent years that microglia have been recognized as an essential component of adult neurogenic niches (Fig. 10.1).

The first evidence for such a role of microglia comes from studies showing that inflammation is detrimental to adult hippocampal neurogenesis, as well as learning and memory, throughout life. Original research from Olle Lindvall and Theo Palmer's groups showed that the number of BrdU-labeled neurons decreases after inflammatory challenge induced by systemic or intrahippocampal administration of lipopolysaccharides (LPS), a component of the wall of Gram-negative bacteria, and this effect is prevented by treatment with indomethacin, a non-steroidal anti-inflammatory drug (NSAID) which inhibits the synthesis of pro-inflammatory prostaglandins (Ekdahl et al. 2003; Monje et al. 2003). Furthermore, LPS also prevents the increase in newborn neurons labeled with BrdU driven by seizures during partial status epilepticus, whereas the anti-inflammatory antibiotic minocycline increases neurogenesis after either partial or generalized status epilepticus (Ekdahl et al. 2003). But in these studies neither LPS nor minocycline affected the proliferation of neuroprogenitors, and their effects seemed to be mostly mediated through a decrease in the survival of newborn cells (Ekdahl et al. 2003; Monje et al. 2003). Indeed, LPS has been shown to increase apoptosis of the newborn cells generated in the hippocampus during normal physiological (or basal) conditions (Sierra et al. 2010). LPS also enhances the synaptic connectivity of the newborn neurons during early stages of their synaptogenesis (Chugh et al. 2013) and prevents their reactivation during spatial exploration, determined as the percentage of BrdU cells co-labeled with the immediate-early gene Arc in situ (Belarbi et al. 2012). A more direct implication of microglia in LPS-induced decreased neurogenesis is that conditioned media from LPS-challenged microglia contain high levels of a pro-inflammatory cytokine, interleukin 6 (IL-6), causing apoptosis of neuroblasts derived from the adult mouse hippocampus in vitro (Monje et al. 2003). In contrast, IL-1β increases the proliferation of cultured human hippocampal embryonic neuroprogenitors, but decreases their differentiation into neurons (Zunszain et al. 2012).

In vivo, the implication of microglia has been revealed by the negative correlation between the numbers of newborn neurons (BrdU+, NeuN+) and microglia expressing ED1, also called CD68 or macrosialin (Ekdahl et al. 2003). The number of ED1-positive microglia also negatively correlated with neurogenesis during inflammation driven by cranial irradiation (Monje et al. 2003). ED1 was originally described as a lysosomal protein over-expressed during inflammatory challenge, and it is commonly used as a proxy for phagocytosis. However, phagocytic microglia do not necessarily express ED1 (Sierra et al. 2010) and reducing ED1 expression with anti-ED1 monoclonal antibodies in cultured macrophages did not impair phagocytosis, and thus the actual role of ED1 in microglia remains elusive (reviewed in Sierra et al. 2013). Furthermore, it is important to notice that none of the above manipulations (LPS, minocycline, etc.) are specific for microglia, also affecting other resident (astrocytes, perivascular, and meningeal macrophages) and infiltrating (monocytes) inflammatory cells. This is an essential notion, because inflammatory cytokines released during LPS challenge (Ji et al. 2007) or even therapeutic levels of cranial irradiation (Morganti et al. 2014) increase the infiltration of monocytes. Furthermore, the relative contributions of microglia versus monocytes in various contexts of physiology and pathology remain widely discussed (Schwartz and Shechter 2010; Prinz et al. 2011). In summary, while the detrimental impact of inflammation on neurogenesis is well-established, the relative roles of the different resident and infiltrating brain inflammatory cells remain to be defined.

10.4.1 Normal Physiological Conditions

In the healthy hippocampus, surveillant microglia were lately shown to be physically intimate with all components of the adult hippocampal neurogenic cascade (Sierra et al. 2010). As microglia are extremely motile cells during normal physiological conditions, continuously responding to neuronal activity, sensory experience, as well as motor learning in vivo (see Chaps. 3, 4 and 9 for further reading), this proximity allows them to actively modulate adult hippocampal neurogenesis. In particular, surveillant microglia were shown to recognise and engulf the newborn cells undergoing apoptosis (between 60 and 80 % of all the newborn cells) in the SGZ (Sierra et al. 2010). This physiological phagocytosis occurs very rapidly (under 1.5 h) and is not disturbed by either normal aging or challenge with LPS (Sierra et al. 2010), indicating an innate capacity of microglia to modulate the neurogenic cascade that is undeterred by inflammation. Indeed, phagocytosis may have functional consequences on hippocampal-dependent learning and memory. Intravenous administration of annexin V, which binds to the phosphatidylserine (PS) receptor present in phagocytes and prevents their recognition of PS exposed on the surface of cells undergoing oxidative stress or apoptosis (Neher et al. 2012), leads to an increased number of apoptotic cells in the SGZ, presumably by blocking phagocytosis (Lu et al. 2011). Both annexin V treatment and reduced expression of ELMO, a protein that promotes internalization of apoptotic cells, resulted in decreased survival of the neuroblasts, ultimately decreasing neurogenesis (Lu et al. 2011). Interestingly, these effects were ascribed to phagocytosis by doublecortin-expressing neuroblasts because exogenously administered apoptotic neuroprogenitors were shown to be engulfed only by these cells (Lu et al. 2011), although the apoptotic newborn cells in the adult SGZ are exclusively phagocytosed by microglia during normal physiological conditions (Sierra et al. 2010). Nonetheless, whether microglia actively select which newborn cells undergo apoptosis and the functional impact of microglial phagocytosis in hippocampal-dependent learning and memory remain to be elucidated with more specific tools.

In addition, it has been hypothesized that anti-inflammatory cytokines produced by phagocytic microglia may further regulate neurogenesis in the healthy mature brain (Sierra et al. 2013). For instance, the anti-inflammatory transforming growth factor beta (TGFβ) is produced by cultured microglia after phagocytosis of apoptotic neurons (De Simone et al. 2003) and is well-known to inhibit the proliferation of SGZ neuroprogenitors in vivo (Buckwalter et al. 2006). Furthermore, cultured microglia adopt a pro-neurogenic and pro-oligodendrogenic phenotype when primed with interleukin 4 (IL-4) or low doses of interferon gamma (IFNγ), two cytokines associated with T helper lymphocytes (Butovsky et al. 2006). This phenotype had a decreased production of tumour necrosis factor α (TNFα) and an increased production of insulin-like growth factor 1 (IGF-1) (Butovsky et al. 2006), which are, respectively, known for regulating negatively (Iosif et al. 2006) and positively (Trejo et al. 2001) neuroprogenitor proliferation in adult hippocampus. Finally, the possibility that microglia control the synaptic integration of the newborn neurons, the ultimate step which determines their contribution to neurogenesis-induced learning and memory, has been recently suggested (Ekdahl 2012) based on the emerging roles of microglia in the activity- and experience-dependent pruning of synapses during normal physiological conditions (see Chap. 9 for further information on this process).

A classic pro-neurogenic paradigm, enriched environment (EE), has also been recently related to microglial involvement with adult hippocampal neurogenesis, learning and memory. In EE, mice are given access to toys, running wheels, and edible treats and are allowed to socially interact within large colonies, to mimic their living in the wild. At the cellular level, EE is also known to regulate excitability, synaptic transmission, and LTP in the dentate gyrus (Irvine et al. 2006). With relation to adult neurogenesis, EE has long been known to increase newborn cell survival, and dentate gyrus size, and enhance performance in the water maze (Kempermann et al. 2004; Nilsson et al. 1999). A similar paradigm, voluntary running, also increases neuroprogenitor proliferation and neurogenesis (van Praag et al. 1999). Importantly, EE is additionally antiapoptotic, neuroprotective (Young et al. 1999), and anti-inflammatory, as it reduces the expression of pro-inflammatory cytokines such as IL-1β and TNFα induced by LPS challenge to the hippocampus (Williamson et al. 2012). EE has also been related to another cytokine, the chemokine fractalkine. Fractalkine, or CX3CL1, is almost exclusively expressed by neurons, either membrane-bound or released, whereas its receptor (CX3CR1) is solely expressed by microglia during normal physiological conditions, forming a unique neuron-microglia signalling unit to control the extent of microglial inflammation and regulate their contribution to normal brain homeostasis (Cardona et al. 2006; Paolicelli et al. 2014). The CX3CR1$^{GFP/GFP}$ mice (CX3CR1$^{-/-}$) are completely deficient in fractalkine signalling due to the replacement of CX3CR1 by a GFP reporter (Jung et al. 2000; Wolf et al. 2013). The CX3CR1$^{GFP/+}$ heterozygous mice are widely used for visualising microglia in two-photon in vivo imaging experiments (Davalos et al. 2005; Nimmerjahn et al. 2005; Fuhrmann et al. 2010; Tremblay 2011), but they were recently shown to have alterations compared to wild-type mice, notably a decreased neurogenesis, impaired performance in hippocampal-dependent memory tasks, and impaired synaptic plasticity (Rogers et al. 2011). In adult (3 months) CX3CR1$^{GFP/GFP}$ mice, compared with heterozygous CX3CR1$^{GFP/+}$ mice, the decreased expression of CX3CR1 resulted in decreased neuroprogenitor proliferation and neuroblasts number in the hippocampus (Bachstetter et al. 2011; Rogers et al. 2011), an effect that was restored by EE (Maggi et al. 2011). Behavioural testing of these mice also revealed an impairment of associative learning and memory, using a standard fear-conditioning paradigm. Even though their freezing behaviour was similar in the training session, the CX3CR1$^{GFP/GFP}$ mice failed to display associative learning, i.e., reduced freezing, when tested 24 h later in the same environment (Rogers et al. 2011). The CX3CR1$^{GFP/GFP}$ mice also displayed deficits in learning the water maze. The detrimental effects of impaired fractalkine signalling on neurogenesis could be mediated by microglial release of the pro-inflammatory cytokine IL-1β, since intrahippocampal infusion of its antagonist IL-1ra prevented the proliferation decrease induced by a CX3CR1 blocking antibody and reversed cognitive dysfunction in the CX3CR1$^{GFP/GFP}$ mice (Rogers et al. 2011) (see Chaps. 4 and 9 for more information on these mice). In contrast, EE still promoted neurogenesis in mice that were null for the IL-1 receptor (Goshen et al. 2009). Although some of the pro-neurogenic effects of EE may be related to a modulation of inflammation, which will be discussed further in the following section, it is unclear whether microglia are directly contributing to the effects of EE on adult neurogenesis. In particular, very little is known about the changes induced by EE in microglia.

One potential mechanism behind the pro-neurogenic actions of EE is that EE triggers an increased surveillance of the hippocampal tissue by T cells, which somehow drive microglia to increase their expression of IGF-1, ultimately resulting in an increased number of newborn neurons in adult rats (Ziv et al. 2006). Whether this interaction between T cells and microglia in EE involves any type of antigen presentation is still unknown, even though EE induced microglia to express the major histocompatibility complex class II (MHC-II). However, antigen presentation in the CNS is thought to be restricted to the meninges and choroid plexus, rather than taking place in the parenchyma where microglia are considered to exclusively reside (Ransohoff and Engelhardt 2012) (see Chap. 5 for further reading). Furthermore, none of the changes induced by EE could be replicated in voluntary running paradigms in adult mice, including T lymphocyte surveillance of the hippocampus, T cells interactions with microglia, and increased expression of IGF-1, IL-1β, and TNFα by microglia (Olah et al. 2009). Finally, a significant correlation between hippocampal proliferation and genetic loci associated to the proportion of circulating CD4+ T lymphocytes was found in a genetically heterogeneous stock of mice (Huang et al. 2010). It is important to note, however, that proliferation does not necessarily imply neurogenesis, as cells other than neuroprogenitors can also proliferate (including astrocytes and microglia), and newly generated cells can undergo apoptosis instead of differentiating into neurons. Nonetheless, T cells seem to be functionally relevant, because nude mice and mice undergoing antibody-based T cell depletion also have impaired performance in the water maze (Kipnis et al. 2004). The underlying mechanisms behind the effects of T lymphocytes on adult neurogenesis are largely unexplored and may not necessarily be related to microglia but to bystander effects of T lymphocytes on other cell types, including the neuroprogenitors themselves. In fact, it has been suggested that T lymphocyte depletion leads to a decreased hippocampal production of brain-derived neurotrophic factor (BDNF) (Wolf et al. 2009), although it has not been addressed whether administration of BDNF could prevent the detrimental effects of T lymphocytes depletion on any stage of the hippocampal neurogenic cascade. Interestingly, BDNF was also found to mediate the pro-neurogenic actions of EE (Rossi et al. 2006), and thus it is possible that the effects of T lymphocytes ascribed to T cell–microglia interactions are in fact mediated by this trophic factor. BDNF has been largely thought to be produced by neurons solely, however recent observations using CX3CR1-CreER mice suggest that mice deficient in microglial BDNF have strong deficits of synaptogenesis in the motor cortex, with functional repercussions on motor learning in adolescent and adult mice (Parkhurst et al. 2013) (see Chap. 9 for further reading), and thus the direct contribution of microglial BDNF to the effects of EE on adult SGZ and SVZ neurogenesis needs to be addressed. Overall, more precise experimental designs are required to directly evaluate the roles of surveillant microglia in the activity- and experience-dependent regulation of adult hippocampal neurogenesis.

In contrast to the body of evidence showing microglial contribution to SGZ neurogenesis, little is known about its role in SVZ neurogenesis (Fig. 10.1). Recent data show that microglia negatively control the size of the neuroprogenitors pool in the embryonic SVZ by means of phagocytosis (Cunningham et al. 2013), but in contrast are necessary for early postnatal SVZ neurogenesis and oligodendrogenesis (Shigemoto-Mogami et al. 2014) (for further reading see Chap. 7). These evidences suggest indeed an overlooked role for microglial phagocytosis in adult SVZ neurogenesis. For instance, it is unknown how the survival of adult newborn cells is regulated in the SVZ/RMS, and whether microglia phagocytose the excess of cells, similar to the hippocampus. The comparison between microglial inflammation in SGZ versus SVZ neurogenesis is not straightforward, however, as different levels of activation have been reported under basal conditions. For instance, SVZ microglia express higher levels of the lymphocyte common antigen CD45 and proliferate at higher rates in the SVZ than in the SGZ and non-neurogenic areas of the adult brain (Goings et al. 2006). Similarly, the vast majority of microglia in the mouse developing SVZ exhibit an amoeboid morphology and express the inducible nitric oxide synthase (iNOS) (Cunningham et al. 2013). In the early postnatal SVZ (postnatal days (P)2–10), microglia are mostly amoeboid as well, and there is a sustained production of pro-inflammatory cytokines, including IL-1β, IL-6, TNFα, and IFNγ (Shigemoto-Mogami et al. 2014), indirectly suggesting a low, ongoing inflammation in the developing and adult SVZ. Further, treating pregnant mice with doxycycline, an antibiotic with anti-inflammatory properties, decreases the percentage of microglia expressing iNOS and IL-1β in the neocortex of developing pups and results in decreased apoptosis, increased number of neuroprogenitors and neurons, and increased thickness of the SVZ at birth (Cunningham et al. 2013). The effects of doxycycline on embryonic neurogenesis are largely attributed to microglia, because their depletion via liposomes filled with chlodronate, a poison, increases the number of neuroprogenitors in the embryonic neocortex (Cunningham et al. 2013). This microglial control of the pool size of neuroprogenitors in the developing SVZ may be related to inflammatory cytokines, phagocytosis, or phagoptosis, a form of cell death associated with inflammation, which triggers a transient exposure of PS on the outer membrane of living cells and leads to their engulfment by microglia (Neher et al. 2012). The effects of microglia promoting early SVZ neurogenesis have also been related to inflammation, as treatment with minocycline decreased the expression of IL-1β, TNFα, IL-6, and IFNγ in the SVZ and concomitantly reduced the number of proliferating cells and the number of neuroblasts in rats during postnatal development (Shigemoto-Mogami et al. 2014). In contrast, adult SGZ neurogenesis in basal conditions seems to be largely independent of inflammation. For instance, neither treatment with the anti-inflammatory drug indomethacin (Monje et al. 2003) nor transgenic knock-down of the IL-1 receptor (Goshen et al. 2009) induce alterations in adult SGZ neurogenesis. Other fundamental differences between hippocampal and SVZ neurogenesis may exist. For instance, conditioned media from a microglial cell line (BV-2) activated with LPS induced apoptosis of differentiated neurospheres derived from the adult hippocampus, an effect mediated by IL-6 (Monje et al. 2003). In contrast, conditioned media from primary microglia activated with LPS induced the neuronal differentiation of neurospheres derived from embryonic (E16) cortices (from SVZ progenitors), an effect mediated by IL-1β and IFNγ possibly acting together (Shigemoto-Mogami et al. 2014). Overall, these results suggest differences in basal inflammation in the developing and adult SVZ and SGZ neurogenic niches which may contribute to the overall role of microglia in neurogenesis during normal physiological conditions. Nonetheless, a formal comparison of the regulation of adult hippocampal and SVZ neurogenesis by microglia and released pro-inflammatory cytokines has not yet been carried out.

The presence of microglia in the physical neurogenic niche of the adult SGZ and SVZ is well established, but the respective roles of microglial phagocytosis and release of soluble mediators such as cytokines and trophic factors, regulating the proliferation of neuroprogenitors, as well as the survival, differentiation, and integration of newborn neurons remain largely unexplored. More importantly, the ultimate contribution of microglia to spatial, associative, and olfactory types of learning and memory in physiological conditions is still a wide open area for future research.

10.4.2 Inflammatory Conditions: Normal Aging and Neurodegenerative Diseases

Inflammation is a common phenomenon during normal aging and neurodegenerative diseases, such as Alzheimer's (AD), and therefore might represent an underlying mechanism explaining the decreased hippocampal neurogenesis and cognitive impairment in these conditions. During aging, microglia express higher levels of pro-inflammatory cytokines, including IL-6, IL-1β, and TNFα ex vivo (Sierra et al. 2007). (see Chap. 13 for further reading.) In parallel, neurogenesis declines with age in mice and humans (Manganas et al. 2007; Encinas et al. 2011), although the decay is more pronounced and occurs later in life in mice than in humans (Spalding et al. 2013). The aging-associated decrease in neurogenesis can be explained as a consequence of the autonomous exhaustion of the quiescent neuroprogenitors, which divide several times in a row and then terminally differentiate into astrocytes (Encinas et al. 2011). In addition, age-related increases in microglial release of cytokines such as IL-1β may further hinder neurogenesis in the aging brain. IL-1β expression is controlled at the transcriptional level by the transcription factor NFκB, and at the post-translational level by the inflammasomes, a group of multimeric proteins which contain the interleukin 1-converting enzyme (ICE, caspase 1) and release the active form of the cytokine (Latz et al. 2013). Treatment with IL-1β decreases hippocampal proliferation in young mice (Koo and Duman 2008), and pharmacological inhibition of ICE was found to partially restore the number of newborn neurons in aged mice without significantly affecting their differentiation rate (Gemma et al. 2007). However, it remains to be determined whether the detrimental effects of IL-1β on neurogenesis in aged mice are specifically exerted on the proliferation of neuroprogenitors, or on the survival of newborn cells.

Another relevant immune molecule for the regulation of neurogenesis during normal aging is the chemokine fractalkine. In the aging hippocampus, fractalkine expression by mature neurons decreases, leading to alterations in neurogenesis that have been linked to an increased IL-1β expression (Bachstetter et al. 2011). In aged rats (22 months old), intracerebroventricular infusion of fractalkine partially rescued the numbers of proliferating neuroprogenitors and doublecortin-expressing neuroblasts (Bachstetter et al. 2011). These effects could be mediated by microglial release of IL-1β since intrahippocampal infusion of its antagonist IL-1ra prevented the proliferation decrease induced by a CX3CR1 blocking antibody, at least in young rats and mice, while reversing the cognitive dysfunction observed in the CX3CR1$^{GFP/GFP}$ mice (Rogers et al. 2011). These evidences suggest a scenario in which aging leads to decreased expression of fractalkine (by unknown mechanisms), which in turn leads to an upregulated production of IL-1β by microglia, and the subsequent reduced proliferation of the few remaining neuroprogenitors in the aging hippocampus.

Inflammatory cytokines, and in particular IL-1β, are over-expressed in the microglial cells associated with amyloid beta (Aβ) plaques in postmortem brain samples of AD patients (Griffin et al. 1989) and transgenic mice modeling the disease (Benzing et al. 1999). In AD, synapses are a main pathogenic substrate (see Siskova and Tremblay 2013), but adult neurogenesis is also severely reduced in most mouse models of AD, possibly due to a decreased proliferation of neuroprogenitors and decreased survival of the newborn cells (reviewed in Sierra et al. (2011)). Several studies have used the anti-inflammatory antibiotic minocycline to study the role of microglia; however, as mentioned before, minocycline acts not only on microglia, but also on astrocytes and monocytes, by reducing expression of the transcription factor NFκB, which is a major controller of inflammation (Bernardino et al. 2009). In mice expressing the human amyloid precursor protein (APP), early treatment with minocycline was shown to improve cognition and reduce the Aβ burden (Seabrook et al. 2006). In another mouse model of AD expressing human APP and a mutated form of presenilin 1 (PS1), which is part of the γ secretase pathway cleaving Aβ, minocycline reduced the levels of pro-inflammatory cytokines and numbers of microglial cells, in addition to restoring neurogenesis and hippocampus-dependent memory deficits, albeit without detectable changes in the Aβ plaques load (Biscaro et al. 2012). Altogether, these evidences suggest that the cognitive decay in AD may be at least partly related to a detrimental effect of inflammation on hippocampal neurogenesis; however, direct evidence for a microglial contribution to these effects is still missing. Furthermore, there is some discordance between the rodent models and the human AD data, and it is unclear whether there is any impairment of neurogenesis in the AD brain at all. For instance, no changes in proliferation were observed in the postmortem hippocampus of presenile AD patients (Boekhoorn et al. 2006), and an increased expression of neuroblast proteins, including doublecortin, was reported (Jin et al. 2004). Several differences may account for the contradictory results between human AD and rodent models, including aging per se. Only a small percentage of AD patients have an early onset (before 65 years of age); however, in most transgenic models of AD, Aβ accumulation and cognitive deficits start to occur when the animals are still young (2–6 months old). As aging induces a cumulative alteration of the function of many physiological systems, including the peripheral and brain immune systems (see Chap. 13 for more information on aging and microglia), and a decline in the number of neuroprogenitors (Encinas et al. 2011), the *status quo* of the brain tissue (including levels of pro-inflammatory cytokines, and number of neuroprogenitors) at the onset of pathology in humans and transgenic mice is appallingly different. Furthermore, a key question is whether the normal age-related loss of neuroprogenitors is accelerated or slowed down in AD, and whether these neuroprogenitors remain at all in the AD brain. In fact, strategic interventions should perhaps focus at preventing the age-related loss of neuroprogenitors in young, healthy subjects, rather than at increasing the neurogenic output of the few neuroprogenitors remaining.

Adult SVZ neurogenesis is profoundly affected by inflammation as well. In experimental acute encephalomyelitis (EAE), a mouse model of multiple sclerosis triggered by immunization against myelin oligodendrocyte glycoprotein (MOG), the proliferation of B and C neuroprogenitors, and the migration of A cells is impaired at the peak of inflammation (Pluchino et al. 2008). EAE mice also have decreased OB neurogenesis accompanied by olfactory dysfunction (Tepavcevic et al. 2011). A possible mediator in EAE-decreased neurogenesis is interleukin 15 (IL-15), which is upregulated by neurons in EAE (Wu et al. 2010), and regulates microglial activation (Gomez-Nicola et al. 2010b). Knocking down IL-15 aggravates motor deficit during EAE (Gomez-Nicola et al. 2010a) and impairs basal SVZ neurogenesis by decreasing proliferation and differentiation (Gomez-Nicola et al. 2011), although whether the two phenomena (i.e., motor deficit and SVZ neurogenesis) are related is unknown. In contrast to the SVZ, SGZ neuroprogenitors show increased proliferation in the acute phase of the disease, although it is unclear whether this increased proliferation results in increased neurogenesis (Giannakopoulou et al. 2013). In both the SVZ and the SGZ, direct evidence for a role of microglia versus other types of inflammatory cells in the decreased neurogenesis induced by EAE-related inflammation is still missing. Further evidence of a microglial-mediated control of OB neurogenesis comes from deafferentiation studies, in which chemical depletion of olfactory sensory neurons leads to a decreased survival of OB newborn cells without affecting SZV proliferation or RMS migration (Alonso et al. 2006). In this context, treatment with minocycline reduces the number of proliferating microglia and microglia-expressing ED1, while restoring OB neurogenesis depleted by deafferentation thus confirming that inflammation is responsible for the depletion of OB neurogenesis after deafferentation (Lazarini and Lledo 2011). In this model, microglial proliferation and expression of ED1 was independent from peripheral immune cells involvement, readily occurring in Rag2 $\gamma c^{-/-}$ mice which lack T and B lymphocytes and natural killer (NK) cells due to their deficiency in the recombinant activating gene 2 (RAG2) and the common cytokine receptor γ chain (Lazarini and Lledo 2011). Unfortunately, the deafferentation-induced defects in OB neurogenesis were not tested in $Rag2^{-/-}$ mice (Lazarini and Lledo 2011), and therefore a direct effect of peripheral immune cells on SVZ neurogenesis in EAE or after deafferentation cannot be ruled out.

In summary, inflammation and particularly release of the pro-inflammatory cytokines IL-6, IL-1β, and IL-15 as well as the anti-inflammatory fractalkine, during normal aging and in experimental models of disease, have strong detrimental effects on hippocampal and OB neurogenesis. The relative contribution of microglia compared to other inflammatory cell types, as well as the actual cellular and molecular targets in the neurogenic cascade, remains an open area of study.

10.5 Conclusion

Inflammation has long been known to be detrimental for adult neurogenesis and is probably one of the underlying mechanisms by which aging and neurodegenerative diseases such as AD and EAE impair adult neurogenesis. Nonetheless, the relative contribution of microglia versus other types of cells involved in the brain inflammatory response, including astrocytes, endothelial cells, meningeal, and perivascular macrophages, as well as circulating monocytes, remains largely unknown at this early stage of investigation. A main problem is the use of non-specific manipulations to either activate (such as LPS) or block (such as indomethacin, doxycycline, or minocyline) the inflammatory response. More specific approaches have been recently developed to manipulate microglia, such as the CX3CR1-CreER mice from Steffen Jung (Yona et al. 2013), although in these mice the CreER construct replaces the endogenous CX3CR1 locus, and a deficient fractalkine signaling leads to several alterations compared to wild-type mice, including decreased neurogenesis, impaired synaptic plasticity, and performance in hippocampal-dependent memory tasks (reviewed in Paolicelli et al. (2014)). Obviously, dissecting the specific role of microglia in adult neurogenic niches will require more sophisticated methods that remain to be developed.

Furthermore, the cellular and molecular mechanisms involved in the modulation of neurogenesis by microglia in health and disease remain only superficially described. In particular, no systematic analysis of microglial behaviour in the SVZ and SGZ niches across development, adulthood, and normal aging has been carried out. As a consequence, our knowledge of the beneficial versus detrimental roles of microglia, and the particular mechanisms involved in each context where neurogenesis has been described, remains fragmented. For instance, among the intriguing differences observed between the two consensus regions and across developmental stages, microglial phagocytosis has been shown to eliminate the newborn cell apoptotic debris in the adult hippocampus and to actively kill the neuroprogenitors in the developing SVZ. Furthermore, IL-1β, IL-6, and fractalkine are emerging as major cytokines controlling the neurogenic cascade in the SGZ, whereas IL-15 seems to be mostly involved in SVZ neurogenesis. Other major players that were proposed are T lymphocytes, at least in the hippocampus, but it remains to be fully addressed whether their actions are mediated at distance, or rather involve direct patrolling of the brain parenchyma and a physical interaction with microglia.

Finally, the ultimate contribution of microglial cells to spatial, associative, and olfactory learning is surprisingly unknown. Importantly, microglia are in a central position to modulate adult neurogenesis: (1) they are physically present within the two main neurogenic niches in the mature brain; (2) they contact all the types of cells participating in the cascade; (3) they are the most motile cell type in the mature brain, and their processes continuously scan the brain parenchyma in relation to the local changes associated with neuronal activity, experience or lifestyle, and disease; (4) they phagocytose neuroprogenitors and newborn cells; (5) they produce trophic factors modulating neuronal and synaptic plasticity; and (6) they orchestrate the brain inflammatory response, release a variety of pro-inflammatory mediators, and coordinate the action of the peripheral immune cells. Thus, microglia have a tremendous potential for modulating adult neurogenic niches in health as much as in diseases. This knowledge of the roles of microglia in neurogenesis will in time be transferred into designing therapeutic interventions aimed at manipulating microglial function to protect and potentiate neurogenesis, and its essential contribution to learning and memory.

Acknowledgement This work was supported by grants from the Banting Research Foundation, Scottish Rite Charitable Foundation of Canada, and Natural Sciences and Engineering Research Council of Canada, as well as a start-up fund from the Université Laval and Centre de recherche du CHU de Québec to Marie-Ève Tremblay, and from the Spanish Ministry of Economy and Competitiveness with FEDER funds (BFU2012-32089) to Amanda Sierra.

References

Alonso M, Viollet C, Gabellec MM, Meas-Yedid V, Olivo-Marin JC, Lledo PM (2006) Olfactory discrimination learning increases the survival of adult-born neurons in the olfactory bulb. J Neurosci 26:10508–10513

Arruda-Carvalho M, Sakaguchi M, Akers KG, Josselyn SA, Frankland PW (2011) Posttraining ablation of adult-generated neurons degrades previously acquired memories. J Neurosci 31: 15113–15127

Bachstetter AD, Morganti JM, Jernberg J, Schlunk A, Mitchell SH, Brewster KW, Hudson CE, Cole MJ, Harrison JK, Bickford PC, Gemma C (2011) Fractalkine and CX 3 CR1 regulate hippocampal neurogenesis in adult and aged rats. Neurobiol Aging 32:2030–2044

Belarbi K, Arellano C, Ferguson R, Jopson T, Rosi S (2012) Chronic neuroinflammation impacts the recruitment of adult-born neurons into behaviorally relevant hippocampal networks. Brain Behav Immun 26:18–23

Belnoue L, Grosjean N, Abrous DN, Koehl M (2011) A critical time window for the recruitment of bulbar newborn neurons by olfactory discrimination learning. J Neurosci 31:1010–1016

Benzing WC, Wujek JR, Ward EK, Shaffer D, Ashe KH, Younkin SG, Brunden KR (1999) Evidence for glial-mediated inflammation in aged APP(SW) transgenic mice. Neurobiol Aging 20:581–589

Bergmann O, Liebl J, Bernard S, Alkass K, Yeung MS, Steier P, Kutschera W, Johnson L, Landen M, Druid H, Spalding KL, Frisen J (2012) The age of olfactory bulb neurons in humans. Neuron 74:634–639

Bernardino AL, Kaushal D, Philipp MT (2009) The antibiotics doxycycline and minocycline inhibit the inflammatory responses to the Lyme disease spirochete Borrelia burgdorferi. J Infect Dis 199:1379–1388

Biscaro B, Lindvall O, Tesco G, Ekdahl CT, Nitsch RM (2012) Inhibition of microglial activation protects hippocampal neurogenesis and improves cognitive deficits in a transgenic mouse model for Alzheimer's disease. Neurodegener Dis 9:187–198

Boekhoorn K, Joels M, Lucassen PJ (2006) Increased proliferation reflects glial and vascular-associated changes, but not neurogenesis in the presenile Alzheimer hippocampus. Neurobiol Dis 24:1–14

Breunig JJ, Arellano JI, Macklis JD, Rakic P (2007) Everything that glitters isn't gold: a critical review of postnatal neural precursor analyses. Cell Stem Cell 1:612–627

Buckwalter MS, Yamane M, Coleman BS, Ormerod BK, Chin JT, Palmer T, Wyss-Coray T (2006) Chronically increased transforming growth factor-beta1 strongly inhibits hippocampal neurogenesis in aged mice. Am J Pathol 169:154–164

Butovsky O, Ziv Y, Schwartz A, Landa G, Talpalar AE, Pluchino S, Martino G, Schwartz M (2006) Microglia activated by IL-4 or IFN-gamma differentially induce neurogenesis and oligodendrogenesis from adult stem/progenitor cells. Mol Cell Neurosci 31:149–160

Cardona AE, Pioro EP, Sasse ME, Kostenko V, Cardona SM, Dijkstra IM, Huang D, Kidd G, Dombrowski S, Dutta R, Lee JC, Cook DN, Jung S, Lira SA, Littman DR, Ransohoff RM (2006) Control of microglial neurotoxicity by the fractalkine receptor. Nat Neurosci 9:917–924

Chugh D, Nilsson P, Afjei SA, Bakochi A, Ekdahl CT (2013) Brain inflammation induces post-synaptic changes during early synapse formation in adult-born hippocampal neurons. Exp Neurol 250:176–188

Crawley JN (2008 Mar 27) Behavioral phenotyping strategies for mutant mice. Neuron 57(6): 809–818. doi:10.1016/j.neuron.2008.03.001

Cunningham CL, Martinez-Cerdeno V, Noctor SC (2013) Microglia regulate the number of neural precursor cells in the developing cerebral cortex. J Neurosci 33:4216–4233

Curtis MA, Kam M, Nannmark U, Anderson MF, Axell MZ, Wikkelso C, Holtas S, van Roon-Mom WM, Bjork-Eriksson T, Nordborg C, Frisen J, Dragunow M, Faull RL, Eriksson PS (2007) Human neuroblasts migrate to the olfactory bulb via a lateral ventricular extension. Science 315:1243–1249

Davalos D, Grutzendler J, Yang G, Kim JV, Zuo Y, Jung S, Littman DR, Dustin ML, Gan WB (2005) ATP mediates rapid microglial response to local brain injury in vivo. Nat Neurosci 8: 752–758

De Simone R, Ajmone-Cat MA, Tirassa P, Minghetti L (2003) Apoptotic PC12 cells exposing phosphatidylserine promote the production of anti-inflammatory and neuroprotective molecules by microglial cells. J Neuropathol Exp Neurol 62:208–216

Deng W, Saxe MD, Gallina IS, Gage FH (2009) Adult-born hippocampal dentate granule cells undergoing maturation modulate learning and memory in the brain. J Neurosci 29: 13532–13542

Dobrossy MD, Drapeau E, Aurousseau C, Le Moal M, Piazza PV, Abrous DN (2003) Differential effects of learning on neurogenesis: learning increases or decreases the number of newly born cells depending on their birth date. Mol Psychiatry 8:974–982
Ekdahl CT (2012) Microglial activation—tuning and pruning adult neurogenesis. Front Pharmacol 3:41
Ekdahl CT, Claasen JH, Bonde S, Kokaia Z, Lindvall O (2003) Inflammation is detrimental for neurogenesis in adult brain. Proc Natl Acad Sci U S A 100:13632–13637
Encinas JM, Enikolopov G (2008) Identifying and quantitating neural stem and progenitor cells in the adult brain. Methods Cell Biol 85:243–272
Encinas JM, Michurina TV, Peunova N, Park JH, Tordo J, Peterson DA, Fishell G, Koulakov A, Enikolopov G (2011) Division-coupled astrocytic differentiation and age-related depletion of neural stem cells in the adult hippocampus. Cell Stem Cell 8:566–579
Eriksson PS, Perfilieva E, Bjork-Eriksson T, Alborn AM, Nordborg C, Peterson DA, Gage FH (1998) Neurogenesis in the adult human hippocampus. Nat Med 4:1313–1317
Fuhrmann M, Bittner T, Jung CK, Burgold S, Page RM, Mitteregger G, Haass C, LaFerla FM, Kretzschmar H, Herms J (2010) Microglial Cx3cr1 knockout prevents neuron loss in a mouse model of Alzheimer's disease. Nat Neurosci 13:411–413
Gage FH (2002) Neurogenesis in the adult brain. J Neurosci 22:612–613
Gage FH, Coates PW, Palmer TD, Kuhn HG, Fisher LJ, Suhonen JO, Peterson DA, Suhr ST, Ray J (1995) Survival and differentiation of adult neuronal progenitor cells transplanted to the adult brain. Proc Natl Acad Sci U S A 92:11879–11883
Ge S, Yang CH, Hsu KS, Ming GL, Song H (2007) A critical period for enhanced synaptic plasticity in newly generated neurons of the adult brain. Neuron 54:559–566
Gemma C, Bachstetter AD, Cole MJ, Fister M, Hudson C, Bickford PC (2007) Blockade of caspase-1 increases neurogenesis in the aged hippocampus. Eur J Neurosci 26:2795–2803
Giannakopoulou A, Grigoriadis N, Bekiari C, Lourbopoulos A, Dori I, Tsingotjidou AS, Michaloudi H, Papadopoulos GC (2013) Acute inflammation alters adult hippocampal neurogenesis in a multiple sclerosis mouse model. J Neurosci Res 91:890–900
Goings GE, Kozlowski DA, Szele FG (2006) Differential activation of microglia in neurogenic versus non-neurogenic regions of the forebrain. Glia 54:329–342
Gomez-Nicola D, Spagnolo A, Guaza C, Nieto-Sampedro M (2010a) Aggravated experimental autoimmune encephalomyelitis in IL-15 knockout mice. Exp Neurol 222:235–242
Gomez-Nicola D, Valle-Argos B, Nieto-Sampedro M (2010b) Blockade of IL-15 activity inhibits microglial activation through the NFkappaB, p38, and ERK1/2 pathways, reducing cytokine and chemokine release. Glia 58:264–276
Gomez-Nicola D, Valle-Argos B, Pallas-Bazarra N, Nieto-Sampedro M (2011) Interleukin-15 regulates proliferation and self-renewal of adult neural stem cells. Mol Biol Cell 22: 1960–1970
Goshen I, Avital A, Kreisel T, Licht T, Segal M, Yirmiya R (2009) Environmental enrichment restores memory functioning in mice with impaired IL-1 signaling via reinstatement of long-term potentiation and spine size enlargement. J Neurosci 29:3395–3403
Gould E, Beylin A, Tanapat P, Reeves A, Shors TJ (1999) Learning enhances adult neurogenesis in the hippocampal formation. Nat Neurosci 2:260–265
Griffin WS, Stanley LC, Ling C, White L, MacLeod V, Perrot LJ, White CL III, Araoz C (1989) Brain interleukin 1 and S-100 immunoreactivity are elevated in Down syndrome and Alzheimer disease. Proc Natl Acad Sci U S A 86:7611–7615
Huang GJ, Smith AL, Gray DH, Cosgrove C, Singer BH, Edwards A, Sims S, Parent JM, Johnsen A, Mott R, Mathis D, Klenerman P, Benoist C, Flint J (2010) A genetic and functional relationship between T cells and cellular proliferation in the adult hippocampus. PLoS Biol 8:e1000561
Imayoshi I, Sakamoto M, Ohtsuka T, Takao K, Miyakawa T, Yamaguchi M, Mori K, Ikeda T, Itohara S, Kageyama R (2008) Roles of continuous neurogenesis in the structural and functional integrity of the adult forebrain. Nat Neurosci 11:1153–1161
Iosif RE, Ekdahl CT, Ahlenius H, Pronk CJ, Bonde S, Kokaia Z, Jacobsen SE, Lindvall O (2006) Tumor necrosis factor receptor 1 is a negative regulator of progenitor proliferation in adult hippocampal neurogenesis. J Neurosci 26:9703–9712

Irvine GI, Logan B, Eckert M, Abraham WC (2006) Enriched environment exposure regulates excitability, synaptic transmission, and LTP in the dentate gyrus of freely moving rats. Hippocampus 16:149–160

Ji KA, Yang MS, Jeong HK, Min KJ, Kang SH, Jou I, Joe EH (2007) Resident microglia die and infiltrated neutrophils and monocytes become major inflammatory cells in lipopolysaccharide-injected brain. Glia 55:1577–1588

Jin K, Peel AL, Mao XO, Xie L, Cottrell BA, Henshall DC, Greenberg DA (2004) Increased hippocampal neurogenesis in Alzheimer's disease. Proc Natl Acad Sci U S A 101:343–347

Jung S, Aliberti J, Graemmel P, Sunshine MJ, Kreutzberg GW, Sher A, Littman DR (2000) Analysis of fractalkine receptor CX(3)CR1 function by targeted deletion and green fluorescent protein reporter gene insertion. Mol Cell Biol 20:4106–4114

Kee N, Teixeira CM, Wang AH, Frankland PW (2007) Preferential incorporation of adult-generated granule cells into spatial memory networks in the dentate gyrus. Nat Neurosci 10:355–362

Kempermann G, Jessberger S, Steiner B, Kronenberg G (2004) Milestones of neuronal development in the adult hippocampus. Trends Neurosci 27:447–452

Kipnis J, Cohen H, Cardon M, Ziv Y, Schwartz M (2004) T cell deficiency leads to cognitive dysfunction: implications for therapeutic vaccination for schizophrenia and other psychiatric conditions. Proc Natl Acad Sci U S A 101:8180–8185

Koo JW, Duman RS (2008) IL-1beta is an essential mediator of the antineurogenic and anhedonic effects of stress. Proc Natl Acad Sci U S A 105:751–756

Latz E, Xiao TS, Stutz A (2013) Activation and regulation of the inflammasomes. Nat Rev Immunol 13:397–411

Lazarini F, Lledo PM (2011) Is adult neurogenesis essential for olfaction? Trends Neurosci 34: 20–30

Leutgeb JK, Leutgeb S, Moser MB, Moser EI (2007) Pattern separation in the dentate gyrus and CA3 of the hippocampus. Science 315:961–966

Lledo PM, Merkle FT, Alvarez-Buylla A (2008) Origin and function of olfactory bulb interneuron diversity. Trends Neurosci 31:392–400

Lu Z, Elliott MR, Chen Y, Walsh JT, Klibanov AL, Ravichandran KS, Kipnis J (2011) Phagocytic activity of neuronal progenitors regulates adult neurogenesis. Nat Cell Biol 13:1076–1083

Maggi L, Scianni M, Branchi I, D'Andrea I, Lauro C, Limatola C (2011) CX(3)CR1 deficiency alters hippocampal-dependent plasticity phenomena blunting the effects of enriched environment. Front Cell Neurosci 5:22

Manganas LN, Zhang X, Li Y, Hazel RD, Smith SD, Wagshul ME, Henn F, Benveniste H, Djuric PM, Enikolopov G, Maletic-Savatic M (2007) Magnetic resonance spectroscopy identifies neural progenitor cells in the live human brain. Science 318:980–985

Marin-Burgin A, Schinder AF (2012) Requirement of adult-born neurons for hippocampus-dependent learning. Behav Brain Res 227:391–399

Monje ML, Toda H, Palmer TD (2003) Inflammatory blockade restores adult hippocampal neurogenesis. Science 302:1760–1765

Morganti JM, Jopson TD, Liu S, Gupta N, Rosi S (2014) Cranial irradiation alters the brain's microenvironment and permits CCR2+ macrophage infiltration. PLoS One 9:e93650

Mouret A, Gheusi G, Gabellec MM, de Chaumont F, Olivo-Marin JC, Lledo PM (2008) Learning and survival of newly generated neurons: when time matters. J Neurosci 28:11511–11516

Neher JJ, Neniskyte U, Brown GC (2012) Primary phagocytosis of neurons by inflamed microglia: potential roles in neurodegeneration. Front Pharmacol 3:27

Nimmerjahn A, Kirchhoff F, Helmchen F (2005) Resting microglial cells are highly dynamic surveillants of brain parenchyma in vivo. Science 308:1314–1318

Nilsson M, Perfilieva E, Johansson U, Orwar O, Eriksson PS (1999) Enriched environment increases neurogenesis in the adult rat dentate gyrus and improves spatial memory. J Neurobiol 39(4):569–578

Olah M, Ping G, De Haas AH, Brouwer N, Meerlo P, Van Der Zee EA, Biber K, Boddeke HW (2009) Enhanced hippocampal neurogenesis in the absence of microglia T cell interaction and microglia activation in the murine running wheel model. Glia 57:1046–1061

Palmer TD, Ray J, Gage FH (1995) FGF-2-responsive neuronal progenitors reside in proliferative and quiescent regions of the adult rodent brain. Mol Cell Neurosci 6:474–486

Paolicelli RC, Bisht K, Tremblay ME (2014) Fractalkine regulation of microglial physiology and consequences on the brain and behaviour. Front Cell Neurosci 8:129

Parkhurst CN, Yang G, Ninan I, Savas JN, Yates JR III, Lafaille JJ, Hempstead BL, Littman DR, Gan WB (2013) Microglia promote learning-dependent synapse formation through brain-derived neurotrophic factor. Cell 155:1596–1609

Plath N, Ohana O, Dammermann B, Errington ML, Schmitz D, Gross C, Mao X, Engelsberg A, Mahlke C, Welzl H, Kobalz U, Stawrakakis A, Fernandez E, Waltereit R, Bick-Sander A, Therstappen E, Cooke SF, Blanquet V, Wurst W, Salmen B, Bosl MR, Lipp HP, Grant SG, Bliss TV, Wolfer DP, Kuhl D (2006) Arc/Arg3.1 is essential for the consolidation of synaptic plasticity and memories. Neuron 52:437–444

Pluchino S, Muzio L, Imitola J, Deleidi M, Alfaro-Cervello C, Salani G, Porcheri C, Brambilla E, Cavasinni F, Bergamaschi A, Garcia-Verdugo JM, Comi G, Khoury SJ, Martino G (2008) Persistent inflammation alters the function of the endogenous brain stem cell compartment. Brain 131:2564–2578

Prinz M, Priller J, Sisodia SS, Ransohoff RM (2011) Heterogeneity of CNS myeloid cells and their roles in neurodegeneration. Nat Neurosci 14:1227–1235

Ransohoff RM, Engelhardt B (2012) The anatomical and cellular basis of immune surveillance in the central nervous system. Nat Rev Immunol 12:623–635

Reynolds BA, Weiss S (1992) Generation of neurons and astrocytes from isolated cells of the adult mammalian central nervous system. Science 255:1707–1710

Richards LJ, Kilpatrick TJ, Bartlett PF (1992) De novo generation of neuronal cells from the adult mouse brain. Proc Natl Acad Sci U S A 89:8591–8595

Rogers JT, Morganti JM, Bachstetter AD, Hudson CE, Peters MM, Grimmig BA, Weeber EJ, Bickford PC, Gemma C (2011) CX3CR1 deficiency leads to impairment of hippocampal cognitive function and synaptic plasticity. J Neurosci 31:16241–16250

Rossi C, Angelucci A, Costantin L, Braschi C, Mazzantini M, Babbini F, Fabbri ME, Tessarollo L, Maffei L, Berardi N, Caleo M (2006) Brain-derived neurotrophic factor (BDNF) is required for the enhancement of hippocampal neurogenesis following environmental enrichment. Eur J Neurosci 24:1850–1856

Sahay A, Scobie KN, Hill AS, O'Carroll CM, Kheirbek MA, Burghardt NS, Fenton AA, Dranovsky A, Hen R (2011) Increasing adult hippocampal neurogenesis is sufficient to improve pattern separation. Nature 472:466–470

Sakamoto M, Imayoshi I, Ohtsuka T, Yamaguchi M, Mori K, Kageyama R (2011) Continuous neurogenesis in the adult forebrain is required for innate olfactory responses. Proc Natl Acad Sci U S A 108:8479–8484

Sanai N, Berger MS, Garcia-Verdugo JM, Alvarez-Buylla A (2007) Comment on "Human neuroblasts migrate to the olfactory bulb via a lateral ventricular extension". Science 318:393, author reply 393

Schwartz M, Shechter R (2010) Systemic inflammatory cells fight off neurodegenerative disease. Nat Rev Neurol 6:405–410

Schellinck HM, Forestell CA, LoLordo VM (2001) A simple and reliable test of olfactory learning and memory in mice. Chem Senses 26(6):663–672

Seabrook TJ, Jiang L, Maier M, Lemere CA (2006) Minocycline affects microglia activation, Abeta deposition, and behavior in APP-tg mice. Glia 53:776–782

Shigemoto-Mogami Y, Hoshikawa K, Goldman JE, Sekino Y, Sato K (2014) Microglia enhance neurogenesis and oligodendrogenesis in the early postnatal subventricular zone. J Neurosci 34:2231–2243

Shors TJ, Miesegaes G, Beylin A, Zhao M, Rydel T, Gould E (2001) Neurogenesis in the adult is involved in the formation of trace memories. Nature 410:372–376

Sierra A, Abiega O, Shahraz A, Neumann H (2013) Janus-faced microglia: beneficial and detrimental consequences of microglial phagocytosis. Front Cell Neurosci 7:6

Sierra A, Encinas JM, Deudero JJ, Chancey JH, Enikolopov G, Overstreet-Wadiche LS, Tsirka SE, Maletic-Savatic M (2010) Microglia shape adult hippocampal neurogenesis through apoptosis-coupled phagocytosis. Cell Stem Cell 7:483–495

Sierra A, Encinas JM, Maletic-Savatic M (2011) Adult human neurogenesis: from microscopy to magnetic resonance imaging. Front Neurosci 5:47

Sierra A, Gottfried-Blackmore AC, McEwen BS, Bulloch K (2007) Microglia derived from aging mice exhibit an altered inflammatory profile. Glia 55:412–424

Šišková Z, Tremblay MÈ (2013) Microglia and synapse: interactions in health and neurodegeneration. Neural Plast 2013:425845. doi:10.1155/2013/425845, Epub 2013 Dec 11

Sousa N, Almeida OF, Wotjak CT (2006) A hitchhiker's guide to behavioral analysis in laboratory rodents. Genes Brain Behav 5(Suppl 2):5–24

Spalding KL, Bergmann O, Alkass K, Bernard S, Salehpour M, Huttner HB, Bostrom E, Westerlund I, Vial C, Buchholz BA, Possnert G, Mash DC, Druid H, Frisen J (2013) Dynamics of hippocampal neurogenesis in adult humans. Cell 153:1219–1227

Tepavcevic V, Lazarini F, Alfaro-Cervello C, Kerninon C, Yoshikawa K, Garcia-Verdugo JM, Lledo PM, Nait-Oumesmar B, Baron-Van Evercooren A (2011) Inflammation-induced subventricular zone dysfunction leads to olfactory deficits in a targeted mouse model of multiple sclerosis. J Clin Invest 121:4722–4734

Trejo JL, Carro E, Torres-Aleman I (2001) Circulating insulin-like growth factor I mediates exercise-induced increases in the number of new neurons in the adult hippocampus. J Neurosci 21:1628–1634

Tremblay ME (2011) The role of microglia at synapses in the healthy CNS: novel insights from recent imaging studies. Neuron Glia Biol 7:67–76

van Praag H, Kempermann G, Gage FH (1999) Running increases cell proliferation and neurogenesis in the adult mouse dentate gyrus. Nat Neurosci 2:266–270

Williamson LL, Chao A, Bilbo SD (2012) Environmental enrichment alters glial antigen expression and neuroimmune function in the adult rat hippocampus. Brain Behav Immun 26:500–510

Wolf SA, Steiner B, Akpinarli A, Kammertoens T, Nassenstein C, Braun A, Blankenstein T, Kempermann G (2009) CD4-positive T lymphocytes provide a neuroimmunological link in the control of adult hippocampal neurogenesis. J Immunol 182:3979–3984

Wolf Y, Yona S, Kim KW, Jung S (2013) Microglia, seen from the CX3CR1 angle. Front Cell Neurosci 7:26

Wu X, Pan W, He Y, Hsuchou H, Kastin AJ (2010) Cerebral interleukin-15 shows upregulation and beneficial effects in experimental autoimmune encephalomyelitis. J Neuroimmunol 223: 65–72

Yamaguchi M, Mori K (2005) Critical period for sensory experience-dependent survival of newly generated granule cells in the adult mouse olfactory bulb. Proc Natl Acad Sci U S A 102:9697–9702

Yona S, Kim KW, Wolf Y, Mildner A, Varol D, Breker M, Strauss-Ayali D, Viukov S, Guilliams M, Misharin A, Hume DA, Perlman H, Malissen B, Zelzer E, Jung S (2013) Fate mapping reveals origins and dynamics of monocytes and tissue macrophages under homeostasis. Immunity 38:79–91

Young D, Lawlor PA, Leone P, Dragunow M, During MJ (1999) Environmental enrichment inhibits spontaneous apoptosis, prevents seizures and is neuroprotective. Nat Med 5:448–453

Ziv Y, Ron N, Butovsky O, Landa G, Sudai E, Greenberg N, Cohen H, Kipnis J, Schwartz M (2006) Immune cells contribute to the maintenance of neurogenesis and spatial learning abilities in adulthood. Nat Neurosci 9:268–275

Zunszain PA, Anacker C, Cattaneo A, Choudhury S, Musaelyan K, Myint AM, Thuret S, Price J, Pariante CM (2012) Interleukin-1beta: a new regulator of the kynurenine pathway affecting human hippocampal neurogenesis. Neuropsychopharmacology 37:939–949

Chapter 11
Neuropathic Pain

Jessica K. Alexander, Simon Beggs, and Michael W. Salter

Abstract It is now widely appreciated that microglia are critical effectors in the development of persistent pain hypersensitivity after traumatic nerve injury. Data to support this view began to emerge in the 1990s, and presently the neuropathic pain literature is replete with research articles describing various roles of microglia in the pathogenesis of pain. With this rapid proliferation of publications, it may be challenging to distill whether there exists a precise microglial function or phenotype that drives pain hypersensitivity. Numerous signaling pathways and countless molecules have been potentially implicated; however, their point(s) of convergence or divergence in injury-induced aberrant nociceptive processing are unclear. In this chapter, we examine key proposed mechanisms of microglial involvement in pain hypersensitivity. We also interrogate the concept of an inflammatory microglia phenotype and peripheral immune cell trafficking to spinal cord dorsal horn induced by nerve injury.

Keywords Microglia • Neuropathic pain • Peripheral nerve injury • Cytokines • Chemokines • Purinergic signaling • Brain-derived neurotrophic factor • Pain hypersensitivity

J.K. Alexander, Ph.D. • S. Beggs, Ph.D.
Program in Neurosciences and Mental Health, Hospital for Sick Children,
686 Bay Street, Toronto, ON, Canada M5G 0A4
e-mail: jessaalexander14@gmail.com; simon.beggs@gmail.com

M.W. Salter, M.D., Ph.D. (✉)
Program in Neurosciences and Mental Health, Hospital for Sick Children,
686 Bay Street, Toronto, ON, Canada M5G 0A4

Department of Physiology, Faculty of Medicine, Centre for the Study of Pain,
University of Toronto, Toronto, ON, Canada M5G 1X8
e-mail: mike.salter@utoronto.ca

M.-È. Tremblay, A. Sierra (eds.), *Microglia in Health and Disease*,
DOI 10.1007/978-1-4939-1429-6_11

Bullet Points

- A role for microglia in neuropathic pain is supported by rodent studies of peripheral nerve injury.
- Microglia exhibit a range of reactions to peripheral nerve injury, including migration, proliferation, cell body hypertrophy, gene expression, and changes in cell surface expression and secretion of proteins.
- Evidence for a predominantly pro-inflammatory phenotype of dorsal horn microglia after peripheral nerve is lacking.
- De novo expression of purinergic P2X4 receptor by microglia represents an essential phenotype for pain hypersensitivity after peripheral nerve injury.

11.1 Basics of Chronic and Neuropathic Pain

Acute nociceptive pain is an evolutionarily conserved and highly adaptive sense. It alerts the body of noxious stimuli to prevent or limit tissue damage. When pain outlasts expected healing of the precipitating cause or injury—regarded as "chronic"—it resigns any known physiological advantage and, rather, is recast as a disease itself. Pain persistence beyond healing indicates a pathological process.

Chronic, or pathological, pain is a major world health problem. It retards personal and societal productivity and generally is inadequately controlled by available pharmacotherapy. Approximately 5–20 % of cases of acute pain lead to chronic pain (Kehlet et al. 2006; Jensen 2005; Jung et al. 2004). According to the World Health Organization, an estimated 20 % of the world population suffers from chronic pain; in the U.S., it affects ~100 million individuals (Gaskin and Richard 2012; Goldberg and McGee 2011). Pain is a leading cause of disability and the most (or second most) common reason to seek physician consultation in the U.S. (Turk and Dworkin 2004), where the financial burden of chronic pain is estimated to be $560–635 billion annually (Gaskin and Richard 2012). Still, it remains an under-recognized and under-treated medical problem.

Beyond significant personal, societal, and economic costs, chronic pain engenders pathological sequelae (Tracey and Bushnell 2009). In recent years, long-term consequences of living with pain have been identified, including loss of brain mass and altered brain region connectivity and function (Cauda et al. 2010; Geha et al. 2008; Malinen et al. 2010). Persistent pain conditions lead to neocortical and/or subcortical (e.g., hippocampal, thalamic) atrophy (Apkarian et al. 2004; Kuchinad et al. 2007; Schmidt-Wilcke et al. 2006) and cognitive deficits. Emotional decision-making, attention, and memory are impaired in individuals with chronic pain (Apkarian et al. 2004; Dick and Rashiq 2007; Sjogren et al. 2005). These apparently downstream effects would be expected to compound personal and economical cost.

11.2 Neuropathic Pain

Perhaps the most devastating type of chronic pain is neuropathic pain (Smith et al. 2007; Torrance et al. 2006). The International Association for the Study of Pain defines neuropathic pain as, "pain arising as a direct consequence of a lesion or disease affecting the somatosensory system" (Treede et al. 2008). As such, diverse conditions and events precipitate neuropathic pain, including stroke, diabetes mellitus, herpes zoster infection, multiple sclerosis, nerve injury, amputation, cancer, and human immunodeficiency virus (HIV); treatments for cancer and HIV also produce neuropathic pain (Bennett 1998; Dworkin et al. 2003). The etiologic diversity can be reduced to fewer, broader classifications of traumatic, chemical, metabolic, and degenerative damage of the peripheral or central nervous system (CNS) (Dworkin et al. 2003). For example, constricted blood flow to distal extremities in diabetes produces an injurious metabolic famine for nerves. Chemotherapeutic agents likely produce direct chemical damage to nerves (Polomano et al. 2001; Xiao and Bennett 2008). A list of human peripheral and central neuropathic pain conditions is shown in Table 11.1.

The affliction of neuropathic pain is such because there is no cure and treatments rarely produce adequate relief. The pain can be evoked or spontaneous and the quality, or descriptor, is variable and includes burning, shooting, stabbing, aching, cold, etc. Lack of sufficient alleviation and/or organic pathophysiologic processes associated with neuropathic pain contribute to poor quality of life, loss of function, and development of comorbidities (e.g., depression, anxiety, insomnia) (Scadding and Koltzenburg 2006). Global population prevalence of neuropathic is thought to be underestimated at 1–2 % (Smith and Torrance 2012); with an aging population and improvements in diagnostics, this estimate is expected to increase (Bennett 1998; Raja and Haythornthwaite 2005).

Table 11.1 Human neuropathic pain conditions (Dworkin et al. 2003)

Clinical neuropathic pain (NP) conditions	
Peripheral NP	Central NP
Diabetes	Spinal cord injury
Herpes zoster (shingles)	Spinal stenosis
Trigeminal neuralgia	HIV
Traumatic nerve injury	Multiple sclerosis
Nerve compression	Parkinson disease
Tumor invasion	Phantom limb
Cancer chemotherapy	Ischemia/stroke
Spinal radiculopathy	Radiation
Carpal tunnel syndrome	
Demyelinating polyradiculopathy	
HIV	
Iatrogenic neuralgias	
Radiation	

Chronic neuropathic pain is distinct from acute pain, not just in terms of duration or etiology. The intractable nature of neuropathic pain is due to mechanisms of substantial nervous system plasticity, which, after sufficient damage, manifests as pathological perturbations to healthy neural transmission. In a properly functioning somatosensory system, pain is the percept of a process termed nociception. Nociception is the process by which specific sets of receptors activated by painful stimuli ("nociceptors," e.g., heat, acid) initiate the propagation of neuronal transmission from the site of stimulation (e.g., skin burn) to centers of conscious awareness in the brain. Transduction of such signals is affected by the microenvironment (e.g., molecular, cellular constituents) at key relay points along the sensory neuro-axis, including the site of injury and at each synapse thereafter. Compared to nociception, in neuropathic pain, the physiologic process of transmitting sensory information (non-painful and painful) is altered at molecular and cellular levels throughout the sensory axis. The mechanisms that underlie the development and maintenance of neuropathic pain are described in this chapter.

11.3 Experimental Models of Neuropathic Pain

Mechanisms of neuropathic pain are largely studied in rodent models. Traumatic peripheral nerve injury is the most commonly utilized rodent model. Methods have been developed to identify and measure pain-like behaviors in rodents that are inferred to reflect the three cardinal symptoms of neuropathic pain: spontaneous pain, and evoked allodynia and hyperalgesia. In humans, a painful response to a typically non-noxious stimulus (e.g., touch, warmth) is called "allodynia." An example of allodynia in acute pain is sensitivity to touch after sunburn. "Hyperalgesia" describes increased pain to a noxious stimulus. Using the same analogy for hyperalgesia, exposure of sunburned skin to hot water is especially painful. Spontaneous pain, or pain without an apparent stimulus, is a primary complaint in human neuropathic pain. After peripheral nerve injury, rodents exhibit spontaneous behavior including altered facial expressions (Sotocinal et al. 2011). Other examples of behavior that imply pain include postural changes that minimize weight bearing, reduction in typical locomotive behavior, and analgesic self-administration (Mogil et al. 2010). It is worthwhile to note, however, that abnormal paw posturing could represent motor effects of nerve injury. Loss of typical sleep or food intake may indicate spontaneous pain and is a correlate of pain in humans. However, altered sleep, appetite, or social interactions may represent experimentally induced physiological changes independent of hypersensitivity. For face validity and thus translational potential, spontaneous measures are preferred but not yet widely used; validating the specificity and improving quantitation of these behaviors is an area of current attention (Sotocinal et al. 2011). For this reason, most behavioral studies report measures of evoked pain-like behaviors.

Stimulus-evoked behavioral response is a common outcome measure in pain research. In rodents, a baseline (normal) threshold of sensitivity (e.g., of the hind paw) is acquired for cold, heat, or tactile stimulation. The threshold at which the animal withdraws its hind paw is interpreted as an uncomfortable, or painful, stimulus. Then, thresholds are assessed after a pain-evoking condition is established (e.g., after nerve injury or inflammation). Reduction in threshold after injury is interpreted as hypersensitivity or pain-like behavior. The validity of this assumption is confirmed by pharmacological reversal of hind paw sensitivity after the development of such conditions (Ossipov et al. 2006).

Peripheral nerve injuries predominate in this field of research; however, models that mimic diverse neuropathic pain states in human also have been developed. Rodent spinal cord injury (Detloff et al. 2008; Hulsebosch et al. 2009; Yezierski 2000), cancer (Schwei et al. 1999), cancer chemotherapy (Chaudhry et al. 1994; Smith et al. 2004), and diabetes (Courteix et al. 1993) models exist. Sciatic nerve injuries are most commonly utilized for several reasons: (1) the procedures are relatively simple and reproducible; (2) unlike systemic treatments, sciatic injury produces direct and easily examined damage; (3) the affected dermatomes include the hind paw, which is amenable to behavioral testing. Dermatomes are the portion of skin innervated by the nerve of a specific spinal root.

11.4 Mechanisms of Neuropathic Pain

Persistent pain may result from pathology at any level of sensory transmission. In most experimental models of neuropathic pain, the peripheral nerve is injured; however, the pathology is not limited to the site of damage. Aberrant relay of signals along neuronal transmission pathways manifest as pathology that is maintained independent of the original injury. The mechanisms that underlie the development and maintenance of neuropathic pain are described in the remainder of this chapter. As neurons are the electrically excitable cell type fundamental to the transmission of nociceptive signals, until the mid-1990s, they served as the sole basis for pain research. Since then, it has become clear that neural transmission is critically modulated by the activity of nearby non-neuronal cells. Microglia, for example, suppress or amplify inhibitory and excitatory neuronal activity. In the case of nerve injury, spinal cord microglia release a factor (brain-derived neurotrophic factor (BDNF), to be described in subsequent sections) that disinhibits lamina I neurons of the dorsal horn, providing a substrate for facilitation of ascending nociceptive transmission (Coull et al. 2005). Projection neurons of lamina I integrate and relay input from nociceptive primary afferents. Hence, it is a network of cellular communications that is a major determinant in pain, and the primary focus here.

11.5 Immune Response to Injury: Inflammation

In most tissue sites, the innate immune response fulfills its purpose; it responds to and defends against infection, prevents damage, and initiates repair of injured or infected tissue (Yeager et al. 2004). The return to homeostasis following nervous system damage, however, is constrained. It appears that the chronic, feed-forward nature of neuropathic pain reflects unresolved immune cell activity and concomitant neuronal sensitization (e.g., abnormal neural-glial interactions). The majority of inflammatory/immune mediators are algogenic (pain eliciting) and sensitize pain transmission pathways by producing ectopic firing at the injury site and sensitization of neural pathways in the CNS. In this regard, the immune response to nervous system injury is mechanistic in the generation and maintenance of pain.

Inflammation, classically, refers to the presence of five clinical signs: tissue swelling, redness, heat, pain, and loss of normal function. The purpose of an inflammatory response is to promote survival against injurious stimuli (e.g., pathogens, trauma). At a cellular and molecular level, inflammation is highly complex and its intricacies are beyond the scope of this discussion (see (Medzhitov 2008) for review and Chap. 5). Broadly, inflammation encompasses processes of plasma protein exudation and leukocyte extravasation and such functions as vasodilation, coagulation, phagocytosis, release of proteases, chemokines, and cytokines. It is important to consider what role "classical" inflammation might serve in the dorsal spinal horn after distal peripheral nerve injury, and more specifically, when the terminology is used, if it is even present.

11.6 Neuroinflammation

When inflammation occurs in neural tissue, it is often referred to as "neuroinflammation." The healthy nervous system is not endowed with cells to support inflammation such as T and B cells. The brain, spinal cord, and peripheral nerve, together with endothelia, must elicit peripherally derived leukocytes to generate classical inflammation (Cui et al. 2000). Thus, "neuroinflammation" describes induction of non-resident immune cell activity in nervous system tissues (Popovich and Longbrake 2008). Microglia are resident cells of the CNS that, under some circumstances, share functions with peripheral macrophages and collaborate with peripherally derived immune cells in neuroinflammation. For example, microglia can release pro-inflammatory cytokines, prostaglandins, proteolytic enzymes, and nitric oxide (NO). Like macrophages, microglia also can present antigen and remove cellular debris by phagocytosis (Streit et al. 1988, 1999; Stoll and Jander 1999; Sierra et al. 2013). The contemporary view of microglia has replaced the notion of binary switching from a singular "resting" state to a singular "activated" state with the concept that microglia continuously survey the CNS and react pleiotropically to stimuli. That is, microglia are capable of responding to specific microenvironments with numerous functional phenotypes. Because microglia are active in both healthy and compromised CNS, "reactive" is used to describe any response to disrupted homeostasis. The reactive state does not imply inflammation or specify function, however.

In the past 20 years, neuroinflammation has gained recognition as a mechanism in the development of neuropathic pain in rodent peripheral nerve injury. Peripheral nerve injury (e.g., sciatic nerve) triggers a spatio-temporally defined inflammatory response in the nerve. The magnitude of the response is determined by the type of nerve damage; ligation injuries show increased measures of inflammation versus non-ligation (e.g., transection, crush) (Cui et al. 2000; Hu et al. 2007). This is intriguing as it suggests that inflammation is not proportional to the damage of endothelia, Schwann cells, and axons (that would be severed in transection), but rather, that the pathology is perhaps protracted and thus cumulative after ligation injuries. In support of this, Fry et al. demonstrated deficient macrophage efflux in ligated nerve versus crush (Fry et al. 2007). Finally, with respect to magnitude, the neuroinflammatory response is moderated with distance from the lesion site (Hu and McLachlan 2003). Sites, time course, and roles of neuroinflammation after peripheral nerve damage are described below.

11.7 Peripheral Nerve and Dorsal Root Ganglion

Common procedures of peripheral nerve damage disrupt the blood-nerve barrier or expose the nerve directly (e.g., by transection). In neuropathic pain models, the lesion site harbors the most inflammation (Sorkin and Schafers 2007). Here, Schwann cells, endothelia, and the few resident immune cells (e.g., mast cells) rapidly respond to damage (Galli et al. 2005; Cui et al. 2000; Zuo et al. 2003; Metcalfe et al. 1997). Products of these early responders, including histamine, chemokines, pro-inflammatory cytokines (e.g., interleukin (IL) 1-beta (IL-1β), tumor necrosis factor-alpha (TNFα)), and proteases increase within an hour post-injury (Uceyler et al. 2007), excite neurons, and aid in the initiation of inflammation via chemoattraction and vasodilation. Schwann cells, not classically considered immune cells, are myelinating resident nerve glia that transiently morph into pro-inflammatory, phagocytic cells (e.g., clearing myelin debris) (Shamash et al. 2002; Bergsteinsdottir et al. 1991; Bolin et al. 1995; Chen et al. 2006; Reichert et al. 1994; Rutkowski et al. 1999; Tofaris et al. 2002; Wagner and Myers 1996). Neutrophils infiltrate as early as 1 h and peak by 24 h (Clatworthy et al. 1995; Perkins and Tracey 2000; Perry et al. 1987), and preemptive neutrophil depletion minimizes pain-like behavior after injury (Perkins and Tracey 2000). Hematogenous macrophages are recruited within 24 h but peak later, approximately 7–28 days post-injury, and pain-like behavior temporally correlates with macrophage infiltration of the nerve (Myers et al. 1996). Neutrophils and macrophages release highly pro-inflammatory substances, including prostaglandins, lipoxygenases, cytokines, reactive oxygen species (ROS), NO, proteases, and adenosine triphosphate (ATP). Importantly, virtually every agent produced by the described cells can exert direct or indirect stimulatory effects on neuron function (Sorkin and Schafers 2007). Neuron sensitization and activation after injury constitutes a principal pain mechanism. Exogenous administration of such agents (e.g., TNFα, prostaglandin E2 (PGE2)) results in pain hypersensitivity (Fukuoka et al. 1994; Jin and Gereau 2006; Ozaktay et al. 2006; Schafers et al. 2003a; von Banchet et al. 2005; Zhang et al. 2002; Obreja et al. 2002, 2005). Immunosuppressive treatment (Clatworthy et al. 1995; Bennett 1999) and macrophage depletion (Barclay et al. 2007; Liu et al. 2000) reduce immune cell infiltration and pain hypersensitivity. Encasing the nerve stump to limit exposure of the nerve to inflammatory cells attenuates autotomy, a behavior indicative of sensory dysfunction (Okuda et al. 2006). Of note, however, macrophage depletion appears to exert modality-specific effects on pain hypersensitivity. Systemic treatment with clodronate liposomes that deplete circulating macrophages and reduce macrophage infiltration of the injured nerve decreases thermal and mechanical hyper-responsivity after nerve injury, but has no effect on injury-induced reduction of mechanical thresholds (Barclay et al. 2007; Rutkowski et al. 2000; Liu et al. 2000).

Nerve inflammation and consequent ectopic neuronal activity contributes substantially to pain behavior; however, factors at the site of injury are not sufficient to maintain behavioral hypersensitivity. Only prophylactic, long-term nerve anesthesia is sufficient to prevent the development of hypersensitivity after nerve injury; after the injury, local nerve anesthesia may reduce but not eliminate hypersensitivity (Suter et al. 2009; Wen et al. 2007).

After injury to the peripheral nerve, immune cells also infiltrate the affected dorsal root ganglion (DRG). The resident glial population, satellite cells, undergo hypertrophy, proliferate, and increase expression of numerous inflammatory and neuron support molecules, including TNFα, nerve growth factor, and neurotrophin-3 (Woodham et al. 1989; Ohtori et al. 2004; Takahashi et al. 2006; Zhang et al. 2000; Zhou et al. 1999). Exposure of DRG neurons to pro-inflammatory cytokines excites nociceptive neurons and produces pain-like behaviors (Fukuoka et al. 1994; Zelenka et al. 2005; von Banchet et al. 2005; Schafers et al. 2003b; Ozaktay et al. 2006). After nerve injury, TNFα and other cytokines are retrogradely transported to the DRG and subsequently to the spinal cord (Shubayev and Myers 2001, 2002). Antidromic shuttling of TNFα, and release into target organs, is also possible (Schafers et al. 2002), suggesting bi-directional pro-inflammatory and neuro-excitatory communication. Activated kinases (e.g., mitogen-activated protein kinase (MAPK)), likewise, can be anterogradely and/or retrogradely transported after nerve injury (Cavalli et al. 2005; Delcroix et al. 2003; Perlson et al. 2005; Reynolds et al. 2001). Suppression of cytokine signaling or downstream intracellular kinase signaling is analgesic (Sommer et al. 2001a, b; Sweitzer et al. 2001b). Relative to peripheral nerve and spinal cord, less attention has been directed at understanding neuroimmune mechanisms of neuropathic pain in DRG.

11.8 Spinal Cord

Blocking peripheral afferent neuronal activity after nerve injury is not sufficient to inhibit hypersensitivity, indicating spinal or supraspinal mechanisms of neuropathic pain. Compared to spinal cord, mechanisms of neuropathic pain are least described in the brain. Peripheral immune cell infiltration in spinal cord is considerably less abundant than in the lesioned nerve. Indeed, only recent data implicate the presence of veritable "neuroinflammation," albeit minimal (Costigan et al. 2009; Sweitzer et al. 2002; Cao and DeLeo 2008). Sciatic nerve injury may produce blood-spinal cord barrier permeability or endothelial reactivity that can enable leukocyte extravasation (Kwan et al. 2008; Gordh et al. 2006; Beggs et al. 2010; Echeverry et al. 2011). Indeed, a small number of T cells enter the spinal cord after nerve injury and T cell-deficient mice exhibit reduced pain behaviors (Costigan et al. 2009; Sweitzer et al. 2002; Cao and DeLeo 2008). Detection of macrophage infiltration is obscured by non-distinct, overlapping expression of markers by microglia (e.g., ionized calcium-binding adaptor molecule 1 (Iba-1), cluster of differentiation (CD) molecules (e.g., CD11b), major histocompatibility complex (MHC) class I & II). When CD45 expression is measured coincident with CD11b, it is a useful tool to distinguish microglia (low CD45 expression) from macrophages (high CD45 expression) (Sedgwick et al. 1991). However, the relative expression of CD45 on spinal cord CD11b+ cells has not been demonstrated after nerve injury. The most obvious immune cell response to peripheral nerve injury is by resident microglia.

11.9 Microglia as Central Sensors of Peripheral Nerve Damage

Every model of traumatic nerve injury—nerve transection, ligation, or crush injury—induces dorsal horn microglial reactivity (DeLeo and Yezierski 2001; Katsura et al. 2006; Raghavendra et al. 2003; Hashizume et al. 2000; Meller et al. 1994) that is characterized by cellular hypertrophy, migration, and proliferation (Tsuda et al. 2004, 2005; Streit et al. 1999). This shift from a distributed arrangement and surveying morphology to a "reactive" phenotype after injury is associated with morphological changes, increased expression of cell surface markers (e.g., CD11b, Iba-1, MHC class II, F4/80), and intracellular signaling molecules (e.g., MAPK, ROS) (Hanisch and Kettenmann 2007). The function of many of these cell surface markers after nerve injury is unknown (Han et al. 2010). The specificity of microglial reactivity to axonal damage is demonstrated by its spatially-restricted reactive zone and its absence after chemical or inflammatory challenge to the nerve (Clark et al. 2007; Honore et al. 2000; Lin et al. 2007). Essentially, microglia react in the zone of injured afferent terminals (Beggs and Salter 2007). In some neuropathic models, microglial reactivity may be observed along the nociceptive pathway (e.g., periaqueductal grey, thalamus, cortex) (Detloff et al. 2008; Banati 2002; Zhao et al. 2007). The extent and diffusion of the microglial reaction may be dependent on injury severity or proximity to the spinal cord (Zhang et al. 2008), but it is not thought to be merely a consequence of exposure to transganglionic degeneration (Beggs et al. 2012; Liu et al. 1998). Rather, microglia are thought to respond to injury-induced neuronal activity or soluble factors.

11.10 Evidence of Spinal Cord Microglial-Mediated Mechanisms of Neuropathic Pain

Microglia play a critical role in neuropathic pain. More than two decades ago, it was observed that there is a prototypical morphological microglial response to nerve injury (Cammermeyer 1965; Gehrmann et al. 1991) that is temporally associated with the development of pain-like behaviors. At that time, and in the past, pain research was dominated by interrogation of neuronal transmission, including a possible role for dorsal horn ATP (Salter et al. 1993). A functional relationship of microglia with pain was suggested by the efficacy of pre-injury treatment with microglial modifying compounds, fluorocitrate (Meller et al. 1994; Milligan et al. 2003; Watkins et al. 1997), minocycline (Hains and Waxman 2006; Hua et al. 2005; Ledeboer et al. 2005b; Narita et al. 2006), and propentofylline (Garry et al. 2005; Sweitzer et al. 2001b), to attenuate hypersensitivity. Although all three agents lack specificity for microglia, these pioneer observations raised the possibility that microglia may have a role in pain.

Microglial morphology changes were, until recently, widely accepted as informative, although the field is now shifting away from it. Reduction of this prototypical response to peripheral nerve injury often has coincided with reduction of hypersensitivity, and hence was considered a meaningful end point. However, this coincident relationship is not faithful. An increasing number of studies report alterations of pain behavior without diminution of microglial reactivity (Colburn et al. 1997; Tozaki-Saitoh et al. 2008; Tsuda et al. 2003, 2009b; Alexander et al. 2012); thus, morphology and expression of cell surface markers alone is an inadequate measure of microglial contribution to neuropathic pain-like behavior. From these data, it is clear that this form of microglial "response" is not sufficient to predict hypersensitivity. An as yet unknown condition is whether the microglial response that accompanies every nerve injury is necessary for the development of hypersensitivity. Importantly, the term "neuroinflammation" should be distinguished from resident glial reactivity (e.g., microglia and astrocytes) that occurs in response to a range of stimuli. It is possible, that in certain circumstances, microglial reactivity (as measured by morphology and common surface markers) is either insignificant, or perhaps even protective/analgesic.

Lack of concordance between the extent of microglial reactivity (by number or proportional area (e.g., morphology)) and hypersensitivity raises appreciation for potential microglial phenotypes in neuropathic pain. If typical markers of reactivity do not consistently co-vary with hypersensitivity, what less visible aspect of microglia is relevant? Microglia exhibit functional diversity that is tailored to the local microenvironment. Discrete phenotypes, or functional states, are not discernible by typical cell surface markers. This limitation or hurdle likely advanced the presumption that microglia, as immune cells, act as classical inflammatory macrophages after nerve injury. Although microglia have demonstrated a broad range of functionality throughout the healthy CNS, microglial function in trauma and disease has focused primarily on inflammatory responses with less attention to alternative phenotypes. Little direct evidence supports this spinal microglial phenotype after distal peripheral nerve injury. For example, it does not appear that nerve injury-reactive microglia produce significant amounts of pro-inflammatory cytokines, activation of the inflammasome, increased MHC class II, or microbicidal activity (see Chap. 5 for further description of these processes). A discrete functional state of spinal dorsal horn microglia critical for neuropathic pain is characterized by de novo expression of the purinergic P2X4 receptor. Converging lines of evidence indicate that this P2X4R+ state is essential for pain hypersensitivity following peripheral nerve damage (Inoue et al. 2003; Ulmann et al. 2008; Trang et al. 2009). The P2X4R+ state does not appear to be pro-inflammatory. These data will be explained in more depth, below.

Microglia might affect neuropathic conditions by releasing pro-algesic factors that sensitize neighboring neuron activity. Conceivably, microglia could be skewed to an anti-inflammatory or other phenotype that blunts abnormal excitatory neuronal transmission. Like Schwann cells and other immune cells after nerve injury, microglia can secrete a range of pro-algesic soluble factors, including pro-inflammatory cytokines (e.g., TNFα, IL-1, IL-6), prostaglandins, NO, ROS, glutamate, and growth factors. Release of such factors by reactive microglia correlates with the development of hypersensitivity, including hyperalgesia and allodynia (Watkins et al. 2001; DeLeo et al. 2004; Ledeboer et al. 2005a; Jin et al. 2003; Coyle 1998; Coull et al. 2005). Indeed, neuropathic pain literature is replete with research articles describing various microglial-derived or -intrinsic factors in the pathogenesis of pain. From this rapidly expanding literature, it is challenging to distill whether there exists a precise microglial function or phenotype that drives pain hypersensitivity. Numerous signaling pathways and molecules have been implicated; however, their point(s) of convergence or divergence in injury-induced aberrant nociceptive processing are unclear. This area of research is incomplete because it is unknown to what extent there is overlapping or sequential release of such molecules. Broad, non-specific microglial inhibitors (e.g., minocycline) likely minimize the overall secretory activated phenotype of the cell, thus inhibiting release of multiple potentially pro-algesic factors. Even specific pathway inhibitors, such as nuclear factor kappa B (NF-κB), MAPK, or mammalian target of rapamycin (mTOR), are non-specific because these pathways converge (e.g., extracellular signal-related kinase (ERK) on TOR or NFκB on p38 MAPK). Furthermore, both the broad microglial inhibitors and specific kinase inhibitors are not cell-specific.

There is a growing list of molecules critical for pain hypersensitivity behaviors in rodent models of persistent pain. How to reconcile the observation that deletion or inhibition of a single factor blocks nerve injury-induced pain behavior in more than one-half dozen distinct cases is difficult, except to assume a chronological succession or interaction among these distinct molecules. In genetically identical mice, after injury, the microenvironment to which microglia are exposed is conceivably the same. Hence, microglia react to a signal, alter cell surface expression, shape, signaling molecules, gene expression, protein production, secretory factors, and ultimately function. At present, gene knockouts for P2X4R, protein kinase C gamma (PKCγ), C-C motif chemokine ligand (CCL) 21 (CCL21), toll-like receptor (TLR) 2 (TLR2), TLR4, C-C motif chemokine receptor (CCR) 2 (CCR2), interferon gamma (IFNγ), matrix metalloproteinase 9 (MMP9), or macrophage migration inhibitory factor (MIF) show protection from neuropathic pain-like behavior after nerve injury (Kawasaki et al. 2008; Zhang et al. 2007; Abbadie et al. 2003; Kim et al. 2007; Tanga et al. 2005; Tsuda et al. 2009a, b; Alexander et al. 2012). These data represent specific gene knockouts in the whole-animal; whereas, gene deletion in microglia (or microglia/macrophages) would specify the role of microglial-derived factors. Several kinases also demonstrate a pivotal role in neuropathic pain-like behavior, including p38 MAPK, ERK MAPK, mTOR, and Src tyrosine kinase (Garry et al. 2005; Ji et al. 1999; Jin et al. 2003; Inoue et al. 2003; Katsura et al. 2006; Obata et al. 2004; Tsuda et al. 2004, 2008b; Zhang et al. 2005).

Elucidation of a core microglial reaction to peripheral nerve injury that is critical for the pathogenesis of pain hypersensitivity and that accounts for a number of the known "critical factors" might be realized in the microglial phenotype characterized by expression of P2X4R. Nerve injury induces de novo expression of P2X4R on microglia that is coincident with the development of hypersensitivity. P2X4R is a purinergic receptor activated by ATP, and, at the time its expression was observed to increase with pain, ATP was being investigated for its role in pathological pain from a neuron-centric standpoint (Salter et al. 1993). When suppression of P2X4R, by pharmacological means or antisense RNA, reversed hypersensitivity, it was presumed to be neuronally mediated. Instead, it was observed that, in the spinal cord, P2X4R is expressed exclusively on microglia and its persistent activation is required to maintain nerve injury-induced hypersensitivity (Tsuda et al. 2003). Lack of hypersensitivity in P2X4R null mice after nerve injury further supports the necessity of this receptor (Ulmann et al. 2008; Tsuda et al. 2009a). Its sufficiency for the development of hypersensitivity is demonstrated by adoptive transfer experiments, wherein P2X4R-stimulated (or control) microglia are intrathecally injected into uninjured, naïve rats. Only P2X4R-stimulated microglia elicit tactile allodynia similar to that seen in neuropathic rats, an effect that is blocked by a pharmacological inhibitor of P2X4R (Tsuda et al. 2003). Taken together, the pharmacological, genetic, and behavioral data provide converging lines of evidence that P2X4R on spinal microglia is both necessary to produce the behavioral manifestations of neuropathic pain arising from peripheral nerve injury and sufficient to convert the response of normally non-painful peripheral inputs from innocuous to nociceptive. For an in-depth review of these data, see (Trang and Salter 2012). These observations form the basis for the concept that the P2X4R+ microglial phenotype is critically involved in the etiology of neuropathic pain.

Identification of the upstream and downstream components of the P2X4R microglial phenotype has revealed that it encompasses a number of other microglial mediators of pain hypersensitivity after nerve injury. Upstream of P2X4R is CCL21, a chemokine released from injured neurons (Biber et al. 2011; de Jong et al. 2005). It triggers microglial expression of P2X4R in vitro and in vivo. Accordingly, nerve-injured mice that lack neuronal CCL21 fail to upregulate P2X4R and do not develop hypersensitivity. Whereas CCL21 alone induces hypersensitivity and P2X4R expression in an uninjured wild-type mouse, it is without effect in a P2X4R null mouse. These data unify what may otherwise have been untested and presumed as two disparate mechanisms for the development of neuropathic pain behavior. Through these experiments, CCL21 was revealed to be an upstream neuronal factor that initiates hypersensitivity via induction of microglial P2X4R. Similarly, IFNγ receptor null mutant mice do not develop tactile allodynia after nerve injury, and IFNγ-mediated allodynia is reduced in P2X4R null mice, indicating upstream interdependence on P2X4R (Tsuda et al. 2009b). CCR2 is a chemokine receptor expressed on microglia (and peripheral macrophages) that is required for the development of neuropathic hypersensitivity (Abbadie et al. 2003; Zhang et al. 2007). Activation of CCR2 promotes microglial expression of P2X4Rs (Toyomitsu et al. 2012). Lyn kinase, a Src family kinase, is activated by injury-induced elevation of fibronectin, and it is necessary for the development of hypersensitivity and P2X4R upregulation after nerve injury (Tsuda et al. 2008b). Fibronectin is an extracellular matrix protein that is increased in the ipsilateral dorsal horn by nerve injury—possibly as a result of blood–brain barrier permeability after nerve injury—and signals through integrin receptors (Nasu-Tada et al. 2006). Inhibition of phosphoinositide 3-kinase—Akt (PI3K-Akt) or ERK signalling, which occur downstream of fibronectin-induced Lyn kinase activation, prevents P2X4R transcription and translation, respectively (Tsuda et al. 2009c). Moreover, fibronectin-induced allodynia is absent in P2X4R null mice (Tsuda et al. 2008a). These data provide another integration point for several critical but distinct molecules (e.g., fibronectin, Lyn kinase, PI3K, ERK) in the induction of hypersensitivity via P2X4R. In summary, significant advances have been made towards understanding the reaction of microglia after nerve injury, where factors critical for neuropathic pain play a role in converting microglia to the P2X4R phenotype.

Although we are gaining an understanding of the microglial reaction to peripheral nerve injury, it is not clear that these cells take on a highly pro-inflammatory phenotype. In the next section, the steps downstream of P2X4R activation leading to changes in neuronal activity are addressed. First, it should be considered that the stimuli that drive P2X4R "polarization" of microglia in turn influence microglial function (i.e., the specific response is programmed by the microenvironment). In macrophages, phagocytosis rapidly increases surface membrane and functional expression of P2X4R; whereas, classical, pro-inflammatory M1 stimuli (e.g., IFNγ + TNFα or lipopolysaccharide (LPS) + IFNγ) reduce surface and functional P2X4R (Stokes and Surprenant 2009). This comparative stimulation approach remains to be tested in microglia, but the macrophage data suggest that the induction of the P2X4R phenotype is not a consequence of a pro-inflammatory microenvironment nor supportive of classical M1 macrophage/microglia functions (e.g., microbicidal activity, IL-12 secretion).

There is little evidence of such a pro-inflammatory microenvironment in the dorsal horn after nerve injury. The detection of a pro-inflammatory cytokine does not imply optimal conditions for inflammation, for example, recruitment of peripheral immune cells or induction of Th1 or Th17 T cell phenotypes (see Chap. 5 for further description of these phenotypes; and Chap. 16 for their roles in multiple sclerosis); the environment in which microglia react and contribute is context-dependent. Cellular function is not predicated on the presence or absence of a single cytokine or chemokine. The context includes interacting cytokines (and other molecules), their form (e.g., total, active), receptor availability, and concentration (Lin and Levison 2009). There are mixed data for ipsilateral dorsal horn cytokine expression after nerve injury (i.e., the "microenvironment"). IL-1β and TNFα protein levels are not consistently increased; whereas IL-6 is elevated in rats and mice (Cao et al. 2009; DeLeo et al. 1997). In contrast, in rats, intrathecal delivery of functional antagonists of IL-1β and TNF α ameliorate pain hypersensitivity after nerve injury

(Sweitzer et al. 2001a). It is worthwhile to acknowledge that the DRG, in addition to the spinal cord, is accessed by compounds administered via the intrathecal route (Wang et al. 2005). Increased cytokine mRNA is observed more consistently than protein (Lee et al. 2004), and cellular-specific measurements after nerve injury are technically complicated due to relatively few cells compared to other conditions (e.g., experimental autoimmune encephalomyelitis, spinal cord injury, stroke). No data appear to support induction of IL-12 or IL-23 (Cao et al. 2009), cytokines that are required for the differentiation of Th1 or Th17 T cells. These are elevated in other CNS conditions and would be widely accepted as demonstrative of inflammatory processes (Becher et al. 2003; Vom Berg et al. 2012). Whole animal deletion of a given cytokine or pro-algesic molecule (e.g., IL-6, IFNγR) is not informative for the role of microglia or CNS involvement, as there is extensive inflammation at the site of the nerve injury that also affects hypersensitivity. It is generally perceived that T cell infiltration of the spinal cord (albeit few cells) and microglial reactivity constitute significant inflammatory processes after nerve injury. It is important to note that site-specific roles of immune cells are difficult to establish (e.g., CNS vs peripheral nervous system), and that the ensemble of immune-glial-neuronal activity in the dorsal horn after nerve injury may not comprise inflammation.

Other factors shown in experiments to be critical for the development of neuropathic pain also fit into the P2X4R schema, downstream. For such molecules that are already established as critical but that lack evidence of interdependence, determining their role in the P2X4R phenotype, offers the possibility of a unified pathway from which to rationally approach targeted therapies. Already, as discussed above, CCL21, CCR2, IFNγ, Lyn kinase, PI3K, and ERK have been tied into the induction of the P2X4R phenotype. How does this phenotype alter neuronal activity to affect pain hypersensitivity?

P2X4R is a purinergic receptor activated by ATP. Its expression is induced after nerve injury, and several key molecules driving its expression are described above. The source of ATP after nerve injury is unknown, although astrocytes are not the source (Foley et al. 2011). The central terminals of large diameter, myelinated primary sensory neurons, are a likely source (Beggs et al. 2012). Stimulation of P2X4R induces conformational changes that permit non-selective cation entry into the cell (Burnstock 2006). Prior to uncovering the intracellular signaling pathways triggered by P2X4R stimulation and the soluble factor that impinges on neuron function, the effect on neuron function was observed. Nerve injury induces disinhibition of nociceptive transmission, wherein $GABA_A$—and glycine-mediated inhibition is weakened (Coull et al. 2003). This disinhibition is caused by downregulation of the principal neuronal potassium-chloride co-transporter, KCC2. The resulting increase in intracellular chloride is sufficient to switch $GABA_A$/glycine responses from inhibition to excitation ("E_{anion} reversal") in some lamina I neurons. Intrathecal administration of P2X4R-stimulated microglia into naïve, uninjured rats mimics the nerve injury-induced depolarizing shift in E_{anion}, unmasking low threshold input to lamina I and causing nociceptive neurons to relay innocuous stimuli, increase output to noxious stimuli, and exhibit spontaneous activity (Coull et al. 2003; Keller et al. 2007). Such neuronal transformations likely form the basis for manifestations of neuropathic pain including allodynia, hyperalgesia, and spontaneous pain.

A P2X4R-dependent microglial-derived soluble factor that increases excitability of nociceptive transmission neurons in the dorsal horn is BDNF (Coull et al. 2005). BDNF, acting through its cognate receptor, trkB, produces the disinhibition and downregulation of KCC2. On P2X4R activation in vitro, calcium influx drives p38 MAPK phosphorylation and subsequent BDNF synthesis and release by cultured microglia (Trang et al. 2009). The role of p38 here may explain the previous observation that microglial p38 is required for the development of pain hypersensitivity after nerve injury (Jin et al. 2003; Tsuda et al. 2004; Zhuang et al. 2007), offering yet another unifying mechanism in microglia–neuron communication by the P2X4R microglial phenotype. P2X4R-driven expression of BDNF by microglia not only mediates pain hypersensitivity after peripheral nerve injury, but also the paradoxical hyperalgesia produced by morphine and other opioids. Indeed, targeted deletion of BDNF in microglia prevents morphine-induced hyperalgesia (Ferrini et al. 2013). Thus, the P2X4R-BDNF-trkB-KCC2 pathway in the dorsal horn nociceptive network is a core signaling cascade for pain hypersensitivity (Fig. 11.1).

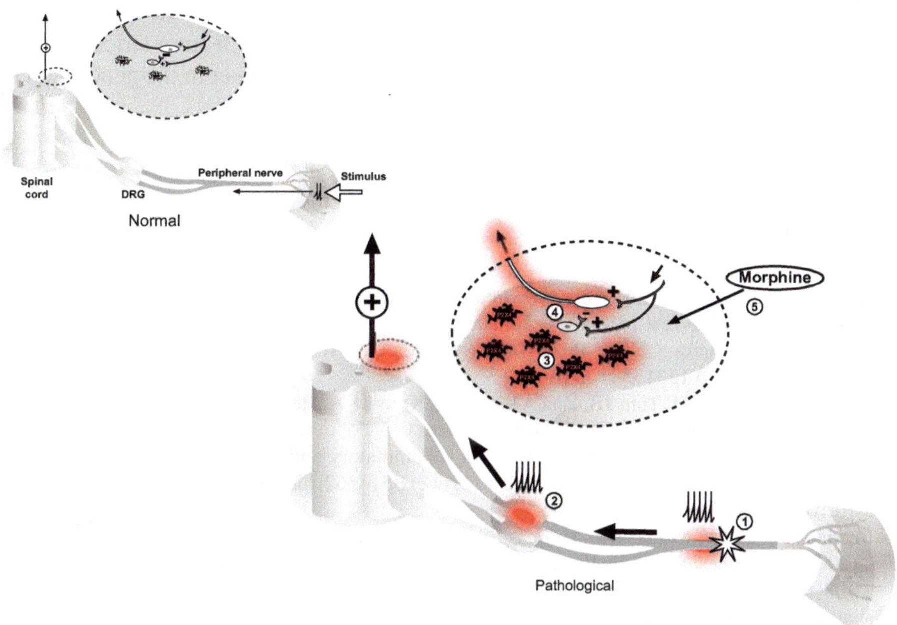

Fig. 11.1 Neuroimmune interactions along the sensory axis. Following peripheral nerve injury (1), release of pro-inflammatory mediators by infiltrating immune cells at the injury site and DRG result in primary afferent sensitization and ectopic discharge (2). In the spinal dorsal horn resident microglia upregulate P2X4 receptors (3), the activation of which leads to the synthesis and release of BDNF. BDNF, signaling through trkB receptors on dorsal horn neurons, induces the downregulation of neuronal KCC2 and consequent decrease in inhibitory tone (4). This core pathway of upregulation of microglial P2X4R expression; synthesis and release of microglial BDNF; downregulation of neuronal KCC2 expression drives the transformation of spinal neuronal output that underlies neuropathic pain hypersensitivity. The same mechanism mediates the paradoxical phenomenon of morphine hyperalgesia (5)

There is value in understanding to what extent the microglial reaction and other immune cell responses reflect veritable dorsal horn inflammation. If not, targeted immunotherapies (e.g., antibodies to cytokines or integrins) may not be useful in the context of neuropathic pain. Blocking inflammation at the level of the nerve does not sustainably reverse pain hypersensitivity, and it may not be the operative mechanism underlying neuronal transformations in the spinal cord. Considerable progress has been made in characterizing constituents of the dorsal spinal cord after peripheral nerve injury, including presence/absence of cells and molecules. What remains is to understand the functional output of these cells, their interactions with other cells, and the impact on the neuronal network and net neuronal transmission. For example, in addition to its presence, what is the role of the T cell in the spinal cord after nerve injury? What is its functional phenotype and how does that alter neuron function?

In summary, potential mechanisms of neuropathic pain may appear complex and numerous but may all feed through a common, core signaling cascade, the heart of which is neuron-microglia-neuron communication. In recent years, the field has been tightening up the use of terminology related to inflammatory processes, as microglial morphological changes do not constitute "inflammation" or "neuroinflammation." Much is to be gained as the field moves away from a strictly neurocentric view and recognizes that microglia and neurons interact in the generation of aberrant neuronal activity that manifests as neuropathic pain.

References

Abbadie C, Lindia JA, Cumiskey AM, Peterson LB, Mudgett JS, Bayne EK, DeMartino JA, MacIntyre DE, Forrest MJ (2003) Impaired neuropathic pain responses in mice lacking the chemokine receptor CCR2. Proc Natl Acad Sci U S A 100(13):7947–7952, doi:10.1073/pnas.1331358100 1331358100 [pii]

Alexander JK, Cox GM, Tian JB, Zha AM, Wei P, Kigerl KA, Reddy MK, Dagia NM, Sielecki T, Zhu MX, Satoskar AR, McTigue DM, Whitacre CC, Popovich PG (2012) Macrophage migration inhibitory factor (MIF) is essential for inflammatory and neuropathic pain and enhances pain in response to stress. Exp Neurol 236(2):351–362, doi:S0014-4886(12)00190-2 [pii] 10.1016/j.expneurol.2012.04.018

Apkarian AV, Sosa Y, Krauss BR, Thomas PS, Fredrickson BE, Levy RE, Harden RN, Chialvo DR (2004) Chronic pain patients are impaired on an emotional decision-making task. Pain 108(1–2):129–136

Banati RB (2002) Brain plasticity and microglia: is transsynaptic glial activation in the thalamus after limb denervation linked to cortical plasticity and central sensitisation? J Physiol Paris 96(3–4):289–299, doi:S0928-4257(02)00018-9 [pii]

Barclay J, Clark AK, Ganju P, Gentry C, Patel S, Wotherspoon G, Buxton F, Song C, Ullah J, Winter J, Fox A, Bevan S, Malcangio M (2007) Role of the cysteine protease cathepsin S in neuropathic hyperalgesia. Pain 130(3):225–234, doi:S0304-3959(06)00664-6 [pii] 10.1016/j.pain.2006.11.017

Becher B, Durell BG, Noelle RJ (2003) IL-23 produced by CNS-resident cells controls T cell encephalitogenicity during the effector phase of experimental autoimmune encephalomyelitis. J Clin Invest 112(8):1186–1191, doi:10.1172/JCI19079 112/8/1186 [pii]

Beggs S, Salter MW (2007) Stereological and somatotopic analysis of the spinal microglial response to peripheral nerve injury. Brain Behav Immun 21(5):624–633

Beggs S, Liu XJ, Kwan C, Salter MW (2010) Peripheral nerve injury and TRPV1-expressing primary afferent C-fibers cause opening of the blood-brain barrier. Mol Pain 6:74, doi:1744-8069-6-74 [pii] 10.1186/1744-8069-6-74

Beggs S, Trang T, Salter MW (2012) P2X4R+microglia drive neuropathic pain. Nat Neurosci 15(8):1068–1073, doi:nn.3155 [pii] 10.1038/nn.3155

Bennett GJ (1998) Neuropathic pain: new insights, new interventions. Hosp Pract (Off Ed) 33(10):95–98, 101–104, 107–110 passim

Bennett GJ (1999) Does a neuroimmune interaction contribute to the genesis of painful peripheral neuropathies? Proc Natl Acad Sci U S A 96(14):7737–7738

Bergsteinsdottir K, Kingston A, Mirsky R, Jessen KR (1991) Rat Schwann cells produce interleukin-1. J Neuroimmunol 34(1):15–23, doi:0165-5728(91)90094-N [pii]

Biber K, Tsuda M, Tozaki-Saitoh H, Tsukamoto K, Toyomitsu E, Masuda T, Boddeke H, Inoue K (2011) Neuronal CCL21 up-regulates microglia P2X4 expression and initiates neuropathic pain development. EMBO J 30(9):1864–1873, doi:emboj201189 [pii] 10.1038/emboj.2011.89

Bolin LM, Verity AN, Silver JE, Shooter EM, Abrams JS (1995) Interleukin-6 production by Schwann cells and induction in sciatic nerve injury. J Neurochem 64(2):850–858

Burnstock G (2006) Purinergic P2 receptors as targets for novel analgesics. Pharmacol Ther 110(3):433–454, doi:S0163-7258(05)00199-3 [pii] 10.1016/j.pharmthera.2005.08.013

Cammermeyer J (1965) Juxtavascular karyokinesis and microglia cell proliferation during retrograde reaction in the mouse facial nucleus. Ergeb Anat Entwicklungsgesch 38:1–22

Cao L, DeLeo JA (2008) CNS-infiltrating CD4+ T lymphocytes contribute to murine spinal nerve transection-induced neuropathic pain. Eur J Immunol 38(2):448–458. doi:10.1002/eji.200737485

Cao L, Palmer CD, Malon JT, De Leo JA (2009) Critical role of microglial CD40 in the maintenance of mechanical hypersensitivity in a murine model of neuropathic pain. Eur J Immunol 39(12):3562–3569. doi:10.1002/eji.200939657

Cauda F, D'Agata F, Sacco K, Duca S, Cocito D, Paolasso I, Isoardo G, Geminiani G (2010) Altered resting state attentional networks in diabetic neuropathic pain. J Neurol Neurosurg Psychiatry 81(7):806–811, doi:jnnp.2009.188631 [pii] 10.1136/jnnp.2009.188631

Cavalli V, Kujala P, Klumperman J, Goldstein LS (2005) Sunday Driver links axonal transport to damage signaling. J Cell Biol 168(5):775–787, doi:jcb.200410136 [pii] 10.1083/jcb.200410136

Chaudhry V, Rowinsky EK, Sartorius SE, Donehower RC, Cornblath DR (1994) Peripheral neuropathy from taxol and cisplatin combination chemotherapy: clinical and electrophysiological studies. Ann Neurol 35(3):304–311. doi:10.1002/ana.410350310

Chen S, Velardez MO, Warot X, Yu ZX, Miller SJ, Cros D, Corfas G (2006) Neuregulin 1-erbB signaling is necessary for normal myelination and sensory function. J Neurosci 26(12):3079–3086, doi:26/12/3079 [pii] 10.1523/JNEUROSCI.3785-05.2006

Clark AK, Gentry C, Bradbury EJ, McMahon SB, Malcangio M (2007) Role of spinal microglia in rat models of peripheral nerve injury and inflammation. Eur J Pain 11(2):223–230, doi:S1090-3801(06)00023-1 [pii] 10.1016/j.ejpain.2006.02.003

Clatworthy AL, Illich PA, Castro GA, Walters ET (1995) Role of peri-axonal inflammation in the development of thermal hyperalgesia and guarding behavior in a rat model of neuropathic pain. Neurosci Lett 184(1):5–8

Colburn RW, DeLeo JA, Rickman AJ, Yeager MP, Kwon P, Hickey WF (1997) Dissociation of microglial activation and neuropathic pain behaviors following peripheral nerve injury in the rat. J Neuroimmunol 79(2):163–175

Costigan M, Moss A, Latremoliere A, Johnston C, Verma-Gandhu M, Herbert TA, Barrett L, Brenner GJ, Vardeh D, Woolf CJ, Fitzgerald M (2009) T-cell infiltration and signaling in the adult dorsal spinal cord is a major contributor to neuropathic pain-like hypersensitivity. J Neurosci 29(46):14415–14422, doi:29/46/14415 [pii] 10.1523/JNEUROSCI.4569-09.2009

Coull JA, Boudreau D, Bachand K, Prescott SA, Nault F, Sik A, De Koninck P, De Koninck Y (2003) Trans-synaptic shift in anion gradient in spinal lamina I neurons as a mechanism of neuropathic pain. Nature 424(6951):938–942

Coull JA, Beggs S, Boudreau D, Boivin D, Tsuda M, Inoue K, Gravel C, Salter MW, De Koninck Y (2005) BDNF from microglia causes the shift in neuronal anion gradient underlying neuropathic pain. Nature 438(7070):1017–1021

Courteix C, Eschalier A, Lavarenne J (1993) Streptozocin-induced diabetic rats: behavioural evidence for a model of chronic pain. Pain 53(1):81–88, doi:0304-3959(93)90059-X [pii]

Coyle DE (1998) Partial peripheral nerve injury leads to activation of astroglia and microglia which parallels the development of allodynic behavior. Glia 23(1):75–83, doi:10.1002/(SICI)1098-1136(199805)23:1<75::AID-GLIA7>3.0.CO;2-3 [pii]

Cui JG, Holmin S, Mathiesen T, Meyerson BA, Linderoth B (2000) Possible role of inflammatory mediators in tactile hypersensitivity in rat models of mononeuropathy. Pain 88(3):239–248, doi:S0304395900003316 [pii]

de Jong EK, Dijkstra IM, Hensens M, Brouwer N, van Amerongen M, Liem RS, Boddeke HW, Biber K (2005) Vesicle-mediated transport and release of CCL21 in endangered neurons: a possible explanation for microglia activation remote from a primary lesion. J Neurosci 25(33):7548–7557, doi:25/33/7548 [pii] 10.1523/JNEUROSCI.1019-05.2005

Delcroix JD, Valletta JS, Wu C, Hunt SJ, Kowal AS, Mobley WC (2003) NGF signaling in sensory neurons: evidence that early endosomes carry NGF retrograde signals. Neuron 39(1):69–84, doi:S0896627303003970 [pii]

DeLeo JA, Yezierski RP (2001) The role of neuroinflammation and neuroimmune activation in persistent pain. Pain 90(1–2):1–6

DeLeo JA, Colburn RW, Rickman AJ (1997) Cytokine and growth factor immunohistochemical spinal profiles in two animal models of mononeuropathy. Brain Res 759(1):50–57

DeLeo JA, Tanga FY, Tawfik VL (2004) Neuroimmune activation and neuroinflammation in chronic pain and opioid tolerance/hyperalgesia. Neuroscientist 10(1):40–52

Detloff MR, Fisher LC, McGaughy V, Longbrake EE, Popovich PG, Basso DM (2008) Remote activation of microglia and pro-inflammatory cytokines predict the onset and severity of below-level neuropathic pain after spinal cord injury in rats. Exp Neurol 212(2):337–347

Dick BD, Rashiq S (2007) Disruption of attention and working memory traces in individuals with chronic pain. Anesth Analg 104(5):1223–1229, tables of contents. doi:104/5/1223 [pii] 10.1213/01.ane.0000263280.49786.f5

Dworkin RH, Backonja M, Rowbotham MC, Allen RR, Argoff CR, Bennett GJ, Bushnell MC, Farrar JT, Galer BS, Haythornthwaite JA, Hewitt DJ, Loeser JD, Max MB, Saltarelli M, Schmader KE, Stein C, Thompson D, Turk DC, Wallace MS, Watkins LR, Weinstein SM (2003) Advances in neuropathic pain: diagnosis, mechanisms, and treatment recommendations. Arch Neurol 60(11):1524–1534

Echeverry S, Shi XQ, Rivest S, Zhang J (2011) Peripheral nerve injury alters blood-spinal cord barrier functional and molecular integrity through a selective inflammatory pathway. J Neurosci 31(30):10819–10828, doi:31/30/10819 [pii] 10.1523/JNEUROSCI.1642-11.2011

Ferrini F, Trang T, Mattioli TA, Laffray S, Del'Guidice T, Lorenzo LE, Castonguay A, Doyon N, Zhang W, Godin AG, Mohr D, Beggs S, Vandal K, Beaulieu JM, Cahill CM, Salter MW, De Koninck Y (2013) Morphine hyperalgesia gated through microglia-mediated disruption of neuronal Cl(-) homeostasis. Nat Neurosci 16(2):183–192, doi:nn.3295 [pii] 10.1038/nn.3295

Foley JC, McIver SR, Haydon PG (2011) Gliotransmission modulates baseline mechanical nociception. Mol Pain 7:93, doi:1744-8069-7-93 [pii] 10.1186/1744-8069-7-93

Fry EJ, Ho C, David S (2007) A role for Nogo receptor in macrophage clearance from injured peripheral nerve. Neuron 53(5):649–662, doi:S0896-6273(07)00108-0 [pii] 10.1016/j.neuron.2007.02.009

Fukuoka H, Kawatani M, Hisamitsu T, Takeshige C (1994) Cutaneous hyperalgesia induced by peripheral injection of interleukin-1 beta in the rat. Brain Res 657(1–2):133–140, doi:0006-8993(94)90960-1 [pii]

Galli SJ, Kalesnikoff J, Grimbaldeston MA, Piliponsky AM, Williams CM, Tsai M (2005) Mast cells as "tunable" effector and immunoregulatory cells: recent advances. Annu Rev Immunol 23:749–786

Garry EM, Delaney A, Blackburn-Munro G, Dickinson T, Moss A, Nakalembe I, Robertson DC, Rosie R, Robberecht P, Mitchell R, Fleetwood-Walker SM (2005) Activation of p38 and p42/44 MAP kinase in neuropathic pain: involvement of VPAC2 and NK2 receptors and mediation by spinal glia. Mol Cell Neurosci 30(4):523–537

Gaskin DJ, Richard P (2012) The economic costs of pain in the United States. J Pain 13(8):715–724, doi:S1526-5900(12)00559-7 [pii] 10.1016/j.jpain.2012.03.009

Geha PY, Baliki MN, Harden RN, Bauer WR, Parrish TB, Apkarian AV (2008) The brain in chronic CRPS pain: abnormal gray-white matter interactions in emotional and autonomic regions. Neuron 60(4):570–581, doi:S0896-6273(08)00748-4 [pii] 10.1016/j.neuron.2008.08.022

Gehrmann J, Monaco S, Kreutzberg GW (1991) Spinal cord microglial cells and DRG satellite cells rapidly respond to transection of the rat sciatic nerve. Restor Neurol Neurosci 2(4):181–198, doi:F2852H884540843H [pii] 10.3233/RNN-1991-245605

Goldberg DS, McGee SJ (2011) Pain as a global public health priority. BMC Public Health 11:770, doi:1471-2458-11-770 [pii] 10.1186/1471-2458-11-770

Gordh T, Chu H, Sharma HS (2006) Spinal nerve lesion alters blood-spinal cord barrier function and activates astrocytes in the rat. Pain 124(1–2):211–221, doi:S0304-3959(06)00291-0 [pii] 10.1016/j.pain.2006.05.020

Hains BC, Waxman SG (2006) Activated microglia contribute to the maintenance of chronic pain after spinal cord injury. J Neurosci 26(16):4308–4317

Han C, Jin J, Xu S, Liu H, Li N, Cao X (2010) Integrin CD11b negatively regulates TLR-triggered inflammatory responses by activating Syk and promoting degradation of MyD88 and TRIF via Cbl-b. Nat Immunol 11(8):734–742, doi:ni.1908 [pii] 10.1038/ni.1908

Hanisch UK, Kettenmann H (2007) Microglia: active sensor and versatile effector cells in the normal and pathologic brain. Nat Neurosci 10(11):1387–1394, doi:nn1997 [pii] 10.1038/nn1997

Hashizume H, DeLeo JA, Colburn RW, Weinstein JN (2000) Spinal glial activation and cytokine expression after lumbar root injury in the rat. Spine 25(10):1206–1217

Honore P, Rogers SD, Schwei MJ, Salak-Johnson JL, Luger NM, Sabino MC, Clohisy DR, Mantyh PW (2000) Murine models of inflammatory, neuropathic and cancer pain each generates a unique set of neurochemical changes in the spinal cord and sensory neurons. Neuroscience 98(3):585–598, doi:S0306-4522(00)00110-X [pii]

Hu P, McLachlan EM (2003) Distinct functional types of macrophage in dorsal root ganglia and spinal nerves proximal to sciatic and spinal nerve transections in the rat. Exp Neurol 184(2):590–605, doi:10.1016/S0014-4886(03)00307-8 S0014488603003078 [pii]

Hu P, Bembrick AL, Keay KA, McLachlan EM (2007) Immune cell involvement in dorsal root ganglia and spinal cord after chronic constriction or transection of the rat sciatic nerve. Brain Behav Immun 21(5):599–616, doi:S0889-1591(06)00347-3 [pii] 10.1016/j.bbi.2006.10.013

Hua XY, Svensson CI, Matsui T, Fitzsimmons B, Yaksh TL, Webb M (2005) Intrathecal minocycline attenuates peripheral inflammation-induced hyperalgesia by inhibiting p38 MAPK in spinal microglia. Eur J Neurosci 22(10):2431–2440

Hulsebosch CE, Hains BC, Crown ED, Carlton SM (2009) Mechanisms of chronic central neuropathic pain after spinal cord injury. Brain Res Rev 60(1):202–213, doi:S0165-0173(08)00148-3 [pii] 10.1016/j.brainresrev.2008.12.010

Inoue K, Koizumi S, Tsuda M, Shigemoto-Mogami Y (2003) Signaling of ATP receptors in glia-neuron interaction and pain. Life Sci 74(2–3):189–197

Jensen TS (2005) From molecules to suffering. Nat Rev Neurosci 6(7):505

Ji R-R, Baba H, Brenner GJ, Woolf CJ (1999) Nociceptive-specific activation of ERK in spinal neurons contributes to pain hypersensitivity. Nat Neurosci 2(12):1114–1119

Jin X, Gereau RW IV (2006) Acute p38-mediated modulation of tetrodotoxin-resistant sodium channels in mouse sensory neurons by tumor necrosis factor-{alpha}. J Neurosci 26(1):246–255. doi:10.1523/jneurosci.3858-05.2006

Jin SX, Zhuang ZY, Woolf CJ, Ji RR (2003) p38 mitogen-activated protein kinase is activated after a spinal nerve ligation in spinal cord microglia and dorsal root ganglion neurons and contributes to the generation of neuropathic pain. J Neurosci 23(10):4017–4022

Jung BF, Johnson RW, Griffin DR, Dworkin RH (2004) Risk factors for postherpetic neuralgia in patients with herpes zoster. Neurology 62(9):1545–1551

Katsura H, Obata K, Mizushima T, Sakurai J, Kobayashi K, Yamanaka H, Dai Y, Fukuoka T, Sakagami M, Noguchi K (2006) Activation of Src-family kinases in spinal microglia contributes to mechanical hypersensitivity after nerve injury. J Neurosci 26(34):8680–8690

Kawasaki Y, Xu ZZ, Wang X, Park JY, Zhuang ZY, Tan PH, Gao YJ, Roy K, Corfas G, Lo EH, Ji RR (2008) Distinct roles of matrix metalloproteases in the early- and late-phase development of neuropathic pain. Nat Med 14(3):331–336
Kehlet H, Jensen TS, Woolf CJ (2006) Persistent postsurgical pain: risk factors and prevention. Lancet 367(9522):1618–1625
Keller AF, Beggs S, Salter MW, De Koninck Y (2007) Transformation of the output of spinal lamina I neurons after nerve injury and microglia stimulation underlying neuropathic pain. Mol Pain 3:27, doi:1744-8069-3-27 [pii] 10.1186/1744-8069-3-27
Kim D, Kim MA, Cho IH, Kim MS, Lee S, Jo EK, Choi SY, Park K, Kim JS, Akira S, Na HS, Oh SB, Lee SJ (2007) A critical role of toll-like receptor 2 in nerve injury-induced spinal cord glial cell activation and pain hypersensitivity. J Biol Chem 282(20):14975–14983
Kuchinad A, Schweinhardt P, Seminowicz DA, Wood PB, Chizh BA, Bushnell MC (2007) Accelerated brain gray matter loss in fibromyalgia patients: premature aging of the brain? J Neurosci 27(15):4004–4007, doi:27/15/4004 [pii] 10.1523/JNEUROSCI.0098-07.2007
Kwan C, Beggs S, Salter M (2008) Changes in blood spinal cord barrier permeability following peripheral nerve injury. Paper presented at the NeuPSIG Satellite on Neuropathic Pain to the Glasgow 2008 World Congress on Pain, London
Ledeboer A, Gamanos M, Lai W, Martin D, Maier SF, Watkins LR, Quan N (2005a) Involvement of spinal cord nuclear factor kappaB activation in rat models of proinflammatory cytokine-mediated pain facilitation. Eur J Neurosci 22(8):1977–1986
Ledeboer A, Sloane EM, Milligan ED, Frank MG, Mahony JH, Maier SF, Watkins LR (2005b) Minocycline attenuates mechanical allodynia and proinflammatory cytokine expression in rat models of pain facilitation. Pain 115(1–2):71–83
Lee HL, Lee KM, Son SJ, Hwang SH, Cho HJ (2004) Temporal expression of cytokines and their receptors mRNAs in a neuropathic pain model. Neuroreport 15(18):2807–2811, doi:00001756-200412220-00024 [pii]
Lin HW, Levison SW (2009) Context-dependent IL-6 potentiation of interferon- gamma-induced IL-12 secretion and CD40 expression in murine microglia. J Neurochem 111(3):808–818, doi:JNC6366 [pii] 10.1111/j.1471-4159.2009.06366.x
Lin T, Li K, Zhang FY, Zhang ZK, Light AR, Fu KY (2007) Dissociation of spinal microglia morphological activation and peripheral inflammation in inflammatory pain models. J Neuroimmunol 192(1–2):40–48, doi:S0165-5728(07)00303-7 [pii] 10.1016/j.jneuroim.2007.09.003
Liu L, Persson JK, Svensson M, Aldskogius H (1998) Glial cell responses, complement, and clusterin in the central nervous system following dorsal root transection. Glia 23(3):221–238, doi:10.1002/(SICI)1098-1136(199807)23:3<221::AID-GLIA5>3.0.CO;2-7 [pii]
Liu T, van Rooijen N, Tracey DJ (2000) Depletion of macrophages reduces axonal degeneration and hyperalgesia following nerve injury. Pain 86(1–2):25–32, doi:S0304-3959(99)00306-1 [pii]
Malinen S, Vartiainen N, Hlushchuk Y, Koskinen M, Ramkumar P, Forss N, Kalso E, Hari R (2010) Aberrant temporal and spatial brain activity during rest in patients with chronic pain. Proc Natl Acad Sci U S A 107(14):6493–6497, doi:1001504107 [pii] 10.1073/pnas.1001504107
Medzhitov R (2008) Origin and physiological roles of inflammation. Nature 454(7203):428–435, doi:nature07201 [pii] 10.1038/nature07201
Meller ST, Dykstra C, Grzybycki D, Murphy S, Gebhart GF (1994) The possible role of glia in nociceptive processing and hyperalgesia in the spinal cord of the rat. Neuropharmacology 33(11):1471–1478
Metcalfe DD, Baram D, Mekori YA (1997) Mast cells. Physiol Rev 77(4):1033–1079
Milligan ED, Twining C, Chacur M, Biedenkapp J, O'Connor K, Poole S, Tracey K, Martin D, Maier SF, Watkins LR (2003) Spinal glia and proinflammatory cytokines mediate mirror-image neuropathic pain in rats. J Neurosci 23(3):1026–1040
Mogil JS, Graham AC, Ritchie J, Hughes SF, Austin JS, Schorscher-Petcu A, Langford DJ, Bennett GJ (2010) Hypolocomotion, asymmetrically directed behaviors (licking, lifting, flinching, and shaking) and dynamic weight bearing (gait) changes are not measures of neuropathic pain in mice. Mol Pain 6:34, doi:1744-8069-6-34 [pii] 10.1186/1744-8069-6-34

Myers RR, Heckman HM, Rodriguez M (1996) Reduced hyperalgesia in nerve-injured WLD mice: relationship to nerve fiber phagocytosis, axonal degeneration, and regeneration in normal mice. Exp Neurol 141(1):94–101, doi:S0014-4886(96)90142-9 [pii] 10.1006/exnr.1996.0142

Narita M, Kuzumaki N, Kaneko C, Hareyama N, Miyatake M, Shindo K, Miyoshi K, Nakajima M, Nagumo Y, Sato F, Wachi H, Seyama Y, Suzuki T (2006) Chronic pain-induced emotional dysfunction is associated with astrogliosis due to cortical delta-opioid receptor dysfunction. J Neurochem 97(5):1369–1378

Nasu-Tada K, Koizumi S, Tsuda M, Kunifusa E, Inoue K (2006) Possible involvement of increase in spinal fibronectin following peripheral nerve injury in upregulation of microglial P2X4, a key molecule for mechanical allodynia. Glia 53(7):769–775. doi:10.1002/glia.20339

Obata K, Yamanaka H, Kobayashi K, Dai Y, Mizushima T, Katsura H, Fukuoka T, Tokunaga A, Noguchi K (2004) Role of mitogen-activated protein kinase activation in injured and intact primary afferent neurons for mechanical and heat hypersensitivity after spinal nerve ligation. J Neurosci 24(45):10211–10222

Obreja O, Schmelz M, Poole S, Kress M (2002) Interleukin-6 in combination with its soluble IL-6 receptor sensitises rat skin nociceptors to heat, in vivo. Pain 96(1–2):57–62, doi:S0304395901004201 [pii]

Obreja O, Biasio W, Andratsch M, Lips KS, Rathee PK, Ludwig A, Rose-John S, Kress M (2005) Fast modulation of heat-activated ionic current by proinflammatory interleukin 6 in rat sensory neurons. Brain 128(Pt 7):1634–1641, doi:awh490 [pii] 10.1093/brain/awh490

Ohtori S, Takahashi K, Moriya H, Myers RR (2004) TNFalpha and TNFalpha receptor type 1 upregulation in glia and neurons after peripheral nerve injury: studies in murine DRG and spinal cord. Spine (Phila Pa 1976) 29(10):1082–1088, doi:00007632-200405150-00006 [pii]

Okuda T, Ishida O, Fujimoto Y, Tanaka N, Inoue A, Nakata Y, Ochi M (2006) The autotomy relief effect of a silicone tube covering the proximal nerve stump. J Orthop Res 24(7):1427–1437. doi:10.1002/jor.20196

Ossipov MH, Lai J, Porreca F (2006) Mechanisms of experimental neuropathic pain: integration from animal models. In: McMahon SB, Koltzenburg M (eds) Wall and Melzack's textbook of pain, 5th edn. Elsevier, Churchill, pp 929–946

Ozaktay AC, Kallakuri S, Takebayashi T, Cavanaugh JM, Asik I, DeLeo JA, Weinstein JN (2006) Effects of interleukin-1 beta, interleukin-6, and tumor necrosis factor on sensitivity of dorsal root ganglion and peripheral receptive fields in rats. Eur Spine J 15(10):1529–1537. doi:10.1007/s00586-005-0058-8

Perkins NM, Tracey DJ (2000) Hyperalgesia due to nerve injury: role of neutrophils. Neuroscience 101(3):745–757, doi:S0306-4522(00)00396-1 [pii]

Perlson E, Hanz S, Ben-Yaakov K, Segal-Ruder Y, Seger R, Fainzilber M (2005) Vimentin-dependent spatial translocation of an activated MAP kinase in injured nerve. Neuron 45(5):715–726, doi:S0896-6273(05)00055-3 [pii] 10.1016/j.neuron.2005.01.023

Perry VH, Brown MC, Gordon S (1987) The macrophage response to central and peripheral nerve injury. A possible role for macrophages in regeneration. J Exp Med 165(4):1218–1223

Polomano RC, Mannes AJ, Clark US, Bennett GJ (2001) A painful peripheral neuropathy in the rat produced by the chemotherapeutic drug, paclitaxel. Pain 94(3):293–304, doi:S0304395901003633 [pii]

Popovich PG, Longbrake EE (2008) Can the immune system be harnessed to repair the CNS? Nat Rev Neurosci 9(6):481–493

Raghavendra V, Tanga F, DeLeo JA (2003) Inhibition of microglial activation attenuates the development but not existing hypersensitivity in a rat model of neuropathy. J Pharmacol Exp Ther 306(2):624–630

Raja SN, Haythornthwaite JA (2005) Combination therapy for neuropathic pain—which drugs, which combination, which patients? N Engl J Med 352(13):1373–1375

Reichert F, Saada A, Rotshenker S (1994) Peripheral nerve injury induces Schwann cells to express two macrophage phenotypes: phagocytosis and the galactose-specific lectin MAC-2. J Neurosci 14(5 Pt 2):3231–3245

Reynolds AJ, Hendry IA, Bartlett SE (2001) Anterograde and retrograde transport of active extracellular signal-related kinase 1 (ERK1) in the ligated rat sciatic nerve. Neuroscience 105(3):761–771, doi:S0306-4522(01)00235-4 [pii]

Rutkowski JL, Tuite GF, Lincoln PM, Boyer PJ, Tennekoon GI, Kunkel SL (1999) Signals for proinflammatory cytokine secretion by human Schwann cells. J Neuroimmunol 101(1):47–60, doi:S0165-5728(99)00132-0 [pii]

Rutkowski MD, Pahl JL, Sweitzer S, van Rooijen N, DeLeo JA (2000) Limited role of macrophages in generation of nerve injury-induced mechanical allodynia. Physiol Behav 71(3–4):225–235, doi:S0031-9384(00)00333-4 [pii]

Salter MW, De Koninck Y, Henry JL (1993) Physiological roles for adenosine and ATP in synaptic transmission in the spinal dorsal horn. Prog Neurobiol 41(2):125–156

Scadding JW, Koltzenburg M (2006) Painful peripheral neuropathies. In: McMahon SB, Koltzenburg M (eds) Wall and Melzack's textbook of pain, 5th edn. Elsevier, Churchill, pp 973–999

Schafers M, Geis C, Brors D, Yaksh TL, Sommer C (2002) Anterograde transport of tumor necrosis factor-alpha in the intact and injured rat sciatic nerve. J Neurosci 22(2):536–545, doi:22/2/536 [pii]

Schafers M, Lee DH, Brors D, Yaksh TL, Sorkin LS (2003a) Increased sensitivity of injured and adjacent uninjured rat primary sensory neurons to exogenous tumor necrosis factor-alpha after spinal nerve ligation. J Neurosci 23(7):3028–3038

Schafers M, Svensson CI, Sommer C, Sorkin LS (2003b) Tumor necrosis factor-alpha induces mechanical allodynia after spinal nerve ligation by activation of p38 MAPK in primary sensory neurons. J Neurosci 23(7):2517–2521

Schmidt-Wilcke T, Leinisch E, Ganssbauer S, Draganski B, Bogdahn U, Altmeppen J, May A (2006) Affective components and intensity of pain correlate with structural differences in gray matter in chronic back pain patients. Pain 125(1–2):89–97, doi:S0304-3959(06)00255-7 [pii] 10.1016/j.pain.2006.05.004

Schwei MJ, Honore P, Rogers SD, Salak-Johnson JL, Finke MP, Ramnaraine ML, Clohisy DR, Mantyh PW (1999) Neurochemical and cellular reorganization of the spinal cord in a murine model of bone cancer pain. J Neurosci 19(24):10886–10897

Sedgwick JD, Schwender S, Imrich H, Dorries R, Butcher GW, ter Meulen V (1991) Isolation and direct characterization of resident microglial cells from the normal and inflamed central nervous system. Proc Natl Acad Sci U S A 88(16):7438–7442

Shamash S, Reichert F, Rotshenker S (2002) The cytokine network of Wallerian degeneration: tumor necrosis factor-alpha, interleukin-1alpha, and interleukin-1beta. J Neurosci 22(8):3052–3060

Shubayev VI, Myers RR (2001) Axonal transport of TNFalpha in painful neuropathy: distribution of ligand tracer and TNF receptors. J Neuroimmunol 114(1–2):48–56, doi:S0165572800004537 [pii]

Shubayev VI, Myers RR (2002) Anterograde TNF alpha transport from rat dorsal root ganglion to spinal cord and injured sciatic nerve. Neurosci Lett 320(1–2):99–101, doi:S0304394002000101 [pii]

Sierra A, Abiega O, Shahraz A, Neumann H (2013) Janus-faced microglia: beneficial and detrimental consequences of microglial phagocytosis. Front Cell Neurosci 7:6. doi:10.3389/fncel.2013.00006

Sjogren P, Christrup LL, Petersen MA, Hojsted J (2005) Neuropsychological assessment of chronic non-malignant pain patients treated in a multidisciplinary pain centre. Eur J Pain 9(4):453–462, doi:S1090-3801(04)00129-6 [pii] 10.1016/j.ejpain.2004.10.005

Smith BH, Torrance N (2012) Epidemiology of neuropathic pain and its impact on quality of life. Curr Pain Headache Rep 16(3):191–198. doi:10.1007/s11916-012-0256-0

Smith SB, Crager SE, Mogil JS (2004) Paclitaxel-induced neuropathic hypersensitivity in mice: responses in 10 inbred mouse strains. Life Sci 74(21):2593–2604, doi:10.1016/j.lfs.2004.01.002 S0024320504000153 [pii]

Smith BH, Torrance N, Bennett MI, Lee AJ (2007) Health and quality of life associated with chronic pain of predominantly neuropathic origin in the community. Clin J Pain 23(2):143–149, doi:10.1097/01.ajp.0000210956.31997.89 00002508-200702000-00005 [pii]

Sommer C, Lindenlaub T, Teuteberg P, Schafers M, Hartung T, Toyka KV (2001a) Anti-TNFneutralizing antibodies reduce pain-related behavior in two different mouse models of painful mononeuropathy. Brain Res 913(1):86–89, doi:S0006-8993(01)02743-3 [pii]

Sommer C, Schafers M, Marziniak M, Toyka KV (2001b) Etanercept reduces hyperalgesia in experimental painful neuropathy. J Peripher Nerv Syst 6(2):67–72

Sorkin L, Schafers M (2007) Immune cells in peripheral nerve. In: DeLeo JA, Sorkin LS, Watkins LR (eds) Immune and glial regulation of pain. IASP Press, Seattle

Sotocinal SG, Sorge RE, Zaloum A, Tuttle AH, Martin LJ, Wieskopf JS, Mapplebeck JC, Wei P, Zhan S, Zhang S, McDougall JJ, King OD, Mogil JS (2011) The Rat Grimace Scale: a partially automated method for quantifying pain in the laboratory rat via facial expressions. Mol Pain 7:55, doi:1744-8069-7-55 [pii] 10.1186/1744-8069-7-55

Stokes L, Surprenant A (2009) Dynamic regulation of the P2X4 receptor in alveolar macrophages by phagocytosis and classical activation. Eur J Immunol 39(4):986–995. doi:10.1002/eji.200838818

Stoll G, Jander S (1999) The role of microglia and macrophages in the pathophysiology of the CNS. Prog Neurobiol 58(3):233–247

Streit WJ, Graeber MB, Kreutzberg GW (1988) Functional plasticity of microglia: a review. Glia 1(5):301–307

Streit WJ, Walter SA, Pennell NA (1999) Reactive microgliosis. Prog Neurobiol 57(6):563–581

Suter MR, Berta T, Gao YJ, Decosterd I, Ji RR (2009) Large A-fiber activity is required for microglial proliferation and p38 MAPK activation in the spinal cord: different effects of resiniferatoxin and bupivacaine on spinal microglial changes after spared nerve injury. Mol Pain 5:53, doi:1744-8069-5-53 [pii] 10.1186/1744-8069-5-53

Sweitzer S, Martin D, DeLeo JA (2001a) Intrathecal interleukin-1 receptor antagonist in combination with soluble tumor necrosis factor receptor exhibits an anti-allodynic action in a rat model of neuropathic pain. Neuroscience 103(2):529–539, doi:S0306-4522(00)00574-1 [pii]

Sweitzer SM, Schubert P, DeLeo JA (2001b) Propentofylline, a glial modulating agent, exhibits antiallodynic properties in a rat model of neuropathic pain. J Pharmacol Exp Ther 297(3):1210–1217

Sweitzer SM, Hickey WF, Rutkowski MD, Pahl JL, DeLeo JA (2002) Focal peripheral nerve injury induces leukocyte trafficking into the central nervous system: potential relationship to neuropathic pain. Pain 100(1–2):163–170

Takahashi N, Kikuchi S, Shubayev VI, Campana WM, Myers RR (2006) TNFalpha and phosphorylation of ERK in DRG and spinal cord: insights into mechanisms of sciatica. Spine (Phila Pa 1976) 31(5):523–529, doi:10.1097/01.brs.0000201305.01522.17 00007632-200603010-00005 [pii]

Tanga FY, Nutile-McMenemy N, DeLeo JA (2005) The CNS role of toll-like receptor 4 in innate neuroimmunity and painful neuropathy. Proc Natl Acad Sci U S A 102(16):5856–5861

Tofaris GK, Patterson PH, Jessen KR, Mirsky R (2002) Denervated Schwann cells attract macrophages by secretion of leukemia inhibitory factor (LIF) and monocyte chemoattractant protein-1 in a process regulated by interleukin-6 and LIF. J Neurosci 22(15):6696–6703, doi:20026699 22/15/6696 [pii]

Torrance N, Smith BH, Bennett MI, Lee AJ (2006) The epidemiology of chronic pain of predominantly neuropathic origin. Results from a general population survey. J Pain 7(4):281–289, doi S1526-5900(05)01029-1 [pii] 10.1016/j.jpain.2005.11.008

Toyomitsu E, Tsuda M, Yamashita T, Tozaki-Saitoh H, Tanaka Y, Inoue K (2012) CCL2 promotes P2X4 receptor trafficking to the cell surface of microglia. Purinergic Signal 8(2):301–310. doi:10.1007/s11302-011-9288-x

Tozaki-Saitoh H, Tsuda M, Miyata H, Ueda K, Kohsaka S, Inoue K (2008) P2Y12 receptors in spinal microglia are required for neuropathic pain after peripheral nerve injury. J Neurosci 28(19):4949–4956, doi:28/19/4949 [pii] 10.1523/JNEUROSCI.0323-08.2008

Tracey I, Bushnell MC (2009) How neuroimaging studies have challenged us to rethink: is chronic pain a disease? J Pain 10(11):1113–1120, doi:S1526-5900(09)00712-3 [pii] 10.1016/j.jpain.2009.09.001

Trang T, Salter MW (2012) P2X4 purinoceptor signaling in chronic pain. Purinergic Signal 8(3):621–628. doi:10.1007/s11302-012-9306-7
Trang T, Beggs S, Wan X, Salter MW (2009) P2X4-receptor-mediated synthesis and release of brain-derived neurotrophic factor in microglia is dependent on calcium and p38-mitogen-activated protein kinase activation. J Neurosci 29(11):3518–3528, doi:29/11/3518 [pii] 10.1523/JNEUROSCI.5714-08.2009
Treede RD, Jensen TS, Campbell JN, Cruccu G, Dostrovsky JO, Griffin JW, Hansson P, Hughes R, Nurmikko T, Serra J (2008) Neuropathic pain: redefinition and a grading system for clinical and research purposes. Neurology 70(18):1630–1635, doi:01.wnl.0000282763.29778.59 [pii] 10.1212/01.wnl.0000282763.29778.59
Tsuda M, Shigemoto-Mogami Y, Koizumi S, Mizokoshi A, Kohsaka S, Salter MW, Inoue K (2003) P2X4 receptors induced in spinal microglia gate tactile allodynia after nerve injury. Nature 424(6950):778–783
Tsuda M, Mizokoshi A, Shigemoto-Mogami Y, Koizumi S, Inoue K (2004) Activation of p38 mitogen-activated protein kinase in spinal hyperactive microglia contributes to pain hypersensitivity following peripheral nerve injury. Glia 45(1):89–95. doi:10.1002/glia.10308
Tsuda M, Inoue K, Salter MW (2005) Neuropathic pain and spinal microglia: a big problem from molecules in "small" glia. Trends Neurosci 28(2):101–107
Tsuda M, Toyomitsu E, Komatsu T, Masuda T, Kunifusa E, Nasu-Tada K, Koizumi S, Yamamoto K, Ando J, Inoue K (2008a) Fibronectin/integrin system is involved in P2X(4) receptor upregulation in the spinal cord and neuropathic pain after nerve injury. Glia 56(5):579–585. doi:10.1002/glia.20641
Tsuda M, Tozaki-Saitoh H, Masuda T, Toyomitsu E, Tezuka T, Yamamoto T, Inoue K (2008b) Lyn tyrosine kinase is required for P2X(4) receptor upregulation and neuropathic pain after peripheral nerve injury. Glia 56(1):50–58. doi:10.1002/glia.20591
Tsuda M, Kuboyama K, Inoue T, Nagata K, Tozaki-Saitoh H, Inoue K (2009a) Behavioral phenotypes of mice lacking purinergic P2X4 receptors in acute and chronic pain assays. Mol Pain 5:28, doi:1744-8069-5-28 [pii] 10.1186/1744-8069-5-28
Tsuda M, Masuda T, Kitano J, Shimoyama H, Tozaki-Saitoh H, Inoue K (2009b) IFN-gamma receptor signaling mediates spinal microglia activation driving neuropathic pain. Proc Natl Acad Sci U S A 106(19):8032–8037, doi:0810420106 [pii] 10.1073/pnas.0810420106
Tsuda M, Toyomitsu E, Kometani M, Tozaki-Saitoh H, Inoue K (2009c) Mechanisms underlying fibronectin-induced up-regulation of P2X4R expression in microglia: distinct roles of PI3K-Akt and MEK-ERK signalling pathways. J Cell Mol Med 13(9B):3251–3259, doi:JCMM719 [pii] 10.1111/j.1582-4934.2009.00719.x
Turk DC, Dworkin RH (2004) What should be the core outcomes in chronic pain clinical trials? Arthritis Res Ther 6(4):151–154, doi:10.1186/ar1196 ar1196 [pii]
Uceyler N, Tscharke A, Sommer C (2007) Early cytokine expression in mouse sciatic nerve after chronic constriction nerve injury depends on calpain. Brain Behav Immun 21(5):553–560
Ulmann L, Hatcher JP, Hughes JP, Chaumont S, Green PJ, Conquet F, Buell GN, Reeve AJ, Chessell IP, Rassendren F (2008) Up-regulation of P2X4 receptors in spinal microglia after peripheral nerve injury mediates BDNF release and neuropathic pain. J Neurosci 28(44): 11263–11268
Vom Berg J, Prokop S, Miller KR, Obst J, Kalin RE, Lopategui-Cabezas I, Wegner A, Mair F, Schipke CG, Peters O, Winter Y, Becher B, Heppner FL (2012) Inhibition of IL-12/IL-23 signaling reduces Alzheimer's disease-like pathology and cognitive decline. Nat Med 18(12):1812–1819, doi:nm.2965 [pii] 10.1038/nm.2965
von Banchet GS, Kiehl M, Schaible HG (2005) Acute and long-term effects of IL-6 on cultured dorsal root ganglion neurones from adult rat. J Neurochem 94(1):238–248, doi:JNC3185 [pii] 10.1111/j.1471-4159.2005.03185.x
Wagner R, Myers RR (1996) Schwann cells produce tumor necrosis factor alpha: expression in injured and non-injured nerves. Neuroscience 73(3):625–629, doi:0306-4522(96)00127-3 [pii]
Wang X, Wang C, Zeng J, Xu X, Hwang PY, Yee WC, Ng YK, Wang S (2005) Gene transfer to dorsal root ganglia by intrathecal injection: effects on regeneration of peripheral nerves. Mol Ther 12(2):314–320, doi:S1525-0016(05)00159-0 [pii] 10.1016/j.ymthe.2005.03.032

Watkins LR, Martin D, Ulrich P, Tracey KJ, Maier SF (1997) Evidence for the involvement of spinal cord glia in subcutaneous formalin induced hyperalgesia in the rat. Pain 71(3):225–235, doi:S0304395997033691 [pii]
Watkins LR, Milligan ED, Maier SF (2001) Spinal cord glia: new players in pain. Pain 93(3):201–205
Wen YR, Suter MR, Kawasaki Y, Huang J, Pertin M, Kohno T, Berde CB, Decosterd I, Ji RR (2007) Nerve conduction blockade in the sciatic nerve prevents but does not reverse the activation of p38 mitogen-activated protein kinase in spinal microglia in the rat spared nerve injury model. Anesthesiology 107(2):312–321, doi:10.1097/01.anes.0000270759.11086.e7 00000542-200708000-00020 [pii]
Woodham P, Anderson PN, Nadim W, Turmaine M (1989) Satellite cells surrounding axotomised rat dorsal root ganglion cells increase expression of a GFAP-like protein. Neurosci Lett 98(1):8–12, doi:0304-3940(89)90364-9 [pii]
Xiao WH, Bennett GJ (2008) Chemotherapy-evoked neuropathic pain: abnormal spontaneous discharge in A-fiber and C-fiber primary afferent neurons and its suppression by acetyl-L-carnitine. Pain 135(3):262–270, doi:S0304-3959(07)00303-X [pii] 10.1016/j.pain.2007.06.001
Yeager MP, Guyre PM, Munck AU (2004) Glucocorticoid regulation of the inflammatory response to injury. Acta Anaesthesiol Scand 48(7):799–813
Yezierski RP (2000) Pain following spinal cord injury: pathophysiology and central mechanisms. Prog Brain Res 129:429–449
Zelenka M, Schafers M, Sommer C (2005) Intraneural injection of interleukin-1beta and tumor necrosis factor-alpha into rat sciatic nerve at physiological doses induces signs of neuropathic pain. Pain 116(3):257–263
Zhang Y, Roslan R, Lang D, Schachner M, Lieberman AR, Anderson PN (2000) Expression of CHL1 and L1 by neurons and glia following sciatic nerve and dorsal root injury. Mol Cell Neurosci 16(1):71–86, doi:10.1006/mcne.2000.0852 S1044-7431(00)90852-2 [pii]
Zhang JM, Li H, Liu B, Brull SJ (2002) Acute topical application of tumor necrosis factor alpha evokes protein kinase A-dependent responses in rat sensory neurons. J Neurophysiol 88(3):1387–1392
Zhang FE, Cao JL, Zhang LC, Zeng YM (2005) Activation of p38 mitogen-activated protein kinase in spinal cord contributes to chronic constriction injury-induced neuropathic pain. Sheng Li Xue Bao 57(5):545–551
Zhang J, Shi XQ, Echeverry S, Mogil JS, De Koninck Y, Rivest S (2007) Expression of CCR2 in both resident and bone marrow-derived microglia plays a critical role in neuropathic pain. J Neurosci 27(45):12396–12406, doi:27/45/12396 [pii] 10.1523/JNEUROSCI.3016-07.2007
Zhang F, Vadakkan KI, Kim SS, Wu LJ, Shang Y, Zhuo M (2008) Selective activation of microglia in spinal cord but not higher cortical regions following nerve injury in adult mouse. Mol Pain 4:15, doi:1744-8069-4-15 [pii] 10.1186/1744-8069-4-15
Zhao P, Waxman SG, Hains BC (2007) Modulation of thalamic nociceptive processing after spinal cord injury through remote activation of thalamic microglia by cysteine cysteine chemokine ligand 21. J Neurosci 27(33):8893–8902
Zhou XF, Deng YS, Chie E, Xue Q, Zhong JH, McLachlan EM, Rush RA, Xian CJ (1999) Satellite-cell-derived nerve growth factor and neurotrophin-3 are involved in noradrenergic sprouting in the dorsal root ganglia following peripheral nerve injury in the rat. Eur J Neurosci 11(5):1711–1722
Zhuang ZY, Kawasaki Y, Tan PH, Wen YR, Huang J, Ji RR (2007) Role of the CX3CR1/p38 MAPK pathway in spinal microglia for the development of neuropathic pain following nerve injury-induced cleavage of fractalkine. Brain Behav Immun 21(5):642–651, doi:S0889-1591(06)00354-0 [pii] 10.1016/j.bbi.2006.11.003
Zuo Y, Perkins NM, Tracey DJ, Geczy CL (2003) Inflammation and hyperalgesia induced by nerve injury in the rat: a key role of mast cells. Pain 105(3):467–479

Chapter 12
Drug Addiction

Xiaohui Wang, Thomas A. Cochran, Mark R. Hutchinson, Hang Yin, and Linda R. Watkins

Abstract Drug addiction is a pervasive worldwide problem characterized by compulsive drug use that continues despite negative consequences and treatment attempts. Historically, the biological basis of drug addiction has focused principally on neuronal activity. However, despite their pivotal role in the underlying pathology of drug addiction, neurons are not the only central nervous system (CNS) component involved. The role of additional cell types, especially the CNS immunocompetent microglial cells, in the development of tolerance and related neuroplastic changes during drug taking, addiction, and withdrawal is also emerging. Within this perspective, this chapter reviews the roles of microglial cells in several aspects of drug addiction and its behavioural consequences, including reward, tolerance, dependence, and withdrawal. The cellular and molecular mechanisms which are particularly recruited will be emphasized. Lastly, we will also summarize the development of pharmacological modulators of microglial activation that offer novel treatment strategies and highlight the need to better understand the roles of microglia in the context of drug addiction.

X. Wang, Ph.D.
Chemical Biology Laboratory, Changchun Institute of Applied Chemistry, Chinese Academy of Sciences, Changchun, Jilin, China

Department of Psychology and the Center for Neuroscience, Department of Chemistry and Biochemistry and BioFrontiers Institute, University of Colorado at Boulder, Boulder, CO, USA
e-mail: xhwang83@gmail.com

T.A. Cochran, B.A. • L.R. Watkins, Ph.D. (✉)
Department of Psychology and the Center for Neuroscience, University of Colorado at Boulder, Boulder, CO, USA
e-mail: thomas.cochran@colorado.edu; linda.watkins@colorado.edu

M.R. Hutchinson, Ph.D.
Discipline of Physiology, School of Medical Sciences, University of Adelaide, Level 5, Medical School South, S508, Frome Road, Adelaide, SA 5005, Australia
e-mail: mark.hutchinson@adelaide.edu.au

H. Yin, Ph.D.
Department of Chemistry and Biochemistry and BioFrontiers Institute, University of Colorado at Boulder, Boulder, CO, USA
e-mail: hubert.yin@colorado.edu

M.-È. Tremblay, A. Sierra (eds.), *Microglia in Health and Disease*,
DOI 10.1007/978-1-4939-1429-6_12

Keywords Microglia • Inflammation • Drug addiction • Toll-like receptor 4 • NMDA receptors • Purinergic signalling • Cytokines • Opioids • Ethanol

Bullet Points

- Microglial cells contribute to several aspects of drug addiction and its behavioural consequences, including reward, tolerance, dependence, and withdrawal.
- The cellular and molecular mechanisms utilized by microglia that contribute to drug addiction include microglial Toll-like receptors, NMDA receptors, and purinergic signaling.
- Targeting microglial activation may lead to the development of pharmacological modulators that seek to alter the behavioural consequences associated with drug addiction.
- A better understanding of microglial involvement in drug addiction is therefore needed.

12.1 Introduction

Drug abuse can lead to drug addiction, which is the compulsive use of a substance, despite its negative effects. Drug addiction represents one of the major medical, social, and economic burdens of human behavior. Abused drugs range from legally available recreational drugs such as alcohol and legally prescribed opioids such as morphine, to illicit "street" and "party" drugs such as heroin, cocaine, cannabis, and methamphetamine (Coller and Hutchinson 2012). In 2011, there were 22.5 million current illicit drug users aged 12 or older in the United States, comprising 8.7 % of the population (Fig. 12.1a) (Substance Abuse and Mental Health Services Administration 2012). The highest rate of illicit drug use was among the 18–20 year olds (23.8 %, Fig. 12.1b). The number of persons with drug dependence or abuse in 2011 was 20.6 million (Fig. 12.1c). Among the people with illicit drug dependence or abuse, marijuana, opioid analgesics, and cocaine were the 3 most used illicit drugs (Fig. 12.1d). These statistics highlight the need for research aimed at characterizing the molecular mechanisms responsible for drug addiction in an effort to develop more effective treatment regimens.

The physiochemical properties of abusive drugs, and the brain structures they affect, are diverse. Consequently, the neurochemical mechanisms responsible for drug addiction vary depending on the drug. However, all drugs of abuse share the common trait that they summate neuronally to produce elevated signaling in the cortico-mesolimbic reward pathway, which behaviorally presents as a rewarding and reinforcing drive after repeated exposure (Coller and Hutchinson 2012). This pathway begins at the ventral tegmental area (VTA), a midbrain cluster of dopaminergic neurons which project principally to a basal forebrain nucleus known as the nucleus accumbens (NAcc). The reward experienced from all addictive drugs has classically

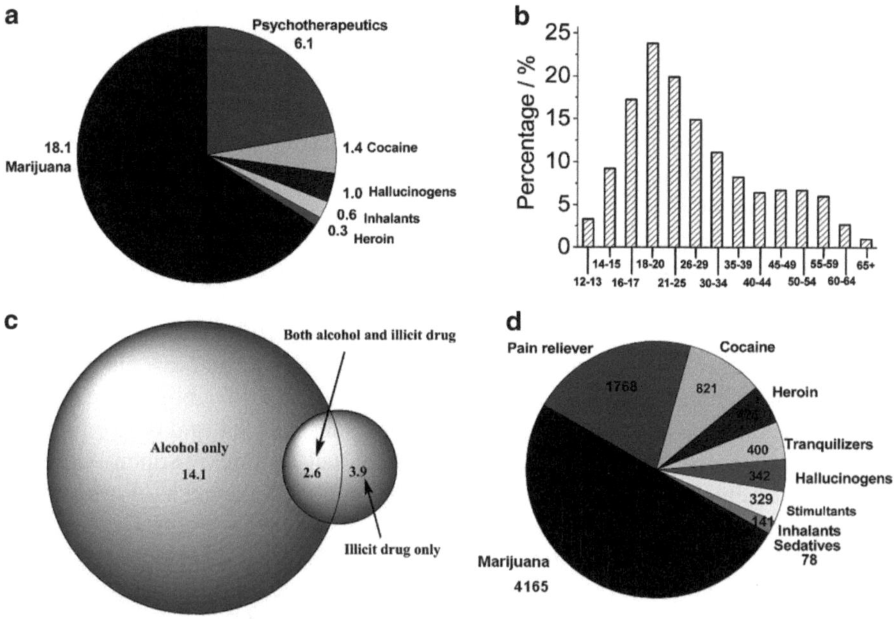

Fig. 12.1 Prevalence of drug addiction. (**a**), illicit drug use among persons aged 12 or older in 2011 in US. Numbers are in millions. Data are derived from 2011 National Survey on Drug Use and Health (NSDUH). (**b**), age distribution of illicit drug use among persons aged 12 or older in 2011 in US. (**c**), drug dependence or abuse in 2011 among persons aged 12 or older. Numbers are in millions. (**d**), specific illicit drug dependence or abuse among Persons aged 12 or older in 2011. Numbers are in thousands

been thought to be associated, at least in part, with increases in extracellular dopamine from the VTA projections to the NAcc (Hyman et al. 2006). In addition to its hedonic effect, dopamine release at the nucleus accumbens core has been reported to be involved with motivation for drug seeking provided by drug learning memory (Li et al. 2011). Increased accumbal dopamine, in addition to activity in mesocortical circuits, has also been associated with the salience for drug use (Goldstein and Volkow 2002). When not exploited by drug use, this reward circuit reinforces behaviors that are essential for survival. Dopamine release at the nucleus accumbens can be triggered by eating or procreating, thereby contributing to reinforcement and ensuring these behavioural processes are repeated (Ikemoto and Panksepp 1999).

In addition to the VTA and NAcc, the amygdala and hippocampus are critical components of the cortico-mesolimbic reward pathway involved in drug addiction. The amygdala consists of nuclei located within the medial temporal lobes acting as a key stress system through interaction with the hypothalamic-pituitary-adrenal (HPA) axis (Lowry 2002). It is hypothesized that environmental stressors emerging from drug dependence and withdrawal activate the central and basolateral amygdala (BLA) and, once activated, create negative emotional states which are capable of motivating drug-seeking behavior (Koob 2009). The BLA also provides input to the ventral subiculum (vSub), a component of the hippocampus that projects to a variety of limbic-related structures and modulates a variety of context-dependent processes involved in drug addiction, including drug sensitization and stress (Groenewegen et al. 1987; Herman and Mueller 2006; Sinha 2001). When activated directly by environmental stressors, or via the BLA, the vSub has been shown to induce a hyperdopaminergic response to amphetamine at the NAcc, and is hypothesized to be a key component responsible for drug sensitization (Belujon and Grace 2011; Lodge and Grace 2008).

The cycle of drug addiction varies greatly depending upon the individual and the type of drugs used. However, it generally involves three key stages: initiation, maintenance, and relapse. Initiation is a strong compulsion to continue drug use which is fueled by the motivation to obtain drug reward and is associated with increased activity in the cortico-mesolimbic dopamine reward pathway and associated neural networks (Coller and Hutchinson 2012). With repeated use, this motivation evolves into maintenance, which is characterized by drug dependence and desire to avoid negative withdrawal symptoms. Should an individual become abstinent, cravings and environmental cues can trigger drug relapse and begin the cycle of drug addiction once more.

Typically, animal models for drug initiation and relapse are measured by a self-administration behavioral paradigm, while the degree of reward can be measured by conditioned place preference (CPP) (Cooper et al. 2008). Briefly, the self-administration behavioral paradigm involves inducing an operant response in an animal by pairing a positive reinforcer with an operant behavior. In a drug self-administration behavioral paradigm, the operant behavior is typically a lever press while the positive reinforcer is the administration of an addictive drug. After several pairings, the animal will continue to elicit the operant behavior in the absence of the reinforcer. In this paradigm, initiation and relapse are measured as the degree of operant behavior exhibited by the animal following conditioning. CPP operates by placing the animal in a box subdivided into two sides having distinct colors and patterns. After a pre-exposure test, animals in the experimental group repeatedly receive a drug of abuse on only one side of the box. This conditions the animal to having a preference for the side where it received the drug. In this way, reward is measured as the increase in the amount of time the animal spent on the side of the box where it received the drug relative to the amount of time spent in that same side prior to drug exposure (Tzschentke 2007).

The initiation, maintenance, and relapse stages of drug addiction involve complex changes in the neurocircuitry of reward and stress, both of which are capable of reinforcing drug-seeking behavior (Kovacs 2012). Consequently, an overwhelming proportion of the research on the molecular and cellular basis of addiction has focused on neuronal activity. Likewise, pharmacological therapies for treating drug addiction have been directed toward molecular pathways in neurons (Miguel-Hidalgo 2009). This research has provided a plethora of knowledge on the neuronal circuits altered in drug addiction and provided valuable insights into how their alterations determine specific aberrant behaviors associated with drug addiction (Miguel-Hidalgo 2009). However, despite their pivotal role in the underlying pathology of drug addiction, neurons are not the only central nervous system (CNS) component regulating neurotransmission. The role of additional cell types, especially the CNS immunocompetent microglial cells, in the development of tolerance and related neuroplastic changes during drug taking, addiction, and withdrawal is also emerging (Hutchinson et al. 2007, 2011; Milligan and Watkins 2009; Watkins et al. 2005, 2009).

Microglia are derived from an erythromyeloid precursor cell of the embryonic hematopoiesis in the yolk sac. As such, they are the resident cells of the innate immune system in the brain (Kierdorf et al. 2013; Neumann and Wekerle 2013). The new role of microglia as neuronal modulators of drug reward has emerged from the opioid's ability to activate microglia and promote the release of pro-inflammatory cytokines (Wang et al. 2012). These cytokines were recently identified as a potential source of microglia-to-neuron signaling stimulating microglia to increase neuronal excitability, thus resulting in a number of pathological conditions. These conditions include maintaining and enhancing pain states, reducing the efficacy of opioid for pain control, and developing analgesic tolerance and opioid withdrawal and dependence (Hutchinson et al. 2007, 2011; Milligan et al. 2009; Watkins et al. 2005, 2009). Since these phenomena are related to the process of addiction, this new role of microglia highlights the need to better understand the mechanisms of drug abuse from a microglial perspective (Coller and Hutchinson 2012). The goal of this chapter is to summarize the cellular and molecular mechanisms utilized by microglia that contribute to drug addiction and review the development of pharmacological microglia modulators that seek to alter the behavioral consequences associated with drug addiction.

12.2 Microglia Mechanisms of Drug Abuse

Microglia cells express markers of macrophage lineage and many receptors related to inflammatory processes (Kettenmann et al. 2011), and become activated when exposed to xenobiotics (Hutchinson et al. 2011). Activation of microglia by addictive drugs results in a change in cell morphology from a quiescent to an activated macrophage-like phenotype and initiates an innate immune response characterized by the release of pro-inflammatory cytokines such as interleukin-1 β (IL-1β), tumor necrosis factor α (TNFα), as well as nitric oxide (NO), within the prefrontal cortex, NAcc, VTA, amygdala, and hippocampus (Hutchinson et al. 2012; Miguel-Hidalgo 2009; Zhang et al. 2012). Pro-inflammatory cytokines down-regulate glutamate transporter (Watkins et al. 2009) and GABA receptor expression (Stellwagen et al. 2005), up-regulate AMPA and NMDA receptor expression and function (De et al. 2003), and enhance neurotransmitter release and synaptic transmission (Beattie et al. 2002; Pascual et al. 2012; Youn et al. 2008). Pro-inflammatory cytokines could therefore amplify the changes in neuronal activity induced by drugs of abuse, at multiple points within the associated circuitry. As the primary immune sentinels of the CNS, microglia engulf and remove apoptotic neurons and cellular debris. Microglia make direct contacts with synaptic elements and could also directly modulate synaptic activity by regulating synaptic numbers through sculpting and

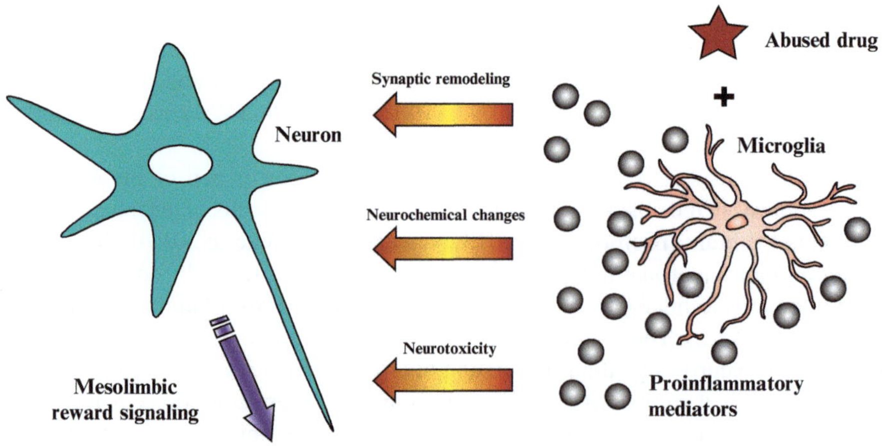

Fig. 12.2 Schematic illustration of the microglia–neuron interactions which could be implicated in drug addiction. Abuse drugs cause microglia activation, leading to a profound pro-inflammatory immune signaling within the reward circuitry of the brain. In particular, pro-inflammatory immune signaling augments the mesolimbic dopamine signaling possibly through increased synaptic remodeling, neurochemical changes in excitatory transmission, and phagocytosis of apoptotic neurons and cellular debris

pruning and by modulating synaptic adhesion molecules and glutamate receptor expression (Ji et al. 2013; Kettenmann et al. 2013; Tremblay and Majewska 2011a; Tremblay et al. 2011b) (see Chap. 9 for further information). Collectively, microglia may modulate neuronal circuits of reward and dependence and alter behavioral responses through synaptic remodeling, neurochemical interaction with excitatory transmission, and phagocytosis of apoptotic neurons and cellular debris (Fig. 12.2) (Coller and Hutchinson 2012; Graeber 2010; Kettenmann et al. 2011; Kovacs 2012), although direct evidence for the role of microglia in remodeling the cortico-mesolimbic reward pathway in the context of drug abuse is still missing. Additionally, it should be noted that the activation of central immune signaling pathways induced by drugs of abuse has only been shown as a complimentary mechanism working in conjunction with previously established neuronal circuits underlying drug reward and dependence. Central immune signaling pathways alone will not produce drug addiction behavioral outcomes (Coller and Hutchinson 2012).

12.2.1 Toll-Like Receptor 4

Accumulating evidences show that morphine can activate microglia and that inhibitors of microglia activation are capable of blocking behavioral measures of morphine reward such as CPP and self-administration in animal models (Hutchinson et al. 2007, 2012; Watkins et al. 2009). While microglia express a variety of "classical" opioid receptor subtypes to which morphine can bind, including δ, κ, and μ-opioid receptors, studies with opioid receptor knockout mice reveal that morphine can act on microglia independently from these receptors. One such study reported that triple opioid receptor knockout mice are still responsive to morphine-induced hyperalgesia (Juni et al. 2007). Furthermore, classical opioid receptors are stereoselective for (−)-opioid agonists (Hutchinson et al. 2010a), yet (+)-opioid agonists, which have no classical opioid receptor activity (Wu et al. 2007), are capable of suppressing (−)-opioid-induced analgesia (Wu et al. 2006). In fact, (+)-opioid agonists are sufficient in themselves for enhancing pain (Hutchinson et al. 2010a). This indicates that (+)-opioid agonists may be capable of inducing microglia activation as well. These findings reveal the existence of a non-classical opioid receptor (Wu et al. 2007) through which morphine might be capable of binding and activating microglia. Subsequently pharmacological blockade and genetic knockout studies have identified the Toll-like receptor 4 (TLR4)/myeloid differentiation factor 2 (MD-2) complex as a non-classical opioid receptor for morphine.

In the CNS, TLR4 is constitutively expressed by microglia, as well as astrocytes (Bsibsi et al. 2002; Lehnardt et al. 2003; Li et al. 2002; Mishra et al. 2006). TLR4 can be activated by microbial associated molecular patterns (MAMP, e.g., lipopolysaccharides (LPS)) (Takeuchi and Akira 2010) and xenobiotic-associated molecular patterns (XAMPs) (Hutchinson et al. 2012). TLR4 works through MyD88 and other adaptor proteins and leads to activation of the IL-1 receptor-associated kinases (IRAKs) and TNF receptor-associated factor-6 (TRAF6), which finally culminates in activation of NFκB and MAPKs and production of pro-inflammatory cytokines (IL-1β, TNFα) as well as NO. (Hameed et al. 2010; Hutchinson et al. 2012; Takeuchi and Akira 2010). Molecular docking and biophysical characterizations show that morphine, morphine-3-glucuronide, methamphetamine, remifentanil, and cocaine bind to the LPS-binding cleft of the TLR4 accessory protein MD-2, and induce pro-inflammatory central immune TLR4 signaling activation (Hutchinson et al. 2007, 2012; Milligan and Watkins 2009).

The p38 mitogen-activated protein kinase (MAPK), which is an important downstream mediator of TLR4 signaling, plays an important role in the acquisition and maintenance of morphine CPP; pharmacological inhibition of p38 by SB203580 in the NAcc microglia blocks morphine-induced CPP (Zhang et al. 2012). Therefore, it is hypothesized that opioids engage inflammatory CNS processes, acting as XAMPs to activate TLR4 signaling, in addition to neuronal targets, thereby potentiating the activity in dopamine reward circuits involved in opioid reward and reinforcement (Hutchinson et al. 2011, 2012).

Acute and chronic alcohol intake also results in region-specific activation of microglia (and astrocytes) in a dose- and time-dependent manner in situ (Coller and Hutchinson 2012; Kovacs 2012; Miguel-Hidalgo 2009). Alcohol induces pro-inflammatory microglia cell activation, including enlargement of their cell bodies and thickening of their processes, as well as over-production of pro-inflammatory factors, such as IL-1β, TNFα, and monocyte chemotactic protein-1 (MCP-1) in prefrontal cortex, cingulate cortex, and VTA (He and Crews 2008; McClain et al. 2011; Miguel-Hidalgo et al. 2002). Recent work in TLR4 genetic knockout mice highlights the essential role of lipid rafts (the microdomains containing an assembly of receptors and glycosphingolipids), TLR4, and their interaction with MD-2 and CD14 accessory proteins in alcohol-induced neuroinflammation (Alfonso-Loeches et al. 2010; Coller and Hutchinson 2012; Kovacs 2012). Ethanol activates primary microglia cells in culture and induces microglia activation in vivo, via MyD88-dependent and MyD88-independent TLR4 signaling pathways (Coller and Hutchinson 2012; Fernandez-Lizarbe et al. 2009; Lewis et al. 2013; Pandey 2012). Alcohol can also trigger a structural association between TLR4/TLR2, thereby contributing to the inflammatory response in microglia during alcohol abuse (Fernandez-Lizarbe et al. 2013). Blocking TLR4 signaling by genetic knock-out of TLR4 or MyD88 or by the non-opioid TLR4 antagonist (+)-naloxone abolish the behavioral responses related to acute and chronic ethanol exposure (Pandey 2012; Pascual et al. 2011; Wu et al. 2012). Taken together, the above evidences suggest that TLR4 is an important molecular target that may play a significant role in the development of alcoholism.

Neurons have been reported to express TLR4 (and the TLR accessory protein MD-2) at low levels (Diogenes et al. 2011; Ferraz et al. 2011), but their functional impact on the mechanisms of drug addiction is unknown (Hutchinson et al. 2012; Okun et al. 2011). This implies that one may be able to selectively target microglia by modulating TLR4 signaling without affecting CNS neurons, which suggests a novel strategy to separate beneficial analgesic effects of opioids from the unwanted tolerance and rewarding side effects. Several novel TLR4 inhibitors are being developed (Wang et al. 2013); if any of them can cross the blood–brain barrier (BBB), they may have efficacy in prevention or attenuation of drug addiction (Coller and Hutchinson 2012). Pharmacological agents that target any of the associated downstream consequences of TLR4 activation may also prove efficacious for the treatment of drug addiction (Coller and Hutchinson 2012).

12.2.2 N-Methyl-D-Aspartate Receptors

The NMDA receptor (NMDAR) family is composed of NMDA R1, NMDA R2A-D, and NMDA R3A-B subunits (Domercq et al. 2013). They form heterotetramers that bind to the major excitatory neurotransmitter, glutamate, and work as voltage-gated ion channels. Accumulating evidences support the involvement of glutamatergic neurotransmission, within the dopaminergic reward circuit, among the mechanisms of drug addiction (Kovacs 2012). Morphine has been shown to up-regulate brain-derived neurotrophic factor (BDNF) in microglia, which up-regulates in turn NMDARs (Kovacs 2012; Ueda and Ueda 2009). Activation of microglia NMDARs triggers the release of pro-inflammatory cytokines (TNFα, IL-1β, IL-6, etc.) (Kaindl et al. 2012), which down-regulates glutamate transporters in astrocyte in vivo and leads to a dysregulation of extracellular glutamate. In vitro microglia also produces quinolinic acid following activation (Espey et al. 1997), which promotes microglia glutamate release through activation of their NMDARs (Kovacs 2012). Down-regulation of glutamate transporters and enhanced glutamate release augment CNS excitotoxicity overall, therefore amplifying the abused drug-induced neuronal activity within the reward circuitry and contributing to opioid withdrawal and dependence (Coller and Hutchinson 2012). In vivo administration of resveratrol, which blocks NMDARs activity and inhibits opioid-induced neuroinflammation in microglia and astrocytes, also attenuates drug tolerance and dependence (Hameed et al. 2010; Tsai et al. 2012). Therefore, NMDARs are considered to play a significant role in drug addiction. However, whether NMDARs activation directly or indirectly controls microglia activation in vivo remains to be determined (Domercq et al. 2013). Therefore, further studies are necessary to characterize in vivo the existence of functional NMDARs in the resident and activated microglia populations, which could respond to the altering glutamate release during drug addiction.

12.2.3 P2X4 Receptor

P2X4 receptor is an ATP-gated ion channel encoded by the P2X purinoceptor 4 gene, which occurs naturally as both homo- and hetero-oligomers (Kaczmarek-Hajek et al. 2012). Microglia express ionotropic purinergic receptor P2X4 (Inoue 2006). Multiple splice variants, often with distinct patterns of expression, have been found for the P2X4 receptor (Kaczmarek-Hajek et al. 2012). The expression of the P2X4 receptor is enhanced in microglia when exposed to morphine (Horvath and DeLeo 2009). Chronic morphine administration induces ATP release in the spinal cord dorsal horn (Horvath et al. 2010), which can activate microglia via P2X4 receptors leading to the release of bioactive substances including cytokines such as TNFα, IL-1β, IL-6, etc. (Horvath et al. 2010; Inoue 2006). Pro-inflammatory factors released from activated microglia contribute to the development of opioid reward and dependence (Coller and Hutchinson 2012; Kovacs 2012). Therefore, modulation of P2X4 receptor expression and activity on microglia may prove an attractive target for the prevention or attenuation of opioid-induced side effects (Horvath and DeLeo 2009; Horvath et al. 2010). However, it should be noted that P2X4 is not exclusively expressed by microglia. Neurons and astrocytes also express P2X4. Further studies are thus required to understand how the microglia-astrocyte and neuron-glia interactions at P2X4 receptors affect the dopaminergic reward circuits and the behaviors of drug abuse.

12.3 Pharmacological Microglia Modulators as a Way to Alter the Behavioral Effect of Drug Abuse

Drug abuse costs the US economy hundreds of billions of dollars in increased healthcare costs, crime, and lost productivity (Mental Health Services Administration 2012). Drug addiction changes the structure of the brain and its functionality, which in turn affect individual personality and behaviors. The impact of addiction can be far reaching. Cardiovascular disease, high blood pressure, diabetes, liver disease, lung disease, stroke, HIV/AIDS, and other health problems can be affected by drug abuse.

Further, drug abuse not only affects the abuser's existence, but also the lives of his/her family and community members. This includes tearing the family apart, loss of employment, failure in school, increasing violence, child abuse, and other crimes. Therefore, there is great need for the development of pharmacological therapies to reverse the behavioral effects of drug abuse. Microglia release neuroinflammatory mediators in response to illicit drugs and have been shown to mediate sensitization and tolerance (Miguel-Hidalgo 2009; Hutchinson et al. 2007, 2011; Milligan and Watkins 2009; Watkins et al. 2009). Therefore, a better understanding of the underlying central immune signaling could provide novel opportunities for developing anti-addiction therapies that specifically target activated microglia (Table 12.1).

Table 12.1 Small molecule modulators of microglia activation which could be used for the treatment of drug addiction

Compound	Structure	Reference
Minocycline	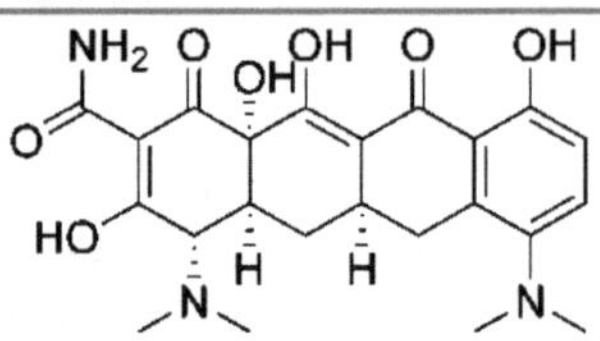	Habibi-Asl et al. (2009), Hutchinson et al. 2008, Zhang et al. (2006)
Ibudilast	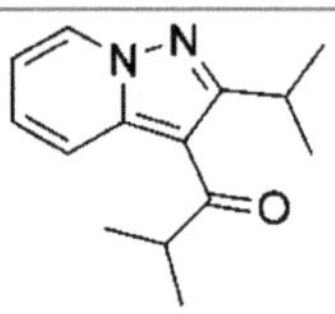	Beardsley et al. (2010), Hutchinson et al. (2009); Ledeboer et al. (2007), Snider et al. (2012, 2013)
(+)-Naloxone	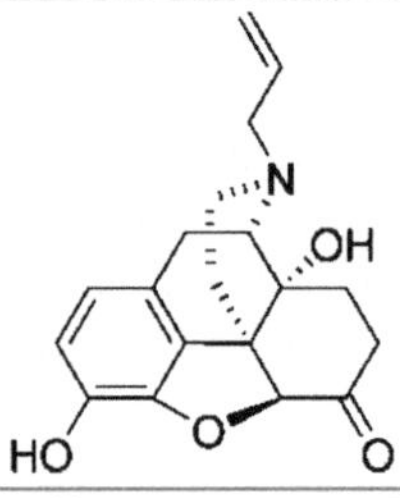	Hutchinson et al. (2012)
Pentoxifylline	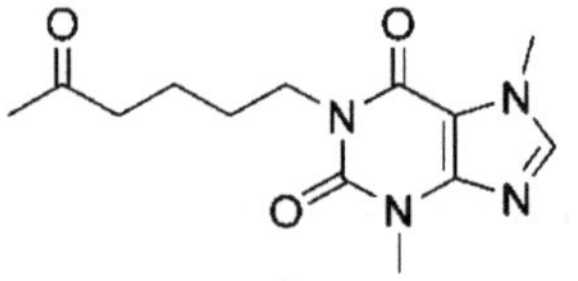	Ciraulo et al. (2005), Yoshikawa et al. (1999)
Propranolol	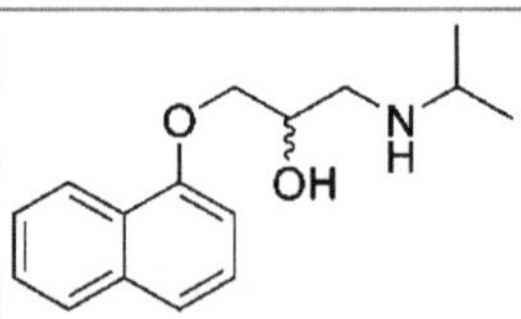	Bernardi et al. (2006); Kampman et al. (2001), Saladin et al. (2013)
Dizocilpine	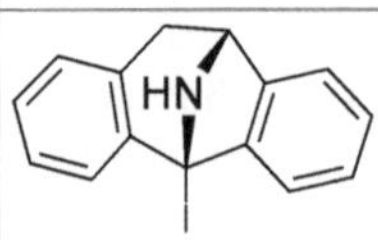	Fan et al. (2012), Hameed et al. (2010), Shu et al. (2008))

12.3.1 Interleukin-10 (IL-10)

Opioids induce microglia activation, which is characterized by the release of pro-inflammatory molecules that contribute, at least in part, to opiate addiction (Hutchinson et al. 2007, 2012; Milligan and Watkins 2009). Among the pro-inflammatory cytokines released by activated microglia are IL-1, IL-6, and TNFα (Watkins et al. 2005). When IL-1, IL-6, and TNFα are injected into the spinal intrathecal (peri-spinal) space of rats in an effort to mimic the presence of cytokines released by activated microglia, each has been found to enhance pain sensitivity, a phenomenon known as hyperalgesia. Additionally, IL-1 and TNFα induce spontaneous and enhanced activity in the responding neurons (Hermann et al. 2001; Reeve et al. 2000). The ability to counter-regulate chronic pain states through morphine-induced microglia activation has set the stage for new research based upon interactions between drugs, microglia, and their effects on neuronal activity. It has been shown that anti-inflammatory factor IL-10 also attenuates morphine craving or relapse and completely knocks out morphine-seeking behavior (Schwarz et al. 2011). Administration of recombinant IL-10 has been found to reduce rats' cravings for morphine (Schwarz et al. 2011). The value of IL-10 therapies has yet to be explored in drug addiction treatment. Currently, Xalud Therapeutics is using XT-101, a microparticle-delivered non-viral gene therapy that drives the production of IL-10 to normalize microglia activity and treat neuropathic pain, multiple sclerosis (MS), and amyotrophic lateral sclerosis (ALS). While these conditions seem unrelated to addiction, this therapeutic approach may reveal the potential of treating a variety of neurological problems, including opioid addiction, by attenuating the neuroinflammatory response generated by activated microglia.

12.3.2 Minocycline

Minocycline is the most lipid-soluble derivative of tetracycline and is known to cross the BBB (Bastos et al. 2012). Minocycline inhibits microglia activation through blockade of NFκB nuclear translocation, inhibition of protein kinase C (PKC), p38 and c-Jun N-terminal kinase (JNK) activities, and suppression of inflammatory cytokines over-production (Bastos et al. 2012). Minocycline suppresses opioid-induced respiratory depression, tolerance, and CPP (Habibi-Asl et al. 2009; Hutchinson et al. 2008; Zhang et al. 2006, 2012). Minocycline was also observed to decrease alcohol consumption in mice thus proposing a new approach for the treatment of alcoholism (Agrawal et al. 2011). Therefore, minocycline may have therapeutic usefulness for reducing addictive behavioral responses associated with drug abuse. However, it should be noted that minocycline also acts on other cell types such as astrocytes, T lymphocytes, and macrophages (Bastos et al. 2012; Brundula et al. 2002; Dutta et al. 2010; Szeto et al. 2011). Most likely the actions of minocycline on peripheral immune cells also play a role in suppressing the unwanted effects of opioids (Bastos et al. 2012).

12.3.3 Ibudilast

Ibudilast (formerly AV-411, now MN-166) attenuates innate immune signaling and can cross the BBB (Ledeboer et al. 2007). It has a complex mechanism of action, including inhibition of phosphodiesterases, TLR4, and macrophage migration inhibitory factor (MIF) (Ledeboer et al. 2007). It reduces pro-inflammatory factor over-production, at least in part by inhibiting TLR4 signaling (Ledeboer et al. 2007). Microglia activation is reduced in a dose-dependent manner by ibudilast with reduction in the levels of LPS-induced NO, reactive oxygen species (ROS), IL-1β, IL-6, and TNFα production, and enhanced production of the anti-inflammatory cytokine IL-10 (Ledeboer et al. 2007). Preclinical data show that ibudilast is well tolerated, is effective via oral administration, and reduces astrocyte and microglia activation markers glial fibrillary acidic protein (GFAP) and cluster of differentiation molecule 11b (CD11b) in vitro (Ledeboer et al. 2007), as well as attenuated morphine and methamphetamine addiction (Beardsley et al. 2010; Hutchinson et al. 2009; Snider et al. 2012, 2013). This CNS-directed anti-inflammatory activity of ibudilast is of potential use in the improved efficacy and safety of opioids and other drugs of abuse such as methamphetamine as it decreases drug reward, dependence, withdrawal, and reinforcement (Snider et al. 2012, 2013).

12.3.4 (+)-Naloxone

(+)-naloxone is the plus enantiomer of the opioid antagonist drug (−)-naloxone. Unlike "normal and natural" (−)-naloxone, (+)-naloxone has no significant affinity for opioid receptors, but instead binds to the TLR4 accessory protein myeloid MD-2 and acts as a selective antagonist of TLR4 (Hutchinson et al. 2008, 2012). Opioids and alcohol activate microglia by TLR4 signaling and chronic use of these drugs causes release of TNFα and IL-1β as well as other downstream effects (Fernandez-Lizarbe et al. 2009, 2013; Hutchinson et al. 2010b, 2012; Lewis et al. 2013; Pandey 2012; Wu et al. 2012). These pro-inflammatory cytokines are thought to contribute to drug tolerance, dependence, and reinforcement by inhibiting TLR4-dependent microglia activation. In addition, it is thought that (+)-naloxone can counteract these effects (Alfonso-Loeches et al. 2010; Hutchinson et al. 2007, 2012). Since (+)-naloxone does not have any apparent binding affinity for the μ-opioid receptor, the desired opioid analgesic effects could be preserved.

12.3.5 Pentoxifylline

Pentoxifylline is a competitive nonselective phosphodiesterase inhibitor which can cross the BBB and reduces cyclic adenosine monophosphate (cAMP) production. This, in turn, decreases release of pro-inflammatory cytokines TNFα and IL-1β and increases production of the anti-inflammatory cytokine IL-10 in cultured microglia (Yoshikawa et al. 1999). Neuroinflammation is a critical component in the development and maintenance of drug abuse (Kovacs 2012). Therefore, pentoxifylline might be used to block drug-induced pro-inflammatory factors and attenuate drug reward and dependence effects (Ciraulo et al. 2005).

12.3.6 *Propranolol*

Cultured microglia express β-adrenergic receptor, which are G-protein-coupled receptors for catecholamines, such as adrenaline and noradrenaline (Tanaka et al. 2002). β-Adrenergic receptor signaling activation increases expression of pro-inflammatory IL-1β, TNFα, and IL-6 in the brain, both at the mRNA and protein levels. Propranolol, a β-adrenergic receptor antagonist, antagonizes pro-inflammatory cytokine production in microglia cells (Wang et al. 2010). Preclinical and clinical evidence shows that propranolol can attenuate cocaine reward, dependence, and withdrawal (Bernardi et al. 2006; Kampman et al. 2001; Saladin et al. 2013). The effect of propranolol on the attenuation of cocaine addiction may be, at least in part, mediated by its suppression of microglia inflammation.

12.3.7 *Dizocilpine (MK-801)*

Dizocilpine (MK-801) is an NMDARs noncompetitive antagonist that reduces pro-inflammatory factors cyclooxygenase-2 (COX-2) and TNFα and prevents microglia activation and neurotoxicity (Thomas and Kuhn 2005). Dizocilpine can reduce methamphetamine-induced activation of microglia in the striatum in vivo (Thomas and Kuhn 2005). Therefore, dizocilpine inhibits addictive behaviors in experimental animals (Fan et al. 2012; Hameed et al. 2010). However, a low dose of dizocilpine (0.05 mg/kg, i.p.) does not impair morphine-induced CPP and delay morphine extinction (Fan et al. 2012). The combined action of a low dose of dizocilpine (0.05 mg/kg, i.p.) and glutamate transporter activation by ceftriaxone (25 mg/kg, i.p.), however, effectively reduces the acquisition of morphine-induced CPP and completely prevents morphine reinstatement (Fan et al. 2012). Dizocilpine injected intrathecally decreases morphine-induced tolerance at the spinal cord level (Hameed et al. 2010; Kest et al. 1993). Dizocilpine's ability to reduce opioid tolerance is, at least in part, attributed to its inhibition of the NMDARs-dependent activation of spinal JNK; this kinase has been shown to be involved in the development of opioid tolerance (Guo et al. 2009).

12.4 Conclusions

In an effort to develop a better understanding of drug addiction, the implication of microglia needs to be considered, given the growing evidence that neuroinflammatory processes generated by these non-neuronal cells profoundly impact reward circuitries. Such processes are capable of modulating neuronal physiology and neuronal circuits of reward and dependence (Fig. 12.2) (Coller and Hutchinson 2012; Kettenmann et al. 2011; Kovacs 2012). Among the CNS immunocompetent cells, resident microglia are emerging as key contributors to the behavioral consequences of drug addiction by promoting neuroinflammation through release of proinflammatory cytokines. This finding invites the development of novel medications that treat drug addiction by modulating microglia activation specifically. In this regard, the small molecule modulators of microglia activation discussed in this chapter have excellent, well-established clinical safety records. Combining opioid agonist activity with the ability to block microglia activation in one small molecule may represent a novel strategy to yield powerful analgesia but with reduced abuse potential. For instance, PTI-609 is a novel agent that activates opioid receptors via a novel binding domain (Burns and Wang 2010) and inhibits the release of proinflammatory cytokines while also activating opioid receptors, thus leading to excellent analgesic action with no apparent potential for addiction (Burns and Wang 2010). This rationale may prove extremely beneficial in the future for the development of powerful but safe analgesics. In closing, it should be noted that microglia modulators also show effects on other CNS immunocompetent cells, such as astrocytes, CNS endothelial cells, and peripheral immune cells such as macrophages, monocytes, and lymphocytes, which can infiltrate the parenchyma in pathological contexts (Coller and Hutchinson 2012). Therefore, further research must also consider the contribution of these other CNS immunocompetent cells and peripheral immune cells to drug addiction.

References

Agrawal RG, Hewetson A, George CM et al (2011) Minocycline reduces ethanol drinking. Brain Behav Immun 25(Suppl 1):S165–S169

Alfonso-Loeches S, Pascual-Lucas M, Blanco AM et al (2010) Pivotal role of TLR4 receptors in alcohol-induced neuroinflammation and brain damage. J Neurosci 30:8285–8295

Bastos LF, de Oliveira AC, Watkins LR et al (2012) Tetracyclines and pain. Naunyn Schmiedebergs Arch Pharmacol 385:225–241

Beardsley PM, Shelton KL, Hendrick E et al (2010) The glial cell modulator and phosphodiesterase inhibitor, AV411 (ibudilast), attenuates prime- and stress-induced methamphetamine relapse. Eur J Pharmacol 637:102–108

Beattie EC, Stellwagen D, Morishita W et al (2002) Control of synaptic strength by glial TNFalpha. Science 295:2282–2285

Belujon P, Grace AA (2011) Hippocampus, amygdala, and stress: interacting systems that affect susceptibility to addiction. Ann N Y Acad Sci 1216:114–121

Bernardi RE, Lattal KM, Berger SP (2006) Postretrieval propranolol disrupts a cocaine conditioned place preference. Neuroreport 17:1443–1447
Brundula V, Rewcastle NB, Metz LM et al (2002) Targeting leukocyte MMPs and transmigration: minocycline as a potential therapy for multiple sclerosis. Brain 125:1297–1308
Bsibsi M, Ravid R, Gveric D et al (2002) Broad expression of Toll-like receptors in the human central nervous system. J Neuropathol Exp Neurol 61:1013–1021
Burns LH, Wang HY (2010) PTI-609: a novel analgesic that binds filamin A to control opioid signaling. Recent Pat CNS Drug Discov 5:210–220
Ciraulo DA, Sarid-Segal O, Knapp CM et al (2005) Efficacy screening trials of paroxetine, pentoxifylline, riluzole, pramipexole and venlafaxine in cocaine dependence. Addiction 100(Suppl 1):12–22
Coller JK, Hutchinson MR (2012) Implications of central immune signaling caused by drugs of abuse: mechanisms, mediators and new therapeutic approaches for prediction and treatment of drug dependence. Pharmacol Ther 134:219–245
Cooper ZD, Truong YN, Shi YG et al (2008) Morphine deprivation increases self-administration of the fast- and short-acting mu-opioid receptor agonist remifentanil in the rat. J Pharmacol Exp Ther 326:920–929
De A, Krueger JM, Simasko SM (2003) Tumor necrosis factor alpha increases cytosolic calcium responses to AMPA and KCl in primary cultures of rat hippocampal neurons. Brain Res 981: 133–142
Diogenes A, Ferraz CC, Akopian AN et al (2011) LPS sensitizes TRPV1 via activation of TLR4 in trigeminal sensory neurons. J Dent Res 90:759–764
Domercq M, Vazquez-Villoldo N, Matute C (2013) Neurotransmitter signaling in the pathophysiology of microglia. Front Cell Neurosci 7:49
Dutta K, Mishra MK, Nazmi A et al (2010) Minocycline differentially modulates macrophage mediated peripheral immune response following Japanese encephalitis virus infection. Immunobiology 215:884–893
Espey MG, Chernyshev ON, Reinhard JF Jr et al (1997) Activated human microglia produce the excitotoxin quinolinic acid. Neuroreport 8:431–434
Fan Y, Niu H, Rizak JD et al (2012) Combined action of MK-801 and ceftriaxone impairs the acquisition and reinstatement of morphine-induced conditioned place preference, and delays morphine extinction in rats. Neurosci Bull 28:567–576
Fernandez-Lizarbe S, Pascual M, Guerri C (2009) Critical role of TLR4 response in the activation of microglia induced by ethanol. J Immunol 183:4733–4744
Fernandez-Lizarbe S, Montesinos J, Guerri C (2013) Ethanol induces TLR4/TLR2 association, triggering an inflammatory response in microglial cells. J Neurochem 126:261–273
Ferraz CC, Henry MA, Hargreaves KM et al (2011) Lipopolysaccharide from Porphyromonas gingivalis sensitizes capsaicin-sensitive nociceptors. J Endod 37:45–48
Goldstein RZ, Volkow ND (2002) Drug addiction and its underlying neurobiological basis: neuroimaging evidence for the involvement of the frontal cortex. Am J Psychiatry 159:1642–1652
Graeber MB (2010) Changing face of microglia. Science 330:783–788
Groenewegen HJ, Vermeulen-Van der Zee E, te Kortschot A et al (1987) Organization of the projections from the subiculum to the ventral striatum in the rat. A study using anterograde transport of Phaseolus vulgaris leucoagglutinin. Neuroscience 23:103–120
Guo RX, Zhang M, Liu W et al (2009) NMDA receptors are involved in upstream of the spinal JNK activation in morphine antinociceptive tolerance. Neurosci Lett 467:95–99
Habibi-Asl B, Hassanzadeh K, Charkhpour M (2009) Central administration of minocycline and riluzole prevents morphine-induced tolerance in rats. Anesth Analg 109:936–942
Hameed H, Hameed M, Christo PJ (2010) The effect of morphine on glial cells as a potential therapeutic target for pharmacological development of analgesic drugs. Curr Pain Headache Rep 14:96–104
He J, Crews FT (2008) Increased MCP-1 and microglia in various regions of the human alcoholic brain. Exp Neurol 210:349–358

Herman JP, Mueller NK (2006) Role of the ventral subiculum in stress integration. Behav Brain Res 174:215–224

Hermann GE, Rogers RC, Bresnahan JC et al (2001) Tumor necrosis factor-alpha induces cFOS and strongly potentiates glutamate-mediated cell death in the rat spinal cord. Neurobiol Dis 8:590–599

Horvath RJ, DeLeo JA (2009) Morphine enhances microglial migration through modulation of P2X4 receptor signaling. J Neurosci 29:998–1005

Horvath RJ, Romero-Sandoval EA, De Leo JA (2010) Inhibition of microglial P2X4 receptors attenuates morphine tolerance, Iba1, GFAP and mu opioid receptor protein expression while enhancing perivascular microglial ED2. Pain 150:401–413

Hutchinson MR, Bland ST, Johnson KW et al (2007) Opioid-induced glial activation: mechanisms of activation and implications for opioid analgesia, dependence, and reward. Scientific World Journal 7:98–111

Hutchinson MR, Zhang Y, Brown K et al (2008) Non-stereoselective reversal of neuropathic pain by naloxone and naltrexone: involvement of toll-like receptor 4 (TLR4). Eur J Neurosci 28: 20–29

Hutchinson MR, Lewis SS, Coats BD et al (2009) Reduction of opioid withdrawal and potentiation of acute opioid analgesia by systemic AV411 (ibudilast). Brain Behav Immun 23:240–250

Hutchinson MR, Lewis SS, Coats BD et al (2010a) Possible involvement of toll-like receptor 4/ myeloid differentiation factor-2 activity of opioid inactive isomers causes spinal proinflammation and related behavioral consequences. Neuroscience 167:880–893

Hutchinson MR, Zhang Y, Shridhar M et al (2010b) Evidence that opioids may have toll-like receptor 4 and MD-2 effects. Brain Behav Immun 24:83–95

Hutchinson MR, Shavit Y, Grace PM et al (2011) Exploring the neuroimmunopharmacology of opioids: an integrative review of mechanisms of central immune signaling and their implications for opioid analgesia. Pharmacol Rev 63:772–810

Hutchinson MR, Northcutt AL, Hiranita T et al (2012) Opioid activation of toll-like receptor 4 contributes to drug reinforcement. J Neurosci 32:11187–11200

Hyman SE, Malenka RC, Nestler EJ (2006) Neural mechanisms of addiction: the role of reward-related learning and memory. Annu Rev Neurosci 29:565–598

Ikemoto S, Panksepp J (1999) The role of nucleus accumbens dopamine in motivated behavior: a unifying interpretation with special reference to reward-seeking. Brain Res Brain Res Rev 31:6–41

Inoue K (2006) The function of microglia through purinergic receptors: neuropathic pain and cytokine release. Pharmacol Ther 109:210–226

Ji K, Akgul G, Wollmuth LP et al (2013) Microglia actively regulate the number of functional synapses. PLoS One 8:e56293

Juni A, Klein G, Pintar JE et al (2007) Nociception increases during opioid infusion in opioid receptor triple knock-out mice. Neuroscience 147:439–444

Kaczmarek-Hajek K, Lorinczi E, Hausmann R et al (2012) Molecular and functional properties of P2X receptors–recent progress and persisting challenges. Purinergic Signal 8:375–417

Kaindl AM, Degos V, Peineau S et al (2012) Activation of microglial N-methyl-D-aspartate receptors triggers inflammation and neuronal cell death in the developing and mature brain. Ann Neurol 72:536–549

Kampman KM, Volpicelli JR, Mulvaney F et al (2001) Effectiveness of propranolol for cocaine dependence treatment may depend on cocaine withdrawal symptom severity. Drug Alcohol Depend 63:69–78

Kest B, Mogil JS, Shamgar BE et al (1993) The NMDA receptor antagonist MK-801 protects against the development of morphine tolerance after intrathecal administration. Proc West Pharmacol Soc 36:307–310

Kettenmann H, Hanisch UK, Noda M et al (2011) Physiology of microglia. Physiol Rev 91:461–553

Kettenmann H, Kirchhoff F, Verkhratsky A (2013) Microglia: new roles for the synaptic stripper. Neuron 77:10–18

Kierdorf K, Erny D, Goldmann T et al (2013) Microglia emerge from erythromyeloid precursors via Pu.1- and Irf8-dependent pathways. Nat Neurosci 16:273–280
Koob GF (2009) Brain stress systems in the amygdala and addiction. Brain Res 1293:61–75
Kovacs KJ (2012) Microglia and drug-induced plasticity in reward-related neuronal circuits. Front Mol Neurosci 5:74
Ledeboer A, Hutchinson MR, Watkins LR et al (2007) Ibudilast (AV-411). A new class therapeutic candidate for neuropathic pain and opioid withdrawal syndromes. Expert Opin Investig Drugs 16:935–950
Lehnardt S, Massillon L, Follett P et al (2003) Activation of innate immunity in the CNS triggers neurodegeneration through a Toll-like receptor 4-dependent pathway. Proc Natl Acad Sci U S A 100:8514–8519
Lewis SS, Hutchinson MR, Zhang Y et al (2013) Glucuronic acid and the ethanol metabolite ethylglucuronide cause toll-like receptor 4 activation and enhanced pain. Brain Behav Immun 30:24–32
Li Y, Ji A, Schäfer MK (2002) Toll-like receptor 4 is expressed by peptidergic presumed nociceptive neurons in rat dorsal root ganglion. Soc Neurosci Abstr 46:19
Li YQ, Xue YX, He YY et al (2011) Inhibition of PKMzeta in nucleus accumbens core abolishes long-term drug reward memory. J Neurosci 31:5436–5446
Lodge DJ, Grace AA (2008) Amphetamine activation of hippocampal drive of mesolimbic dopamine neurons: a mechanism of behavioral sensitization. J Neurosci 28:7876–7882
Lowry CA (2002) Functional subsets of serotonergic neurones: implications for control of the hypothalamic-pituitary-adrenal axis. J Neuroendocrinol 14:911–923
McClain JA, Morris SA, Deeny MA et al (2011) Adolescent binge alcohol exposure induces long-lasting partial activation of microglia. Brain Behav Immun 25(Suppl 1):S120–S128
Substance Abuse and Mental Health Services Administration (2012) Results from the 2011 National Survey on Drug Use and Health: Summary of National Findings, NSDUH Series H-44, HHS Publication No. (SMA) 12-4713. Rockville, MD
Miguel-Hidalgo JJ (2009) The role of glial cells in drug abuse. Curr Drug Abuse Rev 2:72–82
Miguel-Hidalgo JJ, Wei J, Andrew M et al (2002) Glia pathology in the prefrontal cortex in alcohol dependence with and without depressive symptoms. Biol Psychiatry 52:1121–1133
Milligan ED, Watkins LR (2009) Pathological and protective roles of glia in chronic pain. Nat Rev Neurosci 10:23–36
Mishra BB, Mishra PK, Teale JM (2006) Expression and distribution of Toll-like receptors in the brain during murine neurocysticercosis. J Neuroimmunol 181:46–56
Neumann H, Wekerle H (2013) Brain microglia: watchdogs with pedigree. Nat Neurosci 16: 253–255
Okun E, Griffioen KJ, Mattson MP (2011) Toll-like receptor signaling in neural plasticity and disease. Trends Neurosci 34:269–281
Pandey SC (2012) TLR4-MyD88 signalling: a molecular target for alcohol actions. Br J Pharmacol 165:1316–1318
Pascual M, Balino P, Alfonso-Loeches S et al (2011) Impact of TLR4 on behavioral and cognitive dysfunctions associated with alcohol-induced neuroinflammatory damage. Brain Behav Immun 25(Suppl 1):S80–S91
Pascual O, Ben Achour S, Rostaing P et al (2012) Microglia activation triggers astrocyte-mediated modulation of excitatory neurotransmission. Proc Natl Acad Sci U S A 109:E197–E205
Reeve AJ, Patel S, Fox A et al (2000) Intrathecally administered endotoxin or cytokines produce allodynia, hyperalgesia and changes in spinal cord neuronal responses to nociceptive stimuli in the rat. Eur J Pain 4:247–257
Saladin ME, Gray KM, McRae-Clark AL et al (2013) A double blind, placebo-controlled study of the effects of post-retrieval propranolol on reconsolidation of memory for craving and cue reactivity in cocaine dependent humans. Psychopharmacology (Berl) 226:721–737
Schwarz JM, Hutchinson MR, Bilbo SD (2011) Early-life experience decreases drug-induced reinstatement of morphine CPP in adulthood via microglial-specific epigenetic programming of anti-inflammatory IL-10 expression. J Neurosci 31:17835–17847

Shu H, Hayashida M, Huang W et al (2008) The comparison of effects of processed Aconiti tuber, U50488H and MK-801 on the antinociceptive tolerance to morphine. J Ethnopharmacol 117:158–165

Sinha R (2001) How does stress increase risk of drug abuse and relapse? Psychopharmacology (Berl) 158:343–359

Snider SE, Vunck SA, van den Oord EJ et al (2012) The glial cell modulators, ibudilast and its amino analog, AV1013, attenuate methamphetamine locomotor activity and its sensitization in mice. Eur J Pharmacol 679:75–80

Snider SE, Hendrick ES, Beardsley PM (2013) Glial cell modulators attenuate methamphetamine self-administration in the rat. Eur J Pharmacol 701:124–130

Stellwagen D, Beattie EC, Seo JY et al (2005) Differential regulation of AMPA receptor and GABA receptor trafficking by tumor necrosis factor-alpha. J Neurosci 25:3219–3228

Szeto GL, Pomerantz JL, Graham DR et al (2011) Minocycline suppresses activation of nuclear factor of activated T cells 1 (NFAT1) in human CD4+ T cells. J Biol Chem 286:11275–11282

Takeuchi O, Akira S (2010) Pattern recognition receptors and inflammation. Cell 140:805–820

Tanaka KF, Kashima H, Suzuki H et al (2002) Existence of functional beta1- and beta2-adrenergic receptors on microglia. J Neurosci Res 70:232–237

Thomas DM, Kuhn DM (2005) MK-801 and dextromethorphan block microglial activation and protect against methamphetamine-induced neurotoxicity. Brain Res 1050:190–198

Tremblay ME, Majewska AK (2011) A role for microglia in synaptic plasticity? Commun Integr Biol 4:220–222

Tremblay ME, Stevens B, Sierra A et al (2011) The role of microglia in the healthy brain. J Neurosci 31:16064–16069

Tsai RY, Chou KY, Shen CH et al (2012) Resveratrol regulates N-methyl-D-aspartate receptor expression and suppresses neuroinflammation in morphine-tolerant rats. Anesth Analg 115:944–952

Tzschentke TM (2007) Measuring reward with the conditioned place preference (CPP) paradigm: update of the last decade. Addict Biol 12:227–462

Ueda H, Ueda M (2009) Mechanisms underlying morphine analgesic tolerance and dependence. Front Biosci 14:5260–5272

Wang J, Li J, Sheng X et al (2010) Beta-adrenoceptor mediated surgery-induced production of pro-inflammatory cytokines in rat microglia cells. J Neuroimmunol 223:77–83

Wang X, Loram LC, Ramos K et al (2012) Morphine activates neuroinflammation in a manner parallel to endotoxin. Proc Natl Acad Sci U S A 109:6325–6330

Wang X, Smith C, Yin H (2013) Targeting Toll-like receptors with small molecule agents. Chem Soc Rev 42:4859–4866

Watkins LR, Hutchinson MR, Johnston IN et al (2005) Glia: novel counter-regulators of opioid analgesia. Trends Neurosci 28:661–669

Watkins LR, Hutchinson MR, Rice KC et al (2009) The "toll" of opioid-induced glial activation: improving the clinical efficacy of opioids by targeting glia. Trends Pharmacol Sci 30: 581–591

Wu HE, Sun HS, Cheng CW et al (2006) dextro-Naloxone or levo-naloxone reverses the attenuation of morphine antinociception induced by lipopolysaccharide in the mouse spinal cord via a non-opioid mechanism. Eur J Neurosci 24:2575–2580

Wu HE, Hong JS, Tseng LF (2007) Stereoselective action of (+)-morphine over (−)-morphine in attenuating the (−)-morphine-produced antinociception via the naloxone-sensitive sigma receptor in the mouse. Eur J Pharmacol 571:145–151

Wu Y, Lousberg EL, Moldenhauer LM et al (2012) Inhibiting the TLR4-MyD88 signalling cascade by genetic or pharmacological strategies reduces acute alcohol-induced sedation and motor impairment in mice. Br J Pharmacol 165:1319–1329

Yoshikawa M, Suzumura A, Tamaru T et al (1999) Effects of phosphodiesterase inhibitors on cytokine production by microglia. Mult Scler 5:126–133

Youn DH, Wang H, Jeong SJ (2008) Exogenous tumor necrosis factor-alpha rapidly alters synaptic and sensory transmission in the adult rat spinal cord dorsal horn. J Neurosci Res 86: 2867–2875
Zhang L, Kitaichi K, Fujimoto Y et al (2006) Protective effects of minocycline on behavioral changes and neurotoxicity in mice after administration of methamphetamine. Prog Neuropsychopharmacol Biol Psychiatry 30:1381–1393
Zhang XQ, Cui Y, Chen Y et al (2012) Activation of p38 signaling in the microglia in the nucleus accumbens contributes to the acquisition and maintenance of morphine-induced conditioned place preference. Brain Behav Immun 26:318–325

Chapter 13
Aging

Rommy von Bernhardi, Betsi Flores, and Hiroshi Nakanishi

Abstract Microglial cells undergo multiple morphological and immunophenotypic changes during normal aging. Abnormal morphology, which includes fewer and shorter ramifications, beading and spheroid swellings, has been observed particularly in the cerebral cortex, as well as in and around the white matter. In aged animals, microglia express some surface antigens which are not normally present in their young counterparts, in addition to presenting altered motility and phagocytosis. Aged microglia exhibit an aberrant production of pro- and anti-inflammatory mediators, accompanied by an exacerbated inflammatory response to pathological changes, a phenomenon known as microglial "priming." Lysosomal dysfunction and mitochondrial DNA oxidative damage further accumulate in aged microglia, resulting in an increased production of reactive oxygen species. These changes could contribute to mediating the neuronal dysfunction observed during normal aging and facilitate the onset of age-associated cognitive decline, as well as neurodegenerative diseases. In this chapter, we describe microglial aging at the cellular and molecular levels, the implications for diseases, and potential strategies to slow down aging based on preserving lysosomal and mitochondrial function.

Keywords Microglia • Neuroinflammation • Cytokines • Priming • Reactive oxygen species • Oxidative stress • lysosome • Mitochondria • DNA damage • NFκB • Cathepsin B

Bullet Points

- Microglial cells undergo various morphologic and immunophenotipic changes during normal aging.

R. von Bernhardi, M.D., Ph.D. (✉) • B. Flores
Department of Neurology, School of Medicine, Pontificia Universidad Católica de Chile, Marcoleta 391, Santiago, Chile
e-mail: rvonb@med.puc.cl; betsi.flores.c@gmail.com

H. Nakanishi, Ph.D.
Department of Aging Science and Pharmacology, Faculty of Dental Sciences, Kyushu University, Fukuoka 812-8582, Japan
e-mail: nakan@dent.kyushu-u.ac.jp

M.-È. Tremblay, A. Sierra (eds.), *Microglia in Health and Disease*,
DOI 10.1007/978-1-4939-1429-6_13

- Aged microglia show an impaired motility towards damage signals, probably reflecting a decreased surveillance capacity.
- Microglial proliferation and phagocytosis of different types of cargo are affected with aging.
- Aged microglia exhibit an altered expression of cytokines and exacerbated inflammatory response, a phenomenon known as microglial "priming."
- Lysosomal dysfunction and mitochondrial DNA oxidative damage also accumulate in microglia during aging, resulting in the increased production of reactive oxygen species (ROS) and activation of the microglial inflammasome.
- These changes are probably both cell autonomous and the result of an altered aging brain milieu.
- The overall contribution of aging microglia to the onset of cognitive dysfunction and neurodegenerative diseases remains to be experimentally tested.

13.1 Introduction

Aging (or senescence) is a complex process of cumulative changes affecting an organism with the passage of time. Aging increases the vulnerability to death and is a primary risk factor for major human pathologies, including cancer, diabetes, cardiovascular disorders, and neurodegenerative diseases. It is characterized by a progressive loss of physiological integrity, leading to impaired function in various levels and systems of the organism, such as skeletal muscles (Klein et al. 2001; Thompson 2009), cardiovascular, endocrine, respiratory processes (Fadel et al. 2004; Lipsitz 2002; Smith et al. 2005), and central nervous system (CNS) activity (Smith et al. 2005). These changes are generally accompanied by alterations in behavior, associated with increased tremor, loss of balance control, and decreased walking proficiency (Glenn et al. 2004; Lipsitz 2002; Lipsitz and Goldberger 1992). The ability to perform complex dual learning tasks, such as memorizing word lists while walking, also significantly decreases during aging (Lindenberger et al. 2000; Salat et al. 2005), even though many elderly people preserve to a certain degree their cognitive abilities (Shock et al. 1984).

At the cellular level, senescence is characterized by the accumulation of DNA damage, oxidative stress, chronic inflammatory activity, and an imbalance between the levels of pro- and anti-inflammatory cytokines in various tissues, including the brain. Cellular aging is also associated with the shortening of telomeres and the activation of tumor suppressor genes (reviewed in Lopez-Otin et al. 2013). The potentially damaging elements may be produced by the organism itself (e.g., cytokines, radical species, eicosanoids, among other mediators) or derived from the prolonged exposure to physical, chemical, or biological agents (e.g., ionic radiation, pollutants, pathogens; see Chap. 6 for further reading) (Droge and Schipper 2007; Vijg and Campisi 2008). In addition, some responses of the immune system particularly decline with age, increasing the susceptibility to infections and cancer, whereas other responses are exacerbated, facilitating the onset of autoimmune diseases (Yung and Julius 2008). As the blood-brain barrier (BBB) undergoes several changes during aging (Marques et al. 2013), it is possible that aging leads to an increased surveillance of the brain parenchyma by peripheral monocytes and lymphocytes, which could further contribute to the aging process. In fact, the expression levels of chemotactic molecules such as interferon-inducible protein 10 (IP-10) and monocyte chemotactic protein-1 (MCP-1), but also the infiltration of $CD11b^{+}$ $CD45^{high}$ cells identified as monocytes, were shown to be increased ex vivo in hippocampal tissue prepared from aged rats (Blau et al. 2012; Enciu et al. 2013).

The CNS also undergo pronounced structural and functional alteration during normal aging, even in clinically healthy middle-age individuals, i.e., 40–50 years old. In particular, brain weight decreases in the order of 2–3 % per decade after the age of 50 and accelerates in later years to reach 10 % at the age of 80 (Drachman 2006). Using magnetic resonance imaging (MRI) and voxel-based morphometry (VBM), it has been shown that the gray matter volume of prefrontal, parietal, and temporal cortices of the human brain progressively decreases during aging (Courchesne et al. 2000; Ge et al. 2002; Good et al. 2001; Jernigan et al. 2001; Salat et al. 2004; Sowell et al. 2003). The temporal cortex is more affected in the left hemisphere than in the right hemisphere, in agreement with an age-related decline in language functions (Sowell et al. 2003). White matter volume also decreases with age; the process begins later but progresses at a more accelerated rate than in the gray matter (Courchesne et al. 2000; Ge et al. 2002; Jernigan et al. 2001).

Some of these changes are undoubtedly related to cell autonomous alterations in the aging neurons (Jurk et al. 2012) and astrocytes (Sheng et al. 2013), but aging microglia may further contribute to the aging pathology, notably by their production of reactive oxygen species (ROS) and pro-inflammatory cytokines, which could together increase neuronal vulnerability to oxidative stress. An accepted view is that neuroinflammation and oxidative stress collectively induce neuronal dysfunction and degeneration, thus resulting in the decline of motor and cognitive functions during aging (Forster et al. 1996; Navarro et al. 2002). In addition, compromised microglial properties have been proposed to cause impaired reaction to neuronal abnormalities during aging (Streit 2006; Aguzzi et al. 2013; Kettenmann et al. 2013; Conde and Streit 2006b; Siskova and Tremblay 2013). In this chapter, we will discuss the cellular and molecular changes observed in senescent microglia, and their implication for our understanding of the normal aging process, related cognitive dysfunction, and brain diseases.

13.2 Changes in Microglial Morphology, Dynamics, Phagocytosis, and Proliferation During Aging

Several authors have shown that microglial cells display various changes in morphology and functional behavior over the course of normal aging. Morphological characteristics associated with normal aging include fewer and shorter ramifications, excessive beading, and formation of spheroid swellings (Conde and Streit 2006b; Flanary 2005; Streit 2006; Streit et al. 2004). These changes are commonly referred to as microglial cell "dystrophy" (Streit 2006; Streit et al. 2004). Moreover, it was shown that microglia often colocalize with neurodegenerating neurons in the aging brain and display additional deterioration such as higher incidence of clumping, irregular distribution, and accumulation of phagocytic debris (lysosomal lipopigments, cellular elements, vacuoles, and large vesicles), particularly observed in cortical areas, as well as in and near the white matter (Hart et al. 2012; Perry et al. 1993; Tremblay et al. 2012; Hefendehl et al. 2013). Microglial accumulation of phagocytic debris might contribute to reducing their dynamism and impairing their phagocytic capacity as discussed below. However, no systematic quantification of microglial interactions with neurons has been performed to determine the extent of their changes during aging. Furthermore, no study has yet clearly determined the functional consequences of their abnormal morphology.

In terms of functional behavior, live imaging of microglial cells in retinal explants has demonstrated ex vivo that the dynamic responses of senescent microglia to injury also show age-dependent variations (Damani et al. 2011). In particular, young microglia were shown to rapidly increase their motility and number of ramifications when exposed to the nucleotide ATP, an injury-associated signal, or to a laser-induced focal tissue injury. In contrast, aged microglia were less dynamic and ramified, as compared to younger counterparts, and became even less dynamic and ramified in the presence of ATP, resulting in slower responses to a laser-induced injury. Moreover, their migration away from the site of injury was retarded in senescent versus young microglia (Damani et al. 2011). A recent characterization of the changes in microglial morphology and dynamic behavior in vivo using two photon imaging also showed similar age-related processes, such as a shortening of processes, increased soma volume, and loss of homogeneous tissue distribution and surveillance rate, in the cerebral cortex of aged mice. In addition, aged microglia examined in vivo also presented a diminished dynamic response to a laser-induced tissue injury, as in the retina, but their migration was however found to be accelerated (Hefendehl et al. 2013). Together, these findings suggest that the microglial capacity to detect and respond to pathological signs might be compromised in the aging brain, probably resulting both from altered microglial properties, especially their motility, and from the altered brain microenvironment.

This altered motility of microglia also seems to be closely related to additional alterations in their phagocytic capacity. As the brain professional phagocytes, microglia have the capacity to engulf apoptotic cells, myelin and axonal debris, deposits of extracellular proteins including beta amyloid (Aβ), and neurites (reviewed in Sierra et al. 2013). During aging, microglial cells over-express ED1, the rodent equivalent of CD68, a lysosomal protein upregulated during inflammation, which has been associated with phagocytosis (Perry et al. 1993). However, the function of ED1 is unknown since reducing ED1 expression with anti-ED1 monoclonal antibodies in cultured macrophages did not impair phagocytosis (reviewed in Sierra et al. 2013). Acutely isolated microglial cells from aged mice also show a decreased ability to phagocytose Aβ, contrarily to microglia derived from young mice (Floden and Combs 2011). In vitro, the phagocytosis of Aβ was also comparable between microglia derived from young and old mice, but only enhanced by a bacterial lipopolysaccharide (LPS) challenge in microglia from young mice

(Tichauer et al. 2014). In addition, once internalized, the proteolytic degradation of Aβ was shown to be impaired in aged mice, due to deficits in lysosome acidification, as required for proper functioning of the lysosome degradation enzymes (Majumdar et al. 2007). Nevertheless, microglial phagocytosis of newborn apoptotic cells in the adult hippocampus neurogenic niche remained functional at least until 12 months of age (Sierra et al. 2010) (see Chap. 10 for further reading), although no systematic observations were carried out in older ages. Thus, the extent to which microglial phagocytosis is impaired during aging remains to be fully determined. A related process, protein homeostasis or proteostasis, which involves chaperone-mediated protein folding and stability, protein trafficking, protein degradation, and autophagy pathways, is also impaired in aged microglia in vitro, possibly explaining the microglial accumulation of phagocytic debris observed at the ultrastructural level (Tremblay et al. 2012). A major consequence of this declining proteostasis is the aggregation of abnormal proteins, which has been linked to the pathogenesis of neurodegenerative diseases such as Parkinson's disease (PD) and Alzheimer's disease (AD) (Taylor and Dillin 2011).

The proliferative capacity of microglia could also become altered during aging. Microglial cells exhibit an increased density in the facial nucleus upon nerve injury in aged versus young rats, and a similar finding was observed in mouse cerebral cortex during age-related impairment of audition and vision, possibly suggesting a less regulated proliferative response (Conde and Streit 2006a; Tremblay et al. 2012). However, microglial replication is generally considered as being reduced during aging, and there is no significant evidence for an overall increase in microglial density in the aging postmortem human brain (VanGuilder et al. 2011). It has been speculated that a reduced microglial replication could induce the depletion of healthy microglia in the aged brain, shifting the balance towards a more senescent and dysfunctional population (Mosher and Wyss-Coray 2014). Clumped microglia have been observed in the aged cerebral cortex, but their possible colocalization with markers of proliferation was not examined, and alternative interpretations were also proposed, such as a breakdown of the mechanisms which regulate their territorial organization (Tremblay et al. 2012). Additionally, interpreting the increased microglial density observed in some studies is limited by the impossibility to distinguish microglia from monocytes/macrophages derived from the periphery using CX3CR1, ionized calcium-binding adapter molecule 1, and other commonly used immunocytochemical markers. Thus, it remains unclear whether there are regional or species-specific changes in microglial density in the aging brain, and if so, whether they result from the proliferation of resident microglial cells, or the infiltration of peripheral inflammatory cells.

Together, these observations suggest the appearance of dysfunctional microglial phenotypes in the aging brain, which combined with the immunophenotypic changes described below, might contribute to the age-associated neuronal dysfunction and cognitive decline.

13.3 "Priming" of Microglia During Aging

During normal aging, a decreased secretion of the anti-inflammatory cytokine interleukin (IL)-10 has been observed (Ye and Johnson 2001), found to be accompanied by increased levels of pro-inflammatory cytokines, such as tumor necrosis factor α (TNFα) and IL-1β in the CNS (Lukiw 2004; Streit et al. 2004), and IL-6 in plasma (Godbout and Johnson 2004; Ye and Johnson 1999, 2001), using gene expression profiling, flow cytometry, and ELISA in mice and human. Aged microglia studied in situ (Sheng et al. 1998) and isolated ex vivo (Njie et al. 2012; Sierra et al. 2007) also showed increased expression of several pro-inflammatory mediators, but reports regarding the levels of anti-inflammatory cytokines, such as IL-10, are less uniform (Sierra et al. 2007; Ye and Johnson 2001). Furthermore, microglia display a significant upregulation of several Toll-like receptors (TLRs), such as TLR1, TLR2, TLR4, TLR5, and TLR7, as well as an increased expression of the TLR4 co-receptor, CD14, during aging (Letiembre et al. 2007). An age-related alteration in the signal transduction of TLR4 and conspicuous changes in the expression profile of scavenger receptors (SRs) have also been reported (Hickman et al. 2008; Yamamoto et al. 2002). TLRs, CD14, and SRs are pattern recognition receptors (PRRs) that participate in host defense response and phagocytosis of pathogen-associated molecules pattern (PAMPs), as well as damage-associated molecules pattern (DAMPs), which are crucial for the innate immune response. Signaling through these receptors is accompanied by microglial cell activation, including their production of pro-inflammatory mediators, and uptake of pathogens and macromolecules, such as the neurotoxic peptide Aβ. Therefore, changes in the expression profile of these receptors might account for the alterations observed in microglial inflammatory profile during normal aging.

Additionally, increased mRNA levels of transforming growth factor β isoform 1 (TGFβ1) have been measured in microglia from aged mice and rats (Bye et al. 2001; Sierra et al. 2007). TGFβ1 is a potent regulator of cytotoxicity and inflammatory response in the CNS. Its downstream signaling involves members of the Smad family, i.e., intracellular proteins that transduce extracellular signals from TGFβ ligands to the nucleus, thus acting as transcription factors. These proteins are homologs of the *Drosophila* mothers against decapentaplegic (MAD) protein and *Caenorhabditis elegans* SMA protein, named after the gene *Sma* for small body size, acting as mitogen-activated protein kinases (MAPKs), although their activation is highly variable and dependent on the cell type (Schmierer and Hill 2007). TGFβ1 modulates the activation of microglial cells induced by a LPS challenge, by decreasing their production of pro-inflammatory molecules, with the consequence of protecting cultured neurons from neurotoxicity and oxidative stress (Herrera-Molina and von Bernhardi 2005; Hu et al. 1995; Lieb et al. 2003). It has also been demonstrated in culture that this influence of TGFβ1 on microglial activation is regulated in a Smad3-dependent manner (Le et al. 2004; Werner et al. 2000). The TGFβ1 and Smad3 pathways were further linked to the reduction of radical species production induced by inflammatory stimuli and to the induction of Aβ phagocytosis in vitro

(Tichauer and von Bernhardi 2012). Additionally, it has recently been shown that the induction of the Smad3 pathway is decreased in normal aging under inflammatory conditions (Tichauer et al. 2014), which could explain, at least partially, that microglial activation is overall increased in the aging brain, even though microglial expression of TGFβ1 is concomitantly increased ex vivo (Sierra et al. 2007).

The over-production of pro-inflammatory cytokines is associated with a repertoire of symptoms commonly known as sickness behavior, an adaptive response that occurs following exposure to infectious microorganisms, and that is exacerbated during aging (Hart 1988). Upon systemic inflammatory stimulation, aged microglia display an exacerbated inflammatory phenotype, compared with young ones, possibly resulting in enhanced sickness behavior (Combrinck et al. 2002; Cunningham et al. 2005; Godbout et al. 2005; Sierra et al. 2007). In particular, systemic inflammation resulted in an exacerbated production of the pro-inflammatory cytokines IL1-β, IL-6, TNFα ex vivo in aged versus young microglia (Sierra et al. 2007). This exacerbated response to inflammatory challenges is also referred to as microglial "priming," a concept first introduced by Perry and colleagues (Perry 2004; Perry et al. 1993). By definition, primed microglia undergo a phenotypic shift towards a more sensitized state, in which they respond more rapidly and to a greater extent to a secondary "triggering" stimulus than non-primed microglia (Perry 2004; Perry et al. 2003, 2007; Harry 2013). An important question is to what extent aging microglial cells do become intrinsically dysfunctional, versus simply react to the changes in their local aging brain environment. Importantly, aging microglial properties were replicated in a mouse model where only neurons are made senescent. In these mice considered as a model of accelerated senescence, deleting the expression of a nucleotide repair protein (Ercc1) exclusively in forebrain neurons results in decreased neuronal plasticity, progressive neuronal pathology, and learning impairment (Borgesius et al. 2011). Importantly, in spite of not carrying the mutation, microglia also displayed hallmark features of "priming" such as an exaggerated response to peripheral LPS exposure in terms of pro-inflammatory cytokines expression, ROS production, and phagocytosis (Raj et al. 2014). Therefore, the exacerbated response of "primed" microglia to inflammatory stimuli could result, at least partially, from an accumulation of neuronal genotoxic stress, in addition to changes in the expression of TLRs (Letiembre et al. 2007) and other alterations including shortening of telomeres (Flanary and Streit 2004; Flanary et al. 2007). In turn, the exaggerated inflammatory response of microglia could further enhance the neuronal dysfunction and sickness behavior associated with normal aging.

13.4 Aged Microglia and Their Relationship with Neurodegeneration

While it remains unclear whether microglial "priming" could directly result from microglial senescence, aging microglial cells have been linked to several age-related neurodegenerative diseases, including PD, amyotrophic lateral sclerosis (ALS), and AD (von Bernhardi 2007) (for further reading, refer to Chap. 18). Microglial "priming" could not only result in an increased inflammatory response and cytotoxicity, but also in the impairment of microglial neuroprotective functions (see Chap. 5 for further reading). Indeed, aged microglia appear to actively participate to the neuronal damage observed in neurodegenerative diseases, especially through their production of ROS (Block et al. 2007). Thus, inflammation, possibly related to the activity of microglia, has been suggested to contribute to the death of dopaminergic neurons in PD, forebrain neurons in AD, and motor neurons in ALS (Boillee et al. 2006; Mount et al. 2007). In particular, it has been shown that TNFα promotes PD progression (McCoy et al. 2006), whereas the absence of TNF receptor 1 protects against AD- and PD-like disease in mice (He et al. 2007; Sriram et al. 2002). Moreover, administration of the anti-inflammatory derivative of thalidomide, lenalidomide, which was accompanied by a reduced expression of TNFα and IL-1β, was shown to improve motor behavior even after the onset of symptoms and to extend the survival in mouse models of ALS (Neymotin et al. 2009). Nonetheless, microglia are not likely the sole producers of TNFα and IL-1β in these diseases, and more specific experimental approaches are necessary to dissect out the effects of other inflammatory cells, including infiltrating macrophages, at different stages of their time course.

Altered responses of the aging microglia have also been linked to AD. In rhesus monkeys, a microinjection of fibrillar Aβ in the cortex was found to trigger neuronal loss, tau phosphorylation, and microglial cell proliferation in aged but not young adult monkeys. This in vivo observation suggests that Aβ neurotoxicity is a pathological response specific to the aging brain (Geula et al. 1998). Moreover, aged microglia are less capable of phagocytosing (Floden and Combs 2011) and degrading Aβ (Majumdar et al. 2007) than young microglia, as discussed above. Microglial cell reactivity to Aβ and phagocytic activity is further modulated by astrocytes, at least in vitro, whose presence attenuates the cytotoxic response of cultured microglia (DeWitt et al. 1998; von Bernhardi and Ramirez 2001). However, this modulation was not observed in "primed" microglia exposed to Aβ (von Bernhardi and Eugenin 2004), which showed increased cytotoxicity, Aβ precursor protein (APP) synthesis, Aβ aggregation, and impaired uptake and degradation of Aβ as compared with non-activated microglia (Rogers et al. 2002; Ramirez et al. 2008; von Bernhardi et al. 2007). TGFβ1 secreted by hippocampal neurons and astrocytes has been identified as an important modulatory cytokine of microglial activation, attenuating the release of pro-inflammatory mediators (Chen et al. 2002; Herrera-Molina and von Bernhardi 2005; Mittaud et al. 2002; Herrera-Molina et al. 2012) and promoting microglia-mediated Aβ phagocytosis and degradation (Wyss-Coray et al. 2001). It has been recently shown that these effects are mediated by Smad3-dependent mechanisms as described above (Flores and von Bernhardi 2012; Tichauer and von Bernhardi 2012). Interestingly, this signaling pathway is impaired in the brains of AD patients and mouse models, resulting in Aβ accumulation, Aβ-induced neurodegeneration, and neurofibrillary tangle formation (Tesseur et al. 2006; Ueberham et al. 2006), even though TGFβ1 levels are elevated in the cerebrospinal fluid of these patients (Blobe et al. 2000). Therefore, the Smad3 pathway could be considered as a target for therapeutic approaches against AD.

13.5 Increased Mitochondrial DNA Damage and Resultant Over-production of Reactive Oxygen Species and Inflammatory Cytokines by Microglia During Aging

In the aging brain, microglia constitute a primary cellular source of inflammatory molecules and oxidative products (Hayashi et al. 2008; Pawate et al. 2004; Qin et al. 2005). In the hippocampus of aged mice, immunoreactivity for 8-oxo-deoxyguanosine (8-oxo-dG), a major DNA peroxidation product, has been mainly observed in microglial cells, and partially in neurons, but not in astrocytes (Hayashi et al. 2008). Furthermore, this immunoreactivity for 8-oxo-dG mainly colocalized with the cytochrome *b*, a marker of mitochondria, thus suggesting that microglia could accumulate oxidative damage to their mitochondrial DNA (mtDNA) over the course of aging. Mitochondrial DNA (mtDNA) serves to encode components of the mitochondria electron transfer complexes and is highly susceptible to oxidative damage due to its close proximity to ROS generated through the respiratory chain, and its paucity of protective histones and DNA-binding proteins. Accumulation of mtDNA damages during aging results in a reduced expression of the mitochondria electron transfer complexes, especially the complexes I and IV, which contain a relatively large number of mtDNA-encoded subunits. Reduced activity of the complex I further increases the generation of ROS (Corral-Debrinski et al. 1992; Lin et al. 2002), thus forming a vicious cycle in the mitochondria (Kang et al. 2007) (Fig. 13.1).

In parallel, several changes induced by the aging environment, such as an increase in systemic inflammation and BBB permeability, as well as dysfunction, oxidative stress and degeneration of the other resident cells including neurons and astrocytes (Fig. 13.2), could further contribute to the production and release of ROS. In aged animals, several studies have proposed that BBB permeability could increase (Blau et al. 2012; Enciu et al. 2013), and therefore, a possible production of ROS by peripheral immune cells in the aged brain must be considered. Lastly, neuronal cells could be implicated as well, since DNA damage to cortical, hippocampal, and peripheral neurons (from the myenteric plexus) has been found to induce their production of ROS and release of the pro-inflammatory cytokine IL-6 (Jurk et al. 2012). A similar role of human astrocytes was also recently suggested from in vitro observations (Sheng et al. 2013).

The over-production of ROS, through a vicious cycle in the aging mitochondria, might also activate redox-sensitive transcription factor NFκB, implicated in the regulation of immunity, inflammation, and cell death (Adler et al. 2007, 2008), thereby provoking excessive inflammation in the aged brain (Hayashi et al. 2008; Nakanishi and Wu 2009) (Fig. 13.1). Increased NFκB signaling during aging further potentiates the expression of NLRP3, a member of the NLR family of cytosolic pattern recognition receptors that control the activity of caspase-1 by forming multi-protein complexes which are termed inflammasomes. After being activated, the pyrin domain containing-3 protein (NLRP3) recruits the adaptor protein ASC, which in turn binds to pro-caspase-1, leading to its autocatalytic processing and activation. Active caspase-1 cleaves the inactive precursors of two inflammatory

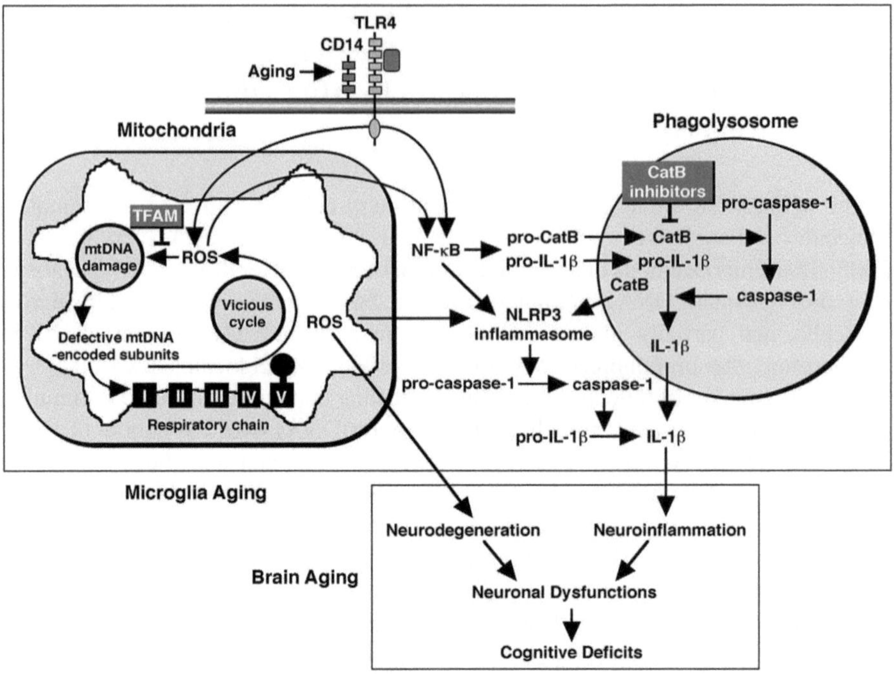

Fig. 13.1 Schematic representation of the "Microglia-Aging" hypothesis. Increased intracellular ROS activate the redox-sensitive NFκB to provoke excessive inflammation in the aged brain. During aging, the NFκB pathway is activated through two different pathways, the mitochondria-derived ROS-dependent pathway and the direct ROS-independent pathway. The activation of the NFκB pathway induces the expression of pro-CatB, pro-IL-1β and NLRP3. CatB and the NLRP3 inflammasome are involved in the activation of pro-caspase-1, an essential enzyme for the proteolytic processing of pro-IL-1β. Therefore, TFAM and CatB inhibitors inhibit the IL-1β-producing pathways in microglia, thereby improving the age-dependent deficits in memory

cytokines, IL-1β and IL-18, into their mature forms (Tschopp and Schroder 2010). NLPR3 has been proposed to be activated by several danger signals, including PAMPs and DAMPs, via three different models: the "ROS model," the "lysosomal rupture model," and the "channel model" (Tschopp and Schroder 2010). In the "channel model," NLRP3 is activated by extracellular ATP released from damaged cells, which binds to the nucleotide receptor P2X7, and triggers a rapid efflux of K^+ and the formation of a pore in the cell membrane, leading to the entry of extracellular factors acting as NLRP3 ligands (including PAMPs and DAMPs). Herein, we will focus on the other two models, which have been associated with the altered function of microglia during aging.

According to the "ROS model" proposed by Tschopp's group, particulate activators of the NLRP3 inflammasome, including asbestos fibers and silica crystals, trigger the generation of short-lived ROS, whereas treatment with various ROS scavengers blocks NLRP3 activation in response to these particulate activators. Monosodium urate crystal and asbestos fiber, which are major causative factors of gout and asbestosis,

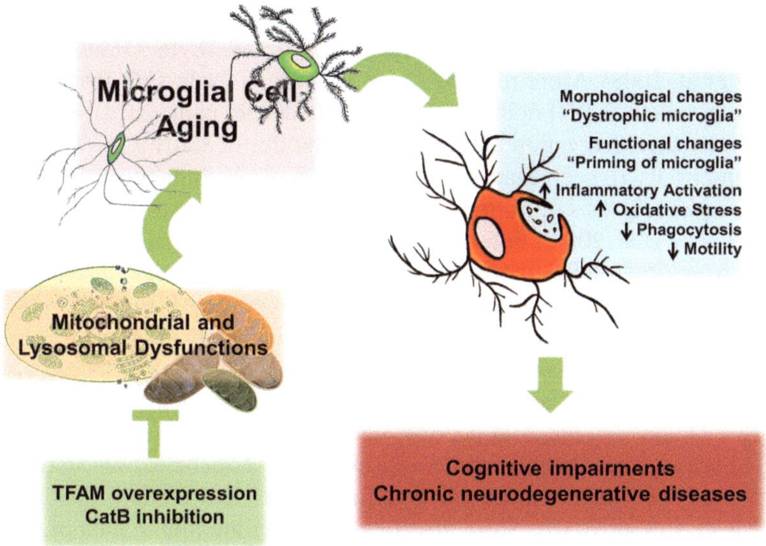

Fig. 13.2 Schematic representation of microglial cell changes during aging. Aged microglia show alterations in their morphology and function, which can lead to neuronal damage and cognitive impairments, facilitating the onset of chronic neurodegenerative diseases. Mitochondrial and lysosomal dysfunctions observed during aging, which are modulated by the over-expression of TFAM and the inhibition of CatB, could represent the cellular basis of these changes. Therefore, TFAM and CatB could be considered as potential targets for the development of pharmacological interventions against aging-associated impairments

respectively, activate the NLRP3 inflammasomes in a ROS-dependent manner (Dostert et al. 2008). Recent studies have demonstrated that autophagic uptake capacity can regulate mitochondrial integrity, ROS production, and subsequent NLRP3 activation (Nakahira et al. 2011; Salminen et al. 2012; Zhou et al. 2011). Furthermore, NLRP3 activation is negatively regulated by autophagy, which plays an important role in clearing the damaged ROS-hypergenerating mitochondria. During aging, the efficiency of autophagic uptake declines and waste materials accumulate within cells (Salminen et al. 2012). Furthermore, the inhibition of autophagy triggers the accumulation of damaged ROS-hypergenerating mitochondria, which augments the activation of the NLRP3 inflammasomes in human macrophages (Zhou et al. 2011). As discussed above, a dysfunction of autophagy has been reported in microglia during aging (reviewed in Wong 2013). Therefore, in addition to the activation of redox-sensitive NFκB, it is also hypothesized that the dominance of ROS-hypergenerating mitochondria, due to the dysfunction of autophagy, could contribute to activating the NLRP3 inflammasome in microglia during aging, leading to excessive production of IL-1β and IL-18 in the aged brain (Fig. 13.1).

On the other hand, according to the "lysosomal rupture model" proposed by Latz's group, the uptake of fibrillar $A\beta_{42}$ or silica crystals by LPS-primed microglia/macrophages causes phagosomal destabilization and lysosomal rupture. The subsequent secretion of cathepsin B (CatB), a typical lysosomal cysteine protease, into the cytoplasm triggers the activation of the NLRP3 inflammasome directly or indirectly, leading to the production and secretion of IL-1β and IL-18 (Halle et al. 2008; Hornung et al. 2008). More recently, CatB was found to directly interact with the leucine-rich-repeat (LRR) domain of NLRP3 (Bruchard et al. 2013). This model is supported by the observation that a specific inhibitor of CatB, CA074Me, significantly inhibits IL-1β secretion from LPS-primed microglia and macrophages following the phagocytosis of fibrillar Aβ and silica crystals, respectively, (Halle et al. 2008; Hornung et al. 2008). Following the phagocytosis of fibrillar Aβ, the secretion of IL-1β from CatB-deficient macrophages is significantly reduced compared with wild-type macrophages (Hornung et al. 2008). Furthermore, NLRP3-deficient mice carrying mutations associated with familial AD demonstrate improvement in spatial memory, a reduced expression of caspase-1 and IL-1β in the brain, and enhanced Aβ clearance (Heneka et al. 2013). Besides fibrillar Aβ and silica crystals, cholesterol crystals and islet amyloid peptide, which are major causative factors of age-related diseases such as atherosclerosis and type 2 diabetes, respectively, also activate the NLRP3 inflammasome in a CatB-dependent manner (Duewell et al. 2010; Masters et al. 2010). More direct evidence on the importance of the "lysosomal rupture model" will require identification of the putative CatB substrates that activate the NLRP3 inflammasome. Phagocytosed particles that are too large to be efficiently cleared are likely to induce the production of ROS on their way to lysosomes. Therefore, the "lysosomal rupture model" could be viewed as part of a more general "ROS model." It is likely that the activation of the NLRP3 inflammasome is more complex, requiring a combination of factors, including enzymatic activity of CatB and ROS activity.

13.6 Preventing or Reversing Microglia Aging by Inhibition of Cathepsin B

In addition to mediating the maturation of pro-caspase-1 through activation of the NLRP3 inflammasome, CatB also directly contributes to the proteolytic maturation of pro-caspase-1. CatB can efficiently cleave pro-caspase-11 in a cell-free system at a neutral pH, but only cleaves pro-caspase-1 at an acidic pH (Vancompernolle et al. 1998). Further cleavage is necessary for the full maturation of pro-caspase-1 after its proteolytic cleavage by CatB, because the fragments generated by CatB cleavage are still larger than the mature caspase-1 (Hentze et al. 2003). This suggests that CatB is involved in the activation of pro-caspase-1 through its direct activation of pro-caspase-11, which in turn activates pro-caspase-1 (Kang et al. 2000). CatB deficiency and selective pharmacological inhibition with CA074Me prevent the activation of precursor forms of IL-1β and IL-18 in microglial cell cultures following treatment with chromogranin A (CGA), a potent activator of microglia, through inhibition of proteolytic maturation of pro-caspase-1 (Terada et al. 2010). CGA does not induce leakage of CatB in microglia (Sun et al. 2012; Wu et al. 2013), but it is known to activate microglia through scavenger receptors class-A (SRA) (Hooper et al. 2009).

CatB-containing enlarged lysosomes are considered to be phagolysosomes formed by the fusion of SRA-mediated phagosomes and primary lysosomes (Sun et al. 2012; Wu et al. 2013). Therefore, pro-caspase-1 and the inactive forms of IL-1β and IL-18 in the cytoplasm may be trapped in phagosomes, which are fused with CatB-containing primary lysosomes to form phagolysosomes and thus degraded rather than being released extracellularly. It is also noted that CatB is increased in the brain during aging (Nakanishi 2003). Therefore, a pharmacological inhibition of CatB could be a potent strategy for slowing brain aging through inhibition of the pro-caspase-1 activation in microglia, and resultant reduction of neuroinflammation (Fig. 13.1).

13.7 Preventing or Reversing Microglia Aging by Elevation of Mitochondrial Transcription Factor-A

Mitochondrial transcription factor-A (TFAM) is a nucleus-encoded protein that binds upstream of the light-strand and heat-strand promoters of mtDNA and induces the transcription of mtDNA (Parisi and Clayton 1991). Therefore, the level of TFAM is a major determinant of the amount of mtDNA (Kanki et al. 2004; Seidel-Rogol and Shadel 2002). In addition to maintaining mtDNA by acting as a transcription factor, TFAM stabilizes mtDNA by forming a nucleoid structure (Kanki et al. 2004). The amounts of both TFAM and mtDNA are significantly increased during aging in the brain and peripheral organs including the liver of rodents (Dinardo et al. 2003; Masuyama et al. 2005). There is growing evidence that mtDNA deficiency and mitochondrial dysfunction play a major role in the development and progression of cardiac failure (Ikeuchi et al. 2005), but whether these changes occur in aging microglia is not known. The over-expression of human TFAM has been shown to prevent the decrease in mtDNA copy number and mitochondrial electron transfer function in a partial myocardial infarction model of a mouse (Ikeuchi et al. 2005). The increased mtDNA copy number observed in hTFAM-transgenic mice could be due to nucleoid formation or the stabilization of mtDNA by hTFAM, because hTFAM is not expected to function as a transcription factor in murine cells (Kang et al. 2007). Therefore, oxidative stress may cause deficiencies of mtDNA, leading to cardiac failure through mitochondrial dysfunction and resultant over-production of ROS.

The increased expression level of hTFAM in HeLa cells effectively reduces ROS generation induced by rotenone, an inhibitor of mitochondrial complex I, and the subsequent nuclear translocation of NFκB, probably through stabilization of mtDNA, which could reduce mitochondrial dysfunction and resultant ROS generation (Corral-Debrinski et al. 1992). Furthermore, hTFAM-transgenic mice exhibit a significant improvement of age-dependent motor and memory impairments, associated with a marked reduction of mtDNA damage and IL-1β production in microglia (Hayashi et al. 2008; Nakanishi and Wu 2009). In addition to the motor and memory functions, sickness behaviors induced by LPS are also affected by aging

(Huang et al. 2008), as discussed above. Increasing evidence suggest that LPS-induced NFκB activation through TLR4-CD14 complex is dependent on the production of ROS (Baeuerle and Henkel 1994; Janssen-Heininger et al. 2000), and a broad range of antioxidants abolish NFκB activation (Blackwell et al. 1996; Zhang et al. 1994). These observations prompted further investigation of the effect of human TFAM over-expression on the age-dependent prolongation of LPS-induced sickness behaviors. In particular, human TFAM-transgenic mice were found to exhibit a significant improvement of age-dependent prolonged sickness behaviors following treatment with LPS, which is closely correlated with attenuation of mtDNA damages and IL-1β expression in microglia (Nakanishi et al. 2011) (Fig. 13.1). Therefore, over-expression of hTFAM could improve the age-dependent memory impairment and prolonged LPS-induced sickness behaviors by ameliorating the mtDNA damage and the resulting redox-regulated inflammatory response.

In the aging brain, there is an impairment of electron transfer in some mitochondrial complex, shifting the intracellular redox balance towards a more oxidized state (Navarro et al. 2002). Aged dysfunctional mitochondria may not respond to sudden increases in ATP demands, which has been speculated to lead to impaired performance in behavioral tests (Navarro et al. 2002). Whether this impairment also occurs in aged microglia is not known, but it is likely that alterations in mitochondrial function result in a decreased microglial motility and phagocytosis, as they are energy-requiring events (Fig. 13.2). However, this hypothesis needs to be experimentally determined. Microglia with highly branched fine processes are now being considered as active players in the normal healthy brain (see Chaps. 3, 4 and 9 for further information). Therefore, the accumulation of damaged ROS-hypergenerating mitochondria in microglia might limit microglial cell defensive behaviors, including motility and phagocytosis, during aging.

13.8 Alternative Strategies for Preventing or Reversing Microglial Cell Aging

There is accumulating evidence that exercise and caloric restriction can play a role in reducing microglial activation and "priming" during aging. In aged animals, small amounts of exercise were found to prevent the infection-induced exaggerated neuroinflammatory response, which is associated with increased cytokine production and increased cognitive deficits (Barrientos et al. 2011). Moreover, voluntary exercise was found to abrogate the age-related "priming" of microglia (Barrientos et al. 2011; Kohman et al. 2013), suggesting that exercise might be an effective intervention to prevent or reverse microglial cell aging. These beneficial effects of exercise may, in part, result from its induction of brain-derived neurotrophic factor (BDNF), which is a potent regulator of synaptic development and plasticity (Barrientos et al. 2011). On the other hand, caloric restriction could also attenuate the age-related activation of microglia, resulting in beneficial effects on neurodegeneration and cognitive decline (Morgan et al. 2007). Caloric restriction has anti-inflammatory and anti-apoptotic properties (Loncarevic-Vasiljkovic et al. 2012). Interestingly, both exercise and caloric restriction were recently shown to promote mitochondrial biogenesis and expression of TFAM in the rat brain (Picca et al. 2012; Zhang et al. 2012). Collectively, both exercise and caloric restriction may effectively slow down the brain aging through preventing or reversing microglial aging. However, their exact underlying mechanisms remain unknown.

13.9 Conclusion

During normal aging, microglia undergo several morphological and functional changes, affecting their neuronal environment and facilitating the appearance of cognitive impairments. Among these changes, increased production of ROS and pro-inflammatory cytokines by microglia and other resident and infiltrating cells, including monocytes, could facilitate the onset of neurodegenerative diseases. Decline of both lysosomal function and mitochondrial DNA damage in these cells results in an exacerbated generation of ROS and pro-inflammatory mediators, which could represent the cellular basis of microglia aging. Therefore, molecules implicated in lysosomal and mitochondrial dysfunction, such as CatB and TFAM, may be considered as potential therapeutic targets. Further research would be necessary, however, to develop effective pharmacological interventions against brain aging. Within this perspective, pharmacological approaches aimed to rejuvenate old microglia in the elderly brain may constitute a promising future research avenue for slowing senescence. Furthermore, non-pharmacological strategies, like exercise and dietary restriction, could promote a healthy aging through their effects on promoting microglial surveillance and physiological functions, while reducing inflammation and ROS production.

References

Adler AS, Sinha S, Kawahara TL, Zhang JY, Segal E, Chang HY (2007) Motif module map reveals enforcement of aging by continual NF-kappaB activity. Genes Dev 21(24):3244–3257

Adler AS, Kawahara TL, Segal E, Chang HY (2008) Reversal of aging by NFkappaB blockade. Cell Cycle 7(5):556–559

Aguzzi A, Barres BA, Bennett ML (2013) Microglia: scapegoat, saboteur, or something else? Science 339(6116):156–161. doi:10.1126/science.1227901

Baeuerle PA, Henkel T (1994) Function and activation of NF-kappa B in the immune system. Annu Rev Immunol 12:141–179. doi:10.1146/annurev.iy.12.040194.001041

Barrientos RM, Frank MG, Crysdale NY, Chapman TR, Ahrendsen JT, Day HE, Campeau S, Watkins LR, Patterson SL, Maier SF (2011) Little exercise, big effects: reversing aging and infection-induced memory deficits, and underlying processes. J Neurosci 31(32):11578–11586. doi:10.1523/JNEUROSCI.2266-11.2011

Blackwell TS, Blackwell TR, Holden EP, Christman BW, Christman JW (1996) In vivo antioxidant treatment suppresses nuclear factor-kappa B activation and neutrophilic lung inflammation. J Immunol 157(4):1630–1637

Blau CW, Cowley TR, O'Sullivan J, Grehan B, Browne TC, Kelly L, Birch A, Murphy N, Kelly AM, Kerskens CM, Lynch MA (2012) The age-related deficit in LTP is associated with changes in perfusion and blood-brain barrier permeability. Neurobiol Aging 33(5):1005. doi:10.1016/j.neurobiolaging.2011.09.035, e23–35
Blobe GC, Schiemann WP, Lodish HF (2000) Role of transforming growth factor beta in human disease. N Engl J Med 342(18):1350–1358. doi:10.1056/NEJM200005043421807
Block ML, Zecca L, Hong JS (2007) Microglia-mediated neurotoxicity: uncovering the molecular mechanisms. Nat Rev Neurosci 8(1):57–69. doi:10.1038/nrn2038
Boillee S, Yamanaka K, Lobsiger CS, Copeland NG, Jenkins NA, Kassiotis G, Kollias G, Cleveland DW (2006) Onset and progression in inherited ALS determined by motor neurons and microglia. Science 312(5778):1389–1392. doi:10.1126/science.1123511
Borgesius NZ, de Waard MC, van der Pluijm I, Omrani A, Zondag GC, van der Horst GT, Melton DW, Hoeijmakers JH, Jaarsma D, Elgersma Y (2011) Accelerated age-related cognitive decline and neurodegeneration, caused by deficient DNA repair. J Neurosci 31(35):12543–12553. doi:10.1523/JNEUROSCI.1589-11.2011
Bruchard M, Mignot G, Derangere V, Chalmin F, Chevriaux A, Vegran F, Boireau W, Simon B, Ryffel B, Connat JL, Kanellopoulos J, Martin F, Rebe C, Apetoh L, Ghiringhelli F (2013) Chemotherapy-triggered cathepsin B release in myeloid-derived suppressor cells activates the Nlrp3 inflammasome and promotes tumor growth. Nat Med 19(1):57–64. doi:10.1038/nm.2999
Bye N, Zieba M, Wreford NG, Nichols NR (2001) Resistance of the dentate gyrus to induced apoptosis during ageing is associated with increases in transforming growth factor-beta1 messenger RNA. Neuroscience 105(4):853–862
Chen S, Luo D, Streit WJ, Harrison JK (2002) TGF-beta1 upregulates CX3CR1 expression and inhibits fractalkine-stimulated signaling in rat microglia. J Neuroimmunol 133(1–2):46–55
Combrinck MI, Perry VH, Cunningham C (2002) Peripheral infection evokes exaggerated sickness behaviour in pre-clinical murine prion disease. Neuroscience 112(1):7–11
Conde JR, Streit WJ (2006a) Effect of aging on the microglial response to peripheral nerve injury. Neurobiol Aging 27(10):1451–1461. doi:10.1016/j.neurobiolaging.2005.07.012
Conde JR, Streit WJ (2006b) Microglia in the aging brain. J Neuropathol Exp Neurol 65(3): 199–203. doi:10.1097/01.jnen.0000202887.22082.6300005072-200603000-00001
Corral-Debrinski M, Horton T, Lott MT, Shoffner JM, Beal MF, Wallace DC (1992) Mitochondrial DNA deletions in human brain: regional variability and increase with advanced age. Nat Genet 2(4):324–329. doi:10.1038/ng1292-324
Courchesne E, Chisum HJ, Townsend J, Cowles A, Covington J, Egaas B, Harwood M, Hinds S, Press GA (2000) Normal brain development and aging: quantitative analysis at in vivo MR imaging in healthy volunteers. Radiology 216(3):672–682
Cunningham C, Wilcockson DC, Campion S, Lunnon K, Perry VH (2005) Central and systemic endotoxin challenges exacerbate the local inflammatory response and increase neuronal death during chronic neurodegeneration. J Neurosci 25(40):9275–9284. doi:10.1523/JNEUROSCI.2614-05.2005
Damani MR, Zhao L, Fontainhas AM, Amaral J, Fariss RN, Wong WT (2011) Age-related alterations in the dynamic behavior of microglia. Aging Cell 10(2):263–276. doi:10.1111/j.1474-9726.2010.00660.x
DeWitt DA, Perry G, Cohen M, Doller C, Silver J (1998) Astrocytes regulate microglial phagocytosis of senile plaque cores of Alzheimer's disease. Exp Neurol 149(2):329–340. doi:10.1006/exnr.1997.6738
Dinardo MM, Musicco C, Fracasso F, Milella F, Gadaleta MN, Gadaleta G, Cantatore P (2003) Acetylation and level of mitochondrial transcription factor A in several organs of young and old rats. Biochem Biophys Res Commun 301(1):187–191
Dostert C, Petrilli V, Van Bruggen R, Steele C, Mossman BT, Tschopp J (2008) Innate immune activation through Nalp3 inflammasome sensing of asbestos and silica. Science 320(5876): 674–677. doi:10.1126/science.1156995
Drachman DA (2006) Aging of the brain, entropy, and Alzheimer disease. Neurology 67(8): 1340–1352. doi:10.1212/01.wnl.0000240127.89601.83

Droge W, Schipper HM (2007) Oxidative stress and aberrant signaling in aging and cognitive decline. Aging Cell 6(3):361–370. doi:10.1111/j.1474-9726.2007.00294.x
Duewell P, Kono H, Rayner KJ, Sirois CM, Vladimer G, Bauernfeind FG, Abela GS, Franchi L, Nunez G, Schnurr M, Espevik T, Lien E, Fitzgerald KA, Rock KL, Moore KJ, Wright SD, Hornung V, Latz E (2010) NLRP3 inflammasomes are required for atherogenesis and activated by cholesterol crystals. Nature 464(7293):1357–1361. doi:10.1038/nature08938
Enciu AM, Gherghiceanu M, Popescu BO (2013) Triggers and effectors of oxidative stress at blood-brain barrier level: relevance for brain ageing and neurodegeneration. Oxid Med Cell Longev 2013:297512. doi:10.1155/2013/297512
Fadel PJ, Barman SM, Phillips SW, Gebber GL (2004) Fractal fluctuations in human respiration. J Appl Physiol 97(6):2056–2064. doi:10.1152/japplphysiol.00657.2004
Flanary B (2005) The role of microglial cellular senescence in the aging and Alzheimer diseased brain. Rejuvenation Res 8(2):82–85. doi:10.1089/rej.2005.8.82
Flanary BE, Streit WJ (2004) Progressive telomere shortening occurs in cultured rat microglia, but not astrocytes. Glia 45(1):75–88. doi:10.1002/glia.10301
Flanary BE, Sammons NW, Nguyen C, Walker D, Streit WJ (2007) Evidence that aging and amyloid promote microglial cell senescence. Rejuvenation Res 10(1):61–74. doi:10.1089/rej.2006.9096
Floden AM, Combs CK (2011) Microglia demonstrate age-dependent interaction with amyloid-beta fibrils. J Alzheimers Dis 25(2):279–293. doi:10.3233/JAD-2011-101014
Flores B, von Bernhardi R (2012) Transforming growth factor beta1 modulates amyloid beta-induced glial activation through the Smad3-dependent induction of MAPK phosphatase-1. J Alzheimers Dis 32(2):417–429. doi:10.3233/JAD-2012-120721
Forster MJ, Dubey A, Dawson KM, Stutts WA, Lal H, Sohal RS (1996) Age-related losses of cognitive function and motor skills in mice are associated with oxidative protein damage in the brain. Proc Natl Acad Sci U S A 93(10):4765–4769
Ge Y, Grossman RI, Babb JS, Rabin ML, Mannon LJ, Kolson DL (2002) Age-related total gray matter and white matter changes in normal adult brain. Part I: volumetric MR imaging analysis. AJNR Am J Neuroradiol 23(8):1327–1333
Geula C, Wu CK, Saroff D, Lorenzo A, Yuan M, Yankner BA (1998) Aging renders the brain vulnerable to amyloid beta-protein neurotoxicity. Nat Med 4(7):827–831
Glenn CF, Chow DK, David L, Cooke CA, Gami MS, Iser WB, Hanselman KB, Goldberg IG, Wolkow CA (2004) Behavioral deficits during early stages of aging in Caenorhabditis elegans result from locomotory deficits possibly linked to muscle frailty. J Gerontol A Biol Sci Med Sci 59(12):1251–1260
Godbout JP, Johnson RW (2004) Interleukin-6 in the aging brain. J Neuroimmunol 147(1–2): 141–144
Godbout JP, Chen J, Abraham J, Richwine AF, Berg BM, Kelley KW, Johnson RW (2005) Exaggerated neuroinflammation and sickness behavior in aged mice following activation of the peripheral innate immune system. FASEB J 19(10):1329–1331. doi:10.1096/fj.05-3776fje
Good CD, Johnsrude IS, Ashburner J, Henson RN, Friston KJ, Frackowiak RS (2001) A voxel-based morphometric study of ageing in 465 normal adult human brains. Neuroimage 14(1 Pt 1):21–36. doi:10.1006/nimg.2001.0786
Halle A, Hornung V, Petzold GC, Stewart CR, Monks BG, Reinheckel T, Fitzgerald KA, Latz E, Moore KJ, Golenbock DT (2008) The NALP3 inflammasome is involved in the innate immune response to amyloid-beta. Nat Immunol 9(8):857–865. doi:10.1038/ni.1636
Harry GJ (2013) Microglia during development and aging. Pharmacol Ther 139(3):313–326. doi:10.1016/j.pharmthera.2013.04.013
Hart BL (1988) Biological basis of the behavior of sick animals. Neurosci Biobehav Rev 12(2):123–137
Hart AD, Wyttenbach A, Perry VH, Teeling JL (2012) Age related changes in microglial phenotype vary between CNS regions: grey versus white matter differences. Brain Behav Immun 26(5):754–765. doi:10.1016/j.bbi.2011.11.006
Hayashi Y, Yoshida M, Yamato M, Ide T, Wu Z, Ochi-Shindou M, Kanki T, Kang D, Sunagawa K, Tsutsui H, Nakanishi H (2008) Reverse of age-dependent memory impairment and

mitochondrial DNA damage in microglia by an overexpression of human mitochondrial transcriptionfactorainmice.JNeurosci28(34):8624–8634.doi:10.1523/JNEUROSCI.1957-08.2008
He P, Zhong Z, Lindholm K, Berning L, Lee W, Lemere C, Staufenbiel M, Li R, Shen Y (2007) Deletion of tumor necrosis factor death receptor inhibits amyloid beta generation and prevents learning and memory deficits in Alzheimer's mice. J Cell Biol 178(5):829–841. doi:10.1083/jcb.200705042
Hefendehl JK, Neher JJ, Suhs RB, Kohsaka S, Skodras A, Jucker M (2013) Homeostatic and injury-induced microglia behavior in the aging brain. Aging Cell 13(1):60–69. doi:10.1111/acel.12149
Heneka MT, Kummer MP, Stutz A, Delekate A, Schwartz S, Vieira-Saecker A, Griep A, Axt D, Remus A, Tzeng TC, Gelpi E, Halle A, Korte M, Latz E, Golenbock DT (2013) NLRP3 is activated in Alzheimer's disease and contributes to pathology in APP/PS1 mice. Nature 493(7434):674–678. doi:10.1038/nature11729
Hentze H, Lin XY, Choi MS, Porter AG (2003) Critical role for cathepsin B in mediating caspase-1-dependent interleukin-18 maturation and caspase-1-independent necrosis triggered by the microbial toxin nigericin. Cell Death Differ 10(9):956–968. doi:10.1038/sj.cdd.44012644401264
Herrera-Molina R, von Bernhardi R (2005) Transforming growth factor-beta 1 produced by hippocampal cells modulates microglial reactivity in culture. Neurobiol Dis 19(1–2):229–236. doi:10.1016/j.nbd.2005.01.003
Herrera-Molina R, Flores B, Orellana JA, von Bernhardi R (2012) Modulation of interferon-gamma-induced glial cell activation by transforming growth factor beta1: a role for STAT1 and MAPK pathways. J Neurochem 123(1):113–123. doi:10.1111/j.1471-4159.2012.07887.x
Hickman SE, Allison EK, El Khoury J (2008) Microglial dysfunction and defective beta-amyloid clearance pathways in aging Alzheimer's disease mice. J Neurosci 28(33):8354–8360. doi:10.1523/JNEUROSCI.0616-08.2008
Hooper C, Fry VA, Sevastou IG, Pocock JM (2009) Scavenger receptor control of chromogranin A-induced microglial stress and neurotoxic cascades. FEBS Lett 583(21):3461–3466. doi:10.1016/j.febslet.2009.09.049
Hornung V, Bauernfeind F, Halle A, Samstad EO, Kono H, Rock KL, Fitzgerald KA, Latz E (2008) Silica crystals and aluminum salts activate the NALP3 inflammasome through phagosomal destabilization. Nat Immunol 9(8):847–856. doi:10.1038/ni.1631
Hu S, Sheng WS, Peterson PK, Chao CC (1995) Cytokine modulation of murine microglial cell superoxide production. Glia 13(1):45–50. doi:10.1002/glia.440130106
Huang Y, Henry CJ, Dantzer R, Johnson RW, Godbout JP (2008) Exaggerated sickness behavior and brain proinflammatory cytokine expression in aged mice in response to intracerebroventricular lipopolysaccharide. Neurobiol Aging 29(11):1744–1753. doi:10.1016/j.neurobiolaging.2007.04.012
Ikeuchi M, Matsusaka H, Kang D, Matsushima S, Ide T, Kubota T, Fujiwara T, Hamasaki N, Takeshita A, Sunagawa K, Tsutsui H (2005) Overexpression of mitochondrial transcription factor a ameliorates mitochondrial deficiencies and cardiac failure after myocardial infarction. Circulation 112(5):683–690. doi:10.1161/CIRCULATIONAHA.104.524835
Janssen-Heininger YM, Poynter ME, Baeuerle PA (2000) Recent advances towards understanding redox mechanisms in the activation of nuclear factor kappaB. Free Radic Biol Med 28(9):1317–1327
Jernigan TL, Archibald SL, Fennema-Notestine C, Gamst AC, Stout JC, Bonner J, Hesselink JR (2001) Effects of age on tissues and regions of the cerebrum and cerebellum. Neurobiol Aging 22(4):581–594
Jurk D, Wang C, Miwa S, Maddick M, Korolchuk V, Tsolou A, Gonos ES, Thrasivoulou C, Saffrey MJ, Cameron K, von Zglinicki T (2012) Postmitotic neurons develop a p21-dependent senescence-like phenotype driven by a DNA damage response. Aging Cell 11(6):996–1004
Kang SJ, Wang S, Hara H, Peterson EP, Namura S, Amin-Hanjani S, Huang Z, Srinivasan A, Tomaselli KJ, Thornberry NA, Moskowitz MA, Yuan J (2000) Dual role of caspase-11 in mediating activation of caspase-1 and caspase-3 under pathological conditions. J Cell Biol 149(3):613–622

Kang D, Kim SH, Hamasaki N (2007) Mitochondrial transcription factor A (TFAM): roles in maintenance of mtDNA and cellular functions. Mitochondrion 7(1–2):39–44. doi:10.1016/j.mito.2006.11.017

Kanki T, Ohgaki K, Gaspari M, Gustafsson CM, Fukuoh A, Sasaki N, Hamasaki N, Kang D (2004) Architectural role of mitochondrial transcription factor A in maintenance of human mitochondrial DNA. Mol Cell Biol 24(22):9823–9834. doi:10.1128/MCB.24.22.9823-9834.2004

Kettenmann H, Kirchhoff F, Verkhratsky A (2013) Microglia: new roles for the synaptic stripper. Neuron 77(1):10–18. doi:10.1016/j.neuron.2012.12.023

Klein CS, Rice CL, Marsh GD (2001) Normalized force, activation, and coactivation in the arm muscles of young and old men. J Appl Physiol 91(3):1341–1349

Kohman RA, Bhattacharya TK, Wojcik E, Rhodes JS (2013) Exercise reduces activation of microglia isolated from hippocampus and brain of aged mice. J Neuroinflammation 10:114. doi:10.1186/1742-2094-10-114

Le Y, Iribarren P, Gong W, Cui Y, Zhang X, Wang JM (2004) TGF-beta1 disrupts endotoxin signaling in microglial cells through Smad3 and MAPK pathways. J Immunol 173(2):962–968

Letiembre M, Hao W, Liu Y, Walter S, Mihaljevic I, Rivest S, Hartmann T, Fassbender K (2007) Innate immune receptor expression in normal brain aging. Neuroscience 146(1):248–254. doi:10.1016/j.neuroscience.2007.01.004

Lieb K, Engels S, Fiebich BL (2003) Inhibition of LPS-induced iNOS and NO synthesis in primary rat microglial cells. Neurochem Int 42(2):131–137, S0197018602000761 [pii]

Lin MT, Simon DK, Ahn CH, Kim LM, Beal MF (2002) High aggregate burden of somatic mtDNA point mutations in aging and Alzheimer's disease brain. Hum Mol Genet 11(2):133–145

Lindenberger U, Marsiske M, Baltes PB (2000) Memorizing while walking: increase in dual-task costs from young adulthood to old age. Psychol Aging 15(3):417–436

Lipsitz LA (2002) Dynamics of stability: the physiologic basis of functional health and frailty. J Gerontol A Biol Sci Med Sci 57(3):B115–B125

Lipsitz LA, Goldberger AL (1992) Loss of "complexity" and aging. Potential applications of fractals and chaos theory to senescence. JAMA 267(13):1806–1809

Loncarevic-Vasiljkovic N, Pesic V, Todorovic S, Popic J, Smiljanic K, Milanovic D, Ruzdijic S, Kanazir S (2012) Caloric restriction suppresses microglial activation and prevents neuroapoptosis following cortical injury in rats. PLoS One 7(5):e37215. doi:10.1371/journal.pone.0037215

Lopez-Otin C, Biasco MA, Partridge L, Serrano M, Kroemer G (2013) The hallmarks of aging. Cell 153(6):1194–1217

Lukiw WJ (2004) Gene expression profiling in fetal, aged, and Alzheimer hippocampus: a continuum of stress-related signaling. Neurochem Res 29(6):1287–1297

Majumdar A, Cruz D, Asamoah N, Buxbaum A, Sohar I, Lobel P, Maxfield FR (2007) Activation of microglia acidifies lysosomes and leads to degradation of Alzheimer amyloid fibrils. Mol Biol Cell 18(4):1490–1496

Marques F, Sousa JC, Sousa N, Palha JA (2013) Blood-brain-barriers in aging and in Alzheimer's disease. Mol Neurodegener 8:38. doi:10.1186/1750-1326-8-38

Masters SL, Dunne A, Subramanian SL, Hull RL, Tannahill GM, Sharp FA, Becker C, Franchi L, Yoshihara E, Chen Z, Mullooly N, Mielke LA, Harris J, Coll RC, Mills KH, Mok KH, Newsholme P, Nunez G, Yodoi J, Kahn SE, Lavelle EC, O'Neill LA (2010) Activation of the NLRP3 inflammasome by islet amyloid polypeptide provides a mechanism for enhanced IL-1beta in type 2 diabetes. Nat Immunol 11(10):897–904. doi:10.1038/ni.1935

Masuyama M, Iida R, Takatsuka H, Yasuda T, Matsuki T (2005) Quantitative change in mitochondrial DNA content in various mouse tissues during aging. Biochim Biophys Acta 1723(1–3):302–308. doi:10.1016/j.bbagen.2005.03.001

McCoy MK, Martinez TN, Ruhn KA, Szymkowski DE, Smith CG, Botterman BR, Tansey KE, Tansey MG (2006) Blocking soluble tumor necrosis factor signaling with dominant-negative tumor necrosis factor inhibitor attenuates loss of dopaminergic neurons in models of Parkinson's disease. J Neurosci 26(37):9365–9375. doi:10.1523/JNEUROSCI.1504-06.2006

Mittaud P, Labourdette G, Zingg H, Guenot-Di Scala D (2002) Neurons modulate oxytocin receptor expression in rat cultured astrocytes: involvement of TGF-beta and membrane components. Glia 37(2):169–177. doi:10.1002/glia.10029

Morgan TE, Wong AM, Finch CE (2007) Anti-inflammatory mechanisms of dietary restriction in slowing aging processes. Interdiscip Top Gerontol 35:83–97. doi:10.1159/000096557

Mosher KI, Wyss-Coray T (2014) Microglial dysfunction in brain aging and Alzheimer's disease. Biochem Pharmacol 88(4):594–604. doi:10.1016/j.bcp.2014.01.008

Mount MP, Lira A, Grimes D, Smith PD, Faucher S, Slack R, Anisman H, Hayley S, Park DS (2007) Involvement of interferon-gamma in microglial-mediated loss of dopaminergic neurons. J Neurosci 27(12):3328–3337. doi:10.1523/JNEUROSCI.5321-06.2007

Nakahira K, Haspel JA, Rathinam VA, Lee SJ, Dolinay T, Lam HC, Englert JA, Rabinovitch M, Cernadas M, Kim HP, Fitzgerald KA, Ryter SW, Choi AM (2011) Autophagy proteins regulate innate immune responses by inhibiting the release of mitochondrial DNA mediated by the NALP3 inflammasome. Nat Immunol 12(3):222–230. doi:10.1038/ni.1980

Nakanishi H (2003) Neuronal and microglial cathepsins in aging and age-related diseases. Ageing Res Rev 2(4):367–381

Nakanishi H, Wu Z (2009) Microglia-aging: roles of microglial lysosome- and mitochondria-derived reactive oxygen species in brain aging. Behav Brain Res 201(1):1–7. doi:10.1016/j.bbr.2009.02.001

Nakanishi H, Hayashi Y, Wu Z (2011) The role of microglial mtDNA damage in age-dependent prolonged LPS-induced sickness behavior. Neuron Glia Biol 7(1):17–23. doi:10.1017/S1740925X1100010X

Navarro A, Sanchez Del Pino MJ, Gomez C, Peralta JL, Boveris A (2002) Behavioral dysfunction, brain oxidative stress, and impaired mitochondrial electron transfer in aging mice. Am J Physiol Regul Integr Comp Physiol 282(4):R985–R992. doi:10.1152/ajpregu.00537.2001

Neymotin A, Petri S, Calingasan NY, Wille E, Schafer P, Stewart C, Hensley K, Beal MF, Kiaei M (2009) Lenalidomide (Revlimid) administration at symptom onset is neuroprotective in a mouse model of amyotrophic lateral sclerosis. Exp Neurol 220(1):191–197. doi:10.1016/j.expneurol.2009.08.028

Njie EG, Boelen E, Stassen FR, Steinbusch HW, Borchelt DR, Streit WJ (2012) Ex vivo cultures of microglia from young and aged rodent brain reveal age-related changes in microglial function. Neurobiol Aging 33(1):195. doi:10.1016/j.neurobiolaging.2010.05.008, e191-112

Parisi MA, Clayton DA (1991) Similarity of human mitochondrial transcription factor 1 to high mobility group proteins. Science 252(5008):965–969

Pawate S, Shen Q, Fan F, Bhat NR (2004) Redox regulation of glial inflammatory response to lipopolysaccharide and interferongamma. J Neurosci Res 77(4):540–551. doi:10.1002/jnr.20180

Perry VH (2004) The influence of systemic inflammation on inflammation in the brain: implications for chronic neurodegenerative disease. Brain Behav Immun 18(5):407–413. doi:10.1016/j.bbi.2004.01.004S0889159104000261

Perry VH, Matyszak MK, Fearn S (1993) Altered antigen expression of microglia in the aged rodent CNS. Glia 7(1):60–67. doi:10.1002/glia.440070111

Perry VH, Newman TA, Cunningham C (2003) The impact of systemic infection on the progression of neurodegenerative disease. Nat Rev Neurosci 4(2):103–112. doi:10.1038/nrn1032nrn1032

Perry VH, Cunningham C, Holmes C (2007) Systemic infections and inflammation affect chronic neurodegeneration. Nat Rev Immunol 7(2):161–167. doi:10.1038/nri2015

Picca A, Fracasso F, Pesce V, Cantatore P, Joseph AM, Leeuwenburgh C, Gadaleta MN, Lezza AM (2012) Age- and calorie restriction-related changes in rat brain mitochondrial DNA and TFAM binding. Age (Dordr) 35(5):1607–1620. doi:10.1007/s11357-012-9465-z

Qin L, Li G, Qian X, Liu Y, Wu X, Liu B, Hong JS, Block ML (2005) Interactive role of the toll-like receptor 4 and reactive oxygen species in LPS-induced microglia activation. Glia 52(1):78–84. doi:10.1002/glia.20225

Raj DDA, Jaarsma D, Holtman IR, Olah M, Ferreira FM, Schaafsma W, Brouwer N, Meijer MM, de Waard MC, van der Pluijm I, Brandt R, Kreft KL, Laman JD, de Haan G, Biber KPH, Hoeijmakers JHJ, Eggen BJL, Boddeke HWGM (2014) Priming of Microglia in a DNA-Repair Deficient Model of Accelerated Aging. Neurobiol Aging 35(9):2147–2160
Ramirez G, Rey S, von Bernhardi R (2008) Proinflammatory stimuli are needed for induction of microglial cell-mediated AbetaPP and Abeta-neurotoxicity in hippocampal cultures. J Alzheimers Dis 15(1):45–59
Rogers J, Strohmeyer R, Kovelowski CJ, Li R (2002) Microglia and inflammatory mechanisms in the clearance of amyloid beta peptide. Glia 40(2):260–269. doi:10.1002/glia.10153
Salat DH, Buckner RL, Snyder AZ, Greve DN, Desikan RS, Busa E, Morris JC, Dale AM, Fischl B (2004) Thinning of the cerebral cortex in aging. Cereb Cortex 14(7):721–730. doi:10.1093/cercor/bhh032bhh032
Salat DH, Tuch DS, Hevelone ND, Fischl B, Corkin S, Rosas HD, Dale AM (2005) Age-related changes in prefrontal white matter measured by diffusion tensor imaging. Ann N Y Acad Sci 1064:37–49. doi:10.1196/annals.1340.009
Salminen A, Ojala J, Kaarniranta K, Kauppinen A (2012) Mitochondrial dysfunction and oxidative stress activate inflammasomes: impact on the aging process and age-related diseases. Cell Mol Life Sci 69(18):2999–3013. doi:10.1007/s00018-012-0962-0
Schmierer B, Hill CS (2007) TGFbeta-SMAD signal transduction: molecular specificity and functional flexibility. Nat Rev Mol Cell Biol 8(12):970–982. doi:10.1038/nrm2297
Seidel-Rogol BL, Shadel GS (2002) Modulation of mitochondrial transcription in response to mtDNA depletion and repletion in HeLa cells. Nucleic Acids Res 30(9):1929–1934
Sheng JG, Mrak RE, Griffin WS (1998) Enlarged and phagocytic, but not primed, interleukin-1 alpha-immunoreactive microglia increase with age in normal human brain. Acta Neuropathol 95(3):229–234
Sheng WS, Hu S, Feng A, Rock RB (2013) Reactive oxygen species from human astrocytes induce functional impairment and oxidative damage. Neurochem Res 38(10):2148–2159
Shock N, Greulich R, Costa P, Andres R, Lakatta E, Arenberg D, Tobin J (1984) Normal human aging: the baltimore longitudinal study of aging. US Department of Health and Human Services, Baltimore
Sierra A, Gottfried-Blackmore AC, McEwen BS, Bulloch K (2007) Microglia derived from aging mice exhibit an altered inflammatory profile. Glia 55(4):412–424. doi:10.1002/glia.20468
Sierra A, Encinas JM, Deudero JJ, Chancey JH, Enikolopov G, Overstreet-Wadiche LS, Tsirka SE, Maletic-Savatic M (2010) Microglia shape adult hippocampal neurogenesis through apoptosis-coupled phagocytosis. Cell Stem Cell 7(4):483–495. doi:10.1016/j.stem.2010.08.014
Sierra A, Abiega O, Shahraz A, Neumann H (2013) Janus-faced microglia: beneficial and detrimental consequences of microglial phagocytosis. Front Cell Neurosci 7:6. doi:10.3389/fncel.2013.00006
Siskova Z, Tremblay ME (2013) Microglia and synapse: interactions in health and neurodegeneration. Neural Plast 2013:425845
Smith RG, Betancourt L, Sun Y (2005) Molecular endocrinology and physiology of the aging central nervous system. Endocr Rev 26(2):203–250. doi:10.1210/er.2002-0017
Sowell ER, Peterson BS, Thompson PM, Welcome SE, Henkenius AL, Toga AW (2003) Mapping cortical change across the human life span. Nat Neurosci 6(3):309–315. doi:10.1038/nn1008nn1008
Sriram K, Matheson JM, Benkovic SA, Miller DB, Luster MI, O'Callaghan JP (2002) Mice deficient in TNF receptors are protected against dopaminergic neurotoxicity: implications for Parkinson's disease. FASEB J 16(11):1474–1476. doi:10.1096/fj.02-0216fje02-0216fje
Streit WJ (2006) Microglial senescence: does the brain's immune system have an expiration date? Trends Neurosci 29(9):506–510. doi:10.1016/j.tins.2006.07.001
Streit WJ, Sammons NW, Kuhns AJ, Sparks DL (2004) Dystrophic microglia in the aging human brain. Glia 45(2):208–212. doi:10.1002/glia.10319
Sun L, Wu Z, Hayashi Y, Peters C, Tsuda M, Inoue K, Nakanishi H (2012) Microglial cathepsin B contributes to the initiation of peripheral inflammation-induced chronic pain. J Neurosci 32(33):11330–11342. doi:10.1523/JNEUROSCI.0677-12.2012

Taylor RC, Dillin A (2011) Aging as an event of proteostasis collapse. Cold Spring Harb Perspect Biol 3:5. doi:10.1101/cshperspect.a004440

Terada K, Yamada J, Hayashi Y, Wu Z, Uchiyama Y, Peters C, Nakanishi H (2010) Involvement of cathepsin B in the processing and secretion of interleukin-1beta in chromogranin A-stimulated microglia. Glia 58(1):114–124. doi:10.1002/glia.20906

Tesseur I, Zou K, Esposito L, Bard F, Berber E, Can JV, Lin AH, Crews L, Tremblay P, Mathews P, Mucke L, Masliah E, Wyss-Coray T (2006) Deficiency in neuronal TGF-beta signaling promotes neurodegeneration and Alzheimer's pathology. J Clin Invest 116(11):3060–3069. doi:10.1172/JCI27341

Thompson LV (2009) Age-related muscle dysfunction. Exp Gerontol 44(1–2):106–111. doi:10.1016/j.exger.2008.05.003

Tichauer JE, von Bernhardi R (2012) Transforming growth factor-beta stimulates beta amyloid uptake by microglia through Smad3-dependent mechanisms. J Neurosci Res 90(10):1970–1980. doi:10.1002/jnr.23082

Tichauer JE, Flores B, Soler B, Eugenin-von Bernhardi L, Ramirez G, von Bernhardi R (2014) Age-dependent changes on TGFbeta1 Smad3 pathway modify the pattern of microglial cell activation. Brain Behav Immun 37:187–196. doi:10.1016/j.bbi.2013.12.018

Tremblay ME, Zettel ML, Ison JR, Allen PD, Majewska AK (2012) Effects of aging and sensory loss on glial cells in mouse visual and auditory cortices. Glia 60(4):541–558. doi:10.1002/glia.22287

Tschopp J, Schroder K (2010) NLRP3 inflammasome activation: the convergence of multiple signalling pathways on ROS production? Nat Rev Immunol 10(3):210–215. doi:10.1038/nri2725

Ueberham U, Ueberham E, Gruschka H, Arendt T (2006) Altered subcellular location of phosphorylated Smads in Alzheimer's disease. Eur J Neurosci 24(8):2327–2334. doi:10.1111/j.1460-9568.2006.05109.x

Vancompernolle K, Van Herreweghe F, Pynaert G, Van de Craen M, De Vos K, Totty N, Sterling A, Fiers W, Vandenabeele P, Grooten J (1998) Atractyloside-induced release of cathepsin B, a protease with caspase-processing activity. FEBS Lett 438(3):150–158, doi:S0014-5793(98)01275-7 [pii]

VanGuilder HD, Bixler GV, Brucklacher RM, Farley JA, Yan H, Warrington JP, Sonntag WE, Freeman WM (2011) Concurrent hippocampal induction of MHC II pathway components and glial activation with advanced aging is not correlated with cognitive impairment. J Neuroinflammation 8:138. doi:10.1186/1742-2094-8-138

Vijg J, Campisi J (2008) Puzzles, promises and a cure for ageing. Nature 454(7208):1065–1071. doi:10.1038/nature07216

von Bernhardi R (2007) Glial cell dysregulation: a new perspective on Alzheimer disease. Neurotox Res 12(4):215–232

von Bernhardi R, Eugenin J (2004) Microglial reactivity to beta-amyloid is modulated by astrocytes and proinflammatory factors. Brain Res 1025(1–2):186–193. doi:10.1016/j.brainres.2004.07.084

von Bernhardi R, Ramirez G (2001) Microglia-astrocyte interaction in Alzheimer's disease: friends or foes for the nervous system? Biol Res 34(2):123–128

von Bernhardi R, Ramirez G, Toro R, Eugenin J (2007) Pro-inflammatory conditions promote neuronal damage mediated by Amyloid Precursor Protein and decrease its phagocytosis and degradation by microglial cells in culture. Neurobiol Dis 26(1):153–164. doi:10.1016/j.nbd.2006.12.006

Werner F, Jain MK, Feinberg MW, Sibinga NE, Pellacani A, Wiesel P, Chin MT, Topper JN, Perrella MA, Lee ME (2000) Transforming growth factor-beta 1 inhibition of macrophage activation is mediated via Smad3. J Biol Chem 275(47):36653–36658. doi:10.1074/jbc.M004536200M004536200

Wong WT (2013) Microglial aging in the healthy CNS: phenotypes, drivers, and rejuvenation. Front Cell Neurosci 7:22

Wu Z, Sun L, Hashioka S, Yu S, Schwab C, Okada R, Hayashi Y, McGeer PL, Nakanishi H (2013) Differential pathways for interleukin-1beta production activated by chromogranin A

and amyloid beta in microglia. Neurobiol Aging 34(12):2715–2725. doi:10.1016/j.neurobiolaging.2013.05.018
Wyss-Coray T, Lin C, Yan F, Yu GQ, Rohde M, McConlogue L, Masliah E, Mucke L (2001) TGF-beta1 promotes microglial amyloid-beta clearance and reduces plaque burden in transgenic mice. Nat Med 7(5):612–618. doi:10.1038/87945
Yamamoto K, Shimokawa T, Yi H, Isobe K, Kojima T, Loskutoff DJ, Saito H (2002) Aging accelerates endotoxin-induced thrombosis: increased responses of plasminogen activator inhibitor-1 and lipopolysaccharide signaling with aging. Am J Pathol 161(5):1805–1814, doi:S0002-9440(10)64457-4 [pii]
Ye SM, Johnson RW (1999) Increased interleukin-6 expression by microglia from brain of aged mice. J Neuroimmunol 93(1–2):139–148
Ye SM, Johnson RW (2001) An age-related decline in interleukin-10 may contribute to the increased expression of interleukin-6 in brain of aged mice. Neuroimmunomodulation 9(4):183–192
Yung RL, Julius A (2008) Epigenetics, aging, and autoimmunity. Autoimmunity 41(4):329–335. doi:10.1080/08916930802024889
Zhang H, Spapen H, Nguyen DN, Benlabed M, Buurman WA, Vincent JL (1994) Protective effects of N-acetyl-L-cysteine in endotoxemia. Am J Physiol 266(5 Pt 2):H1746–H1754
Zhang Q, Wu Y, Zhang P, Sha H, Jia J, Hu Y, Zhu J (2012) Exercise induces mitochondrial biogenesis after brain ischemia in rats. Neuroscience 205:10–17. doi:10.1016/j.neuroscience.2011.12.053
Zhou R, Yazdi AS, Menu P, Tschopp J (2011) A role for mitochondria in NLRP3 inflammasome activation. Nature 469(7329):221–225. doi:10.1038/nature09663

Part III
Their Pathological Implications
Conditions Impacting on the Quality of Life

Chapter 14
Neurodevelopmental and Neuropsychiatric Disorders

Akira Monji, Izumi Maezawa, Yoshito Mizoguchi, Takahiro A. Kato, and Lee-Way Jin

Abstract The etiology of neuropsychiatric disorders such as autism spectrum disorders (ASDs) and schizophrenia remains unclear. However, many aspects of their neuropathology were recently reported to be closely associated with microglial dysfunction. Microglia, which are the major players of innate immunity in the central nervous system, respond rapidly to pathological changes, even minor ones, and contribute directly to neuroinflammation by producing various cytokines and free radicals. Recent studies revealed that microglia become activated over the course of ASDs and schizophrenia, using brain neuroimaging and postmortem analyzes. Recent studies have also shown inhibitory effects of some antipsychotics on the release of inflammatory cytokines and free radicals from activated microglia, causing synaptic and white matter abnormalities as seen in ASDs and schizophrenia postmortem brains. In addition, recent evidence strongly suggests a neurodevelopmental role for microglia in regulating the formation/function of neuronal circuits by their phagocytic activity and structural interactions with synapses. In Rett syndrome (RTT) particularly, microglia become dysfunctional and neurotoxic, thus contributing

A. Monji, M.D., Ph.D. (✉) • Y. Mizoguchi, M.D., Ph.D.
Department of Psychiatry, Faculty of Medicine, Saga University Hospital, Nabeshima 5-1-1, Saga 849-8501, Japan
e-mail: amonji@hf.rim.or.jp; ymizo@pk2.so-net.ne.jp

I. Maezawa, Ph.D.
Department of Pathology and Laboratory Medicine, School of Medicine, University of California Davis, MIND Institute, 2805 50th Street, Sacramento, CA 95817, USA
e-mail: imaezawa@ucdavis.edu

T.A. Kato, M.D., Ph.D.
Department of Neuropsychiatry, Graduate School of Medical Sciences, Kyushu University, Maidashi 3-1-1, Fukuoka 812-8582, Japan
e-mail: takahiro@npsych.med.kyushu-u.ac.jp

L.-W. Jin, M.D., Ph.D. (✉)
Department of Pathology and Laboratory Medicine, School of Medicine, University of California Davis, MIND Institute, 2805 50th Street, Sacramento, CA 95817, USA
e-mail: lee-way.jin@ucdmc.ucdavis.edu

M.-È. Tremblay, A. Sierra (eds.), *Microglia in Health and Disease*,
DOI 10.1007/978-1-4939-1429-6_14

to abnormal brain development. Populating the brain of RTT mice with wild-type microglia was also found to arrest the disease, indicating an essential role of microglia in regulating the neurodevelopmental trajectory. In summary, emerging evidence indicates that microglia are closely related to the progression and outcome of ASDs and schizophrenia. Understanding microglial pathology may shed new light on the most promising therapeutic strategies for ASDs and schizophrenia, among other neuropsychiatric disorders.

Keywords Microglia • Schizophrenia • Autism spectrum disorders • Rett syndrome • Neuroinflammation • Cytokines • Free radicals • Phagocytosis • Antipsychotics

Bullet Points

- An accumulating body of evidence points to the significance of neuroinflammatory processes in the etiology of autism spectrum disorders (ASDs) and schizophrenia.
- In the central nervous system, their neuropathology is closely associated with microglial dysfunction, including release of pro-inflammatory and neurotoxic mediators, and impaired phagocytosis.
- Antipsychotics, minocycline, non-steroidal anti-inflammatory drugs, and inhibitors of the mammalian target of rapamycin (mTOR) signaling pathway inhibit release of pro-inflammatory and neurotoxic mediators from microglia and cells of the peripheral immune system.
- Controlling microglial activation and rescuing their normal physiological properties may shed new light on the therapeutic strategies for ASDs and schizophrenia.

14.1 Introduction

Research into neuropsychiatric disorders has so far been "neuronocentric". However, several categories of neuropsychiatric disorders show no overt alterations of brain architecture, leaving neuropathologists and neuroanatomists clueless. Schizophrenia, for example, was famously named "the graveyard of neuropathologists" due to its lack of useful pathological features for diagnosis, both at the anatomical or microscopic levels (Iritani 2007). Similarly, while many neuroimaging and neuropathological abnormalities have been described for ASDs, there is no widely validated structural abnormality allowing for clinicopathological correlation (Pickett and London 2005; Casanova 2007; Polsek et al. 2011). However, recent neurochemical and brain imaging studies have moved microglia into the center of attention. A few radiotracers were developed to evaluate changes in microglial function in vivo using positron emission tomography (PET), including [11C]-(R)-PK11195, a ligand of peripheral benzodiazepine receptors which predominantly binds to morphologically "activated" microglia (Benavides et al. 1988; Pike et al. 1993). The availability of well-preserved postmortem brain samples and several established microglial histological markers in recent years has also allowed neurochemical evaluations of microglia in brain pathologies.

At the experimental biology level, innovative molecular genetic and biophotonic approaches are now enabling investigators to visualize the dynamic behavior of microglial cells in vivo, while introducing microglia-targeted gene deletions and ablations. Therefore, we are at the brink of making fundamental discoveries regarding microglial contributions to neuropsychiatric disorders. In this chapter, we will review two prominent categories of neuropsychiatric disorders, ASDs and schizophrenia, to exemplify the pathological roles of microglia, either acting as primary instigators due to genetic defects, or responding aberrantly to prenatal, perinatal, or postnatal insults specifically associated with the disease, at the interphase between environmental alterations and the resulting brain responses.

14.2 Autism Spectrum Disorders

Autism is a complex developmental disability which was first described by Leo Kanner in 1943. Although initially considered a rare infantile disorder, recent surveys estimated that as many as one in 80 children in the United States are affected by some form of ASDs, an alarming statistics that could mount to an "autism epidemic" (Rice 2009; Williams 2012). Individuals with autism fail to properly share emotions or understand how others feel, have problems in verbal and non-verbal communications (such as eye contact or smiling), show repetitive behaviors, and obsessively follow certain routines. The term ASDs is used to include heterogeneous disorders grouped together based on similar features of atypical development in socialization, communication, and behavior. These conditions include autistic disorder (also called "classic" autism), Asperger syndrome, and pervasive developmental disorder-not otherwise specified (PDD-NOS). Children with Asperger syndrome or PDD-NOS have fewer diagnostic symptoms and milder impairment compared with classic autism. The clinical heterogeneity of ASDs is perhaps the result of etiological heterogeneity; it is now generally agreed upon that autism is the consequence of complex interplays between heritable genetic factors and environmental factors influencing individual's epigenome (Geschwind 2011; Hallmayer et al. 2011; LaSalle 2011). Symptoms of ASDs typically are present before 3 years of age and often are accompanied by abnormalities in cognitive functioning, learning, attention, and sensory processing (Yeargin-Allsopp et al. 2003).

There currently is no cure for autism. Although there are a growing number of proposed and actually administered treatments for patients with autism (many of them without much justification), behavioral intervention programs remain the treatment of choice (Vismara and Rogers 2010). Major challenges for finding a cure include the etiological heterogeneity and the lack of consistent and reliable genetic or biologic diagnostic markers for accurate classification and early diagnosis of ASDs. The behavioral disabilities can be attributed to abnormal functions of synapses, which normally show remarkable plasticity throughout life. Functional imaging studies have established autism as a disorder of under-connectivity among the brain regions participating in cortical networks (Minshew and Keller 2010).

Studies on the genetic etiology of autism have also uncovered genes that regulate synaptic functions (Geschwind 2008; Betancur et al. 2009). However, these rare genetic variants, which by themselves alone might be sufficient to cause autism, only account for 10–20 % of autism cases. The more common neurobiological basis of autism seems to be a complex combination of common genetic variants, epigenetic regulation, environmental factors, glial cell abnormalities, aberrant developmental and possibly adult neurogenesis, blood factors (for example, auto-antibodies), to name just a few. These risk factors are interrelated, and it has been argued previously that microglia and associated neuroinflammation play an important role in mediating the heterogeneous effects of various insults, especially environmental ones, on the neurodevelopmental trajectory leading to autism (Maezawa et al. 2011).

14.2.1 Immune Dysregulation in ASDs

Perinatal exposures, obstetric complications, and environmental toxins are all known to impose risks to autism, thus pointing to an immune/inflammatory etiology (Goines et al. 2011a, b), similar to schizophrenia as discussed below. Indeed, systemic immune dysregulation is an inherent component of ASDs. For example, many patients with ASDs have food allergies and allergy-like symptoms without positive objective test results. The rather common presentation of gastrointestinal problems is often attributed to dietary allergy. Patients with ASD frequently have inflammatory bowel disorders of various severities, showing lymphoid hyperplasia. Peripheral blood profiles often present irregularities including decreased numbers of B and T lymphocytes, reduced lymphocytic response to stimulation, increased numbers of monocytes, abnormal cytokine profiles, abnormal immunoglobulin levels, and increased myeloid dendritic cell densities (Goines et al. 2011a, b; Breece et al. 2013). Numerous studies of serum cytokines demonstrated lower levels of transforming growth factor β (TGFβ) and higher levels of macrophage inhibitory factor, leptin, interleukin 1β (IL-1β), IL-6, interferon γIFNγ, and IL-12 in various age groups of patients with ASDs (Goines et al. 2011a, b; Goines and Ashwood 2013). Skewed cytokine levels were also observed in brain and gastrointestinal specimens, in circulating immune cells isolated from ASD subjects, and in amniotic fluid samples from mothers giving birth to a child with ASD. Although ASD patients rarely present defined auto-immune disorders, several groups have independently discovered auto-antibodies against various components of the central nervous system (CNS). These antibodies have diverse brain-specific targets, such as myelin basic protein (MBP), brain serotonin receptors, brain endothelial cell proteins, brain-derived neurotrophic factor, neurofilament proteins, heat shock proteins, Purkinje cell proteins, cerebellar Golgi cell proteins, and proteins in caudate nucleus, hypothalamus, and thalamus (Goines et al. 2011a, b). How the above-described aberrant immune activities contribute to the core symptoms and comorbidities of ASDs is unknown, but they are clear signs of immune dysregulation that likely promotes neuroinflammation.

14.2.2 ASDs and Microglial Activation

The pioneer work of Vargas and colleagues and subsequent studies revealed an active neuroinflammatory phenotype for microglia in the postmortem brains of patients with autism (Vargas et al. 2005; Zimmerman et al. 2005; Chez et al. 2007; Morgan et al. 2010). Marked changes in microglial morphology, accompanied by a unique profile of pro-inflammatory cytokines were seen in the cerebral cortex, white matter, and cerebellum. Two cytokines in particular, the pro-inflammatory macrophage chemoattractant protein 1 (MCP-1) and the anti-inflammatory TGFβ, both derived from microglia and astrocytes, were the most prevalent cytokines in autism brain tissues. Cerebrospinal fluid (CSF) samples from patients with autism also showed an increase in MCP-1 (Vargas et al. 2005) and TNFα (Chez et al. 2007). The presence of high levels of MCP-1 and TGFβ is indicative of a chronic inflammatory state involving a cascade of pro- and anti-inflammatory responses in autism. The implications of these findings could be highly significant when interpreted in the context of the above-described systemic immune dysregulation in autism. These findings could provide a link between the peripheral immune abnormalities and the neurotoxic events in the brain. For example, activated microglia were observed to be intimately associated with Purkinje cells undergoing apoptosis in cerebellar organotypic cultures during normal development. This could be consistent with a role for microglia in developmentally regulated neuronal death by promoting Purkinje cell apoptosis (Marin-Teva et al. 2004) (see Chap. 7 for further reading), an important physiological activity which could be impaired in autism. Microglial abnormality could also result from CNS or peripheral immune signals, such as auto-antibodies as discussed above (Wills et al. 2009; Goines et al. 2011a, b), or from peripheral chemokines/cytokines (such as IL-1β, IL-6, and TNFα) up-regulated in autism (Gupta et al. 1998; Jyonouchi et al. 2001; Vargas et al. 2005). The chemotactic/phagocytic activity of microglia could also be impaired, further aggravating the symptoms by insufficient clearance of cellular debris (Derecki et al. 2013). (see discussion below in Sect. 2.3.) Excessive microglia activation in vivo in young adults (age 18–31) affected by ASDs was confirmed by Suzuki and colleagues using PET with [11C]-(R)-PK11195. In this study, ASD brain regions showing increased binding of the radiotracer included the cerebellum, midbrain, pons, fusiform gyri, and the anterior cingulate and orbitofrontal cortices. The most prominent increase was observed in the cerebellum (Suzuki et al. 2013).

14.2.3 The Central Role of Microglia in Rett Syndrome, A Syndromic ASD

RTT is one of the "syndromic" ASDs. These are a group of genetically diverse neurodevelopmental disorders with high penetrance of ASD diagnosis (Levitt and Campbell 2009) that include fragile X (FXS), Rett, Angelman, Prader-Willi,

15q duplication, Timothy, and Smith-Lemli-Opitz syndromes, as well as neurofibromatosis and tuberous sclerosis complex (TSC). Although they represent only a minority of children with autism, studying these disorders is likely to lead to important clues about the pathogenetic pathways of autism considering that they are relatively homogeneous compared to classic autism and show clear genetic and metabolic abnormalities that are amenable to animal modeling (Levitt and Campbell 2009). For example, autism occurs in approximately 30 % of FXS cases, and PDD-NOS in an additional 30 % of FXS cases (Hagerman et al. 2010). Similarly, abnormal expression of methyl CpG-binding protein 2 (MeCP2), the protein that is deficient in patients with RTT due to loss-of-function mutations of the X-linked MECP2, also significantly contributes to the development of autism, due to epigenetic dysregulation (Lasalle and Yasui 2009). Reduced MeCP2 expression has been observed in 79 % of autism brain samples, correlating with aberrant methylation of the MECP2 promoter in male autism samples (Nagarajan et al. 2006).

RTT primarily affects young girls ($Mecp2^{-/+}$ with mosaic expression of MeCP2 due to X-chromosome inactivation), who develop normally until 6–18 months of age, at which time they start a progressive loss of neurodevelopmental milestones. This regressive period is characterized by deceleration of brain growth, loss of motor skills (including purposeful hand movements), ataxia, loss of vocalization skills, loss of cognitive capability, onset of autistic features, seizures, and respiratory dysfunction. This regressive course is strikingly similar to that of regressive autism (Zoghbi 2003). Both RTT and regressive autism share the neuropathological basis of dendritic and synaptic abnormalities (Zoghbi 2003; Chahrour and Zoghbi 2007; Shepherd and Katz 2011), suggesting that MeCP2 dysregulation could also play a role in promoting the regressive course of autism.

Remarkably, microglia play an essential role in RTT, as demonstrated by mouse model studies. First, Maezawa and Jin identified a neurotoxic activity from highly pure MeCP2-deficient microglia derived from a line of Mecp2-null mice and later identified the activity as glutamate, a major excitatory neurotransmitter acting as an excitatory neurotoxin when in excess. MeCP2-deficient microglia produce five times more glutamate than wild-type microglia, in part due to a higher expression of the mitochondrial enzyme glutaminase which generates glutamate from glutamine (Maezawa and Jin 2010). The over-produced glutamate was also demonstrated to be toxic to neurons, similar to what has been shown in several neurodegenerative disorders. In the context of neurodevelopmental disorders, these findings may additionally point to a possible detrimental effect of elevated glutamate on microglial physiological properties, including their motility and phagocytosis activity, as well as neuronal and synaptic activity. (see Chaps. 3, 4 and 9 for further reading.) Microglia processes make brief and direct contacts with synaptic elements (dendritic spines, axon terminals) in adolescent and adult cerebral cortex in vivo, at regular frequencies which may depend on neuronal activity (Wake et al. 2009; Tremblay et al. 2010), and presumably on activity-dependent glutamate release, considering that microglia express several ionotropic and metabotropic glutamate receptors. (see Chap. 3 for more information.) In the larval zebrafish, microglial processes preferentially navigate toward and subsequently make bulbous contact with neurons exhibiting high levels of spontaneous activity. Increasing neuronal activity by glutamate uncaging also steers microglial processes toward the uncaging site while inducing the formation of bulbous endings (Li et al. 2012). Pharmacological studies in retinal explants further revealed that microglial morphological parameters (such as size and complexity of dendritic morphology) and process motility were increased by ionotropic glutamatergic neurotransmission and decreased by ionotropic GABAergic neurotransmission (Fontainhas et al. 2011). Interestingly, the net consequence of microglia-neuron contact during activity-dependent glutamate release was a reduction in neuronal activity (Li et al. 2012). Thus, microglia could perform a delicate regulation of neuronal activity, using glutamate as an essential signaling molecule. Within this perspective, the "hyperactive" phenotype of RTT microglia might disturb normal homeostasis by over-producing glutamate, thus contributing to the synaptic dysfunction in RTT; however, this hypothesis remains to be experimentally determined.

Another unexplored aspect is the particular mechanism by which glutamate over-production might affect microglial phagocytic activity (Domercq et al. 2013). In a recent study by Derecki and colleagues, a bone marrow transplant experiment that successfully engrafted microglia-like cells from donor wild-type mice into the brains of recipient Mecp2-null mice was conducted, leading to the surprising discovery that RTT-like pathology could be halted in these mice. The null mice receiving transplant showed remarkable improvement in several RTT-like phenotypes including apnea, tremors, and locomotor performance, and their lifespan was extended from 8 to 10 weeks to nearly 1 year. This study further revealed that primary Mecp2-null microglia have a striking impairment in their phagocytic ability, while pharmacological inhibition of phagocytic activity, by using annexin V to block phosphatidylserine residues on apoptotic targets, diminished the beneficial effects of the transplant (Derecki et al. 2012). The exact microglial mechanism implicated in RTT remains to be determined, but this study showing that simply providing healthy myeloid cells rescues a severe brain pathology strongly supports a central role for systemic and brain immune function in ASDs, and possibly other neurodevelopmental disorders affecting synaptic connectivity. The evidence for an emerging role of microglial phagocytosis of synapses in neurodevelopment is discussed in details in Chap. 9 (Tremblay et al. 2010; Paolicelli et al. 2011; Schafer et al. 2012).

14.3 Schizophrenia

Schizophrenia is a severe neuropsychiatric disorder affecting about 1 % of the world population. The onset of full-blown schizophrenia is typically in late adolescence or early adulthood, with characteristic symptoms which are both positive (delusions, hallucinations, and disorganized speech) and negative (affective flattening, alogia, and avolition). Cognitive functioning, working memory, learning, attention, and

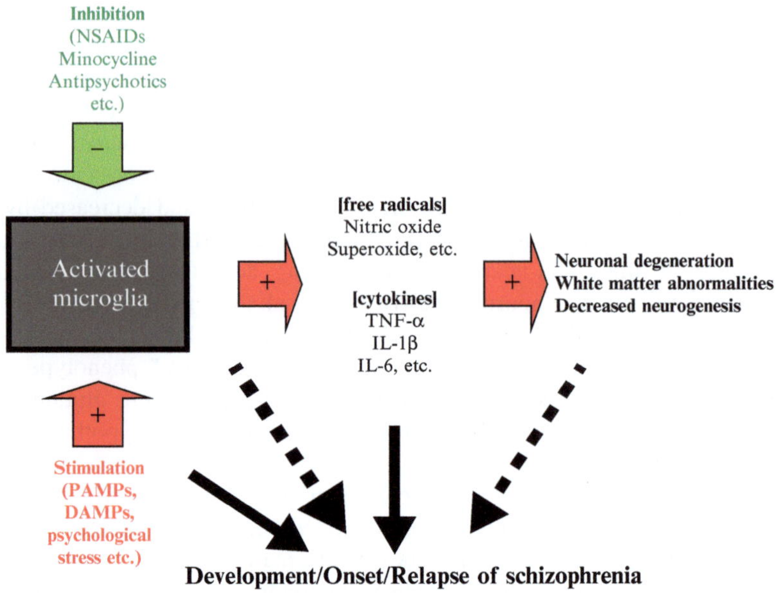

Fig. 14.1 *The microglial hypothesis of schizophrenia*. Several lines of evidence suggest that microglia might be involved in schizophrenia. (see main text for details.) Microglia activated by PAMPs and DAMPs (pathogen- and danger-associated molecular patterns), as well as psychosocial stress produce free radicals and pro-inflammatory cytokines, which contribute to neuronal degeneration and white matter abnormality, in addition to reducing neurogenesis. Together, these alterations may contribute to the development, onset, or relapse of schizophrenia. Microglial inflammatory activity is reduced by NSAIDs (non-steroidal anti-inflammatory drugs), antibiotics such as minocyline, and antipsychotic drugs. Direct evidence is shown by the *solid arrows*, whereas hypotheses to be tested are shown by the *dashed arrows*

sensory processing also become impaired, as in autism, sometimes causing great emotional distress. In addition to severely disrupting the life of the patient and their family, schizophrenia incurs a great cost to society in terms of years lost and treatment-related expenses. Many recent neuroimaging studies using magnetic resonance imaging (MRI) have provided evidence for a progressive brain atrophy over the course of schizophrenia (Davis et al. 2003; Kumra et al. 2005; Salisbury et al. 2007; Hulshoff Pol and Kahn 2008; Womack et al. 2011), thus suggesting neurodevelopmental and neurodegenerative aspects to the disease, which are together resulting in cognitive impairment (Lieberman 1999; Perez-Neri et al. 2006). Increased serum and CSF levels of the calcium-binding protein S100B, a neuroinflammation marker produced mainly by astrocytes that is associated with the destruction of brain tissue in various forms of neurodegeneration, have been reported in patients showing negative symptoms (Schmitt et al. 2005). In this section we will review the roles of neuroinflammation and microglial cells particularly in apoptosis and white matter disorders in the pathophysiology of schizophrenia and discuss the potential therapeutic strategies aimed at targeting microglia (Fig. 14.1).

14.3.1 Neuroinflammation and Schizophrenia

Jurius Wagner-Jauregg, who was awarded the Nobel Prize in Medicine in 1927, first proposed the use of fever induction as a method for treating mental diseases. His well-known "pyrotherapy" might have instigated a field of investigation regarding this involvement of the immune system in schizophrenia. An accumulating body of evidence points to the significance of neuroinflammation in schizophrenia, characterized by an increased serum concentration of several pro-inflammatory cytokines including IL-2, IL-6, and IL-8 (Lin et al. 1998; Zhang et al. 2004; Drzyzga et al. 2006; Potvin et al. 2008; Meyer 2011; Miller et al. 2011). A DNA microarray study also revealed increased expression of genes related to immune functions in the dorsolateral prefrontal cortex (involved in sustaining attention and working memory) of schizophrenic patients (Arion et al. 2007), while another investigation of the same brain region revealed changes in the molecular substrates of schizophrenia from early to chronic stages of disease progression (Narayan et al. 2008). In the latter study, short-term illness was particularly associated with a gene expression profile characterized by the disruption of gene transcription, metal-binding, RNA expression, and vesicle-mediated transport, whereas long-term illness was associated with inflammation, stimulus–response, and immune functions (Narayan et al. 2008). Recent genome-wide studies in schizophrenia additionally revealed an association with genetic markers within the major histocompatibility complex (MHC), which is involved in antigen presentation (see Chap. 5 for additional information) (Stefansson et al. 2009; Jia et al. 2010). Peripheral inflammatory responses were further linked to variations in the number of circulating monocytes and T-cells in the transcriptome of schizophrenic patients (Drexhage et al. 2010). Moreover, recent reports have shown elevated levels of IL-1β in the CSF of drug-naïve patients (Soderlund et al. 2009), as well as blood lymphocyte abnormalities in drug-naïve first episode psychosis (Miller et al. 2013), thus suggesting immune dysfunction throughout the brain and periphery. With two additional studies demonstrating increased levels of inflammatory markers in schizophrenia (Dean et al. 2013; Fillman et al. 2013), including TNF receptor 1 mRNA (Dean et al. 2013), these observations together support an important immunological aspect of the disease neuropathology.

Several epidemiologic studies demonstrate a significant impact of exposures, obstetric complications, and environmental toxins during perinatal development on the risk of developing schizophrenia, thus pointing to an immune/inflammatory etiology as in autism. Infection with influenza, rubella, herpes simplex virus-type2, or toxoplasma gondii during the perinatal period of life acts as a vulnerability factor for late-life alterations in the production of pro-inflammatory cytokines, accompanied by marked changes in cognitive and affective behavior throughout the lifespan. Two case-controlled studies revealed an association between elevated serum levels of maternal TNFα or IL-8 and increased risk of developing schizophrenia in the offspring (Ashdown et al. 2006; Nawa and Takei 2006; Deverman and Patterson 2009; Ellman and Susser 2009; Brown and Derkits 2010). Other known risk factors for schizophrenia such as malnutrition and psychological stress during prenatal development are similarly accompanied by upregulation of pro-inflammatory cytokines in the serum of the mother (Deverman and Patterson 2009). Evidence from recent animal studies suggests that most viral infections do not appear to cross the placenta; therefore, their teratogenic influence might be more related to maternal, foetal, and/or placental responses to infection.

Schizophrenic patients have a reduced lifespan, with cardiovascular disorders being the main cause of death. In addition, these patients have a high prevalence of type 2 diabetes mellitus, which is a particularly strong risk factor for cardiovascular mortality, i.e., a risk equivalent to myocardial infarction, and is linked to neurodegenerative diseases. Diabetes mellitus is a metabolic syndrome characterized by high blood sugar that is related to an inflammatory state in the fat tissue. In recent years, there has been an exponential increase in the schizophrenia literature discussing the high prevalence of type 2 diabetes and pre-diabetic states including metabolic syndrome. These results are very important because insulin resistance has been observed in antipsychotic-naïve schizophrenic patients. In addition, second-generation antipsychotics such as olanzapine and clozapine have been associated with the development of metabolic syndrome which is a cluster of the most dangerous heart attack risk factors, and type 2 diabetes (Meyer and Stahl 2009; Devaraj et al. 2010; Nielsen et al. 2010). Beumer and colleagues recently suggested that the increased serum levels of cytokines, chemokines, and adipokines in schizophrenic patients may not only be related to the disease itself, but also to the metabolic syndrome (Beumer et al. 2012). Lastly, C-reactive protein (CRP) is a pentameric protein that is generated largely in the liver and secreted in the blood, providing a reliable marker of chronic inflammation in the body. In the blood of schizophrenic patients, CRP levels were found to be increased and further associated with the cardiovascular risk factors especially in patients treated with antipsychotics of the second generation which will be discussed below (Dieset et al. 2012; Dickerson et al. 2013).

14.3.2 Schizophrenia and Microglial Activation

To account for these immunological changes in schizophrenia, Bilbo and colleagues hypothesized that a subset of microglial cells could permanently remain in a "primed" state following perinatal infection, causing exaggerated release of pro-inflammatory cytokines by the "primed" microglia upon subsequent immune challenge (Bilbo and Schwarz 2009; Bland et al. 2010). Many infections can induce symptoms of mental illness, but even though most symptoms generally disappear after recovery from the acute illness, some symptoms may persist if the acute infection becomes chronic. Microbes, which have the ability to permanently reside in the body, could permanently disturb brain functions. In the prefrontal cortex of schizophrenic patients, an increased prevalence of DNA from the infecting bacteria

Chlamydophila causing microglial activation was indeed reported (Fellerhoff and Wank 2011), suggesting that chronically infected microglia may permanently impair normal brain functions thus represent a mechanism causing illness in schizophrenia. Using an analog of viral double-stranded RNA (polynsinic-polycytidylic acid sodium salt [poly I:C]), a recent study further revealed that maternal infection during embryogenesis contributes to exacerbating microglial activation in the offspring (Juckel et al. 2011).

Recent animal studies have additionally shown that psychological stress induces changes in microglial morphology suggestive of an altered function. For example, microglia become more numerous and display hyper-ramified processes in the hippocampus and prefrontal cortex after repeated restraint stress, and they lose their processes in the hippocampus, prefrontal cortex, and amygdala following repeated social defeat (Frank et al. 2007; Schiavone et al. 2009; Tynan et al. 2010; Wohleb et al. 2011; Hinwood et al. 2012). A combined exposure to prenatal immune challenge and variable, unpredictable stress during peripubertal development was lastly shown to induce a synergy of pathological effects on adult behavior, notably the impairment of associative learning, accompanied by microglial activation, including increased numbers of primary processes and branch points in hippocampus (Giovanoli et al. 2013).

Hypoglutamatergic states and impaired *N*-methyl-D-aspartate (NMDA) signaling are considered as main contributors to the pathophysiology of schizophrenia. NMDA antagonists such as phencyclidine (PCP), ketamine, and MK-801 are commonly used for modeling schizophrenia in experimental animals (Javitt 2010). However, all three NMDA antagonists are known to induce microglial activation in rodent brains (Nakki et al. 1995, 1996). Interestingly, an increased microglial density is also suggested by postmortem observations, at least in a subpopulation of individuals with schizophrenia (Bayer et al. 1999; Radewicz et al. 2000; Steiner et al. 2006; Schnieder and Dwork 2011). Highly elevated microglial cell numbers were measured in the anterior cingulate cortex and mediodorsal thalamus of schizophrenic patients who committed suicide during acute psychosis (Steiner et al. 2006). By using [11C] (R)-PK11195 to visualize microglial activation in a non-invasive manner, researchers have recently reported an increased microglial activation in the grey matter or hippocampus of patients with schizophrenia (van Berckel et al. 2008a; Doorduin et al. 2009). These PET studies demonstrated that activated microglia are present in the early stages of disease putatively associated with ongoing neuronal injury (van Berckel et al. 2008a, b), but also in patients just recovering from a psychotic state in which neuroinflammation has been proposed to be exacerbated (Doorduin et al. 2009). Another recent PET study in chronic schizophrenia has shown a positive correlation between levels of the [11C] DAA1106 radiotracer, binding to the peripheral benzodiazepin receptor similarly to [11C] (R)-PK11195, and the positive symptoms scores as well as the illness duration in schizophrenic patients (Takano et al. 2010). Nevertheless, it remains unclear whether microglial activation is neuroprotective or detrimental in schizophrenia.

14.3.3 Schizophrenia and Apoptosis

Structural brain abnormalities are extensively and consistently documented in schizophrenia. Longitudinal studies of brain structure using high-resolution MRI found a progressive reduction of brain volumes over time, following the course of illness and treatment outcome in schizophrenic patients (Davis et al. 2003; Kumra et al. 2005; Salisbury et al. 2007). A recent review proposed that brain tissue progressively decreases, while lateral ventricle volume increases, in chronically ill patients with schizophrenia, until over 20 years after onset of the symptoms (Hulshoff Pol and Kahn 2008). In fact, multiple lines of evidence combine to implicate an increased susceptibility for apoptotic death in the pathophysiology of schizophrenia. Indeed, the well-characterized reduction in neuronal and astrocytic numbers, and the decreased volumes of neuropil, is considered as especially caused by the loss of synaptic elements. Neuroimaging evidence of progressive gray matter loss has been observed early over the course of disorder, as described above, and make apoptosis a plausible mechanism explaining the neurodegenerative aspect of schizophrenia. The activation of apoptotic processes rapidly can lead to neuronal death. However, recent experimental data also indicate that sublethal apoptotic activity can trigger the selective elimination of synapses (Jarskog et al. 2005; Glantz et al. 2006). Pro-inflammatory cytokines such as TNFα are well characterized as mediators of oxidative stress, inducing apoptosis of human cortical neuron and oligodendrocytes in vitro (Medina et al. 2002; Buntinx et al. 2004). In addition, nitric oxide (NO) has been reported to directly induce neuronal apoptosis mediated by pro-inflammatory cytokines (Palluy and Rigaud 1996; Hu et al. 1997). Further, the interaction between NO and superoxide anion ($O2^-$), which can be generated by activated microglia, forms the highly toxic peroxynitrite ($ONOO^-$) that is well known for triggering apoptotic cell death. High levels of NO and TNFα could affect synaptogenesis, synaptic plasticity, and neuronal connectivity, as well as composition of synaptic membranes (Sunico et al. 2005; Stellwagen and Malenka 2006). In light of these observations, alteration in the synaptic organization of the brain is now emerging as one of the key features of schizophrenia (Roberts et al. 2005). Several postmortem studies found changes in the density of dendritic spines within brain regions showing the greatest indices of gray matter loss in schizophrenia, particularly the dorsolateral prefrontal cortex, superior temporal gyrus, and hippocampus. These results reveal a strong association between the region-specific loss of gray matter and reduced density of dendritic spines, respectively, measured in vivo using high-resolution MRI and in situ using light microscopy analyzes of postmortem samples, with the hypoactivity measured in vivo using functional MRI (fMRI) (Penzes et al. 2011). Therefore, it is becoming clear that a synaptic pathology coexists with the inflammatory response in schizophrenic patients. Nonetheless, direct evidence that microglial NO and pro-inflammatory cytokines are responsible for the loss of synapses in these patients is still missing.

14.3.4 Schizophrenia and White Matter Disorders

Neuroimaging studies using diffusion tensor imaging (DTI), high-resolution MRI, and fMRI have shown that first-episode schizophrenic patients also have a significant reduction of white matter volume resulting in abnormal brain functional connectivity (Price et al. 2006; Schlosser et al. 2007; Lee et al. 2013; Wang et al. 2013). A reduced density and compromised morphology of oligodendroglia as well as signs of deviant myelination are also evident in postmortem brains (Uranova et al. 2004, 2007; Bernstein et al. 2009). Combined with the evidence showing dysregulation of the myelination-related genes, a disruption of oligodendrocytic function is strongly suspected in schizophrenia (McCullumsmith et al. 2007). While microglial activation in the CNS has been implicated in the pathogenesis of white matter disorders, recently it has been reported that cytotoxicity to oligodendrocytes could be mediated through microglial release of molecules related to free radicals such as NO and peroxynitrite (Merrill et al. 1993; Li et al. 2005), and pro-inflammatory cytokines such as TNFα and IFNγ (Buntinx et al. 2004). In addition, TNFα has been shown to compromise the growth of oligodendrocytes and the expression of mRNA for MBP in vitro (Cammer and Zhang 1999), and to inhibit the survival and proliferation of oligodendrocytic progenitors, and their subsequent differentiation into mature myelinating phenotypes in vitro (Feldhaus et al. 2004). These results are intriguing because Mittelbronn and colleagues previously demonstrated that microglial density differs in the normal adult human brain by up to one order of magnitude between regions, with significantly more microglia observed in white matter than in gray matter (Mittelbronn et al. 2001). Therefore, it is possible that microglial dysfunction in schizophrenic patients leads to alterations in oligodendrocytic function resulting in aberrant brain connectivity.

14.4 Treatments Aiming at Controlling Neuroinflammation

Although ASDs and schizophrenia are two apparently disparate groups of disorders, developing at different stages of the lifespan, they share several clinical characteristics such as a pronounced impairment of social cognition (King and Lord 2011; Sugranyes et al. 2011; Lugnegard et al. 2013). A recent complex network and computational analysis suggests that genetic variations associated with ASDs and schizophrenia occur in the same molecular pathways and functional domains (Cristino et al. 2014). From the above discussion, it is clear that abnormal microglial activation or intrinsic abnormalities, possibly resulting from maternal stress, infections, and obstetric complications, are also a shared feature of these disorders. Taming microglia-associated neuroinflammation as a therapeutic approach has been extensively studied in schizophrenia, but only recently proposed or tested for ASDs.

The effect of antipsychotics on microglial activation

drugs	microglia	stimulant	production of cytokines	NO production	Journals
Flupentixol **Trifluperidol**	Primary Culture	LPS	TNF-α: Inhibited TNF-α: Inhibited	Inhibited Inhibited	Kowalski J et al.(2005) Pol.J.Pharmacol.
Chlorpromazine **Loxapine**	Primary Culture	LPS	IL-1β& IL-2 Inhibited IL-1β& IL-2 Inhibited		Labuzek K et al.(2005) Eur.Neuropsychopharmacol.
Haloperidol **Clozapine** **Olanzapine**	Cell Line (N 9)	LPS		Not Inhibited Not Inhibited Inhibited	Hou Y et al.(2006) Prog Neuro-Psychopharmacology & Biol Psychiatry
Haloperidol **Risperidone**	Cell Line (6-3)	**IFN-Y**	IL-1β, IL-6 & TNF-α:Inhibited IL-1β, IL-6 & TNF-α:Inhibited	Inhibited(+–) Inhibited(++)	Kato T et al.(2007) Schizophrenia Res
Quetiapine **Perospirone** **Ziprasidone**	Cell Line (6-3)	**IFN-Y**	TNF-α: Inhibited TNF-α: Inhibited TNF-α: Activated	Inhibited(++) Inhibited(++) Inhibited(++)	Bian Q et al.(2008) Prog Neuro-Psychopharmacology & Biol Psychiatry
Aripiprazole	Cell Line (6-3)	**IFN-Y**	TNF-α: Inhibited	Inhibited(++)	Kato T et al.(2008) J Neurochem
	Primary Culture	**LPS**		Inhibited(++)	

First generation antipsychotics are Flupentixol, Trifluperidol, Chlorpromazine, Loxapine, and Haloperidol while **second generation antipsychotics** are Clozapine, Riperidone, Quetiapine, Perospirone, Ziprasidone, and Aripiprazole.
Aripirrazole is a dopamine-partial agonist while all the rest are dopamine-antagonists.

Fig. 14.2 *The effect of antipsychotics on microglial activation*. First-generation antipsychotics are Flupentixol, Trifluperidol, Chlorpromazine, Loxapine, and Haloperidol, while second-generation antipsychotics are Clozapine, Riperidone, Quetiapine, Perospirone, Ziprasidone, and Aripiprazole. Aripiprazole is a dopamine-partial agonist while all the rest are dopamine-antagonists

14.4.1 Anti-inflammatory Strategy for Schizophrenia

Second-generation, also named atypical, antipsychotics such as olanzapine and risperidone are becoming standard drugs for alleviating symptoms in schizophrenia because of their reduced adverse effects compared with first-generation antipsychotics, and increased effectiveness against negative symptoms. These antipsychotics are acting in various manners on the brain neuromodulatory systems, having dopamine D2 receptors antagonistic activity for most of them, and partial D2 receptors agonist activity in the case of aripiprazole, in addition to binding to receptors for other neuromodulators such as serotonin (Lieberman et al. 2005). Certain second-generation drugs also seem to exert positive effects on neuronal cell growth and survival by increasing the levels of growth factors, at least in vitro (Lu and Dwyer 2005), and some of these neuroprotective effects could be mediated through their pharmacological influence on the immune system, including most notably resident microglia as described below and summarized in Fig. 14.2.

In particular, in vitro studies comparing the effects of several second-generation antipsychotics on microglia have shown that flupentixol and trifluperidol reduce microglial secretion of TNFα and NO (Kowalski et al. 2003), and flupentixol, trifluperidol, chlorpromazine, and loxapine reduce microglial secretion of IL-1β and IL-2 (Kowalski et al. 2004; Labuzek et al. 2005). Olanzapine also inhibits microglial release of NO, contrarily to the first-generation antipsychotics haloperidol and clozapine (Hou et al. 2006). Furthermore, the second-generation antipsychotic risperidone inhibits microglial production of NO, IL-1β, IL-6, and TNFα following their stimulation with IFNγ, contrarily to the effects of first-generation haloperidol (Kato et al. 2007). Moreover, the serum levels of IL-2 and IFN-γ, and their production by peripheral blood mononuclear cells (PBMC) stimulated with the mitogen phytohemagglutinin (PHA), were reported to be significantly higher in schizophrenia (Cazzullo et al. 2001). More recently, the same inhibitory effects on IFN-γ-induced microglial activation were demonstrated for the second-generation antipsychotics perospirone and quetiapine (Bian et al. 2008). Sugino and colleagues further revealed that clozapine, olanzapine, and risperidone, but not haloperidol, suppress production of the pro-inflammatory cytokines TNFα and IL-6 and up-regulate secretion of the anti-inflammatory cytokine IL-10 in lipopolysaccharide (LPS)-treated mice. This study has also shown that clozapine alone can suppress poly I:C-induced inflammation in mice (Sugino et al. 2009). Similarly, risperidone was recently reported to normalize LPS-induced changes, by decreasing expression of IL-1β, TNFα, and the activity of inducible pro-inflammatory nitric oxide synthase (NOS) and cyclooxygenase (COX), while increasing activity of anti-inflammatory pathways involving deoxyprostaglandins and peroxisome proliferator-activated receptor γ (PPARγ) (MacDowell et al. 2013).

Even though aripiprazole is a partial dopamine D2 receptor agonist, contrarily to the other second-generation drugs discussed above, it was similarly found to inhibit the generation of NO and TNFα from IFNγ-activated microglia in vitro (Kato et al. 2008). Aripiprazole was also recently demonstrated to inhibit the production of superoxide through the nicotinamide dinucleotide phosphate (NADPH) oxidase system (NOX) (McKeith 2006) in phorbol-myristate-acetate (PMA)-stimulated microglia in vitro (Kato et al. 2011). These results are very intriguing because social isolation has been reported to cause behavioral and pathological alterations resembling the symptoms of schizophrenia by increasing oxidative stress in rats, particularly microglial release of superoxide derived from the NOX system in vivo (Schiavone et al. 2009). Additionally, the loss of fast-spiking, parvalbumin (PV)-positive interneurons found in the prefrontal cortex following rodent administration of ketamine, as a model of schizophrenia, was shown to be mediated by increased levels of oxidative stress through the NOX system. Repetitive adult exposure to ketamine has also been reported to increase the levels of IL-6 in the brain of rodents, through activation of the NOX system, leading to a loss of the GABAergic phenotype in PV-interneurons and to a decreased inhibitory activity in the prefrontal cortex (Behrens et al. 2007; Behrens and Sejnowski 2009; Powell et al. 2012).

Together these studies suggest that some antipsychotics may have a potentially useful therapeutic influence on patients with schizophrenia by reducing inflammatory reactions, which might be implicated in the apoptotic process and white matter abnormalities described above (Lieberman et al. 2005). Surprisingly, not only antipsychotics with dopamine D2 receptor antagonism such as olanzapine and risperidol but also the dopamine D2 receptor partial agonist aripiprazole was found to exert anti-inflammatory effects by causing inhibition of microglial activation.

However, recent reports have also shown that antipsychotics can increase the production of pro-inflammatory cytokines, either in patients treated with clozapine or olanzapine, two antipsychotics associated with fever induction, or in human blood stimulated with toxic shock syndrome toxin (TSST-1) supplemented with antipsychotics in vitro (Kluge et al. 2009; Himmerich et al. 2011). Other recent reports have additionally demonstrated the possible antipsychotic effect of minocycline, a potent but non-specific antibiotic associated with the inhibition of microglial activation. In these studies, adjunctive therapy of minocycline with antipsychotics was found to be beneficial (Miyaoka et al. 2007, 2008; Levkovitz et al. 2010; Kelly et al. 2011), predominantly improving the negative symptoms which are usually resistant to antipsychotics (Chaudhry et al. 2012).

A putative mechanism contributing to the neurodevelopmental, neurodegenerative, and cognitive deficits associated with schizophrenia is microglial secretion of neurotrophic factors, especially of brain-derived neurotrophic factor (BDNF) which is a potent regulator of synaptic development and plasticity (Weickert et al. 2003). Recently, BDNF was shown to induce a sustained mobilization of intracellular calcium in primary microglial cells derived from postnatal rats and to further reduce the release of NO from activated microglia stimulated with INFγ (Mizoguchi et al. 2009). These observations suggest that BDNF could induce neuroprotection by steering microglia towards an anti-inflammatory phenotype. Additionally, a recent study has shown that an α7 nicotinic acetylcholine receptor agonist (α7nAChR) induces a sustained elevation of intracellular calcium, while suppressing the release of TNFα from LPS-activated microglia in vitro (Suzuki et al. 2006). These results are particularly relevant because α7nAChR agonists have recently emerged as a promising new treatment for improving cognitive dysfunction in schizophrenic patients (Lieberman et al. 2008). Lastly, dietary omega-3 fatty acids, which can inhibit microglial activation, were recently reported to reduce the rate of progression towards a first psychosis episode (Amminger et al. 2010; Lu et al. 2010). Based on the above results, elucidating the changes in microglial phenotype over the course of schizophrenia pathogenesis, including negative and positive symptoms, could help to discover new targets for the prevention and treatment of schizophrenia.

In this regard, immunomodulatory drugs such as cyclooxygenase-2 (COX-2) inhibitors were recently acknowledged for their beneficial effects on the symptoms of schizophrenia. A randomized, double-blind, placebo-controlled trial revealed that adjuvant therapy with the non-steroidal anti-inflammatory agents (NSAID) acetylsalicylic acid, commonly known as aspirin, can reduce the symptoms of schizophrenia (Muller et al. 2005, 2010; Akhondzadeh et al. 2007; Laan et al. 2010). A recent meta-analysis on the use of NSAIDs in schizophrenia additionally proposed NSAID augmentation as a potentially useful strategy for treating both the positive and negative symptoms, also considering the additional benefits of aspirin on reducing cardiac and cancer mortality (Sommer et al. 2012). Within this perspective, immunosuppressive or immunomodulatory drugs might prove to be beneficial, at least for the treatment of acute schizophrenia (Knight et al. 2007), but again, direct evidence in support of this hypothesis is missing.

14.4.2 Anti-inflammatory Strategies for ASDs

Second-generation antipsychotics, now standard drugs for schizophrenia, are also first-line pharmacological agents for the treatment of irritability and associated behaviors in children with autism (Chavez et al. 2007; McDougle et al. 2008; Politte and McDougle 2014). Currently, aripiprazole and risperidone are the only approved medications for treating irritability in autistic disorder by the American Food and Drug Administration (McCracken et al. 2002; Owen et al. 2009; Ghanizadeh et al. 2014). It would be interesting to determine if the microglial modulating actions of aripiprazole and risperidone described above also influence the pathophysiology of autism.

As mentioned above, minocycline, a potent but non-specific inhibitor of microglial activation, was reported to be beneficial for the treatment of schizophrenia when administered in conjunction with antipsychotics. Minocycline also provides significant functional benefits to patients with FXS (Paribello et al. 2010; Dansie et al. 2013; Leigh et al. 2013). However, a pilot open-label trial of minocycline in patients with autism and regressive features showed only negligible clinical improvements. Following treatment, only the chemokine CXCL8 (IL-8) was significantly different while no significant changes in chemokines such as MCP-1 or cytokines such as TNFα, CD40L, IL-6, IFNγ, and IL-1β were observed in CSF or serum (Pardo et al. 2013). Possible explanations for this apparent lack of anti-inflammatory effects of minocycline in autism may include the relatively late age of intervention (mean age 7.19 years; range 3–12 years) and unique pathways of microglial abnormalities in autism.

Other immunomodulatory reagents mentioned for schizophrenia in the above section may also be beneficial to children with autism. A COX2 inhibitor, celecoxib, was recently used as adjunctive treatment to risperidone in children with autism in a randomized, double-blind, placebo-controlled trial. It was found that the combination of risperidone and celecoxib was superior to risperidone alone in treating irritability, social withdrawal, and stereotypy of children with autism (Asadabadi et al. 2013). The anti-oxidant *N*-acetylcysteine (NAC) was also reported to effectively reduce irritability in patients with ASDs as an augmentation therapy adjunctive to resperidone in a randomized, double-blind, placebo-controlled trial (Ghanizadeh and Moghimi-Sarani 2013).

One potential anti-inflammatory approach is through the mTOR (mammalian target of rapamycin) pathway. mTOR is a key regulator of protein synthesis, and itself a serine/threonine kinase activated by the phosphoinositide-3 kinase (PI3K) pathway. The proteins encoded by TSC1/TSC2, NF1, and PTEN genes, the mutations of which cause syndromic ASDs such as TSC, neurofibromatosis type 1, and PTEN harmatoma tumor syndromes (Zafeiriou et al. 2013), act as negative regulators of mTOR complex 1. In addition, the mTOR downstream pathway is upregulated in patients with FXS, a frequent cause of autism. As discussed above, the pathological pathways of these genetic syndromes may converge upon common pathogenesis of ASDs (Zafeiriou et al. 2013). Activation of the mTOR pathway was found to cause proliferation and survival of mast cells and inflammatory microglia in vitro (Shang et al. 2013; Theoharides 2013; Theoharides et al. 2013). Mast cells, although best known for their role in allergy and anaphylaxis in the periphery, reside also in the brain and are able to release various mediators activating microglia. They serve a potentially important peripheral immune signaling link to the brain in an inflammatory setting (Skaper et al. 2012, 2014) and, in the context of the above mTOR-activating mutations of ASDs, could aggravate microglia-mediated neuroinflammation (Skaper et al. 2012). Since mTOR inhibitors have shown efficacy in tuberous sclerosis patients and experimental animals, they could also have an efficacy in patients with autism (Ehninger and Silva 2011; Curatolo and Moavero 2012; Sahin 2012; Wang and Doering 2013). Two open-label pilot trials of an anti-inflammatory mTOR inhibitor called luteolin found encouraging improvement in several outcome domains as well as gastrointestinal and allergy symptoms in patients with ASDs (Theoharides et al. 2012; Taliou et al. 2013).

Although our attention here is on microglia, we need to keep in mind that the CNS is by no means immune-privileged as previously thought. Cells instrumental in peripheral immune regulation such as dendritic cells and mast cells also reside in the brain parenchyma (Skaper et al. 2012; Colton 2013), and various leukocytes can migrate into the brain through discreet and highly regulated portals in immune surveillance to help initiate neuroinflammation (Kivisakk et al. 2003; Galea et al. 2007; Reboldi et al. 2009). Therefore, the pharmacological actions of the above described antipsychotics, minocycline, NSAIDs, and mTOR inhibitors should be considered in the broader context of immune interactions between CNS microglia (and even neurons and astrocytes) and cells of the peripheral immune system. In this spirit, the systemic immune dysregulation and aberrant microglia activation seen in autism (see Sects. 2.1 and 2.2) are interrelated, and anti-inflammatory strategies to treat allergic gastrointestinal symptoms, for example, could also be effective in ameliorating neuroinflammation in ASDs.

14.5 Conclusions

In many aspects, the neuropathology of schizophrenia and ASDs is closely associated with neuroinflammation, especially microglial activation. As described above, their respective anti-inflammation therapies are also strikingly similar. Our understanding about the connection between microglial abnormalities and neuropsychiatric disorders is at its infant stage, but we foresee the tremendous potential of therapeutic approaches targeting microglia. This anticipation is based on two considerations. First, microglia quickly respond to various stimuli with an amazingly dynamic behavioral repertoire, therefore these immune cells are highly transmutable for therapeutic interventions. Second, our view of the brain being "immune privileged" has fundamentally changed in recent years, with the realization that the systemic immune system is able to modulate multiple brain functions (Lynch and Mills 2012), in part through interactions of peripheral immune cells such as dendritic cells and mast cells with their CNS counterparts, such as microglia. Therefore, approaches to modify the neuroinflammatory environment that is conducive to neuropsychiatric disorders can be launched peripherally. Peripheral immune cells are relatively straightforward to isolate and manipulate ex vivo, and they can enter the CNS via endogenous mechanisms (Derecki et al. 2013), providing a foundation for the development of novel cell-based therapies or "vaccines". To achieve this goal, a fundamental question that needs to be investigated in depth is how peripheral immune modulation of microglial physiological properties affects normal brain function and dysregulation in diseases. In our view, ASDs and schizophrenia provide excellent opportunities for gaining such understanding of the complex interactions between CNS and systemic immune system, a gateway to a true "mind-body" problem.

References

Akhondzadeh S, Tabatabaee M, Amini H, Abhari SAA, Abbasi SH, Behnam B (2007) Celecoxib as adjunctive therapy in schizophrenia: a double-blind, randomized and placebo-controlled trial. Schizophr Res 90:179–185

Amminger GP, Schafer MR, Papageorgiou K, Klier CM, Cotton SM, Harrigan SM, Mackinnon A, McGorry PD, Berger GE (2010) Long-chain omega-3 fatty acids for indicated prevention of psychotic disorders: a randomized, placebo-controlled trial. Arch Gen Psychiatry 67:146–154

Arion D, Unger T, Lewis DA, Levitt P, Mirnics K (2007) Molecular evidence for increased expression of genes related to immune and chaperone function in the prefrontal cortex in schizophrenia. Biol Psychiatry 62:711–721

Asadabadi M, Mohammadi MR, Ghanizadeh A, Modabbernia A, Ashrafi M, Hassanzadeh E, Forghani S, Akhondzadeh S (2013) Celecoxib as adjunctive treatment to risperidone in children with autistic disorder: a randomized, double-blind, placebo-controlled trial. Psychopharmacology (Berl) 225:51–59

Ashdown H, Dumont Y, Ng M, Poole S, Boksa P, Luheshi GN (2006) The role of cytokines in mediating effects of prenatal infection on the fetus: implications for schizophrenia. Mol Psychiatry 11:47–55

Bayer TA, Buslei R, Havas L, Falkai P (1999) Evidence for activation of microglia in patients with psychiatric illnesses. Neurosci Lett 271:126–128

Behrens MM, Sejnowski TJ (2009) Does schizophrenia arise from oxidative dysregulation of parvalbumin-interneurons in the developing cortex? Neuropharmacology 57:193–200

Behrens MM, Ali SS, Dao DN, Lucero J, Shekhtman G, Quick KL, Dugan LL (2007) Ketamine-induced loss of phenotype of fast-spiking interneurons is mediated by NADPH-oxidase. Science 318:1645–1647

Benavides J, Cornu P, Dennis T, Dubois A, Hauw JJ, MacKenzie ET, Sazdovitch V, Scatton B (1988) Imaging of human brain lesions with an omega 3 site radioligand. Ann Neurol 24:708–712

Bernstein HG, Steiner J, Bogerts B (2009) Glial cells in schizophrenia: pathophysiological significance and possible consequences for therapy. Expert Rev Neurother 9:1059–1071

Betancur C, Sakurai T, Buxbaum JD (2009) The emerging role of synaptic cell-adhesion pathways in the pathogenesis of autism spectrum disorders. Trends Neurosci 32:402–412

Beumer W, Drexhage RC, De Wit H, Versnel MA, Drexhage HA, Cohen D (2012) Increased level of serum cytokines, chemokines and adipokines in patients with schizophrenia is associated with disease and metabolic syndrome. Psychoneuroendocrinology 37:1901–1911

Bian Q, Kato T, Monji A, Hashioka S, Mizoguchi Y, Horikawa H, Kanba S (2008) The effect of atypical antipsychotics, perospirone, ziprasidone and quetiapine on microglial activation induced by interferon-gamma. Prog Neuropsychopharmacol Biol Psychiatry 32:42–48

Bilbo SD, Schwarz JM (2009) Early-life programming of later-life brain and behavior: a critical role for the immune system. Front Behav Neurosci 3:14

Bland ST, Beckley JT, Young S, Tsang V, Watkins LR, Maier SF, Bilbo SD (2010) Enduring consequences of early-life infection on glial and neural cell genesis within cognitive regions of the brain. Brain Behav Immun 24:329–338

Breece E, Paciotti B, Nordahl CW, Ozonoff S, Van de Water JA, Rogers SJ, Amaral D, Ashwood P (2013) Myeloid dendritic cells frequencies are increased in children with autism spectrum disorder and associated with amygdala volume and repetitive behaviors. Brain Behav Immun 31:69–75

Brown AS, Derkits EJ (2010) Prenatal infection and schizophrenia: a review of epidemiologic and translational studies. Am J Psychiatry 167:261–280

Buntinx M, Moreels M, Vandenabeele F, Lambrichts I, Raus J, Steels P, Stinissen P, Ameloot M (2004) Cytokine-induced cell death in human oligodendroglial cell lines: I Synergistic effects of IFN-gamma and TNF-alpha on apoptosis. J Neurosci Res 76:834–845

Cammer W, Zhang H (1999) Maturation of oligodendrocytes is more sensitive to TNF alpha than is survival of precursors and immature oligodendrocytes. J Neuroimmunol 97:37–42

Casanova MF (2007) The neuropathology of autism. Brain Pathol 17:422–433

Cazzullo CL, Sacchetti E, Galluzzo A, Panariello A, Colombo F, Zagliani A, Clerici M (2001) Cytokine profiles in drug-naive schizophrenic patients. Schizophr Res 47:293–298

Chahrour M, Zoghbi HY (2007) The story of Rett syndrome: from clinic to neurobiology. Neuron 56:422–437

Chaudhry IB, Hallak J, Husain N, Minhas F, Stirling J, Richardson P, Dursun S, Dunn G, Deakin B (2012) Minocycline benefits negative symptoms in early schizophrenia: a randomised double-blind placebo-controlled clinical trial in patients on standard treatment. J Psychopharmacol 26:1185–1193

Chavez B, Chavez-Brown M, Sopko MA Jr, Rey JA (2007) Atypical antipsychotics in children with pervasive developmental disorders. Paediatr Drugs 9:249–266

Chez MG, Dowling T, Patel PB, Khanna P, Kominsky M (2007) Elevation of tumor necrosis factor-alpha in cerebrospinal fluid of autistic children. Pediatr Neurol 36:361–365

Colton CA (2013) Immune heterogeneity in neuroinflammation: dendritic cells in the brain. J Neuroimmune Pharmacol 8:145–162

Cristino AS, Williams SM, Hawi Z, An JY, Bellgrove MA, Schwartz CE, Costa LD, Claudianos C (2014) Neurodevelopmental and neuropsychiatric disorders represent an interconnected molecular system. Mol Psychiatry 19(3):294–301

Curatolo P, Moavero R (2012) mTOR inhibitors in tuberous sclerosis complex. Curr Neuropharmacol 10:404–415

Dansie LE, Phommahaxay K, Okusanya AG, Uwadia J, Huang M, Rotschafer SE, Razak KA, Ethell DW, Ethell IM (2013) Long-lasting effects of minocycline on behavior in young but not adult Fragile X mice. Neuroscience 246:186–198

Davis KL, Stewart DG, Friedman JI, Buchsbaum M, Harvey PD, Hof PR, Buxbaum J, Haroutunian V (2003) White matter changes in schizophrenia: evidence for myelin-related dysfunction. Arch Gen Psychiatry 60:443–456

Dean B, Gibbons AS, Tawadros N, Brooks L, Everall IP, Scarr E (2013) Different changes in cortical tumor necrosis factor-alpha-related pathways in schizophrenia and mood disorders. Mol Psychiatry 18:767–773

Derecki NC, Cronk JC, Lu Z, Xu E, Abbott SB, Guyenet PG, Kipnis J (2012) Wild-type microglia arrest pathology in a mouse model of Rett syndrome. Nature 484:105–109

Derecki NC, Cronk JC, Kipnis J (2013) The role of microglia in brain maintenance: implications for Rett syndrome. Trends Immunol 34:144–150

Devaraj S, Dasu MR, Jialal I (2010) Diabetes is a proinflammatory state: a translational perspective. Expert Rev Endocrinol Metab 5:19–28

Deverman BE, Patterson PH (2009) Cytokines and CNS development. Neuron 64:61–78

Dickerson F, Stallings C, Origoni A, Vaughan C, Khushalani S, Yang SJ, Yolken R (2013) C-reactive protein is elevated in schizophrenia. Schizophr Res 143:198–202

Dieset I, Hope S, Ueland T, Bjella T, Agartz I, Melle I, Aukrust P, Rossberg JI, Andreassen OA (2012) Cardiovascular risk factors during second generation antipsychotic treatment are associated with increased C-reactive protein. Schizophr Res 140:169–174

Domercq M, Vazquez-Villoldo N, Matute C (2013) Neurotransmitter signaling in the pathophysiology of microglia. Front Cell Neurosci 7:49

Doorduin J, de Vries EFJ, Willemsen ATM, de Groot JC, Dierckx RA, Klein HC (2009) Neuroinflammation in schizophrenia-related psychosis: a PET study. J Nucl Med 50: 1801–1807

Drexhage RC, Knijff EM, Padmos RC, van der Heul-Nieuwenhuijzen L, Beumer W, Versnel MA, Drexhage HA (2010) The mononuclear phagocyte system and its cytokine inflammatory networks in schizophrenia and bipolar disorder. Expert Rev Neurother 10:59–76

Drzyzga L, Obuchowicz E, Marcinowska A, Herman ZS (2006) Cytokines in schizophrenia and the effects of antipsychotic drugs. Brain Behav Immun 20:532–545

Ehninger D, Silva AJ (2011) Rapamycin for treating Tuberous sclerosis and Autism spectrum disorders. Trends Mol Med 17:78–87

Ellman LM, Susser ES (2009) The promise of epidemiologic studies: neuroimmune mechanisms in the etiologies of brain disorders. Neuron 64:25–27

Feldhaus B, Dietzel ID, Heumann R, Berger R (2004) Effects of interferon-gamma and tumor necrosis factor-alpha on survival and differentiation of oligodendrocyte progenitors. J Soc Gynecol Investig 11:89–96

Fellerhoff B, Wank R (2011) Increased prevalence of Chlamydophila DNA in post-mortem brain frontal cortex from patients with schizophrenia. Schizophr Res 129:191–195

Fillman SG, Cloonan N, Catts VS, Miller LC, Wong J, McCrossin T, Cairns M, Weickert CS (2013) Increased inflammatory markers identified in the dorsolateral prefrontal cortex of individuals with schizophrenia. Mol Psychiatry 18:206–214

Fontainhas AM, Wang MH, Liang KJ, Chen S, Mettu P, Damani M, Fariss RN, Li W, Wong WT (2011) Microglial morphology and dynamic behavior is regulated by ionotropic glutamatergic and GABAergic neurotransmission. PLoS One 6:14

Frank MG, Baratta MV, Sprunger DB, Watkins LR, Maier SF (2007) Microglia serve as a neuroimmune substrate for stress-induced potentiation of CNS pro-inflammatory cytokine responses. Brain Behav Immun 21:47–59

Galea I, Bechmann I, Perry VH (2007) What is immune privilege (not)? Trends Immunol 28:12–18

Geschwind DH (2008) Autism: many genes, common pathways? Cell 135:391–395

Geschwind DH (2011) Genetics of autism spectrum disorders. Trends Cogn Sci 15:409–416

Ghanizadeh A, Moghimi-Sarani E (2013) A randomized double blind placebo controlled clinical trial of *N*-acetylcysteine added to risperidone for treating autistic disorders. BMC Psychiatry 13:196

Ghanizadeh A, Sahraeizadeh A, Berk M (2014) A head-to-head comparison of aripiprazole and risperidone for safety and treating autistic disorders, a randomized double blind clinical trial. Child Psychiatry Hum Dev 45(2):185–192

Giovanoli S, Engler H, Engler A, Richetto J, Voget M, Willi R, Winter C, Riva MA, Mortensen PB, Feldon J, Schedlowski M, Meyer U (2013) Stress in puberty unmasks latent neuropathological consequences of prenatal immune activation in mice. Science 339:1095–1099

Glantz LA, Gilmore JH, Lieberman JA, Jarskog LF (2006) Apoptotic mechanisms and the synaptic pathology of schizophrenia. Schizophr Res 81:47–63

Goines PE, Ashwood P (2013) Cytokine dysregulation in autism spectrum disorders (ASD): possible role of the environment. Neurotoxicol Teratol 36:67–81

Goines P, Haapanen L, Boyce R, Duncanson P, Braunschweig D, Delwiche L, Hansen R, Hertz-Picciotto I, Ashwood P, Van de Water J (2011a) Autoantibodies to cerebellum in children with autism associate with behavior. Brain Behav Immun 25:514–523

Goines P, Zimmerman A, Ashwood P, Van de Water J (2011b) The immune system, autoimmunity, allergy, and autism spectrum disorders. In: Amaral DG, Dawson G, Geschwind DH (eds) Autism spectrum disorders. Oxford University Press, New York, pp 395–419

Gupta S, Aggarwal S, Rashanravan B, Lee T (1998) Th1- and Th2-like cytokines in CD4+ and CD8+ T cells in autism. J Neuroimmunol 85:106–109

Hagerman R, Hoem G, Hagerman P (2010) Fragile X and autism: intertwined at the molecular level leading to targeted treatments. Mol Autism 1:12

Hallmayer J, Cleveland S, Torres A, Phillips J, Cohen B, Torigoe T, Miller J, Fedele A, Collins J, Smith K, Lotspeich L, Croen LA, Ozonoff S, Lajonchere C, Grether JK, Risch N (2011) Genetic heritability and shared environmental factors among twin pairs with autism. Arch Gen Psychiatry 68(11):1095–1102

Himmerich H, Schonherr J, Fulda S, Sheldrick AJ, Bauer K, Sack U (2011) Impact of antipsychotics on cytokine production in-vitro. J Psychiatr Res 45:1358–1365

Hinwood M, Morandini J, Day TA, Walker FR (2012) Evidence that microglia mediate the neurobiological effects of chronic psychological stress on the medial prefrontal cortex. Cereb Cortex 22:1442–1454

Hou Y, Wu CF, Yang JY, He X, Bi XL, Yu L, Guo T (2006) Effects of clozapine, olanzapine and haloperidol on nitric oxide production by lipopolysaccharide-activated N9 cells. Prog Neuropsychopharmacol Biol Psychiatry 30:1523–1528

Hu S, Peterson PK, Chao CC (1997) Cytokine-mediated neuronal apoptosis. Neurochem Int 30:427–431

Hulshoff Pol HE, Kahn RS (2008) What happens after the first episode? A review of progressive brain changes in chronically ill patients with schizophrenia. Schizophr Bull 34:354–366

Iritani S (2007) Neuropathology of schizophrenia: a mini review. Neuropathology 27:604–608

Jarskog LF, Glantz LA, Gilmore JH, Lieberman JA (2005) Apoptotic mechanisms in the pathophysiology of schizophrenia. Prog Neuropsychopharmacol Biol Psychiatry 29:846–858

Javitt DC (2010) Glutamatergic theories of schizophrenia. Isr J Psychiatry Relat Sci 47:4–16

Jia PL, Wang LL, Meltzer HY, Zhao ZM (2010) Common variants conferring risk of schizophrenia: a pathway analysis of GWAS data. Schizophr Res 122:38–42

Juckel G, Manitz MP, Brune M, Friebe A, Heneka MT, Wolf RJ (2011) Microglial activation in a neuroinflammational animal model of schizophrenia—a pilot study. Schizophr Res 131:96–100

Jyonouchi H, Sun S, Le H (2001) Proinflammatory and regulatory cytokine production associated with innate and adaptive immune responses in children with autism spectrum disorders and developmental regression. J Neuroimmunol 120:170–179

Kato T, Monji A, Hashioka S, Kanba S (2007) Risperidone significantly inhibits interferon–gamma-induced microglial activation in vitro. Schizophr Res 92:108–115

Kato T, Mizoguchi Y, Monji A, Horikawa H, Suzuki SO, Seki Y, Iwaki T, Hashioka S, Kanba S (2008) Inhibitory effects of aripiprazole on interferon–gamma-induced microglial activation via intracellular Ca^{2+} regulation in vitro. J Neurochem 106:815–825

Kato TA, Monji A, Yasukawa K, Mizoguchi Y, Horikawa H, Seki Y, Hashioka S, Han YH, Kasai M, Sonoda N, Hirata E, Maeda Y, Inoguchi T, Utsumi H, Kanba S (2011) Aripiprazole inhibits superoxide generation from phorbol-myristate-acetate (PMA)-stimulated microglia in vitro: implication for antioxidative psychotropic actions via microglia. Schizophr Res 129(2–3):172–182

Kelly DL, Vyas G, Richardson CM, Koola M, McMahon RP, Buchanan RW, Wehring HJ (2011) Adjunct minocycline to clozapine treated patients with persistent schizophrenia symptoms. Schizophr Res 133:257–258

King BH, Lord C (2011) Is schizophrenia on the autism spectrum? Brain Res 1380:34–41

Kivisakk P, Mahad DJ, Callahan MK, Trebst C, Tucky B, Wei T, Wu L, Baekkevold ES, Lassmann H, Staugaitis SM, Campbell JJ, Ransohoff RM (2003) Human cerebrospinal fluid central memory CD4+ T cells: evidence for trafficking through choroid plexus and meninges via P-selectin. Proc Natl Acad Sci U S A 100:8389–8394

Kluge M, Schuld A, Schacht A, Himmerich H, Dalal MA, Wehmeier PM, Hinze-Selch D, Kraus T, Dittmann RW, Pollmacher T (2009) Effects of clozapine and olanzapine on cytokine systems are closely linked to weight gain and drug-induced fever. Psychoneuroendocrinology 34:118–128

Knight JG, Menkes DB, Highton J, Adams DD (2007) Rationale for a trial of immunosuppressive therapy in acute schizophrenia. Mol Psychiatry 12:424–431

Kowalski J, Labuzek K, Herman ZS (2003) Flupentixol and trifluperidol reduce secretion of tumor necrosis factor-alpha and nitric oxide by rat microglial cells. Neurochem Int 43:173–178

Kowalski J, Labuzek K, Herman ZS (2004) Flupentixol and trifluperidol reduce interleukin-1 beta and interleukin-2 release by rat mixed glial and microglial cell cultures. Pol J Pharmacol 56:563–570

Kumra S, Ashtari M, Cervellione KL, Henderson I, Kester H, Roofeh D, Wu J, Clarke T, Thaden E, Kane JM, Rhinewine J, Lencz T, Diamond A, Ardekani BA, Szeszko PR (2005) White matter abnormalities in early-onset schizophrenia: a voxel-based diffusion tensor imaging study. J Am Acad Child Adolesc Psychiatry 44:934–941

Laan W, Grobbee DE, Selten JP, Heijnen CJ, Kahn RS, Burger H (2010) Adjuvant aspirin therapy reduces symptoms of schizophrenia spectrum disorders: results from a randomized, double-blind, placebo-controlled trial. J Clin Psychiatry 71:520–527

Labuzek K, Kowalski J, Gabryel B, Herman ZS (2005) Chlorpromazine and loxapine reduce interleukin-1beta and interleukin-2 release by rat mixed glial and microglial cell cultures. Eur Neuropsychopharmacol 15:23–30

LaSalle JM (2011) A genomic point-of-view on environmental factors influencing the human brain methylome. Epigenetics 6:862–869

Lasalle JM, Yasui DH (2009) Evolving role of MeCP2 in Rett syndrome and autism. Epigenomics 1:119–130

Lee SH, Kubicki M, Asami T, Seidman LJ, Goldstein JM, Mesholam-Gately RI, McCarley RW, Shenton ME (2013) Extensive white matter abnormalities in patients with first-episode schizophrenia: a Diffusion Tensor Imaging (DTI) study. Schizophr Res 143:231–238

Leigh MJ, Nguyen DV, Mu Y, Winarni TI, Schneider A, Chechi T, Polussa J, Doucet P, Tassone F, Rivera SM, Hessl D, Hagerman RJ (2013) A randomized double-blind, placebo-controlled trial of minocycline in children and adolescents with fragile x syndrome. J Dev Behav Pediatr 34:147–155

Levitt P, Campbell DB (2009) The genetic and neurobiologic compass points toward common signaling dysfunctions in autism spectrum disorders. J Clin Invest 119:747–754

Levkovitz Y, Mendlovich S, Riwkes S, Braw Y, Levkovitch-Verbin H, Gal G, Fennig S, Treves I, Kron S (2010) A double-blind, randomized study of minocycline for the treatment of negative and cognitive symptoms in early-phase schizophrenia. J Clin Psychiatry 71:138–149

Li J, Baud O, Vartanian T, Volpe JJ, Rosenberg PA (2005) Peroxynitrite generated by inducible nitric oxide synthase and NADPH oxidase mediates microglial toxicity to oligodendrocytes. Proc Natl Acad Sci U S A 102:9936–9941

Li Y, Du XF, Liu CS, Wen ZL, Du JL (2012) Reciprocal regulation between resting microglial dynamics and neuronal activity in vivo. Dev Cell 23:1189–1202

Lieberman JA (1999) Is schizophrenia a neurodegenerative disorder? A clinical and neurobiological perspective. Biol Psychiatry 46:729–739

Lieberman JA, Tollefson GD, Charles C, Zipursky R, Sharma T, Kahn RS, Keefe RS, Green AI, Gur RE, McEvoy J, Perkins D, Hamer RM, Gu H, Tohen M (2005) Antipsychotic drug effects on brain morphology in first-episode psychosis. Arch Gen Psychiatry 62:361–370

Lieberman JA, Javitch JA, Moore H (2008) Cholinergic agonists as novel treatments for schizophrenia: the promise of rational drug development for psychiatry. Am J Psychiatry 165:931–936

Lin A, Kenis G, Bignotti S, Tura GJB, De Jong R, Bosmans E, Pioli R, Altamura C, Scharpe S, Maes M (1998) The inflammatory response system in treatment-resistant schizophrenia: increased serum interleukin-6. Schizophr Res 32:9–15

Lu XH, Dwyer DS (2005) Second-generation antipsychotic drugs, olanzapine, quetiapine, and clozapine enhance neurite outgrowth in PC12 cells via PI3K/AKT, ERK, and pertussis toxin-sensitive pathways. J Mol Neurosci 27:43–64

Lu DY, Tsao YY, Leung YM, Su KP (2010) Docosahexaenoic acid suppresses neuroinflammatory responses and induces heme oxygenase-1 expression in BV-2 microglia: implications of antidepressant effects for omega-3 fatty acids. Neuropsychopharmacology 35:2238–2248

Lugnegard T, Hallerback MU, Hjarthag F, Gillberg C (2013) Social cognition impairments in Asperger syndrome and schizophrenia. Schizophr Res 143:277–284

Lynch MA, Mills KH (2012) Immunology meets neuroscience—opportunities for immune intervention in neurodegenerative diseases. Brain Behav Immun 26:1–10
MacDowell KS, Garcia-Bueno B, Madrigal JL, Parellada M, Arango C, Mico JA, Leza JC (2013) Risperidone normalizes increased inflammatory parameters and restores anti-inflammatory pathways in a model of neuroinflammation. Int J Neuropsychopharmacol 16:121–135
Maezawa I, Jin LW (2010) Rett syndrome microglia damage dendrites and synapses by the elevated release of glutamate. J Neurosci 30:5346–5356
Maezawa I, Calafiore M, Wulff H, Jin LW (2011) Does microglial dysfunction play a role in autism and Rett syndrome? Neuron Glia Biol 7:85–97
Marin-Teva JL, Dusart I, Colin C, Gervais A, van Rooijen N, Mallat M (2004) Microglia promote the death of developing Purkinje cells. Neuron 41:535–547
McCracken JT, McGough J, Shah B, Cronin P, Hong D, Aman MG, Arnold LE, Lindsay R, Nash P, Hollway J, McDougle CJ, Posey D, Swiezy N, Kohn A, Scahill L, Martin A, Koenig K, Volkmar F, Carroll D, Lancor A, Tierney E, Ghuman J, Gonzalez NM, Grados M, Vitiello B, Ritz L, Davies M, Robinson J, McMahon D (2002) Risperidone in children with autism and serious behavioral problems. N Engl J Med 347:314–321
McCullumsmith RE, Gupta D, Beneyto M, Kreger E, Haroutunian V, Davis KL, Meador-Woodruff JH (2007) Expression of transcripts for myelination-related genes in the anterior cingulate cortex in schizophrenia. Schizophr Res 90:15–27
McDougle CJ, Stigler KA, Erickson CA, Posey DJ (2008) Atypical antipsychotics in children and adolescents with autistic and other pervasive developmental disorders. J Clin Psychiatry 69(Suppl 4):15–20
McKeith IG (2006) Consensus guidelines for the clinical and pathologic diagnosis of dementia with Lewy bodies (DLB): report of the Consortium on DLB International Workshop. J Alzheimers Dis 9:417–423
Medina S, Martinez M, Hernanz A (2002) Antioxidants inhibit the human cortical neuron apoptosis induced by hydrogen peroxide, tumor necrosis factor alpha, dopamine and beta-amyloid peptide 1–42. Free Radic Res 36:1179–1184
Merrill JE, Ignarro LJ, Sherman MP, Melinek J, Lane TE (1993) Microglial cell cytotoxicity of oligodendrocytes is mediated through nitric oxide. J Immunol 151:2132–2141
Meyer U (2011) Anti-inflammatory signaling in schizophrenia. Brain Behav Immun 25:1507–1518
Meyer JM, Stahl SM (2009) The metabolic syndrome and schizophrenia. Acta Psychiatr Scand 119:4–14
Miller BJ, Buckley P, Seabolt W, Mellor A, Kirkpatrick B (2011) Meta-analysis of cytokine alterations in schizophrenia: clinical status and antipsychotic effects. Biol Psychiatry 70:663–671
Miller BJ, Gassama B, Sebastian D, Buckley P, Mellor A (2013) Meta-analysis of lymphocytes in schizophrenia: clinical status and antipsychotic effects. Biol Psychiatry 73:993–999
Minshew NJ, Keller TA (2010) The nature of brain dysfunction in autism: functional brain imaging studies. Curr Opin Neurol 23:124–130
Mittelbronn M, Dietz K, Schluesener HJ, Meyermann R (2001) Local distribution of microglia in the normal adult human central nervous system differs by up to one order of magnitude. Acta Neuropathol 101:249–255
Miyaoka T, Yasukawa R, Yasuda H, Hayashida M, Inagaki T, Horiguchi J (2007) Possible antipsychotic effects of minocycline in patients with schizophrenia. Prog Neuropsychopharmacol Biol Psychiatry 31:304–307
Miyaoka T, Yasukawa R, Yasuda H, Hayashida M, Inagaki T, Horiguchi J (2008) Minocycline as adjunctive therapy for schizophrenia: an open-label study. Clin Neuropharmacol 31:287–292
Mizoguchi Y, Monji A, Kato T, Seki Y, Gotoh L, Horikawa H, Suzuki SO, Iwaki T, Yonaha M, Hashioka S, Kanba S (2009) Brain-derived neurotrophic factor induces sustained elevation of intracellular Ca^{2+} in rodent microglia. J Immunol 183:7778–7786
Morgan JT, Chana G, Pardo CA, Achim C, Semendeferi K, Buckwalter J, Courchesne E, Everall IP (2010) Microglial activation and increased microglial density observed in the dorsolateral prefrontal cortex in autism. Biol Psychiatry 68:368–376
Muller N, Riedel M, Schwarz MJ, Engel RR (2005) Clinical effects of COX-2 inhibitors on cognition in schizophrenia. Eur Arch Psychiatry Clin Neurosci 255:149–151

Muller N, Krause D, Dehning S, Musil R, Schennach-Wolff R, Obermeier M, Moller HJ, Klauss V, Schwarz MJ, Riedel M (2010) Celecoxib treatment in an early stage of schizophrenia: results of a randomized, double-blind, placebo-controlled trial of celecoxib augmentation of amisulpride treatment. Schizophr Res 121:118–124
Nagarajan RP, Hogart AR, Gwye Y, Martin MR, LaSalle JM (2006) Reduced MeCP2 expression is frequent in autism frontal cortex and correlates with aberrant MECP2 promoter methylation. Epigenetics 1:e1–e11
Nakki R, Koistinaho J, Sharp FR, Sagar SM (1995) Cerebellar toxicity of phencyclidine. J Neurosci 15:2097–2108
Nakki R, Nickolenko J, Chang J, Sagar SM, Sharp FR (1996) Haloperidol prevents ketamine- and phencyclidine-induced HSP70 protein expression but not microglial activation. Exp Neurol 137:234–241
Narayan S, Tang B, Head SR, Gilmartin TJ, Sutcliffe JG, Dean B, Thomas EA (2008) Molecular profiles of schizophrenia in the CNS at different stages of illness. Brain Res 1239:235–248
Nawa H, Takei N (2006) Recent progress in animal modeling of immune inflammatory processes in schizophrenia: implication of specific cytokines. Neurosci Res 56:2–13
Nielsen J, Skadhede S, Correll CU (2010) Antipsychotics associated with the development of type 2 diabetes in antipsychotic-naive schizophrenia patients. Neuropsychopharmacology 35:1997–2004
Owen R, Sikich L, Marcus RN, Corey-Lisle P, Manos G, McQuade RD, Carson WH, Findling RL (2009) Aripiprazole in the treatment of irritability in children and adolescents with autistic disorder. Pediatrics 124:1533–1540
Palluy O, Rigaud M (1996) Nitric oxide induces cultured cortical neuron apoptosis. Neurosci Lett 208:1–4
Paolicelli RC, Bolasco G, Pagani F, Maggi L, Scianni M, Panzanelli P, Giustetto M, Ferreira TA, Guiducci E, Dumas L, Ragozzino D, Gross CT (2011) Synaptic pruning by microglia is necessary for normal brain development. Science 333:1456–1458
Pardo CA, Buckley A, Thurm A, Lee LC, Azhagiri A, Neville DM, Swedo SE (2013) A pilot open-label trial of minocycline in patients with autism and regressive features. J Neurodev Disord 5:9
Paribello C, Tao L, Folino A, Berry-Kravis E, Tranfaglia M, Ethell IM, Ethell DW (2010) Open-label add-on treatment trial of minocycline in fragile X syndrome. BMC Neurol 10:91
Penzes P, Cahill ME, Jones KA, VanLeeuwen JE, Woolfrey KM (2011) Dendritic spine pathology in neuropsychiatric disorders. Nat Neurosci 14:285–293
Perez-Neri I, Ramirez-Bermudez J, Montes S, Rios C (2006) Possible mechanisms of neurodegeneration in schizophrenia. Neurochem Res 31:1279–1294
Pickett J, London E (2005) The neuropathology of autism: a review. J Neuropathol Exp Neurol 64:925–935
Pike VW, Halldin C, Crouzel C, Barre L, Nutt DJ, Osman S, Shah F, Turton DR, Waters SL (1993) Radioligands for PET studies of central benzodiazepine receptors and PK (peripheral benzodiazepine) binding sites—current status. Nucl Med Biol 20:503–525
Politte LC, McDougle CJ (2014) Atypical antipsychotics in the treatment of children and adolescents with pervasive developmental disorders. Psychopharmacology (Berl) 231(6):1023–1036
Polsek D, Jagatic T, Cepanec M, Hof PR, Simic G (2011) Recent developments in neuropathology of autism spectrum disorders. Transl Neurosci 2:256–264
Potvin S, Stip E, Sepehry AA, Gendron A, Bah R, Kouassi E (2008) Inflammatory cytokine alterations in schizophrenia: a systematic quantitative review. Biol Psychiatry 63:801–808
Powell SB, Sejnowski TJ, Behrens MM (2012) Behavioral and neurochemical consequences of cortical oxidative stress on parvalbumin-interneuron maturation in rodent models of schizophrenia. Neuropharmacology 62:1322–1331
Price G, Cercignani M, Bagary MS, Barnes TR, Barker GJ, Joyce EM, Ron MA (2006) A volumetric MRI and magnetization transfer imaging follow-up study of patients with first-episode schizophrenia. Schizophr Res 87:100–108

Radewicz K, Garey LJ, Gentleman SM, Reynolds R (2000) Increase in HLA-DR immunoreactive microglia in frontal and temporal cortex of chronic schizophrenics. J Neuropathol Exp Neurol 59:137–150

Reboldi A, Coisne C, Baumjohann D, Benvenuto F, Bottinelli D, Lira S, Uccelli A, Lanzavecchia A, Engelhardt B, Sallusto F (2009) C–C chemokine receptor 6-regulated entry of TH-17 cells into the CNS through the choroid plexus is required for the initiation of EAE. Nat Immunol 10:514–523

Rice C (2009) Prevalence of autism spectrum disorders—autism and developmental disabilities monitoring network, United States, 2006 MMWR. Surveillance Summaries, pp 1–20

Roberts RC, Roche JK, Conley RR (2005) Synaptic differences in the postmortem striatum of subjects with schizophrenia: a stereological ultrastructural analysis. Synapse 56:185–197

Sahin M (2012) Targeted treatment trials for tuberous sclerosis and autism: no longer a dream. Curr Opin Neurobiol 22:895–901

Salisbury DF, Kuroki N, Kasai K, Shenton ME, McCarley RW (2007) Progressive and interrelated functional and structural evidence of post-onset brain reduction in schizophrenia. Arch Gen Psychiatry 64:521–529

Schafer DP, Lehrman EK, Kautzman AG, Koyama R, Mardinly AR, Yamasaki R, Ransohoff RM, Greenberg ME, Barres BA, Stevens B (2012) Microglia sculpt postnatal neural circuits in an activity and complement-dependent manner. Neuron 74:691–705

Schiavone S, Sorce S, Dubois-Dauphin M, Jaquet V, Colaianna M, Zotti M, Cuomo V, Trabace L, Krause KH (2009) Involvement of NOX2 in the development of behavioral and pathologic alterations in isolated rats. Biol Psychiatry 66:384–392

Schlosser RG, Nenadic I, Wagner G, Gullmar D, von Consbruch K, Kohler S, Schultz CC, Koch K, Fitzek C, Matthews PM, Reichenbach JR, Sauer H (2007) White matter abnormalities and brain activation in schizophrenia: a combined DTI and fMRI study. Schizophr Res 89:1–11

Schmitt A, Bertsch T, Henning U, Tost H, Klimke A, Henn FA, Falkai P (2005) Increased serum S100B in elderly, chronic schizophrenic patients: negative correlation with deficit symptoms. Schizophr Res 80:305–313

Schnieder TP, Dwork AJ (2011) Searching for neuropathology: gliosis in schizophrenia. Biol Psychiatry 69:134–139

Shang YC, Chong ZZ, Wang SH, Maiese K (2013) Tuberous sclerosis protein 2 (TSC2) modulates CCN4 cytoprotection during apoptotic amyloid toxicity in microglia. Curr Neurovasc Res 10:29–38

Shepherd GM, Katz DM (2011) Synaptic microcircuit dysfunction in genetic models of neurodevelopmental disorders: focus on Mecp2 and Met. Curr Opin Neurobiol 21:827–833

Skaper SD, Giusti P, Facci L (2012) Microglia and mast cells: two tracks on the road to neuroinflammation. FASEB J 26:3103–3117

Skaper SD, Facci L, Giusti P (2014) Mast cells, glia and neuroinflammation: partners in crime? Immunology 141:314–327

Soderlund J, Schroder J, Nordin C, Samuelsson M, Walther-Jallow L, Karlsson H, Erhardt S, Engberg G (2009) Activation of brain interleukin-1 beta in schizophrenia. Mol Psychiatry 14:1069–1071

Sommer IE, de Witte L, Begemann M, Kahn RS (2012) Nonsteroidal anti-inflammatory drugs in schizophrenia: ready for practice or a good start? A meta-analysis. J Clin Psychiatry 73:414–419

Stefansson H, Ophoff RA, Steinberg S, Andreassen OA, Cichon S, Rujescu D, Werge T, Pietilainen OP, Mors O, Mortensen PB, Sigurdsson E, Gustafsson O, Nyegaard M, Tuulio-Henriksson A, Ingason A, Hansen T, Suvisaari J, Lonnqvist J, Paunio T, Borglum AD, Hartmann A, Fink-Jensen A, Nordentoft M, Hougaard D, Norgaard-Pedersen B, Bottcher Y, Olesen J, Breuer R, Moller HJ, Giegling I, Rasmussen HB, Timm S, Mattheisen M, Bitter I, Rethelyi JM, Magnusdottir BB, Sigmundsson T, Olason P, Mason G, Gulcher JR, Haraldsson M, Fossdal R, Thorgeirsson TE, Thorsteinsdottir U, Ruggeri M, Tosato S, Franke B, Strengman E, Kiemeney LA, Melle I, Djurovic S, Abramova L, Kaleda V, Sanjuan J, de Frutos R, Bramon E, Vassos E, Fraser G, Ettinger U, Picchioni M, Walker N, Toulopoulou T, Need AC, Ge D, Yoon JL,

Shianna KV, Freimer NB, Cantor RM, Murray R, Kong A, Golimbet V, Carracedo A, Arango C, Costas J, Jonsson EG, Terenius L, Agartz I, Petursson H, Nothen MM, Rietschel M, Matthews PM, Muglia P, Peltonen L, St Clair D, Goldstein DB, Stefansson K, Collier DA, Stefansson H, Ophoff RA, Steinberg S, Andreassen OA, Cichon S, Rujescu D, Werge T, Pietilainen OP, Mors O, Mortensen PB, Sigurdsson E, Gustafsson O, Nyegaard M, Tuulio-Henriksson A, Ingason A, Hansen T, Suvisaari J, Lonnqvist J, Paunio T, Borglum AD, Hartmann A, Fink-Jensen A, Nordentoft M, Hougaard D, Norgaard-Pedersen B, Bottcher Y, Olesen J, Breuer R, Moller HJ, Giegling I, Rasmussen HB, Timm S, Mattheisen M, Bitter I, Rethelyi JM, Magnusdottir BB, Sigmundsson T, Olason P, Mason G, Gulcher JR, Haraldsson M, Fossdal R, Thorgeirsson TE, Thorsteinsdottir U, Ruggeri M, Tosato S, Franke B, Strengman E, Kiemeney LA, Melle I, Djurovic S, Abramova L, Kaleda V, Sanjuan J, de Frutos R, Bramon E, Vassos E, Fraser G, Ettinger U, Picchioni M, Walker N, Toulopoulou T, Need AC, Ge D, Yoon JL, Shianna KV, Freimer NB, Cantor RM, Murray R, Kong A, Golimbet V, Carracedo A, Arango C, Costas J, Jonsson EG, Terenius L, Agartz I, Petursson H, Nothen MM, Rietschel M, Matthews PM, Muglia P, Peltonen L, St Clair D, Goldstein DB, Stefansson K, Collier DA (2009) Common variants conferring risk of schizophrenia. Nature 460(7256):744–747

Steiner J, Mawrin C, Ziegeler A, Bielau H, Ullrich O, Bernstein HG, Bogerts B (2006) Distribution of HLA-DR-positive microglia in schizophrenia reflects impaired cerebral lateralization. Acta Neuropathol 112:305–316

Stellwagen D, Malenka RC (2006) Synaptic scaling mediated by glial TNF-alpha. Nature 440:1054–1059

Sugino H, Futamura T, Mitsumoto Y, Maeda K, Marunaka Y (2009) Atypical antipsychotics suppress production of proinflammatory cytokines and up-regulate interleukin-10 in lipopolysaccharide-treated mice. Prog Neuropsychopharmacol Biol Psychiatry 33:303–307

Sugranyes G, Kyriakopoulos M, Corrigall R, Taylor E, Frangou S (2011) Autism spectrum disorders and schizophrenia: meta-analysis of the neural correlates of social cognition. PLoS One 6:e25322

Sunico CR, Portillo F, Gonzalez-Forero D, Moreno-Lopez B (2005) Nitric-oxide-directed synaptic remodeling in the adult mammal CNS. J Neurosci 25:1448–1458

Suzuki T, Hide I, Matsubara A, Hama C, Harada K, Miyano K, Andra M, Matsubayashi H, Sakai N, Kohsaka S, Inoue K, Nakata Y (2006) Microglial alpha7 nicotinic acetylcholine receptors drive a phospholipase C/IP3 pathway and modulate the cell activation toward a neuroprotective role. J Neurosci Res 83:1461–1470

Suzuki K, Sugihara G, Ouchi Y, Nakamura K, Futatsubashi M, Takebayashi K, Yoshihara Y, Omata K, Matsumoto K, Tsuchiya KJ, Iwata Y, Tsujii M, Sugiyama T, Mori N (2013) Microglial activation in young adults with autism spectrum disorder. JAMA Psychiatry 70:49–58

Takano A, Arakawa R, Ito H, Tateno A, Takahashi H, Matsumoto R, Okubo Y, Suhara T (2010) Peripheral benzodiazepine receptors in patients with chronic schizophrenia: a PET study with [11C]DAA1106. Int J Neuropsychopharmacol 13:943–950

Taliou A, Zintzaras E, Lykouras L, Francis K (2013) An open-label pilot study of a formulation containing the anti-inflammatory flavonoid luteolin and its effects on behavior in children with autism spectrum disorders. Clin Ther 35:592–602

Theoharides TC (2013) Is a subtype of Autism an allergy of the brain? Clin Ther 35:584–591

Theoharides TC, Asadi S, Panagiotidou S (2012) A case series of a luteolin formulation (Neuroprotek (R)) in children with autism spectrum disorders. Int J Immunopathol Pharmacol 25:317–323

Theoharides TC, Asadi S, Patel AB (2013) Focal brain inflammation and autism. J Neuroinflammation 10:46

Tremblay ME, Lowery RL, Majewska AK (2010) Microglial interactions with synapses are modulated by visual experience. PLoS Biol 8:e1000527

Tynan RJ, Naicker S, Hinwood M, Nalivaiko E, Buller KM, Pow DV, Day TA, Walker FR (2010) Chronic stress alters the density and morphology of microglia in a subset of stress-responsive brain regions. Brain Behav Immun 24:1058–1068

Uranova NA, Vostrikov VM, Orlovskaya DD, Rachmanova VI (2004) Oligodendroglial density in the prefrontal cortex in schizophrenia and mood disorders: a study from the Stanley Neuropathology Consortium. Schizophr Res 67:269–275

Uranova NA, Vostrikov VM, Vikhreva OV, Zimina IS, Kolomeets NS, Orlovskaya DD (2007) The role of oligodendrocyte pathology in schizophrenia. Int J Neuropsychopharmacol 10:537–545

van Berckel BN, Bossong MG, Boellaard R, Kloet R, Schuitemaker A, Caspers E, Luurtsema G, Windhorst AD, Cahn W, Lammertsma AA, Kahn RS (2008a) Microglia activation in recent-onset schizophrenia: a quantitative (R)-[C-11]PK11195 positron emission tomography study. Biol Psychiatry 64:820–822

van Berckel BNM, Bossong MG, Boellaard R, Kloet R, Schuitemaker A, Caspers E, Luurtsema G, Windhorst AD, Cahn W, Lammertsma AA, Kahn RS (2008b) Microglia activation in recent onset schizophrenia: a quantitative (R)-[C-11]PK11195 study. Neuroimage 41:T119–T119

Vargas DL, Nascimbene C, Krishnan C, Zimmerman AW, Pardo CA (2005) Neuroglial activation and neuroinflammation in the brain of patients with autism. Ann Neurol 57:67–81

Vismara LA, Rogers SJ (2010) Behavioral treatments in autism spectrum disorder: what do we know? Annu Rev Clin Psychol 6(6):447–468

Wake H, Moorhouse AJ, Jinno S, Kohsaka S, Nabekura J (2009) Resting microglia directly monitor the functional state of synapses in vivo and determine the fate of ischemic terminals. J Neurosci 29:3974–3980

Wang HS, Doering LC (2013) Reversing autism by targeting downstream mTOR signaling. Front Cell Neurosci 7:28

Wang Q, Cheung C, Deng W, Li M, Huang C, Ma X, Wang Y, Jiang L, Sham PC, Collier DA, Gong Q, Chua SE, McAlonan GM, Li T (2013) White-matter microstructure in previously drug-naive patients with schizophrenia after 6 weeks of treatment. Psychol Med 43:2301–2309

Weickert CS, Hyde TM, Lipska BK, Herman MM, Weinberger DR, Kleinman JE (2003) Reduced brain-derived neurotrophic factor in prefrontal cortex of patients with schizophrenia. Mol Psychiatry 8:592–610

Williams SC (2012) Genetics: searching for answers. Nature 491:S4–S6

Wills S, Cabanlit M, Bennett J, Ashwood P, Amaral DG, Van de Water J (2009) Detection of autoantibodies to neural cells of the cerebellum in the plasma of subjects with autism spectrum disorders. Brain Behav Immun 23:64–74

Wohleb ES, Hanke ML, Corona AW, Powell ND, Stiner LM, Bailey MT, Nelson RJ, Godbout JP, Sheridan JF (2011) β-Adrenergic receptor antagonism prevents anxiety-like behavior and microglial reactivity induced by repeated social defeat. J Neurosci 31:6277–6288

Womack KB, Diaz-Arrastia R, Aizenstein HJ, Arnold SE, Barbas NR, Boeve BF, Clark CM, DeCarli CS, Jagust WJ, Leverenz JB, Peskind ER, Turner RS, Zamrini EY, Heidebrink JL, Burke JR, DeKosky ST, Farlow MR, Gabel MJ, Higdon R, Kawas CH, Koeppe RA, Lipton AM, Foster NL (2011) Temporoparietal hypometabolism in frontotemporal lobar degeneration and associated imaging diagnostic errors. Arch Neurol 68:329–337

Yeargin-Allsopp M, Rice C, Karapurkar T, Doernberg N, Boyle C, Murphy C (2003) Prevalence of autism in a US metropolitan area. JAMA 289:49–55

Zafeiriou DI, Ververi A, Dafoulis V, Kalyva E, Vargiami E (2013) Autism spectrum disorders: the quest for genetic syndromes. Am J Med Genet B Neuropsychiatr Genet 162B:327–366

Zhang XY, Zhou DF, Cao LY, Zhang PY, Wu GY, Shen YC (2004) Changes in serum interleukin-2,-6, and-8 levels before and during treatment with risperidone and haloperidol: relationship to outcome in schizophrenia. J Clin Psychiatry 65:940–947

Zimmerman AW, Jyonouchi H, Comi AM, Connors SL, Milstien S, Varsou A, Heyes MP (2005) Cerebrospinal fluid and serum markers of inflammation in autism. Pediatr Neurol 33:195–201

Zoghbi HY (2003) Postnatal neurodevelopmental disorders: meeting at the synapse? Science 302:826–830

Chapter 15
Human Immunodeficiency Virus

Daniel F. Marker, Shao-Ming Lu, and Harris A. Gelbard

Abstract Microglia have been known to be the resident immune cells in the brain for quite some time, but their role as pathologic players in neurological infectious disease has only recently been studied. For example, microglial activation has recently been shown to be a strong driver of neurological impairment in human immunodeficiency virus (HIV)-associated neurocognitive disorders (HAND). Studying the interactions between microglia and synaptic networks during chronic HIV infection is pivotal to understanding why normal synaptic communication is impaired and is the etiologic substrate for HIV-associated neurocognitive deficits, despite successful control of viral replication by combination antiretroviral therapy (cART). In this chapter we focus on the phenomenology of these interactions and suggest that upregulation of mixed lineage kinase type 3 (MLK3) and leucine-rich repeat kinase 2 (LRRK2) activity plays a pivotal role in microglial activation and synaptic dysfunction that give rise to HAND. We further discuss the rationale for targeting pathologic activation of these kinases in creating disease-modifying strategies for the treatment of HAND.

Keywords Microglia • Human immunodeficiency virus • Human immunodeficiency virus-associated neurocognitive disorder • Mixed lineage kinase type 3 • Leucine-rich repeat kinase 2 • Inflammation • Phagocytosis • Synapses

Bullet Points

- Microglial activation is a strong driver of neurological impairment in human immunodeficiency virus (HIV)-associated neurocognitive disorders (HAND).
- HIV infection of perivascular macrophage and microglia alters their secretome, including pro-inflammatory cytokines and chemokines.
- HIV infection enhances macrophage and microglial phagocytosis of synaptic elements.
- Mixed lineage kinase type 3 (MLK3) and leucine-rich repeat kinase 2 (LRRK2) are key enzymes involved in this neuroinflammatory response to HIV.

D.F. Marker, Ph.D. • S.-M. Lu, Ph.D. • H.A. Gelbard, M.D., Ph.D. (✉)
Center for Neural Development and Disease, Department of Neurology
(Child Neurology Division), University of Rochester Medical Center,
Box 645, 601 Elmwood Avenue, Rochester, NY 14642, USA
e-mail: daniel_marker@urmc.rochester.edu; Pat_Lu@URMC.Rochester.edu;
Harris_Gelbard@URMC.Rochester.edu

M.-È. Tremblay, A. Sierra (eds.), *Microglia in Health and Disease*,
DOI 10.1007/978-1-4939-1429-6_15

15.1 Introduction

In 1960, there was a single article entitled "Reactional aspects of the microglia in spontaneous granulomatous encephalitis" (de Amorim 1960); by the middle of 2013, a PubMed search cross-referencing central nervous system (CNS) infections and microglia yielded 1,572 references. This chapter will focus on a single type of infection of microglia with the lentivirus, human immunodeficiency virus (HIV), to describe the remarkably pleiotropic effects on normal neuroimmune signaling between microglia and neurons that are hijacked during infection. The first report of microglial infection with HIV-1 (Price et al. 1988) in 1988 has been followed by over 300 studies that highlight the diversity of roles that microglia can play in both infectious disease and neuroimmune signaling in the CNS. The disease complex known as HIV-associated neurocognitive disorders (HAND) has continued to inflict people living with HIV infection, despite combination antiretroviral therapy (cART) achieving an efficacious reduction of productive viral infection. HAND is thought to affect over 50 % of people living with HIV-1 in the USA, and with a worldwide incidence of approximately 40,000,000 people, continues to be a problem of pandemic importance. Thus, this chapter will emphasize why the microglial reservoir in the CNS remains such a compelling target for investigation and disease-modifying therapies in people living with HIV and HAND.

15.2 Human Immunodeficiency Virus-Associated Neurocognitive Disorder: Clinical Presentation

HIV is a lentivirus that primarily infects and kills $CD4^+$ T-cells, leading to the eventual destruction of the adaptive immune response. $CD4^+$ T-cells, also known as helper T-cells, are responsible for facilitating cell-mediated immunity through the secretion of cytokines that promote cytotoxic T-cell function and B-cell maturation. The CD4 surface glycoprotein is a co-receptor that plays an important role in cell to cell interactions between $CD4^+$ T-cells and antigen presenting cells. In addition to infecting $CD4^+$ T-cells, HIV can infect other subsets of cells that have lower levels of CD4 expression. These cell types include monocytes and perivascular macrophages, and specifically in the CNS, microglia and astrocytes. Primary HIV infection is characterized by four stages: (1) an asymptomatic incubation phase of up to 4 weeks in duration during which the virus divides exponentially, (2) an acute infection stage, in which the patient's immune system begins to combat the virus and the patient may experience constitutional symptoms including fever, fatigue, lymphadenopathy, etc., for up to 2 weeks, (3) an asymptomatic latency stage that can last for years during which time there is a progressive loss of the $CD4^+$ T-cell population leading to, (4) the acquired immunodeficiency syndrome (AIDS), when the patient is at severe risk of opportunistic infection, cancer, and death (Ellis et al. 2009; Kahn and Walker 1998). Early HIV infection can also present as an acute meningoencephalitis, which is of particular relevance to HIV-associated brain disease, as it suggests very early viral entry into the brain (Li et al. 2002; Pascual-Lozano et al. 2005; del Saz et al. 2008).

Today, HIV can be effectively treated, but not cured, through drug regimens known as cART. Because of the high viral mutation rate imparted by HIV's error-prone reverse transcriptase, it rapidly develops resistance to single drug regimens, making multiple drugs that target multiple steps in the viral life cycle necessary for effective long-term treatment (Ellis et al. 2009). With cART treatment, there is significant reduction in HIV-associated morbidity and mortality (Palella et al. 1998) and significant improvement in immune function (Gulick et al. 1997). Unfortunately, while modern cART is able to suppress viral replication to undetectable levels (fewer than 50 viral RNA copies per milliliter of blood) for years, it is not a curative treatment. It is unable to eradicate the virus for a number of reasons, including: (1) HIV's ability to integrate into the cellular genomes of long lived cell types and stem cell populations (McNamara et al. 2012), (2) its ability to enter and emerge from long-term latent phases without productive viral production, and (3) its ability to infect cells in cellular reservoirs that are protected from many antiretroviral drugs by endothelial barriers, such as the testes and the brain (Valcour et al. 2011).

Along with the progressive destruction of the peripheral adaptive immune system, HIV also infects cells in the CNS, where it can cause a spectrum of disorders known today as HAND. Untreated, HIV infection in the brain can lead to acute encephalitis, rapid onset dementia, and death (Boisse et al. 2008). By significantly decreasing viral load, the development of cART has greatly reduced the prevalence of these severe HIV sequelae (Dore et al. 2003; Mirza and Rathore 2012). Recent large-scale studies of cART-treated HIV-infected patients have shown, however, that mild to moderate neurologic impairment persists despite effective antiviral treatment. For instance, the "CNS HIV antiretroviral therapy effects research" (CHARTER) study prospectively and longitudinally examined a population of greater than 1,500 HIV-infected individuals with various comorbidities and differing rates of adherence to cART for neuropsychological impairment. They found that over 40 % of patients with undetectable viral load and minimal other comorbidities still experienced either mild or moderate neuropsychological deficit (Heaton et al. 2010). This study confirmed the findings of a number of previous smaller studies that demonstrated that cART treatment alone could not eliminate HAND (Marra et al. 2009; Cysique et al. 2004).

Comparison of the CHARTER study to comparable pre-cART studies using more contemporary operational criteria for HAND (Antinori 2007) demonstrated that patients with asymptomatic HIV infection in the post-cART era actually have significantly higher rates of HAND than those in the pre-cART era [36 % post-cART vs. 25 % pre-cART (Heaton et al. 2011)]. This increase reflects the fact that as the lifespan of HIV-infected individuals increases under antiretroviral treatment, CNS disease continues to progress despite treatment (Heaton et al. 2010, 2011; Mothobi and Brew 2012). It is therefore necessary to develop adjunctive therapies with targets other than the viral life cycle to slow, stop, or reverse the progression of HAND, such as normalizing pathologic microglia-neuron interactions, as will be discussed later in the chapter.

15.3 Neuropathologic Basis of HAND

HIV enters the brain during the initial incubation phase of infection, either by directly crossing the blood-brain barrier (BBB) or primarily through the entry of infected patrolling peripheral monocytes and T-cells (Kure et al. 1990; Anderson et al. 2002). Once in the brain, the virus can productively infect brain-resident microglia and perivascular monocytes (Ho et al. 1985; Koenig et al. 1986) via CD4 receptors and CCR3, CCR5, and CXCR4 chemokine receptors. These immune cell surface receptors are normally involved in cell to cell communication and chemotaxis through various soluble factors known as chemokines. Interestingly, people with a double allelic loss of CCR5 are for all practical purposes immune to HIV infection. One of the pathologic hallmarks of HIV infection in the CNS is microgliosis, which is an increase in the total microglial number and volume, and is seen in a number of models of CNS injury. The presence of the virus activates microglia-inducing morphologic changes that include an increase in number and thickness of cell processes, as demonstrated in autopsy samples of patients infected with HIV (Gelman 1993). Additionally, HIV infection in the brain drives increased production of inflammatory cytokines, such as tumor necrosis factor alpha (TNFα) and interleukin 1 beta (IL-1β), and chemokines, such as monocyte chemoattractant protein 1 (MCP-1), by brain-resident microglia and astrocytes, based on analysis of cerebrospinal fluid (CSF) from HIV-infected individuals and in situ hybridization analysis of autopsy samples (Mintz et al. 1989; Wesselingh et al. 1993). In vitro cellular models and in vivo animal models based on this human subject data have further shown that this neuroinflammatory response can lead to BBB destruction (Kanmogne et al. 2007) and peripheral monocyte recruitment (Clay et al. 2007; Chaudhuri et al. 2008), which further drives viral invasion of the CNS. The virus can also nonproductively infect astrocytes, with expression of early phase regulatory gene products (Ranki et al. 1995), but without production of virions. Regulatory gene products can in turn upregulate chemokine gradients to further attract peripheral monocytes into the CNS (Weiss et al. 1999), which in turn may amplify HAND progression (Churchill et al. 2009). HIV does not directly infect oligodendrocytes or neurons, as these cells lack the necessary CD4 and cofactor expression.

In the early 1990s, there was a significant push to discover the CNS pathologic correlate of HAND severity. Postmortem tissue analysis from the brains of patients with HIV infection consistently demonstrated the presence of infected monocytes, multi-nucleated giant cells (multiple activated monocyte lineage cells that merge into a single soma), gliosis, and diffuse loss of white matter (Boisse et al. 2008). Patients with advanced infection also had damage to synaptodendritic architecture in frontal cortex (Everall et al. 1999; Masliah et al. 1992) and significant neuronal loss (Everall et al. 1991, 1993). Of all these immune and neuronal-related pathologic findings, only synaptodendritic density (Everall et al. 1994; Glass et al. 1993; Masliah et al. 1997) and total monocyte/microglia number (Glass et al. 1995) actually correlated with neurocognitive impairment. Surprisingly, loss of neuronal cell bodies (i.e., neuronal death) did not correlate with neurologic disease severity (Everall et al. 1991, 1994). As the synaptodendritic arbor is plastic with the ability to regenerate following injury, whereas neuronal loss is permanent, these findings helped to explain the reversible nature of HAND when patients were started on cART. The phenomenon of dendritic regeneration has been particularly well studied in in vivo stroke models, with damaged dendritic arbors reconstituting in minutes to hours after reperfusion (Murphy et al. 2008; Zhang et al. 2005). The studies of dendritic damage in patients with HIV agreed with earlier findings in Alzheimer's dementia (DeKosky and Scheff 1990; Terry et al. 1991); synapse loss or damage is a major correlate of neurologic disease severity. It was also interesting that the total number of monocyte lineage cells present in the CNS correlated with disease severity, whereas the number of HIV-infected monocyte cells did not. This suggests that the host immune response to the viral presence may play a greater role in disease progression than the virus itself. In hindsight, these data support the findings of the CHARTER study; that viral suppression by cART alone does not halt the progression of HAND.

15.4 Molecular Basis of Synaptic Damage in HAND: Direct Viral Protein Toxicity

HIV damages neuronal architecture through two main mechanisms: (1) direct neuronal toxicity of the HIV proteins and, (2) immune cell activation and subsequent damage from soluble mediators or direct cellular actions. In terms of direct viral neurotoxicity, the two most well-studied HIV proteins are the envelope glycoprotein 120 (gp120) and the trans activator of transcription (Tat).

The envelope protein gp120 is encoded by the HIV *env* gene and, along with gp41, forms a heterotrimeric spike, which is located on the outer envelope of the virus, and is essential for the virus to bind CD4 and fuse with the target cell (Gallo et al. 2003; Zhu et al. 2008). In the brain, gp120 causes neuronal damage through two potential mechanisms. First, gp120 inhibits the production of brain-derived neurotrophic factor (BDNF) in cultured neurons by decreasing the levels of the proteolytic enzyme furin, which cleaves pro-BDNF into its active form (Bachis et al. 2012). This decrease in active BDNF leads to shortening and simplification of neurite arbors (Bachis et al. 2012) and could potentially be the cause of the caspase-dependent cell death seen in in vivo models of gp120 exposure (Acquas et al. 2004). This cell death can be reversed by treatment with active BDNF and subsequent anti-apoptotic signaling through its receptor TrkB (Bachis et al. 2003; Mocchetti and Bachis 2004). gp120 can also cause calcium-dependent excitotoxic death in neurons by causing increased production of arachidonic acid and sensitization of the *N*-methyl-D-aspartate receptor (NMDAR) as well as accelerating the surface localization of NMDAR in vulnerable synapses (Dreyer et al. 1990; Xu et al. 2011; Muller et al. 1996; Ushijima et al. 1995).

The HIV Tat gene is encoded in the transactivator region of the HIV genome and is essential for transcription of the HIV long terminal repeat promoter and full-length viral gene transcription (Campbell and Loret 2009). Tat also changes the expression of host genes directly by binding to nuclear factor κβ (NFκβ) enhancer sequences (Dandekar et al. 2004). NFκβ is a transcription factor that is present in nearly all cell types and controls important cellular functions such as response pathways to cytokine signals, inflammatory mediators, and immune mediators.

HIV Tat is of particular interest in HAND for two reasons: (1) it is continually produced in the CNS by cellular reservoirs that are likely to include microglia, perivascular macrophages, and astrocytes even in the presence of cART (Cowley et al. 2011; Li et al. 2009), and (2) it is secreted from infected cells and, because of its protein transduction domain, Tat can enter any cell type (Campbell and Loret 2009). This means that Tat can directly enter neurons and alter gene transcription and interfere with normal function. Once secreted from infected cells in the CNS, Tat can directly induce neuronal damage in a number of ways. Tat has been shown to interact with the low-density lipoprotein receptor-related protein (LRP), resulting in Tat uptake into neurons (Liu et al. 2000). In the process of this binding and uptake, Tat prevents the normal transport function of LRP, inducing metabolic imbalance, neuronal dysfunction, and apoptosis. Further studies demonstrated that Tat treatment of neuronal cultures induced a complex of LRP, NMDAR, post-synaptic density 95 (PSD95), and neuronal nitric oxide synthase (nNOS) (Eugenin et al. 2007). In this complex, LRP potentiated calcium influx through the NMDAR, activating nNOS, leading to the production of radical oxygen species (ROS) and subsequent neuronal apoptosis. Blockade of LRP (preventing Tat uptake), NMDAR (preventing calcium influx), or nNOS (preventing ROS generation) all protected the neurons from Tat-induced apoptosis. This process has since been demonstrated in in vitro models of Tat-induced reversible synapse loss prior to irreversible induction neuronal apoptosis (Kim et al. 2008; Shin and Thayer 2012).

Tat can also interfere directly with normal microtubule dynamics in neurons and other cell types (de Mareuil et al. 2005; Butler et al. 2011). Cultured neurons exposed to Tat experienced rapid destabilization of the cytoskeleton. Specifically, Tat rapidly caused proteasome-dependent destruction of microtubule-associated protein 2 (Map-2), a cytoskeletal protein found only in neuronal dendrites, destabilizing the neurite structure (Aprea et al. 2006). These results are of particular interest because dystrophic Map-2 staining is a pathologic hallmark of CNS tissue from HAND patients (Aprea et al. 2006; Bellizzi et al. 2005; Everall et al. 1999; Masliah et al. 1997). Finally, this Tat-induced microtubule disruption has been linked to apoptosis, through a mitochondrial caspase-dependent mechanism (Giacca 2005).

Overall, these studies provide compelling evidence that HIV proteins can directly damage neuronal architecture by interfering with neurotrophic signaling, damaging cytoskeletal components, or inducing excitotoxicity. HIV Tat is of particular interest in the modern cART era, as it is an early response gene that is continually produced and released from infected cells even in the presence of effective cART, whereas gp120 requires full viral particle production to exit infected cells and exert direct and indirect (i.e., stimulation of pro-inflammatory mediators from mononuclear cells) neurotoxic effects.

15.5 Molecular Basis of Synaptic Damage in HAND Revisited: Immune Cell Activation and Interaction with Neuronal Elements

With the failure of treatment strategies aimed at directly preventing HIV neurotoxicity, there has been a renewed focus on the role of the host immune response in HAND. HIV infection rapidly leads to alteration in cytokine production by glia and immune cells in the infected CNS, notably inducing TNFα and IL-1β (Mintz et al. 1989; Wesselingh et al. 1993). These factors are produced by infected astrocytes, monocytes, and microglia, as well as uninfected cells that are exposed to HIV viral proteins. Pathologic levels of TNFα can inhibit glutamate uptake by astrocytes, leading to a build-up of glutamate at the synapse and eventual excitotoxic cell death (Fine et al. 1996). Work with rodent brain slice models has demonstrated that even slightly elevated levels of TNFα can interfere with the development of early phase long-term potentiation (LTP), a process thought to be the molecular basis of learning and memory (Tancredi et al. 1992; Xiong et al. 1999; Butler et al. 2004). In rodent models, IL-1β damages neurons by inducing astrocytic production of TNFα and inducible nitric oxide synthase (iNOS), leading to ROS production, in addition to further increasing the activation of microglia and monocytes (Brabers and Nottet 2006; Zhao et al. 2001). The role of IL-6, another important immune cytokine that is significantly upregulated in the brains of patients with HIV, is less clear, with potentially both neuroprotective and neurodegenerative effects (Erta et al. 2012).

HIV infection also results in immune-mediated increases in the production of pro-inflammatory lipids including arachidonic acid and platelet-activating factor (PAF) (Gelbard et al. 1994; Genis et al. 1992). PAF is of particular interest in HAND because it plays a physiologic role in the development of LTP, being produced by post-synaptic neurons as a result of NMDAR activation and acting pre-synaptically to increase glutamate release (Wieraszko et al. 1993; Kato et al. 1994). However, patients with HAND have significantly elevated levels of PAF expression in their CSF when compared to HIV seronegative patients or patients infected with HIV devoid of neurologic impairment, indicating a potential role specifically in the development of HAND (Gelbard et al. 1994). Further work demonstrated that pathologic levels of PAF produce dendritic spine loss and dendritic beading in cultured neurons, two pathologic hallmarks of HAND (Bellizzi et al. 2005). This dendritic beading is associated with impaired development of LTP and occurs in a calcium-dependent manner, thus indicating a potential excitotoxic mechanism. These findings are further supported by autopsy studies of brain tissue from HAND patients, where there is strong expression of PAF receptor on neurons that have beaded dendrites and dendritic spine loss (Bellizzi et al. 2005).

There have been several randomized controlled clinical trials based on the mechanisms of viral neurotoxicity listed in the previous section, including a number that have progressed to Phase II studies of efficacy. The first trials were based on protecting neurons by scavenging or reducing the production of ROS, as this was shown to have neuroprotective effects on Tat-exposed neurons in vitro (Eugenin et al. 2007).

The most promising study examined the use of the FDA approved drug selegiline, a monoamine oxidase B (MAO-B) inhibitor that is used to increase dopamine availability and reduce the symptoms of early Parkinson's disease (PD) or depression. Aside from its MAO-B inhibition, selegiline also increases ROS scavenging (Carrillo et al. 1994) and protects neurons from death after axotomy or neurotoxin exposure in vivo (Salo and Tatton 1992; Tatton 1993), possibly by stimulating neurotrophic support. These mechanisms of action seemed to counteract many of the effects caused by gp120 and Tat exposure, which are outlined in the previous section. Despite apparent initial success in a small randomized controlled trial (Sacktor et al. 2000), selegiline for the treatment of HAND failed in Phase II clinical trials due to lack of efficacy (Evans et al. 2007; Schifitto et al. 2007). Other trials with antioxidants that demonstrated neuroprotection in preclinical models similarly failed due to lack of efficacy (The Dana Consortium on the Therapy of HIV Dementia and Related Cognitive Disorders 1997; Clifford et al. 2002), suggesting that ROS reduction alone does not prevent the progression of HAND or that these treatments could not restore redox homeostasis in critical intracellular compartments such as mitochondria or rough endoplasmic reticulum of vulnerable neurons.

Other trials attempted to treat HAND by blocking the NMDAR-dependent excitotoxicity caused by exposure to HIV viral proteins. Because of its importance in excitatory neuronal signaling throughout the CNS, it can be difficult to antagonize NMDAR signaling without adverse effects. Therefore, these studies used the drug memantine, a low affinity non-competitive antagonist of glutamate and NMDAR signaling (Chen et al. 1992; Chen and Lipton 1997). Memantine is generally well tolerated and has proven to be marginally effective in slowing the progression of moderate to severe Alzheimer's disease in multiple large-scale clinical trials (Reisberg et al. 2006; Tariot et al. 2004). The relative success of memantine in Alzheimer's disease, coupled with memantine's protection in various models of HAND (Nath et al. 2000; Toggas et al. 1996), led to a placebo-controlled open-label Phase II trial of memantine in patients with HAND. While the drug was well tolerated, there was no significant improvement seen in patients with HAND treated with memantine compared to placebo control in any neurologic measures (Zhao et al. 2010).

To date, there has been no successful clinical therapy based on targets that influence the direct toxicity of HIV products to neurons. These findings demonstrate that while there is no doubt that HIV itself is toxic to neurons and interferes with neuronal function, this toxicity may not be the main driving force for the progression of HAND, especially in the cART era. Recent evidence indicates that the pathologic substrate of HAND may be better explained by direct interaction between microglia, invasive peripheral immune cells, and synaptic elements (Lu et al. 2011). A single exposure to HIV Tat in the CNS of mice leads to a rapid transient influx of neutrophils and a sustained influx of peripheral monocytes, as well as sustained microglial activation as determined by persistently altered morphology, in the form of a simpler, thicker process arbor and possible increase in cell volume, weeks after exposure. The invading immune cells and activated microglia directly contact and engulf neuronal components, potentially leading to neuronal process simplification through phagocytosis (Lu et al. 2011; Marker et al. 2013). Specifically, studies with in vivo two-photon microscopy show sustained direct contact and apparent engulfment of intact neuronal elements by microglia. Microglial phagocytosis of neuronal elements was confirmed at the ultrastructural level using immuno-electron microscopy, with a significant increase in neuronal inclusions in monocyte lineage cells 24 h and 7 days after Tat exposure. Interestingly, axonal components seemed most susceptible to phagocytosis in vivo, a finding that is supported by in vitro microglia/neuronal co-culture experiments (Marker et al. 2013). While these results are very intriguing, it is still unclear whether microglia phagocytosis of neuronal components is a driving force in the pathology of HAND, or if it is merely a response to neuronal death caused through some other mechanism. Further work will be needed in this area to determine causality.

Activated microglia may also damage neurons directly through the local pathologic release of matrix metalloproteinases (MMPs). Physiologically, MMPs in the CNS are produced by neurons, astrocytes, and microglia, and they are important for synaptic plasticity as required for learning and memory (Nagy et al. 2006; Dziembowska and Wlodarczyk 2012). In HIV infection, however, MMP activity can disrupt the BBB integrity, facilitating peripheral immune invasion into the CNS and furthering CNS HIV infection. MMP9, specifically, is upregulated in the brains of patients with severe HAND in autopsy immunohistological studies, particularly in areas close to the vasculature and activated monocyte lineage cells (Ghorpade et al. 2001). Further in vitro studies based on these findings showed that HIV Tat exposure leads to MMP9 production and release from cultured astrocytes in a MAPK-dependent manner (Ju et al. 2009). Additionally, widespread MMP activity can result in the cleavage of extracellular adhesion proteins and basement membranes, resulting in the loss of neuronal processes (Conant et al. 2012; Ghorpade et al. 2001; Ju et al. 2009).

Finally, there is also a complicated interplay between the invading immune cells and resident microglia, with widespread total engulfment of invading peripheral neutrophils by distal microglial processes, as well as partial neutrophil engulfment of microglia (Lu et al. 2011). The significance and molecular underpinnings of these cross lineage phagocytic interactions currently remain unclear, although it is possible that microglial engulfment of peripheral cells acts as a mechanism to limit the extent of the peripheral immune reaction. Immune cell activation, recruitment, and cytokine production all play important roles in HAND progression. The current evidence supports a scenario in which HIV exposure causes production of inflammatory mediators and peripheral immune chemoattractants, such as MCP-1 in microglia and astrocytes (Weiss et al. 1999), which drives peripheral monocyte and neutrophil invasion into the inflamed CNS. This could potentially lead to a positive feedback cycle in which HIV exposure leads to increase in inflammatory mediators, which in turn attracts more peripheral monocytes to the CNS, in turn leading to more inflammation. In this scenario, inflammatory cytokine production would serve to initiate and perpetuate direct immune cell damage to the neuronal arbor, even in the absence of active viral replication. It also presents a unique opportunity for treatment, with the potential for a therapeutic agent to break the inflammatory cycle and allow for neuronal recovery, or to directly prevent pathologic immune cell interactions with neurons.

15.6 Mixed Lineage Kinase 3: A Target That Mediates Neuroinflammation

One promising target in the treatment of neuroinflammatory disease is mixed lineage kinase 3 (MLK3, aka MAP3K11). MLK3 is a serine/threonine mitogen-associated protein kinase kinase kinase (MAPKKK) (Gallo et al. 1994), which controls a signaling cascade that activates the downstream MAP kinase Jun-n-terminal kinase (JNK), a pivotal kinase in the control of the innate immune response (Arthur and Ley 2013; Tibbles et al. 1996). MLK3 is expressed at basal levels in peripheral immune cells (Handley et al. 2007), microglia (Wang et al. 2010), and neurons (Maroney et al. 2001), among other cell types. In neurons, MLK3 is upregulated and phosphorylated in response to trophic factor withdrawal, ROS exposure, and ultra violet radiation, three classical activators of JNK (Xu et al. 2001; Lotharius et al. 2005). This increase in MLK3 activity results in JNK-dependent c-Jun upregulation and activation, upregulation and activation of many pro-apoptotic genes and proteins, and eventual neuronal apoptosis (Dhanasekaran and Reddy 2008). Additionally, JNK activation has been strongly implicated in the pathology of Alzheimer's disease, PD, and other neurodegenerative disease states (Borsello and Forloni 2007).

MLK3 plays a significant role in HIV infection and HAND progression. MLK3 was identified in a genome-wide screen as a potent enhancer of HIV infection (Nguyen et al. 2007). The same study demonstrated that MLK3 over-expression increased Tat-dependent gene transcription at the HIV long terminal repeat, the major promoter for HIV replication, sevenfold over control, and this increase was dependent on the MLK3 protein having an intact kinase region. Furthermore, genetic silencing of MLK3 inhibited viral replication in vitro (Nguyen et al. 2007). The mechanism by which MLK3 increased viral replication appears to be mediated through NFκβ, which can be activated by MLK3 (Hehner et al. 2000) and directly enhances viral transcription at the LTR (Yang et al. 1999). Additionally, HIV Tat and gp120 have been shown to induce autophosphorylation of MLK3 (Sui et al. 2006). HIV Tat- and gp120-induced neuronal apoptosis appears to be dependent on MLK3, as both genetic ablation and small molecule inhibition of MLK3 are neuroprotective in vitro (Bodner et al. 2002; Sui et al. 2006). Finally, Tat- and gp120-induced MLK3 activation in monocytes results in increased production of TNFα (Sui et al. 2006). Some of these relationships are depicted in the cartoon in Fig. 15.1. In conclusion, these studies indicate that HIV proteins increase the activity of MLK3 as a mechanism to enhance viral replication in an NFκβ-dependent manner. This increase in MLK3 activity is both directly neurotoxic and neuroinflammatory and may therefore be a major contributor to HAND.

15.7 Leucine-Rich Repeat Kinase 2: An Emerging Target in Neuroinflammation

Another promising kinase target recently implicated in neuroinflammation is the leucine-rich repeat kinase 2 (LRRK2). LRRK2 was discovered through genetic screening in families with autosomal-dominant PD (Funayama et al. 2002; Zimprich et al. 2004), and the most common mutation, G2019S, was further identified as a major cause of sporadic PD as well (Goldwurm et al. 2005). LRRK2 is a large GTPase with many functional domains, including a MAPKKK-like kinase domain (Zimprich et al. 2004). Many of the mutations in LRRK2 associated with PD are gain of function mutations, meaning they increase phosphorylation and activation of the LRRK2 kinase at baseline. LRRK2 is expressed both in neurons and non-neuronal cell populations in the brain; however, this study did not include co-localization markers for microglia (Han et al. 2008). Although direct substrates of LRRK2 have yet to be conclusively identified, there is significant evidence that mutant LRRK2 can lead to the phosphorylation and activation of MAP kinase kinases (MAPKKs) 3/6 and 7, which are directly upstream of MAPKs p38 and JNK, respectively (Gloeckner et al. 2009; Hsu et al. 2010; Chen et al. 2012).

Although the bulk of the current research has centered on LRRK2's function in neurons in PD, there has been a recent focus on its role in the immune system, particularly on its role in neuroinflammation. To date, these studies have found that LRRK2 is expressed by many peripheral immune cells, in particular peripheral monocytes, and is upregulated in response to microbial antigen presentation (Hakimi et al. 2011). LRRK2 is also involved in the maturation and differentiation of peripheral monocytes (Thevenet et al. 2011). Finally, increased LRRK2 protein levels were demonstrated in LPS-stimulated microglia, both in vitro and in vivo (Kim et al. 2012; Moehle et al. 2012). Further, LRRK2 kinase inhibition or genetic knockdown lessened microglial cytokine release and ROS production in response to LPS exposure (Kim et al. 2012; Moehle et al. 2012). LRRK2 has been recently linked to HAND, as recent studies further suggest that LRRK2 activation in microglia is instrumental for Tat-mediated neuroinflammation as well as phagocytosis of neuronal elements, which is in part due to the presence of phosphatidylserine groups on axons (Marker et al. 2012). Some of these relationships are depicted in the cartoon in Fig. 15.1.

15.8 Conclusions

The failure of anti-excitotoxic therapies for the treatment of HAND has led to a re-examination of the existing paradigms related to its neuropathogenesis. In particular, the notion that innate and adaptive immune responses shape normal relationships between microglia and synaptic architecture/function and that presence of the lentivirus HIV can disrupt the homeostasis between microglia and synapses by altering these relationships has profound consequences for the understanding of the

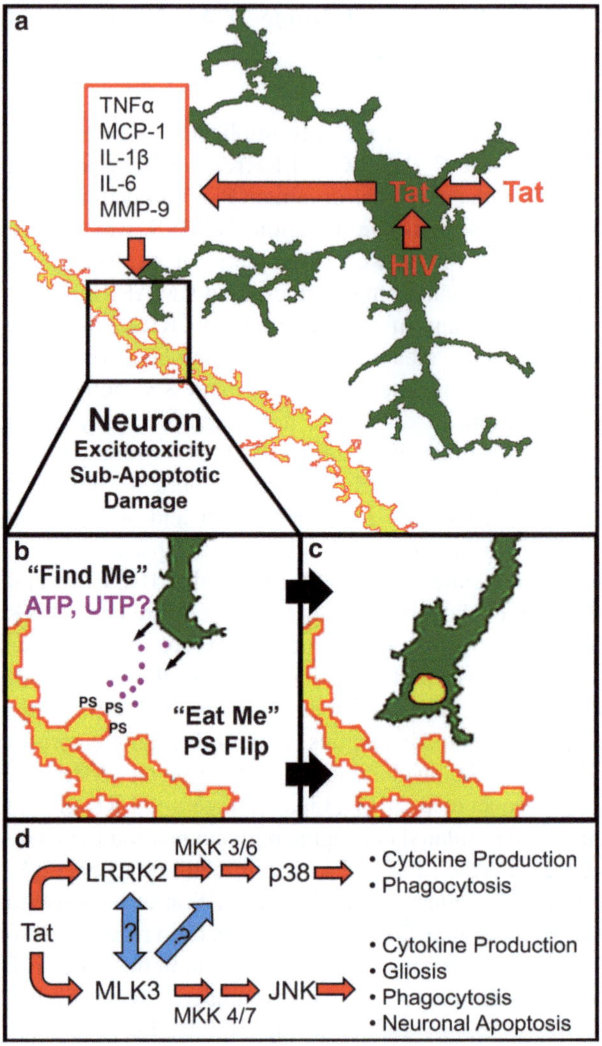

Fig. 15.1 Working models of pathologic synaptic pruning, mixed lineage kinase type 3 (MLK3) signaling, and leucine-rich repeat kinase 2 (LRRK2) signaling in HAND. (**a**) Diagram summarizing the effects of Tat on microglia and neurons. Tat exposure stimulates microglia to release many inflammatory cytokines that can damage nearby neuronal structures. Tat also directly damages neuronal structures and induces excitotoxic events. (**b**) Diagram showing hypothesized direct interaction between microglia and neurons in an inflammatory environment. Sub-apoptotic neuronal damage caused by HIV protein exposure or cytokines results in release of nucleotides, which are potent chemoattractants for microglia (Gyoneva et al. 2009), and local membrane exposure of phosphatidylserine (PS) in synaptic elements, which is a classic signal for phagocytosis (Ravichandran 2011). (**c**) This process results in activated microglia "stripping" neuronal synaptic components. (**d**) Signaling diagram showing the proposed signaling pathways for LRRK2 and MLK3 in microglia and neurons exposed to HIV Tat

pathogenesis of HAND and its treatment. Identification of pathologic ingestion of synaptic elements by microglia is a first step in understanding which key molecular entities may be involved in this phenomenon; investigating signaling through kinase "control hubs" for inflammation such as MLK3 and LRRK2 is the second step in devising rational strategies to restore homeostasis between microglia and synapses. In doing this, a framework for moving from a position of treating symptoms to an approach for modification of the course of disease in conditions such as HAND can be constructed.

Acknowledgments This work was funded by NIH P30 AI078498 (H.A.G.), PO1 MH64570 (H.A.G.), NIH RO1 MH56838 (H.A.G.), NIH UL1 RR024160 (H.A.G.), NIH T32 GM07356 (D.F.M.), NIH T32 AI049815 (D.F.M.), NIH F30 MH095664 (D.F.M.), and the Geoffrey Waisdorp Pediatric Neurology Fund (H.A.G.).

References

Acquas E, Bachis A, Nosheny RL, Cernak I, Mocchetti I (2004) Human immunodeficiency virus type 1 protein gp120 causes neuronal cell death in the rat brain by activating caspases. Neurotox Res 5(8):605–615

Anderson E, Zink W, Xiong H, Gendelman HE (2002) HIV-1-associated dementia: a metabolic encephalopathy perpetrated by virus-infected and immune-competent mononuclear phagocytes. J Acquir Immune Defic Syndr 31(Suppl 2):S43–S54

Antinori A, Arendt G, Becker JT, Brew BJ, Byrd DA, Cherner M, Clifford DB, Cinque P, Epstein LG, Goodkin K, Gisslen M, Grant I, Heaton RK, Joseph J, Marder K, Marra CM, McArthur JC, Nunn M, Price RW, Pulliam L, Robertson KR, Sacktor N, Valcour V, Wojna VE (2007) Updated research nosology for HIV-associated neurocognitive disorders. Neurology 69:1789–1799

Aprea S, Del Valle L, Mameli G, Sawaya BE, Khalili K, Peruzzi F (2006) Tubulin-mediated binding of human immunodeficiency virus-1 Tat to the cytoskeleton causes proteasomal-dependent degradation of microtubule-associated protein 2 and neuronal damage. J Neurosci 26(15):4054–4062. doi:10.1523/JNEUROSCI.0603-06.2006

Arthur JS, Ley SC (2013) Mitogen-activated protein kinases in innate immunity. Nat Rev Immunol 13(9):679–692. doi:10.1038/nri3495

Bachis A, Avdoshina V, Zecca L, Parsadanian M, Mocchetti I (2012) Human immunodeficiency virus type 1 alters brain-derived neurotrophic factor processing in neurons. J Neurosci 32(28):9477–9484. doi:10.1523/JNEUROSCI.0865-12.2012

Bachis A, Major EO, Mocchetti I (2003) Brain-derived neurotrophic factor inhibits human immunodeficiency virus-1/gp120-mediated cerebellar granule cell death by preventing gp120 internalization. J Neurosci 23(13):5715–5722

Bellizzi MJ, Lu SM, Masliah E, Gelbard HA (2005) Synaptic activity becomes excitotoxic in neurons exposed to elevated levels of platelet-activating factor. J Clin Invest 115(11):3185–3192. doi:10.1172/JCI25444

Bodner A, Maroney AC, Finn JP, Ghadge G, Roos R, Miller RJ (2002) Mixed lineage kinase 3 mediates gp120IIIB-induced neurotoxicity. J Neurochem 82(6):1424–1434

Boisse L, Gill MJ, Power C (2008) HIV infection of the central nervous system: clinical features and neuropathogenesis. Neurol Clin 26(3):799–819. doi:10.1016/j.ncl.2008.04.002

Borsello T, Forloni G (2007) JNK signalling: a possible target to prevent neurodegeneration. Curr Pharm Des 13(18):1875–1886

Brabers NA, Nottet HS (2006) Role of the pro-inflammatory cytokines TNF-alpha and IL-1beta in HIV-associated dementia. Eur J Clin Invest 36(7):447–458. doi:10.1111/j.1365-2362.2006.01657.x
Butler MP, O'Connor JJ, Moynagh PN (2004) Dissection of tumor-necrosis factor-alpha inhibition of long-term potentiation (LTP) reveals a p38 mitogen-activated protein kinase-dependent mechanism which maps to early-but not late-phase LTP. Neuroscience 124(2):319–326. doi:10.1016/j.neuroscience.2003.11.040
Butler TR, Smith KJ, Self RL, Braden BB, Prendergast MA (2011) Neurodegenerative effects of recombinant HIV-1 Tat(1-86) are associated with inhibition of microtubule formation and oxidative stress-related reductions in microtubule-associated protein-2(a, b). Neurochem Res 36(5):819–828. doi:10.1007/s11064-011-0409-2
Campbell GR, Loret EP (2009) What does the structure-function relationship of the HIV-1 Tat protein teach us about developing an AIDS vaccine? Retrovirology 6:50. doi:10.1186/1742-4690-6-50
Carrillo MC, Kitani K, Kanai S, Sato Y, Miyasaka K, Ivy GO (1994) (-)deprenyl increases activities of superoxide dismutase and catalase in certain brain regions in old male mice. Life Sci 54(14):975–981
Chaudhuri A, Yang B, Gendelman HE, Persidsky Y, Kanmogne GD (2008) STAT1 signaling modulates HIV-1-induced inflammatory responses and leukocyte transmigration across the blood-brain barrier. Blood 111(4):2062–2072. doi:10.1182/blood-2007-05-091207
Chen CY, Weng YH, Chien KY, Lin KJ, Yeh TH, Cheng YP et al (2012) (G2019S) LRRK2 activates MKK4-JNK pathway and causes degeneration of SN dopaminergic neurons in a transgenic mouse model of PD. Cell Death Differ 19(10):1623–1633. doi:10.1038/cdd.2012.42
Chen HS, Lipton SA (1997) Mechanism of memantine block of NMDA-activated channels in rat retinal ganglion cells: uncompetitive antagonism. J Physiol 499(Pt 1):27–46
Chen HS, Pellegrini JW, Aggarwal SK, Lei SZ, Warach S, Jensen FE et al (1992) Open-channel block of N-methyl-D-aspartate (NMDA) responses by memantine: therapeutic advantage against NMDA receptor-mediated neurotoxicity. J Neurosci 12(11):4427–4436
Churchill MJ, Wesselingh SL, Cowley D, Pardo CA, McArthur JC, Brew BJ et al (2009) Extensive astrocyte infection is prominent in human immunodeficiency virus-associated dementia. Ann Neurol 66(2):253–258. doi:10.1002/ana.21697
Clay CC, Rodrigues DS, Ho YS, Fallert BA, Janatpour K, Reinhart TA et al (2007) Neuroinvasion of fluorescein-positive monocytes in acute simian immunodeficiency virus infection. J Virol 81(21):12040–12048. doi:10.1128/JVI.00133-07
Clifford DB, McArthur JC, Schifitto G, Kieburtz K, McDermott MP, Letendre S et al (2002) A randomized clinical trial of CPI-1189 for HIV-associated cognitive-motor impairment. Neurology 59(10):1568–1573
Conant K, Lim ST, Randall B, Maguire-Zeiss KA (2012) Matrix metalloproteinase dependent cleavage of cell adhesion molecules in the pathogenesis of CNS dysfunction with HIV and methamphetamine. Curr HIV Res 10(5):384–391
Cowley D, Gray LR, Wesselingh SL, Gorry PR, Churchill MJ (2011) Genetic and functional heterogeneity of CNS-derived tat alleles from patients with HIV-associated dementia. J Neurovirol 17(1):70–81. doi:10.1007/s13365-010-0002-5
Cysique LA, Maruff P, Brew BJ (2004) Prevalence and pattern of neuropsychological impairment in human immunodeficiency virus-infected/acquired immunodeficiency syndrome (HIV/AIDS) patients across pre- and post-highly active antiretroviral therapy eras: a combined study of two cohorts. J Neurovirol 10(6):350–357. doi:10.1080/13550280490521078
Dandekar DH, Ganesh KN, Mitra D (2004) HIV-1 Tat directly binds to NFkappaB enhancer sequence: role in viral and cellular gene expression. Nucleic Acids Res 32(4):1270–1278. doi:10.1093/nar/gkh289
de Amorim M (1960) Reactional aspects of the microglia in spontaneous granulomatous encephalitis. Arch De Vecchi Anat Patol 31:1–9
de Mareuil J, Carre M, Barbier P, Campbell GR, Lancelot S, Opi S et al (2005) HIV-1 Tat protein enhances microtubule polymerization. Retrovirology 2:5. doi:10.1186/1742-4690-2-5
DeKosky ST, Scheff SW (1990) Synapse loss in frontal cortex biopsies in Alzheimer's disease: correlation with cognitive severity. Ann Neurol 27(5):457–464. doi:10.1002/ana.410270502

del Saz SV, Sued O, Falco V, Aguero F, Crespo M, Pumarola T et al (2008) Acute meningoencephalitis due to human immunodeficiency virus type 1 infection in 13 patients: clinical description and follow-up. J Neurovirol 14(6):474–479. doi:10.1080/13550280802195367

Dhanasekaran DN, Reddy EP (2008) JNK signaling in apoptosis. Oncogene 27(48):6245–6251. doi:10.1038/onc.2008.301

Dore GJ, McDonald A, Li Y, Kaldor JM, Brew BJ, National HIVSC (2003) Marked improvement in survival following AIDS dementia complex in the era of highly active antiretroviral therapy. AIDS 17(10):1539–1545. doi:10.1097/01.aids.0000076282.54156.c3

Dreyer EB, Kaiser PK, Offermann JT, Lipton SA (1990) HIV-1 coat protein neurotoxicity prevented by calcium channel antagonists. Science 248(4953):364–367

Dziembowska M, Wlodarczyk J (2012) MMP9: a novel function in synaptic plasticity. Int J Biochem Cell Biol 44(5):709–713. doi:10.1016/j.biocel.2012.01.023

Ellis RJ, Calero P, Stockin MD (2009) HIV infection and the central nervous system: a primer. Neuropsychol Rev 19(2):144–151. doi:10.1007/s11065-009-9094-1

Erta M, Quintana A, Hidalgo J (2012) Interleukin-6, a major cytokine in the central nervous system. Int J Biol Sci 8(9):1254–1266. doi:10.7150/ijbs.4679

Eugenin EA, King JE, Nath A, Calderon TM, Zukin RS, Bennett MV et al (2007) HIV-tat induces formation of an LRP-PSD-95- NMDAR-nNOS complex that promotes apoptosis in neurons and astrocytes. Proc Natl Acad Sci U S A 104(9):3438–3443. doi:10.1073/pnas.0611699104

Evans SR, Yeh TM, Sacktor N, Clifford DB, Simpson D, Miller EN et al (2007) Selegiline transdermal system (STS) for HIV-associated cognitive impairment: open-label report of ACTG 5090. HIV Clin Trials 8(6):437–446. doi:10.1310/hct0806-437

Everall IP, Glass JD, McArthur J, Spargo E, Lantos P (1994) Neuronal density in the superior frontal and temporal gyri does not correlate with the degree of human immunodeficiency virus-associated dementia. Acta Neuropathol 88(6):538–544

Everall IP, Heaton RK, Marcotte TD, Ellis RJ, McCutchan JA, Atkinson JH et al (1999) Cortical synaptic density is reduced in mild to moderate human immunodeficiency virus neurocognitive disorder. HNRC Group. HIV Neurobehavioral Research Center. Brain Pathol 9(2):209–217

Everall IP, Luthert PJ, Lantos PL (1991) Neuronal loss in the frontal cortex in HIV infection. Lancet 337(8750):1119–1121

Everall IP, Luthert PJ, Lantos PL (1993) Neuronal number and volume alterations in the neocortex of HIV infected individuals. J Neurol Neurosurg Psychiatry 56(5):481–486

Fine SM, Angel RA, Perry SW, Epstein LG, Rothstein JD, Dewhurst S et al (1996) Tumor necrosis factor alpha inhibits glutamate uptake by primary human astrocytes. Implications for pathogenesis of HIV-1 dementia. J Biol Chem 271(26):15303–15306

Funayama M, Hasegawa K, Kowa H, Saito M, Tsuji S, Obata F (2002) A new locus for Parkinson's disease (PARK8) maps to chromosome 12p11.2-q13.1. Ann Neurol 51(3):296–301

Gallo KA, Mark MR, Scadden DT, Wang Z, Gu Q, Godowski PJ (1994) Identification and characterization of SPRK, a novel src-homology 3 domain-containing proline-rich kinase with serine/threonine kinase activity. J Biol Chem 269(21):15092–15100

Gallo SA, Finnegan CM, Viard M, Raviv Y, Dimitrov A, Rawat SS et al (2003) The HIV Env-mediated fusion reaction. Biochim Biophys Acta 1614(1):36–50

Gelbard HA, Nottet HS, Swindells S, Jett M, Dzenko KA, Genis P et al (1994) Platelet-activating factor: a candidate human immunodeficiency virus type 1-induced neurotoxin. J Virol 68(7):4628–4635

Gelman BB (1993) Diffuse microgliosis associated with cerebral atrophy in the acquired immunodeficiency syndrome. Ann Neurol 34(1):65–70. doi:10.1002/ana.410340112

Genis P, Jett M, Bernton EW, Boyle T, Gelbard HA, Dzenko K et al (1992) Cytokines and arachidonic metabolites produced during human immunodeficiency virus (HIV)-infected macrophage-astroglia interactions: implications for the neuropathogenesis of HIV disease. J Exp Med 176(6):1703–1718

Ghorpade A, Persidskaia R, Suryadevara R, Che M, Liu XJ, Persidsky Y et al (2001) Mononuclear phagocyte differentiation, activation, and viral infection regulate matrix metalloproteinase expression: implications for human immunodeficiency virus type 1-associated dementia. J Virol 75(14):6572–6583. doi:10.1128/JVI.75.14.6572-6583.2001

Giacca M (2005) HIV-1 Tat, apoptosis and the mitochondria: a tubulin link? Retrovirology 2:7. doi:10.1186/1742-4690-2-7

Glass JD, Fedor H, Wesselingh SL, McArthur JC (1995) Immunocytochemical quantitation of human immunodeficiency virus in the brain: correlations with dementia. Ann Neurol 38(5): 755–762. doi:10.1002/ana.410380510

Glass JD, Wesselingh SL, Selnes OA, McArthur JC (1993) Clinical-neuropathologic correlation in HIV-associated dementia. Neurology 43(11):2230–2237

Gloeckner CJ, Schumacher A, Boldt K, Ueffing M (2009) The Parkinson disease-associated protein kinase LRRK2 exhibits MAPKKK activity and phosphorylates MKK3/6 and MKK4/7, in vitro. J Neurochem 109(4):959–968. doi:10.1111/j.1471-4159.2009.06024.x

Goldwurm S, Di Fonzo A, Simons EJ, Rohe CF, Zini M, Canesi M et al (2005) The G6055A (G2019S) mutation in LRRK2 is frequent in both early and late onset Parkinson's disease and originates from a common ancestor. J Med Genet 42(11):e65. doi:10.1136/jmg.2005.035568

Gulick RM, Mellors JW, Havlir D, Eron JJ, Gonzalez C, McMahon D et al (1997) Treatment with indinavir, zidovudine, and lamivudine in adults with human immunodeficiency virus infection and prior antiretroviral therapy. N Engl J Med 337(11):734–739. doi:10.1056/NEJM199709113371102

Gyoneva S, Orr AG, Traynelis SF (2009) Differential regulation of microglial motility by ATP/ADP and adenosine. Parkinsonism Relat Disord 15(Suppl 3):S195–S199. doi:10.1016/S1353-8020(09)70813-2

Hakimi M, Selvanantham T, Swinton E, Padmore RF, Tong Y, Kabbach G et al (2011) Parkinson's disease-linked LRRK2 is expressed in circulating and tissue immune cells and upregulated following recognition of microbial structures. J Neural Transm 118(5):795–808. doi:10.1007/s00702-011-0653-2

Han BS, Iacovitti L, Katano T, Hattori N, Seol W, Kim KS (2008) Expression of the LRRK2 gene in the midbrain dopaminergic neurons of the substantia nigra. Neurosci Lett 442(3):190–194. doi:10.1016/j.neulet.2008.06.086

Handley ME, Rasaiyaah J, Chain BM, Katz DR (2007) Mixed lineage kinases (MLKs): a role in dendritic cells, inflammation and immunity? Int J Exp Pathol 88(2):111–126. doi:10.1111/j.1365-2613.2007.00531.x

Heaton RK, Clifford DB, Franklin DR Jr, Woods SP, Ake C, Vaida F et al (2010) HIV-associated neurocognitive disorders persist in the era of potent antiretroviral therapy: CHARTER Study. Neurology 75(23):2087–2096. doi:10.1212/WNL.0b013e318200d727

Heaton RK, Franklin DR, Ellis RJ, McCutchan JA, Letendre SL, Leblanc S et al (2011) HIV-associated neurocognitive disorders before and during the era of combination antiretroviral therapy: differences in rates, nature, and predictors. J Neurovirol 17(1):3–16. doi:10.1007/s13365-010-0006-1

Hehner SP, Hofmann TG, Ushmorov A, Dienz O, Wing-Lan Leung I, Lassam N et al (2000) Mixed-lineage kinase 3 delivers CD3/CD28-derived signals into the IkappaB kinase complex. Mol Cell Biol 20(7):2556–2568

Ho DD, Rota TR, Schooley RT, Kaplan JC, Allan JD, Groopman JE et al (1985) Isolation of HTLV-III from cerebrospinal fluid and neural tissues of patients with neurologic syndromes related to the acquired immunodeficiency syndrome. N Engl J Med 313(24):1493–1497. doi:10.1056/NEJM198512123132401

Hsu CH, Chan D, Greggio E, Saha S, Guillily MD, Ferree A et al (2010) MKK6 binds and regulates expression of Parkinson's disease-related protein LRRK2. J Neurochem 112(6):1593–1604. doi:10.1111/j.1471-4159.2010.06568.x

Ju SM, Song HY, Lee JA, Lee SJ, Choi SY, Park J (2009) Extracellular HIV-1 Tat up-regulates expression of matrix metalloproteinase-9 via a MAPK-NF-kappaB dependent pathway in human astrocytes. Exp Mol Med 41(2):86–93

Kahn JO, Walker BD (1998) Acute human immunodeficiency virus type 1 infection. N Engl J Med 339(1):33–39. doi:10.1056/NEJM199807023390107

Kanmogne GD, Schall K, Leibhart J, Knipe B, Gendelman HE, Persidsky Y (2007) HIV-1 gp120 compromises blood-brain barrier integrity and enhances monocyte migration across blood-brain barrier: implication for viral neuropathogenesis. J Cereb Blood Flow Metab 27(1): 123–134. doi:10.1038/sj.jcbfm.9600330

Kato K, Clark GD, Bazan NG, Zorumski CF (1994) Platelet-activating factor as a potential retrograde messenger in CA1 hippocampal long-term potentiation. Nature 367(6459):175–179. doi:10.1038/367175a0

Kim B, Yang MS, Choi D, Kim JH, Kim HS, Seol W et al (2012) Impaired inflammatory responses in murine Lrrk2-knockdown brain microglia. PLoS One 7(4):e34693. doi:10.1371/journal.pone.0034693

Kim HJ, Martemyanov KA, Thayer SA (2008) Human immunodeficiency virus protein Tat induces synapse loss via a reversible process that is distinct from cell death. J Neurosci 28(48):12604–12613. doi:10.1523/JNEUROSCI.2958-08.2008

Koenig S, Gendelman HE, Orenstein JM, Dal Canto MC, Pezeshkpour GH, Yungbluth M et al (1986) Detection of AIDS virus in macrophages in brain tissue from AIDS patients with encephalopathy. Science 233(4768):1089–1093

Kure K, Lyman WD, Weidenheim KM, Dickson DW (1990) Cellular localization of an HIV-1 antigen in subacute AIDS encephalitis using an improved double-labeling immunohistochemical method. Am J Pathol 136(5):1085–1092

Li CM, Lee YY, Ho YR (2002) Acute meningoencephalitis as initial presentation of human immunodeficiency virus infection: report of two cases. J Microbiol Immunol Infect 35(3):195–198

Li W, Li G, Steiner J, Nath A (2009) Role of Tat protein in HIV neuropathogenesis. Neurotox Res 16(3):205–220. doi:10.1007/s12640-009-9047-8

Liu Y, Jones M, Hingtgen CM, Bu G, Laribee N, Tanzi RE et al (2000) Uptake of HIV-1 tat protein mediated by low-density lipoprotein receptor-related protein disrupts the neuronal metabolic balance of the receptor ligands. Nat Med 6(12):1380–1387. doi:10.1038/82199

Lotharius J, Falsig J, van Beek J, Payne S, Dringen R, Brundin P et al (2005) Progressive degeneration of human mesencephalic neuron-derived cells triggered by dopamine-dependent oxidative stress is dependent on the mixed-lineage kinase pathway. J Neurosci 25(27):6329–6342. doi:10.1523/JNEUROSCI.1746-05.2005

Lu SM, Tremblay ME, King IL, Qi J, Reynolds HM, Marker DF et al (2011) HIV-1 Tat-induced microgliosis and synaptic damage via interactions between peripheral and central myeloid cells. PLoS One 6(9):e23915. doi:10.1371/journal.pone.0023915

Marker DF, Puccini JM, Mockus TE, Barbieri J, Lu SM, Gelbard HA (2012) LRRK2 kinase inhibition prevents pathological microglial phagocytosis in response to HIV-1 Tat protein. J Neuroinflammation 9:261. doi:10.1186/1742-2094-9-261

Marker DF, Tremblay ME, Puccini JM, Barbieri J, Gantz Marker MA, Loweth CJ et al (2013) The new small-molecule mixed-lineage kinase 3 inhibitor URMC-099 is neuroprotective and anti-inflammatory in models of human immunodeficiency virus-associated neurocognitive disorders. J Neurosci 33(24):9998–10010. doi:10.1523/JNEUROSCI.0598-13.2013

Maroney AC, Finn JP, Connors TJ, Durkin JT, Angeles T, Gessner G et al (2001) Cep-1347 (KT7515), a semisynthetic inhibitor of the mixed lineage kinase family. J Biol Chem 276(27):25302–25308. doi:10.1074/jbc.M011601200

Marra CM, Zhao Y, Clifford DB, Letendre S, Evans S, Henry K et al (2009) Impact of combination antiretroviral therapy on cerebrospinal fluid HIV RNA and neurocognitive performance. AIDS 23(11):1359–1366. doi:10.1097/QAD.0b013e32832c4152

Masliah E, Ge N, Morey M, DeTeresa R, Terry RD, Wiley CA (1992) Cortical dendritic pathology in human immunodeficiency virus encephalitis. Lab Invest 66(3):285–291

Masliah E, Heaton RK, Marcotte TD, Ellis RJ, Wiley CA, Mallory M et al (1997) Dendritic injury is a pathological substrate for human immunodeficiency virus-related cognitive disorders. HNRC Group. The HIV Neurobehavioral Research Center. Ann Neurol 42(6):963–972. doi:10.1002/ana.410420618

McNamara LA, Ganesh JA, Collins KL (2012) Latent HIV-1 infection occurs in multiple subsets of hematopoietic progenitor cells and is reversed by NF-kappaB activation. J Virol 86(17):9337–9350. doi:10.1128/JVI.00895-12

Mintz M, Rapaport R, Oleske JM, Connor EM, Koenigsberger MR, Denny T et al (1989) Elevated serum levels of tumor necrosis factor are associated with progressive encephalopathy in children with acquired immunodeficiency syndrome. Am J Dis Child 143(7):771–774

Mirza A, Rathore MH (2012) Human immunodeficiency virus and the central nervous system. Semin Pediatr Neurol 19(3):119–123. doi:10.1016/j.spen.2012.02.007

Mocchetti I, Bachis A (2004) Brain-derived neurotrophic factor activation of TrkB protects neurons from HIV-1/gp120-induced cell death. Crit Rev Neurobiol 16(1–2):51–57

Moehle MS, Webber PJ, Tse T, Sukar N, Standaert DG, DeSilva TM et al (2012) LRRK2 inhibition attenuates microglial inflammatory responses. J Neurosci 32(5):1602–1611. doi:10.1523/JNEUROSCI.5601-11.2012

Mothobi NZ, Brew BJ (2012) Neurocognitive dysfunction in the highly active antiretroviral therapy era. Curr Opin Infect Dis 25(1):4–9. doi:10.1097/QCO.0b013e32834ef586

Muller WE, Pergande G, Ushijima H, Schleger C, Kelve M, Perovic S (1996) Neurotoxicity in rat cortical cells caused by N-methyl-D-aspartate (NMDA) and gp120 of HIV-1: induction and pharmacological intervention. Prog Mol Subcell Biol 16:44–57

Murphy TH, Li P, Betts K, Liu R (2008) Two-photon imaging of stroke onset in vivo reveals that NMDA-receptor independent ischemic depolarization is the major cause of rapid reversible damage to dendrites and spines. J Neurosci 28(7):1756–1772. doi:10.1523/JNEUROSCI.5128-07.2008

Nagy V, Bozdagi O, Matynia A, Balcerzyk M, Okulski P, Dzwonek J et al (2006) Matrix metalloproteinase-9 is required for hippocampal late-phase long-term potentiation and memory. J Neurosci 26(7):1923–1934. doi:10.1523/JNEUROSCI.4359-05.2006

Nath A, Haughey NJ, Jones M, Anderson C, Bell JE, Geiger JD (2000) Synergistic neurotoxicity by human immunodeficiency virus proteins Tat and gp120: protection by memantine. Ann Neurol 47(2):186–194

Nguyen DG, Yin H, Zhou Y, Wolff KC, Kuhen KL, Caldwell JS (2007) Identification of novel therapeutic targets for HIV infection through functional genomic cDNA screening. Virology 362(1):16–25. doi:10.1016/j.virol.2006.11.036

Palella FJ Jr, Delaney KM, Moorman AC, Loveless MO, Fuhrer J, Satten GA et al (1998) Declining morbidity and mortality among patients with advanced human immunodeficiency virus infection. HIV Outpatient Study Investigators. N Engl J Med 338(13):853–860. doi:10.1056/NEJM199803263381301

Pascual-Lozano AM, Salvador-Aliaga A, Coret-Ferrer F, Lainez-Andres JM (2005) Subacute uveomeningoencephalitis as the presenting symptom of acute infection by the human immunodeficiency virus. Rev Neurol 41(5):273–276

Price RW, Brew B, Sidtis J, Rosenblum M, Scheck AC, Cleary P (1988) The brain in AIDS: central nervous system HIV-1 infection and AIDS dementia complex. Science 239(4840):586–592

Ranki A, Nyberg M, Ovod V, Haltia M, Elovaara I, Raininko R et al (1995) Abundant expression of HIV Nef and Rev proteins in brain astrocytes in vivo is associated with dementia. AIDS 9(9):1001–1008

Ravichandran KS (2011) Beginnings of a good apoptotic meal: the find-me and eat-me signaling pathways. Immunity 35(4):445–455

Reisberg B, Doody R, Stoffler A, Schmitt F, Ferris S, Mobius HJ (2006) A 24-week open-label extension study of memantine in moderate to severe Alzheimer disease. Arch Neurol 63(1):49–54. doi:10.1001/archneur.63.1.49

Sacktor N, Schifitto G, McDermott MP, Marder K, McArthur JC, Kieburtz K (2000) Transdermal selegiline in HIV-associated cognitive impairment: pilot, placebo-controlled study. Neurology 54(1):233–235

(1997) Safety and tolerability of the antioxidant OPC-14117 in HIV-associated cognitive impairment. The Dana Consortium on the Therapy of HIV Dementia and Related Cognitive Disorders. Neurology 49(1):142–146

Salo PT, Tatton WG (1992) Deprenyl reduces the death of motoneurons caused by axotomy. J Neurosci Res 31(2):394–400. doi:10.1002/jnr.490310223

Schifitto G, Zhang J, Evans SR, Sacktor N, Simpson D, Millar LL et al (2007) A multicenter trial of selegiline transdermal system for HIV-associated cognitive impairment. Neurology 69(13):1314–1321. doi:10.1212/01.wnl.0000268487.78753.0f
Shin AH, Thayer SA (2012) Human immunodeficiency virus-1 protein Tat induces excitotoxic loss of presynaptic terminals in hippocampal cultures. Mol Cell Neurosci 54C:22–29. doi:10.1016/j.mcn.2012.12.005
Sui Z, Fan S, Sniderhan L, Reisinger E, Litzburg A, Schifitto G et al (2006) Inhibition of mixed lineage kinase 3 prevents HIV-1 Tat-mediated neurotoxicity and monocyte activation. J Immunol 177(1):702–711
Tancredi V, D'Arcangelo G, Grassi F, Tarroni P, Palmieri G, Santoni A et al (1992) Tumor necrosis factor alters synaptic transmission in rat hippocampal slices. Neurosci Lett 146(2):176–178
Tariot PN, Farlow MR, Grossberg GT, Graham SM, McDonald S, Gergel I et al (2004) Memantine treatment in patients with moderate to severe Alzheimer disease already receiving donepezil: a randomized controlled trial. JAMA 291(3):317–324. doi:10.1001/jama.291.3.317
Tatton WG (1993) Selegiline can mediate neuronal rescue rather than neuronal protection. Mov Disord 8(Suppl 1):S20–S30
Terry RD, Masliah E, Salmon DP, Butters N, DeTeresa R, Hill R et al (1991) Physical basis of cognitive alterations in Alzheimer's disease: synapse loss is the major correlate of cognitive impairment. Ann Neurol 30(4):572–580. doi:10.1002/ana.410300410
Thevenet J, Pescini Gobert R, Hooft van Huijsduijnen R, Wiessner C, Sagot YJ (2011) Regulation of LRRK2 expression points to a functional role in human monocyte maturation. PLoS One 6(6):e21519
Tibbles LA, Ing YL, Kiefer F, Chan J, Iscove N, Woodgett JR et al (1996) MLK-3 activates the SAPK/JNK and p38/RK pathways via SEK1 and MKK3/6. EMBO J 15(24):7026–7035
Toggas SM, Masliah E, Mucke L (1996) Prevention of HIV-1 gp120-induced neuronal damage in the central nervous system of transgenic mice by the NMDA receptor antagonist memantine. Brain Res 706(2):303–307
Ushijima H, Nishio O, Klocking R, Perovic S, Muller WE (1995) Exposure to gp120 of HIV-1 induces an increased release of arachidonic acid in rat primary neuronal cell culture followed by NMDA receptor-mediated neurotoxicity. Eur J Neurosci 7(6):1353–1359
Valcour V, Sithinamsuwan P, Letendre S, Ances B (2011) Pathogenesis of HIV in the central nervous system. Curr HIV/AIDS Rep 8(1):54–61. doi:10.1007/s11904-010-0070-4
Wang MJ, Huang HY, Chen WF, Chang HF, Kuo JS (2010) Glycogen synthase kinase-3beta inactivation inhibits tumor necrosis factor-alpha production in microglia by modulating nuclear factor kappaB and MLK3/JNK signaling cascades. J Neuroinflammation 7:99. doi:10.1186/1742-2094-7-99
Weiss JM, Nath A, Major EO, Berman JW (1999) HIV-1 Tat induces monocyte chemoattractant protein-1-mediated monocyte transmigration across a model of the human blood-brain barrier and up-regulates CCR5 expression on human monocytes. J Immunol 163(5):2953–2959
Wesselingh SL, Power C, Glass JD, Tyor WR, McArthur JC, Farber JM et al (1993) Intracerebral cytokine messenger RNA expression in acquired immunodeficiency syndrome dementia. Ann Neurol 33(6):576–582. doi:10.1002/ana.410330604
Wieraszko A, Li G, Kornecki E, Hogan MV, Ehrlich YH (1993) Long-term potentiation in the hippocampus induced by platelet-activating factor. Neuron 10(3):553–557
Xiong H, Zeng YC, Zheng J, Thylin M, Gendelman HE (1999) Soluble HIV-1 infected macrophage secretory products mediate blockade of long-term potentiation: a mechanism for cognitive dysfunction in HIV-1-associated dementia. J Neurovirol 5(5):519–528
Xu H, Bae M, Tovar-y-Romo LB, Patel N, Bandaru VV, Pomerantz D et al (2011) The human immunodeficiency virus coat protein gp120 promotes forward trafficking and surface clustering of NMDA receptors in membrane microdomains. J Neurosci 31(47):17074–17090. doi:10.1523/JNEUROSCI.4072-11.2011
Xu Z, Maroney AC, Dobrzanski P, Kukekov NV, Greene LA (2001) The MLK family mediates c-Jun N-terminal kinase activation in neuronal apoptosis. Mol Cell Biol 21(14):4713–4724. doi:10.1128/MCB.21.14.4713-4724.2001

Yang X, Chen Y, Gabuzda D (1999) ERK MAP kinase links cytokine signals to activation of latent HIV-1 infection by stimulating a cooperative interaction of AP-1 and NF-kappaB. J Biol Chem 274(39):27981–27988

Zhang S, Boyd J, Delaney K, Murphy TH (2005) Rapid reversible changes in dendritic spine structure in vivo gated by the degree of ischemia. J Neurosci 25(22):5333–5338. doi:10.1523/JNEUROSCI.1085-05.2005

Zhao ML, Kim MO, Morgello S, Lee SC (2001) Expression of inducible nitric oxide synthase, interleukin-1 and caspase-1 in HIV-1 encephalitis. J Neuroimmunol 115(1–2):182–191

Zhao Y, Navia BA, Marra CM, Singer EJ, Chang L, Berger J et al (2010) Memantine for AIDS dementia complex: open-label report of ACTG 301. HIV Clin Trials 11(1):59–67. doi:10.1310/hct1101-59

Zhu P, Winkler H, Chertova E, Taylor KA, Roux KH (2008) Cryoelectron tomography of HIV-1 envelope spikes: further evidence for tripod-like legs. PLoS Pathog 4(11):e1000203. doi:10.1371/journal.ppat.1000203

Zimprich A, Biskup S, Leitner P, Lichtner P, Farrer M, Lincoln S et al (2004) Mutations in LRRK2 cause autosomal-dominant parkinsonism with pleomorphic pathology. Neuron 44(4):601–607. doi:10.1016/j.neuron.2004.11.005

Chapter 16
Multiple Sclerosis

Yoshifumi Sonobe and Akio Suzumura

Abstract Multiple sclerosis (MS) is an inflammatory disease characterized by demyelination and axonal degeneration in the central nervous system (CNS). Although MS is considered an autoimmune disease against myelin antigens, its pathogenesis still remains unclear. Microglia are macrophage-like cells in the CNS which play a critical role in innate immunity, in addition to activating pathways associated with adaptive immunity. Microglia produce pro-inflammatory and anti-inflammatory mediators, including cytokines and chemokines, and phagocytose various types of cellular debris. In MS, microglia critically contribute to the inflammatory milieu, but also participate in disrupting the blood–brain barrier integrity, thus inducing the migration of various types of immune cells such as T and B lymphocytes, macrophages, and neutrophils into the CNS. In this disease, microglia may additionally behave as antigen-presenting cells and function as effector cells causing demyelination and axonal degeneration. However, recent evidence also indicates that microglia could play a beneficial role in remyelination and neuroprotection in MS. In this chapter, we will discuss about microglial involvement in MS, with an emphasis on the experimental autoimmune encephalomyelitis (EAE) animal model and describe the cellular and molecular mechanisms which could be specifically implicated in the pathogenesis.

Keywords Microglia • Inflammation • Cytokine • Chemokine • Blood–brain barrier • Antigen presentation • Demyelination • Neurodegeneration • Multiple sclerosis • Experimental autoimmune encephalomyelitis

Bullet Points

- Microglia, macrophage-like cells in the central nervous system (CNS), play a substantial role in the pathogenesis of multiple sclerosis (MS).
- In MS, microglia contribute to the development of neuroinflammation by producing both pro-inflammatory and anti-inflammatory mediators.

Y. Sonobe, Ph.D. • A. Suzumura, M.D., Ph.D. (✉)
Department of Neuroimmunology, Research Institute of Environmental Medicine, Nagoya University, Furo-cho, Chikusa-ku, Nagoya, Aichi 464-8601, Japan
e-mail: sonobe@riem.nagoya-u.ac.jp; suzumura@riem.nagoya-u.ac.jp

M.-È. Tremblay, A. Sierra (eds.), *Microglia in Health and Disease*,
DOI 10.1007/978-1-4939-1429-6_16

- Microglia also promote the migration of peripheral immune cells into the CNS and might additionally behave as antigen-presenting cells in MS.
- Microglia could lastly influence by their effector functions the demyelination and axonal degeneration observed in MS.

16.1 Introduction

Multiple sclerosis (MS) is a chronic demyelinating disease of the central nervous system (CNS) resulting from an autoimmune response against myelin antigens. It affects approximately 2.5 million people worldwide, with a predominance in women (ratio of females to males, 2:1). The disease is characterized by a progressive loss of neurological functions caused by the destruction of axonal myelin sheaths throughout the brain and spinal cord white matter. The loss of myelin translates into clinical symptoms ranging from paralysis, muscle spasms, and optic neuritis, to neuropathic pain. Pathological features of MS lesions include increased blood–brain barrier (BBB) permeability, axonal degeneration, glial scar formation, and the prevalence of peripheral immune cells such as T and B lymphocytes, macrophages, and neutrophils within the CNS (Williams et al. 2007; Jadidi-Niaragh and Mirshafiey 2011). The etiology of MS is still unclear. Genetic factors like variations in the HLA-DRB1 gene coding for the major histocompatibility complex (MHC) class II complex (DRB1-9 beta chain), and the IL-7R gene coding for the interleukin (IL)-7 receptor, as well as environmental factors such as exposure to the Epstein-Barr virus, low levels of vitamin D, and smoking, have been associated with an increased risk of developing MS (Haines et al. 1996; Sawcer et al. 1996; Teutsch et al. 2003; Oreja-Guevara et al. 2014). Approximately 85 % of MS patients repeatedly undergo relapse followed by partial or complete recovery periods (or remissions), a form of the disease which is termed relapsing-remitting MS. In more than 50 % of the relapse-remitting MS patients, the disease progressively worsens with minor remissions, reaching a stage of secondary progressive MS. In the remaining 10–15 % of MS patients, however, the disease only advances without remission, a form of the disease which is termed primary progressive MS (Thompson et al. 1997; Haines et al. 2011).

16.1.1 Inflammatory Mechanisms in MS/EAE

Experimental autoimmune encephalomyelitis (EAE) is commonly used as an animal model of MS, being similarly associated with axonal degeneration and chronic demyelination, primarily in the spinal cord, resulting in tail and hindlimb paralysis. However, similar to MS, the disease symptoms reflect the anatomical location of the inflammatory lesions and may also include emotional instability, sensory loss, ataxia, muscle weakness, and spasms. EAE is generally induced in rodents by immunization with myelin peptides, such as myelin basic protein (MBP), myelin proteolipid protein (PLP), and myelin oligodendrocyte glycoprotein (MOG), emulsified in an adjuvant (typically complete Freund's adjuvant) to enhance the immune response. EAE is also produced by adoptive transfer of myelin-reactive T lymphocytes expressing CD4 glycoproteins on their surface ($CD4^+$ T cells), isolated from mice immunized with myelin peptides, and further stimulated in vitro with myelin peptides. Depending on the nature of the antigens, and on the background of the animals, an acute stretch of EAE, a relapsing-remitting form, or chronic EAE can be induced (Rangachari and Kuchroo 2013).

In both EAE and MS, the infiltration of T and B lymphocytes, macrophages, and neutrophils is pronounced around the demyelinating lesions in situ, within the perivascular space and/or parenchyma. In addition, oligoclonal IgGs are commonly detected in the cerebrospinal fluid of EAE mice and MS patients (Mehta et al. 1985; Tomioka and Matsui 2014), thus suggesting the presence of an immune response in the CNS. Although the initiation mechanisms of EAE still remain unclear, they were shown to be mediated by Th effector T cells, a phenotype of CD4+ cells resulting from their activation by antigen-presenting cells (APCs) (Montero et al. 2004; reviewed in Kawakami et al. 2012) (see Chap. 5 for further reading on antigen presentation).

16.1.2 Microglia in MS/EAE

It is usually difficult to discriminate microglia from infiltrated macrophages in the postmortem brains of MS patients. However, several lines of evidence have suggested that microglia could play a pivotal role in mediating neuroinflammation in MS. Microglia are macrophage-like cells that reside in the CNS and contribute in various manners to maintaining CNS integrity. In the inflamed CNS, microglia can also function as immunocompetent cells, particularly involved with the production of inflammatory mediators and/or the presentation of antigens, depending on the context (Ransohoff et al. 2003; Tran and Miller 2003; Raivich and Banati 2004; Chastain et al. 2011).

During the development of EAE and MS, microglia display several signs of 'activation' at the morphological and gene expression levels. For instance, microglia have larger cell bodies, accumulate around lesions sites, and show immunoreactivity for MHC class II and CD68, a lysosomal marker also named 'macrosialin' in mouse that is upregulated during inflammation (Minagar et al. 2002; Jack et al. 2005; Marik et al. 2007; also see Chap. 10 for additional reading about CD68). In the postmortem brains of progressive MS patients, demyelination and neuronal damage reportedly correlate with an increased density and clustering of CD68 or MHC class II positive cells, a pathological feature commonly referred to as 'microglial nodules' (Prineas et al. 2001; Singh et al. 2013). In addition, microglia have been proposed to behave as APCs in MS. They were shown to express MHC class II, display antigens on their cellular surface, and colocalize with $CD4^+$ cells before the onset of EAE, and the infiltration of myeloid cells in bone marrow chimera in vivo (Ponomarev et al. 2005). These observations suggest a possible role for microglia in the activation of T cells in MS, or in their reactivation following antigen presentation in the periphery (see Chap. 5 for more information on both processes), although direct evidence remains to be shown. In addition, microglia and macrophages contained phagocytosed myelin debris around the white matter demyelinating lesions in MS postmortem samples, highlighting their possible involvement, whether detrimental or beneficial, to the demyelination and axonal degeneration (Tanaka et al. 1975; Bauer et al. 1994; also see Napoli and Neumann 2010).

Moreover, it has been reported that preventing microglial activation could repress the development of EAE in vivo (Heppner et al. 2005). In particular, Heppner and colleagues have generated a mouse model expressing the herpes simplex virus thymidine kinase (HSVTK) specifically in microglia/macrophages, under the CD11b promoter, thus rendering these cells susceptible to ganciclovir cytotoxicity. Following transplantation of wild-type bone-marrow cells, to spare the peripheral myeloid cell population from ganciclovir treatment, and the subsequent peripheral injection of ganciclovir, microglial transformation to amoeboid morphologies was found to be arrested, a phenomenon referred to as "microglial paralysis", which resulted in delayed EAE onset and reduced clinical score (Heppner et al. 2005). These findings strongly suggested that microglia could play a significant role in the pathogenesis of EAE and MS. The focus of this chapter is on microglial implication in multiple immunological aspects of the disease pathogenesis, including antigen presentation, inflammation, demyelination, and neurotoxicity.

16.2 Microglia as Antigen-Presenting Cells

Dendritic cells (DCs), which are monocyte-derived cells considered as 'professional' APCs, are often encountered in the leptomeninges and white matter lesions of MS patients (Ganguly et al. 2013; Nuyts et al. 2013). However, microglial cells could also behave as APCs in MS as will be discussed below (Smith et al. 1998) (see Chap. 5 for further reading). After the phagocytosis of antigens, such as myelin peptides, APCs become engaged in antigen presentation through MHC class II signaling to $CD4^+$ cells, which express the cognate T cell receptor, leading to their activation (or reactivation) and differentiation into various Th effector T cell subsets, such as the pro-inflammatory, encephalitogenic Th1 and Th17 cells. Costimulatory molecules such as CD80 and CD86 on APCs, or the CD40 ligand (CD40L) expressed on T cells further contribute to activating Th cells via CD28 (member of the B7 family), or to activating APCs via CD40, to promote cellular expansion and survival. Conversely, the costimulatory molecules-programmed cell death-ligand 1 (PD-L1) and PD-L2 suppress Th cell activation by acting on their programmed cell death 1 (PD-1) receptor (Keir et al. 2008; Elgueta et al. 2009).

During normal physiological conditions, microglia express undetectable to low levels of MHC class II molecules (Wong et al. 1984; Suzumura et al. 1987) and constitutively express low levels of CD80 and high levels of CD86 (Satoh et al. 1995; Dangond et al. 1997). Over the course of EAE, however, the expression of MHC class II, CD80 and CD86, was found to be upregulated in CD11b$^+$CD45^{low} microglial cells, isolated from the brains of EAE mice by flow cytometry (Ponomarev et al. 2005; Murphy et al. 2010). Expression of MHC class II is also induced in cultured microglia/macrophages upon stimulation with interferon (IFN)-γ (Suzumura et al. 1987). This pro-inflammatory cytokine produced by Th1 cells, T cells expressing the CD8 glycoprotein (CD8$^+$ T cells), macrophages, DCs, and microglial cells in culture, is well known for promoting immune responses against viral and bacterial infection, as well as the development of tumors (Munder et al. 1998; Kawanokuchi et al. 2006; Vremec et al. 2007). Additionally, treatment of cultured microglial cells with the supernatant from IFNγ-producing Th1 cell lines, specific for MBP, particularly induced microglial expression of MHC class II, CD80, CD86, and CD40 in vitro (Seguin et al. 2003). The binding of CD40 to CD40L induces APCs to produce pro-inflammatory mediators such as TNF-α, IL-6, and IL-12 in vitro (Aloisi et al. 1999; Rezai-Zadeh et al. 2008). When cultured in the presence of IFNγ-stimulated microglial cells, the proliferative capacity of Th cells is additionally increased and accompanied by an enhanced production of IL-2 and IFNγ in vitro (Aloisi et al. 1998). On the other hand, IFNγ also induces the expression of PD-L1 on microglia, while suppressing Th cell activation and the production of IFNγ in vitro (Magnus et al. 2005). Together, these findings suggest that IFNγ-stimulated microglial cells may not only activate Th cells via the induction of CD80 and CD86, or CD40, but also suppress T cell activation via the induction of PD-L1 expression, at least in vitro.

The cytokine granulocyte macrophage colony-stimulating factor (GM-CSF), which is secreted by T cells, as well as astrocytes and macrophages in vitro (Ohno et al. 1990; Shi et al. 2006), may also play a critical role in the induction of APCs functions in microglia (Matyszak et al. 1999). During EAE, MHC class II expression is considerably reduced ex vivo in microglial cells derived from GM-CSF-deficient mice, compared to wild-type mice (Ponomarev et al. 2007; Codarri et al. 2011), suggesting that GM-CSF regulates microglial expression of MHC class II. Interestingly, GM-CSF also promotes DCs-like properties in microglia ex vivo and in vivo, enhancing their expression of the DCs marker CD11c, as well as MHC class II, CD80, and CD86 (Schermer and Humpel 2002; Li et al. 2011). GM-CSF-stimulated microglia also have the ability to induce the proliferation and reactivation of CD4$^+$ T cells in vitro (Fischer et al. 1993; Aloisi et al. 2000). In GM-CSF-stimulated microglia, however, the levels of MHC class II are lower than in DCs and associated with reduced Th cells proliferation (Lambert et al. 2008), in agreement with the view that microglia have limited antigen-presenting capacity (Ransohoff and Engelhardt 2012). Therefore, GM-CSF-stimulated microglia could be used for stimulating adaptive immune responses, such as antigen presentation, in EAE and MS, albeit with a limited capacity as compared to monocyte-derived DCs.

16.3 Microglia as Inflammatory Cells

Activated microglial cells produce a variety of cytokines and monokines (i.e., cytokines mainly produced by monocytes/macrophages) involved in mediating neuroinflammation. In particular, the secretion of IL-1β, IL-6, and tumour necrosis factor α (TNFα) (Fig. 16.1) is upregulated in cultured microglia upon direct contact with MBP-primed T cells (Dasgupta et al. 2005), while their production of TNFα, IL-6, and IL-12 is enhanced by antigen presentation to CD4+ T cells via the CD40-CD40L signaling pathway in vitro (Rezai-Zadeh et al. 2008; Aloisi et al. 1998). These findings suggest that microglial interactions with T cells could influence their contribution to the neuroinflammatory milieu in EAE and MS, specifically by modulating their release of TNFα, IL-1β, IL-6, and IL-12 as discussed below.

1. *TNFα* is produced in the MS/EAE brain by microglia as well as macrophages (Renno et al. 1995) and functions as a pro-inflammatory mediator in the CNS, by inducing the production of chemokines (or **chemo**tactic cyto**kines**; mediating

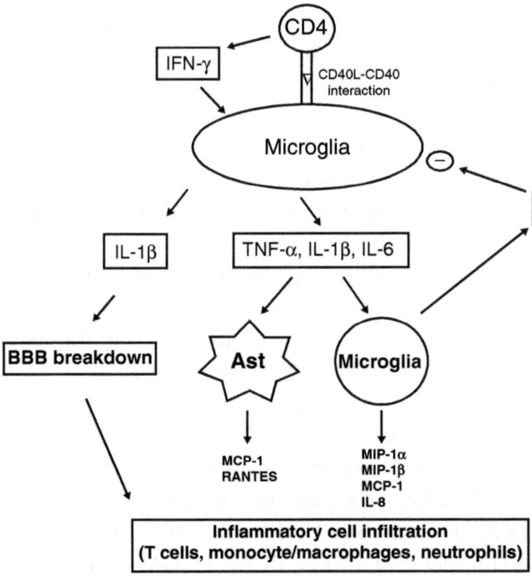

Fig. 16.1 The role of microglia in neuroinflammation. Microglia produce monokines including tumor necrosis factor (TNF)-α, interleukin (IL)-1β, and IL-6 when stimulated with CD40 ligand (CD40L) expressed on activated T cells and/or T cell-derived cytokines. In addition, activated microglia produce a variety of chemokines and induce inflammatory cell infiltration. Monokines activate microglia and astrocytes to induce the production of chemokines, leading to the infiltration of inflammatory cells including T cells, monocytes/macrophages, and neutrophils. IL-1β further disrupts blood–brain barrier (BBB) and induces the production of the chemokine monocyte chemoattractant protein (MCP)-1, which contributes to the infiltration of inflammatory cells. However, TNF-α also induces microglial production of IL-10 via the TNF receptor (TNFR)2, thus exerting anti-inflammatory responses as well

the attraction of their responsive cells) such as IL-8, macrophage inflammatory protein (MIP)-1α and MIP-1β in cultured human microglia (Ehrlich et al. 1998; McManus et al. 1998), and the production of monocyte chemoattractant protein (MCP)-1 and regulated on activation, normal T cell expressed and secreted (RANTES) in cultured rat astrocytes (Guo et al. 1998). MIP-1α, MIP-1β, MCP-1, and RANTES are chemotactic for T cells, macrophages, and microglia. Interestingly, in mice devoid of CNS expression of TNF receptor 1 (TNFR1), the recruitment of macrophages and granulocytes is reduced over the course of EAE, induced using MOG-reactive T cells. The levels of MCP-1 and MIP-2, i.e., a chemotactic factor for neutrophils produced by macrophages and microglia, were also found to be reduced in the CNS of these mice (Gimenez et al. 2006). These findings suggest that TNFα could enhance neuroinflammation in EAE and MS by acting on the infiltration of inflammatory cells from the periphery, and the induction of additional chemokines in microglia or astrocytes. Consistently, transgenic mice over-expressing TNFα were shown to spontaneously develop an inflammatory demyelinating disease characterized by the activation of astrocytes and microglial cells, together with the infiltration of $CD4^+$ and $CD8^+$ T cells into the meninges and CNS parenchyma (Probert et al. 1995). TNFα-deficient mice also displayed a delayed EAE onset, but similar to higher levels of EAE severity were observed in later phases of the disease (Kassiotis et al. 1999), thus suggesting that TNFα could exert distinct roles, either detrimental or beneficial, depending on the stage of disease progression.

The functions of TNFα are exerted via either TNFR1 or TNFR2 expressed on various types of cells (Dopp et al. 1997; Baker and Reddy 1998; Tracey et al. 2008; Martin et al. 2014). In the healthy CNS, these receptors are found on neurons, astrocytes, and oligodendrocytes (Yang et al. 2002; Kuno et al. 2006; Faustman and Davis 2013). In EAE/MS, the infiltrated lymphocytes, neutrophils, macrophages, and MHC class II positive cells additionally express TNFR1 and TNFR2 around EAE lesions (Kahn et al. 1999), while oligodendrocytes express TNFR1 around MS lesions (Probert et al. 2000). TNFR1, but not TNFR2, contains a death domain. The affinity of TNFα for TNFR1 is significantly greater than for TNFR2 (Grell et al. 1998). In previous reports, TNFR1-deficient mice were found to be resistant to MOG-induced EAE, whereas TNFR2-deficient mice displayed enhanced $CD4^+$ and $F4/80^+$ cells infiltration in vivo, together with an exacerbated EAE outcome (Eugster et al. 1999; Suvannavejh et al. 2000). In addition, TNFR2 stimulation promotes microglial expression of the anti-inflammatory cytokine IL-10 in vitro (Veroni et al. 2010). The findings suggest that TNFα could mediate neuroinflammation in MS and EAE, through the activation of TNFR1, and anti-inflammatory responses via TNFR2.

2. *IL-1β* is also detected in microglial cells and macrophages during EAE (Bauer et al. 1993; Cash et al. 1994). MBP-primed T cells induce the production of IL-1β from murine microglial cells and macrophages in vitro (Dasgupta et al. 2005). Microinjection of IL-1β into the CNS reportedly increases BBB permeability, accompanied by a pronounced recruitment of neutrophils (Ferrari et al. 2004). In addition, IL-1β induces the expression of genes favoring blood vessel

plasticity, such as the hypoxia-inducible factor 1α (HIF-1α) and its target, vascular endothelial growth factor (VEGF)-A, in cultured human astrocytes. This results in increased BBB permeability through downregulation of the tight junction proteins claudin 5 and occludin in endothelial cells (Argaw et al. 2012). In a mouse model of traumatic brain injury, IL-1β similarly induces the invasion of neutrophils and T cells into the CNS (Clausen et al. 2009). In vitro studies also demonstrate that IL-1β induces MCP-1, IL-8, MIP-1α, and MIP-1β expression in microglial cells (Calvo et al. 1996; Ehrlich et al. 1998; McManus et al. 1998), and that of MCP-1 and RANTES in astrocytes (Hayashi et al. 1995; Barnes et al. 1996). Thus, microglia-derived IL-1β could induce disruption of the BBB, release of chemoattractant mediators from microglia and astrocytes, and infiltration of peripheral inflammatory cells into the CNS. However, the particular involvement of microglia-derived IL-1β as a neuroinflammatory mediator in EAE and MS remains to be tested experimentally.

3. *IL-6* is also a potent inducer of microglia-mediated neuroinflammation. In EAE, IL-6 is produced by microglial cells as well as T cells and macrophages (Diab et al. 1997; Wlodarczyk et al. 2014). Many reports have suggested that IL-6 plays an inflammatory role in the pathogenesis of EAE (Erta et al. 2012). Neuron-targeted expression of IL-6 has been shown to induce reactive astrogliosis and microglial activation in transgenic mice in vivo (Fattori et al. 1995). Accordingly, IL-6 also induces MCP-1 mRNA expression by rat microglia in vitro (Calvo et al. 1996). In transgenic mice where the production of IL-6 is restricted to the cerebellum, MOG-induced EAE additionally resulted in the activation of infiltrated macrophages and microglia, accompanied by severe ataxia, enhanced cerebellum infiltration of neutrophils and B cells, and expression of RANTES (or CCL-5), MCP-5 (or CCL-12), and TNFα in situ (Quintana et al. 2009), suggesting its involvement in EAE and MS.
4. *IL-12*: IFNγ induces the expression of CD40 on microglia (Aloisi et al. 1998), while binding of CD40 to its ligand CD40L induces microglial production of IL-12p70 in vitro (Aloisi et al. 1999). This heterodimeric cytokine, which is composed of the IL-12p35 and IL-12p40 subunits, is a crucial differentiation factor for pro-inflammatory Th1 cells, which can trigger inflammatory responses and activate APCs and cytotoxic T cells to attack their target cells (Knutson and Disis 2005). In addition, microglial cells stimulated with IFNγ in conjunction with the Toll-like receptor (TLR) 4 ligand, bacterial lipopolysaccharide (LPS), produced IL-12p70 and IL-23 in vitro (Suzumura et al. 1998; Sonobe et al. 2005). IL-23 is a heterodimer consisting of p19 and the IL-12p40 subunit. It induces the development of IL-17-producing Th17 cells, another type of pro-inflammatory $CD4^+$ cells, which play a crucial role in the pathogenesis of EAE (Langrish et al. 2005). However, more recent studies have suggested that mice deficient in the IL-12p35 subunit are fully susceptible to EAE (Becher et al. 2002), indicating that factors other than IL-12p70 might be crucial for the induction of EAE. In contrast, p19-deficient mice are reportedly resistant to EAE (Cua et al. 2003), suggesting that IL-23 production is more critical than IL-12p70 production in the CNS. It is important to note that mice in which the IL-23

subunit IL-23p40 is devoid of CNS expression have decreased EAE severity, indicating that IL-23 produced by CNS-resident cells is important for the development of EAE (Becher et al. 2003). Moreover, expression of the IL-23p19 subunit was observed in APCs including microglia and macrophages localized around the lesion sites in MS postmortem brains (Li et al. 2007). Because IFNγ is also involved with the production of IL-23, IFNγ could regulate the differentiation of both Th1 and Th17 cells.

Taken together, these findings suggest that three main monokines secreted by microglia, namely TNF-α, IL-1β, and IL-6, could synergistically 'activate' glial cells and promote infiltration of peripheral immune cells in the CNS, as observed in MS and EAE, while IL-12 and IL-23 could mediate differentiation of the encephalitogenic Th cells. Nonetheless, the specific contribution of microglial cells to the release of these monokines, at different stages of MS and EAE, and their ultimate impact on the disease pathogenesis remain unknown.

16.4 Microglia in Demyelination

Microglial cells are recruited to areas of demyelination in MS/EAE, where they transform their morphology and actively proliferate. Activated phenotypes are also observed in rodents upon feeding with dietary cuprizone, a copper chelator which causes demyelination of the corpus callosum (Remington et al. 2007; Groebe et al. 2009). Recent findings suggest that microglia could be directly involved in the mechanisms of demyelination, by their release of excitotoxic glutamate, reactive oxygen species, pro-inflammatory cytokines, nitric oxide, and mediators of apoptosis as described below (see Fig. 16.2).

The expression levels of glutaminase, an enzyme that converts glutamine into glutamate, were shown to be increased in microglial cells localized within the active MS lesions in situ (Werner et al. 2001), suggesting a possible role of excitotoxic glutamate released from microglia in the demyelination process. LPS-activated microglia also increase the extracellular glutamate levels and induce the death of oligodendrocytes in vitro (Domercq et al. 2007). IL-1β reportedly induces apoptosis in cultures of oligodendrocytes, astrocytes and microglia, and this effect is blocked pharmacologically by applying antagonists of the AMPA/kainate glutamate receptors (Takahashi et al. 2003). The combined findings suggest that glutamate released from microglia could be directly involved in demyelination during EAE and MS, by inducing toxicity against oligodendrocytes, although this hypothesis remains to be tested.

In the brains of MS patients, DNA oxidation and lipid peroxidation are mainly observed in the nucleus and cytoplasm of oligodendrocytes, respectively, thus suggesting an ongoing state of oxidative stress (Haider et al. 2011). Accordingly, the expression levels of several enzymes controlling the respiratory burst, including the nicotinamide adenine dinucleotide phosphate-oxidase (NADPH) catalytic subunits p91phox, p22phox, and p47phox, were found to be upregulated in MS lesions

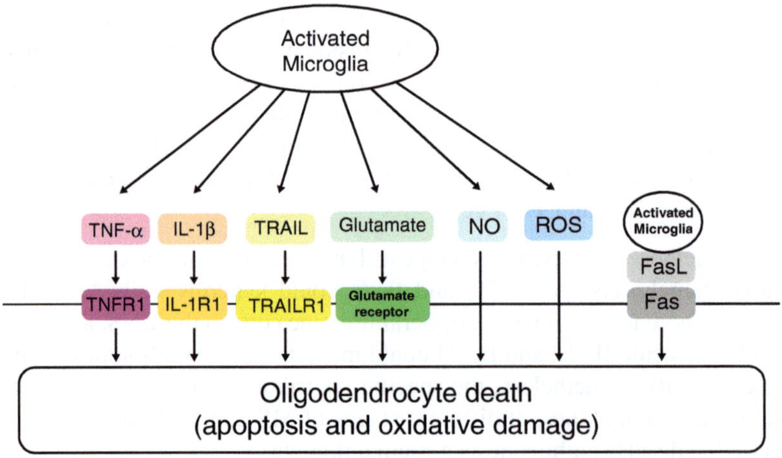

Fig. 16.2 The role of microglia in demyelination. Activated microglia produce soluble factors including TNF-α, IL-1β, TNF-related apoptosis-inducing ligand (TRAIL), glutamate, nitric oxide (NO), and reactive oxygen species (ROS), which damage oligodendrocytes via the induction of apoptosis and/or oxidative stress. Alternatively, Fas ligand (FasL) expressed on activated microglia interacts with Fas on oligodendrocytes and induces apoptosis of oligodendrocytes. Abbreviations as in Fig. 16.1

(Fischer et al. 2012). IFNγ- or GM-CSF-stimulated microglial cells reportedly produce reactive oxygen species (ROS) in vitro (Hu et al. 1995; Smith et al. 1998), showing cytotoxicity against cultured oligodendrocytes (Schreibelt et al. 2007), thus suggesting a possible role for microglial production of ROS in demyelination.

Microglial release of TNFα and IL-1β could be specifically involved in this process. It has been reported that microglia-secreted TNFα induces the death of oligodendrocytes and their progenitor cells in vitro (Zajicek et al. 1992; Pang et al. 2010). Oligodendrocyte-specific ablation of TNFR1 also attenuates the clinical signs of EAE, suggesting that oligodendrocytic TNFR1 could be involved in demyelination (Hovelmeyer et al. 2005). On the other hand, using the cuprizone-induced demyelinating model, Arnett and colleagues revealed that TNFR2, but not TNFR1, is critical for the regeneration of oligodendrocytes in vivo (Arnett et al. 2001). These observations suggest that microglia-derived TNFα could mediate the death of oligodendrocytes via TNFR1, and their regeneration through TNFR2.

In previous reports, IL-1β was also reported to cause demyelination in vivo and in vitro, while IL-1β-stimulated microglia exerted an increased oxidative activity in vitro (Smith et al. 1998), suggesting that IL-1β could damage oligodendrocytes through the release of ROS. In addition, IL-1β degrades intracellular sphingomyelin to ceramide and induces the apoptosis of oligodendrocytes in vitro (Brogi et al. 1997). Sphingomyelin, which mainly consists of ceramide and phosphocholine, is a component of the myelin sheath. Accordingly, administration of the IL-1 receptor antagonist suppressed EAE in rats, while mice deficient in the IL-1 receptor 1 (IL-1R1) showed ameliorated symptoms of EAE (Martin and Near 1995; Sutton et al. 2009). These findings suggest that IL-1β could contribute to demyelination in MS and EAE via the induction of oligodendrocyte apoptosis. On the other hand, IL-1β-deficient mice undergoing the cuprizone-induced demyelination failed to remyelinate, following withdrawal of the dietary cuprizone, suggesting that IL-1β additionally contributes to remyelination (Mason et al. 2001). This failure of remyelination also appears to correlate with a lack of insulin-like growth factor-1 (IGF-1) production by microglia and astrocytes, as their mRNA levels were shown to be reduced in brain extracts from IL-1β-deficient mice (Mason et al. 2001). Since IGF-1 is also required for the differentiation of precursor cells into mature oligodendrocytes, it has been speculated that IL-1β could play a crucial role in remyelination through the induction of microglial and/or astrocytic IGF-1 (Mason et al. 2001). Thus, it is possible that microglial IL-1β is similarly required in MS for remyelination.

Furthermore, in vitro studies have shown that microglia stimulated with IFNγ and/or LPS produce nitric oxide (NO) and the TNF-related apoptosis-inducing ligand (TRAIL) (Chao et al. 1992; Zielasek et al. 1992; Genc et al. 2003). NO reportedly causes single-stranded DNA breaks and mitochondrial damage in oligodendrocytes in vitro (Mitrovic et al. 1994), suggesting another mechanism by which microglia could contribute to demyelination in MS and EAE. In addition, TRAIL, which is a member of the death-signaling molecule family, reportedly induces human oligodendrocyte apoptosis via TRAILR1 in vitro (Matysiak et al. 2002). Thus, IFNγ might also promote demyelination via the production of NO and TRAIL by microglia.

In MS lesions, microglia, infiltrated macrophages, and T lymphocytes were lastly found to express the Fas ligand (FasL), which belongs to the TNF family and induces apoptosis upon binding to its receptor FasR (Dowling et al. 1996; D'Souza et al. 1996). Fas ligation with an anti-Fas antibody or the Fas ligand induced oligodendrocyte cell membrane lysis and subsequent cellular death in vitro (D'Souza et al. 1996). In addition, mice lacking Fas in oligodendrocytes exclusively were reported to be partially resistant to EAE (Hovelmeyer et al. 2005). Further research will test the direct contribution of microglia-derived glutamate, ROS, TNFα, IL-1β, IGF-1, NO, TRAIL, and FasL in the processes of oligodendrocyte apoptosis and regeneration, as well as concomitant demyelination and remyelination in MS.

16.5 Microglia in Neurodegeneration

MS has long been considered as a chronic inflammatory demyelinating disease of the CNS. However, several lines of evidence also suggest that axonal degeneration in MS and EAE could occur independently from demyelination. Even though axonal degeneration is widespread in the corpus callosum in EAE (Mangiardi et al. 2011), the underlying molecular mechanisms still remain unclear. Howell and colleagues proposed that microglial cells could be involved, since neurodegeneration is accompanied by the presence of activated microglia in MS postmortem brains, showing thicker and shorter processes than observed in control microglia (Howell et al. 2010). The evidence from in vivo and in vitro studies that activated microglia could cause 'inflammation-induced neurodegeneration' in MS and EAE, via the release of NO, ROS, glutamate, and various pro-inflammatory cytokines, will be reviewed in the following section.

Neuronal degeneration is induced in cultures of microglia and neurons upon stimulation with LPS and IFNγ and reduced by the addition of NOS inhibitors (Chao et al. 1992). In addition, heat shock protein (HSP)60 induces microglial production of NO via TLR4, in addition to causing extensive axonal loss and neuronal death in vitro (Lehnardt et al. 2008). Lipoteichoic acid (LPA), an agonist of TLR2 derived from staphylococcus aureus, which is involved in pathogen recognition and has been reported to exacerbate both EAE and MS (Gambuzza et al., 2011), also promotes the production of NO and superoxide by microglial cells. LPA additionally induces neuronal death, and this process is blocked by an iNOS inhibitor and a peroxynitrite scavenger in vitro, suggesting a direct detrimental contribution of microglia-derived NO to neuronal death (Kinsner et al., 2005).

In the EAE model, Nikić and colleagues revealed that ROS and reactive nitrogen species (RNS) released from activated macrophages/microglia induce focal axonal degeneration, using two-photon in vivo imaging (Nikić et al. 2011). Microglial expression of p47 phox, a cytosolic subunit of NADPH oxidase, was further shown to be upregulated in the active lesions of MS postmortem brains (Fischer et al. 2012). Microglia stimulated with the nucleotide ATP also released superoxides via P2X7 receptor, a purinergic receptor for ATP, and elicited toxicity against cultured neurons (Mead et al. 2012). It has been shown that blockade of the P2X7 receptor ameliorates EAE (Matute et al. 2007). Interestingly, microglia activated with thrombin, which induces blood coagulation via converting fibrinogen to fibrin, induce oxidative stress resulting in hippocampal neuronal cell death (Choi et al. 2005). Prothrombin kringle-2, a domain of prothrombin distinct from thrombin, also induces the loss of cortical neurons and this effect is partially inhibited by a NADPH oxidase inhibitor (Won et al. 2009). In addition, the leakage of fibrinogen from blood vessels has been shown to activate microglia and further induce axonal damage in EAE, using two-photon in vivo imaging (Davalos et al. 2012) (see Chap. 4 for additional information on these observations). It is possible that these blood-derived proteins enter the CNS through damaged BBB, thus activating microglia and inducing neurodegeneration through the release of ROS. Furthermore, high mobility group box chromosomal protein 1 (HMGB1), a chromatin-associated nuclear protein, has been detected in active lesions of MS and EAE (Andersson et al. 2008). It reportedly induces p47 phox membrane translocation and microglial production of ROS via MAC1, which consists of integrin alpha M and beta 2 (Gao et al. 2011). These findings strongly suggest that ROS produced by microglia, possibly activated by ATP or thrombin, contribute to axonal and neurodegeneration in MS and EAE.

Moreover, microglial stimulation with Chromogranin A, a marker of neurodegeneration that is released from damaged neurons and elevated in the CSF of MS patients (Stoop et al. 2008), induced their production of NO and glutamate in vitro. The conditioned medium from Chromogranin A-stimulated microglia also killed rat cerebellar granule cells via caspase-3-dependent apoptosis, blocked with an ionotropic glutamate receptor antagonist, thus suggesting that microglia induce neuronal death via glutamate release (Kingham et al. 1999). In addition, in vitro studies have suggested that LPS and TNFα could induce neurotoxicity through the release of glutamate from activated microglia in vitro (Takeuchi et al. 2006; Yawata et al. 2008).

Neuroinflammation, including microglial TNFα production, is associated with neurodegeneration in EAE (Centonze et al. 2009). Thus, it is possible that microglia-derived glutamate exerts toxicity against neurons in EAE and MS.

16.6 Conclusion

In EAE and MS, microglial cells function as APCs, albeit possibly at lower levels than professional APCs such as DCs (Fig. 16.3). IFNγ and GM-CSF induce expression of MHC class II and costimulatory molecules of the B7 family in microglia,

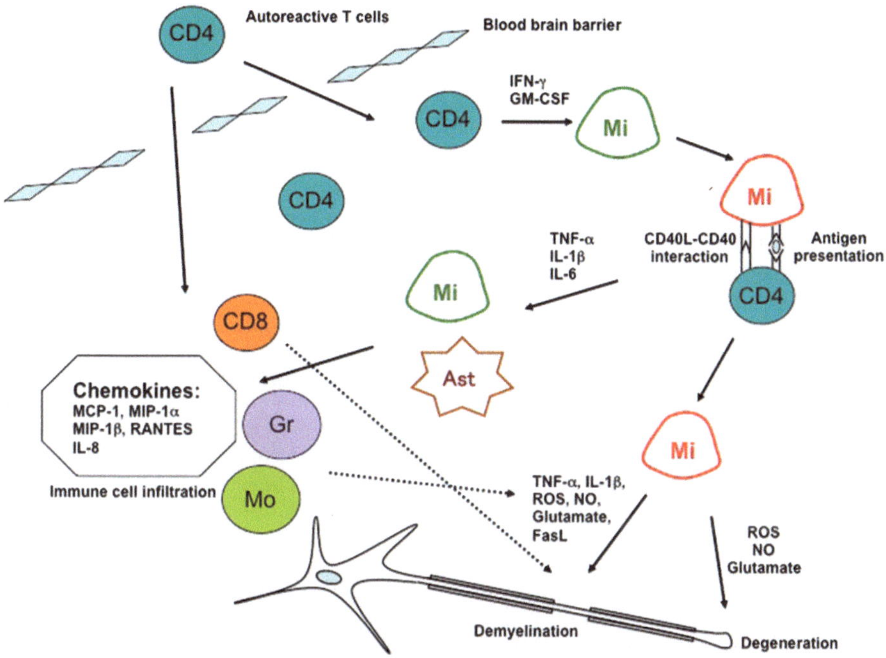

Fig. 16.3 The role of microglia in neuroinflammatory diseases. Microglia are activated by CD4+ T cells via CD40L–CD40 interactions and/or soluble factors. In some cases, activated microglia express major histocompatibility complex (MHC) class II and costimulatory molecules including CD80 and CD86, and possibly behave as antigen presenting cells (APCs) that restimulate infiltrating CD4+ cells. In another cases, activated microglia produce monokines such as TNF-α, IL-1β, and IL-6, which contribute to the activation of astrocytes and bystander microglial cells. Astrocytes and microglia activated by the monokines produce chemokines including monocyte chemoattractant protein (MCP)-1, macrophage inflammatory protein (MIP)-1α, MIP-1β, regulated on activation, normal T cell expressed and secreted (RANTES), and IL-8 which induce chemotaxis of various immune cells including monocytes/macrophages, neutrophils, CD4+, and CD8+ T cells. Activated microglial cells could also induce demyelination via FasL-Fas interactions, and the production of TNF-α, IL-1β, NO, glutamate, and ROS. Moreover, activated microglia could further induce neurodegeneration via the production of NO, glutamate, and ROS. Abbreviations as in Figs. 16.1 and 16.2

activate (or reactivate) myelin-specific T cells, and enhance neuroinflammation. In addition, microglia are activated by monokines such as TNFα, IL-1β, and IL-6 early in disease. Activated microglia also produce monokines, which in turn act on microglia and astrocytes to further enhance neuroinflammation via the production of cytokines and chemokines. Microglia-derived inflammatory monokines such as, IL-1β and TNFα, and degenerative factors such as ROS, NO, and glutamate could additionally induce demyelination. Furthermore, microglia-derived degenerative factors, such as ROS, NO, and glutamate, could cause inflammation-induced neurodegeneration. Thus, microglial cells could be considered as conductors that orchestrate a plethora of neuroinflammatory phenomena involved in the pathogenesis of EAE and MS. However, many of the discussed studies were performed in vitro. Further investigation is needed to clarify the direct contribution of microglia versus the other inflammatory cells in this exciting field.

References

Aloisi F, Ria F, Penna G et al (1998) Microglia are more efficient than astrocytes in antigen processing and in Th1 but not Th2 cell activation. J Immunol 160(10):4671–4680

Aloisi F, Penna G, Polazzi E et al (1999) CD40–CD154 interaction and IFNgamma are required for IL-12 but not prostaglandin E2 secretion by microglia during antigen presentation to Th1 cells. J Immunol 162(3):1384–1391

Aloisi F, De Simone R, Columba-Cabezas S et al (2000) Functional maturation of adult mouse resting microglia into an APC is promoted by granulocyte-macrophage colony-stimulating factor and interaction with Th1 cells. J Immunol 164(4):1705–1712

Andersson A, Covacu R, Sunnemark D et al (2008) Pivotal advance: HMGB1 expression in active lesions of human and experimental multiple sclerosis. J Leukoc Biol 84(5):1248–1255

Argaw AT, Asp L, Zhang J et al (2012) Astrocyte-derived VEGF-A drives blood–brain barrier disruption in CNS inflammatory disease. J Clin Invest 122(7):2454–2468

Arnett HA, Mason J, Marino M et al (2001) TNF alpha promotes proliferation of oligodendrocyte progenitors and remyelination. Nat Neurosci 4(11):1116–1122

Baker SJ, Reddy EP (1998) Modulation of life and death by the TNF receptor superfamily. Oncogene 17(25):3261–3270

Barnes DA, Huston M, Holmes R et al (1996) Induction of RANTES expression by astrocytes and astrocytoma cell lines. J Neuroimmunol 71(1–2):207–214

Bauer J, Berkenbosch F, Van Dam AM et al (1993) Demonstration of interleukin-1 beta in Lewis rat brain during experimental allergic encephalomyelitis by immunocytochemistry at the light and ultrastructural level. J Neuroimmunol 48(1):13–21

Bauer J, Sminia T, Wouterlood FG et al (1994) Phagocytic activity of macrophages and microglial cells during the course of acute and chronic relapsing experimental autoimmune encephalomyelitis. J Neurosci Res 38(4):365–375

Becher B, Durell BG, Noelle RJ (2002) Experimental autoimmune encephalitis and inflammation in the absence of interleukin-12. J Clin Invest 110(4):493–497

Becher B, Durell BG, Noelle RJ (2003) IL-23 produced by CNS-resident cells controls T cell encephalitogenicity during the effector phase of experimental autoimmune encephalomyelitis. J Clin Invest 112(8):1186–1191

Brogi A, Strazza M, Melli M et al (1997) Induction of intracellular ceramide by interleukin-1 beta in oligodendrocytes. J Cell Biochem 66(4):532–541

Calvo CF, Yoshimura T, Gelman M et al (1996) Production of monocyte chemotactic protein-1 by rat brain macrophages. Eur J Neurosci 8(8):1725–1734

Cash El, Minty A, Ferrara P, Caput D, Fradelizi D, Rott O (1994) Macrophage-inactivating IL-13 suppresses experimental autoimmune encephalomyelitis in rats. J Immunol 153(9):4258–4267
Centonze D, Muzio L, Rossi S et al (2009) Inflammation triggers synaptic alteration and degeneration in experimental autoimmune encephalomyelitis. J Neurosci 29(11):3442–3452
Chao CC, Hu S, Molitor TW et al (1992) Activated microglia mediate neuronal cell injury via a nitric oxide mechanism. J Immunol 149(8):2736–2741
Chastain EM, Duncan DS, Rodgers JM et al (2011) The role of antigen presenting cells in multiple sclerosis. Biochim Biophys Acta 1812(2):265–274
Choi SH, Lee DY, Kim SU et al (2005) Thrombin-induced oxidative stress contributes to the death of hippocampal neurons in vivo: role of microglial NADPH oxidase. J Neurosci 25(16):4082–4090
Clausen F, Hanell A, Bjork M et al (2009) Neutralization of interleukin-1beta modifies the inflammatory response and improves histological and cognitive outcome following traumatic brain injury in mice. Eur J Neurosci 30(3):385–396
Codarri L, Gyulveszi G, Tosevski V et al (2011) RORgammat drives production of the cytokine GM-CSF in helper T cells, which is essential for the effector phase of autoimmune neuroinflammation. Nat Immunol 12(6):560–567
Cua DJ, Sherlock J, Chen Y et al (2003) Interleukin-23 rather than interleukin-12 is the critical cytokine for autoimmune inflammation of the brain. Nature 421(6924):744–748
D'Souza SD, Bonetti B, Balasingam V et al (1996) Multiple sclerosis: Fas signaling in oligodendrocyte cell death. J Exp Med 184(6):2361–7230
Dangond F, Windhagen A, Groves CJ et al (1997) Constitutive expression of costimulatory molecules by human microglia and its relevance to CNS autoimmunity. J Neuroimmunol 76(1–2):132–138
Dasgupta S, Jana M, Liu X et al (2005) Myelin basic protein-primed T cells of female but not male mice induce nitric-oxide synthase and proinflammatory cytokines in microglia: implications for gender bias in multiple sclerosis. J Biol Chem 280(38):32609–32617
Davalos D, Ryu JK, Merlini M et al (2012) Fibrinogen-induced perivascular microglial clustering is required for the development of axonal damage in neuroinflammation. Nat Commun 3:1227
Diab A, Zhu J, Xiao BG et al (1997) High IL-6 and low IL-10 in the central nervous system are associated with protracted relapsing EAE in DA rats. J Neuropathol Exp Neurol 56(6):641–650
Domercq M, Sanchez-Gomez MV, Sherwin C et al (2007) System xc- and glutamate transporter inhibition mediates microglial toxicity to oligodendrocytes. J Immunol 178(10):6549–6556
Dopp JM, Mackenzie-Graham A, Otero GC et al (1997) Differential expression, cytokine modulation, and specific functions of type-1 and type-2 tumor necrosis factor receptors in rat glia. J Neuroimmunol 75(1–2):104–112
Dowling P, Shang G, Raval S et al (1996) Involvement of the CD95 (APO-1/Fas) receptor/ligand system in multiple sclerosis brain. J Exp Med 184(4):1513–1518
Ehrlich LC, Hu S, Sheng WS et al (1998) Cytokine regulation of human microglial cell IL-8 production. J Immunol 160(4):1944–1948
Elgueta R, Benson MJ, de Vries VC et al (2009) Molecular mechanism and function of CD40/CD40L engagement in the immune system. Immunol Rev 229(1):152–172
Erta M, Quintana A, Hidalgo J et al (2012) Interleukin-6, a major cytokine in the central nervous system. Int J Biol Sci 8(9):1254–1266
Eugster HP, Frei K, Bachmann R et al (1999) Severity of symptoms and demyelination in MOG-induced EAE depends on TNFR1. Eur J Immunol 29(2):626–632
Fattori E, Lazzaro D, Musiani P et al (1995) IL-6 expression in neurons of transgenic mice causes reactive astrocytosis and increase in ramified microglial cells but no neuronal damage. Eur J Neurosci 7(12):2441–2449
Faustman DL, Davis M (2013) TNF receptor 2 and disease: autoimmunity and regenerative medicine. Front Immunol 4:478
Ferrari CC, Depino AM, Prada F et al (2004) Reversible demyelination, blood–brain barrier breakdown, and pronounced neutrophil recruitment induced by chronic IL-1 expression in the brain. Am J Pathol 165(5):1827–1837

Fischer HG, Nitzgen B, Germann T et al (1993) Differentiation driven by granulocyte-macrophage colony-stimulating factor endows microglia with interferon-gamma-independent antigen presentation function. J Neuroimmunol 42(1):87–95

Fischer MT, Sharma R, Lim JL et al (2012) NADPH oxidase expression in active multiple sclerosis lesions in relation to oxidative tissue damage and mitochondrial injury. Brain 135(Pt 3): 886–899

Gambuzza M, Licata N, Palella E et al (2011) Targeting Toll-like receptors: emerging therapeutics for multiple sclerosis management. J Neuroimmunol 239(1–2):1–12

Ganguly D, Haak S, Sisirak V et al (2013) The role of dendritic cells in autoimmunity. Nat Rev Immunol 13(8):566–577

Gao HM, Zhou H, Zhang F et al (2011) HMGB1 acts on microglia Mac1 to mediate chronic neuroinflammation that drives progressive neurodegeneration. J Neurosci 31(3):1081–1092

Genc S, Kizildag S, Genc K et al (2003) Interferon gamma and lipopolysaccharide upregulate TNF-related apoptosis-inducing ligand expression in murine microglia. Immunol Lett 85(3):271–274

Gimenez MA, Sim J, Archambault AS et al (2006) A tumor necrosis factor receptor 1-dependent conversation between central nervous system-specific T cells and the central nervous system is required for inflammatory infiltration of the spinal cord. Am J Pathol 168(4):1200–1209

Grell M, Wajant H, Zimmermann G et al (1998) The type 1 receptor (CD120a) is the high-affinity receptor for soluble tumor necrosis factor. Proc Natl Acad Sci U S A 95(2):570–575

Groebe A, Clarner T, Baumgartner W et al (2009) Cuprizone treatment induces distinct demyelination, astrocytosis, and microglia cell invasion or proliferation in the mouse cerebellum. Cerebellum 8(3):163–174

Guo H, Jin YX, Ishikawa M et al (1998) Regulation of beta-chemokine mRNA expression in adult rat astrocytes by lipopolysaccharide, proinflammatory and immunoregulatory cytokines. Scand J Immunol 48(5):502–508

Haider L, Fischer MT, Frischer JM et al (2011) Oxidative damage in multiple sclerosis lesions. Brain 134(Pt 7):1914–1924

Haines JL, Ter-Minassian M, Bazyk A et al (1996) A complete genomic screen for multiple sclerosis underscores a role for the major histocompatability complex. The Multiple Sclerosis Genetics Group. Nat Genet 13(4):469–471

Haines JD, Inglese M, Casaccia P (2011) Axonal damage in multiple sclerosis. Mt Sinai J Med 78(2):231–243

Hayashi M, Luo Y, Laning J et al (1995) Production and function of monocyte chemoattractant protein-1 and other beta-chemokines in murine glial cells. J Neuroimmunol 60(1–2):143–150

Heppner FL, Greter M, Marino D et al (2005) Experimental autoimmune encephalomyelitis repressed by microglial paralysis. Nat Med 11(2):146–152

Hovelmeyer N, Hao Z, Kranidioti K et al (2005) Apoptosis of oligodendrocytes via Fas and TNF-R1 is a key event in the induction of experimental autoimmune encephalomyelitis. J Immunol 175(9):5875–5884

Howell OW, Rundle JL, Garg A et al (2010) Activated microglia mediate axoglial disruption that contributes to axonal injury in multiple sclerosis. J Neuropathol Exp Neurol 69(10):1017–1033

Hu S, Sheng WS, Peterson PK et al (1995) Cytokine modulation of murine microglial cell superoxide production. Glia 13(1):45–50

Jack C, Ruffini F, Bar-Or A et al (2005) Microglia and multiple sclerosis. J Neurosci Res 81(3):363–373

Jadidi-Niaragh F, Mirshafiey A (2011) Th17 cell, the new player of neuroinflammatory process in multiple sclerosis. Scand J Immunol 74(1):1–13

Kahn MA, Dopp JM, Liva S et al (1999) Temporal kinetics and cellular phenotype of TNF p55/p75 receptors in experimental allergic encephalomyelitis. J Neuroimmunol 95(1–2):19–34

Kassiotis G, Pasparakis M, Kollias G et al (1999) TNF accelerates the onset but does not alter the incidence and severity of myelin basic protein-induced experimental autoimmune encephalomyelitis. Eur J Immunol 29(3):774–780

Kawakami N, Bartholomäus I, Pesic M et al (2012) An autoimmunity odyssey: how autoreactive T cells infiltrate into the CNS. Immunol Rev 248(1):140–55

Kawanokuchi J, Mizuno T, Takeuchi H et al (2006) Production of interferon-gamma by microglia. Mult Scler 12(5):558–564

Keir ME, Butte MJ, Freeman GJ et al (2008) PD-1 and its ligands in tolerance and immunity. Annu Rev Immunol 26:677–704

Kingham PJ, Cuzner ML, Pocock JM (1999) Apoptotic pathways mobilized in microglia and neurones as a consequence of chromogranin A-induced microglial activation. J Neurochem 73(2):538–547

Kinsner A, Pilotto V, Deininger S et al (2005) Inflammatory neurodegeneration induced by lipoteichoic acid from Staphylococcus aureus is mediated by glia activation, nitrosative and oxidative stress, and caspase activation. J Neurochem 95(4):1132–1143

Knutson KL, Disis ML (2005) Tumor antigen-specific T helper cells in cancer immunity and immunotherapy. Cancer Immunol Immunother 54(8):721–728

Kuno R, Yoshida Y, Nitta A (2006) The role of TNF-alpha and its receptors in the production of NGF and GDNF by astrocytes. Brain Res 1116(1):12–18

Lambert C, Desbarats J, Arbour N et al (2008) Dendritic cell differentiation signals induce anti-inflammatory properties in human adult microglia. J Immunol 181(12):8288–8297

Langrish CL, Chen Y, Blumenschein WM et al (2005) IL-23 drives a pathogenic T cell population that induces autoimmune inflammation. J Exp Med 201(2):233–240

Lassmann H (2010) Axonal and neuronal pathology in multiple sclerosis: what have we learnt from animal models. Exp Neurol 225(1):2–8

Lassmann H, van Horssen J (2011) The molecular basis of neurodegeneration in multiple sclerosis. FEBS Lett 585(23):3715–3723

Lehnardt S, Schott E, Trimbuch T et al (2008) A vicious cycle involving release of heat shock protein 60 from injured cells and activation of toll-like receptor 4 mediates neurodegeneration in the CNS. J Neurosci 28(10):2320–2331

Li Y, Chu N, Hu A et al (2007) Increased IL-23p19 expression in multiple sclerosis lesions and its induction in microglia. Brain 130(Pt 2):490–501

Li H, Sonobe Y, Tabata H et al (2011) Tumor necrosis factor-α promotes granulocyte-macrophage colony-stimulating factor-stimulated microglia to differentiate into competent dendritic cell-like antigen-presenting cells. Clin Exp Neuroimmunol 2(1):1–11

Magnus T, Schreiner B, Korn T et al (2005) Microglial expression of the B7 family member B7 homolog 1 confers strong immune inhibition: implications for immune responses and autoimmunity in the CNS. J Neurosci 25(10):2537–2546

Mangiardi M, Crawford DK, Xia X et al (2011) An animal model of cortical and callosal pathology in multiple sclerosis. Brain Pathol 21(3):263–278

Marik C, Felts PA, Bauer J et al (2007) Lesion genesis in a subset of patients with multiple sclerosis: a role for innate immunity? Brain 130(Pt 11):2800–2815

Martin D, Near SL (1995) Protective effect of the interleukin-1 receptor antagonist (IL-1ra) on experimental allergic encephalomyelitis in rats. J Neuroimmunol 61(2):241–245

Martin EM, Remke A, Pfeifer E et al (2014) TNFR2 maintains adequate IL-12 production by dendritic cells in inflammatory responses by regulating endogenous TNF levels. Innate Immun. doi:10.1177/1753425913506949

Mason JL, Suzuki K, Chaplin DD et al (2001) Interleukin-1beta promotes repair of the CNS. J Neurosci 21(18):7046–7052

Matute C, Torre I, Perez-Cerda F et al (2007) P2X(7) receptor blockade prevents ATP excitotoxicity in oligodendrocytes and ameliorates experimental autoimmune encephalomyelitis. J Neurosci 27(35):9525–9233

Matysiak M, Jurewicz A, Jaskolski D et al (2002) TRAIL induces death of human oligodendrocytes isolated from adult brain. Brain 125(Pt 11):2469–2480

Matyszak MK, Denis-Donini S, Citterio S et al (1999) Microglia induce myelin basic protein-specific T cell anergy or T cell activation, according to their state of activation. Eur J Immunol 29(10):3063–3076

McManus CM, Brosnan CF, Berman JW (1998) Cytokine induction of MIP-1 alpha and MIP-1 beta in human fetal microglia. J Immunol 160(3):1449–1455

Mead EL, Mosley A, Eaton S et al (2012) Microglial neurotransmitter receptors trigger superoxide production in microglia; consequences for microglial–neuronal interactions. J Neurochem 121(2):287–301

Mehta PD, Patrick BA, Wisniewski HM (1985) Specificity of oligoclonal IgG bands in sera from chronic relapsing experimental allergic encephalomyelitis guinea pigs. J Immunol 134(4): 2338–2342

Minagar A, Shapshak P, Fujimura R, Eisdorfer C et al (2002) The role of macrophage/microglia and astrocytes in the pathogenesis of three neurologic disorders: HIV-associated dementia, Alzheimer disease, and multiple sclerosis. J Neurol Sci 202(1–2):13–23

Mitrovic B, Ignarro LJ, Montestruque S et al (1994) Nitric oxide as a potential pathological mechanism in demyelination: its differential effects on primary glial cells in vitro. Neuroscience 61(3):575–585

Montero E, Nussbaum G, Kaye JF et al (2004) Regulation of experimental autoimmune encephalomyelitis by $CD4^+$, $CD25^+$ and $CD8^+$ T cells: analysis using depleting antibodies. J Autoimmun 23(1):1–7

Munder M, Mallo M, Eichmann K et al (1998) Murine macrophages secrete interferon gamma upon combined stimulation with interleukin (IL)-12 and IL-18: a novel pathway of autocrine macrophage activation. J Exp Med 187(12):2103–2108

Murphy AC, Lalor SJ, Lynch MA et al (2010) Infiltration of Th1 and Th17 cells and activation of microglia in the CNS during the course of experimental autoimmune encephalomyelitis. Brain Behav Immun 24(4):641–651

Napoli I, Neumann H (2010) Protective effects of microglia in multiple sclerosis. Exp Neurol 225(1):24–28

Nikić I, Merkler D, Sorbara C et al (2011) A reversible form of axon damage in experimental autoimmune encephalomyelitis and multiple sclerosis. Nat Med 17(4):495–499

Nuyts AH, Lee WP, Bashir-Dar R et al (2013) Dendritic cells in multiple sclerosis: key players in the immunopathogenesis, key players for new cellular immunotherapies? Mult Scler 19(8):995–1002

Ohno K, Suzumura A, Sawada M et al (1990) Production of granulocyte/macrophage colony-stimulating factor by cultured astrocytes. Biochem Biophys Res Commun 169(2):719–724

Oreja-Guevara C, Wiendl H, Kieseier BC et al (2014) Specific aspects of modern life for people with multiple sclerosis: considerations for the practitioner. Ther Adv Neurol Disord 7(2): 137–149

Pang Y, Campbell L, Zheng B et al (2010) Lipopolysaccharide-activated microglia induce death of oligodendrocyte progenitor cells and impede their development. Neuroscience 166(2): 464–475

Ponomarev ED, Shriver LP, Maresz K et al (2005) Microglial cell activation and proliferation precedes the onset of CNS autoimmunity. J Neurosci Res 81(3):374–389

Ponomarev ED, Shriver LP, Maresz K et al (2007) GM-CSF production by autoreactive T cells is required for the activation of microglial cells and the onset of experimental autoimmune encephalomyelitis. J Immunol 178(1):39–48

Prineas JW, Kwon EE, Cho ES et al (2001) Immunopathology of secondary-progressive multiple sclerosis. Ann Neurol 50(5):646–657

Probert L, Akassoglou K, Pasparakis M et al (1995) Spontaneous inflammatory demyelinating disease in transgenic mice showing central nervous system-specific expression of tumor necrosis factor alpha. Proc Natl Acad Sci U S A 92(24):11294–11298

Probert L, Eugster HP, Akassoglou K et al (2000) TNFR1 signalling is critical for the development of demyelination and the limitation of T-cell responses during immune-mediated CNS disease. Brain 123(Pt 10):2005–2019

Quintana A, Muller M, Frausto RF et al (2009) Site-specific production of IL-6 in the central nervous system retargets and enhances the inflammatory response in experimental autoimmune encephalomyelitis. J Immunol 183(3):2079–2088

Raivich G, Banati R (2004) Brain microglia and blood-derived macrophages: molecular profiles and functional roles in multiple sclerosis and animal models of autoimmune demyelinating disease. Brain Res Brain Res Rev 46(3):261–281

Rangachari M, Kuchroo K (2013) Using EAE to better understand principles of immune function and autoimmune pathology. J Autoimmun 45:31–39

Ransohoff RM, Engelhardt B (2012) The anatomical and cellular basis of immune surveillance in the central nervous system. Nat Rev Immunol 12(9):623–635

Ransohoff RM, Kivisakk P, Kidd G (2003) Three or more routes for leukocyte migration into the central nervous system. Nat Rev Immunol 3(7):569–581

Remington LT, Babcock AA, Zehntner SP et al (2007) Microglial recruitment, activation, and proliferation in response to primary demyelination. Am J Pathol 170(5):1713–1724

Renno T, Krakowski M, Piccirillo C et al (1995) TNF-alpha expression by resident microglia and infiltrating leukocytes in the central nervous system of mice with experimental allergic encephalomyelitis. Regulation by Th1 cytokines. J Immunol 154(2):944–953

Rezai-Zadeh K, Ehrhart J, Bai Y et al (2008) Apigenin and luteolin modulate microglial activation via inhibition of STAT1-induced CD40 expression. J Neuroinflammation 5:41

Satoh J, Lee YB, Kim SU (1995) T-cell costimulatory molecules B7-1 (CD80) and B7-2 (CD86) are expressed in human microglia but not in astrocytes in culture. Brain Res 704(1):92–96

Sawcer S, Jones HB, Feakes R et al (1996) A genome screen in multiple sclerosis reveals susceptibility loci on chromosome 6p21 and 17q22. Nat Genet 13(4):464–468

Schermer C, Humpel C (2002) Granulocyte macrophage-colony stimulating factor activates microglia in rat cortex organotypic brain slices. Neurosci Lett 328(2):180–184

Schreibelt G, van Horssen J, van Rossum S et al (2007) Therapeutic potential and biological role of endogenous antioxidant enzymes in multiple sclerosis pathology. Brain Res Rev 56(2):322–330

Seguin R, Biernacki K, Prat A et al (2003) Differential effects of Th1 and Th2 lymphocyte supernatants on human microglia. Glia 42(1):36–45

Shi Y, Liu CH, Roberts AI et al (2006) Granulocyte-macrophage colony-stimulating factor (GM-CSF) and T-cell responses: what we do and don't know. Cell Res 16(2):126–133

Singh S, Metz I, Armor S et al (2013) Microglial nodules in early multiple sclerosis white matter are associated with degenerating axons. Acta Neuropathol 125(4):595–608

Smith ME, van der Maesen K, Somera FP (1998) Macrophage and microglial responses to cytokines in vitro: phagocytic activity, proteolytic enzyme release, and free radical production. J Neurosci Res 54(1):68–78

Sonobe Y, Yawata I, Kawanokuchi J et al (2005) Production of IL-27 and other IL-12 family cytokines by microglia and their subpopulations. Brain Res 1040(1–2):202–207

Stoop MP, Dekker LJ, Titulaer MK et al (2008) Multiple sclerosis-related proteins identified in cerebrospinal fluid by advanced mass spectrometry. Proteomics 8(8):1576–1585

Sutton CE, Lalor SJ, Sweeney CM et al (2009) Interleukin-1 and IL-23 induce innate IL-17 production from gammadelta T cells, amplifying Th17 responses and autoimmunity. Immunity 31(2):331–341

Suvannavejh GC, Lee HO, Padilla J et al (2000) Divergent roles for p55 and p75 tumor necrosis factor receptors in the pathogenesis of MOG(35-55)-induced experimental autoimmune encephalomyelitis. Cell Immunol 205(1):24–33

Suzumura A, Mezitis SG, Gonatas NK et al (1987) MHC antigen expression on bulk isolated macrophage-microglia from newborn mouse brain: induction of Ia antigen expression by gamma-interferon. J Neuroimmunol 15(3):263–278

Suzumura A, Sawada M, Takayanagi T (1998) Production of interleukin-12 and expression of its receptors by murine microglia. Brain Res 787(1):139–142

Takahashi JL, Giuliani F, Power C et al (2003) Interleukin-1beta promotes oligodendrocyte death through glutamate excitotoxicity. Ann Neurol 53(5):588–595

Takeuchi H, Jin S, Wang J et al (2006) Tumor necrosis factor-alpha induces neurotoxicity via glutamate release from hemichannels of activated microglia in an autocrine manner. J Biol Chem 281(30):21362–21368

Tanaka R, Iwasaki Y, Koprowski H (1975) Ultrastructural studies of perivascular cuffing cells in multiple sclerosis brain. Am J Pathol 81(3):467–478

Teutsch SM, Booth DR, Bennetts BH et al (2003) Identification of 11 novel and common single nucleotide polymorphisms in the interleukin-7 receptor-alpha gene and their associations with multiple sclerosis. Eur J Hum Genet 11(7):509–515

Thompson AJ, Polman CH, Miller DH et al (1997) Primary progressive multiple sclerosis. Brain 120(Pt6):1085–1096

Tomioka R, Matsui M (2014) Biomarkers for multiple sclerosis. Intern Med 53(5):361–365

Tracey D, Klareskog L, Sasso EH et al (2008) Tumor necrosis factor antagonist mechanisms of action: a comprehensive review. Pharmacol Ther 117(2):244–279

Tran PB, Miller RJ (2003) Chemokine receptors: signposts to brain development and disease. Nat Rev Neurosci 4(6):444–455

Veroni C, Gabriele L, Canini I et al (2010) Activation of TNF receptor 2 in microglia promotes induction of anti-inflammatory pathways. Mol Cell Neurosci 45(3):234–244

Vremec D, O'Keeffe M, Hochrein H et al (2007) Production of interferons by dendritic cells, plasmacytoid cells, natural killer cells, and interferon-producing killer dendritic cells. Blood 109(3):1165–1173

Werner P, Pitt D, Raine CS (2001) Multiple sclerosis: altered glutamate homeostasis in lesions correlates with oligodendrocyte and axonal damage. Ann Neurol 50(2):169–180

Williams A, Piaton G, Lubetzki C (2007) Astrocytes—friends or foes in multiple sclerosis? Glia 55(13):1300–1312

Wlodarczyk A, Løbner M, Cédile O et al (2014) Comparison of microglia and infiltrating CD11c+ cells as antigen presenting cells for T cell proliferation and cytokine response. J Neuroinflammation 11:57

Won SY, Choi SH, Jin BK (2009) Prothrombin kringle-2-induced oxidative stress contributes to the death of cortical neurons in vivo and in vitro: role of microglial NADPH oxidase. J Neuroimmunol 214(1–2):83–92

Wong GH, Bartlett PF, Clark-Lewis I et al (1984) Inducible expression of H-2 and Ia antigens on brain cells. Nature 310(5979):688–691

Yang L, Lindholm K, Konishi Y et al (2002) Target depletion of distinct tumor necrosis factor receptor subtypes reveals hippocampal neuron death and survival through different signal transduction pathways. J Neurosci 22(8):3025–3032

Yawata I, Takeuchi H, Doi Y et al (2008) Macrophage-induced neurotoxicity is mediated by glutamate and attenuated by glutaminase inhibitors and gap junction inhibitors. Life Sci 82(21–22):1111–1116

Zajicek JP, Wing M, Scolding NJ et al (1992) Interactions between oligodendrocytes and microglia. A major role for complement and tumour necrosis factor in oligodendrocyte adherence and killing. Brain 115(Pt 6):1611–1631

Zielasek J, Tausch M, Toyka KV et al (1992) Production of nitrite by neonatal rat microglial cells/brain macrophages. Cell Immunol 141(1):111–120

Chapter 17
Ischemia and Stroke

Raffaela Cipriani, Maria Domercq, and Carlos Matute

Abstract Ischemic stroke is a complex brain pathology caused by an interruption of blood supply to the brain. It is associated with neurological deficits (paralysis, memory disturbances, and thinking, language, and movement deficits), which reflect the localization and the size of the compromised brain area, and that are the manifestation of complex cellular events triggered by energy depletion. The major pathogenic events involved are excitotoxicity, peri-infarct neuronal depolarization, inflammation, oxidative stress, and apoptosis. Inflammation plays a prominent role, worsening the injury in the early phase and influencing post-stroke recovery in the late phase. Activated microglia are one of the most important cellular components of post-stroke inflammation, appearing from the first few hours after hypoxic–ischemic injury and persisting for days and weeks thereafter. Activation of microglia could be the major cause of delayed neuronal cell death: microglial harmful actions range from the release of reactive oxygen species (ROS) and nitric oxide (NO) to pro-inflammatory cytokines and matrix metalloproteinases (MMPs). In addition, microglia responses including phagocytosis of cell debris, remodeling of the extracellular matrix as well as the release of cytokines and trophic factors can improve the outcome of stroke. This knowledge opens a dynamic and possible time-dependent role of microglia in the brain response to ischemic injury.

In this chapter, we will discuss the nature of the inflammatory response in the hypoxic–ischemic brain, the contribution of microglia to injury and to regenerative processes after stroke, and finally, how cerebral ischemia directly affect microglial functions and survival.

Keywords Microglia • Cerebral ischemia • Post-ischemic inflammation • Pro-inflammatory cytokines • Trophic factor • Metalloproteinases • Phagocytosis • Reactive oxygen species

---

R. Cipriani, Ph.D. • M. Domercq, Ph.D. • C. Matute, Ph.D. (✉)
Department of Neurosciences, CIBERNED and Achucarro Basque Center for Neuroscience, University of Basque Country EHU/UPV, Leioa 48940, Spain
e-mail: raffaelacipriani@gmail.com; maria.domercq@ehu.es; carlos.matute@ehu.es

M.-È. Tremblay, A. Sierra (eds.), *Microglia in Health and Disease*,
DOI 10.1007/978-1-4939-1429-6_17

Bullet Points

- Stroke is a complex pathology caused by a disturbance in the blood supply to the brain; a sequence of pathological events rapidly occurs in the compromised brain area, resulting in dysfunction and death of brain cells and neurological deficits; the progression of the damage is dynamic in time and space and depends on the severity of the insult and/or reperfusion efficiency.
- The cellular and subcellular events responsible for the onset and progression of ischemic neuronal death include excitotoxicity, peri-infarct neuronal depolarization, post-ischemic inflammation, oxidative stress, and apoptosis.
- The inflammatory response in ischemic brain involves resident immune cells, like microglia, as well as infiltrating leukocytes, in particular neutrophils, macrophages, and lymphocytes.
- Microglia play a fundamental role in developing brain reaction to ischemic insults.
- Microglial responses in cerebral ischemia are complex. They range from induced neurotoxicity to neuroprotection, and depend on the severity of ischemic stress, the damage signals released, the duration/timing of the insult, the microenvironment, and the interaction with other brain cells.
- Deleterious effects of microglia in ischemic brain injury are mainly associated with production of reactive oxygen species (ROS), and release of pro-inflammatory cytokines and matrix metalloproteinases.
- The beneficial roles of microglia after stroke are associated with the release of growth factors and anti-inflammatory cytokines, the clearance of debris, and the promotion of late tissue recovery and regeneration.
- Stroke directly affects microglial functions and survival.
- Knowledge and control of the time course of the beneficial and deleterious effects of microglia will be key to the development of therapeutic drugs acting on microglia.

17.1 Introduction

Stroke is a devastating neurological condition derived from the permanent or transient interruption of the blood flow. It has an important impact on global health, being one of the leading causes of death and chronic adult disability worldwide (Green 2008; Moskowitz et al. 2010), and the resulting sensorimotor and cognitive impairment deeply affects the quality of life of stroke survivors. Although the severity of the neurological deficits depends on the brain area involved, stroke patients typically experience varying degrees of motor and neuropsychological deficits. Thus, attention, memory, language, and visuospatial and motor ability are generally affected, and the clinical situation is complicated by emotional deficits and mood disorders, particularly depression, which negatively influence the rehabilitative capacity (Chen et al. 2013; Cumming et al. 2013). The main risk factors for stroke occurrence are hypertension, atherosclerosis, diabetes, hypercholesterolemia, a family history of cerebrovascular diseases, aging, male sex, and Hispanic or Black race (Lo et al. 2003; Moskowitz et al. 2010).

Up to 85 % of all strokes are of ischemic origin as a consequence of blood clotting, in most cases affecting one or more cerebral arteries and drastically reducing or blocking cerebral blood flow (CBF) in the corresponding territory. *Thrombotic stroke* occurs when a clot is locally formed in an artery, while *embolic stroke* is typically the consequence of a clot formed at a distant site within brain blood vessels or somewhere else (Moskowitz et al. 2010). The majority of human strokes result from an occlusion of the middle cerebral artery (Green 2008). Less frequently (<15 %), stroke is caused by hemorrhage or cardiac arrest (Dirnagl et al. 1999; Krafft et al. 2012; Moskowitz et al. 2010). *Hemorrhagic stroke* may be due to intracerebral or subarachnoid hemorrhage.

In spite of the progression in the knowledge of stroke pathophysiology, the only treatment available up to now for patients with ischemic stroke is thrombolysis (i.e., the dissolution of a blood clot) with recombinant tissue plasminogen activator (rt-PA), which promotes reperfusion and restores normal CBF. The main limitation of this therapy is the short therapeutic window (<4.5 h) and a higher risk of symptomatic hemorrhagic complications (Zhang et al. 2012).

Hemorrhagic stroke is the least treatable type (Hwang et al. 2011) and surgery may be needed to reduce hematoma and repair the source of bleeding. Palliative treatment includes anticonvulsants to control seizures, corticosteroids, or diuretics to reduce swelling, painkillers, and anti-hypertensive drugs. However, current surgical and medical therapies are extremely limited in efficacy (Arai et al. 2011) and the prognosis remains poor (Krafft et al. 2012).

The difficulty in finding effective therapeutic strategies and/or translating successful treatments from experimentation to the clinic is largely due to the complex sequence of events that occur in the ischemic brain and that evolve over time and space (Dirnagl et al. 1999).

In this section we provide an overview on the processes that lead to ischemic injury, with particular emphasis on the inflammatory mechanisms.

17.1.1 Overview on the Pathophysiology of Stroke

Oxidative phosphorylation is the main source of energy in the central nervous system (CNS). The brain has a high metabolic consumption of oxygen and glucose and low energy, which makes this organ especially vulnerable to ischemic insults (Dirnagl et al. 1999; Moskowitz et al. 2010). Normal CBF is approximately 50–60 mL/100 g/min and when the CBF goes below 20 mL/100 g/min neurons stop their normal activity and an electrical silence rapidly occurs. CBF lower than 10 mL/100 mg/min causes irreversible neuronal injury (Jaffer et al. 2011). In focal stroke, the central region or *core* of the insult shows a drastic reduction of CBF and neurons die within minutes (Dirnagl et al. 1999). Between the *core* and the normal brain, the CBF deficit is lower, residual perfusion persists due to collateral blood vessels, and partial energy metabolism is maintained. This area, called *penumbra*, is on the edge between life and death: it could survive for a certain time, but, without treatment, the tissue in the penumbra also becomes progressively damaged (Astrup et al. 1977; Lo 2008a). The penumbra is therefore a potentially salvageable tissue, and the target of most neuroprotective strategies.

Progression of injury and cell death continues over time depending on many factors including the vulnerability of the particular brain region affected, its cellular constituents, and the extent of residual perfusion (Dirnagl et al. 1999; Moskowitz et al. 2010). For example, in focal cerebral ischemia, tissue damage is located in the vascular territory of the brain artery affected, whereas in global cerebral ischemia, as seen after cardiac arrest, lesions appear in selectively vulnerable brain regions such as the hippocampus, the caudate, thalamus, neocortex, and cerebellum (Back et al. 2004; Lo 2008a; Lo et al. 2003). On the other hand, many pathological events that represent a neurotoxic component at early stages may set in the late post-ischemic period the conditions for tissue remodelling and recovery, and this complexity increases the difficulty of defining a generalized therapeutic strategy (Lo 2008a, b).

Immediately after vascular occlusion, affected neurons are unable to maintain ion homeostasis and their normal transmembrane ionic gradient, resulting in permanent anoxic depolarization and irreparable tissue damage. In turn, depolarization induces the release of excitatory amino acids, particularly glutamate (Glu), which accumulates into the extracellular space. The absence of energy-dependent synaptic reuptake also contributes to this accumulation. Excitotoxicity is the result of Glu receptor over-stimulation and leads ultimately to necrotic or apoptotic cell death depending on the intensity of the stimulus and the metabolic state of the neurons (Lipton 1999; Moskowitz et al. 2010). Over-activation of Glu ionotropic receptors results in excessive Na^+ and H_2O passive influx and massive Ca^{2+} entry into neurons. Excessive H_2O causes cell swelling, loss of membrane integrity, and cell demise, generally leading to brain edema which negatively affects reperfusion (Dirnagl et al. 1999; Lipton 1999). In turn, Ca^{2+} overload in the cytoplasm activates proteases, lipases, and nucleases that rapidly degrade membranes and proteins, leading to cell death by necrosis. Moreover, Ca^{2+} leads to generation of reactive oxygen species (ROS) via the activation of enzymes (like phospholipase A2, cyclooxygenase, nitric oxide synthase/NOS) and/or to primary mitochondria alterations (Lo et al. 2003). ROS directly damage lipids, proteins, nucleic acid, and carbohydrates; in addition, they may react with NO to produce highly toxic peroxynitrites. ROS impair mitochondrial function leading to generation of further ROS and to the release of pro-apoptotic factors such as cytochrome C (Lipton 1999; Moskowitz et al. 2010). Moreover, ROS contribute to the onset of the inflammatory response after ischemia-reperfusion.

Typically, the ischemic core rapidly undergoes necrosis as a consequence of bioenergetic collapse, whereas in the penumbra residual blood flow allows the preservation of neuronal resting membrane potentials, although not enough to generate action potentials: in this area, neuronal death develops more slowly as a result of

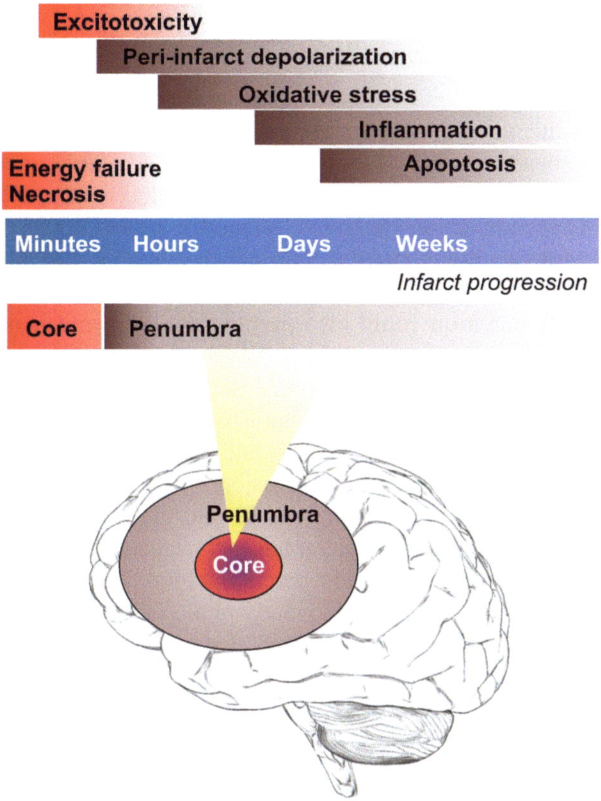

Fig. 17.1 Overview of the pathobiological mechanisms involved in ischemic injury. The ischemic cascade comprises a series of deleterious events that evolve over time and space. Acute damage occurs in the ischemic core, where neurons rapidly die from the total energy failure and excitotoxicity. In the ischemic penumbra, more delayed events could extend the damage to this affected but potentially salvageable tissue. Among these events, post-ischemic inflammation has a prominent role, with microglia, the immune cells of the Central Nervous System, representing its most important player

excitotoxicity and secondary events, like peri-infarct depolarization, oxidative stress, post-ischemic inflammation, and apoptosis (Dirnagl et al. 1999) (Fig. 17.1).

On the other hand, restoring CBF triggers deleterious pathways associated with delayed progression of the injury, ROS, and inflammation being the most prominent (Green 2008). Although thrombolysis of acute stroke patients is currently the main effective therapy, the so-called "reperfusion injury", a phenomenon referring to a cascade of deleterious biochemical processes triggered by reperfusion, must be taken into account as highlighted in animal models of transient cerebral ischemia (Aronowski et al. 1997; Pan et al. 2007). Multiple pathological processes are involved including oxidative and nitrosative stress, leukocyte infiltration, platelet and complement activation, and breakdown of the blood–brain barrier (BBB) (Lo et al. 2003; Pan et al. 2007).

17.1.2 Nature of the Inflammatory Response in the Post-ischemic Brain

Focal cerebral ischemia and subsequent reperfusion are associated with an inflammatory response headed initially by resident brain cells, mainly microglia, as well as by the infiltration of peripheral blood immune cells, neutrophils, monocytes/macrophages, and lymphocytes (Kriz and Lalancette-Hébert 2009). The inflammatory response starts few hours after CBF failure, persists for days after stroke and, at least early on, is associated with the secondary progression of the ischemic damage. However, inflammation could also participate in the mechanisms of tissue repair and brain function recovery, especially in the late post-ischemic period (Kriz 2006; Kriz and Lalancette-Hébert 2009; Moskowitz et al. 2010). In this paragraph, we provide a short overview of the mechanisms of post-ischemic inflammation, and in the next section we will focus on the role played by microglial cells.

Inflammation after brain ischemia-reperfusion is the result of events occurring at vascular, perivascular, and parenchymal levels and involves a wide array of cell types (leucocytes, perivascular macrophages, resident microglia, astrocytes) (Iadecola and Anrather 2011). As the cerebral blood vessels are the first ones affected by a vascular occlusion, the very early phase of the inflammatory response starts at the intravascular compartment. Thus, hypoxia, production of ROS (especially in the case of reperfusion injury), and changes in shear stress are among the main factors priming the activation of platelets (i.e., the non-nucleated blood components implicated in the formation of blood clots), endothelial cells, and the complement system, which is a proteolytic cascade composed of several subunits eventually leading to cell lysis, enhancing phagocytosis (opsonisation), and mediating chemotactic recruitment of inflammatory cells (Iadecola and Anrather 2011; Yanamadala and Friedlander 2010). These events induce the formation of fibrin, a protein involved in blood clotting. Fibrin is formed from fibrinogen by the protease thrombin, and then polymerised to form a mesh that entangles platelets generally forming a clot. In the case of cerebral ischemia, intravascular formation of fibrin, and trapping platelets, and leukocytes results in microvascular occlusions (Iadecola and Anrather 2011), while P-selectin expressed on the membrane of platelets and endothelial cells bursts the production of pro-inflammatory mediators (Iadecola and Anrather 2011).

At the vascular compartment, NO derived from endothelial nitric oxide synthase (eNOS) is a potent vasodilator and sustains blood flow in the ischemic territory (Iadecola et al. 1994; Lo 2008b). Oxidative stress can lead to a reduction of NO produced by eNOS, and to a constriction of pericytes (contractile cells that wrap around the endothelial cells of capillaries and venules), further increasing microvascular occlusions and aggravating ischemic injury (Yemisci et al. 2009). The permeability of BBB is also altered by ROS, inflammatory mediators, proteases, and downregulation of junctional proteins.

In the perivascular compartment, different components of the immune system are involved in the brain response to cerebral ischemia. Thus, mast cells within the cerebral microvasculature release their cytoplasmic granules which contain vasoactive mediators such as tumor necrosis factor α (TNFα), histamine, heparin, and proteases (Lindsberg et al. 2010). Also, perivascular macrophages in the perivascular compartment release pro-inflammatory cytokines such as TNFα and interleukin-1β (IL-1β) (Konsman et al. 2007). These vasoactive compounds and pro-inflammatory cytokines act on the basal membrane further contributing to damage of the BBB and to the expression of endothelial adhesion molecules (e.g., intercellular adhesion molecule-1/ICAM-1, P and E-selectin). These molecules, along with integrins, drive the infiltration of leukocytes (neutrophils, lymphocytes, and monocytes) into the brain parenchyma, thus promoting brain edema, prolonged extravasation, and hemorrhage (Iadecola and Anrather 2011; Konsman et al. 2007; Lindsberg et al. 2010).

The first cells that adhere to the endothelium and cross the vascular wall to enter in the brain parenchyma are neutrophils, followed by monocytes-macrophages and, 1–2 days after stroke, by lymphocytes (Dirnagl et al. 1999; Moskowitz et al. 2010).

In the brain parenchyma, injured neurons burst the inflammatory response driven by microglia which in the first hours after stroke became activated and undistinguishable from circulating macrophages (Schilling et al. 2003). Microglia accumulate at the site of injury producing pro-inflammatory cytokines and toxic mediators, and phagocytosing cell debris (see below). In addition, several T cell subtypes modulate secondary infarct progression (Iadecola and Anrather 2011).

All the above cellular reactions result in early inflammatory-induced cell death along with cerebral infarction progression. The main effectors of cell death in inflammation include direct damage induced by oxidative and nitrosative stress, activated complement system, activation of FasL- and TNF-mediated apoptosis, perforin cytolysis, and release of the apoptosis inducer granzyme by activated microglia and infiltrating immune cells (Iadecola and Anrather 2011). At later stages of stroke, the inflammatory response helps to restore tissue homeostasis by removing dead cells and debris by microglia and infiltrated macrophages, producing anti-inflammatory cytokines such as transforming growth factor β(TGF-β) and inteleukin-10 (IL-10), and growth factors such as insulin-like growth factor-1(IGF-1), vascular endothelial growth factor (VEGF), which act in concert with metalloproteinases (MMPs) to reorganize and repair damaged tissue.

17.2 Microglia in Cerebral Ischemia

Microglia are cells of myeloid origin derived from primitive hematopoiesis in the fetal yolk sac and take up residence in the brain during early fetal development (Ginhoux et al. 2010; Saijo and Glass 2011). Microglia are considered as the resident macrophage population of the CNS, where they maintain surveillance and continuously scan the tissue with their processes, ready to act when any type of disturbance occurs (Hanisch and Kettenmann 2007). Microglia are the most versatile cell population in the brain. Depending on the state of activation, type, and intensity of the stimulus, microglia undergo a plethora of morphological and biochemical transformations to mediate the appropriate responses (Hanisch and Kettenmann 2007; Prinz et al. 2011).

Microglia play a critical role in the inflammatory response that occurs after cerebral ischemia. Minutes after an ischemic lesion resident microglia shift their activity state from a highly ramified phenotype to a more amoeboid morphology, proliferate, and change the pattern of surface marker expression, becoming indistinguishable from macrophages. In this reactive state, microglia migrate to and accumulate in the infarct core and in the penumbra, where they drive the inflammatory response and lymphocytes infiltration, closely affecting later recovery too. Infiltration of reactive microglia in the ipsilateral hemisphere occurs already at 12–24 h after experimental ischemia and is followed by neutrophils and macrophages invasion (Schilling et al. 2003; Weinstein et al. 2010). Positron emission tomography (PET) and magnetic resonance imaging (MRI) visualization techniques have confirmed the presence of activated microglia and macrophages in the peri-infarct zone during the subacute phase of stroke in vivo in humans (Weinstein et al. 2010).

Microglia may persist in an activated state for weeks after initial injury, with a peak in proliferation at 48–72 h (Denes et al. 2007; Lalancette-Hébert et al. 2007).

It is generally assumed that activation of microglial cells in cerebral ischemia is neurotoxic and exacerbates ischemic injury. However, microglial responses are highly variable and depend on the nature, strength, and duration of the insult, ranging from the production of pro-inflammatory cytokines (TNFα, IL-1β), ROS, and NO, to the production of neurotrophic factors and anti-inflammatory cytokines (glial cell line-derived neurotrophic factor/GDNF, brain-derived neurotrophic factor/BDNF, basic fibroblast growth factor/bFGF, IGF-1, TGF-β, VEGF) (Lai and Todd 2008; Kriz and Lalancette-Hébert 2009).

As in the case of macrophages, two activated phenotypes of microglia have been identified: a "classically activated" (also called M1) and an "alternatively activated" (also called M2) (Hu et al. 2012; Michelucci et al. 2009; Perego et al. 2011). This suggests that microglia can adopt different functional programs in response to specific microenvironmental cues. In vitro stimulation with lipopolysaccharide and interferon-γ (IFNγ) promotes the acquisition of a M1 phenotype in microglia/macrophages, typically associated with the release of pro-inflammatory mediators. In contrast, interleukin (IL)-4 and IL-10 induce a M2 phenotype that possesses neuroprotective and anti-inflammatory properties, with enhanced phagocytic activity and reduced production of inflammatory mediators (Durafourt et al. 2011; Michelucci et al. 2009; Olah et al. 2011). However, the phenotype-specific roles of microglia in ischemic brain injury are poorly characterized. It seems that microglia/macrophages at the ischemic lesion present specific phenotypes with distinct spatial and temporal features (Perego et al. 2011), with local microglia and newly recruited macrophages showing the M2 phenotype at the early stages of ischemic stroke and gradually transforming into the M1 phenotype at the site of injury (Hu et al. 2012). This is consistent with the idea that phagocytosis of tissue debris and dying cells, and secretion of pro-survival/neurotrophic factors, become relevant to facilitate and/or initiate the recovery processes and tissue repair.

Microglia are equipped with different receptors to react to homeostatic perturbations. In brain ischemia, microglia sense molecules released from damaged or dead neurons. Among them, purines (ATP and UTP) released into the extracellular space as a result of injury stimulate purinergic receptors present on the microglial membrane (Iadecola and Anrather 2011; Yenari et al. 2010). Purines transmit their signal through two different classes of receptors, P1 (for adenosine) and P2 (for ATP, UTP and ADP, UDP). ATP receptors (P2 receptors) are divided into ionotropic P2X (P2X1-7) and metabotropic P2Y (P2Y1,2,4,6,11-14) receptors (Abbracchio et al. 2009).

Major subtypes of purinergic receptors expressed by microglia include P2X4, P2X7, PXY1, P2Y2, and P2Y12 receptors (Haynes et al. 2006; Pocock and Kettenmann 2007), having important roles in activation, movement, and microglial response to injury. Thus, P2X7 receptors in microglia, whose expression is upregulated after brain injury, stimulate proliferation and release of cytokines like IL-1β, TNF-α (Brough et al. 2002; Monif et al. 2009; Suzuki et al. 2004), while short-term exposure to exogenous ATP reduces their phagocytic capacity through P2X7 receptor activation (Fang et al. 2009). Inhibition of P2X7 receptor reduces microglia-induced inflammation and brain infarct in global (Chu et al. 2012) and focal models of cerebral ischemia in rats (Melani et al. 2006). P2X4, as well as P2Y12, are involved in ATP-induced microglial chemotaxis in vitro (Ohsawa et al. 2007). P2X4 expression is upregulated in reactive microglia in vitro, after oxygen-glucose deprivation (OGD) in rat organotypic brain slices, and in vivo, after cerebral ischemia in gerbil (Cavaliere et al. 2003), suggesting an involvement of P2X4 activation in the modulation of microglial cell migration under ischemic conditions. In turn, activation of P2Y12 is associated with microglial process extension and migration and this is the main receptor through which extracellular purines promote directed microglial movement following CNS injury, as microglia derived from P2Y12 knockout mice are unable to polarize, migrate, or extend processes towards nucleotides in vitro or in vivo (Haynes et al. 2006), and in vitro stimulation of microglia with ATP regulates process extension through the integrin-β1 activation (Ohsawa et al. 2010). However, a specific role for microglial P2Y12 in cerebral ischemia remains to be elucidated.

In addition, neurotransmitters released after ischemia and reperfusion may counteract microglial inflammatory response and reduce cytokines, ROS, and NO production. In fact, microglia express receptors for most neurotransmitters including Glu, GABA, purines (as mentioned above), and dopamine in vitro (Pocock and Kettenmann 2007). Typically, neurotransmitters suppress microglial activation (Biber et al. 2007) and it appears that the local release of neurotransmitter during normal brain function may be a mechanism by which microglia can sense the proper functioning of neurons. Thus, activation of microglia through GABA, adrenergic, adenosine, or cannabinoid receptors suppresses cytokine release and the neurotoxic phenotype (Biber et al. 2007; Pocock and Kettenmann 2007). However, Glu has a more complex role since stimulation of ionotropic Glu receptors (iGluR) induces microglial proliferation, morphological changes, and release of IL-1β, TNFα, NO, and ATP (Biber et al. 2007; Pocock and Kettenmann 2007; Yenari et al. 2010), while metabotropic Glu receptors (mGluRs) can transform microglia into a neuroprotective (via group III mGluRs) or neurotoxic (via group II mGluRs, particularly mGluR2) phenotype (Pocock and Kettenmann 2007). In models of cerebral ischemia, specific types of iGluRs are expressed in reactive microglial cells after ischemia (Gottlieb and Matute 1997), and a neuroprotective effect for a mGluR5 agonist has been demonstrated after acute administration in experimental stroke (Kohara et al. 2008).

Moreover, several intracellular molecules (e.g., hyaluronic acid, fibronectin, heat shock protein 60 (HSP60), and high-mobility group box 1) that are released from dying neurons or generated by the action of lytic enzymes activate Toll-like receptors (TLRs) signalling pathway, a family of transmembrane proteins related to recognition and defence from microorganisms. In particular, activation of TLR4 leads to upregulation of an array of proinflammatory genes in microglia (Iadecola and Anrather 2011; Kilic et al. 2008; Yenari et al. 2010).

Cell-to-cell interaction between neurons and microglia keep the latter in a quiescent state in normal tissue. After neuronal death, these interactions are lost and disrupted signalling causes microglia to become fully reactive (Biber et al. 2007; Hanisch and Kettenmann 2007). This is the case of CD200 and CX3CL1/fractalkine. CD200, a cell surface glycoprotein belonging to an immunoglobulin superfamily constitutively expressed on the neuronal membrane in healthy conditions interacts with its receptor CD200R on microglia, enforcing a resting phenotype (Hoek et al. 2000). As demonstrated in vivo using antibodies against the CD200 receptor or CD200-deficient mice, disruption of this interaction promotes a microglial inflammatory response in several types of experimental neurological disorders like experimental autoimmune encephalomyelitis and facial nerve axotomy (Biber et al. 2007), or in transient middle cerebral artery occlusion (MCAO) in rats (Matsumoto et al. 2007). The chemokine CX3CL1 plays a similar role. CX3CL1 is constitutively expressed by neurons and is synthesized as a transmembrane glycoprotein, from which a soluble chemokine can be proteolytically released. The interaction with its receptor CX3CR1, expressed only by microglia in the healthy CNS, is important for neuron-microglia communication. It has been shown that CX3CL1 reduces microglia neurotoxicity and, consequently, neuronal damage both in vitro and in in vivo models of neuropathologies and brain inflammation (Cardona et al. 2006; Mizuno et al. 2003; Zujovic et al. 2000). However, the role of CX3CL1 in cerebral ischemia appears more complex: in transient focal cerebral ischemia, genetic deletion of CX3CL1 ($CX3CL1^{-/-}$ mice) reduces post-ischemic brain injury (Soriano et al. 2002) and CX3CR1 deficiency ($CX3CR1^{GFP/GFP}$ mice, in which the CX3CR1 gene was replaced by GFP) significantly attenuates ischemic and BBB damage, IL-1β, and TNFα release (Dénes et al. 2008). On the other hand, when CX3CL1 is given exogenously a long-term protective effect on the ischemic lesion is observed (Cipriani et al. 2011), suggesting a pivotal role of the constitutive CX3CL1/CX3CR1 signalling in modulating microglial response during cerebral ischemia.

Once exposed to injury-released signals, microglia acquire an effector phenotype and upregulate the production of inflammatory mediators (TNFα, iNOS, IL-6) and proteases (MMP3, MMP9) (Dirnagl et al. 1999; Yenari et al. 2010). NFkB in particular promotes the transcription of many genes involved in inflammation, including TNFα, ICAM-1, cyclooxygenase-2 (COX-2), iNOS, and IL-6 and drives the morphological transformation of microglia into their effector phenotype (Yenari et al. 2010).

17.2.1 Deleterious Effects of Microglia in Ischemic Brain Injury

Deleterious effects of microglial activation in brain ischemia are typically associated with the M1 phenotype. Activated microglia migrate to the site of injury, proliferate, and release cytotoxic substances such as pro-inflammatory cytokines (i.e., IL-1β and TNFα) (Barone et al. 1997; Rothwell et al. 1997), ROS (superoxide, hydrogen peroxide, nitric oxide), MPPs, and Glu (Yenari et al. 2010) (Fig. 17.2).

Superoxide is primarily produced by microglia via NAPDH oxidase (NOX) (Brown and Neher 2010; Chen et al. 2009; Surace and Block 2012). The main

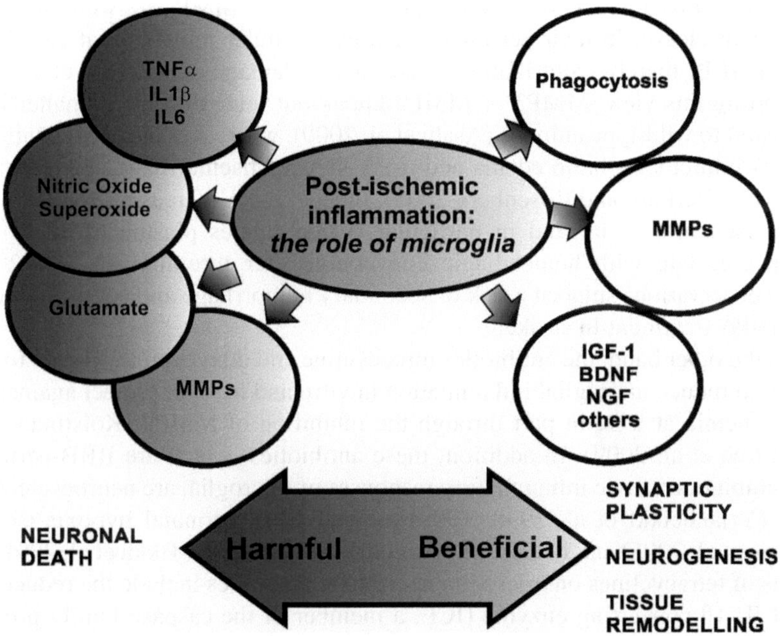

Fig. 17.2 Summary of microglial-mediated responses in cerebral ischemia. Microglial responses after cerebral ischemia show both neurotoxic and neuroprotective effects. This dichotomy could result from different factors influencing the shift from one microglial phenotype to another; among these, the severity of neuronal injury and the temporal factor are crucial, with the cytotoxic aspect being more important in the acute phase of the ischemic injury and the trophic component in the late phase of recovery

cytotoxic role of superoxide derives from the reaction with other molecules to produce other ROS, peroxynitrite and hydroxyl radicals, which are highly toxic to neurons. Accordingly, early treatment with NOX inhibitors as apocynin has a protective effect against experimental cerebral ischemia-reperfusion (Chen et al. 2009; Yenari et al. 2010) and knock-out mice lacking the gp91 subunit of NOX2 present a smaller infarct in cerebral ischemia compared with wild-type littermates (Walder et al. 1997). NOX-derived superoxide could exacerbate BBB injury directly by acting on its constituents, astrocytes and endothelial cells, since inhibiting NOX superoxide production with apocynin preserves BBB constituents in vitro and reduces BBB disruption and hemorrhage in vivo (Yenari et al. 2006).

During cerebral ischemia, inducible nitric oxide synthase (iNOS) is upregulated in activated microglia resulting in an increased production of NO (Brown and Neher 2010; Iadecola et al. 1995; Yenari et al. 2010). Reacting with superoxide, NO leads to the formation of peroxynitrite, a molecular species which is highly reactive and toxic for neurons (Iadecola et al. 1995), and directly inhibits mitochondrial respiration (Mander et al. 2005). As in the case of NOX, pharmacological inhibition of iNOS with aminoguanidine and iNOS genetic deletion in mice reduces infarct size (Iadecola et al. 1995; Zhao et al. 2000).

Microglia are the major source of MMPs released following cerebral ischemia, especially MMP3 and MMP9 (del Zoppo et al. 2007). MMPs are well known proteases that cleave, in their activated state, extracellular proteins that can directly disrupt BBB, thereby contributing to secondary damage (del Zoppo et al. 2007). Supporting this view, MMP3 or MMP9 knock-out mice display a smaller infarct compared to wild-type animals (Asahi et al. 2000), whereas acute MMPs inhibition reduces infarct size, brain edema and rt-PA-induced haemorrhage in experimental stroke (Pfefferkorn and Rosenberg 2003). Indeed, rt-PA increases levels of MMPs in the ischemic brain, and in particular it upregulates plasma MMP-9 levels, which correlate with hemorrhagic conversion after thrombolysis (Lo 2008a). These observations suggest a risk of secondary hemorrhage and edema associated with rt-PA treatment in stroke.

On the other hand, the antibiotics minocycline and doxycycline, known for their ability to reduce microglial inflammation in vitro and in vivo, protect against cerebral ischemia at least in part through the inhibition of MMP9 (Koistinaho et al. 2005; Lee et al. 2009). In addition, these antibiotics, which are BBB-permeable and inhibit neurotoxic inflammatory responses of microglia, are neuroprotective in adult (Yrjänheikki et al. 1998, 1999) as well as in perinatal hypoxia-ischemia (Jantzie et al. 2005) and in cell culture models of cell death (Tikka et al. 2001). The effects of tetracyclines on microglia neurotoxic responses include the reduced levels of IL-1β-converting enzyme (ICE, a member of the caspase-family proteases that cleaves pro-IL-1β, an inactive precursor, to IL-1β) and iNOS (Yrjänheikki et al. 1998), the inhibition of caspases (Chen et al. 2000; Jantzie et al. 2005) and COX-2 expression, and of prostaglandin E2 production (Yrjänheikki et al. 1999). However, non-specific effects of minocycline, acting on other cell types such as astrocytes and macrophages, should be taken into account (Jin et al. 2014; Liu and Yang, 2012).

Finally, microglia could exacerbate excitotoxic damage through the release of Glu in the post-ischemic period, thus adding to the increase in Glu extracellular levels caused by the damage to neurons and the failure of reuptake mechanisms (Takeuchi et al. 2008).

17.2.2 Beneficial Effects of Microglia After Stroke

Microglia response in cerebral ischemia can also help restoring tissue homeostasis and repair (Fig. 17.2). Thus, microglia can support neuronal survival through the clearance of cell debris by phagocytosis, and the production of neurotrophic factors and anti-inflammatory cytokines, in order to promote tissue recovery and regeneration at the lesion site.

Proliferating resident microglia are crucial for a better outcome in cerebral ischemia since their ablation increases apoptotic neurons and decreases the levels of IGF-1 (Lalancette-Hébert et al. 2007). In addition, exogenous microglia injected into the blood stream (Imai et al. 2007) or intracerebrally (Kitamura et al. 2004) can increase post-stroke neuronal survival by producing BDNF and GDNF.

In addition to secreting growth factors, microglia can also be protective by engulfing activated and neurotoxic neutrophils, as shown in organotypic brain slices subjected to OGD (Neumann et al. 2008). Thus, the rapid increase in mononuclear phagocytes numbers at the site of injury during the first 3 days as a consequence of microglial proliferation negatively correlates with the extent of ischemic brain damage, as demonstrated in a model of transient MCAO in mice using MRI combined with blood-born macrophages labeling techniques to discriminate microglia proliferation from macrophage infiltration (Denes et al. 2007). This amelioration in ischemia-reperfusion injury appears to be due to phagocytosis of infiltrating neutrophils and to attenuated inflammation (Denes et al. 2007). On the other hand, beneficial aspects of phagocytic activity of microglia are also associated with the removal of necrotic/apoptotic cells and debris in the ischemic tissue (Denes et al. 2007; Schilling et al. 2005). Phagocytosis of dead cells by microglia promotes the production of immunomodulatory cytokines, such as TGF-β, that could suppress inflammation by inhibiting helper T-cell responses and promoting regulatory T-cell development, and IL-10, that has neuroprotective and anti-inflammatory properties (Iadecola and Anrather 2011). These cytokines can facilitate the resolution of inflammation and brain repair, as well as exert direct neuroprotective effects in cerebral ischemia (Grilli et al. 2000; Lehrmann et al. 1998). Other cytokines can also have a protective role in ischemia. This is the case of TNFα, a pro-inflammatory cytokine whose main source in the brain are resident-activated microglia and infiltrating leucocytes (Lambertsen et al. 2009). TNFα is commonly associated with neurotoxicity and a detrimental stroke outcome. However, recent evidence suggests a dual role for this cytokine. In particular, microglial-derived TNFα reduces the size of cortical infarcts and behavioral deficits, and modulates inflammatory responses after focal cerebral ischemia in mice (Lambertsen et al. 2009). The protective role of TNFα involves the activation of TNF-p55 receptor as infarction increases in TNF-p55 receptor knockout mice compared with TNF-p75 receptor knockout and wild-type mice (Lambertsen et al. 2009).

Finally, microglia have an important role in neurogenesis and neuroplasticity after cerebral ischemia. Ischemic stroke caused by MCAO triggers an extensive and long-lasting cell proliferation in the rat subventricular zone (SVZ), and the newly formed neuroblasts migrate into the damaged striatum producing new striatal neurons (Thored et al. 2009). In this model, the production of neuroblasts is accompanied by a long-lasting accumulation of reactive microglia in the SVZ. These microglia show a ramified morphology and a proneurogenic phenotype associated with an upregulated expression of IGF-1 which influences neurogenesis and suggests a supportive role of microglia in the continuous neuroblast production (Thored et al. 2009). Moreover, activated microglia and reactive astrocytes in the ischemic area express monocyte chemoattractant protein-1, which promotes the migration of neuroblasts to the damaged area (Yan et al. 2007).

Conversely, post-ischemic inhibition of the inflammatory response by minocycline reduces neurogenesis (Kim et al. 2009). Moreover, following focal cerebral ischemia in rats microglia have a supportive role in brain neuroplasticity possibly through the production of BDNF. Indeed, pharmacological inhibition of the inflammatory response with an inhibitor of the poly(ADP-ribose) polymerase-1 (PARP-1), which contributes to the activation of NFkB, is associated with a long-term reduction in the expression of two neuronal plasticity proteins, synaptophysin (marker of synaptogenesis) and GAP-43 (marker of neuritogenesis), as well as with lower levels of BDNF secretion (Madinier et al. 2009). These studies support the hypothesis of an involvement of microglia in neurogenesis and synaptic plasticity after stroke (For more detailed information on microglial role in neurogenesis and synaptic plasticity, see Chaps. 9 and 10).

Taken together, these studies indicate that the beneficial functions of microglia depend on a plethora of factors, far from being an all-or-none phenomenon. Several different "activation states" exist in which microglia may selectively upregulate neuroprotective and/or neurotoxic effectors (Czeh et al. 2011). The severity of neuronal injury determines the type of molecules released by damaged tissue which ultimately determines microglial responses (Gomes-Leal 2012; Lai and Todd 2008).

Post-ischemic changes over time are also relevant for determining the detrimental or protective edge of microglia (Kriz 2006). Thus, acute overactivation of microglia results in an initial harmful response, and the subsequent resolution of inflammation may shift microglia to a pro-survival phenotype, the M2 phenotype, oriented towards the recovery and regeneration of the tissue. At this stage, microglia present enhanced phagocytic activity (Hu et al. 2012), a reduced production of inflammatory mediators, and the capacity of releasing neurotrophic factors, such as GDNF, BDNF, bFGF, IGF-1 TGF-β, and VEGF (Xing et al. 2012). A paradigmatic example that supports this idea is the role of MMPs in cerebral ischemia. MMPs are known to promote injury of BBB by degrading neurovascular matrix. Early MMPs inhibition confers neuroprotection, while late inhibition may conversely mediate beneficial plasticity and remodelling during stroke recovery. VEGF generated through MMP-9 processing of pro-VEGF and matrix-bound VEGF is responsible, at least in part, since inhibition of MMPs reduces the amount of endogenous VEGF in peri-infarct cortex at 14 days after transient MCAO (Zhao et al. 2006).

17.2.3 Altered Functions and Damage to Microglia After Stroke

The present debate about the role of microglia in stroke is focused on whether its depletion or introduction in vivo or ex vivo improves or worsens tissue damage and the neurological consequences of stroke. However, few studies have analyzed how energy-limiting conditions affect microglia functions and survival directly. In response to microdamage in a capillary, microglia rapidly outgrow larger processes and migrate to the site of injury to prevent damage expansion (Davalos et al. 2005; Haynes et al. 2006; Hines et al. 2009). This outgrowth involves functional purinergic P2Y12 receptors and chloride channels and is also actin-dependent (Haynes et al. 2006; Hines et al. 2009). Actin polymerization in motile cells is an ATP-dependent process (Atkinson et al. 2004) and so, the early spatiotemporal changes in microglial dynamics after ischemia are dramatically affected by CBF (Masuda et al. 2011). Indeed, after focal and global ischemia the dynamics of microglia in the penumbra area are strongly correlated with capillary blood flow measured within 10 μm of microglial somata. In contrast, changes in blood flow around distal microglial processes (>30 μm from somata) had no effect on microglial dynamics (Masuda et al. 2011). Similarly, in an ex vivo model of ischemia, microglia mobility was severely reduced during OGD (Eyo and Dailey 2012). However, there are still a reduced number of microglia cells that maintain their motility capacity intact during the ischemic insult (Masuda et al. 2011; Eyo and Dailey 2012). Restoration of blood flow or reperfusion leads to partial or total recovery of microglia motility depending on the nature of the insult. In ex vivo slices, a transient OGD (2 h) caused a prolonged alteration of microglial behavior, with a 50 % reduction of motility and migration capacity (Eyo and Dailey 2012).

In the normal brain, two-photon in vivo imaging recently revealed that microglia monitor the functional status of synapses and make transient connections with synapses (approximately 5 min long) resulting in microglial process retraction (Tremblay et al. 2010; Wake et al. 2009). Microglial motility towards synapses is modulated by neuronal activity and experience (Tremblay et al. 2010; Wake et al. 2009). After transient cerebral ischemia, the duration of the microglia-synapse contacts in the penumbra is prolonged (approximately 1 h), and these contacts were sometimes followed by the complete disappearance of the presynaptic bouton (Wake et al. 2009). However, the number of microglia process extension–retraction events is severely blunted by 76 % in the peri-infarct region after ischemic stroke and 24-h reperfusion in mice (Morrison and Filosa 2013). In addition, the surveillance function of microglia depends on their extensive arborization of dynamic processes. While the withdrawal of microglial processes is generally considered a hallmark of microglial activation and transformation into a phagocytic phenotype, as occurring in ischemia, such morphological manifestation of microglial activation requires some levels of residual blood flow (Masuda et al. 2011) and is more evident during reperfusion (Morrison and Filosa 2013). Microglial process de-ramification is driven by upregulation of the G_s-coupled A_{2A} receptor (Orr et al. 2009) concomitant with downregulation of the G_i-coupled $P2Y_{12}$ receptor following stroke or LPS stimulation (Haynes et al. 2006). However, whether such morphological changes represent an increase in phagocytic activity during ischemia remains to be determined. A recent study in hippocampal tissue slices demonstrates that the microglia surviving the ischemic insult (OGD) can engulf dead cells as revealed by time lapse microscopy (Eyo and Dailey 2012).

Since microglia are necessary to phagocytose dead cells and to promote recovery, damage to microglia could affect the stroke outcome. Pioneer studies have already suggested that cultured microglia are vulnerable to ischemia (Lyons and Kettenmann 1998; Yenari and Giffard 2001), despite the fact that cultured cells developed a metabolic switch making them more resistant to the insult (Chock and Giffard 2005). After permanent MCAO some microglia remained intact in the ischemic core 24 h after the insult (Mabuchi et al. 2000; Wen et al. 2004), suggesting that these cells may survive for some time under permanent ischemia. However, transient MCAO in neonates induces a 40 % decrease in microglia cell number in the ischemic core relative to the penumbra (Faustino et al. 2011). Clear evidences about the effects of stroke on microglial viability in vivo are lacking because it is difficult to estimate microglial cell death due to the concomitant proliferation of surviving microglia and the infiltration of circulating monocytic cells expressing very similar markers. However, using ex vivo acute brain slices, it was demonstrated that OGD, not hypoxia, induces clear damage to microglia (i.e., nearly 44 % of microglial cell death after 6 h OGD) and that microglial vulnerability increases dramatically with age (Eyo and Dailey 2012). Different mechanisms of microglia cell death have been described including apoptosis, autophagy (Arroyo et al. 2013), pyroptosis, programmed cell death triggered by caspase-1 activation by the inflammasome, and necroptosis (Fricker et al. 2013). However, little is known about the signals that regulate microglial cell death in ischemia and its impact on the outcome of stroke. One of the signals that regulate microglial cell death after OGD in acute neonatal slices is ATP through the activation of the low-affinity P2X7 receptor (Eyo et al. 2013).

17.3 Conclusions

In recent years, we have witnessed a tremendous growth of knowledge about the cellular and biochemical processes underlying tissue damage in cerebral ischemia. In parallel, a large variety of neuroprotectants have proven to be beneficial in animal models of stroke, but this potential therapeutic armamentarium has failed in clinical studies with the exception of rt-PA. However, combined therapy targeting several mechanisms with novel drugs during appropriate time windows may result in better treatment of stroke. Some novel concepts may be instrumental in developing new effective drugs. In particular, pharmacological interventions aimed at restoring normal communication at the "neurovascular unit", which includes dynamic interactions between neurons, endothelial cells, vascular smooth muscle, associated tissue matrix proteins, and glial cells, will be crucial for halting the progression of ischemic injury (Lo et al. 2003; Moskowitz et al. 2010; del Zoppo 2009). Post-ischemic inflammation has a profound impact on the integrity of neurovascular unit, and microglia, the main resident immune effectors, are a promising target for the treatment of stroke. However, the biphasic microglial response after ischemia makes therapeutic approaches difficult. Because of this dual response, it is most important to evaluate drug efficacy at both the initial and later phases post-stroke and to take into account the extent of vascular occlusion (Kriz 2006). Current research into microglia-based therapies endeavours a selective suppression of the deleterious effects of microglial activation without affecting mechanisms of neuronal circuits repair and remodelling. Very interesting in this respect is the use of specific viral vector for targeted gene delivery to microglial cells (Balcaitis et al. 2005).

Minocycline, a BBB-permeable tetracycline antibiotic, is commonly used as strong inhibitor of microglia polarization to the proinflammatory state M1 without influencing its apparent M2 phenotype (Kobayashi et al. 2013). These properties of minocycline make it a good candidate to tame microglial reactions towards a more benign response following stroke and warrant further investigations. Indeed, recent early phase clinical trials have shown minocycline to be safe and potentially effective in acute ischemic stroke, alone or in combination with rt-PA (Fagan et al. 2011). Other pharmacological agents modulating microglial function, such as antagonists of TLRs, are also under investigation (Weinstein et al. 2010).

As demonstrated in experimental models of cerebral ischemia, exogenous macrophages can migrate into the brain parenchyma and protect against neurodegeneration (Imai et al. 2007; Kitamura et al. 2004, 2005; Tanaka et al. 2003). Therefore, administration of bone-marrow-derived monocytes/microglia might be a potential tool for cell or gene therapy in the treatment of stroke. In addition, transplantation of bone marrow-derived cells has shown positive results in preclinical models of cerebral ischemia in part by modulation of post-ischemic inflammatory responses (Brenneman et al. 2010; Ohtaki et al. 2008; Ramos et al. 2013).

Acknowledgments Work in our laboratory is supported by MINECO, CIBERNED, UPV/EHU, Gobierno Vasco, and ERANET-NEURON (NanoStroke).

References

Abbracchio MP, Burnstock G, Verkhratsky A, Zimmermann H (2009) Purinergic signalling in the nervous system: an overview. Trends Neurosci 32:19–29

Arai K, Lok J, Guo S, Hayakawa K, Xing C, Lo EH (2011) Cellular mechanisms of neurovascular damage and repair after stroke. J Child Neurol 26:1193–1198

Aronowski J, Strong R, Grotta JC (1997) Reperfusion injury: demonstration of brain damage produced by reperfusion after transient focal ischemia in rats. J Cereb Blood Flow Metab 17:1048–1056

Arroyo DS, Soria JA, Gaviglio EA, Garcia-Keller C, Cancela LM, Rodriguez-Galan MC, Wang JM, Iribarren P (2013) Toll-like receptor 2 ligands promote microglial cell death by inducing autophagy. FASEB J 27:299–312

Asahi M, Asahi K, Jung JC, del Zoppo GJ, Fini ME, Lo EH (2000) Role for matrix metalloproteinase 9 after focal cerebral ischemia: effects of gene knockout and enzyme inhibition with BB-94. J Cereb Blood Flow Metab 20:1681–1689

Astrup J, Symon L, Branston NM, Lassen NA (1977) Cortical evoked potential and extracellular K+ and H+ at critical levels of brain ischemia. Stroke 8:51–57

Atkinson SJ, Hosford MA, Molitoris BA (2004) Mechanism of actin polymerization in cellular ATP depletion. J Biol Chem 279:5194–5199

Back T, Hemmen T, Schüler OG (2004) Lesion evolution in cerebral ischemia. J Neurol 251:388–397

Balcaitis S, Weinstein JR, Li S, Chamberlain JS, Möller T (2005) Lentiviral transduction of microglial cells. Glia 50:48–55

Barone FC, Arvin B, White RF, Miller A, Webb CL, Willette RN, Lysko PG, Feuerstein GZ (1997) Tumor necrosis factor-alpha. A mediator of focal ischemic brain injury. Stroke 28:1233–1244

Biber K, Neumann H, Inoue K, Boddeke HWGM (2007) Neuronal "On" and "Off" signals control microglia. Trends Neurosci 30:596–602

Brenneman M, Sharma S, Harting M, Strong R, Cox CS, Aronowski J, Grotta JC, Savitz SI (2010) Autologous bone marrow mononuclear cells enhance recovery after acute ischemic stroke in young and middle-aged rats. J Cereb Blood Flow Metab 30:140–149

Brough D, Le Feuvre RA, Iwakura Y, Rothwell NJ (2002) Purinergic (P2X7) receptor activation of microglia induces cell death via an interleukin-1-independent mechanism. Mol Cell Neurosci 19:272–280

Brown GC, Neher JJ (2010) Inflammatory neurodegeneration and mechanisms of microglial killing of neurons. Mol Neurobiol 41:242–247

Cardona AE, Pioro EP, Sasse ME, Kostenko V, Cardona SM, Dijkstra IM, Huang D, Kidd G, Dombrowski S, Dutta R, Lee J-C, Cook DN, Jung S, Lira SA, Littman DR, Ransohoff RM (2006) Control of microglial neurotoxicity by the fractalkine receptor. Nat Neurosci 9:917–924

Cavaliere F, Florenzano F, Amadio S, Fusco FR, Viscomi MT, D'Ambrosi N, Vacca F, Sancesario G, Bernardi G, Molinari M, Volontè C (2003) Up-regulation of p2x2, p2x4 receptor and ischemic cell death: prevention by p2 antagonists. Neuroscience 120:85–98

Chen M, Ona VO, Li M, Ferrante RJ, Fink KB, Zhu S, Bian J, Guo L, Farrell LA, Hersch SM, Hobbs W, Vonsattel JP, Cha JH, Friedlander RM (2000) Minocycline inhibits caspase-1 and caspase-3 expression and delays mortality in a transgenic mouse model of Huntington disease. Nat Med 6:797–801

Chen H, Song YS, Chan PH (2009) Inhibition of NADPH oxidase is neuroprotective after ischemia-reperfusion. J Cereb Blood Flow Metab 29:1262–1272

Chen C, Leys D, Esquenazi A (2013) The interaction between neuropsychological and motor deficits in patients after stroke. Neurology 80:S27–34

Chock VY, Giffard RG (2005) Development of neonatal murine microglia in vitro: changes in response to lipopolysaccharide and ischemia-like injury. Pediatr Res 57:475–480

Chu K, Yin B, Wang J, Peng G, Liang H, Xu Z, Du Y, Fang M, Xia Q, Luo B (2012) Inhibition of P2X7 receptor ameliorates transient global cerebral ischemia/reperfusion injury via modulating inflammatory responses in the rat hippocampus. J Neuroinflammation 9:69

Cipriani R, Villa P, Chece G, Lauro C, Paladini A, Micotti E, Perego C, De Simoni M-G, Fredholm BB, Eusebi F, Limatola C (2011) CX3CL1 is neuroprotective in permanent focal cerebral ischemia in rodents. J Neurosci 31:16327–16335

Cumming TB, Marshall RS, Lazar RM (2013) Stroke, cognitive deficits, and rehabilitation: still an incomplete picture. Int J Stroke 8:38–45

Czeh M, Gressens P, Kaindl AM (2011) The yin and yang of microglia. Dev Neurosci 33:199–209

Davalos D, Grutzendler J, Yang G, Kim JV, Zuo Y, Jung S, Littman DR, Dustin ML, Gan WB (2005) ATP mediates rapid microglial response to local brain injury in vivo. Nat Neurosci 8:752–758

Del Zoppo GJ (2009) Relationship of neurovascular elements to neuron injury during ischemia. Cerebrovasc Dis 27(Suppl 1):65–76

Del Zoppo GJ, Milner R, Mabuchi T, Hung S, Wang X, Berg GI, Koziol JA (2007) Microglial activation and matrix protease generation during focal cerebral ischemia. Stroke 38:646–651

Denes A, Vidyasagar R, Feng J, Narvainen J, McColl BW, Kauppinen RA, Allan SM (2007) Proliferating resident microglia after focal cerebral ischaemia in mice. J Cereb Blood Flow Metab 27:1941–1953

Dénes A, Ferenczi S, Halász J, Környei Z, Kovács KJ (2008) Role of CX3CR1 (fractalkine receptor) in brain damage and inflammation induced by focal cerebral ischemia in mouse. J Cereb Blood Flow Metab 28:1707–1721

Dirnagl U, Iadecola C, Moskowitz MA (1999) Pathobiology of ischaemic stroke: an integrated view. Trends Neurosci 22:391–397

Durafourt BA, Lambert C, Johnson TA, Blain M, Bar-Or A, Antel JP (2011) Differential responses of human microglia and blood-derived myeloid cells to FTY720. J Neuroimmunol 230:10–16

Eyo U, Dailey ME (2012) Effects of oxygen–glucose deprivation on microglial mobility and viability in developing mouse hippocampal tissues. Glia 60:1747–1760

Eyo UB, Miner SA, Ahlers KE, Wu L-J, Dailey ME (2013) P2X7 receptor activation regulates microglial cell death during oxygen–glucose deprivation. Neuropharmacology 73C:311–319

Fagan SC, Cronic LE, Hess DC (2011) Minocycline development for acute ischemic stroke. Transl Stroke Res 2:202–208

Fang K-M, Yang C-S, Sun SH, Tzeng S-F (2009) Microglial phagocytosis attenuated by short-term exposure to exogenous ATP through P2X receptor action. J Neurochem 111:1225–1237

Faustino JV, Wang X, Johnson CE, Klibanov A, Derugin N, Wendland MF, Vexler ZS (2011) Microglial cells contribute to endogenous brain defenses after acute neonatal focal stroke. J Neurosci 31:12992–13001

Fricker M, Vilalta A, Tolkovsky AM, Brown GC (2013) Caspase inhibitors protect neurons by enabling selective necroptosis of inflamed microglia. J Biol Chem 288:9145–9152

Ginhoux F, Greter M, Leboeuf M, Nandi S, See P, Gokhan S, Mehler MF, Conway SJ, Ng LG, Stanley ER, Samokhvalov IM, Merad M (2010) Fate mapping analysis reveals that adult microglia derive from primitive macrophages. Science 330:841–845

Gomes-Leal W (2012) Microglial physiopathology: how to explain the dual role of microglia after acute neural disorders? Brain Behav 2:345–356

Gottlieb M, Matute C (1997) Expression of ionotropic glutamate receptor subunits in glial cells of the hippocampal CA1 area following transient forebrain ischemia. J Cereb Blood Flow Metab 17:290–300

Green AR (2008) Pharmacological approaches to acute ischaemic stroke: reperfusion certainly, neuroprotection possibly. Br J Pharmacol 153:S325–S338

Grilli M, Barbieri I, Basudev H, Brusa R, Casati C, Lozza G, Ongini E (2000) Interleukin-10 modulates neuronal threshold of vulnerability to ischaemic damage. Eur J Neurosci 12:2265–2272

Hanisch U-K, Kettenmann H (2007) Microglia: active sensor and versatile effector cells in the normal and pathologic brain. Nat Neurosci 10:1387–1394

Haynes SE, Hollopeter G, Yang G, Kurpius D, Dailey ME, Gan W-B, Julius D (2006) The P2Y12 receptor regulates microglial activation by extracellular nucleotides. Nat Neurosci 9:1512–1519

Hines DJ, Hines RM, Mulligan SJ, MacVicar BA (2009) Microglia processes block the spread of damage in the brain and require functional chloride channels. Glia 57:1610–1618

Hoek RM, Ruuls SR, Murphy CA, Wright GJ, Goddard R, Zurawski SM, Blom B, Homola ME, Streit WJ, Brown MH, Barclay AN, Sedgwick JD (2000) Down-regulation of the macrophage lineage through interaction with OX2 (CD200). Science 290:1768–1771

Hu X, Li P, Guo Y, Wang H, Leak RK, Chen S, Gao Y, Chen J (2012) Microglia/macrophage polarization dynamics reveal novel mechanism of injury expansion after focal cerebral ischemia. Stroke 43:3063–3070

Hwang BY, Appelboom G, Ayer A, Kellner CP, Kotchetkov IS, Gigante PR, Haque R, Kellner M, Connolly ES (2011) Advances in neuroprotective strategies: potential therapies for intracerebral hemorrhage. Cerebrovasc Dis 31:211–222

Iadecola C, Anrather J (2011) The immunology of stroke: from mechanisms to translation. Nat Med 17:796–808

Iadecola C, Pelligrino DA, Moskowitz MA, Lassen NA (1994) Nitric oxide synthase inhibition and cerebrovascular regulation. J Cereb Blood Flow Metab 14(2):175–192

Iadecola C, Zhang F, Xu X (1995) Inhibition of inducible nitric oxide synthase ameliorates cerebral ischemic damage. Am J Physiol 268:R286–292

Imai F, Suzuki H, Oda J, Ninomiya T, Ono K, Sano H, Sawada M (2007) Neuroprotective effect of exogenous microglia in global brain ischemia. J Cereb Blood Flow Metab 27:488–500

Jaffer H, Morris VB, Stewart D, Labhasetwar V (2011) Advances in stroke therapy. Drug Deliv Transl Res 1(6):409–419

Jantzie LL, Cheung P-Y, Todd KG (2005) Doxycycline reduces cleaved caspase-3 and microglial activation in an animal model of neonatal hypoxia–ischemia. J Cereb Blood Flow Metab 25:314–324

Jin WJ, Feng SW, Feng Z, Lu SM, Qi T, Qian YN (2014) Minocycline improves postoperative cognitive impairment in aged mice by inhibiting astrocytic activation. Neuroreport 25(1):1–6

Kilic U, Kilic E, Matter CM, Bassetti CL, Hermann DM (2008) TLR-4 deficiency protects against focal cerebral ischemia and axotomy-induced neurodegeneration. Neurobiol Dis 31:33–40

Kim BJ, Kim M-J, Park J-M, Lee S-H, Kim Y-J, Ryu S, Kim YH, Yoon B-W (2009) Reduced neurogenesis after suppressed inflammation by minocycline in transient cerebral ischemia in rat. J Neurol Sci 279:70–75

Kitamura Y, Takata K, Inden M, Tsuchiya D, Yanagisawa D (2004) Intracerebroventricular injection of microglia protects against focal brain ischemia. J Pharmacol Sci 94(2):203–206

Kitamura Y, Yanagisawa D, Inden M, Takata K, Tsuchiya D, Kawasaki T, Taniguchi T, Shimohama S (2005) Recovery of focal brain ischemia-induced behavioral dysfunction by intracerebroventricular injection of microglia. J Pharmacol Sci 97(2):289–293. Epub 2005 Jan 30

Kobayashi K, Imagama S, Ohgomori T, Hirano K, Uchimura K, Sakamoto K, Hirakawa A, Takeuchi H, Suzumura A, Ishiguro N, Kadomatsu K (2013) Minocycline selectively inhibits M1 polarization of microglia. Cell Death Dis 4:e525

Kohara A, Takahashi M, Yatsugi S-I, Tamura S, Shitaka Y, Hayashibe S, Kawabata S, Okada M (2008) Neuroprotective effects of the selective type 1 metabotropic glutamate receptor antagonist YM-202074 in rat stroke models. Brain Res 1191:168–179

Koistinaho M, Malm TM, Kettunen MI, Goldsteins G, Starckx S, Kauppinen RA, Opdenakker G, Koistinaho J (2005) Minocycline protects against permanent cerebral ischemia in wild type but not in matrix metalloprotease-9-deficient mice. J Cereb Blood Flow Metab 25:460–467

Konsman JP, Drukarch B, Van Dam A-M (2007) (Peri)vascular production and action of pro-inflammatory cytokines in brain pathology. Clin Sci (Lond) 112:1–25

Krafft PR, Bailey EL, Lekic T, Rolland WB, Altay O, Tang J, Wardlaw JM, Zhang JH, Sudlow CLM (2012) Etiology of stroke and choice of models. Int J Stroke 7:398–406

Kriz J (2006) Inflammation in ischemic brain injury: timing is important. Crit Rev Neurobiol 18(1–2):145–157

Kriz J, Lalancette-Hébert M (2009) Inflammation, plasticity and real-time imaging after cerebral ischemia. Acta Neuropathol 117:497–509

Lai AY, Todd KG (2008) Differential regulation of trophic and proinflammatory microglial effectors is dependent on severity of neuronal injury. Glia 56(3):259–270

Lalancette-Hébert M, Gowing G, Simard A, Weng YC, Kriz J (2007) Selective ablation of proliferating microglial cells exacerbates ischemic injury in the brain. J Neurosci 27:2596–2605

Lambertsen KL, Clausen BH, Babcock AA, Gregersen R, Fenger C, Nielsen HH, Haugaard LS, Wirenfeldt M, Nielsen M, Dagnaes-Hansen F, Bluethmann H, Faergeman NJ, Meldgaard M, Deierborg T, Finsen B (2009) Microglia protect neurons against ischemia by synthesis of tumor necrosis factor. J Neurosci 29:1319–1330

Lee H, Park J-W, Kim S-P, Lo EH, Lee S-R (2009) Doxycycline inhibits matrix metalloproteinase-9 and laminin degradation after transient global cerebral ischemia. Neurobiol Dis 34:189–198

Lehrmann E, Kiefer R, Christensen T, Toyka KV, Zimmer J, Diemer NH, Hartung HP, Finsen B (1998) Microglia and macrophages are major sources of locally produced transforming growth factor-beta1 after transient middle cerebral artery occlusion in rats. Glia 24:437–448

Lindsberg PJ, Strbian D, Karjalainen-Lindsberg M-L (2010) Mast cells as early responders in the regulation of acute blood–brain barrier changes after cerebral ischemia and hemorrhage. J Cereb Blood Flow Metab 30:689–702

Lipton P (1999) Ischemic cell death in brain neurons. Physiol Rev 79:1431–1568

Liu D, Yang PS (2012) Minocycline hydrochloride nanoliposomes inhibit the production of TNF-α in LPS-stimulated macrophages. Int J Nanomedicine 7:4769–4775

Lo EH (2008a) Experimental models, neurovascular mechanisms and translational issues in stroke research. Br J Pharmacol 153:S396–S405

Lo EH (2008b) A new penumbra: transitioning from injury into repair after stroke. Nat Med 14:497–500

Lo EH, Dalkara T, Moskowitz MA (2003) Mechanisms, challenges and opportunities in stroke. Nat Rev Neurosci 4:399–415

Lyons SA, Kettenmann H (1998) Oligodendrocytes and microglia are selectively vulnerable to combined hypoxia and hypoglycemia injury in vitro. J Cereb Blood Flow Metab 18:521–530

Mabuchi T, Kitagawa K, Ohtsuki T, Kuwabara K, Yagita Y, Yanagihara T, Hori M, Matsumoto M (2000) Contribution of microglia/macrophages to expansion of infarction and response of oligodendrocytes after focal cerebral ischemia in rats. Stroke 31:1735–1743

Madinier A, Bertrand N, Mossiat C, Prigent-Tessier A, Beley A, Marie C, Garnier P (2009) Microglial involvement in neuroplastic changes following focal brain ischemia in rats. PLoS One 4:e8101

Mander P, Borutaite V, Moncada S, Brown GC (2005) Nitric oxide from inflammatory-activated glia synergizes with hypoxia to induce neuronal death. J Neurosci Res 79:208–215

Masuda T, Croom D, Hida H, Kirov SA (2011) Capillary blood flow around microglial somata determines dynamics of microglial processes in ischemic conditions. Glia 59:1744–1753

Matsumoto H, Kumon Y, Watanabe H, Ohnishi T, Takahashi H, Imai Y, Tanaka J (2007) Expression of CD200 by macrophage-like cells in ischemic core of rat brain after transient middle cerebral artery occlusion. Neurosci Lett 418:44–48

Melani A, Amadio S, Gianfriddo M, Vannucchi MG, Volontè C, Bernardi G, Pedata F, Sancesario G (2006) P2X7 receptor modulation on microglial cells and reduction of brain infarct caused by middle cerebral artery occlusion in rat. J Cereb Blood Flow Metab 26:974–982

Michelucci A, Heurtaux T, Grandbarbe L, Morga E, Heuschling P (2009) Characterization of the microglial phenotype under specific pro-inflammatory and anti-inflammatory conditions: effects of oligomeric and fibrillar amyloid-beta. J Neuroimmunol 210:3–12

Mizuno T, Kawanokuchi J, Numata K, Suzumura A (2003) Production and neuroprotective functions of fractalkine in the central nervous system. Brain Res 979:65–70

Monif M, Reid CA, Powell KL, Smart ML, Williams DA (2009) The P2X7 receptor drives microglial activation and proliferation: a trophic role for P2X7R pore. J Neurosci 29:3781–91

Morrison HW, Filosa JA (2013) A quantitative spatiotemporal analysis of microglia morphology during ischemic stroke and reperfusion. J Neuroinflammation 10:4

Moskowitz MA, Lo EH, Iadecola C (2010) The science of stroke: mechanisms in search of treatments. Neuron 67:181–198

Neumann J, Sauerzweig S, Rönicke R, Gunzer F, Dinkel K, Ullrich O, Gunzer M, Reymann KG (2008) Microglia cells protect neurons by direct engulfment of invading neutrophil granulocytes: a new mechanism of CNS immune privilege. J Neurosci 28:5965–5975

Ohsawa K, Irino Y, Nakamura Y, Akazawa C, Inoue K, Kohsaka S (2007) Involvement of P2X4 and P2Y12 receptors in ATP-induced microglial chemotaxis. Glia 55:604–616

Ohsawa K, Irino Y, Sanagi T, Nakamura Y, Suzuki E, Inoue K, Kohsaka S (2010) P2Y12 receptor-mediated integrin-beta1 activation regulates microglial process extension induced by ATP. Glia 58:790–801

Ohtaki H, Ylostalo JH, Foraker JE, Robinson AP, Reger RL, Shioda S, Prockop DJ (2008) Stem/progenitor cells from bone marrow decrease neuronal death in global ischemia by modulation of inflammatory/immune responses. Proc Natl Acad Sci U S A 105:14638–14643

Olah M, Biber K, Vinet J, Boddeke HWGM (2011) Microglia phenotype diversity. CNS Neurol Disord Drug Targets 10:108–118

Orr AG, Orr AL, Li X-J, Gross RE, Traynelis SF (2009) Adenosine A2A receptor mediates microglial process retraction. Nat Neurosci 12:872–878

Pan J, Konstas A-A, Bateman B, Ortolano GA, Pile-Spellman J (2007) Reperfusion injury following cerebral ischemia: pathophysiology, MR imaging, and potential therapies. Neuroradiology 49:93–102

Perego C, Fumagalli S, De Simoni M-G (2011) Temporal pattern of expression and colocalization of microglia/macrophage phenotype markers following brain ischemic injury in mice. J Neuroinflammation 8:174

Pfefferkorn T, Rosenberg GA (2003) Closure of the blood–brain barrier by matrix metalloproteinase inhibition reduces rtPA-mediated mortality in cerebral ischemia with delayed reperfusion. Stroke 34:2025–2030

Pocock JM, Kettenmann H (2007) Neurotransmitter receptors on microglia. Trends Neurosci 30:527–535

Prinz M, Priller J, Sisodia SS, Ransohoff RM (2011) Heterogeneity of CNS myeloid cells and their roles in neurodegeneration. Nat Neurosci 14:1227–1235

Ramos AB, Vasconcelos-Dos-Santos A, Lopes de Souza SA, Rosado-de-Castro PH, Barbosa da Fonseca LM, Gutfilen B, Cintra WM, Mendez-Otero R (2013) Bone-marrow mononuclear cells reduce neurodegeneration in hippocampal CA1 layer after transient global ischemia in rats. Brain Res 1522:1–11

Rothwell NJ, Loddick SA, Stroemer P (1997) Interleukins and cerebral ischaemia. Int Rev Neurobiol 40:281–298

Saijo K, Glass CK (2011) Microglial cell origin and phenotypes in health and disease. Nat Rev Immunol 11:775–787

Schilling M, Besselmann M, Leonhard C, Mueller M, Ringelstein EB, Kiefer R (2003) Microglial activation precedes and predominates over macrophage infiltration in transient focal cerebral ischemia: a study in green fluorescent protein transgenic bone marrow chimeric mice. Exp Neurol 183:25–33

Schilling M, Besselmann M, Müller M, Strecker JK, Ringelstein EB, Kiefer R (2005) Predominant phagocytic activity of resident microglia over hematogenous macrophages following transient focal cerebral ischemia: an investigation using green fluorescent protein transgenic bone marrow chimeric mice. Exp Neurol 196:290–297

Soriano SG, Amaravadi LS, Wang YF, Zhou H, Yu GX, Tonra JR, Fairchild-Huntress V, Fang Q, Dunmore JH, Huszar D, Pan Y (2002) Mice deficient in fractalkine are less susceptible to cerebral ischemia-reperfusion injury. J Neuroimmunol 125:59–65

Surace MJ, Block ML (2012) Targeting microglia-mediated neurotoxicity: the potential of NOX2 inhibitors. Cell Mol Life Sci 69:2409–2427

Suzuki T, Hide I, Ido K, Kohsaka S, Inoue K, Nakata Y (2004) Production and release of neuroprotective tumor necrosis factor by P2X7 receptor-activated microglia. J Neurosci 24:1–7

Takeuchi H, Jin S, Suzuki H, Doi Y, Liang J, Kawanokuchi J, Mizuno T, Sawada M, Suzumura A (2008) Blockade of microglial glutamate release protects against ischemic brain injury. Exp Neurol 214:144–146

Tanaka R, Komine-Kobayashi M, Mochizuki H, Yamada M, Furuya T, Migita M, Shimada T, Mizuno Y, Urabe T (2003) Migration of enhanced green fluorescent protein expressing bone marrow-derived microglia/macrophage into the mouse brain following permanent focal ischemia. Neuroscience 117:531–539

Thored P, Heldmann U, Gomes-Leal W, Gisler R, Darsalia V, Taneera J, Nygren JM, Jacobsen S-EW, Ekdahl CT, Kokaia Z, Lindvall O (2009) Long-term accumulation of microglia with proneurogenic phenotype concomitant with persistent neurogenesis in adult subventricular zone after stroke. Glia 57:835–849

Tikka T, Fiebich BL, Goldsteins G, Keinanen R, Koistinaho J (2001) Minocycline, a tetracycline derivative, is neuroprotective against excitotoxicity by inhibiting activation and proliferation of microglia. J Neurosci 21:2580–2588

Tremblay M-È, Lowery RL, Majewska AK (2010) Microglial interactions with synapses are modulated by visual experience. PLoS Biol 8:e1000527

Wake H, Moorhouse AJ, Jinno S, Kohsaka S, Nabekura J (2009) Resting microglia directly monitor the functional state of synapses in vivo and determine the fate of ischemic terminals. J Neurosci 29:3974–3980

Walder CE, Green SP, Darbonne WC, Mathias J, Rae J, Dinauer MC, Curnutte JT, Thomas GR (1997) Ischemic stroke injury is reduced in mice lacking a functional NADPH oxidase. Stroke 28:2252–2258

Weinstein JR, Koerner IP, Möller T (2010) Microglia in ischemic brain injury. Future Neurol 5:227–246

Wen TC, Rogido M, Genetta T, Sola A (2004) Permanent focal cerebral ischemia activates erythropoietin receptor in the neonatal rat brain. Neurosci Lett 355:165–168

Xing C, Arai K, Lo EH, Hommel M (2012) Pathophysiologic cascades in ischemic stroke. Int J Stroke 7:378–385

Yan Y-P, Sailor KA, Lang BT, Park S-W, Vemuganti R, Dempsey RJ (2007) Monocyte chemoattractant protein-1 plays a critical role in neuroblast migration after focal cerebral ischemia. J Cereb Blood Flow Metab 27:1213–1224

Yanamadala V, Friedlander RM (2010) Complement in neuroprotection and neurodegeneration. Trends Mol Med 16:69–76

Yemisci M, Gursoy-Ozdemir Y, Vural A, Can A, Topalkara K, Dalkara T (2009) Pericyte contraction induced by oxidative-nitrative stress impairs capillary reflow despite successful opening of an occluded cerebral artery. Nat Med 15:1031–1037

Yenari MA, Giffard RG (2001) Ischemic vulnerability of primary murine microglial cultures. Neurosci Lett 298:5–8

Yenari MA, Xu L, Tang XN, Qiao Y, Giffard RG (2006) Microglia potentiate damage to blood–brain barrier constituents: improvement by minocycline in vivo and in vitro. Stroke 37:1087–1093

Yenari MA, Kauppinen TM, Swanson RA (2010) Microglial activation in stroke: therapeutic targets. Neurotherapeutics 7:378–391

Yrjänheikki J, Keinänen R, Pellikka M, Hökfelt T, Koistinaho J (1998) Tetracyclines inhibit microglial activation and are neuroprotective in global brain ischemia. Proc Natl Acad Sci U S A 95:15769–15774

Yrjänheikki J, Tikka T, Keinänen R, Goldsteins G, Chan PH, Koistinaho J (1999) A tetracycline derivative, minocycline, reduces inflammation and protects against focal cerebral ischemia with a wide therapeutic window. Proc Natl Acad Sci U S A 96:13496–13500

Zhang L, Zhang ZG, Chopp M (2012) The neurovascular unit and combination treatment strategies for stroke. Trends Pharmacol Sci 33:415–422

Zhao X, Haensel C, Araki E, Ross ME, Iadecola C (2000) Gene-dosing effect and persistence of reduction in ischemic brain injury in mice lacking inducible nitric oxide synthase. Brain Res 872:215–218

Zhao B-Q, Wang S, Kim H-Y, Storrie H, Rosen BR, Mooney DJ, Wang X, Lo EH (2006) Role of matrix metalloproteinases in delayed cortical responses after stroke. Nat Med 12:441–445

Zujovic V, Benavides J, Vigé X, Carter C, Taupin V (2000) Fractalkine modulates TNF-alpha secretion and neurotoxicity induced by microglial activation. Glia 29:305–315

Chapter 18
Neurodegenerative Diseases

Diego Gomez-Nicola and V. Hugh Perry

Abstract The role of microglial cells in neurodegenerative conditions is key to understanding the development and progression of the innate immune response during brain pathology. Although past and present research efforts have provided clues for understanding the contribution of microglia to neurodegeneration, the future offers new and exciting opportunities to study and modulate microglial biology in the degenerating brain. In this chapter we will summarize the main findings defining the role of microglia in neurodegenerative diseases, both in experimental animal models of disease and in studies with human brain tissue samples. We will also review the technical limitations to the study of microglia in neurodegenerative diseases and discuss the possible further lines of research to be pursued. In summary, we aim to provide a comprehensive picture of the role of microglial cells in the development, progression, and possible treatment of neurodegenerative diseases, to help build on the recent progress in this exciting field of neuroscience.

Keywords Microglia • Macrophages • Inflammation • Chronic neurodegeneration • Alzheimer's disease • Parkinson's disease • Amyotrophic lateral sclerosis • Prion disease • Huntington's disease

Bullet Points

- Evidence from genome-wide association and neuroimaging studies supports that inflammation is a driver of neurodegenerative diseases, rather than a consequence of ongoing pathology.
- Microglial activation is influenced by systemic inflammation, accelerating the progression of chronic neurodegeneration.
- The study of microglial biology in human samples of patients with chronic neurodegeneration is crucial, as current experimental models fail to reproduce all the pathological features observed in humans.

D. Gomez-Nicola, Ph.D. (✉) • V.H. Perry, Ph.D.
Centre for Biological Sciences, Southampton General Hospital, University of Southampton, South Laboratory and Pathology Block, MP840, Tremona Road, Southampton, Hampshire SO166YD, UK
e-mail: d.gomez-nicola@soton.ac.uk

M.-È. Tremblay, A. Sierra (eds.), *Microglia in Health and Disease*,
DOI 10.1007/978-1-4939-1429-6_18

- The composition and dynamics of the macrophages of the brain (microglia, perivascular macrophages, and meningeal macrophages) is not fully understood in neurodegenerative diseases, supporting further research in the field.

18.1 Introduction. Role of Microglia in Brain Pathology: Friend or Foe?

Microglial cells are active sensors of the disturbances in their microenvironment, capable of elaborating versatile responses to brain pathology (Kreutzberg 1996). These responses are neither linear, compartmentalized or binary, but plastic and multifaceted, i.e., finely tuned to the nature of the stimulus, the molecular repertoire that is engaged, and the prior state of the macrophage (Gordon 2003; Ransohoff and Perry 2009). When defining microglial activation during neurodegeneration, it is crucial to understand the temporal and spatial context of these diseases. Neurodegenerative diseases inexorably progress with the death of a slowly increasing number of neurons over years or even decades, to regional and cellular differential extents, determining the loss of disease-specific brain functions (memory, motor control, etc.). The slow degeneration of the neuronal components (synapses, axons, soma, and myelin sheath) may continue over many years, providing a stimulus that must lead to adaptive changes in the surrounding microglia. The adaptive changes may be then influenced by other comorbidities and systemic influences that communicate with the brain (Perry et al. 2007) (Fig. 18.1).

A recurrent interest in the studies of microglial involvement in neuropathology is the dichotomy between their contributions to neurodegeneration versus neuroprotection. The current literature has extensively reviewed this issue, presenting microglial cells as "friend or foe" or a "double-edged sword", trying to understand the determinants of the positive versus negative microglial contributions to brain pathology, with the goal of minimizing the harmful and favouring the beneficial (Crutcher et al. 2006; Popovich and Longbrake 2008). Achieving this ambitious objective becomes particularly difficult when taking into account the status of the immune-privilege of the brain that defines and tightly controls innate and acquired immune responses, and the influence of pathological processes from peripheral organs. For example, data from a genetically determined inflammatory demyelinating metabolic disorder, X-linked adrenoleukodystrophy, suggest a direct influence of the activation of microglia in the cerebral form of this disease (Eichler et al. 2008). Supporting the detrimental contribution of microglia to neurodegenerative diseases is the repeated observation that non-steroidal anti-inflammatory drugs (NSAIDs) delay the progression of Alzheimer's disease (AD) (Vlad et al. 2008) and Parkinson's disease (PD) (Chen et al. 2005), suggesting a major effect of prostaglandins and cyclooxygenase (COX) activation (Cunningham 2013). However, at the molecular level, the assumption that morphologically activated microglia show a classical pro-inflammatory phenotype, associated with the most detrimental effects during neurodegeneration, is no longer valid. On one hand, studies with transgenic models of AD report low gene expression of the pro-inflammatory cytokine interleukin-1β (IL-1β) or inducible nitric oxide synthase

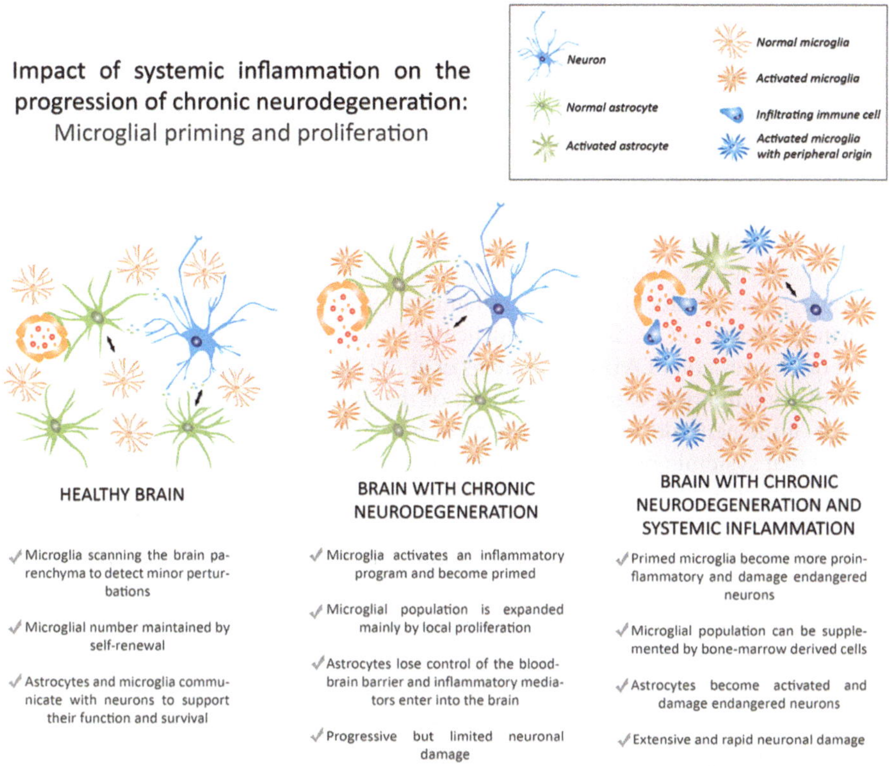

Fig. 18.1 Impact of systemic inflammation on the progression of chronic neurodegeneration: microglial priming and proliferation. Schematic representation of the cross-talk of microglial cells with neurons and astrocytes in the healthy brain (*left*), during chronic neurodegeneration (*middle*) and when chronic neurodegeneration is combined with a systemic inflammatory event (*right*). The legend for the different cell types and phenotypes is provided in the *top right corner*. Figure adapted from Gomez-Nicola and Perry (2014)

(iNOS), with low levels of protein production (Schwab et al. 2010). On the other, both mouse models and human cases of AD have been characterized to have limited pro-inflammatory polarization, but a consistent upregulation of anti-inflammatory markers, associated with potentially beneficial effects of microglia (Cunningham 2013).

Therefore, a detailed and pathology-specific definition of the microglial response needs to be completed, to fully understand the roles of microglia during neurodegeneration.

18.2 Alzheimer's Disease

AD is a chronic neurodegenerative disease and the most common form of dementia in Western countries. The clinical manifestations of AD include memory deficits, mood and behaviour changes, and disorientation. The pathological substrate of AD includes the generation of two abnormal structures: extracellular plaques or deposits of beta-amyloid (Aβ) protein and intracellular tangles of Tau protein. Despite a long-standing interest in the inflammatory response in AD, and the extensive research focused on understanding the role of microglia in this disease, the scientific community has failed to shed a clear and uniform light onto their contribution to the disease (Akiyama et al. 2000; Heneka and O'Banion 2007; Ransohoff and Perry 2009). The neuropathology of AD shows a robust innate immune response characterized by the presence of activated microglia, with increased or de novo expression of diverse macrophage antigens (Akiyama et al. 2000; Edison et al. 2008), and at least in some cases production of pro-inflammatory cytokines (Dickson et al. 1993; Fernandez-Botran et al. 2011). Evidence indicates that NSAIDs protect from the onset or progression of AD (Hoozemans et al. 2011), which is suggestive of the idea that inflammation is a causal component of the disease rather than simply a consequence of the neurodegeneration. There is, however, a growing body of evidence to suggest that systemic inflammation may interact with the innate immune response in the brain to act as a 'driver' of disease progression and exacerbate symptoms (Holmes et al. 2009, 2011) (Fig. 18.1). Studies in animal models show evidence of interactions between systemic inflammation and inflammation in the brain, and importantly provide biologically plausible mechanisms for its contribution to the progression of neurodegeneration. The impact of systemic inflammation means that any neuropathological studies on the inflammatory response in the AD brain must take into account systemic comorbidities that may influence the microglia phenotype.

The definition of the brain inflammatory profile of AD shows contrasting ideas in the literature, probably arising from the heterogeneity of the human postmortem samples and the difficult application of detection methods (for review see (Boche et al. 2013)). For example, AD has been associated with a pro-inflammatory phenotype, characterized by expression of interleukin IL-1β and complement proteins, with a direct association with Aβ plaques in human samples (Griffin et al. 1989, 1995; McGeer et al. 1989). By contrast, an upregulation of genes linked to an anti-inflammatory phenotype, arginase 1 or the transforming growth factor β (TGFβ), has been associated with AD in human samples and mouse models (Wang et al. 2003; Colton et al. 2006). However, the link with inflammation seems clear, as highlighted by a recent study on the gene signature of ageing and AD, using microarray technology (Cribbs et al. 2012). These results support the notion that an activation of the innate inflammatory response in microglia is a prelude to the subsequent development of AD (Cribbs et al. 2012). Furthermore, studies on incipient AD (iAD) post-mortem samples show a strong correlation between genes associated with the microglial response and the progression to AD (Blalock et al. 2004). The concept of the interconnection of AD and the innate immune response is further supported by evidence from genome-wide association studies (GWAS) implicating genes involved in innate immunity (Lambert et al. 2009). Recent studies link genetic variants of TREM2, a protein regulating the activation and phagocytic functions of myeloid cells, with the risk of developing AD (Guerreiro et al. 2013; Jonsson et al. 2013). TREM2 has been described to have a balancing role between phagocytic and pro-inflammatory microglial activities and is expressed in microglia around the plaques (Frank et al. 2008). Similarly, dysregulation of the complement system in humans has been associated with AD (McGeer and McGeer 2002; Lambert et al. 2009) and may influence the priming of microglia, defined as the conditioning of microglial response by a primary stimulus which results in an exaggerated response to a secondary challenge (Cunningham 2013). These promising studies are opening new avenues into the understanding of the impact of the innate immune response in AD, while supporting the need for future exploration.

The morphological activation of microglia is evident in transgenic mouse models of AD, which reproduce the deposition of Aβ (LaFerla and Oddo 2005; Perry et al. 2007; Jucker 2010), but their associated cytokine profile is by no means clear, with the changes in expression level compounded by the various detection methods (for review about research methods to study microglial biology see (Ransohoff and Perry 2009)). An alternative approach to pinpoint the contribution of microglia to the progression of AD is to study their impact on the Aβ plaque load. The plaque burden in AD increases with age, in both mouse models and human patients, indicating the rather ineffective phagocytic activity of microglia. Aβ deposits have been shown to exert a potent chemoattractant activity on microglia, although their removal by phagocytosis has not been clearly evidenced in vivo (for review see (Sierra et al. 2013)) (see Chap. 4 for additional reading). However, the removal of Aβ can be improved by further activation of microglia with bacterial lipopolysaccharides (LPS) challenge (Herber et al. 2004) or the induction of IL-1β (Shaftel et al. 2007). Microglial activation in neurodegeneration is accompanied by an increase in their density, yet further amplifying the effects of systemic inflammation on the brain. In addition, other brain macrophages, perivascular macrophages (PVMs) and meningeal macrophages (MMs), play a critical role in signalling from the periphery to the brain.

A significant body of literature suggests that bone marrow-derived macrophages (BMCs) infiltrate the AD brain, playing a leading role in the removal of Aβ, therefore complementing the poor phagocytic activity of microglia (Simard and Rivest 2006; Simard et al. 2006) (Fig. 18.2). The relative contribution of BMCs to the PVMs, MMs, or microglial pool is a matter of intense debate, and recent studies support a minor or even absent contribution of BMCs to the microglial population in a mouse model of AD (Mildner et al. 2011). Although BMCs recruitment has been demonstrated in experimental models with a complete, partial, or no detectable blood-brain barrier (BBB) disruption (Davoust et al. 2008), several studies point to in situ microglial proliferation as the mechanism regulating microglial turnover, with little or no contribution from circulating progenitors (Lawson et al. 1992; Prinz and Mildner 2011). Recent studies support the notion that microglia are maintained and function independently of BMCs in health (Ginhoux et al. 2010) and disease, as evidenced by models of demyelination, neurodegeneration, or axotomy (Ajami et al. 2007; Mildner et al. 2007, 2011). But analyzing PVMs, MMs, and microglial proliferation under pathological conditions with widespread chronic neurodegeneration is critical for understanding how innate inflammation contributes to the onset and progression of disease (Gomez-Nicola et al. 2014). Recent studies have highlighted the ability of PVMs to clear Aβ in experimental models of

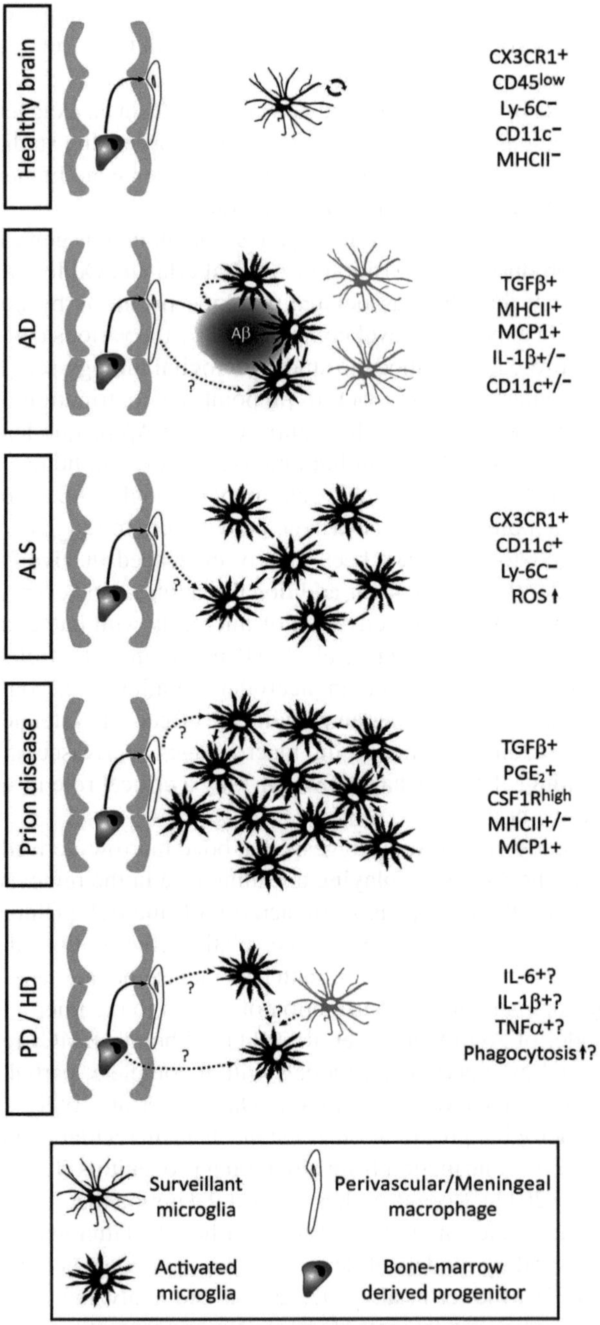

Fig. 18.2 Microglial cells constantly scan the parenchyma of the central nervous system (CNS), seeking to detect changes in the functional or structural integrity and maintain homeostasis. In a normal brain, the microglial population is maintained by self-renewal, while the perivascular

AD (Mildner et al. 2011) and show the need for a better understanding of the differential contribution of BMCs, MMs, and PVMs to the expansion of the microglial population, thereby providing a key link with systemic inflammation (Fig. 18.2).

Although proliferation was assumed to be responsible for the increased number of microglial cells observed in AD samples, direct evidence of proliferating microglial cells (Ki67, nuclear protein associated with the cell cycle, expression in Iba1+ microglia/macrophages) was reported only recently, together with the upregulation of the transcription factor PU.1 and the mitogen interleukin-34 (IL-34), key components of the pathway regulating microglial proliferation (Gomez-Nicola et al. 2013). Another determinant of microglial proliferation, colony-stimulating factor 1 receptor (CSF1R), has also been found to be upregulated in microglial cells from human postmortem samples of AD, indicating a prominent activity of this pathway (Akiyama et al. 1994). The expansion of the microglial population has been consistently documented in transgenic mouse models of AD, mainly accumulating around the plaques (Frautschy et al. 1998; Bolmont et al. 2008). However, direct evidence of microglial proliferation (incorporation of BrdU, a synthetic nucleoside analogue of thymidine, into Iba1+ cells; see Chap. 10 for more information) was only recently reported, suggesting direct effects of the plaque microenvironment on the regulation of microglial mitogenesis (Kamphuis et al. 2012). Therefore, these studies pinpoint the importance of controlling microglial proliferation during AD, offering new avenues for the regulation of the innate immune response in the brain.

18.3 Parkinson's Disease

PD is a chronic neurodegenerative disease characterised by tremor, rigidity, and slowness of movement. The pathological basis of PD includes the death of neurons in the substantia nigra pars compacta (SNpc) and the subsequent loss of dopaminergic tone in addition to a more widespread loss of neurons. The activation of the innate immune response during the progression of PD has been evidenced by the presence of morphologically activated microglia in human postmortem samples

Fig. 18.2 (continued) macrophages (PVMs) can be renewed by bone-marrow-derived progenitors. In Alzheimer's disease (AD) microglia proliferate and accumulate around Aβ plaques, participating in the removal of the misfolded protein, at which the PVMs are more efficient players. In AD, microglia are expanded without a contribution of circulating progenitors. Microglia are expanded and activated during the course of amyotrophic lateral sclerosis (ALS), without a contribution from circulating progenitors. In prion disease, the microglial population is expanded dramatically by local proliferation, being primed to give an exaggerated response to systemic inflammatory events. Little evidence is available regarding the expansion/renewal of the microglial population during Parkinson's or Huntington's diseases, or the dominant inflammatory phenotype. For all the considered neurodegenerative diseases, little evidence is available on the possible contribution of PVMs to the expansion/renewal of the microglial population. Figure adapted from Gomez-Nicola and Perry (2014)

(McGeer et al. 1988), and in vivo PET imaging, showing increased binding of the ligand [11C]PK-11195, considered as mainly labeling 'activated' microglia, without any correlation with the clinical symptoms, thus potentially dissociating microglial activation from the progression of the disease (Gerhard et al. 2006). Some studies suggest that microglia have a pro-inflammatory activation phenotype in PD, which is potentially driving neuronal injury (Mogi et al. 1994; Hunot et al. 1996), although no mechanistic study has yet addressed microglial contribution to the disease progression. The fact that PD has a late onset and that most studies analysed end-stage samples, representing a brain that has been suffering from the disease for many years, complicates the interpretation. Ageing alone has an impact on the phenotype of microglia, and systemic comorbidities, which can influence the microglial physiology, have not been taken into account in the previous studies focusing on PD (for review see (Perry 2012)). The clinical course of PD is often associated with other comorbidities, like chronic constipation or aspiration pneumonia, driving a peripheral inflammatory response that might impact the brain microglial responses and the progression of PD (for review see (Perry 2012); Fig. 18.1).

A significant body of knowledge regarding the role of microglial in PD comes from the study of experimental animal models. However, the current models fail to accurately reflect all aspects of the neuropathology of PD as described in humans. PD is characterised by a slowly evolving degeneration of the SNpc dopaminergic neurons, an aspect not replicated in the rodent models using either neurotoxic toxins or inflammatory challenges. The intracerebral use of neurotoxins, most commonly 6-hydroxydopamine (6OHDA), 1-methyl-4-phenyl-1,2,3,6-tetrahydropyridine (MPTP), or rotenone, provides a rapid degeneration (within a few days) of the SNpc dopaminergic neurons. Microglial activation has been described in the 6OHDA and MPTP models of PD (McGeer et al. 2003; Walsh et al. 2011), although limited information is available regarding the inflammatory phenotype of these cells, contrarily to their morphological features which have been described in details. However, gene-expression studies defining the inflammatory profile in these animal models of PD are numerous, but limited evidence is available regarding the protein expression levels and the particular roles of these molecules in the disease (Hirsch and Hunot 2009). Studies modulating microglial activity with minocycline, an antibiotic having anti-inflammatory actions, provided contrasting results, model-dependent, about the contribution of innate inflammation to the acute neurodegeneration of dopaminergic neurons in the SNpc (Wu et al. 2002; Sriram et al. 2006). Systemic inflammation, induced by IL-1β administration, was shown to impact the survival of dopaminergic neurons in the 6OHDA model, providing a clear evidence of the influence of immune-to-brain communication on the progression of PD (Pott Godoy et al. 2008). Even though the studies with toxin models shed some light onto the understanding of microglial reaction during neurodegeneration in the SNpc, the translation of these findings into neuroprotective or regenerative strategies appears premature and requires further understanding of the complexity of the innate immune response in postmortem samples from well-characterised PD patients.

Additionally, the generation of transgenic mouse models of PD, based on the identification of genes linked to familial PD, also provides a promising approach to model chronic neurodegeneration (Dawson et al. 2010). Transgenic over-expression of α-synuclein, a presynaptic neuronal protein that is linked genetically and neuropathologically to PD, leads to microglial activation and production of tumour necrosis factor alpha (TNF-α) in the SNpc (Su et al. 2008), although little neuronal death is observed. The injection of inflammatory agents, such as LPS, into the SNpc induces the selective loss of dopaminergic neurons, mimicking the neuropathology of PD, albeit acutely (Herrera et al. 2000). The LPS-induced neuronal death involves TNF-α, indicating a direct contribution of microglial activation (McCoy et al. 2006). Both transgenic and inflammatory models of PD can capture aspects of the disease, but fail to provide a comprehensive picture in which to address the roles of the innate immune response, in the context of a slowly evolving neurodegenerative condition. To summarize, the contribution of microglial cells to the onset or progression of PD is not yet established. Further research into the effect of systemic comorbidities and in refining the experimental animal models will help to understand the roles of the innate immune response in PD.

18.4 Amyotrophic Lateral Sclerosis

The pathogenesis of ALS involves the gain-of-function of a mutant protein, with approximately 5–10 % of the cases having a familial inheritable form. ALS is characterised by the progressive loss of motor neurons in the spinal cord, the brain stem, and the motor cortex, caused by increased levels of reactive oxygen species (ROS), leading to an increased muscle weakness and atrophy, with the malfunction of the muscles controlling respiratory functions being the most common cause of death. A significant proportion of the familial ALS subjects have mutations in the enzyme superoxide dismutase-1 (SOD-1), facilitating the modelling of the disease in experimental animals. Microglial cells produce pro-inflammatory molecules and ROS in the regions showing motor neurons degeneration, suggesting an impact on the progression of the disease (Troost et al. 1990; Clement et al. 2003). The modulation of microglial activity in ALS, either by removing the specific downregulatory influence of CX3CL1, mainly produced by neurons and signalling through microglial CX3CR1 (Cardona et al. 2006) or by lowering the levels of mutant SOD-1 in microglia (Boillee et al. 2006), indicates a detrimental role of these cells in the pathogenesis. These results were independently confirmed in microglia-devoid PU.1−/− mice, using a repopulation method with SOD-1-expressing bone-marrow cells, thus leading to a shortening of mouse survival (Beers et al. 2006). A recent report supports the detrimental role of microglia in the pathology of ALS, evidencing that microglia induce motor neuron death via activation of the NFκB pathway (Frakes et al. 2014). Microglial cells carrying the G93A SOD-1 mutation show exaggerated responses to stimulation with LPS and interferon-γ (IFNγ), associated with the activity of C/EBPβ, a transcription factor which regulates genes such as TNF-α, IL-1β, or iNOS (Valente et al. 2012). These results are suggestive of a priming effect on microglia, also taking place in ALS, which is supported by experiments using repeated systemic dosing with LPS on mice carrying the G37R SOD-1 mutation (Nguyen et al. 2004). Other approaches, aiming at removing the contribution of the microglial population from the equation, by either the transgenic expression of thymidine kinase (TK), leading to the "suicide" of proliferating CD11b+ cells (Gowing et al. 2008) (see also Chap. 6 for further discussion of this method) or by the administration of the non-specific blocker of mitosis Ara-C (Audet et al. 2012), indicated a neutral or benign role of microglia in ALS. However, the methods used in these studies lead to a massive and uncontrolled death of microglia, or to a switch of microglial phenotype following treatment with Ara-C (Gomez-Nicola et al. 2013), which does not provide a 'physiologically silent' way to address the contribution of microglial cells in the context of on-going neurodegeneration. Additionally, the activation of the TK transgene in CD11b-TK mice is achieved by administration of ganciclovir: this agent was recently identified to have a potent anti-proliferative impact on microglia during brain pathology (Ding et al. 2014). Other alternative approaches boosting the intrinsic proliferative activity of microglia with recombinant CSF1 have supported a detrimental role of microglia in the pathophysiology of ALS (Gowing et al. 2009), although these experiments also affected the contribution from CSF1-responsive peripheral cells.

In conclusion, the picture of the microglial contribution seems much clearer in ALS than in other neurodegenerative diseases, but further research is necessary to address the specific involvement of the different macrophage populations (peripheral, perivascular, microglia) (Fig. 18.2), in order to understand the mechanisms underlying neurodegeneration. Once again it is particularly important to precisely define the microglial phenotype, and other aspects of the innate inflammatory response, at different stages of disease evolution in mouse models as well as in clinically characterised postmortem human brain samples.

18.5 Prion Disease

Prion diseases, also known as transmissible spongiform encephalopathies (TSEs), are a group of rare progressive neurodegenerative disorders which show similar temporal and neuropathological profiles in both humans and animals models, characterised by depression and cognitive impairments followed by problems with motor control. The pathological substrate of prion diseases is the misfolding of the prion protein (PrP), leading to spongiform changes and neuronal loss. Murine models of prion disease are tractable laboratory models showing a slowly progressing chronic and fatal neurodegeneration, mimicking human prion diseases and presenting features in common with other neurodegenerative conditions. Prion-diseased brains show large numbers of microglia with a morphologically activated phenotype (Perry et al. 2010) and a cytokine profile similar to that of AD (Perry et al. 2002; Cunningham et al. 2003), with low levels of pro-inflammatory cytokines but high levels of TGF-β and prostaglandin E2 (PGE2), associated with a phagocytosing phenotype (Perry et al. 2002), although limited abilities for microglial removal of the misfolded prion protein have been described (Hughes et al. 2010). Following systemic inflammatory challenge the microglia are switched to adopt a pro-inflammatory phenotype (Cunningham et al. 2007, 2009), exacerbating the acute symptoms and accelerating the disease progression. Microglia in chronic neurodegeneration were therefore proposed to be 'primed' by the on-going pathology and then switched by systemic inflammation to produce tissue damaging inflammatory mediators (Perry et al. 2007) (Fig. 18.1). Previous studies in prion disease suggested that microglia arose from bone marrow precursors (BMPs), as evidenced by the use of bone marrow chimeras (Priller et al. 2006). However, recent studies have shown that the expansion of the microglial population during prion disease is maintained by local proliferation in a mouse model of prion disease (Gomez-Nicola et al. 2013) (Fig. 18.2). The expansion of the pool of parenchymal microglial cells is independent from the recruitment of circulating monocytes, while the population of PVMs is expanded by infiltrated cells, in a CCR2-dependent manner (Gomez-Nicola et al. 2014). Furthermore, microglial proliferation in prion disease is maintained by the activity of the CSF1R signaling pathway, and specific antagonism of the receptor, using either blocking antibodies or the selective CSF1R inhibitor GW2580, highlights the contribution of microglial cells as detrimental to the disease (Gomez-Nicola et al. 2013). A reduction in the numbers of proliferating microglia delayed the onset of behavioural deficits and modestly extended the time to terminal disease. The components of this mitogenic pathway are common to the human prion disease (variant Creutzfeldt-Jakob disease; vCJD) and to AD, suggesting that common pathways are controlling microglial proliferation and activation in chronic neurodegeneration (Gomez-Nicola et al. 2013). The analysis of the experimental models of prion disease offers an attractive perspective for the future, as they exhibit the main pathological features observed in many human neurodegenerative conditions (prion disease, AD, PD): protein misfolding, synaptic dysfunction, neurodegeneration, and an innate inflammatory reaction (for review see (Ransohoff and Perry 2009)).

18.6 Huntington's Disease

Huntington's disease (HD) is an inherited disorder characterized by the progressive degeneration of medium spiny striatal GABAergic interneurons, leading to a wide spectrum of clinical symptoms including impairment of movement control, cognitive deficits, and psychiatric symptoms. Progressive morphological activation of microglia and increase in their number has been evidenced in human brain from early pre-symptomatic stages of HD, suggestive of a causative role for these cells in the disease (Sapp et al. 2001; Tai et al. 2007). Binding studies of PK11195 using PET imaging in HD patients suggests that microglial activation correlates with the severity of the disease (Pavese et al. 2006), leading to the suggestion that they might provide a useful diagnostic tool to predict the disease onset (Politis et al. 2011). Recent studies further suggest that microglial activation during HD is a cell-autonomous mechanism, as mutant Huntingtin directly promotes the activation of microglial cells (Crotti et al. 2014).

Microglial activation can be exacerbated by systemic LPS administration in a mouse model of HD, however, without any impact on the neurological symptoms (Franciosi et al. 2012) (Fig. 18.1). The impact of systemic inflammatory events is clear during the progression of HD, as peripheral myeloid cells have been shown to produce altered levels of inflammatory cytokines (Trager and Tabrizi 2013; Trager et al. 2014). A detrimental contribution of microglia in HD has been suggested, through complement-mediated neuronal damage, although supporting mechanistic evidence is limited (Singhrao et al. 1999). Other in vitro studies have evidenced microglial proliferation and pro-inflammatory activation in HD, suggesting a reparative role in the removal of dysfunctional neurites at early and middle stages of the pathology (Kraft et al. 2012). Microglia have been shown to present defective chemotactic responses, in parallel with a reduced migration of immune cells, in a mouse model of HD (Kwan et al. 2012). The current evidence supports the idea that microglial cells are activated in HD, but the question of whether the innate immune response is a bystander consequence or whether they have a direct effect on the disease progression is still a matter of debate which needs further research (for review see (Moller 2010)).

18.7 Concluding Remarks

The role of microglia in neurodegenerative diseases and their contribution to the neuropathology is still a matter of intense debate. Although a significant effort has been directed towards unveiling the phenotype and activity of microglia in different models of experimental neurodegeneration, correlation with the human diseases is still to be determined. There is a need for improving the experimental models to better reflect the molecular neuropathology and inflammatory processes seen in human neurodegenerative diseases, in order to achieve a full understanding of the multi-faceted nature of chronic neurodegeneration. A more precise definition of the correlation between the clinical and molecular pathology and the systemic comorbidities in well-defined cohorts of patients will help us to understand the role of the innate immune response. With an increasing awareness of the potential importance of the innate immune response in neurodegenerative diseases, further research is needed in order to bridge the gap from bench to bedside, and finally understand the beautiful complexity of their cellular effectors, including particularly, the microglia.

References

Ajami B, Bennett JL, Krieger C, Tetzlaff W, Rossi FM (2007) Local self-renewal can sustain CNS microglia maintenance and function throughout adult life. Nat Neurosci 10:1538–1543

Akiyama H, Nishimura T, Kondo H, Ikeda K, Hayashi Y, McGeer PL (1994) Expression of the receptor for macrophage colony stimulating factor by brain microglia and its upregulation in brains of patients with Alzheimer's disease and amyotrophic lateral sclerosis. Brain Res 639:171–174

Akiyama H, Barger S, Barnum S, Bradt B, Bauer J, Cole GM et al (2000) Inflammation and Alzheimer's disease. Neurobiol Aging 21:383–421

Audet JN, Gowing G, Paradis R, Soucy G, Julien JP (2012) Ablation of proliferating cells in the CNS exacerbates motor neuron disease caused by mutant superoxide dismutase. PLoS One 7:e34932

Beers DR, Henkel JS, Xiao Q, Zhao W, Wang J, Yen AA et al (2006) Wild-type microglia extend survival in PU.1 knockout mice with familial amyotrophic lateral sclerosis. Proc Natl Acad Sci U S A 103:16021–16026

Blalock EM, Geddes JW, Chen KC, Porter NM, Markesbery WR, Landfield PW (2004) Incipient Alzheimer's disease: microarray correlation analyses reveal major transcriptional and tumor suppressor responses. Proc Natl Acad Sci U S A 101:2173–2178

Boche D, Perry VH, Nicoll JA (2013) Review: activation patterns of microglia and their identification in the human brain. Neuropathol Appl Neurobiol 39:3–18

Boillee S, Yamanaka K, Lobsiger CS, Copeland NG, Jenkins NA, Kassiotis G et al (2006) Onset and progression in inherited ALS determined by motor neurons and microglia. Science 312:1389–1392

Bolmont T, Haiss F, Eicke D, Radde R, Mathis CA, Klunk WE et al (2008) Dynamics of the microglial/amyloid interaction indicate a role in plaque maintenance. J Neurosci 28:4283–4292

Cardona AE, Pioro EP, Sasse ME, Kostenko V, Cardona SM, Dijkstra IM et al (2006) Control of microglial neurotoxicity by the fractalkine receptor. Nat Neurosci 9:917–924

Chen H, Jacobs E, Schwarzschild MA, McCullough ML, Calle EE, Thun MJ et al (2005) Nonsteroidal antiinflammatory drug use and the risk for Parkinson's disease. Ann Neurol 58:963–967

Clement AM, Nguyen MD, Roberts EA, Garcia ML, Boillee S, Rule M et al (2003) Wild-type nonneuronal cells extend survival of SOD1 mutant motor neurons in ALS mice. Science 302:113–117

Colton CA, Mott RT, Sharpe H, Xu Q, Van Nostrand WE, Vitek MP (2006) Expression profiles for macrophage alternative activation genes in AD and in mouse models of AD. J Neuroinflammation 3:27

Cribbs DH, Berchtold NC, Perreau V, Coleman PD, Rogers J, Tenner AJ et al (2012) Extensive innate immune gene activation accompanies brain aging, increasing vulnerability to cognitive decline and neurodegeneration: a microarray study. J Neuroinflammation 9:179

Crotti A, Benner C, Kerman BE, Gosselin D, Lagier-Tourenne C, Zuccato C et al (2014) Mutant Huntingtin promotes autonomous microglia activation via myeloid lineage-determining factors. Nat Neurosci 17:513–521

Crutcher KA, Gendelman HE, Kipnis J, Perez-Polo JR, Perry VH, Popovich PG et al (2006) Debate: "is increasing neuroinflammation beneficial for neural repair?". J Neuroimmune Pharmacol 1:195–211

Cunningham C (2013) Microglia and neurodegeneration: the role of systemic inflammation. Glia 61:71–90

Cunningham C, Deacon R, Wells H, Boche D, Waters S, Diniz CP et al (2003) Synaptic changes characterize early behavioural signs in the ME7 model of murine prion disease. Eur J Neurosci 17:2147–2155

Cunningham C, Campion S, Teeling J, Felton L, Perry VH (2007) The sickness behaviour and CNS inflammatory mediator profile induced by systemic challenge of mice with synthetic double-stranded RNA (poly I:C). Brain Behav Immun 21:490–502

Cunningham C, Campion S, Lunnon K, Murray CL, Woods JF, Deacon RM et al (2009) Systemic inflammation induces acute behavioral and cognitive changes and accelerates neurodegenerative disease. Biol Psychiatry 65:304–312

Davoust N, Vuaillat C, Androdias G, Nataf S (2008) From bone marrow to microglia: barriers and avenues. Trends Immunol 29:227–234

Dawson TM, Ko HS, Dawson VL (2010) Genetic animal models of Parkinson's disease. Neuron 66:646–661

Dickson DW, Lee SC, Mattiace LA, Yen SH, Brosnan C (1993) Microglia and cytokines in neurological disease, with special reference to AIDS and Alzheimer's disease. Glia 7:75–83

Ding Z, Mathur V, Ho PP, James ML, Lucin KM, Hoehne A et al (2014) Antiviral drug ganciclovir is a potent inhibitor of microglial proliferation and neuroinflammation. J Exp Med 211:189–198

Edison P, Archer HA, Gerhard A, Hinz R, Pavese N, Turkheimer FE et al (2008) Microglia, amyloid, and cognition in Alzheimer's disease: an [11C](R)PK11195-PET and [11C]PIB-PET study. Neurobiol Dis 32:412–419

Eichler FS, Ren JQ, Cossoy M, Rietsch AM, Nagpal S, Moser AB et al (2008) Is microglial apoptosis an early pathogenic change in cerebral X-linked adrenoleukodystrophy? Ann Neurol 63:729–742

Fernandez-Botran R, Ahmed Z, Crespo FA, Gatenbee C, Gonzalez J, Dickson DW et al (2011) Cytokine expression and microglial activation in progressive supranuclear palsy. Parkinsonism Relat Disord 17(9):683–688

Frakes AE, Ferraiuolo L, Haidet-Phillips AM, Schmelzer L, Braun L, Miranda CJ et al (2014) Microglia induce motor neuron death via the classical NF-kB pathway in amyotrophic lateral sclerosis. Neuron 81:1009–1023

Franciosi S, Ryu JK, Shim Y, Hill A, Connolly C, Hayden MR et al (2012) Age-dependent neurovascular abnormalities and altered microglial morphology in the YAC128 mouse model of Huntington disease. Neurobiol Dis 45:438–449

Frank S, Burbach GJ, Bonin M, Walter M, Streit W, Bechmann I et al (2008) TREM2 is upregulated in amyloid plaque-associated microglia in aged APP23 transgenic mice. Glia 56:1438–1447

Frautschy SA, Yang F, Irrizarry M, Hyman B, Saido TC, Hsiao K et al (1998) Microglial response to amyloid plaques in APPsw transgenic mice. Am J Pathol 152:307–317

Gerhard A, Pavese N, Hotton G, Turkheimer F, Es M, Hammers A et al (2006) In vivo imaging of microglial activation with [11C](R)-PK11195 PET in idiopathic Parkinson's disease. Neurobiol Dis 21:404–412

Ginhoux F, Greter M, Leboeuf M, Nandi S, See P, Gokhan S et al (2010) Fate mapping analysis reveals that adult microglia derive from primitive macrophages. Science 330:841–845

Gomez-Nicola D, Perry VH (2014) Microglial dynamics and role in the healthy and diseased brain: a paradigm of functional plasticity. Neuroscientist [2014 Apr 10, Epub ahead of print]

Gomez-Nicola D, Fransen NL, Suzzi S, Perry VH (2013) Regulation of microglial proliferation during chronic neurodegeneration. J Neurosci 33:2481–2493

Gomez-Nicola D, Schetters ST, Perry VH (2014) Differential role of CCR2 in the dynamics of microglia and perivascular macrophages during prion disease. Glia 62(7):1041–1052

Gordon S (2003) Alternative activation of macrophages. Nat Rev Immunol 3:23–35

Gowing G, Philips T, Van Wijmeersch B, Audet JN, Dewil M, Van Den Bosch L et al (2008) Ablation of proliferating microglia does not affect motor neuron degeneration in amyotrophic lateral sclerosis caused by mutant superoxide dismutase. J Neurosci 28:10234–10244

Gowing G, Lalancette-Hebert M, Audet JN, Dequen F, Julien JP (2009) Macrophage colony stimulating factor (M-CSF) exacerbates ALS disease in a mouse model through altered responses of microglia expressing mutant superoxide dismutase. Exp Neurol 220:267–275

Griffin WS, Stanley LC, Ling C, White L, MacLeod V, Perrot LJ et al (1989) Brain interleukin 1 and S-100 immunoreactivity are elevated in Down syndrome and Alzheimer disease. Proc Natl Acad Sci U S A 86:7611–7615

Griffin WS, Sheng JG, Roberts GW, Mrak RE (1995) Interleukin-1 expression in different plaque types in Alzheimer's disease: significance in plaque evolution. J Neuropathol Exp Neurol 54:276–281

Guerreiro R, Wojtas A, Bras J, Carrasquillo M, Rogaeva E, Majounie E et al (2013) TREM2 variants in Alzheimer's disease. N Engl J Med 368:117–127

Heneka MT, O'Banion MK (2007) Inflammatory processes in Alzheimer's disease. J Neuroimmunol 184:69–91

Herber DL, Roth LM, Wilson D, Wilson N, Mason JE, Morgan D et al (2004) Time-dependent reduction in Abeta levels after intracranial LPS administration in APP transgenic mice. Exp Neurol 190:245–253

Herrera AJ, Castano A, Venero JL, Cano J, Machado A (2000) The single intranigral injection of LPS as a new model for studying the selective effects of inflammatory reactions on dopaminergic system. Neurobiol Dis 7:429–447

Hirsch EC, Hunot S (2009) Neuroinflammation in Parkinson's disease: a target for neuroprotection? Lancet Neurol 8:382–397

Holmes C, Cunningham C, Zotova E, Woolford J, Dean C, Kerr S et al (2009) Systemic inflammation and disease progression in Alzheimer disease. Neurology 73:768–774

Holmes C, Cunningham C, Zotova E, Culliford D, Perry VH (2011) Proinflammatory cytokines, sickness behavior, and Alzheimer disease. Neurology 77:212–218

Hoozemans JJ, Veerhuis R, Rozemuller JM, Eikelenboom P (2011) Soothing the inflamed brain: effect of non-steroidal anti-inflammatory drugs on Alzheimer's disease pathology. CNS Neurol Disord Drug Targets 10:57–67

Hughes MM, Field RH, Perry VH, Murray CL, Cunningham C (2010) Microglia in the degenerating brain are capable of phagocytosis of beads and of apoptotic cells, but do not efficiently remove PrPSc, even upon LPS stimulation. Glia 58:2017–2030

Hunot S, Boissiere F, Faucheux B, Brugg B, Mouatt-Prigent A, Agid Y et al (1996) Nitric oxide synthase and neuronal vulnerability in Parkinson's disease. Neuroscience 72:355–363

Jonsson T, Stefansson H, Steinberg S, Jonsdottir I, Jonsson PV, Snaedal J et al (2013) Variant of TREM2 associated with the risk of Alzheimer's disease. N Engl J Med 368:107–116

Jucker M (2010) The benefits and limitations of animal models for translational research in neurodegenerative diseases. Nat Med 16:1210–1214

Kamphuis W, Orre M, Kooijman L, Dahmen M, Hol EM (2012) Differential cell proliferation in the cortex of the APPswePS1dE9 Alzheimer's disease mouse model. Glia 60:615–629

Kraft AD, Kaltenbach LS, Lo DC, Harry GJ (2012) Activated microglia proliferate at neurites of mutant huntingtin-expressing neurons. Neurobiol Aging 33(621):e17–e33

Kreutzberg GW (1996) Microglia: a sensor for pathological events in the CNS. Trends Neurosci 19:312–318

Kwan W, Trager U, Davalos D, Chou A, Bouchard J, Andre R et al (2012) Mutant huntingtin impairs immune cell migration in Huntington disease. J Clin Invest 122:4737–4747

LaFerla FM, Oddo S (2005) Alzheimer's disease: Abeta, tau and synaptic dysfunction. Trends Mol Med 11:170–176

Lambert JC, Heath S, Even G, Campion D, Sleegers K, Hiltunen M et al (2009) Genome-wide association study identifies variants at CLU and CR1 associated with Alzheimer's disease. Nat Genet 41:1094–1099

Lawson LJ, Perry VH, Gordon S (1992) Turnover of resident microglia in the normal adult mouse brain. Neuroscience 48:405–415

McCoy MK, Martinez TN, Ruhn KA, Szymkowski DE, Smith CG, Botterman BR et al (2006) Blocking soluble tumor necrosis factor signaling with dominant-negative tumor necrosis factor inhibitor attenuates loss of dopaminergic neurons in models of Parkinson's disease. J Neurosci 26:9365–9375

McGeer PL, McGeer EG (2002) The possible role of complement activation in Alzheimer disease. Trends Mol Med 8:519–523

McGeer PL, Itagaki S, Boyes BE, McGeer EG (1988) Reactive microglia are positive for HLA-DR in the substantia nigra of Parkinson's and Alzheimer's disease brains. Neurology 38:1285–1291

McGeer PL, Akiyama H, Itagaki S, McGeer EG (1989) Activation of the classical complement pathway in brain tissue of Alzheimer patients. Neurosci Lett 107:341–346

McGeer PL, Schwab C, Parent A, Doudet D (2003) Presence of reactive microglia in monkey substantia nigra years after 1-methyl-4-phenyl-1,2,3,6-tetrahydropyridine administration. Ann Neurol 54:599–604

Mildner A, Schmidt H, Nitsche M, Merkler D, Hanisch UK, Mack M et al (2007) Microglia in the adult brain arise from Ly-6ChiCCR2+ monocytes only under defined host conditions. Nat Neurosci 10:1544–1553

Mildner A, Schlevogt B, Kierdorf K, Bottcher C, Erny D, Kummer MP et al (2011) Distinct and non-redundant roles of microglia and myeloid subsets in mouse models of Alzheimer's disease. J Neurosci 31:11159–11171

Mogi M, Harada M, Kondo T, Riederer P, Inagaki H, Minami M et al (1994) Interleukin-1 beta, interleukin-6, epidermal growth factor and transforming growth factor-alpha are elevated in the brain from Parkinsonian patients. Neurosci Lett 180:147–150

Moller T (2010) Neuroinflammation in Huntington's disease. J Neural Transm 117:1001–1008
Nguyen MD, D'Aigle T, Gowing G, Julien JP, Rivest S (2004) Exacerbation of motor neuron disease by chronic stimulation of innate immunity in a mouse model of amyotrophic lateral sclerosis. J Neurosci 24:1340–1349
Pavese N, Gerhard A, Tai YF, Ho AK, Turkheimer F, Barker RA et al (2006) Microglial activation correlates with severity in Huntington disease: a clinical and PET study. Neurology 66:1638–1643
Perry VH (2012) Innate inflammation in Parkinson's disease. Cold Spring Harb Perspect Med 2:a009373
Perry VH, Cunningham C, Boche D (2002) Atypical inflammation in the central nervous system in prion disease. Curr Opin Neurol 15:349–354
Perry VH, Cunningham C, Holmes C (2007) Systemic infections and inflammation affect chronic neurodegeneration. Nat Rev Immunol 7:161–167
Perry VH, Nicoll JA, Holmes C (2010) Microglia in neurodegenerative disease. Nat Rev Neurol 6:193–201
Politis M, Pavese N, Tai YF, Kiferle L, Mason SL, Brooks DJ et al (2011) Microglial activation in regions related to cognitive function predicts disease onset in Huntington's disease: a multimodal imaging study. Hum Brain Mapp 32:258–270
Popovich PG, Longbrake EE (2008) Can the immune system be harnessed to repair the CNS? Nat Rev Neurosci 9:481–493
Pott Godoy MC, Tarelli R, Ferrari CC, Sarchi MI, Pitossi FJ (2008) Central and systemic IL-1 exacerbates neurodegeneration and motor symptoms in a model of Parkinson's disease. Brain 131:1880–1894
Priller J, Prinz M, Heikenwalder M, Zeller N, Schwarz P, Heppner FL et al (2006) Early and rapid engraftment of bone marrow-derived microglia in scrapie. J Neurosci 26:11753–11762
Prinz M, Mildner A (2011) Microglia in the CNS: immigrants from another world. Glia 59:177–187
Ransohoff RM, Perry VH (2009) Microglial physiology: unique stimuli, specialized responses. Annu Rev Immunol 27:119–145
Sapp E, Kegel KB, Aronin N, Hashikawa T, Uchiyama Y, Tohyama K et al (2001) Early and progressive accumulation of reactive microglia in the Huntington disease brain. J Neuropathol Exp Neurol 60:161–172
Schwab C, Klegeris A, McGeer PL (2010) Inflammation in transgenic mouse models of neurodegenerative disorders. Biochim Biophys Acta 1802:889–902
Shaftel SS, Kyrkanides S, Olschowka JA, Miller JN, Johnson RE, O'Banion MK (2007) Sustained hippocampal IL-1 beta overexpression mediates chronic neuroinflammation and ameliorates Alzheimer plaque pathology. J Clin Invest 117:1595–1604
Sierra A, Abiega O, Shahraz A, Neumann H (2013) Janus-faced microglia: beneficial and detrimental consequences of microglial phagocytosis. Front Cell Neurosci 7:6
Simard AR, Rivest S (2006) Neuroprotective properties of the innate immune system and bone marrow stem cells in Alzheimer's disease. Mol Psychiatry 11:327–335
Simard AR, Soulet D, Gowing G, Julien JP, Rivest S (2006) Bone marrow-derived microglia play a critical role in restricting senile plaque formation in Alzheimer's disease. Neuron 49:489–502
Singhrao SK, Neal JW, Morgan BP, Gasque P (1999) Increased complement biosynthesis by microglia and complement activation on neurons in Huntington's disease. Exp Neurol 159:362–376
Sriram K, Miller DB, O'Callaghan JP (2006) Minocycline attenuates microglial activation but fails to mitigate striatal dopaminergic neurotoxicity: role of tumor necrosis factor-alpha. J Neurochem 96:706–718
Su X, Maguire-Zeiss KA, Giuliano R, Prifti L, Venkatesh K, Federoff HJ (2008) Synuclein activates microglia in a model of Parkinson's disease. Neurobiol Aging 29:1690–1701
Tai YF, Pavese N, Gerhard A, Tabrizi SJ, Barker RA, Brooks DJ et al (2007) Microglial activation in presymptomatic Huntington's disease gene carriers. Brain 130:1759–1766

Trager U, Tabrizi SJ (2013) Peripheral inflammation in neurodegeneration. J Mol Med (Berl) 91:673–681

Trager U, Andre R, Lahiri N, Magnusson-Lind A, Weiss A, Grueninger S et al (2014) HTT-lowering reverses Huntington's disease immune dysfunction caused by NFkappaB pathway dysregulation. Brain 137:819–833

Troost D, Van den Oord JJ, Vianney de Jong JM (1990) Immunohistochemical characterization of the inflammatory infiltrate in amyotrophic lateral sclerosis. Neuropathol Appl Neurobiol 16:401–410

Valente T, Mancera P, Tusell JM, Serratosa J, Saura J (2012) C/EBPbeta expression in activated microglia in amyotrophic lateral sclerosis. Neurobiol Aging 33:2186–2199

Vlad SC, Miller DR, Kowall NW, Felson DT (2008) Protective effects of NSAIDs on the development of Alzheimer disease. Neurology 70:1672–1677

Walsh S, Finn DP, Dowd E (2011) Time-course of nigrostriatal neurodegeneration and neuroinflammation in the 6-hydroxydopamine-induced axonal and terminal lesion models of Parkinson's disease in the rat. Neuroscience 175:251–261

Wang G, Zhang Y, Chen B, Cheng J (2003) Preliminary studies on Alzheimer's disease using cDNA microarrays. Mech Ageing Dev 124:115–124

Wu DC, Jackson-Lewis V, Vila M, Tieu K, Teismann P, Vadseth C et al (2002) Blockade of microglial activation is neuroprotective in the 1-methyl-4-phenyl-1,2,3,6-tetrahydropyridine mouse model of Parkinson disease. J Neurosci 22:1763–1771

Chapter 19
Spinal Cord and Brain Trauma

Samuel David and Phillip G. Popovich

Abstract Microglia are the first responders to central nervous system (CNS) trauma. They are thought to be a central player in initiating and orchestrating a cascade of changes that lead to an elaborate inflammatory response at the site of injury. This inflammatory response also includes the influx of peripheral immune cells into the injured CNS. In this chapter we will focus our attention on the microglial response to spinal cord injury, but where possible we will include discussion of microglial responses after traumatic brain injury. We will discuss the differences in the early and late responses of microglia to CNS injury; the signaling molecules, cytokines and other factors that modulate their responses, the evidence for their beneficial and detrimental effects, and the effects of their activation at the epicenter of the injury and in sites distal to the injury. Attention will also be focused on the evidence of microglial changes in chronic spinal cord and brain trauma. We will highlight the preclinical evidence for targeting some aspects of the microglial response to treat spinal cord and brain trauma and take a look at some future directions to pursue.

Keywords Microglia • Spinal cord injury • Traumatic brain injury • Purinergic signaling • Damage-associated molecular patterns • Toll-like receptors • Cytokines • Free radicals • Wallerian degeneration • Regenerative sprouting

Bullet Points

- Microglia respond within minutes by sending processes towards the injury.
- The early microglial response to micro-lesions is protective and mediated by ATP-purinergic receptor, damage-associated molecular patterns (DAMPs)-Toll-like receptors (TLRs) signaling, and other factors.

S. David, Ph.D. (✉)
Center for Research in Neuroscience, The Research Institute of the McGill University Health Center, 1650 Cedar Avenue, Montreal, QC, Canada H3G 1A4
e-mail: sam.david@mcgill.ca

P.G. Popovich, Ph.D.
Department of Neuroscience, Center for Brain and Spinal Cord Repair, The Ohio State University Wexner Medical Center, 694 Biomedical Research Tower 460 W 12th Avenue, Columbus, OH 43210, USA
e-mail: phillip.popovich@osumc.edu; Popovich.2@osu.edu

M.-È. Tremblay, A. Sierra (eds.), *Microglia in Health and Disease*,
DOI 10.1007/978-1-4939-1429-6_19

- DAMP-TLR signaling in microglia leads to pro-inflammatory cytokine expression.
- Microglia, like peripheral macrophages entering the injured spinal cord, become polarized into cytotoxic M1 and non-cytotoxic M2 states.
- M1 polarization is predominant in the injured spinal cord.
- Microglia respond very slowly in areas of CNS white matter undergoing Wallerian degeneration the reasons for which are not fully understood.
- Microglia in grey matter regions far removed from the lesion site also respond.
- Microglia play a role in synaptic pruning and reorganization of circuits after regenerative sprouting.
- Chronic activation of microglia can contribute to pain and functional disability.
- Chronic microglial activation generates free radicals and other toxic mediators that can impair neurological function even years after injury.

19.1 Introduction

Microglia, which are the resident tissue macrophages of the central nervous system (CNS), are derived from the yolk sac and populate the CNS during early embryonic development (Ginhoux et al. 2010). They are widely distributed throughout the CNS (Lawson et al. 1990; Vela et al. 1995). Their number and distribution pattern varies between different regions of the CNS and also within the same CNS region, e.g., in the cerebellum they are more numerous in the deep nuclei as compared to the molecular layer or granule cell layer (Lawson et al. 1990; Vela et al. 1995). In general they appear to be more abundant in the grey than the white matter (Lawson et al. 1990). Microglia, which are located in the CNS parenchyma, have a ramified morphology in the normal, uninjured CNS. In contrast, macrophages in the leptomeninges and choroid plexus, which share many antigenic markers with microglia, are rounded or ameboid in shape (David and Kroner 2011; Perry et al. 1985). The shape and extent of microglial ramification is also region-specific, with those in the cerebral cortex and hippocampus having extensively branched morphologies, while those in the white matter have long processes and slender-shaped cell bodies (Lawson et al. 1990). The overall shape of these cells appears to reflect the cytoarchitecture of the region in which they are found. Microglia in lower vertebrates such as zebrafish also have a similar branching shape (Sieger et al. 2012). In some grey matter regions of the mammalian CNS, microglia are so abundant that their processes almost touch each other to give the appearance of "tiling". This relatively close distribution means that microglia in these regions would not need to move far in response to perturbations, unlike those in the developing zebrafish that migrate to regions of damage (Sieger et al. 2012).

The characteristic branching morphology of microglia gives the appearance that these cells are surveying and monitoring the CNS tissue. Two-photon imaging of microglia in real-time in the normal mouse cerebral cortex and spinal cord white matter revealed rapid extension and retraction of microglial processes from 2 to 8 μm within seconds to minutes (Davalos et al. 2005; Dibaj et al. 2010) without movement of the cell soma (Dibaj et al. 2010). These data indicate that the cytoplasmic processes of these cells are dynamic and play an active role in constantly monitoring their microenvironment. Microglia are quick to respond to any type of perturbation of the CNS, whether it be injury or disease (discussed below). Although they often respond to CNS alterations by retracting their processes, such morphological changes are not an essential feature of all microglial responses as phenotype switching in terms of cytokine expression can be induced without morphological changes as seen in brains of mice with prion disease after peripheral lipopolysaccharide challenge (Perry et al. 2007). Microglia in the normal uninjured brain can also be identified by the expression of various cell surface and intracellular antigens including CD11b, $CD45^{low}$, $CX3CR1^{high}$, $Ly6C^{low}$, $Gr\text{-}1^{low}$ (David and Kroner 2011). Against this backdrop, damage or injury to the CNS causes rapid changes in microglial morphology, and expression of cell surface or intracellular antigens and synthesis/release of cytokines. Like inflammation elsewhere, the inflammatory response initiated and mediated by microglia after CNS injury is largely meant to control infection and initiate wound healing. This response, which is inherently protective as it ensures survival against sepsis, can also exacerbate tissue damage and worsen recovery. In this chapter we will discuss some of the salient features of the microglial response to CNS injury and its implications for CNS repair, with greater focus on spinal cord injury (SCI) and the relationship of microglia to macrophages derived from blood monocytes during the onset of intraspinal inflammation.

19.2 Early Responses: Minutes to Hours After Injury

Morphological and phenotypic changes at the lesion epicenter: Extreme physical forces including blunt trauma, penetrating bone or projectiles (e.g., knife, bullet), or rapid acceleration/deceleration (e.g., blast, whiplash) causes shearing of axons, dendrites, and blood vessels. At the primary site of injury, bleeding and necrotic and apoptotic cell death occur creating a "hot zone" that is often referred to as the injury epicenter. Micro-lesions in the brain and spinal cord generated by either high-powered laser pulses or with a glass micropipette and then imaged in real-time by two-photon live imaging showed striking and unexpectedly rapid changes in motility of microglial processes. The first paper using this approach showed that instead of process retraction, microglia respond within 10–15 min by extending their cytoplasmic processes towards the lesion (Davalos et al. 2005). By 20–60 min these microglial processes form a dense network tightly around the lesion (Davalos et al. 2005). Other groups have reported similar responses in the cerebral cortex in slice cultures (Hines et al. 2009) and in vivo in spinal cord white matter microglia (Dibaj et al. 2010). The studies on the spinal cord laser lesions showed that microglia located 50–100 μm from the lesion respond by extending their processes towards the lesion, while at the same time process retraction occurred from regions of the cell facing away from the lesion (Dibaj et al. 2010; Hines et al. 2009). The processes that were retracted were restored after ~40 min (Hines et al. 2009). In experiments done in the spinal cord, some microglia soma in white matter were found to migrate towards the lesion within 2 h, sometimes along degenerating axons (Dibaj et al. 2010). A consistent finding is that these microglial responses, regardless of the CNS region, create within minutes a dense meshwork of microglial processes that wall-off the lesion border. Blocking or preventing this response in slice cultures by selective laser ablation of surrounding microglia, preventing actin polymerization, or chloride channel blocking causes the lesion to expand three to fourfold (Hines et al. 2009), indicating that this early and rapid microglial response within the first hour after injury is beneficial and protective. It makes sense that this first and earliest response of microglia would be protective, perhaps by participating in lesion containment or by rapidly surveying regions where the likelihood of pathogen entry is enhanced (e.g., regions of bleeding or tissue barrier compromise). In contrast to these small, well-defined laser or micropipette lesions, contusion injuries to spinal cord or brain cause more extensive damage and it is not known whether microglia respond similarly to small or large lesions, mostly because comparative real-time studies of this type have not been completed.

Molecular control of the rapid microglial response to injury: Glia-derived ATP, and damage-associated molecular patterns (DAMPs) from dying and necrotic cells and plasma proteins, most notably fibrinogen, are among the most prominent early microglia activating factors (Davalos et al. 2012; Kigerl and Popovich 2009; Popovich and Longbrake 2008). The rapid microglial process extension towards cortical laser or traumatic micro-lesions was shown to be mediated by ATP, as it could be inhibited by an ATP-degrading enzyme, or be induced by injection of ATP into the cortex (Davalos et al. 2005). The injury-induced response but not normal resting state motility of microglial processes is mediated by $P2Y_{12}$ G-protein-coupled purinergic receptors (Haynes et al. 2006). In addition, the injury response but not the resting state motility of microglial processes also requires functional volume-sensitive chloride channels (Hines et al. 2009). However, both types of process extension require actin polymerization (Hines et al. 2009). Interestingly, microglial process motility in the normal brain and injury-induced response is mediated by ATP released from astrocytes and can be prevented by blocking connexin hemichannels expressed in astrocytes (Davalos et al. 2005). These findings indicate functional communication between astrocytes and microglia that mediate microglial behavior in the normal and injured CNS. Nitric oxide gradients at the lesion also contribute to microglial process extension and soma migration, which can be augmented by tissue ATP (Dibaj et al. 2010). In the developing zebrafish brain, where microglia are more sparsely distributed, laser lesions induce microglia located at a distance to sense signals originating from the lesion and migrate towards the site of damage (Sieger et al. 2012). These authors have shown that glutamate released by damaged cells induce influx of intracellular calcium (Ca^{2+}) and the spread of Ca^{2+} waves in astrocytes that leads to release of ATP that is sensed by microglia located at a distance via the $P2Y_{12}$ receptors (Sieger et al. 2012). Such mechanisms may also operate in regions of the mammalian CNS in which microglia are sparsely distributed such as the white matter in the brain (Lawson et al. 1990) and spinal cord (Carlson et al. 1998).

DAMPs-TLR-mediated responses: As microglia extend processes towards the site of injury, they are likely to interact with intracellular molecules and fragments of extracellular molecules (ECM) generated from damaged cells and tissue. These DAMPs can signal microglia via binding and activation of various pattern recognition receptors (PRRs). In response to DAMPs that are liberated by sterile injury, such as in non-penetrating, contusion or compression type CNS injury, or pathogens that enter the CNS, microglial PRRs respond in a similar manner with the ultimate goal of preventing infection and promoting tissue repair. Toll-like receptors (TLRs) are one of the major classes of PRRs through which DAMPs signal microglial responses. Some of the known DAMPs include intracellular molecules released after injury such as heat-shock proteins (Hsp 22, 60, 70 and 72), some members of the S100 family of calcium-binding proteins, mRNA and single-stranded RNA (ssRNA); fragments of ECM molecules fibronectin, tenascin, versican, heparan sulfate, and hyaluronic acid; and serum amyloid (Piccinini and Midwood 2010). Except for mRNA and ssRNA, the other DAMPs listed signal via TLR2 and TLR4. Activation of TLR2 and 4 via the MyD88-dependent pathway leads to NFκB activation and expression of pro-inflammatory cytokines such as interleukin 1 β (IL-1β), TNFα, as well as expression of cyclooxygenase (COX), inducible nitric oxide synthase, and matrix metalloproteinases, all of which are increased after CNS injury. In a model of autoimmune demyelination, DAMP-mediated activation of microglia via TLR4 and integrin receptors (CD11b/CD18) results in the production of inflammatory chemokines and destructive oxidative bursts (Davalos et al. 2012; Smiley et al. 2001). In vivo two-photon imaging of the spinal cord in experimental autoimmune encephalomyelitis (EAE) showed microglia clusters co-localized to zones of fibrinogen leakage in the perivascular space (Davalos et al. 2012). These changes preceded axonal injury and demyelination, which could be inhibited by blocking reactive oxygen species (ROS) or deleting the fibrinogen binding site in microglia CD11b receptors (Davalos et al. 2012). A similar response is likely to occur after SCI or traumatic brain injury (TBI) since fibrinogen, albumin, and other as yet to be defined factors in blood persist at and near the epicenter for extended periods of time (days to weeks post-injury).

The binding of DAMPs to TLRs induces inflammatory signaling. When such signals persist or when active resolution of inflammation is impaired, chronic inflammation and pathology ensue resulting in conditions such as rheumatoid arthritis, and inflammatory lung or bowel disease (Piccinini and Midwood 2010). Spinal cord injuries result in an immediate and abundant availability of DAMPs and robust induction of TLRs (Kigerl et al. 2007). DAMP-TLR signaling likely continues chronically after SCI as the period of secondary tissue damage can extend for weeks or months, and as components of extracellular matrix molecules involved in tissue repair can also act as DAMPs. In addition, resolution of inflammation is impaired in injured spinal cord (Pruss et al. 2011). After spinal cord contusion injury in mice, there is increased expression of TLR1, 2, 5, and 7 mRNA immediately

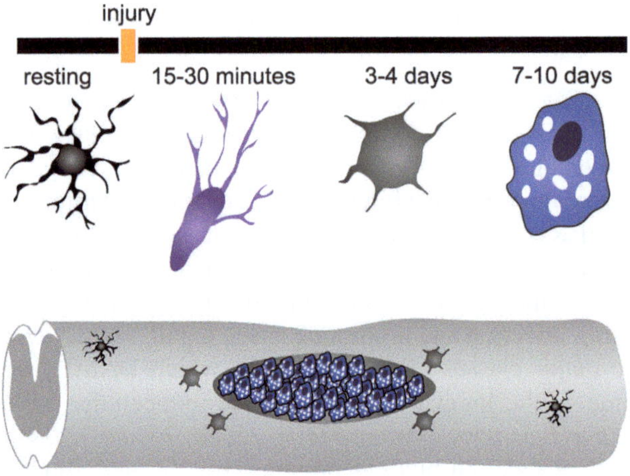

Fig. 19.1 *Upper*: Morphology of microglia prior to injury (resting state) and changes in their morphology over time after SCI. *Lower*: The *bottom* part of the figure is a representation of the spatial distribution of these cells in the contused spinal cord about 2 weeks after contusion injury. "*Resting*" *microglia*: CD11b, CD45^{low}, CX3CR1^{high}, Ly6C^{low}, Gr-1^{low}. "*Activated*" *microglia* (*and newly infiltrating monocytes*) *at and near lesion*: CD11b, CD45^{high}, CX3CR1^{low}, CD68, MHC class II, Ly6C^{high}, Gr-1^{low}. *Microglia located at a distance from lesion*: CD11b, CD45, CXCR3, CCL21, IL-6, MHC class II

after spinal cord contusion injury (1–14 days after SCI), in addition to increased MyD88 and NFκB (Kigerl et al. 2007). Interestingly, mice lacking TLR2 and TLR4 showed a subtle but noticeable worsening of locomotor control, unusual patterns of myelin and axon pathology, and a poorly formed glial limitans after SCI (Kigerl et al. 2007). In wild-type mice after spinal cord contusion injury, macrophages that appear "foamy" due to the breakdown of phagocytosed material (Fig. 19.1) often remain clustered in the central core of the lesion within a well-defined glial boundary. In contrast, TLR-deficient SCI mice had smaller clusters of phagocytic macrophages scattered in the ventral white matter beyond the normal extent of the glial limitans (Kigerl et al. 2007). This distribution of macrophages in TLR null mice may be due to failure of normal wound healing in the CNS that includes the formation of a glia limitans. The reformation of the glia limitans prevents immune cells from entering damaged CNS tissue (Bush et al. 1999). These findings suggest that DAMP-TLR signaling immediately after SCI prior to the formation of the glia limitans may be protective. Whether this is due to reparative programs that are activated in microglia or macrophages by TLR signaling or to the effects of TLR signaling in astrocytes which also express TLR2 and 4 (Kigerl et al. 2007) is not clear at present. Furthermore, the effects of prolonged DAMP-TLR signaling beyond the acute period after SCI cannot be known from these studies because signaling mutants or conventional knockout mice were used. Studies using conditional deletion TLR2 and 4 in microglia/macrophages after the first week after injury may help to better define the role of TLR signaling in these cell types after SCI.

Expression of pro-inflammatory cytokines: DAMP-TLR signaling can be expected to lead to the early expression of IL-1β and TNFα in microglia after injury. In fact, increased expression of these cytokines is seen soon after spinal cord contusion injury (Pineau and Lacroix 2007; Rice et al. 2007). The IL-1β mRNA expressing cells increase in number 6 h after contusion injury, peak at 12 h then return to normal levels by 48 h (Pineau and Lacroix 2007). At the peak of expression this cytokine is expressed mainly by microglia and astrocytes (Pineau and Lacroix 2007). IL-1 receptor antagonist (IL-1ra) is also up-regulated after SCI but at much lower levels than IL-1β (Pan et al. 2002). Intrathecal treatment with IL-1ra for the first 3 days after SCI in rats prevented apoptosis and caspase-3 expression (Nesic et al. 2001) and improved locomotor recovery (Zong et al. 2012). After TBI, activated microglia also increased expression of IL-1β (Bye et al. 2007). Following brain ischemic injury, treatment with IL-1ra reduces infarct volume (Loddick and Rothwell 1996; Relton et al. 1996), while treatment with exogenous IL-1β worsens brain injury (Lawrence et al. 1998; Loddick and Rothwell 1996). Furthermore, neutralizing IL-1β with function-blocking antibodies reduces tissue damage and improves cognition after brain injury (Clausen et al. 2011).

TNFα mRNA expressing cells detected by in-situ hybridization are seen within 15 min after spinal cord contusion injury, reach peak expression at 1 h, and return to baseline levels after 2 days. In addition, there is a second period of TNFα expression between 14 and 28 days (the maximum time examined) (Pineau and Lacroix 2007). During the period immediately after injury, TNFα is expressed by microglia rather than invading monocytes because of the rapid expression within 1 h, which precedes entry of monocytes (Pineau and Lacroix 2007). In addition to microglia, astrocytes, oligodendrocytes, and neurons also expressed TNFα at this early period after injury (Pineau and Lacroix 2007). During the second peak of expression at 2–4 weeks after injury, microglia and macrophages appear to be the main cells expressing TNFα (Pineau and Lacroix 2007). In the injured human spinal cord, TNFα protein is also detected early at 1–3 h after injury in microglia and neurons (Yang et al. 2005). There is also evidence that IL-1β and TNFα produced by microglia are associated with white matter damage around the ventricles in brain hypoxia injury (Deng et al. 2011). Blocking TNFα with a monoclonal antibody (Infliximab) or a TNF receptor–Fc fusion protein (Etanercept) improves histological and locomotor outcomes after SCI (Genovese et al. 2006, 2008).

Is the early microglial response beneficial or detrimental: The two-photon imaging studies showing that microglial ablation or prevention of microglial process extension towards the site of lesion enlarged lesion volume (Hines et al. 2009) indicate that this early response is beneficial. The latter work was done with micro-lesions. Whether microglial process extension occurs after large lesions such as spinal cord contusion injury is not known. Furthermore, if microglial process extension was to occur to any extent after contusion injuries, it would be important to assess if it is beneficial in limiting the size of the lesion. The early DAMP-TLR signaling response, which serves to alert the body of tissue damage, is thought to be involved in restoring tissue homeostasis and effect repair. The evidence obtained from conventional TLR knockout mice indicates that lack of TLR2 and 4 results in noticeable worsening of locomotor control, and a poorly formed glial limitans at the injury site with spreading of clusters of phagocytic macrophages beyond the normal extent of the glial limitans that surround the lesion core (Kigerl et al. 2007). The expression of the pro-inflammatory cytokines IL-1β and TNFα by injured CNS tissue has been shown in a variety of experiments to be detrimental and contributing to secondary damage.

19.3 Late Responses that Occur Days and Weeks After Injury

Phenotypic and functional polarization of microglia/macrophages: Within 2–3 days after SCI, circulating monocytes infiltrate the injury epicenter where they differentiate into tissue macrophages, i.e., monocyte-derived macrophages (MDMs). Within a few days after SCI microglia begin to retract their processes and become phagocytic at the site of lesion (Fig. 19.1). Within 1 week post-injury, routine phenotypic or morphological criteria cannot be used to distinguish between macrophages derived from activated microglia or MDMs. In general, a large proportion of macrophage/microglia that are phagocytic (foamy-looking cells) migrate into a cluster at the very center of the lesion, while activated microglia with short processes are seen in the surrounding areas (Fig. 19.1). As these macrophage/microglia accumulate, factors in the extracellular milieu will modulate their phenotype with discrete effects on neuron survival and axon growth (Kigerl et al. 2009; Stout et al. 2005). For example, cytokines and TLR agonists (e.g., DAMPs) can promote the differentiation of classically activated “M1” or alternatively activated “M2” macrophages. The canonical in vitro model for stimulating M1 differentiation is exposure of naive myeloid cells to LPS and IFNγ or, to promote M2 differentiation, cells are stimulated with IL-4 (Gordon and Taylor 2005). M1 macrophages were originally described as potent microbicidal effector cells; a function associated with release of inflammatory cytokines and ROS. M2 macrophages were found to have enhanced phagocytic capacity and release cytokines and growth factors that promote revascularization and tissue formation (Gordon and Taylor 2005). Both types of macrophages can phagocytose dying cells and tissue debris and their phagocytic activity is likely influenced by the tissue environment. SCI and TBI induce a heterogeneous macrophage response that can be defined by the increased expression of M1 and M2 phenotypic markers (Hsieh et al. 2013; Kigerl et al. 2009). After SCI, M1 macrophages dominate at the injury epicenter and nearby penumbra, overwhelming a comparatively smaller number and transient M2 macrophage response (Kigerl et al. 2009). M1 macrophages likely contribute to neurotoxicity and axonal “die-back”. Depletion or inhibition of the acute macrophage response, which are likely to be M1 macrophages, is neuroprotective and reduces axonal retraction at the site of injury (Blight 1994; Gris et al. 2004; Horn et al. 2008; Popovich et al. 1999). In vitro, M2 macrophages can promote long-distance axon growth without causing neurotoxicity (Kigerl et al. 2009). Therefore, in theory, any manipulation that biases the endogenous response toward an M2 phenotype could limit tissue damage and/or enhance tissue repair and perhaps axonal growth/regeneration. In support of this hypothesis, recent data show that post-injury administration of substance P, neural stem cells, granulocyte-colony-stimulating factor (G-CSF), mesenchymal stem cells or acidic-fibroblast growth factor laced peripheral nerve grafts improve recovery and/or promote repair in the injured spinal cord, and in each case, these interventions enhanced intraspinal M2 macrophage reactions (Guo et al. 2013; Jiang et al. 2012; Nakajima et al. 2012). A direct effect of M2 macrophages on mediating tissue repair or improving functional recovery has not been proved unequivocally in SCI or TBI models.

Phenotypic changes: Morphological transformation of microglia is accompanied by de novo expression of new membrane and cytoplasmic proteins. Since microglial phenotype varies throughout the CNS (de Haas et al. 2008), the phenotypic changes induced by injury, whether the injury is to brain or spinal cord, will vary as a function of injury location. For example, in brain and spinal cord, microglia express CD11b, CD40, CD45, CD80, CD86, F4/80, TREM-2b, CXCR3, and CCR9, albeit at relatively different levels indicating increased migration, phagocytic activity, antigen presentation, T and B cell interactions, inflammatory responses, and others. CD11b and CD45 are expressed by both microglia and MDMs, but overall levels of CD45 are reduced in microglia. Differential expression of CD45 can be detected using flow cytometry and is often used to distinguish between microglia and MDMs, but such differences cannot be distinguished by immunofluorescence in tissue sections. Microglia constitutively express various surface antigens common to other cells of the myeloid cell lineage including CX3CR1 and complement receptors. Activated microglia increase expression of these proteins and begin to express other antigens including major histocompatibility complex class II antigens (MHC class II). After SCI and TBI in rats, mice, non-human primates and humans, MHC class II expression is increased primarily on microglia found within white matter undergoing degeneration distal to the site of injury called, Wallerian degeneration (Kigerl et al. 2006; Popovich et al. 1993; Schmitt et al. 2000; Watanabe et al. 1999). There is no obvious relationship between MHC class II expression and demyelination in these regions, signifying diverse roles for MHC class II in microglia.

Mechanisms and functional implications of microglial activation distal to the injury epicenter: Axonal degeneration and plasticity also occur at sites distal to the epicenter, either as a result of physical severing of axons or diffuse axonal injury. Severed axonal segments located distal to the epicenter, whether they are located in spinal cord, brain, or peripheral nerve, undergo Wallerian degeneration. Wallerian degeneration is an active process that invariably transforms the phenotype of microglia. Several factors produced by Wallerian degenerating axons or their surrounding glia activate microglia. Complement proteins, members of the phospholipase A_2 superfamily of enzymes, galectins, and PRRs (e.g., TLRs), among others, all have been implicated in microglial activation and/or recruitment of MDMs (Boivin et al. 2007; Gaudet et al. 2009; Lopez-Vales et al. 2008; Mietto et al. 2013). Although microglia and MDMs phagocytose degenerating axons and myelin debris, whether these cells also cause collateral damage to intact axons or directly promote regeneration of injured axons remains controversial (Gaudet et al. 2011). Clearly, effective regeneration of injured peripheral nerve requires inflammatory signaling in macrophages (Boivin et al. 2007) that promote rapid clearance of myelin and axonal debris, and the secretion of growth promoting factors. In the CNS, the macrophage/microglial response that underlie Wallerian degeneration is significantly delayed (George and Griffin 1994; Vargas and Barres 2007). This delay in Wallerian degeneration is thought to contribute to the failure of axon regeneration in the CNS.

Microglial activation also occurs in grey matter regions remote from the site of injury and can have pathological consequences. For example, after a mid-thoracic SCI (e.g., T9), remote activation of microglia occurs over a period of several weeks post-injury in lumbar spinal cord (~10 segments below the level of injury), sensory relay nuclei in the brainstem (e.g., cuneate nucleus), and the ventral posterolateral (VPL) nucleus of the thalamus (Detloff et al. 2008; Hains and Waxman 2006; Koshinaga and Whittemore 1995). The microglia in these regions remain process-bearing (Fig. 19.1) but display other immunohistochemical evidence of activation. These responses occur within regions that process and relay sensory information, including pain. After SCI, activated microglia are implicated in the onset and progression of neuropathic pain (Hulsebosch et al. 2009; Tsuda et al. 2005) (see Chap. 11). A novel mechanism for transynaptic microglial activation has been described in which injured neurons increase synthesis of CCL21, a chemokine that is packaged into vesicles and is transported to presynaptic terminals and activates microglia via interactions with CXCR3 (de Jong et al. 2005). Transynaptic increases in CCL21 after SCI may also regulate microglial contributions to neuropathic pain. In a model of mid-thoracic SCI, CCL21 levels increase in neurons and microglia below the level of injury and in the VPL of the thalamus (Zhao et al. 2007b). Blocking axonal transmission rostral to the injury site or locally applying anti-CCL21 antibodies reduces CCL21 in the VPL, impairs microglial activation, and attenuates neuropathic pain (Zhao et al. 2007b). Cortical contusion injuries also cause remote microglia activation in cerebellum and spinal cord (Czeiter et al. 2008; Fukuda et al. 1996). These findings suggest that delay in microglial activation in white matter tracts undergoing Wallerian degeneration distal to the site of injury may contribute to the lack of regeneration in the CNS, while rapid and prolonged activation of microglia in grey matter regions distal to the injury site contribute to the generation of neuropathic pain.

Implications of chronic microglia and macrophage activation: Microglia and macrophages within the epicenter and in regions distal to the injury site remain activated for months or even years post-injury. Immunohistochemical analysis of postmortem human SCI specimens reveal activated phagocytic macrophage/microglia in degenerating white matter distal to the injury epicenter where they cluster or assume rounded phagocytic morphologies. These cells increase *de novo* expression of MHC class II glycoproteins or the lysosomal protein CD68, which may be functionally related to phagocytosis (Fleming et al. 2006; Schmitt et al. 2000). Identical responses are evident in white matter of SCI rats and mice several weeks after injury (Kigerl et al. 2006; Popovich et al. 1997). SCI also causes chronic microglial activation in remote regions of spinal cord gray matter and in the thalamus (Detloff et al. 2008; Zhao et al. 2007a, b). Expression of inflammatory signaling proteins including p38 mitogen-activated protein kinase (MAPK) and phosphorylated extracellular signal regulated kinase 1/2 (pERK1/2) increases in microglia in lumbar spinal cord gray matter, several segments below the level of a thoracic SCI (Detloff et al. 2008; Zhao et al. 2007a). pERK1/2 is an upstream regulator of prostaglandin E2 (PGE2), a lipid inflammatory mediator that is produced downstream of pERK1/2 and COX. Pharmacological blocking of PGE2 release in chronically injured spinal cord lowers the firing threshold for dorsal horn neurons resulting in aberrant sensory function including the onset of neuropathic pain (Zhao et al. 2007a). The neuronal chemokine CCL21 which activates microglia is increased in the dorsal horn of the spinal cord and VPL of the thalamus after SCI and was shown to induce pain-related behaviors (Zhao et al. 2007b). IL-6 levels are increased 3 weeks and later in regions caudal to the site of SCI and may be involved in the maintenance of pain (Detloff et al. 2008). These findings suggest that microglial activation in regions far removed from the site of SCI can contribute to pain and functional disability.

Interestingly, blockade of the C5a fragment of complement, a potent microglia activator and leukocyte chemoattractant, impairs recovery of locomotor function and exacerbates pathology at the site of injury when given beginning at 2 weeks post-injury (Beck et al. 2010). These data may indicate a role for complement in regulating inflammatory-mediated repair processes or, alternatively, the importance of complement in signaling microglia to participate in the surveillance and maintenance of spinal circuitry. Microglial phagocytosis of synapses is triggered by synaptic complement proteins expressed by neurons during postnatal development (Schafer et al. 2012) (see Chap. 9). If microglia function is impaired, the formation and refinement of new circuits resulting from regenerative sprouting or therapeutic intervention could be adversely affected (Paolicelli et al. 2011; Schafer et al. 2012; Stephan et al. 2012). In the adult CNS, microglia strip and phagocytose axon-somatic synapses on facial motor neurons after nerve injury (Kreutzberg 1993).

Chronic activation of microglia or macrophages may predispose the injured brain and spinal cord to excess oxidative stress and inflammatory signaling. Gene expression profiling studies show that a subset of inflammatory genes expressed early after SCI or TBI persist for up to 6 months post-injury and are translated into proteins that may have pathogenic significance (Byrnes et al. 2006). For example, $p22^{Phox}$ and $gp91^{Phox}$ are components of NADPH oxidase and are increased early after injury but expression is maintained for up to 6 weeks (latest time examined). NADPH oxidase produces ROS which in turn activates NFκB and MAPK-dependent inflammatory signaling. Chronic activation of microglial NADPH oxidase can cause neurotoxicity in vivo (Lull and Block 2010). Inhibition of NADPH oxidase during the first week post-contusive SCI with diphenylene iodonium reduces the normally high levels of NADPH oxidase, as well as other inflammatory genes and proteins, and is associated with a decrease in lesion volume (Byrnes et al. 2011).

Chronic oxidative stress also produces a pool of lipid inflammatory mediators that persist in chronically injured spinal cord. Oxidative metabolism of arachidonic acid (AA) yields leukotrienes (LTs) and prostaglandins (PGs). LTs and PGs are elevated in injured rat spinal cord for up to 9 months post-injury (Dulin et al. 2013). These inflammatory mediators are potent activators of microglia and macrophages and can elicit inflammatory signaling that lowers the threshold for neuronal depolarization and can result in neuropathic pain. Licofelone, a novel inhibitor of COX and 5-lipoxygenase (5-LOX), i.e., the enzymes responsible for LT and PG synthesis, normalized the oxidative and inflammatory microenvironment associated with chronic SCI and reduced post-injury allodynia. Licofelone did not improve motor recovery (Dulin et al. 2013).

Despite evidence that acute neuroinflammatory cascades are distinct in injured brain and spinal cord (Batchelor et al. 2008; Schnell et al. 1999), microglia and macrophage reactions in the chronic lesion environment of traumatically injured brain are remarkably similar to those described above for injured spinal cord. In non-human primates, lesions in the primary motor cortex elicit microglial activation for up to 1 year post-injury; microglia in cortex exhibit an activated phenotype ($CD68^{+}$) with prolonged expression of brain-derived neurotrophic factors (BDNF) and tyrosine receptor kinase B (TrkB) receptors evident in cortical and spinal cord microglia (Nagamoto-Combs et al. 2007). Activated microglia were detected in subcortical grey and white matter in humans 11 months or longer after TBI using positron emission tomography (PET) for the translocator protein (TSPO), which is found in high levels in the mitochondria of activated microglia (Ramlackhansingh et al. 2011). They found that the magnitude of microglial activation in thalamus correlated with cognitive impairment but not structural brain damage (Ramlackhansingh et al. 2011). In addition, in a post-mortem study of humans with single TBI, activated microglia detected by immunohistochemistry for CR3-43 and CD68 were seen in 28 % of cases in which there was also evidence of white matter damage (Johnson et al. 2013). Although correlative, these data indicate that chronically activated microglia could contribute to damage and long-term neurologic impairment. If true, microglia-specific therapeutic intervention may still be effective for several years post-injury.

19.4 Future Directions

Microglia and MDMs are derived from distinct myeloid precursors with discrete transcriptional requirements (Ginhoux et al. 2010; Kierdorf et al. 2013); monocytes develop from hematopoietic stem cells in bone marrow, a process that requires the transcription factor Myb (Schulz et al. 2012). Conversely, microglial development from yolk sac myeloid progenitors is Myb-independent (Schulz et al. 2012). Additional cell-specific differences are likely to be discovered and will yield new genetic tools that will allow scientists to independently manipulate these distinct myeloid cell lineages, further refining our knowledge about how these myeloid cell subsets contribute to CNS injury and repair.

How aging affects microglia and CNS repair is also an important experimental and clinical variable for future consideration, especially since aging affects microglial function (Streit 2006) (see Chap. 13). Aging can affect hematopoiesis, and activation of these and other innate immune cells may affect macrophage and microglial function (Kovacs et al. 2009; Shaw et al. 2010). Aging has become an area of intense preclinical study and a significant clinical concern after SCI and TBI, especially since the effects of injury may accelerate aging of some organ systems such as the cardiovascular, endocrine, and musculoskeletal (Hitzig et al. 2011). There is little known about how aging affects microglia and MDMs after SCI, although this is likely to be an area of research that will receive increased attention (Kumamaru et al. 2012).

Why microglia and macrophages remain activated in chronically injured brain and spinal cord is also unclear. It is probable that specific "on" signals including ATP, DAMPs, degenerating axons, myelin debris, and cytokines persist indefinitely or reappear slowly over time creating a pool of ligands that can stimulate microglia and macrophages (Biber et al. 2007). Chronic activation may also be explained by the loss of "off" signals including CD200L or CX3CL1. These proteins are normally found on or are released by healthy neurons and effectively inhibit or silence microglia effector functions (Biber et al. 2007). The presence of "off" signals indicates that the intact nervous system actively regulates microglial function. However, after SCI or TBI, inflammation does not appear to be self-limiting (Pruss et al. 2011).

It seems intuitive that limiting chronic activation of microglia and MDMs would yield better outcomes after SCI or TBI. However, a more in-depth understanding about the functional heterogeneity of these cell types in both the acute and chronic lesion microenvironment is needed. As treatment and care for SCI and TBI individuals improve, lifespan is increasing. Thus, the consequences of an aging microglia repertoire also need to be considered in order to optimize CNS repair.

References

Batchelor PE, Tan S, Wills TE, Porritt MJ, Howells DW (2008) Comparison of inflammation in the brain and spinal cord following mechanical injury. J Neurotrauma 25:1217–1225

Beck KD, Nguyen HX, Galvan MD, Salazar DL, Woodruff TM, Anderson AJ (2010) Quantitative analysis of cellular inflammation after traumatic spinal cord injury: evidence for a multiphasic inflammatory response in the acute to chronic environment. Brain 133:433–447

Biber K, Neumann H, Inoue K, Boddeke HW (2007) Neuronal 'On' and 'Off' signals control microglia. Trends Neurosci 30:596–602

Blight AR (1994) Effects of silica on the outcome from experimental spinal cord injury: implication of macrophages in secondary tissue damage. Neuroscience 60:263–273

Boivin A, Pineau I, Barrette B, Filali M, Vallieres N, Rivest S, Lacroix S (2007) Toll-like receptor signaling is critical for Wallerian degeneration and functional recovery after peripheral nerve injury. J Neurosci 27:12565–12576

Bush TG, Puvanachandra N, Horner CH, Polito A, Ostenfeld T, Svendsen CN, Mucke L, Johnson MH, Sofroniew MV (1999) Leukocyte infiltration, neuronal degeneration, and neurite outgrowth after ablation of scar-forming, reactive astrocytes in adult transgenic mice. Neuron 23:297–308

Bye N, Habgood MD, Callaway JK, Malakooti N, Potter A, Kossmann T, Morganti-Kossmann MC (2007) Transient neuroprotection by minocycline following traumatic brain injury is associated with attenuated microglial activation but no changes in cell apoptosis or neutrophil infiltration. Exp Neurol 204:220–233

Byrnes KR, Garay J, Di Giovanni S, De Biase A, Knoblach SM, Hoffman EP, Movsesyan V, Faden AI (2006) Expression of two temporally distinct microglia-related gene clusters after spinal cord injury. Glia 53:420–433

Byrnes KR, Washington PM, Knoblach SM, Hoffman E, Faden AI (2011) Delayed inflammatory mRNA and protein expression after spinal cord injury. J Neuroinflammation 8:130

Carlson SL, Parrish ME, Springer JE, Doty K, Dossett L (1998) Acute inflammatory response in spinal cord following impact injury. Exp Neurol 151:77–88

Clausen F, Hanell A, Israelsson C, Hedin J, Ebendal T, Mir AK, Gram H, Marklund N (2011) Neutralization of interleukin-1beta reduces cerebral edema and tissue loss and improves late cognitive outcome following traumatic brain injury in mice. Eur J Neurosci 34:110–123

Czeiter E, Pal J, Kovesdi E, Bukovics P, Luckl J, Doczi T, Buki A (2008) Traumatic axonal injury in the spinal cord evoked by traumatic brain injury. J Neurotrauma 25:205–213

Davalos D, Grutzendler J, Yang G, Kim JV, Zuo Y, Jung S, Littman DR, Dustin ML, Gan WB (2005) ATP mediates rapid microglial response to local brain injury in vivo. Nat Neurosci 8:752–758

Davalos D, Ryu JK, Merlini M, Baeten KM, Le Moan N, Petersen MA, Deerinck TJ, Smirnoff DS, Bedard C, Hakozaki H, Gonias Murray S, Ling JB, Lassmann H, Degen JL, Ellisman MH, Akassoglou K (2012) Fibrinogen-induced perivascular microglial clustering is required for the development of axonal damage in neuroinflammation. Nat Commun 3:1227

David S, Kroner A (2011) Repertoire of microglial and macrophage responses after spinal cord injury. Nat Rev Neurosci 12:388–399

de Haas AH, Boddeke HW, Biber K (2008) Region-specific expression of immunoregulatory proteins on microglia in the healthy CNS. Glia 56:888–894

de Jong EK, Dijkstra IM, Hensens M, Brouwer N, van Amerongen M, Liem RS, Boddeke HW, Biber K (2005) Vesicle-mediated transport and release of CCL21 in endangered neurons: a possible explanation for microglia activation remote from a primary lesion. J Neurosci 25:7548–7557

Deng YY, Lu J, Ling EA, Kaur C (2011) Role of microglia in the process of inflammation in the hypoxic developing brain. Front Biosci (Schol Ed) 3:884–900

Detloff MR, Fisher LC, McGaughy V, Longbrake EE, Popovich PG, Basso DM (2008) Remote activation of microglia and pro-inflammatory cytokines predict the onset and severity of below-level neuropathic pain after spinal cord injury in rats. Exp Neurol 212:337–347

Dibaj P, Nadrigny F, Steffens H, Scheller A, Hirrlinger J, Schomburg ED, Neusch C, Kirchhoff F (2010) NO mediates microglial response to acute spinal cord injury under ATP control in vivo. Glia 58:1133–1144

Dulin JN, Karoly ED, Wang Y, Strobel HW, Grill RJ (2013) Licofelone modulates neuroinflammation and attenuates mechanical hypersensitivity in the chronic phase of spinal cord injury. J Neurosci 33:652–664

Fleming JC, Norenberg MD, Ramsay DA, Dekaban GA, Marcillo AE, Saenz AD, Pasquale-Styles M, Dietrich WD, Weaver LC (2006) The cellular inflammatory response in human spinal cords after injury. Brain 129:3249–3269

Fukuda K, Aihara N, Sagar SM, Sharp FR, Pitts LH, Honkaniemi J, Noble LJ (1996) Purkinje cell vulnerability to mild traumatic brain injury. J Neurotrauma 13:255–266

Gaudet AD, Leung M, Poirier F, Kadoya T, Horie H, Ramer MS (2009) A role for galectin-1 in the immune response to peripheral nerve injury. Exp Neurol 220:320–327

Gaudet AD, Popovich PG, Ramer MS (2011) Wallerian degeneration: gaining perspective on inflammatory events after peripheral nerve injury. J Neuroinflammation 8:110

Genovese T, Mazzon E, Crisafulli C, Di Paola R, Muia C, Bramanti P, Cuzzocrea S (2006) Immunomodulatory effects of etanercept in an experimental model of spinal cord injury. J Pharmacol Exp Ther 316:1006–1016

Genovese T, Mazzon E, Crisafulli C, Di Paola R, Muia C, Esposito E, Bramanti P, Cuzzocrea S (2008) TNF-alpha blockage in a mouse model of SCI: evidence for improved outcome. Shock 29:32–41

George R, Griffin JW (1994) Delayed macrophage responses and myelin clearance during Wallerian degeneration in the central nervous system: the dorsal radiculotomy model. Exp Neurol 129:225–236

Ginhoux F, Greter M, Leboeuf M, Nandi S, See P, Gokhan S, Mehler MF, Conway SJ, Ng LG, Stanley ER, Samokhvalov IM, Merad M (2010) Fate mapping analysis reveals that adult microglia derive from primitive macrophages. Science 330:841–845

Gordon S, Taylor PR (2005) Monocyte and macrophage heterogeneity. Nat Rev Immunol 5:953–964

Gris D, Marsh DR, Oatway MA, Chen Y, Hamilton EF, Dekaban GA, Weaver LC (2004) Transient blockade of the CD11d/CD18 integrin reduces secondary damage after spinal cord injury, improving sensory, autonomic, and motor function. J Neurosci 24:4043–4051

Guo Y, Zhang H, Yang J, Liu S, Bing L, Gao J, Hao A (2013) Granulocyte colony-stimulating factor improves alternative activation of microglia under microenvironment of spinal cord injury. Neuroscience 238:1–10

Hains BC, Waxman SG (2006) Activated microglia contribute to the maintenance of chronic pain after spinal cord injury. J Neurosci 26:4308–4317

Haynes SE, Hollopeter G, Yang G, Kurpius D, Dailey ME, Gan WB, Julius D (2006) The P2Y12 receptor regulates microglial activation by extracellular nucleotides. Nat Neurosci 9:1512–1519

Hines DJ, Hines RM, Mulligan SJ, Macvicar BA (2009) Microglia processes block the spread of damage in the brain and require functional chloride channels. Glia 57:1610–1618

Hitzig SL, Eng JJ, Miller WC, Sakakibara BM (2011) An evidence-based review of aging of the body systems following spinal cord injury. Spinal Cord 49:684–701

Horn KP, Busch SA, Hawthorne AL, van Rooijen N, Silver J (2008) Another barrier to regeneration in the CNS: activated macrophages induce extensive retraction of dystrophic axons through direct physical interactions. J Neurosci 28:9330–9341

Hsieh CL, Kim CC, Ryba BE, Niemi EC, Bando JK, Locksley RM, Liu J, Nakamura MC, Seaman WE (2013) Traumatic brain injury induces macrophage subsets in the brain. Eur J Immunol 43:2010–2022

Hulsebosch CE, Hains BC, Crown ED, Carlton SM (2009) Mechanisms of chronic central neuropathic pain after spinal cord injury. Brain Res Rev 60:202–213

Jiang MH, Chung E, Chi GF, Ahn W, Lim JE, Hong HS, Kim DW, Choi H, Kim J, Son Y (2012) Substance P induces M2-type macrophages after spinal cord injury. Neuroreport 23:786–792

Johnson VE, Stewart JE, Begbie FD, Trojanowski JQ, Smith DH, Stewart W (2013) Inflammation and white matter degeneration persist for years after a single traumatic brain injury. Brain 136:28–42

Kierdorf K, Erny D, Goldmann T, Sander V, Schulz C, Perdiguero EG, Wieghofer P, Heinrich A, Riemke P, Holscher C, Muller DN, Luckow B, Brocker T, Debowski K, Fritz G, Opdenakker G, Diefenbach A, Biber K, Heikenwalder M, Geissmann F, Rosenbauer F, Prinz M (2013) Microglia emerge from erythromyeloid precursors via Pu.1- and Irf8-dependent pathways. Nat Neurosci 16:273–280

Kigerl KA, Gensel JC, Ankeny DP, Alexander JK, Donnelly DJ, Popovich PG (2009) Identification of two distinct macrophage subsets with divergent effects causing either neurotoxicity or regeneration in the injured mouse spinal cord. J Neurosci 29:13435–13444

Kigerl KA, Lai W, Rivest S, Hart RP, Satoskar AR, Popovich PG (2007) Toll-like receptor (TLR)-2 and TLR-4 regulate inflammation, gliosis, and myelin sparing after spinal cord injury. J Neurochem 102:37–50

Kigerl KA, McGaughy VM, Popovich PG (2006) Comparative analysis of lesion development and intraspinal inflammation in four strains of mice following spinal contusion injury. J Comp Neurol 494:578–594

Kigerl KA, Popovich PG (2009) Toll-like receptors in spinal cord injury. Curr Top Microbiol Immunol 336:121–136

Koshinaga M, Whittemore SR (1995) The temporal and spatial activation of microglia in fiber tracts undergoing anterograde and retrograde degeneration following spinal cord lesion. J Neurotrauma 12:209–222

Kovacs EJ, Palmer JL, Fortin CF, Fulop T Jr, Goldstein DR, Linton PJ (2009) Aging and innate immunity in the mouse: impact of intrinsic and extrinsic factors. Trends Immunol 30: 319–324

Kreutzberg GW (1993) Dynamic changes in motoneurons during regeneration. Restor Neurol Neurosci 5:59–60

Kumamaru H, Saiwai H, Ohkawa Y, Yamada H, Iwamoto Y, Okada S (2012) Age-related differences in cellular and molecular profiles of inflammatory responses after spinal cord injury. J Cell Physiol 227:1335–1346

Lawrence CB, Allan SM, Rothwell NJ (1998) Interleukin-1beta and the interleukin-1 receptor antagonist act in the striatum to modify excitotoxic brain damage in the rat. Eur J Neurosci 10:1188–1195

Lawson LJ, Perry VH, Dri P, Gordon S (1990) Heterogeneity in the distribution and morphology of microglia in the normal adult mouse brain. Neuroscience 39:151–170

Loddick SA, Rothwell NJ (1996) Neuroprotective effects of human recombinant interleukin-1 receptor antagonist in focal cerebral ischaemia in the rat. J Cereb Blood Flow Metab 16:932–940

Lopez-Vales R, Navarro X, Shimizu T, Baskakis C, Kokotos G, Constantinou-Kokotou V, Stephens D, Dennis EA, David S (2008) Intracellular phospholipase A(2) group IVA and group VIA play important roles in Wallerian degeneration and axon regeneration after peripheral nerve injury. Brain 131:2620–2631

Lull ME, Block ML (2010) Microglial activation and chronic neurodegeneration. Neurotherapeutics 7:354–365

Mietto BS, Jurgensen S, Alves L, Pecli C, Narciso MS, Assuncao-Miranda I, Villa-Verde DM, de Souza Lima FR, de Menezes JR, Benjamim CF, Bozza MT, Martinez AM (2013) Lack of galectin-3 speeds Wallerian degeneration by altering TLR and pro-inflammatory cytokine expressions in injured sciatic nerve. Eur J Neurosci 37:1682–1690

Nagamoto-Combs K, McNeal DW, Morecraft RJ, Combs CK (2007) Prolonged microgliosis in the rhesus monkey central nervous system after traumatic brain injury. J Neurotrauma 24:1719–1742

Nakajima H, Uchida K, Guerrero AR, Watanabe S, Sugita D, Takeura N, Yoshida A, Long G, Wright KT, Johnson WE, Baba H (2012) Transplantation of mesenchymal stem cells promotes an alternative pathway of macrophage activation and functional recovery after spinal cord injury. J Neurotrauma 29:1614–1625

Nesic O, Xu GY, McAdoo D, High KW, Hulsebosch C, Perez-Pol R (2001) IL-1 receptor antagonist prevents apoptosis and caspase-3 activation after spinal cord injury. J Neurotrauma 18:947–956

Pan JZ, Ni L, Sodhi A, Aguanno A, Young W, Hart RP (2002) Cytokine activity contributes to induction of inflammatory cytokine mRNAs in spinal cord following contusion. J Neurosci Res 68:315–322

Paolicelli RC, Bolasco G, Pagani F, Maggi L, Scianni M, Panzanelli P, Giustetto M, Ferreira TA, Guiducci E, Dumas L, Ragozzino D, Gross CT (2011) Synaptic pruning by microglia is necessary for normal brain development. Science 333:1456–1458

Perry VH, Cunningham C, Holmes C (2007) Systemic infections and inflammation affect chronic neurodegeneration. Nat Rev Immunol 7:161–167

Perry VH, Hume DA, Gordon S (1985) Immunohistochemical localization of macrophages and microglia in the adult and developing mouse brain. Neuroscience 15:313–326

Piccinini AM, Midwood KS (2010) DAMPening inflammation by modulating TLR signalling. Mediators Inflamm 2010. DOI: 10.1155/2010/672395

Pineau I, Lacroix S (2007) Proinflammatory cytokine synthesis in the injured mouse spinal cord: multiphasic expression pattern and identification of the cell types involved. J Comp Neurol 500:267–285

Popovich PG, Guan Z, Wei P, Huitinga I, van Rooijen N, Stokes BT (1999) Depletion of hematogenous macrophages promotes partial hindlimb recovery and neuroanatomical repair after experimental spinal cord injury. Exp Neurol 158:351–365

Popovich PG, Longbrake EE (2008) Can the immune system be harnessed to repair the CNS? Nat Rev Neurosci 9:481–493

Popovich PG, Streit WJ, Stokes BT (1993) Differential expression of MHC class II antigen in the contused rat spinal cord. J Neurotrauma 10:37–46

Popovich PG, Wei P, Stokes BT (1997) Cellular inflammatory response after spinal cord injury in Sprague-Dawley and Lewis rats. J Comp Neurol 377:443–464

Pruss H, Kopp MA, Brommer B, Gatzemeier N, Laginha I, Dirnagl U, Schwab JM (2011) Non-resolving aspects of acute inflammation after spinal cord injury (SCI): indices and resolution plateau. Brain Pathol 21:652–660

Ramlackhansingh AF, Brooks DJ, Greenwood RJ, Bose SK, Turkheimer FE, Kinnunen KM, Gentleman S, Heckemann RA, Gunanayagam K, Gelosa G, Sharp DJ (2011) Inflammation after trauma: microglial activation and traumatic brain injury. Ann Neurol 70:374–383

Relton JK, Martin D, Thompson RC, Russell DA (1996) Peripheral administration of interleukin-1 receptor antagonist inhibits brain damage after focal cerebral ischemia in the rat. Exp Neurol 138:206–213

Rice T, Larsen J, Rivest S, Yong VW (2007) Characterization of the early neuroinflammation after spinal cord injury in mice. J Neuropathol Exp Neurol 66:184–195

Schafer DP, Lehrman EK, Kautzman AG, Koyama R, Mardinly AR, Yamasaki R, Ransohoff RM, Greenberg ME, Barres BA, Stevens B (2012) Microglia sculpt postnatal neural circuits in an activity and complement-dependent manner. Neuron 74:691–705

Schmitt AB, Buss A, Breuer S, Brook GA, Pech K, Martin D, Schoenen J, Noth J, Love S, Schroder JM, Kreutzberg GW, Nacimiento W (2000) Major histocompatibility complex class II expression by activated microglia caudal to lesions of descending tracts in the human spinal cord is not associated with a T cell response. Acta Neuropathol 100:528–536

Schnell L, Fearn S, Klassen H, Schwab ME, Perry VH (1999) Acute inflammatory responses to mechanical lesions in the CNS: differences between brain and spinal cord. Eur J Neurosci 11:3648–3658

Schulz C, Gomez Perdiguero E, Chorro L, Szabo-Rogers H, Cagnard N, Kierdorf K, Prinz M, Wu B, Jacobsen SE, Pollard JW, Frampton J, Liu KJ, Geissmann F (2012) A lineage of myeloid cells independent of Myb and hematopoietic stem cells. Science 336:86–90

Shaw AC, Joshi S, Greenwood H, Panda A, Lord JM (2010) Aging of the innate immune system. Curr Opin Immunol 22:507–513

Sieger D, Moritz C, Ziegenhals T, Prykhozhij S, Peri F (2012) Long-range Ca^{2+} waves transmit brain-damage signals to microglia. Dev Cell 22:1138–1148

Smiley ST, King JA, Hancock WW (2001) Fibrinogen stimulates macrophage chemokine secretion through toll-like receptor 4. J Immunol 167:2887–2894

Stephan AH, Barres BA, Stevens B (2012) The complement system: an unexpected role in synaptic pruning during development and disease. Annu Rev Neurosci 35:369–389

Stout RD, Jiang C, Matta B, Tietzel I, Watkins SK, Suttles J (2005) Macrophages sequentially change their functional phenotype in response to changes in microenvironmental influences. J Immunol 175:342–349

Streit WJ (2006) Microglial senescence: does the brain's immune system have an expiration date? Trends Neurosci 29:506–510

Tsuda M, Inoue K, Salter MW (2005) Neuropathic pain and spinal microglia: a big problem from molecules in "small" glia. Trends Neurosci 28:101–107

Vargas ME, Barres BA (2007) Why is Wallerian degeneration in the CNS so slow? Annu Rev Neurosci 30:153–179

Vela JM, Dalmau I, Gonzalez B, Castellano B (1995) Morphology and distribution of microglial cells in the young and adult mouse cerebellum. J Comp Neurol 361:602–616

Watanabe T, Yamamoto T, Abe Y, Saito N, Kumagai T, Kayama H (1999) Differential activation of microglia after experimental spinal cord injury. J Neurotrauma 16:255–265

Yang L, Jones NR, Blumbergs PC, Van Den Heuvel C, Moore EJ, Manavis J, Sarvestani GT, Ghabriel MN (2005) Severity-dependent expression of pro-inflammatory cytokines in traumatic spinal cord injury in the rat. J Clin Neurosci 12:276–284

Zhao P, Waxman SG, Hains BC (2007a) Extracellular signal-regulated kinase-regulated microglia-neuron signaling by prostaglandin E2 contributes to pain after spinal cord injury. J Neurosci 27:2357–2368

Zhao P, Waxman SG, Hains BC (2007b) Modulation of thalamic nociceptive processing after spinal cord injury through remote activation of thalamic microglia by cysteine cysteine chemokine ligand 21. J Neurosci 27:8893–8902

Zong S, Zeng G, Wei B, Xiong C, Zhao Y (2012) Beneficial effect of interleukin-1 receptor antagonist protein on spinal cord injury recovery in the rat. Inflammation 35:520–526

Index

M.-È. Tremblay, A. Sierra (eds.), *Microglia in Health and Disease*,
DOI 10.1007/978-1-4939-1429-6

O

T

MIX
Papier aus verantwortungsvollen Quellen
Paper from responsible sources
FSC® C105338

If you have any concerns about our products,
you can contact us on
ProductSafety@springernature.com

In case Publisher is established outside the EU,
the EU authorized representative is:
Springer Nature Customer Service Center GmbH
Europaplatz 3, 69115 Heidelberg, Germany

Printed by Libri Plureos GmbH
in Hamburg, Germany